FARADAY DISCUSSIONS
NO. 114   1999

The Surface Science of Metal Oxides

The Faraday Division
The Royal Society of Chemistry
London

Organising Committee
Professor G. Thornton (*Chairman*)
Dr R. G. Egdell
Professor M. J. Gillan
Professor B. E. Hayden
Professor C. R. A. Catlow
Professor Dr H.-J. Freund
Professor N. M. Harrison

ISBN: 0-85404-884-7
ISSN: 0301-7249

Typeset by Santype International Ltd., Netherhampton Road, Salisbury, Wiltshire and printed and bound in Great Britain by Whitstable Litho Printers Ltd.

A General Discussion

on

The Surface Science of Metal Oxides

1st, 2nd and 3rd September, 1999

A General Discussion on The Surface Science of Metal Oxides was held at St. Martin's College, Ambleside, UK on 1st, 2nd and 3rd September, 1999.

Contents

1   Introductory Lecture: Oxide surfaces
    Hans-Joachim Freund

33 *Ab initio* calculations on the $Al_2O_3(0001)$ surface
    Iskander Batyrev, Ali Alavi and **Michael W. Finnis**

45 Ultrathin alumina film Al-sublattice structure, metal island nucleation at terrace point defects, and how hydroxylation affects wetting
    D. R. Jennison and **A. Bogicevic**

53 Electronic properties, structure and adsorption at vanadium oxide: density functional theory studies
    K. Hermann, M. Witko and **R. Druzinic**

67 The growth of vanadium oxide on alumina and titania single crystal surfaces
    Robert J. Madix, Jürgen Biener, Marcus Bäumer and **Andreas Dinger**

85 General Discussion

105 A study of the electronic, magnetic, structural and dynamic properties of low-dimensional NiO on MgO(100) surfaces
    William C. Mackrodt, Claudine Noguera and **Neil L. Allan**

129 Scanning tunnelling microscopy on the growth and structure of NiO(100) and CoO(100) thin films
    Ina Sebastian, Thomas Bertrams, Klaus Meinel and **Henning Neddermeyer**

141 Structure determination of molecular adsorbates on oxide surfaces using scanned-energy mode photoelectron diffraction
    M. Polcik, R. Lindsay, P. Baumgärtel, R. Terborg, O. Schaff, S. Kulkarni, A. M. Bradshaw, R. L. Toomes and **D. P. Woodruff**

157 What can we learn on the structure and morphology of metal oxide/metal interfaces by measurement of X-ray crystal truncation rods *in situ*, during growth
    G. Renaud, O. Robach and **A. Barbier**

173 Mg clusters on MgO surfaces: study of the nucleation mechanism with MIES and *ab initio* calculations
    L. N. Kantorovich, A. L. Shluger, P. V. Sushko, J. Günster, P. Stracke, D. W. Goodman and **V. Kempter**

195 A microcalorimetric study of the heat of adsorption of copper on well-defined oxide thin film surfaces: MgO(100), p(2 × 1) oxide on Mo(100) and disordered W oxide
    Jeffrey T. Ranney, David E. Starr, Jana E. Musgrove, Dan J. Bald and **Charles T. Campbell**

209 Cu atoms and clusters on regular and defect sites of the SiO_2 surface. Electronic structure and properties from first principle calculations
    Gianfranco Pacchioni, Nuria Lopez and **Francesc Illas**

This journal is © The Royal Society of Chemistry 2000

223	General Discussion
245	Oxygen-induced restructuring of rutile $TiO_2(110)$: formation mechanism, atomic models, and influence on surface chemistry **Min Li, Wilhelm Hebenstreit, Ulrike Diebold, Michael A. Henderson** and **Dwight R. Jennison**
259	The selective adsorption and kinetic behaviour of molecules on $TiO_2(110)$ observed by STM and NC-AFM **Yasuhiro Iwasawa, Hiroshi Onishi, Ken-ichi Fukui, Shushi Suzuki** and **Takehiko Sasaki**
267 ■	Scanning tunnelling microscopy studies of the reactivity of the $TiO_2(110)$ surface: Re-oxidation and the thermal treatment of metal nanoparticles **R. A. Bennett, P. Stone** and **M. Bowker**
279	Oxygen-induced morphological changes of Ag nanoclusters supported on $TiO_2(110)$ **Xiaofeng Lai, Todd P. St.Clair** and **D. Wayne Goodman**
285	First principles simulations of titanium oxide clusters and surfaces **Tristan Albaret, Fabio Finocchi** and **Claudine Noguera**
305 ■	The influence of soft vibrational modes on our understanding of oxide surface structure **N. M. Harrison, X.-G. Wang, J. Muscat** and **M. Scheffler**
313	The chemistry of methanol on the $TiO_2(110)$ surface: the influence of vacancies and coadsorbed species **Michael A. Henderson, Sary Otero-Tapia** and **Miguel E. Castro**
331	General Discussion
351	Metal oxides: O^{2-} chemistry and dynamical effects on oxide reactivity **Luciano Triguero, Stefano de Carolis, Micael Baudin, Mark Wójcik, Kersti Hermansson, Martin A. Nygren** and **Lars G. M. Pettersson**
363	Structure and reactivity of iron oxide surfaces **Sh. K. Shaikhutdinov, Y. Joseph, C. Kuhrs, W. Ranke** and **W. Weiss**
381	Atomistic simulation of oxide surfaces and their reactivity with water **S. C. Parker, N. H. de Leeuw** and **S. E. Redfern**
395	Theory of $PbTiO_3$, $BaTiO_3$, and $SrTiO_3$ surfaces **B. Meyer, J. Padilla** and **David Vanderbilt**
407	Electronic structure and surface reactivity of $La_{1-x}Sr_xCoO_3$ **Wendy R. Flavell, Andrew G. Thomas, Jane Hollingworth, Samantha Warren, Sarah C. Grice, Patricia M. Dunwoody, Caroline E. J. Mitchell, Peter G. D. Marr, David Teehan, Stuart Downes, Elaine A. Seddon, Vinod R. Dhanak, Kichizo Asai, Yoshihiko Koboyashi** and **Nobuyoshi Yamada**
421	QM investigations on perovskite-structured transition metal oxides: bulk, surfaces and interfaces **F. Corà** and **C. R. A. Catlow**
443	General Discussion
461	Concluding remarks **T. E. Madey**
467	List of Posters
471	List of Participants
475	Index of Contributors

■ Electronic supplementary information is available on http://www.rsc.org/esi
See article for further information.

Introductory Lecture
Oxide surfaces

Hans-Joachim Freund

Fritz-Haber-Institut der Max-Planck-Gesellschaft, Department of Chemical Physics, Faradayweg 4-6, D-14195 Berlin, Germany

Received 6th September 1999

Oxides have gained increasing interest in surface science during recent years because of their important role in applications. In the first part of the lecture we review the current knowledge on morphology and structure of surfaces of bulk single crystals as well as oxide films. The interaction of oxide surfaces with molecules is thoroughly discussed and the role of defects on adsorption is highlighted. In a further part, structure and morphology of deposited aggregates on clean and modified substrates are discussed. Such systems may serve as models for heterogeneous catalysts. Electronic structure as a function of the size of the deposited particle is studied, as well as size dependent adsorption properties and reactivities.

Introduction

The bulk properties of simple binary oxides are well understood and there are excellent reviews and text books available treating the various physical aspects.[1–5] In sharp contrast to the situation encountered for the bulk properties rather little is known about the surfaces of oxides, even the most simple ones. Only recently, if compared with the thirty years of surface science that have passed by,[6] researchers have started to study the surface science of oxides. There is a very useful book that marks a first milestone in this effort entitled "The surface science of oxides" by V. E. Henrich and P. A. Cox.[7] Since the publication of this book several reviews have appeared which have covered the field up to the present date.[8–15] It is understood that there are classes of oxides exhibiting external and internal surfaces, *i.e.* zeolites and meso-porous materials which are technologically very important. The present lecture will not discuss these even though some of the aspects which are dwelled upon here could be applied to those materials. We refer the reader to the paper of Thomas summarizing his Introductory Lecture of *Faraday Discussion* no. 105 where he discusses some aspects of this field as well.[16]

The present lecture has been organized as follows. In the first part we discuss several aspects of the geometric and electronic structure of clean oxide surfaces as determined by a variety of experimental methods. We show examples of surfaces terminating bulk single crystals as well as surfaces of epitaxial oxide films. This part is followed by examples attempting to illustrate some of the principles governing the interaction of molecules with oxide surfaces. A short comparison between the situation encountered on single crystalline surfaces with microcrystalline surfaces is included in order to demonstrate the influence of defects on the adsorption properties. The third part is dedicated to the interaction of metals with oxide surfaces and the study of deposited metal aggregates, including adsorption and reaction of molecules on such systems. Such composites represent model systems for heterogeneous catalysts and allow us to try to bridge the material's gap between single crystal metal surfaces and real catalysts.[8–15,17–19] When a physical chemist talks about catalysis the situation is similar to a mathematician trying to convince engineers that what he does is of use for them. G. H. Hardy's *A Mathematician's Apology*, 1940, contains many useful thoughts on this problem.[20] One is:

> It is one of the first duties of a professor, for example, in any subject, to exaggerate a little both the importance of his subject and his own importance in it. A man who is always asking 'Is what I do worth while?' and 'Am I the right person to do it?' will always be ineffective himself and a discouragement to others. He must shut his eyes a little and think a little more of his subject and himself than they deserve.

Structure and adsorption on clean oxide surfaces

The preparation of a clean oxide surface is a rather difficult task. Several strategies have been followed.[7,21,22]

The most straightforward strategy is UHV *in situ* cleavage, which, however, only leads to good results in certain cases, such as MgO, NiO, ZnO, SrTiO$_3$, *etc.*[13] Some very interesting materials such as Al$_2$O$_3$, SiO$_2$, TiO$_2$, *etc.* are hard to cleave.[7] A disadvantage with respect to experimental investigations of cleaved bulk single crystal insulators is their low conductivity. An alternative way of bulk single crystal surface preparation is *ex situ* cutting and polishing followed by an *in situ* treatment by sputtering and subsequent annealing in oxygen. Through such a process a sufficient number of defects is created in the near surface region and in the bulk to support conductivity of the material. This leads to a situation where electron spectroscopies as well as STM can be applied.[7]

Single crystalline oxide surfaces may also be prepared *via* the growth of thin oxide films on single crystal metal supports.[11,21,22] To such systems all surface science tools can be applied without further problems. If the oxide film is supposed to represent the bulk situation special care has to be taken in the control of film thickness. Also, if adsorption and reactivity studies are intended the continuity of the film has to be guaranteed. There are several examples in the literature where this has been achieved.[10,11,23]

Probably, the best studied clean oxide surfaces are the TiO$_2$(100) and TiO$_2$(110) surfaces.[7,21,24] A STM image of the clean (1 × 1) TiO$_2$(110) surface taken by Diebold and her group[25] is shown in Fig. 1. It is noteworthy that one of the first atomically resolved images of this surface was reported by Thornton and his group.[26,27] The inset shows a ball and stick model of the surface.

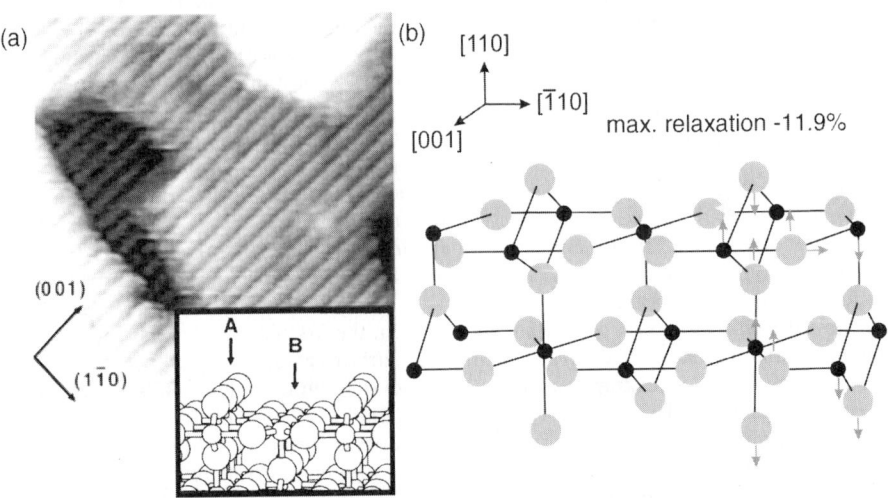

Fig. 1 Structure of the TiO$_2$(110) (1 × 1) surface as determined *via* STM (a, reproduced from ref. 25) and grazing incidence X-ray scattering (b, adapted from ref. 32).

There is now accumulating evidence from theoretical modeling of the tunneling conditions, but also from adsorbate studies using molecules which are assumed to bind to the exposed Ti-sites, that the bright rows represent Ti atoms. Iwasawa and his group[28–31] have successfully used formic acid in such a study, and showed in line with the theoretical predictions, and counter-intuitive with respect to topological arguments, that the Ti ions are imaged as bright lines and the oxygen rows as dark lines. Taking the resolvable interatomic distances within the surface layer the values correspond to the structure of the charge neutral truncation of the stoichiometric (110) surface.[32] Interatomic distances normal to the surface, however, are substantially different from the bulk values as is revealed by X-ray scattering experiments.[32] The top layer six-fold coordinated Ti atoms move outward and the five-fold-coordinated Ti atoms inward. This leads to a rumpling of 0.3 ± 0.1 Å. The rumpling repeats itself in the second layer down with an amplitude of about half of that in the top layer. Bond length variations range from 11.3% contraction to 9.3% expansion. These strong relaxations are not untypical for oxide surfaces and had been theoretically predicted for quite a while.[33]

The relaxations are particularly pronounced for the so-called charge-neutralized polar surfaces.[34–36] There are several experimental results,[37–40] basically corroborating the theoretical predictions although the quantitative agreement is not always good.[41–44] Specifically, the (0001) surfaces of corundum-type materials such as Al_2O_3,[41,42] Cr_2O_3[43] and Fe_2O_3[44] have been studied with X-ray diffraction, quantitative LEED as well as with STM and theoretical methods. Fig. 2 reminds the reader briefly of the fact that a polar surface (*e.g.* (111) orientation for a rock-salt structure) exhibits, if bulk terminated, a diverging surface potential due to the missing compensation of the interlayer dipole moments, as is nicely discussed in Noguera's book.[36] Consequently, polar surfaces reconstruct and/or relax substantially, while non-polar surfaces often exhibit much less pronounced relaxations although, as shown above for TiO_2 the degree of relaxation is substantial. Fig. 3 shows the results of structural determinations for the three related

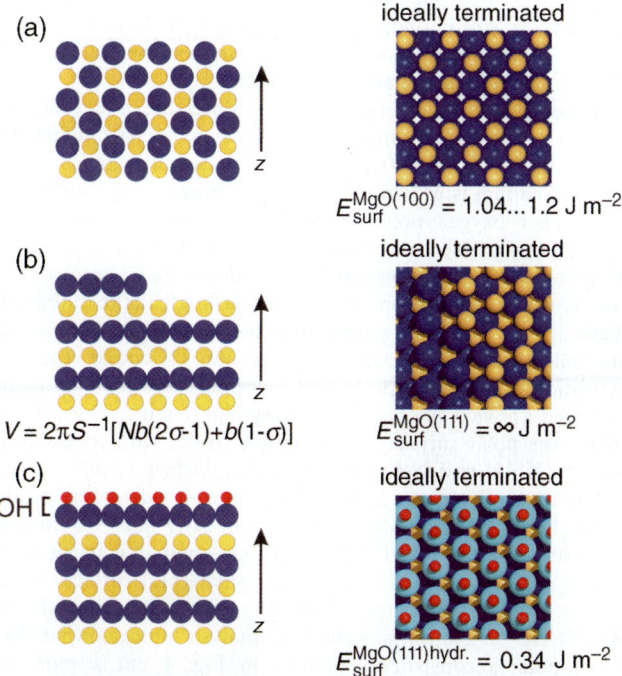

Fig. 2 Schematic representation (side and top views) of the structure of a non-polar (a, MeO(100), an unreconstructed polar (b, MeO(111)) and hydroxylated polar (c, MeO(111) adsorbate stabilized surface) surface of a rock-salt type crystal. The energies given refer to MgO.[177,178] (*V*, surface potential; *S*, area of surface unit cell; *N*, number of layers; *b*, interlayer spacing; σ, charge on surface layer relative to a layer in the bulk.)

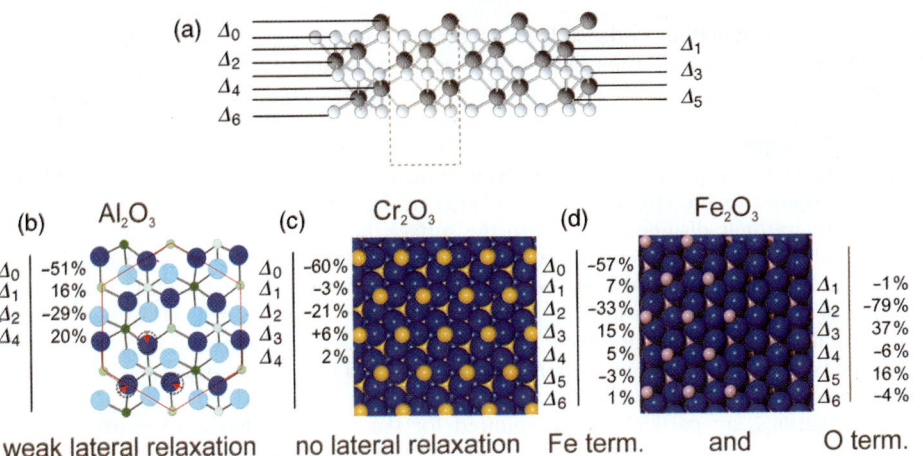

Fig. 3 Experimental data on the structure of corundum-type depolarized (0001) surfaces (side and top views). Adapted from (b) ref. 33; (c) ref. 39, and (d) ref. 44.

systems $Al_2O_3(0001)$, $Cr_2O_3(0001)$ and $Fe_2O_3(0001)$ as addressed above. In all cases a stable structure in UHV is the metal ion terminated surface retaining only half of the number of metal ions in the surface as compared to a full buckled layer of metal ions within the bulk. The interlayer distances are very strongly relaxed down to several layers below the surface. The perturbation of the structure due to the presence of the surface in oxides is considerably more pronounced than in metals, where the interlayer relaxations are typically of the order of a few percent.[45] The absence of the screening charge in a dielectric material such as an oxide contributes to this effect considerably. It has recently been pointed out[46] that oxide structures may not be as rigid as one might think judged on the relatively stiff phonon spectrum in the bulk. In fact, at the surface the phonon spectrum may become soft so that the geometric structure becomes rather flexible, and thus also very much dependent on the presence of adsorbed species.

Bulk oxide stoichiometries depend strongly on oxygen pressure, a fact that has been recognized for a long time.[47] So do oxide surfaces, structures and stoichiometries, a fact that has been shown again in a recent study on the $Fe_2O_3(0001)$ surface by the Scheffler and Schlögl groups.[44] In fact, if a Fe_2O_3 single crystalline film is grown in low oxygen pressure, the surface is metal terminated while growth under higher oxygen pressures leads to a complete oxygen termination.[44] This surface would be formally unstable on the basis of the electrostatic arguments presented above. However, calculations by the Scheffler group[44] have shown that a strong rearrangement of the electron distribution as well as relaxation between the layers leads to stabilization of the system. STM images by Weiss and co-workers[44] corroborate the coexistence of oxygen and iron terminated layers and thus indicate that stabilization must occur. Of course, there is need for further structural characterization. The idea of polar and non-polar surfaces only really holds in its simplest version as presented above if the material is very highly ionic. Thus, the most extreme cases to look at are perhaps the polar surfaces of the simple oxides with rock-salt structure[48] such as MgO and NiO, i.e. MgO(111) and NiO(111). Recently, Barbier et al.[49] have succeeded in preparing a single crystal NiO(111) surface, and to characterize it via grazing incidence X-ray diffraction (GIXD)! As was shown earlier for the case of thin NiO films of different crystallographic orientations, i.e. NiO(100)[50] and NiO(111),[51,52] a surface prepared in air or under residual gas pressure exhibits a p(1 × 1) structure while the clean polar (111) surfaces are reconstructed. The p(2 × 2) reconstruction originally reported for the thin film system has also been found for the bulk single crystal surfaces.[48,49] An initial structural analysis indicated that the actual structure is not the expected octopolar reconstruction shown in Fig. 4 but a more complicated one.[48] However, more recent investigations[53] of more carefully prepared bulk single crystal surfaces reveal that a stoichiometric surface actually reconstructs according to the octopolar scheme.[36,54] The small (100) terminated pyramids are oxygen terminated. Very recently, NiO(111) films grown on Au(111), which were initially studied by Neddermeyer and his group,[55] have been investigated

O termination
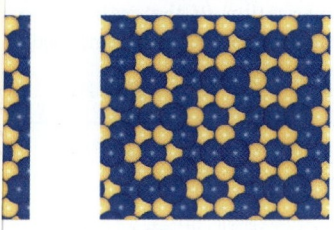

...lar reconstruction of a polar rock-salt (111) surface in oxygen

by GIXD.[56] The p(2 × 2) reconstruction was again corroborated, but the structural analysis undertaken up to now would seem to favor a structure where oxygen as well as Ni-terminated octopols, possibly arranged on adjacent terraces constitute the surface layer. Both, the bulk single crystal surfaces as well as the NiO(111) film surfaces grown on Au(111) exhibit high degrees of surface order. This is probably one reason why these surfaces do not quickly restructure upon exposure to water, while NiO(111) films grown on Ni(111) do reconstruct to form a hydroxy terminated NiO(111) surface.[51,52] A microscopic mechanism would involve massive material transport across the surface, which is the more unfavorable the better the order, and may therefore be kinetically hindered on well ordered single crystals. We would like to note at this point that the interaction of water with polar oxide surfaces is a topic of general interest in geochemical and environmental issues[57] as well as in catalysis. With respect to the latter, Papp *et al.* have found indications that NiO catalysts prepared with preferential (111) crystallographic orientation by topotactical dehydration of $Ni(OH)_2$ do show the highest activity towards $DeNO_x$-reactions after the last monolayer of H_2O has been desorbed.[58,59] Already in 1977, Derouane and co-workers[60] had theoretically analyzed on the basis of energetic considerations that real crystallites must be terminated partly by polar surfaces whose charge has been reduced *via* OH adsorption.

It is thus evident that the general study of the interaction of molecules with oxide surfaces represents an interesting and important field of study. In the following we are going to discuss several examples in order to bring out certain aspects of the bonding and interaction of molecules with oxide surfaces. This will be discussed in comparison with the adsorption of molecules on metal surfaces.

Before we proceed to a more detailed discussion of the binding of adsorbed molecules, a few remarks concerning the electronic structure of oxide surfaces are appropriate.

Early on, Hüfner and co-workers investigated the electronic structure of transition metal bulk samples and a great deal has been learned.[61,62] There have also been attempts to investigate the surfaces of these materials with respect to electronic structure. Qualitatively, it was expected that, due to the high ionicity of some compounds, there are pronounced surface effects. Photoelectron spectroscopy has been used to experimentally verify these expectations through the detection of chemically shifted core levels. However, the shifts are not large enough to be detectable due to the relatively complex satellite structure accompanying metal core ionization.[62] Eventually, this was also rationalized *via* more sophisticated quantitative calculations which showed that there are several compensating contributions rendering the surface fields only slightly different from the bulk. Applying techniques allowing for higher energy resolution have then clearly demonstrated surface effects. In the Merz group[63] and our laboratory[64] electron energy loss spectroscopy (EELS) in the regime of electronic excitations has been used to identify excitations in the surface layer. Fig. 5 shows a set of spectra taken on Ni(100) surfaces. The lowest trace has been taken on a clean single crystal. The broad features peaking at 4–5 eV correspond to charge-transfer excitations crossing the band gap of the insulating NiO bulk. In the gap there are narrow features due to excitations within the d-electron state manifold of the open shell Ni^{2+} ions. As the excitation energy within this manifold increases the number of states increases, so that near the charge-transfer band those states overlap and lead to the monotonous increase of intensity in this energy region. Most of the optically allowed transitions have been spectroscopically observed by transmission spectroscopy of bulk samples.[65–67] Fromme and Kisker have recently performed spin-polarized EELS measurements, which allow an assignment of the spin character of all states *via* the control of spin-polarization in the scattering conditions.[68,69] The assignment and a spin-polarization measurement have been superimposed on the spectra. The important point here is

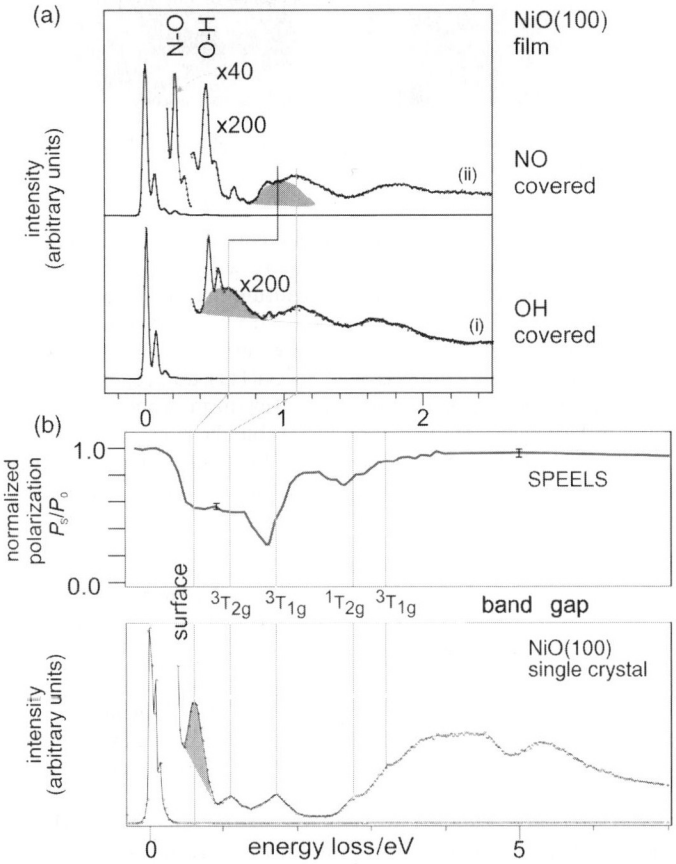

Fig. 5 Electron energy loss spectra of NiO(100) surfaces. (a) Adsorbate covered NiO(100) films, (i) defects OH saturated, before NO adsorption, (ii) after adsorption of NO. (b) Clean NiO(100) surfaces, UHV-cleaved single crystal. The assignment of the features according to theory is given and supported by spin-polarized measurements (adapted from ref. 68 and 69).

that there are additional spectroscopic features, most pronounced at 0.6 eV excitation energy which are not due to excitations in the bulk but rather in the surface layer. This can be experimentally demonstrated by an adsorption study.[64] Those excitations localized in the surface should be most strongly affected by adsorbed species. The experiment has been performed on a thin film sample because the surface has to be cooled to adsorb an appreciable amount of NO in this case. It is very obvious that the peak at 0.6 eV is influenced. In fact, it is shifted towards the position of a feature originating from an excitation in the bulk. In passing we note that the NiO(100) film has been treated with water before the experiments had been performed, in order to saturate the defects with hydroxy groups *via* dissociative adsorption of water. The vibrational losses caused by the hydroxys are clearly visible before NO adsorption took place. NO adsorption then induces yet a further vibrational loss at lower loss energy. What is the nature of the surface excitation and why is it different from the bulk excitation? Staemmler and his group have performed *ab initio* cluster calculations,[64,70] the result of which can be summarized as follows: Due to the localization of the Ni-d-electrons it is sufficient to consider a single Ni^{2+} ion within its octahedral coordination sphere if we consider the situation encountered in the bulk (see Fig. 6). The ground state is a ^3E state with two unpaired d-electrons in the e_g orbitals of the ligand field split set of d-orbitals. The first excited state results from an excitation from the completely filled t_{2g} subset into the partly filled e_g subset giving rise to an excitation located near 1 eV. There are many more higher excited states, some of which are assigned according to the work of Fromme *et al.*[68,69] In the surface, however, one of the coordinated oxygen ions is missing and the symmetry of the local Ni^{2+} site is reduced to C_{4v}. Consequently, the degenerate e_g subset is split. The t_{2g} subset is also

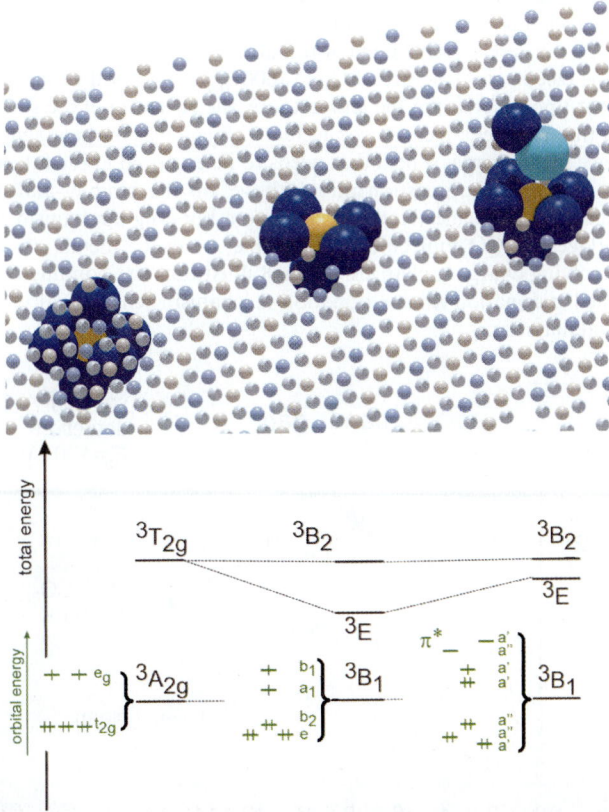

Fig. 6 Correlation of structural data with electronically excited states on NiO(100). Upper panel: (left) coordination of a Ni ion in the bulk, (middle) coordination of a Ni ion in the clean (100) surface as well as (right) in the case of adsorbed NO. Lower panel: orbital diagram and total energies from cluster calculations.[64,70]

split, but this effect is not so important. The d-orbital of the former e_g subset pointing along the Ni–O axis has lost part of its destabilizing interaction and, consequently, its energy decreases. The calculation shows that still both orbitals in the former e_g subset are singly occupied after reduction of symmetry. Therefore, the first excited state in a Ni^{2+} surface ion is at lower energy than in the bulk, as also revealed by the experiment. If an NO molecule is now coordinated to the Ni^{2+} surface ion, the energetic position of the orbital is raised again (similar to the presence of the sixth oxygen ion), effectively moving the excitation energy back close to the bulk position. It is clear from these results that the surface effect on the excitation energies is of the order of 0.4 eV, and thus the above mentioned lack of evidence from other techniques can be rationalized.

So far we have discussed the localized metal-ion states. Are there also surface modifications onto the charge-transfer states? The answer is yes! Unfortunately, NiO is not a good example to support this experimentally. The electronic excitations of the $Cr_2O_3(0001)$ surface, on the other hand, show the effect very clearly.[71–74] Fig. 7 shows the EELS spectrum[72–74] of this surface at low temperature. Again, a thin film has been used. The sharp features in the band gap are excitations within the manifold of d-orbitals. A detailed discussion[74] has shown that the excitations are characteristic of surface Cr ions with three d-electrons. However, the Cr ions do not carry a net charge of 3+ as expected (and found for the bulk Cr ions) but rather of 2+ charge due to strong hybridization with the neighboring oxygen atoms. When we now perform EELS measurements after adsorption of CO_2, not only the d-excitations are influenced but also the very intense feature near 3.8 eV. Again, on the basis of cluster calculations performed in the Staemmler group,[71] it has been possible to assign this intense feature to a surface charge-transfer excitation at the band gap, which is shifted to lower energy as compared with the corresponding excitation in the bulk (see Fig. 7). The decrease in energy is reasonable because the Cr d-orbitals have been lowered, the more open surface structure (see Fig. 3), and the coordination of the Cr ions to only three oxygen ions allows for better charge separation than in the bulk. In NiO the effect is less pronounced because the surface Ni ions are still five-fold coordinated.

The analysis of the electronic structure of $Cr_2O_3(0001)$, as discussed so far, has been performed at 90 K. We note that upon increasing the temperature, the structure of the surface changes,[74] and we have speculated that these changes are connected with changes in the magnetic structure of the surface. Oxide surface magnetism is a field that needs to be explored in the future.

At this point we return to the question raised above, on the binding and interaction of molecules with oxide surfaces. Molecules bind to oxides *via* a bonding mechanism considerably different from metal surfaces. A CO molecule, for example, binds to metals *via* chemical bonds of varying strength involving charge exchanges.[75] Fig. 8 illustrates the bonding of CO to a Ni-metal atom *via* the so-called σ-donation/π-back-donation mechanism schematically, and on the basis of a one electron orbital diagram.

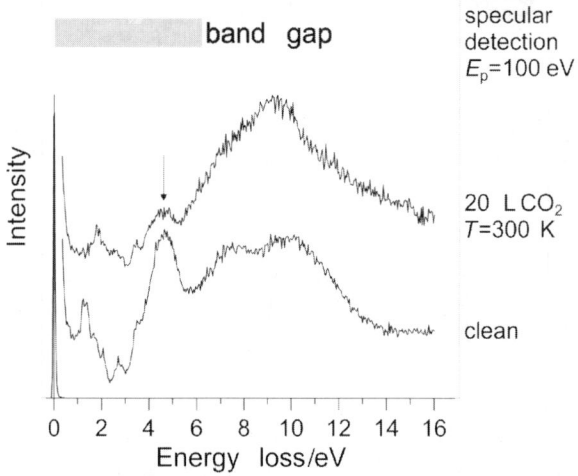

Fig. 7 EELS spectra of the clean and adsorbate covered $Cr_2O_3(0001)$ surface.[72–74]

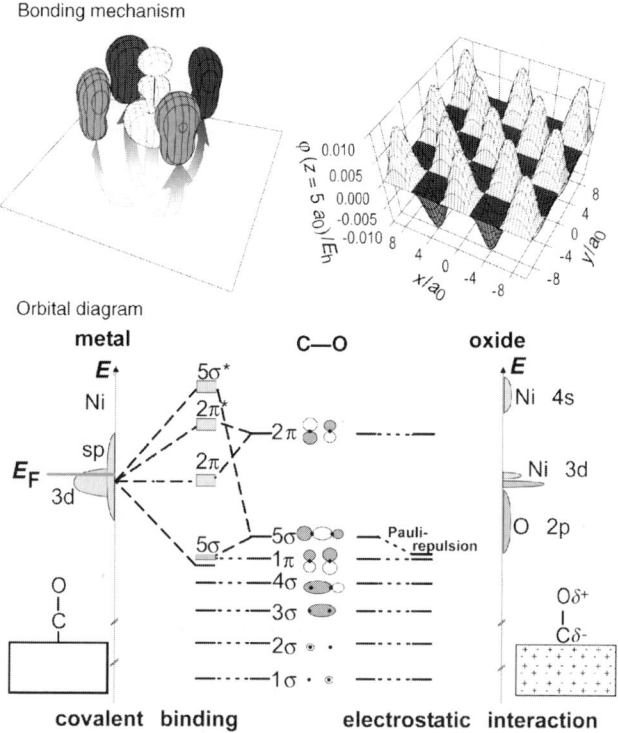

Fig. 8 Orbital diagram for the bonding of CO to Ni-metal (left) and to Ni-oxide (right).

The σ- and π-interactions lead to a relative shift of those σ- and π-orbitals involved in the bond with respect to those orbitals not involved. The diagram reflects this *via* the correlation lines. This may be contrasted by the electrostatically dominated interaction between a CO molecule and a Ni ion in nickel oxide.[70,76] There is a noticeable σ-repulsion between the CO carbon lone pair and the oxide leading to a similar shift of the CO 5σ-orbital as in the case of the metal atom. However, there is no or little π-back-donation so that the CO π-orbitals are not modified.[11,77] Conceptually, the situation is transparent and one would expect that a detailed calculation reveals the differences quantitatively. However, as it turns out the description by *ab initio* calculations is very much involved and today a full account cannot be given.[78] Theoretically (Table 1)[70,78–89] the prediction is that CO as well as NO bind very weakly to NiO.[78] The predicted binding energy of CO is of the order of 0.1 eV and it is expected to be similar to CO binding to MgO(100), *i.e.* the influence of the Ni d-electrons should be negligible.[78]

To shed light on this problem it was necessary to perform thermal desorption measurements on cleaved single crystal surfaces, being the surfaces with the least number of defects.[90] In Figs. 9 and 10 TDS data for CO and NO on vacuum-cleaved NiO(100) are compared with data for thin NiO(100) films grown by oxidation of Ni(100). At temperatures of 30 and 56 K multilayer desorption for CO and NO, respectively, shows up. The pronounced features at higher temperatures correspond to desorption of the respective adsorbate at (sub)monolayer coverage. In the case of the CO adsorbate at 34 K desorption of the second layer is found and the states at 45 and 145 K for CO and NO, respectively, are due to adsorption on defects as concluded from data obtained from ion bombarded surfaces (not shown here). It is obvious that for both adsorbates the thin film data and the data of the cleaved samples agree well, in particular for NiO(100) the thin film data are comparable to those from the more perfect surfaces of the cleaved samples. The higher defect density of the thin film surfaces leads to small, but clearly visible additional peaks in the TDS data which show up as shoulders near to the main peak, for example in the NO spectra. Nevertheless, the general shapes of the thin film spectra of both adsorbates are very similar to those of the cleaved samples.

Table 1 Table of literature data for adsorption of CO and NO on NiO(100) and MgO(100)[a]

Author	System	Method	Adsorption energy eV
Pacchioni and Bagus[79]	CO/NiO(100)	Ab initio cluster calculation	0.24
Klüner and Freund[80]	NO/NiO(100)	Ab initio cluster calculation, BSSE correction	≈0
Pöhlchen and Staemmler[70]	CO/NiO(100)	Ab initio cluster calculation, BSSE correction	0.03 to 0.1
Cappus et al.[81]	CO/NiO(100)/Ni(100)	TDS, Redhead	0.32
Vesecky et al.[82]	CO/NiO(100)/Ni(100)	IRS, Clausius–Clapeyron	0.45
Staemmler[83]	NO/NiO(100)	Ab initio cluster calculation, BSSE correction	0.1
Pöhlchen[84]	NO/NiO(100)	Ab initio cluster calculation, BSSE correction	<0.23
Kuhlenbeck et al.[85]	NO/NiO(100)/Ni(100) and NO/NiO(100)	TDS, Redhead	0.52
Nygren and Pettersson[78]	CO/MgO(100)	Ab initio cluster calculation, BSSE correction	0.08
Chen et al.[86]	CO/MgO(100)	DFT	0.28
Neyman et al.[87]	CO/MgO(100)	DFT, BSSE correction	0.11
He et al.[88]	CO/MgO(100)/Mo(100)	IRS, Clausius–Clapeyron, TDS, Redhead	0.43 0.46
Furuyama et al.[89]	CO/MgO powder	IRS, Clausius–Clapeyron	0.15 to 0.17

[a] BSSE, basis set superposition error; TDS, thermal desorption spectroscopy; DFT, density functional theory; IRS, infrared spectroscopy; Clausius–Clapeyron, evaluation of pressure and temperature dependent IR intensities with the Clausius–Clapeyron equation; Redhead, evaluation of TDS data with the Redhead equation.[179]

For CO the shift of the peak maximum with increasing coverage indicates that at higher coverage repulsive lateral interaction comes into play which may lead to occupation of energetically less favorable sites. This is not the case for the NO adsorbate which may be attributed to smaller lateral interactions and to the higher adsorption energy which makes adsorption more site specific, and thus may inhibit compression of the layer involving site changes.

TDS data for NO and CO on vacuum-cleaved MgO(100), for comparison, are plotted in Figs. 11 and 12. Multilayer desorption is found at 29 and 56 K for CO and NO, respectively. The small features around 45 and 100 K are likely due to defect adsorption since they saturate at rather low coverage. Desorption from layers with small coverage is found at 57 and 84 K for CO and NO, respectively. The data for CO and NO on NiO(100) have been evaluated using the leading edge method as well as a complete analysis. Details of the procedures may be found in ref. 91 and 92. Both methods determine the heat of adsorption as a function of the coverage of molecules already on the surface. The results of the evaluation are shown in Figs. 13 and 14. Both graphs exhibit a trend which is generally to be expected for laterally interacting adsorbate layers: the adsorption energy decreases with increasing coverage. At coverages near to 1 monolayer the energies converge towards the multilayer values (0.09 and 0.18 eV for CO and NO, respectively[93]). At low coverage the lateral interactions are most likely small so that the corresponding adsorption energies may be compared with theoretical results since in the calculations lateral interactions have not been considered. As indicated in Figs. 13 and 14, the low coverage adsorption energies are 0.30 and 0.57 eV for CO and NO, respectively.

The low coverage adsorption energies for CO and NO on NiO(100) and MgO(100) are compiled in Table 2. According to theory the interaction of the adsorbates with MgO(100) and NiO(100) are expected to be similar since the bonding should be mainly electrostatic in nature[78] (the electric fields at the surfaces of NiO(100) and MgO(100) are similar). However, according to Table 2 the bonding energies are considerably different, with the higher values being obtained for NiO(100). Covalent interactions involving the Ni 3d-electrons may play a role for the adsorbate–substrate interaction which does not show up in the calculations published so far.

As far as it concerns the basis set superposition error (BSSE) corrected calculations listed in Table 1 which are expected to yield qualitatively better results as compared to the non-corrected

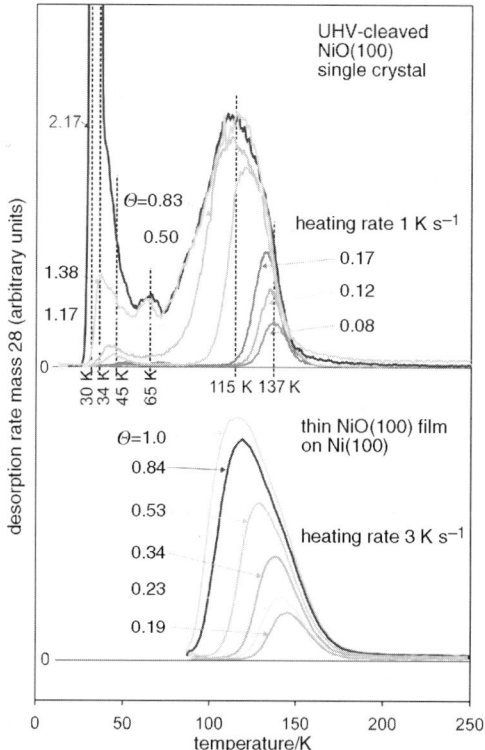

Fig. 9 Thermal desorption spectra of CO on NiO(100) cleaved in vacuum (upper part) and CO on a thin NiO(100) film grown by oxidation of Ni(100) (lower part). The mass spectrometer was set to mass 28 (CO). CO doses are given relative to the dose needed to prepare a monolayer.

calculations, it appears that the theoretical results for adsorption on MgO(100) are in general in line with our experimental results, whereas a similarly favorable comparison can not be made for NiO(100). It appears necessary to re-investigate the role of the Ni 3d-electrons in future theoretical studies.

The adsorption of CO on MgO has been thoroughly investigated by Heidberg and his group using IR-spectroscopy[94] and by Weiss and co-workers using helium scattering spectroscopy.[95] They have clearly demonstrated that CO develops ordered phases on the cleavage planes and that order and spectroscopic properties depend on the quality of the prepared surfaces. From their experiments the influence of the presence of surface defects on adsorption properties is very obvious but a quantitative evaluation based on the number and the nature of the defects has not been reported.

The quantitative evaluation of defects is a well defined but hard to tackle problem for future studies that has to be taken on by our research community. Water adsorption is an example that lends itself to a study of the influence of defects, because at lower coverage the (100) cleavage planes of MgO and NiO do not dissociate water, while the presence of defects does induce water

Table 2 Compilation of low-coverage bonding energies for NO and CO on NiO(100) and MgO(100) obtained in this work

	NiO(100)	MgO(100)
CO	0.30 eV	0.14 eV
NO	0.57 eV	0.22 eV

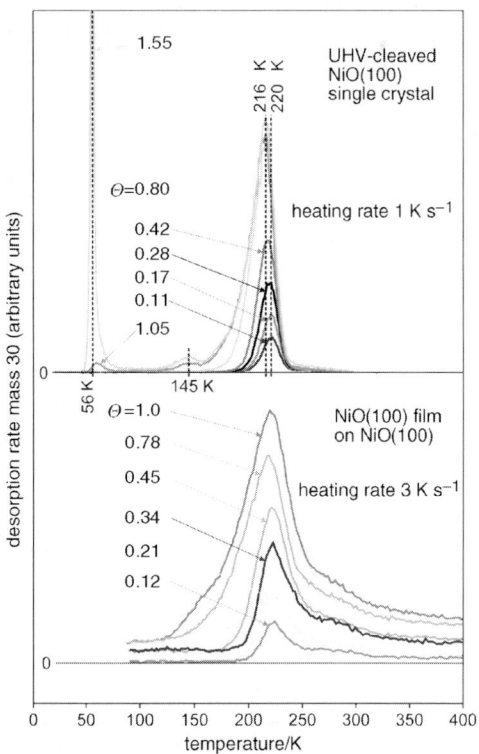

Fig. 10 Thermal desorption spectra of NO on NiO(100) cleaved in vacuum (upper part) and NO on a thin NiO(100) film grown by oxidation of Ni(100) (lower part). The mass spectrometer was set to mass 30 (NO). NO doses are given relative to the dose needed to prepare a monolayer.

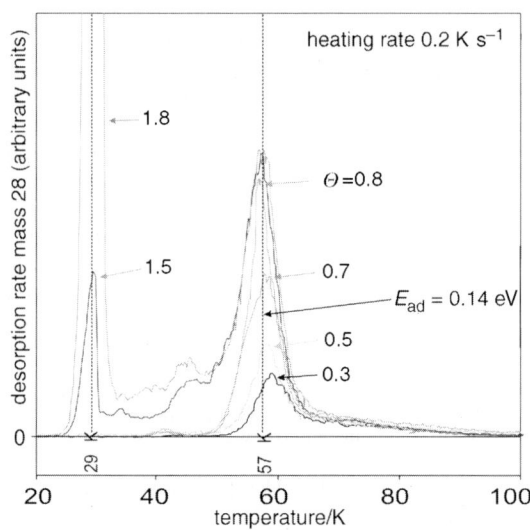

Fig. 11 Thermal desorption spectra of CO on MgO(100) cleaved in UHV. The mass spectrometer was set to mass 28 (CO). CO doses are given relative to the dose needed for the preparation of a monolayer.

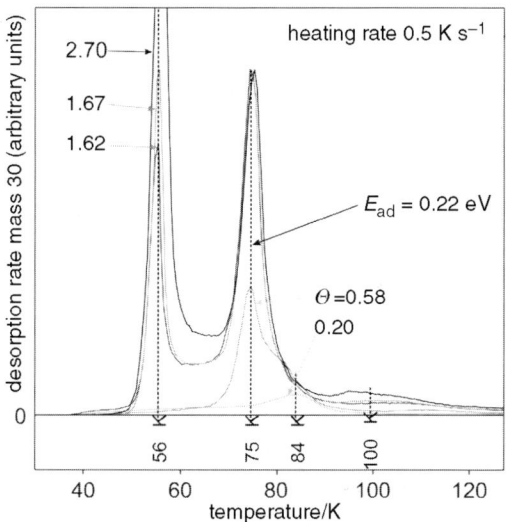

Fig. 12 Thermal desorption spectra of NO on MgO(100) cleaved in UHV. The mass spectrometer was set to mass 30 (NO). NO doses are given relative to the dose needed for the preparation of a monolayer.

dissociation. This can be seen in TDS spectra of H_2O from (100) rock-salt type surfaces. Fig. 15 shows results for H_2O desorption from MgO(100) and NiO(100).[96] The most pronounced features in the spectra are due to condensed water layers at lowest desorption temperature and the conversion of a compact layer (with c(4 × 2) periodicity in the case of MgO) to the monolayer which desorbs at 225 K (240 K for MgO(100).[97,98] The difference in desorption temperature between

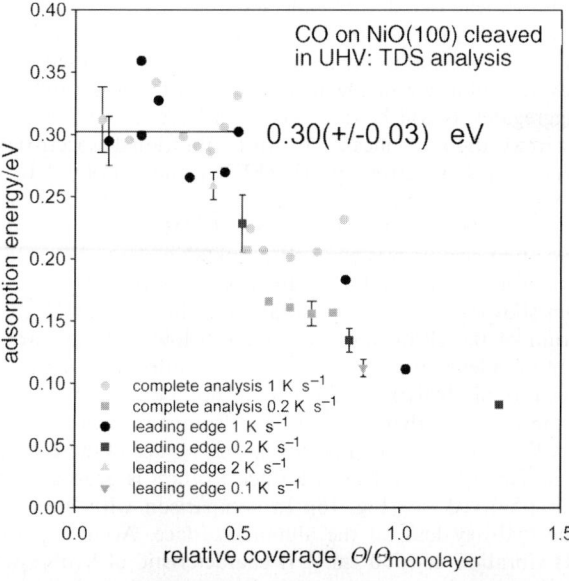

Fig. 13 Adsorption energy of CO on NiO(100) cleaved in vacuum as a function of coverage. The data have been determined from TDS spectra like those shown in Fig. 9 (upper part) using the leading edge method and complete analysis. TDS data taken with heating rates of 0.1, 0.2, 1 and 2 K s^{-1} have been used.

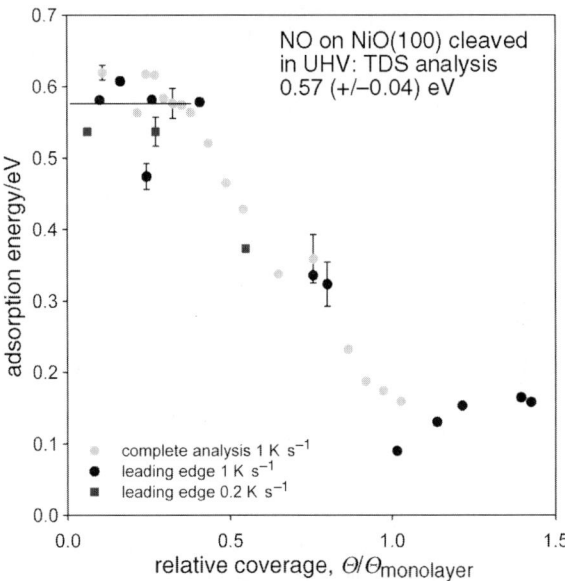

Fig. 14 Adsorption energy of NO on NiO(100) cleaved in vacuum as a function of coverage. The data have been determined from TDS spectra like those shown in Fig. 10 (upper part) using the leading edge method and complete analysis. TDS data taken with heating rates of 0.2 and 1 K s^{-1} have been used.

MgO(100) and NiO(100) seems to be characteristic for the H$_2$O substrate interaction. Most of that information is lost when we create defects *via* sputtering. Thermal desorption is now observed up to relatively high temperatures and the features are broad. Which kind of defects and how many have been created is not yet known. A combination of various techniques to characterize the defects by probe molecule adsorption together with infrared spectroscopy, EPR and electron spectroscopies may in the future lead to a deeper understanding.

Dissociative adsorption of water on oxide surfaces can also be used in a preparative way, namely to chemically modify the surface by hydroxylation. We have used this technique for a thin alumina film to study the influence of the presence of hydroxy groups on the nucleation and growth of metallic aggregates as will be discussed later.[99] At this point we show in Fig. 16 the result of such a hydroxylation as measured with vibrational spectroscopies, such as high resolution electron energy loss spectroscopy (HREELS) and FTIR.[100] In the case of the thin alumina film on NiAl(110) it was impossible to hydroxylate the oxide just by water dissociation, while on a similar film on NiAl(100)[101] formation of OH from dissociative H$_2$O adsorption occurs. The clean oxide film surface was exposed to metallic aluminium and then the aluminium was hydrolyzed *via* water adsorption to form a hydroxy overlayer.[99,100] In Fig. 16 at the bottom an HREELS spectrum showing the hydroxy vibration at 465 meV (3750 cm^{-1}) is plotted atop a corresponding spectrum of the clean film. The peaks below 120 meV are due to the alumina phonons,[102] which are broadened through hydroxylation influencing surface order. The observed hydroxy loss coincides very nicely with the FTIR absorption observed for the same system. In this case more water was adsorbed so that a broad band from water clusters is seen also. The sharp extra band at 3705 cm^{-1} is due to free OH groups at the surface of these water clusters,[103] as they are known from the surface of ice. In fact, if a thick ice film is grown on the alumina film this particular vibration is observed (see Fig. 16). In comparison with literature data[104] it is now possible to assign the hydroxy loss on the alumina surface. According to a review article by Knözinger[104] an OH-vibration at 3750 cm^{-1} is characteristic of hydroxys bridging aluminium ions both in octahedral, or one in an octahedral and one in a tetrahedral site. We mention that on alumina films grown on a different NiAl substrate[101] other types of OH species may be formed as was shown by Hemminger's group. Therefore, it is conceivable that the influence of the nature of

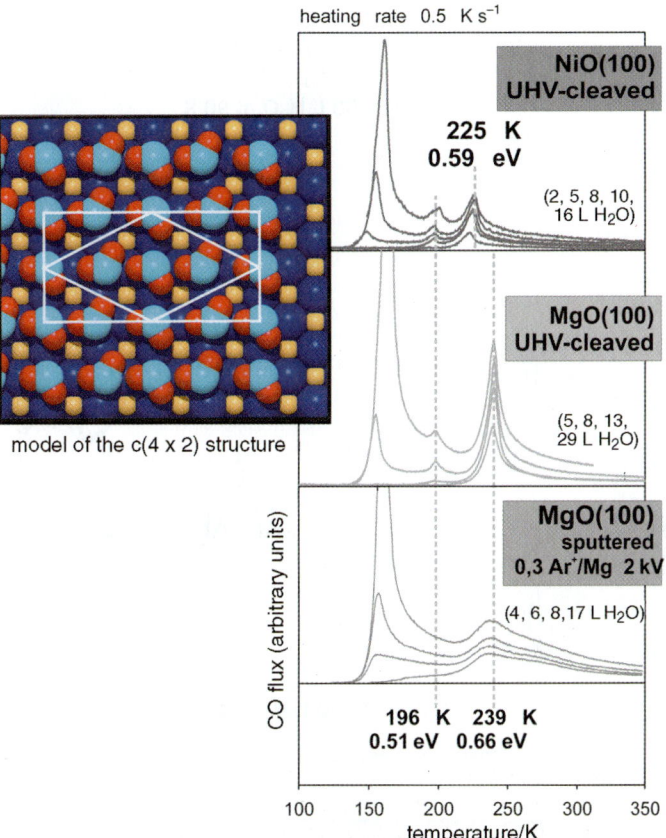

Fig. 15 Thermal desorption spectra of H_2O on UHV-cleaved MgO(100) and NiO(100). A schematic representation of the c (4 × 2) structure is included (reproduced from Heidberg, Redlich and Wetter, *Ber. Bunsenges. Phys. Chem.*, 1995, **99**, 1333). For comparison a thermal desorption spectrum from MgO(100) after creation of defects *via* sputtering is shown.

the hydroxy species modifying the surface on the interaction with additional adsorbates, *i.e.* metal deposits, could be investigated.

Before we move on to discuss the properties of metals on oxides we would like to briefly discuss the adsorption of CO_2 on oxides as an example of a molecular adsorbate system with more degrees of freedom.

TDS spectra indicate[105,106] that there are more weakly and less weakly bound CO_2 species on a Cr_2O_3(0001) surface. We have studied the nature of those species by various techniques including infrared spectroscopy. Fig. 17 shows several sets of IR spectra. The pair of sharp bands around 2300 cm^{-1} can easily be assigned to the more weakly bound CO_2 with only a slightly distorted structure as compared with the gas phase species. By a combination of isotopically labeling the adsorbed CO_2 (shift of frequencies) as well as the oxide layer (no shift of CO_2 bands) we have demonstrated that the single band centered around 1400 cm^{-1} is due to the presence of a carboxylate species, *i.e.* a bent anionic CO_2 species, and not, as perhaps expected, to a carbonate.[72] The bands between 1610 and 1700 cm^{-1} are missing because of the applicability of surface selection rules in thin film systems. This means, all non-totally symmetric bands are suppressed in intensity. A quick comparison with CO_2 adsorption on chromia microcrystalline material as shown in Fig. 17 indicates the presence of the bands between 1610 and 1700 cm^{-1} as expected for adsorption on a bulk dielectric material. It is remarkable how similar the thin film data are in comparison with the microcrystalline material. This has been discussed in detail by Zecchina's group.[107] Also, the

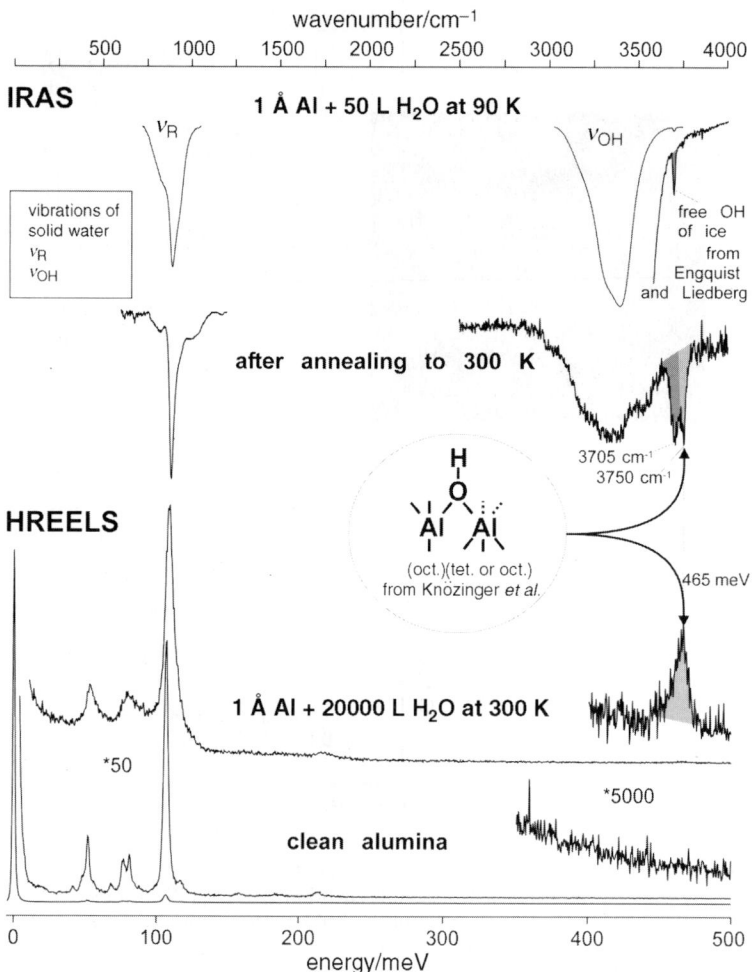

Fig. 16 Fourier transform IR spectra (IRAS) and electron energy loss spectra (HREELS) of a clean and OH(+H$_2$O)-covered alumina film.

response of the two systems with respect to preadsorption of oxygen is very similar. In fact, as shown in Fig. 17, CO$_2$ adsorption in the form of the less weakly bound CO$_2^-$ is fully suppressed on the thin film system and very strongly attenuated for the microcrystalline system. This indicates that CO$_2$ occupies the chromium sites, because we know that oxygen from the gas phase adsorbs on the chromium ions. As we remarked above, ELS[73] and XPS[108] spectra of the Cr$_2$O$_3$(0001) surface have been used to deduce that the Cr-ions in the surface are in a low oxidation state, i.e. Cr^{2+} as opposed to chromium ions in the near surface and bulk regions. It is therefore not surprising that such a surface provides electrons to adsorbed molecules, leading to electron transfer as, for example, documented by the formation of O$_2^-$ and CO$_2^-$. The low valence state of the Cr surface ions also has consequences in other reactions, such as the polymerization of ethene which has been studied on Cr$_2$O$_3$(0001),[109] and in connection with other, more realistic model studies.[110]

A field that has not been investigated in any detail as far as well characterized single crystal oxide surfaces are concerned is connected with photoinduced chemical reactions on larger molecules. Photoinduced desorption of CO and NO from oxides has been studied extensively[111–114] but the reactivity of larger molecules has not. Yates and his group have reported such studies on

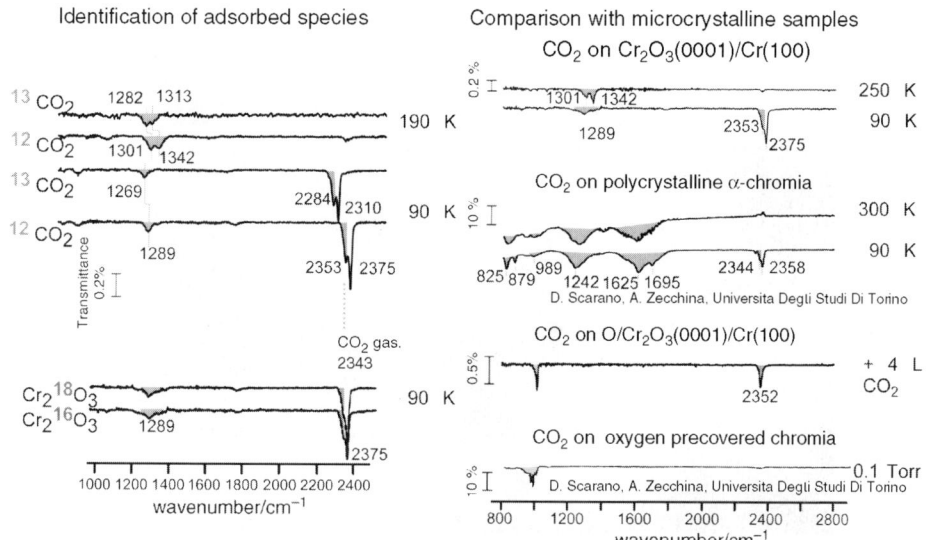

Fig. 17 IRAS spectra of CO_2 adsorbed on $Cr_2O_3(0001)$ surfaces and on polycrystalline chromia. Left panel: IRAS spectra at different surface temperatures and with isotopically labelled CO_2 as well as Cr_2O_3. Right panel: adsorption of CO_2 after pre-adsorption of oxygen.

powder samples, *i.e.* Rh complexes deposited on Al_2O_3 powder and very interesting results concerning C–H bond activation have been reported.[115] We refer to the literature for details,[115] and note that this should be considered as a new promising area in connection with single crystalline systems.

Metals on oxides

So far, we have considered the clean oxide surface and its reactivity. In the following we will modify the oxide surface by deposition of metal onto the surface. This represents a route towards the preparation and characterization of more complex model systems in heterogeneous catalysis in order to bridge the so-called materials gap.

Over past years several strategies have been followed along this route. Very early on small metal particles have been put onto oxide bulk single crystal surfaces, particularly MgO, and characterized by transmission electron microscopy (TEM). Poppa has been the pioneer in this field,[18] and the very important contributions to the field have been recently reviewed by Henry, who himself was involved in the early TEM measurements.[14] While these efforts where mainly aimed at preparing small well defined particles, another strategy has been followed by Møller and his group[116–119] as well as Madey and co-workers[19] by trying to prepare thin metal films on bulk oxide single crystals, such as $TiO_2(110)$ surfaces. As mentioned above the advent of scanning tunneling microscopy has had a substantial influence on the understanding of the structure of clean oxide surfaces. Several groups[120–122] have started to investigate metal deposition on TiO_2 surfaces. Interesting initial results concerning metal particle migration, and oxide migration onto the metal particles (the so-called SMSI effect) have been obtained.[121,122] Particularly well suited for the application of scanning tunneling microscopy are metal particles deposited onto thin film oxide surfaces.[8,10,11,14] Goodman's group, for example, made major contributions to this field early on.[10] In Fig. 18 we show the result of a STM study from our laboratory. The left panel shows the clean alumina surface as imaged by a scanning tunneling microscope.[123] The surface is well ordered and there are several kinds of defects on the surface: Firstly, the reflection domain boundaries between the two growth domains of $Al_2O_3(0001)$ on the NiAl(110) surface, the substrate on which the film is grown *via* a well established oxidation recipe.[102] Secondly, there are anti-phase domain boundaries within the reflection domains, and in addition, there are point defects which are not resolved in the images. The image does not change dramatically after

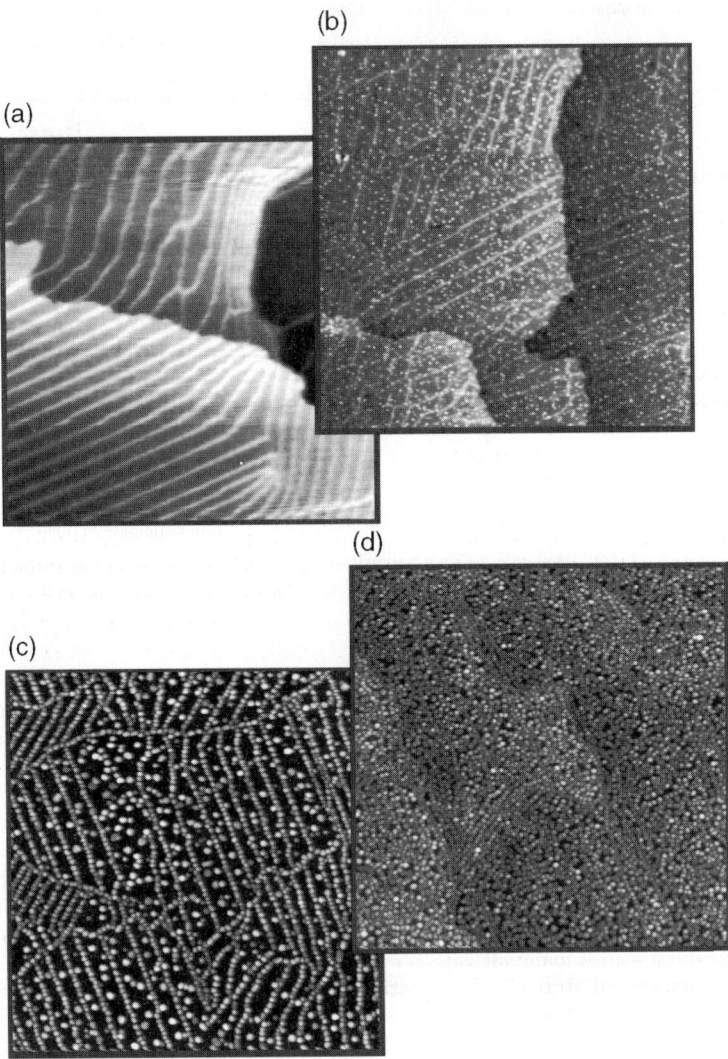

Fig. 18 Scanning tunneling images (3000 × 3000 Å^2, Al$_2$O$_3$/NiAl(110), u_{tip} = 8 V, I = 0.8 nA). (a) Clean alumina film, (b) after deposition of 0.1 Å of Rh at 90 K, (c) after deposition of 2 Å of Rh at 300 K, and (d) after deposition of 2 Å of Rh at 300 K on hydroxylated substrate onto the pre-hydroxylated alumina film.

hydroxylating the film, a procedure we had mentioned above.[99] The additional panels show STM images of rhodium deposits on the clean surface at low temperature, and at room temperature,[15,124] as well as an image after deposition of Rh at room temperature on a hydroxylated substrate.[125] Please note, that the amount deposited onto the hydroxylated surface is equivalent to the amount deposited onto the clean alumina surface at room temperature. Upon deposition of Rh from the metal vapor onto the clean surface at low temperature small particles nucleate on the point defects of the substrate and a narrow distribution of sizes of particles is formed. If the deposition of Rh is carried out at room temperature the mobility of Rh atoms is considerably higher compared with low temperature so that nucleation at the pronounced line defects of the substrate becomes dominant. Consequently, all the material nucleates on the reflection domain and anti-phase domain boundaries. The particles have a relatively uniform size given by the amount of deposited material. If the same amount of material is deposited onto a hydroxylated surface the particles are considerably smaller and distributed across the entire surface showing that hydroxylation leads to higher metal dispersion.[15,99] The thermal behavior of the deposits is important with respect to studies of chemical reactivity because the ensemble of particles may undergo morphological changes adopting their equilibrium shape which could be different with and without the presence of a reactive gas phase. In the present case detailed studies have been undertaken on the particles deposited onto the clean substrate and less detailed studies for the deposit on the hydroxylated surface. As a result of these studies it is known that the morphology of the ensemble is not altered within a temperature window from 90 to approximately 450 K. The window is extended to even higher temperatures on the hydroxylated substrate. Above the upper temperature limit the particles tend to agglomerate and also start to diffuse through the film into the metal substrate underneath.[15]

Studying this agglomeration process is an interesting subject in itself and research in this direction is only starting.[15] A more basic aspect, of course, would be a study of metal atom diffusion on oxide substrates. The obvious method to perform such a study is the STM.[126] However, in contrast to diffusion studies on metal surfaces, similar studies on oxide surfaces have not been reported. On the other hand, field ion microscopy studies on metal atom diffusion on oxide films are under way and a first estimate of activation energies for diffusion has been reported.[127] It is obvious that the area of diffusion studies will considerably profit from atomic resolution, once it is obtained routinely for deposited aggregates on oxide surfaces. While for TiO_2 and very few other oxide substrates atomic resolution may be obtained routinely, there are very few studies on deposited metal particles where atomic resolution has been reported.[128] The first report for an atomically resolved image of a Pd metal cluster on MoS_2 was reported by Henry and his group.[128] A joint effort between Besenbacher and our group[129] has led to atomically resolved images of Pd aggregates deposited on a thin alumina film. Fig. 19 shows such an image of an aggregate of about 80 Å in width. The particle is obviously crystalline and exposes on its top facet, the (111) Pd surface. Also, the (111) facets on the side, typical for a cuboctahedral particle, can be discerned. The small (100) facets predicted *via* equilibrium shape considerations on the basis of the Wulff-construction could not be atomically resolved. If we, however, apply the concept of the Wulff-construction, we may deduce the metal surface interaction energy.[129] The basic equation is

$$W_{adh} = \gamma_{oxide} + \gamma_{metal} - \gamma_{interface} \quad (1)$$

Provided the surface energies (γ_{metal}) of the various crystallographic planes of the metal are known,[130] a relative work of adhesion (W_{adh}) may be defined.[129] We find 2.9 ± 0.2 J m^{-2} which is still rather different with respect to recent calculations by Jennison *et al.*[131] where metal adsorption energies (1.05 J m^{-2}) have been calculated on a defect free thin alumina film. It is not unlikely that this discrepancy is connected with the rather complicated nucleation and growth behavior of the aggregates involving defects in the substrate.

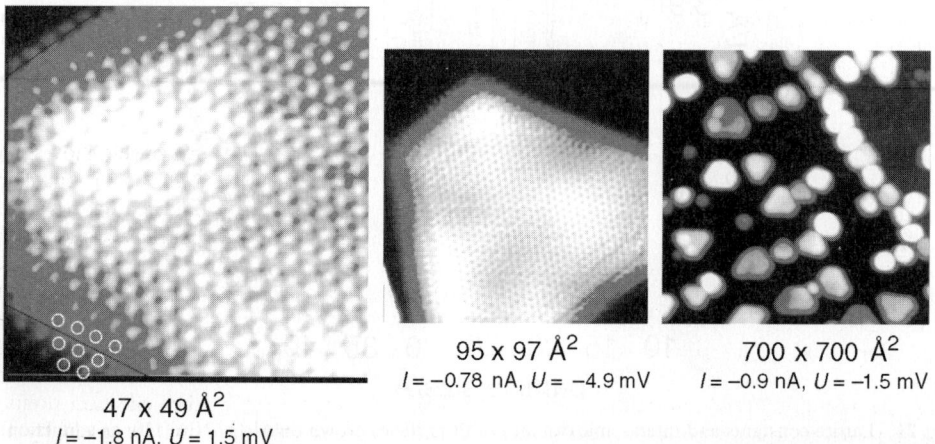

47 × 49 Å²
I = −1.8 nA, *U* = 1.5 mV

95 × 97 Å²
I = −0.78 nA, *U* = −4.9 mV

700 × 700 Å²
I = −0.9 nA, *U* = −1.5 mV

Fig. 19 Scanning tunneling images at atomic resolution of Pd aggregates grown on an alumina film.[129]

While STM reveals the surface structure of deposited particles, their internal structure, in particular as a function of size, is not easily accessible through STM. In this connection TEM studies on the same model systems can be of help.[132] Fig. 20 shows a schematic drawing of a sample. After growing the film and deposition of the particles, the sample is ion-milled from the back so that a small hole is finally formed. In this way, a wedge is obtained which is thin enough for the imaging process. A positive side effect of this procedure is the fact that also the unsupported film next to the edge can be studied.[133] This opens the opportunity to judge whether the metal substrate has any structural effect on the deposits. On the basis of numerous high resolution TEM images and a subsequent analysis of the Moiré periodicities, it has been possible to calculate the lattice constants as a function of particle size.[132] The corresponding plot is depicted in Fig. 21 and indeed proves that the atomic distances continuously decrease to 90% of the bulk value at a cluster size of 10 Å. On the other hand, the lattice constant approaches the Pt bulk value already at a diameter of 30 Å. This effect has also been detected for Ta and Pd clusters on the thin alumina film, but it seems to be less pronounced in these cases.[134–136]

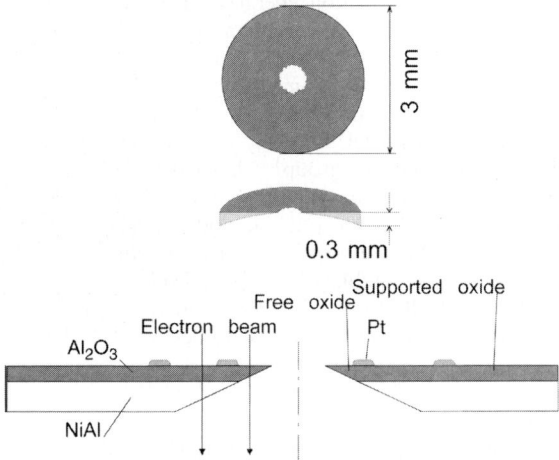

Fig. 20 Schematic drawing of a sample prepared for transmission electron microscopy (sample milling technique).

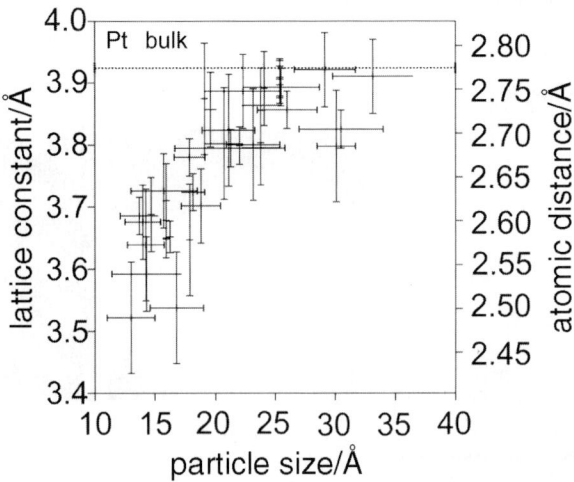

Fig. 21 Lattice constants and interatomic distance of Pt particles grown on Al$_2$O$_3$/NiAl(110) as a function of their size (the ends of the horizontal bars represent the width and the length of the particular clusters, respectively, while the vertical bars are error bars).

Of course, the electronic structure of deposited metal aggregates reflects to a certain extent the geometric structure and *vice versa*. The electronic structure, which will be discussed next, has been investigated using various methods including photoemission, X-ray absorption and scanning tunneling microscopy. One particularly interesting aspect in connection with aggregates is the size dependence of the electronic structure in relation with adsorption and reactivity.

Starting from an atomic level diagram, Fig. 22 shows how such a level diagram develops when more and more atoms are agglomerated to form an aggregate and finally a solid with a periodic lattice. Upon formation of an aggregate from equivalent atoms the atomic levels are split into molecular orbitals many of which are degenerate if the symmetry of the system is high. The splittings are characteristic for the intermolecular interactions. Depending on the interaction strength the split levels derived from a given atomic orbital start to energetically overlap with levels derived from other atomic orbitals. As long as the system has molecular character there is an energy gap left between occupied and unoccupied levels, in contrast to the situation encountered for an infinite periodic metallic solid as represented on the right hand side of the figure, where there is no longer a gap between occupied and unoccupied levels. It is not hard to envision now, that as we slowly enlarge the number of atoms in an agglomerate, the gap between occupied and unoccupied orbitals effectively vanishes. It effectively vanishes if the gap energy decreases to a value close to kT. In this situation the changes in the electronic structure would be responsible for an insulator(molecule)–metal transition. The question arises: how many atoms are necessary to induce such a transition? There are several reports in the literature claiming numbers ranging from 20 to several hundred atoms to be necessary.[135,137–149] In this connection, there is one very interesting extrapolation deduced from spectroscopic measurements of the gap of inorganic carbonyl cluster compounds as a function of the metal cluster size. It is shown in Fig. 23[139] and stems from the Longoni group in Bologna. The extrapolation would suggest that 70 atoms are sufficient to close the gap. Comparatively, we have studied deposited clusters of varying size with a combination of photoelectron spectroscopy and X-ray absorption.[141,147,149] Both, the naked as well as the CO covered aggregates, have been studied. Without discussing the results we only state that the extrapolation on the CO covered clusters yields a vanishing gap just below 100 metal

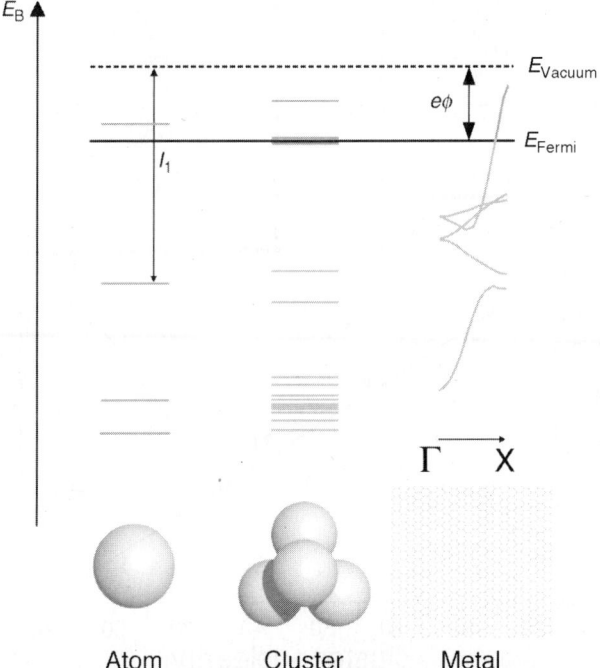

Fig. 22 Diagram illustrating the transition from an atom to a metal. (E_B, binding energy; I_1, first ionisation energy; e, electron charge; ϕ, workfunction; Γ, X, symmetry points in the Brillouin zone.)

atoms. It appears from the STM images that such a situation is reached when the diameter of the aggregate decreases down below 25 Å diameter and a height of 15–20 Å. The aggregate of this size contains 75–100 atoms, a size which well correlates with the extrapolation on the spectroscopic data of carbonyl cluster compounds in Fig. 23, as well as with our results. We take this as a strong indication that at least for the carbon monoxide covered cluster a non-metal-to-metal transition occurs in the vicinity of such a size. Goodman and his group have used scanning tunneling microscopy to investigate the electronic structure of aggregates deposited on oxides.[10,150–152] Fig. 24 shows typical current–voltage curves for some aggregate sizes, *i.e.* Au on TiO_2(110).[150] While the large particles do not exhibit a plateau near $I = V = 0$, the smaller clusters do show the behavior expected for a system with a gap. However, the discrete structures observed for other systems, *i.e.* nanoparticles on graphite[152] and related substrates[153] are not found. The authors report on indications that it is particularly the second layer in the gold aggregates that is responsible for the non-metal-to-metal transition. Au is an interesting low temperature CO oxidation catalyst and the STM findings are important to understand the size specificity of the reaction.

Before we take a closer look at the reactivities of deposited particles we will briefly discuss adsorption properties, as observed mainly through the probe molecule CO. The technique to study CO adsorption is Fourier transform infrared spectroscopy (FTIR) because it provides the resolution to differentiate between various adsorbed species. Again the thin film based systems are particularly well suited because the metallic support of the oxide films acts as a mirror at infrared frequencies. It is, however, also possible to perform such experiments on surfaces of bulk dielectrics as was shown by the Hayden group.[154,155]

Goodman and his group were active in this field early on[152] and have published an interesting study of CO adsorption on Pd aggregates on Al_2O_3 films. The results have been interpreted as characteristic for the adsorption of CO on different facets of the small crystalline aggregates. While this interpretation does not take into account adsorption on the various defect sites of the

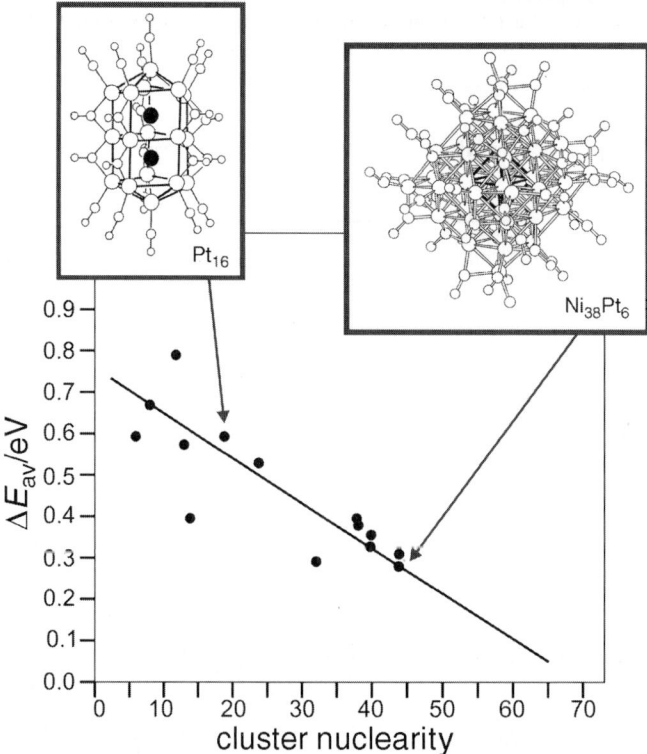

Fig. 23 Electronic excitation of lowest energy for several cluster compounds as a function of metal atoms in the cluster (ΔE_{av} is the energy gap for cluster compounds).

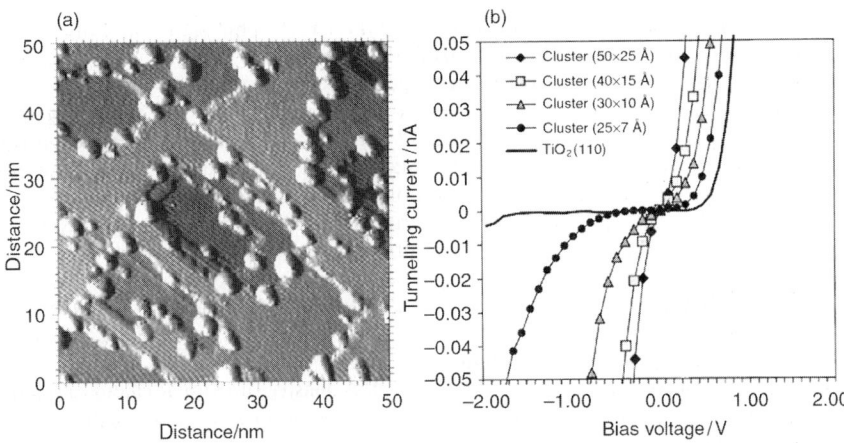

Fig. 24 Current–voltage (b) recorded for Au clusters of various sizes deposited onto a TiO$_2$(110) surface. A typical STM picture of the system is shown in (a). (Adapted from ref. 151.)

aggregates, which has been pointed out in a more recent study,[156] the data are indicative of the potential of the tool for the study of size dependent absorption studies. We have recently prepared metal deposits on well ordered alumina films at lower temperature including liquid He temperatures, *i.e.* in the range of 50 to 90 K substrate temperature,[157,158] in order to determine the IR characteristics of specific sites.

The infrared spectrum taken from a Rh deposit prepared and saturated with CO at 90 K (average particle size: nine atoms) is displayed in Fig. 25 (left, top corresponding to the spectrum in the middle on the right). The most prominent feature in the stretching region of terminally

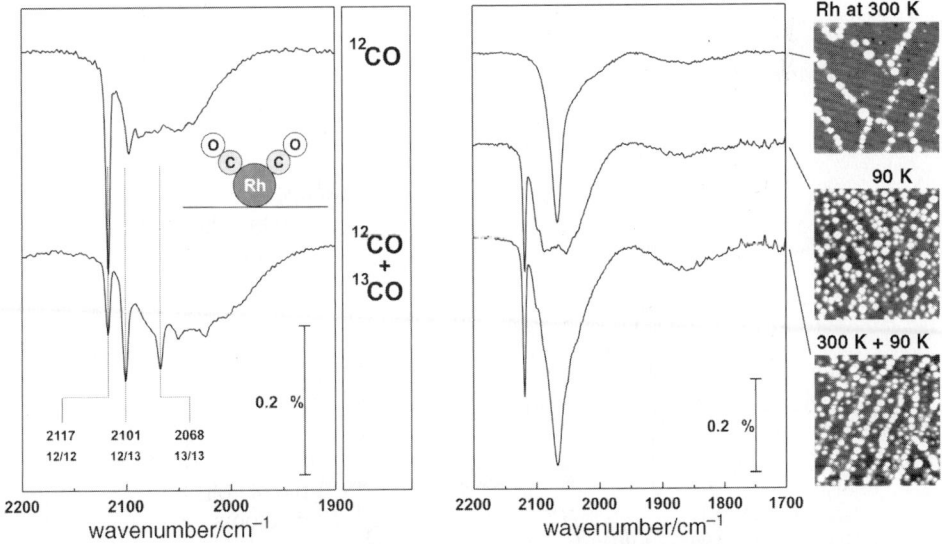

Fig. 25 (Left) Infrared spectra taken after deposition of 0.028 ML Rh on Al$_2$O$_3$ and subsequent saturation with ^{12}CO (top) and an approximately equimolar mixture of ^{12}CO and ^{13}CO (bottom) at 90 K. The isotopic compositions giving rise to the three dicarbonyl bands are indicated below the corresponding wavenumbers. Average particle size: nine atoms. (Right) Infrared spectra recorded after CO saturation of Rh deposits at 90 K, along with corresponding room temperature STM images (500 Å × 500 Å). Top: 0.057 ML Rh deposited at 300 K. Middle: 0.057 ML Rh deposited at 90 K. Bottom: 0.057 ML Rh deposited at 300 K, followed by the same exposure at 90 K.

Faraday Discuss., 1999, **114**, 1–31

bound CO molecules is a sharp, intense band at 2117 cm^{-1}. This signal has previously been shown to arise from isolated Rh bound to oxide defects.[157] Both the number of adsorbed CO molecules and the nature of the defect site remained unclear. Features at lower frequencies are assigned to molecules on Rh aggregates. In order to get insight into the stoichiometry of the Rh–carbonyl species giving rise to the band at 2117 cm^{-1}, isotopic mixing experiments have been carried out. These experiments allowed us to unambiguously assign this band to a Rh(CO)$_2$ species. Large particles prepared at 300 K deposition temperature, residing on line defects, do not exhibit the Rh(CO)$_2$ band (Fig. 25, right, top spectrum). However, if the spectra are recorded after saturation of the line defects at 300 K, and then further metal is deposited at 90 K (Fig. 25, right bottom spectrum) the Rh(CO)$_2$ band is found, indicating that this species resides in point defect sites. By further reducing the size of the particles we have also identified RhCO and species carrying more than three CO molecules.[158]

In summary, however, we may conclude that several different types of Rh particles are responsible for the observed infrared features. Presently, density functional calculations on small Rh carbonyls are in progress.[159] Calculated vibrational frequencies of such systems may help to identify the species present on the alumina film.

These studies on small Rh particles have been extended to include neighboring elements in the periodic table. Infrared spectra recorded after deposition of comparable amounts of Pd, Rh, and Ir and subsequent CO saturation at 90 K are displayed in Fig. 26. We note differences in the low wavenumber region, where vibrational frequencies of molecules in multiple coordinated sites are located. As on single crystals, the population of such sites is highest on Pd,[160,161] while no such CO is observed on Ir.[162,163]

The differences in the region of terminally bound CO, however, are much more pronounced. In the case of Ir, several distinct features are observed. In analogy to the Rh(CO)$_2$ band at 2117 cm^{-1}, the sharp signal at 2107 cm^{-1} may be attributed to Ir(CO)$_2$ species *via* isotopic mixture experiments (not shown). Bands with similar frequencies have been assigned to the symmetric stretch of Ir$^+$(CO)$_2$ on technical Ir/Al$_2$O$_3$ catalysts (2107–2090 cm^{-1})[164] and on the iridium-loaded zeolite H-ZSM-5 (2104 cm^{-1}).[165]

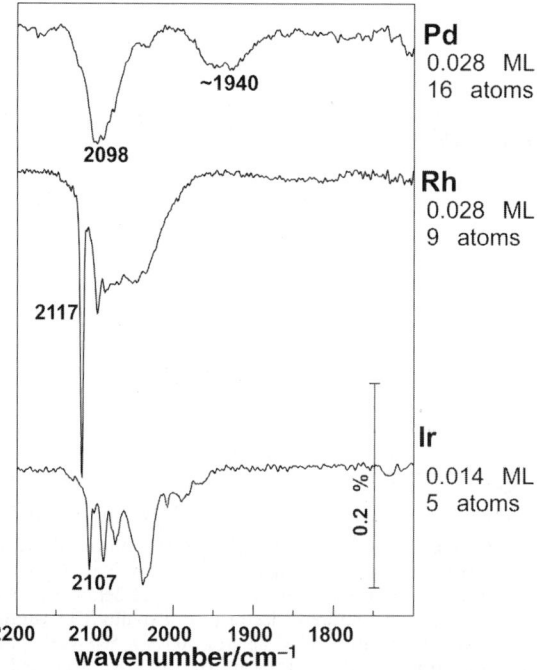

Fig. 26 Infrared spectra of Pd, Ir, and Rh deposited at 90 K and saturated with CO at the same temperature.

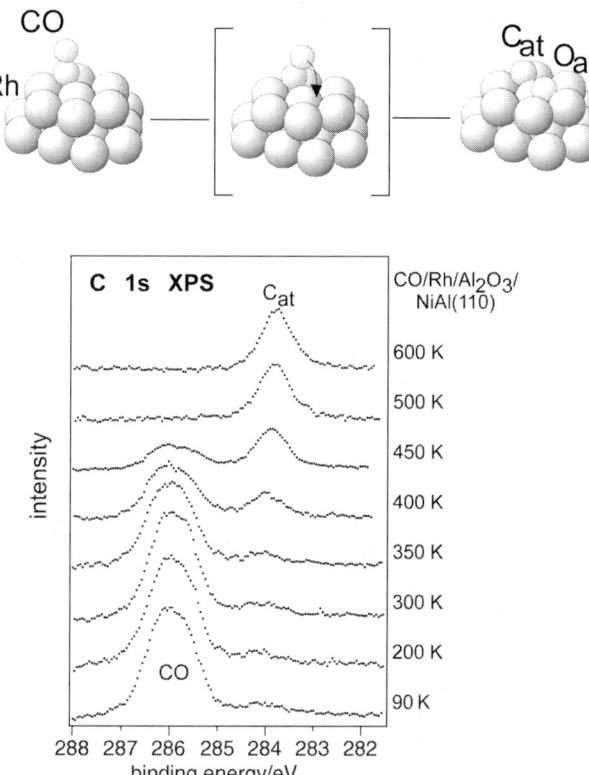

Fig. 27 CO dissociation on Rh/Al$_2$O$_3$/NiAl(110): representative series of C 1s spectra taken after CO saturation at 90 K and heating to the indicated temperatures (data acquisition at 90 K).

By contrast, no signs of atomically disperse Pd or of structurally well-defined aggregates are observed. Indeed, the infrared spectrum is rather similar to that observed on much larger, disordered Pd aggregates.[156] At the same metal exposure, the Pd particles are found to be larger than the Rh aggregates by room temperature STM.

Our observations show that infrared spectra of adsorbed CO provide valuable information on the size of metal nanoparticles, as has been long recognized in the catalysis related literature. In the nucleation regime the metal particle size increases from Ir across Rh to Pd, implying the opposite trend in metal oxide interaction strength, *i.e.* Pd < Rh < Ir.

The literature contains several adsorption studies, see for example,[166] employing other probe molecules such as hydrocarbons but here also reaction comes into play which renders the situation even more complicated.

In the following final section several simple chemical reactions of O$_2$ and CO on small aggregates are addressed.

A simple reaction is the dissociative adsorption of oxygen on small particles. Pd aggregates as shown in Fig. 19 can be imaged at atomic resolution after a dosage to saturation with molecular oxygen from the gas phase.[167] On the side facets the corrugation due to the presence of adsorption of oxygen can be identified. A doubled periodicity corresponding to a p(2 × 2) structure can be identified. This structure is very similar to the p(2 × 2) structure observed after dissociative oxygen adsorption on Pd(111).[168] We therefore conclude that a similar situation is encountered in the case of the deposited aggregates. The p(2 × 2) structure interestingly appears on the different facets at different tunneling conditions. When the oxygen covered Pd aggregates are exposed to carbon monoxide the reactivity of the different facets appear to be different in the sense that the oxygen adsorbate structure is lost on the various facets at variable temperatures and exposures. It will be interesting in the future to study these effects in more detail.

CO oxidation at low temperatures and as a function of size on TiO_2 supported gold aggregates has been studied by the Goodman group.[151] They find a marked size effect of the catalytic activity which correlates with the original observations by the Haruta group[169] for Au on large area titania catalysts. The aggregates near 35 Å size show the maximum activity.

In the future it will be important to perform kinetic measurements for such reactions under well defined conditions including UHV and ambient environments. A very good example has recently been reported by Henry and his group.[170]

Similar to the studies of CO oxidation on Au aggregates reported above, we have undertaken a detailed study of CO dissociation as a function of Rh particle size.[171-173] For various cluster sizes we have taken C 1s photoelectron spectra as a function of sample temperature. An example is shown in Fig. 27 covering a temperature range where we know that the morphology of the ensemble of aggregates does not change. At low temperature the signal typical for molecular CO is observed. Near 400 K a second signal appears indicating the dissociation of CO into carbon and oxygen atoms. At 500 K all molecular CO has been either dissociated or desorbed. The dissociation probability is then given by the relative ratio of the molecular and the atomic C 1s signal. This is plotted in Fig. 28 as a function of the aggregate size. Also included in the figure are data for very small aggregates, where it has been shown that CO dissociation is negligible.[174] This is also true for closed packed single crystal surfaces[175] which would apply to the far end of infinitely large aggregate sizes. If, however, we look at data gained for stepped Rh surfaces then there is a probability for CO dissociation.[176] At an intermediate aggregate size of 200–300 atoms the dissociation probabilities are maximal.

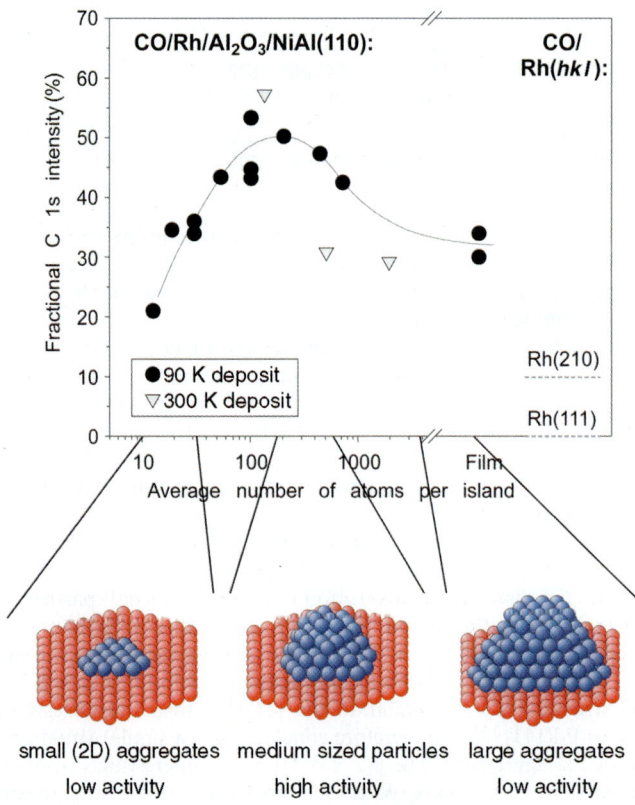

Fig. 28 CO dissociation activity on Rh particles deposited on Al_2O_3/NiAl(110) as determined by XPS. According to ref. 171 and 172 the dissociation activity also passes a maximum for the 300 K deposits, *i.e.* the activity decreases in the regime of small particle sizes reflecting the behavior of the 90 K deposits.

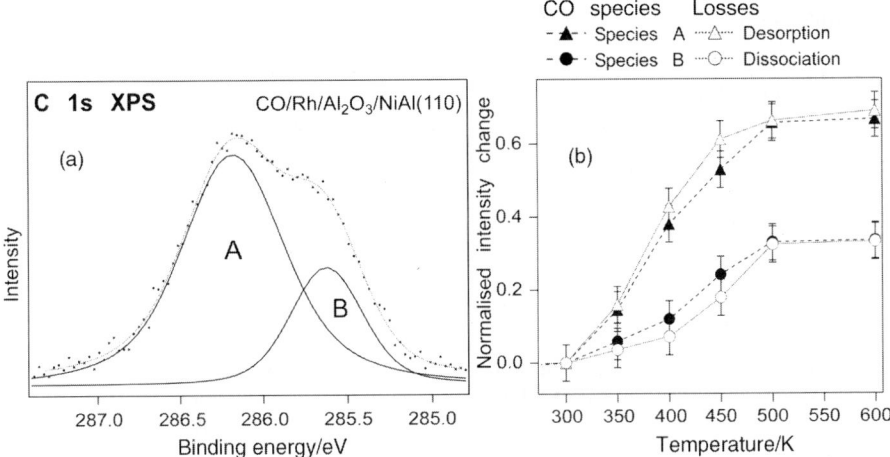

Fig. 29 (a) C 1s spectra of CO adsorbed on Rh particles after saturation at 90 K. (b) Intensity changes for the components A and B as well as the intensity losses due to dissociation and desorption as a function of the annealing temperature (average particle size: $\approx 10^4$ atoms).

Although electronic effects cannot be completely excluded as a reason for the onset of the dissociation phenomenon for small particles, an explanation on the basis of structural properties of the system seems more likely.

Since the Rh deposits are basically disordered, it is easily imaginable that aggregates of medium size exhibit a maximum defect density in terms of steps, kinks and other low coordinate surface atoms. Smaller units should contain less defects, in particular if they are still two-dimensional. In addition to that, spatial constraints may play a role here as well (accommodation of C and O on adjacent sites, see Fig. 27). At high exposures, the step density is reduced due to coalescence processes. For deposition at 300 K, the observed tendency to form crystalline aggregates in the high coverage regime is another factor contributing to a lower defect density. This is consistent with the observation that the dissociation activity declines much faster in this case (see Fig. 28).

An interesting detail concerning the dissociation process has been discovered by a closer inspection of the C 1s emission of the molecularly adsorbed CO.[172] As demonstrated in Fig. 29 (a), the peak actually consists of two components, denoted A and B. If the fraction of the total intensity found for component B after heating to 300 K is compared to the fraction of CO finally dissociating (see Fig. 28), it turns out that the species giving rise to B can be regarded as a kind of dissociation precursor. In fact, the evolution of these two quantities as a function of the particle size is identical, i.e. both pass a maximum at the same point.[172] At 90 K, however, this is not yet the case. Here, the relative intensity step which causes a shift of intensity from component A to component B, i.e. an increase of the B species which is most pronounced for the medium-sized particles. Interestingly, this conversion is irreversible. Cooling down to 90 K does not lead to an intensity redistribution again.

The conclusion that B is indeed a dissociation precursor is additionally corroborated by Fig. 29 (b) showing the intensity changes for the A and B peaks as well as the losses which result either from desorption or dissociation.[173] Unambiguously, the desorption curve follows the curve for component A, whereas the dissociation curve mimics the development of the component B. Unfortunately, the results allow no further statement as to the nature of the A and B species. It can be assumed, however, that the B species is connected with CO adsorbed on defects. Based on the fact that higher coordinated CO species give rise to lower C 1s binding energies (see above), it may be furthermore speculated whether B is associated with CO in a higher coordination as compared to the A species.

The field of investigations of chemical reactivity as a function of aggregate size is in full development and there are more exciting results at the horizon.

Concluding remarks

After 30 years of surface science, which have seen an enormous development of methods and instrumentation to be applied to the investigation of solid surfaces, the field is now ready to tackle questions of rather complex natures. Naturally, so far metal surfaces have been the focus of attention in surface science and this will continue to be the case. However, also for such systems the complexity of problems is constantly increasing, in particular if molecular adsorbates and self-organized systems are considered. Metal oxide surfaces have received some attention in the recent past and the study of such systems as well as more complex metal–metal-oxide composite systems will, or perhaps has already defined a direction in surface science that promises to reveal interesting results of fundamental interest as well as of appeal towards applications. It is good to see that surface science is very healthy and alive and it has never been farther away from fatality.

I am grateful to my co-workers, present and past, who have contributed to the results presented and whose names appear in the list of references. Also, I am happy to thank many colleagues for stimulating discussions and collaboration. Special thanks go to Ralph Wichtendahl for his help with transparencies and figures.

Over the years many funding agencies and also the private sector have supported our work: Deutsche Forschungsgemeinschaft, Bundesministerium für Bildung und Forschung, Ministerium für Wissenschaft und Forschung des Landes Nordrhein-Westfalen, Fonds der Chemischen Industrie, German–Israeli Foundation, European Union, NEDO International Joint Research Grant on Photon and Electron Controlled Surface Processes, Hoechst Celanese, and Synetix, a member of the ICI group, through their Strategy Research Fund.

References

1 P. A. Cox, *Transition Metal Oxides. An Introduction to their Electronic Structure and Properties*, Clarendon Press, Oxford, 1992.
2 D. A. Johnson, *Some Thermodynamic Aspects of Inorganic Chemistry*, Cambridge University Press, Cambridge, 1982.
3 *Non-stoichiometric Oxides*, ed. O. T. Sorensen, Academic Press, New York, 1981.
4 A. Hamnett and J. B. Goodenough, in *Binary transition metal oxides*, ed. O. Madelung, Springer, 1984.
5 J. F. Owen, K. J. Teegarden and H. R. Shanks, *Phys. Rev. B*, 1978, **18**, 3827.
6 *Surface Science: The First Thirty Years*, ed. C. B. Duke, Elsevier, Amsterdam, 1994.
7 V. E. Henrich and P. A. Cox, *The Surface Science of Metal Oxides*, Cambridge University Press, Cambridge, 1994.
8 C. T. Campbell, *Surf. Sci. Rep.*, 1997, **27**, 1.
9 C. T. Campbell, *Curr. Opin. Solid State Mater. Sci.*, 1998, **3**, 439.
10 D. W. Goodman, *Surf. Rev. Lett.*, 1995, **2**, 9.
11 H.-J. Freund, *Angew. Chem. Int. Ed. Engl.*, 1997, **36**, 452.
12 H.-J. Freund, *Phys. Status Solidi B*, 1995, **192**, 407.
13 H.-J. Freund, H. Kuhlenbeck and V. Staemmler, *Rep. Prog. Phys.*, 1996, **59**, 283.
14 C. R. Henry, *Surf. Sci. Rep.*, 1998, **31**, 231.
15 M. Bäumer and H.-J. Freund, *Prog. Surf. Sci.*, 1999, **61**, 127.
16 J. M. Thomas, *Faraday Discuss.*, 1996, **105**, 1.
17 G. Ertl and H.-J. Freund, *Phys. Today*, 1999, **52**, 32.
18 H. Poppa, *Catal. Rev. Sci. Eng.*, 1993, **35**, 359.
19 U. Diebold, J.-M. Pan and T. E. Madey, *Surf. Sci.*, 1995, **331–333**, 845.
20 G. H. Hardy, *A Mathematician's Apology*, Cambridge University Press, 1940.
21 *Adsorption on Ordered Surfaces of Ionic Solids and Thin Films*, ed. H.-J. Freund and E. Umbach, Springer, Heidelberg, 1993.
22 G. H. Vurens, M. Salmeron and G. A. Somorjai, *Prog. Surf. Sci.*, 1989, **32**, 333.
23 P. L. J. Gunter, J. W. H. Niemantsverdriet, F. H. Ribeiro and G. A. Somorjai, *Catal. Rev. Sci. Eng.*, 1997, **39**, 77.
24 D. A. Bonnell, *Prog. Surf. Sci.*, 1998, **57**, 187.
25 U. Diebold, J. F. Anderson, K.-O. Ng and D. Vanderbilt, *Phys. Rev. Lett.*, 1996, **77**, 1322.
26 P. W. Murray, F. M. Leibsle, C. A. Muryn, H. J. Fisher, C. F. J. Flipse and G. Thornton, *Phys. Rev. Lett.*, 1994, **72**, 689.
27 P. W. Murray, N. G. Condon and G. Thornton, *Phys. Rev. B*, 1995, **51**, 10989.
28 H. Onishi and Y. Iwasawa, *Surf. Sci.*, 1994, **313**, L783.
29 H. Onishi and Y. Iwasawa, *Chem. Phys. Lett.*, 1994, **226**, 111.

30 H. Onishi, K. Fukui and Y. Iwasawa, *Bull. Chem. Soc. Jpn.*, 1995, **68**, 2447.
31 H. Onishi and Y. Iwasawa, *Phys. Rev. Lett.*, 1996, **76**, 791.
32 G. Charlton, P. B. Howes, C. L. Nicklin, P. Steadman, J. S. G. Taylor, C. A. Muryn, S. P. Harte, J. Mercer, R. McGrath, D. Norman, T. S. Turner, and G. Thornton, *Phys. Rev. Lett.*, 1997, **78**, 495.
33 G. Renaud, *Surf. Sci. Rep.*, 1998, **32**, 1.
34 P. W. Tasker, *Adv. Ceram.*, 1984, **10**, 176.
35 W. C. Mackrodt, R. J. Davey, I. N. Black and R. Docherty, *J. Cryst. Growth*, 1987, **82**, 441.
36 C. Noguera, *Physics and Chemistry at Oxide Surfaces*, Cambridge University Press, 1996.
37 P. Guénard, G. Renaud, A. Barbier and M. Gantier-Soyer, *Surf. Rev. Lett.*, 1998, **5**, 321.
38 F. Rohr, M. Bäumer, H.-J. Freund, J. A. Mejias, V. Staemmler, S. Müller, L. Hammer and K. Heinz, *Surf. Sci.*, 1997, **372**, L 291.
39 R. Rohr, M. Bäumer, H.-J. Freund, J. A. Mejias, V. Staemmler, S. Muller, L. Hammer and K. Heinz, *Surf. Sci.*, 1997, **389**, 391.
40 W. Weiss, *Surf. Sci.*, 1997, **377–379**, 943.
41 L. Manassidis and M. J. Gillan, *Surf. Sci.*, 1993, **285**, L517.
42 L. Manassidis and M. J. Gillan, *J. Am. Ceram. Soc.*, 1994, **77**, 335.
43 C. Rebhein, N. M. Harrison and A. Wander, *Phys. Rev. B*, 1996, **54**, 14066.
44 X.-G. Wang, W. Weiss, S. K. Shaikhutdinov, M. Ritter, M. Petersen, F. Wagner, R. Schloegl and M. Scheffler, *Phys. Rev. Lett.*, 1998, **81**, 1038.
45 K. Heinz, *Surf. Sci.*, 1984, **299**, 433.
46 N. M. Harrison, X.-G. Wang, M. Muscat and M. Scheffler, *Faraday Discuss.*, 1999, **114**, 305.
47 H. Schmalzried, *Chemical Kinetics of Solids*, VCH, Weinheim, 1995.
48 D. Wolf, *Phys. Rev. Lett.*, 1992, **68**, 3315.
49 A. Barbier and G. Renauld, *Surf. Sci.*, 1997, **392**, L15.
50 D. Cappus, C. Xu, D. Ehrlich, B. Dillmann, C. A. Ventrice Jr., K. Al-Shamery, H. Kuhlenbeck and H.-J. Freund, *Chem. Phys.*, 1993, **177**, 533.
51 F. Rohr, K. Wirth, J. Libuda, D. Cappus, M. Bäumer and H.-J. Freund, *Surf. Sci.*, 1994, **315**, L977.
52 D. Cappus, M. Hassel, E. Neuhaus, M. Heber, F. Rohr and H.-J. Freund, *Surf. Sci.*, 1995, **337**, 268.
53 A. Barbier, *Proc. 1st Int. Workshop Oxide Surf. (IWOX1)*, Elmau, Germany, 1999.
54 R. Lacman, *Colloq. Int. CNRS*, 1965, **152**, 195.
55 C. A. Ventrice Jr., T. Bertrams, H. Hannemann, A. Brodde and H. Neddermeyer, *Phys. Rev. B*, 1994, **49**, 1773.
56 A. Barbier, C. Mocuta, H. Kuhlenbeck, K. Peters, B. Richter and G. Renaud, in preparation.
57 W. H. Casey, H. R. Westrich and G. W. Arnold, *Geochim. Cosmochim. Acta*, 1988, **53**, 2795.
58 H. Papp and B. Egersdörfer, personal communication.
59 B. Egersdörfer, PhD Thesis, Bochum, 1993.
60 J. G. Fripiat, A. A. Lucas, J. M. André and E. G. Derouane, *Chem. Phys.*, 1977, **21**, 101.
61 S. Hüfner, P. Steiner, I. Sander, M. Neumann and S. Witzel, *Z. Phys. B*, 1991, **83**, 185.
62 S. Hüfner, *Photoelectron Spectroscopy: Principles and Applications*, Springer-Verlag, Berlin, Heidelberg, 1995.
63 A. Gorschlüter and H. Merz, *Phys. Rev. B*, 1994, **49**, 17293.
64 A. Freitag, V. Staemmler, D. Cappus, C. A. Ventrice, Jr., K. Al-Shamery, H. Kuhlenbeck and H.-J. Freund, *Chem. Phys. Lett.*, 1993, **210**, 10.
65 R. Newman and R. M. Chrenko, *Phys. Rev.*, 1959, **114**, 1507.
66 R. J. Powell, Report no. 5220-1 (unpublished), Stanford Electronics Laboratory, Stanford, 1967.
67 D. Adler and J. Feinleib, *Phys. Rev. B*, 1970, **2**, 3112.
68 B. Fromme, C. Koch, R. Deussen and E. Kisker, *Phys. Rev. Lett.*, 1995, **75**, 693.
69 B. Fromme, M. Möller, T. Anschütz, C. Bethke and E. Kisker, *Phys. Rev. Lett.*, 1996, **77**, 1548.
70 M. Pöhlchen and V. Staemmler, *J. Chem. Phys.*, 1992, **97**, 2583.
71 J. A. Mejias, V. Staemmler and H.-J. Freund, *J. Phys. Condens. Matter*, in press.
72 H. Kuhlenbeck, C. Xu, B. Dillmann, M. Haßel, B. Adam, D. Ehrlich, S. Wohlrab, H.-J. Freund, U. A. Ditzinger, H. Neddermeyer, M. Neuber and M. Neumann, *Ber. Bunsen-ges. Phys. Chem.*, 1992, **96**, 15.
73 C. Xu, B. Dillmann, H. Kuhlenbeck and H.-J. Freund, *Phys. Rev. Lett.*, 1991, **67**, 3551.
74 M. Bender, D. Ehrlich, I. N. Yakovkin, F. Rohr, M. Bäumer, H. Kuhlenbeck, H. J. Freund and V. Staemmler, *J. Phys. Condens. Matter*, 1995, **7**, 5289.
75 M. Henzler and W. Göpel, *Oberflächenphysik des Festkörpers*, Teubner–Verlag, Stuttgart, 1991.
76 K. M. Neyman, G. Pacchioni and N. Rösch, in *Recent Developments and Applications of Modern Density Functional Theory and Computational Chemistry*, ed. J. M. Seminario, Elsevier, Amsterdam, 1996, p. 569.
77 M. Bäumer, J. Libuda and H.-J. Freund, in *Chemisorption and Reactivity on Supported Clusters and Thin Films*, ed. R. M. Lambert and G. Pacchioni, Kluwer Academic Press, Dordrecht, 1997, p. 61.
78 M. A. Nygren and L. G. M. Pettersson, *J. Chem. Phys.*, 1996, **105**, 9339.
79 G. Pacchioni and P. S. Bagus, in *Adsorption on Ordered Surfaces of Ionic Solids and Thin Films*, ed. H.-J. Freund and E. Umbach, Springer Verlag, Berlin, 1993, p. 180.
80 T. Klüner, personal communication.
81 D. Cappus, J. Klinkmann, H. Kuhlenbeck and H.-J. Freund, *Surf. Sci.*, 1995, **325**, L 421.

82 S. M. Vesecky, X. Xu and D. W. Goodman, *J. Vac. Sci. Technol. A*, 1994, **12**, 2114.
83 V. Staemmler, in *Adsorption on Ordered Surfaces of Ionic Solids and Thin Films*, ed. H.-J. Freund and E. Umbach, Springer Verlag, Berlin, 1993, p. 169.
84 M. Pöhlchen, PhD Thesis, Ruhr-Universität, Bochum, 1992.
85 H. Kuhlenbeck, G. Odörfer, R. Jaeger, G. Illing, M. Menges, T. Mull, H.-J. Freund, M. Pöhlchen, V. Staemmler, S. Witzel, C. Scharfschwerdt, K. Wennemann, T. Liedtke and M. Neumann, *Phys. Rev. B*, 1991, **43**, 1969.
86 L. Chen, R. Wu, N. Kioussis and Q. Zhang, *Chem. Phys. Lett.*, 1998, **290**, 255.
87 K. M. Neyman, S. P. Ruzankin and N. Rösch, *Chem. Phys. Lett.*, 1995, **246**, 546.
88 J.-W. He, C. A. Estrada, J. S. Corneille, M.-C. Wu and D. W. Goodman, *Surf. Sci.*, 1992, **261**, 164.
89 S. Furuyama, H. Fuji, M. Kawamura and T. Morimoto, *J. Phys. Chem.*, 1978, **82**, 1028.
90 R. Wichtendahl, M. Rodriguez-Rodrigo, U. Härtel, H. Kuhlenbeck and H.-J. Freund, *Phys. Status Solidi A*, 1999, **173**, 93.
91 H. Pfnür, P. Feulner and D. Menzel, *J. Chem. Phys.*, 1983, **79**, 4613.
92 C. N. Chittenden, E. D. Pylant, A. L. Schwaner and J. M. White, *Thermal Desorption and Mass Spectrometry*, ed. A. T. Hubbard, CRC Press, Boca Raton, FL, 1995.
93 H. Schlichting and D. Menzel, *Rev. Sci. Instrum.*, 1993, **64**, 2013.
94 J. Heidberg, M. Kandel, D. Meine and U. Wildt, *Surf. Sci.*, 1995, **333**, 1467.
95 R. Gerlach, A. Glebov, G. Lange, J. P. Toennies and W. Weiss, *Surf. Sci.*, 1995, **331–333**, 1490.
96 R. Wichtendahl, PhD Thesis, Freie Universität, Berlin, 1999.
97 M. J. Stirniman, C. Huang, R. C. Smith, J. A. Joyce and B. D. Kay, *J. Chem. Phys.*, 1996, **105**, 1295.
98 C. Xu and D. W. Goodman, *Chem. Phys. Lett.*, 1997, **L65**, 341.
99 J. Libuda, M. Frank, A. Sandell, S. Andersson, P. A. Brühwiler, M. Bäumer, N. Mårtensson and H.-J. Freund, *Surf. Sci.*, 1997, **384**, 106.
100 K. Wolter and M. Frank, personal communication.
101 J. Hemminger, personal communication.
102 R. M. Jaeger, H. Kuhlenbeck, H.-J. Freund, M. Wuttig, W. Hoffmann, R. Franchy and H. Ibach, *Surf. Sci.*, 1991, **259**, 235.
103 I. Engquist and B. Liedberg, *J. Phys. Chem.*, 1996, **100**, 20089.
104 H. Knözinger and P. Ratnasaniy, *Catal. Rev. Sci. Eng.*, 1978, **17**, 31.
105 D. Ehrlich, PhD Thesis, Bochum, 1995.
106 O. Seiferth, K. Wolter, B. Dillmann, G. Klivenyi, H.-J. Freund, D. Scarano and A. Zecchina, *Surf. Sci.*, 1999, **421**, 176.
107 A. Zecchina, D. Scarano, S. Bordiga, G. Ricchiardi, G. Spoto and F. Geobaldo, *Catal. Today*, 1996, **27**, 403.
108 C. Xu, M. Haßel, H. Kuhlenbeck and H.-J. Freund, *Surf. Sci.*, 1991, **258**, 23.
109 I. Hemmerich, F. Rohr, O. Seiferth, B. Dillmann and H.-J. Freund, *Z. Phys. Chem.*, 1997, **202**, 31.
110 P. C. Thüne and J. W. Niemantsverdriet, *Isr. J. Chem.*, 1998, **38**, 385.
111 M. Menges, B. Baumeister, K. Al-Shamery, H.-J. Freund, C. Fischer and P. Andresen, *J. Chem. Phys.*, 1994, **101(4)**, 3318.
112 I. Beauport, K. Al-Shamery and H.-J. Freund, *Chem. Phys. Lett.*, 1996, **256**, 641.
113 T. Klüner, H.-J. Freund, V. Staemmler and R. Kosloff, *Phys. Rev. Lett.*, 1998, **80**, 5208.
114 K. Al-Shamery, *Appl. Phys. A*, 1996, **63**, 509.
115 E. A. Wovchko and J. Yates, Jr., *Langmuir*, 1999, **15**, 3506.
116 M. C. Wu and P. J. Møller, *Surf. Sci.*, 1989, **221**, 250.
117 P. J. Møller and M. C. Wu, *Surf. Sci.*, 1989, **224**, 265.
118 M. C. Wu and P. J. Møller, *Surf. Sci.*, 1990, **235**, 228.
119 P. J. Møller and J. Nerlov, *Surf. Sci.*, 1993, **307–9**, 591.
120 P. W. Murray, J. Shen, N. G. Condon, S. J. Peng and G. Thornton, *Surf. Sci.*, 1997, **380**, L455.
121 P. Stone, R. A. Bennett and M. Bowker, *New J. Phys.*, 1998, **1**, 8.
122 U. Diebold, *Proc. 1st Int. Workshop Oxide Surf. (IWOX1)*, Elmau, Germany, 1999.
123 J. Libuda, F. Winkelmann, M. Bäumer, H.-J. Freund, T. Bertrams, H. Neddermeyer and K. Müller, *Surf. Sci.*, 1994, **318**, 61.
124 S. Stempel, M. Bäumer and H.-J. Freund, *Surf. Sci.*, 1998, **402–404**, 424.
125 M. Heemeier, PhD Thesis, Berlin, in preparation.
126 T. R. Linderoth, S. Horch, E. Laensgaard, I. Stensgaard and F. Besenbacher, *Surf. Sci.*, 1998, **404**, 308.
127 N. Ernst, B. Duncombe, G. Bozdech, M. Naschitzki and H.-J. Freund, *Ultramicroscopy*, 1999, **79**, 231.
128 A. Piednoir, E. Pernot, S. Granjeand, A. Humbert, C. Chapon and C. R. Henry, *Surf. Sci.*, 1997, **391**, 19.
129 K. H. Hansen, T. Warren, S. Stempel, E. Laegsgaard, M. Bäumer, H.-J. Freund, F. Besenbacher and I. Stensgaard, *Phys. Rev. Lett.*, in the press.
130 M. Methfessel, D. Hennig and M. Scheffler, *Phys. Rev. B*, 1992, **46**, 4816.
131 A. Bogicevic and D. R. Jennison, *Phys. Rev. Lett.*, 1999, **82**, 4050.
132 M. Klimenkov, S. Nepijko, H. Kuhlenbeck, M. Bäumer, R. Schlögl and H.-J. Freund, *Surf. Sci.*, 1997, **391**, 27.
133 M. Klimenkov, S. Nepijko, H. Kuhlenbeck and H.-J. Freund, *Surf. Sci.*, 1997, **385**, 66.

134 S. Nepijko, M. Klimenkov, H. Kuhlenbeck, D. Zemylanov, D. Herein, R. Schlögl and H.-J. Freund, *Surf. Sci.*, 1998, **413**, 192.
135 *Clusters and Colloids: From Theory to Applications*, ed. G. Schmid, VCH, Weinheim, 1994.
136 S. A. Nepijko, M. Klimenkov, M. Adelt, H. Kuhlenbeck, R. Stoğl and H.-J. Freund, *Langmuir*, 1999, **15**, 5309.
137 *Metal Clusters*, ed. W. Ekardt, John Wiley, Chichester, 1999.
138 *Transition Metal Clusters*, ed. B. F. G. Johnson, John Wiley, Chichester, 1980.
139 F. F. de Biani, C. Femoni, M. C. Iapalucci, G. Longoni, P. Zanello and A. Ceriotti, *Inorg. Chem.*, 1999, **38**(16), 3721.
140 G. K. Wertheim, S. B. DiCenzo and D. N. E. Buchanan, *Phys. Rev. B*, 1986, **33**, 5384.
141 A. Sandell, J. Libuda, P. Brühwiler, S. Andersson, A. Maxwell, M. Bäumer, N. Mårtensson and H.-J. Freund, *J. Electron Spectrosc. Relat. Phenom.*, 1995, **76**, 301.
142 R. Unwin and A. M. Bradshaw, *Chem. Phys. Lett.*, 1978, **58**, 58.
143 G. K. Wertheim, *Z. Phys. D*, 1989, **12**, 319.
144 S.-T. Lee, G. Apai, M. G. Mason, R. Benbow and Z. Hurych, *Phys. Rev. B*, 1981, **23**, 505.
145 V. de Gouveia, B. Bellamy, Y. Hadj Romdhane, A. Mason and M. Che, *Z. Phys. D*, 1989, **12**, 587.
146 C. Kuhrt and M. Harsdorff, *Surf. Sci.*, 1991, **245**, 173.
147 A. Sandell, J. Libuda, P. A. Brühwiler, S. Andersson, A. J. Maxwell, M. Bäumer, N. Mårtensson and H.-J. Freund, *J. Vac. Sci. Technol. A*, 1996, **14**, 1546.
148 O. D. Häberlen, S.-C. Chung, M. Stener and N. Rösch, *J. Chem. Phys.*, 1997, **106**, 5189.
149 A. Sandell, J. Libuda, P. A. Brühwiler, S. Andersson, M. Bäumer, A. J. Maxwell, N. Mårtensson and H.-J. Freund, *Phys. Rev. B*, 1997, **55**, 7233.
150 D. W. Goodman, *J. Vac. Sci. Technol. A*, 1996, **14**, 1526.
151 M. Valden and D. W. Goodman, *Science*, 1998, **281**, 1647.
152 D. R. Rainer and D. W. Goodman, in *NATO ASI*, ed. G. Pacchioni and R. M. Lambert, Kluwer, Dordrecht, 1997, p. 27.
153 M. Adelt, S. Nepijko, W. Drachsel and H.-J. Freund, *Chem. Phys. Lett.*, 1998, **291**, 425.
154 J. Evans, B. Hayden, F. Mosselman and A. Murray, *Surf. Sci.*, 1992, **279**, L159.
155 J. Evans, B. Hayden, F. Mosselman and A. Murray, *Surf. Sci.*, 1994, **301**, 61.
156 K. Wolter, O. Seiferth, H. Kuhlenbeck, M. Bäumer and H.-J. Freund, *Surf. Sci.*, 1998, **399**, 190.
157 M. Frank, R. Kühnemuth, M. Bäumer and H.-J. Freund, *Surf. Sci.*, 1999, **427/428**, 288.
158 M. Frank, R. Kühnemuth, M. Bäumer and H.-J. Freund, *Surf. Sci.*, submitted.
159 T. Mineva and N. Russo, personal communication.
160 P. Uvdal, P.-A. Karlsson, C. Nyberg, S. Andersson and N. V. Richardson, *Surf. Sci.*, 1988, **202**, 167.
161 T. Giessel, O. Schaff, C. J. Hirschmugl, V. Fernandez, K. M. Schindler, A. Theobald, S. Bao, R. Lindsay, W. Berndt, A. M. Bradshaw, C. Baddeley, A. F. Lee, R. M. Lambert and D. P. Woodruff, *Surf. Sci.*, 1998, **406**, 90.
162 G. Kisters, J. G. Chen, S. Lehwald and H. Ibach, *Surf. Sci.*, 1991, **245**, 65.
163 J. Lauterbach, R. W. Boyle, M. Schick, W. J. Mitchell, B. Meng and W. H. Weinberg, *Surf. Sci.*, 1996, **350**, 32.
164 F. Solymosi, É. Novák and A. Molnár, *J. Phys. Chem.*, 1990, **94**, 7250.
165 T. W. Voskobojnikov, E. S. Shpiro, H. Landmesser, N. I. Jaeger and G. Schulz-Ekloff, *J. Mol. Catal. A: Chem.*, 1996, **104**, 299.
166 C. DeLaCruz and N. Sheppard, *J. Chem. Soc. Faraday Trans.*, 1997, **93**, 3569.
167 K. Højrup Hansen, S. Stempel, E. Laegsgaard, M. Bäumer, F. Besenbacher and I. Stensgaard, personal communication.
168 H. Conrad, G. Ertl and J. Küppers, *Surf. Sci.*, 1978, **76**, 323.
169 M. Haruta, *Catal. Today*, 1997, **36**, 153.
170 L. Piccolo, C. Becker and C. R. Henry, *Euro. Phys. J.*, in the press.
171 M. Frank, S. Andersson, J. Libuda, S. Stempel, A. Sandell, B. Brena, A. Giertz, P. A. Brühwiler, M. Bäumer, N. Mårtensson and H.-J. Freund, *Chem. Phys. Lett.*, 1997, **279**, 92.
172 S. Andersson, M. Frank, A. Sandell, A. Giertz, B. Brena, P. A. Brühwiler, N. Mårtensson, J. Libuda, M. Bäumer and H.-J. Freund, *J. Chem. Phys.*, 1998, **108**, 2967.
173 S. Andersson, M. Frank, A. Sandell, J. Libuda, B. Brena, A. Giertz, P. A. Brühwiler, M. Bäumer, N. Mårtensson and H.-J. Freund, *Vacuum*, 1998, **49**, 167.
174 M. P. Irion, personal communication.
175 J. T. Yates, E. D. Williams and W. H. Weinberg, *Surf. Sci.*, 1980, **91**, 562.
176 M. Rebholz, R. Prins and N. Kruse, *Surf. Sci.*, 1991, **259**, L791.
177 K. Refson, R. A. Wogelius, D. G. Fraser, M. C. Payne, M. H. Lee and V. Milman, *Phys. Rev. B*, 1995, **52**, 10823.
178 U. Birkenheuer, J. C. Boettger and N. Rösch, *J. Chem. Phys.*, 1994, **100**, 6826.
179 P. A. Redhead, *Vacuum*, 1962, **12**, 203.

Paper 9/07182B

Ab initio calculations on the Al$_2$O$_3$(0001) surface

Iskander Batyrev, Ali Alavi and Michael W. Finnis*

Atomistic Simulation Group, School of Mathematics and Physics, The Queen's University of Belfast, Belfast, UK BT7 1NN

Received 26th April 1999

We calculate using a density functional pseudopotential method the atomic and electronic structure of the (0001) surface of α-alumina (Al$_2$O$_3$). The material is studied in the form of a slab with periodic boundary conditions, containing up to eight layers of the stoichiometric Al$_2$O$_3$ units. Five different terminations of the surface are calculated, representing different surface excesses of oxygen, and their free energies are estimated as a function of oxygen partial pressure. Internal relaxations of the atomic positions are obtained. The aluminium terminated surface, which is stoichiometric, has the lowest surface energy for a wide range of oxygen pressures.

1. Introduction

The (0001) surface of α-Al$_2$O$_3$ (corundum or sapphire) has been studied experimentally and theoretically for many years since it is a widely used substrate for many kinds of thin films and an archetypal wide band-gap ionic oxide. This surface terminates a layered structure, in which each oxygen plane in the bulk has an associated aluminium plane at a distance 0.838 Å above and below it, forming a stoichiometric triple layer. The most convenient unit cell is a rhombohedral prism comprising six such (0001) oxygen planes separated by the associated pairs of aluminium planes. The cell contains just one atom in each Al plane and three atoms in each O plane. In the bulk structure the oxygen planes are separated by 2.166 Å and form a hexagonal lattice with ABABAB ... stacking. Their positions are slightly laterally distorted from ideal hexagonal sites. The Al atoms occupy two-thirds of the octahedral holes in the oxygen sublattice, at positions which alternate above and below the centres of these holes. The unoccupied octahedral holes are themselves stacked on a face-centred cubic lattice, abcabc ... A C_{3v} symmetry axis passes through each Al atom and through the centres of the unoccupied octahedral sites. Fig. 1 shows an elevation of the structure.

Only recently, with the advent of grazing incidence X-ray diffraction, has it become possible to compare structural predictions for the (0001) surface directly with experiment. Renaud has published a comprehensive review of the experimental and theoretical results.[1] The normally observed structure is (1 × 1), for which one can postulate three possible bulk terminations: single Al, double Al or O. The observed surface is believed to be terminated by a single Al plane, as in Fig. 1 in which this plane is labelled Al1. This termination is stoichiometric, that is, it exhibits no surface excess of Al or O. The top layers relax strongly; best estimates of the interlayer relaxations of the top four layers from experiment[2] are −51%, 16%, −29% and +20%.

Manassidis *et al.*[3] were the first to use density functional theory (DFT) within the local density approximation (LDA) for exchange and correlation, to calculate the energy and relaxation of this surface. They used a pseudopotential approach and modelled the surface with a slab containing

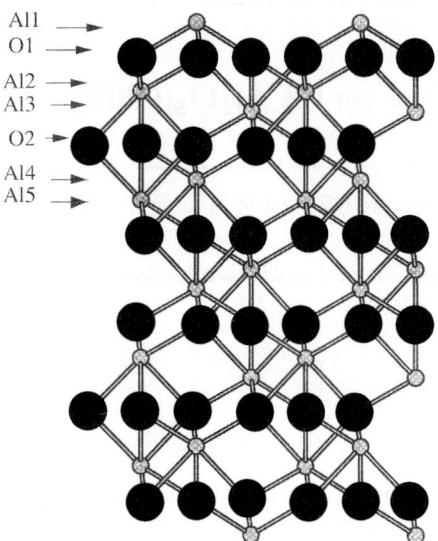

Fig. 1 View of the corundum lattice terminated at the (0001) surface by the Al plane labelled Al1. This is the stoichiometric termination ($\Gamma_O = 0$) and the main one observed experimentally.

three layers of oxygen. The most striking feature of the surface is the large inward relaxation of the surface Al plane, which in this calculation reduces the first interplanar spacing by 85%. Their results were later confirmed with a similar calculation by Kruse et al.[4] A recent DFT calculation by Verdozzi and co-workers[5] again confirmed the large surface relaxation; they used slabs containing up to 18 layers of oxygen, eliminating any uncertainty as to the effect of finite slab thickness. Their use of a thicker slab also enabled them to predict the interlayer spacings of deeper layers. For the first four layers they found relaxations of −87%, +3%, −42%, +19%.

Puchin et al.[6] have also calculated the atomic and electronic structure of the stoichiometric, Al-terminated surface modelled as a slab containing 3 layers of oxygen. They used the Hartree–Fock (HF) method (embodied in the CRYSTAL code) with an 85-11G/8-411G basis set for each atom. The surface Al layer was found to relax inwards by 68%, somewhat less than the DFT calculations, (the difference being 0.016 nm), but significantly more than the previous HF results (48%) using a smaller basis.[7,8] Puchin et al. went on to calculate theoretical UV photoelectron spectra (UPS) and metastable impact electron spectra (MIES), and compared them with experiment, concluding that the model gave a good description of the surface resolved density of states (DOS) in the valence band, and even that the relaxation was more consistent with these experimental data than the somewhat larger DFT relaxation.

Calculations of the atomic and electronic structure of the (0001) surface have been made with a non-self-consistent tight-binding model by Godin and LaFemina.[9] They also reported a very large (>90%) relaxation of the surface Al layer, which they attributed to the formation of almost planar sp^2 bonds between the surface Al and the O triangle on which it sits. However, this large relaxation was also found in the earliest studies with classical ionic interatomic potentials,[10] which of course have no representation of chemical bonding, whether sp^2 or sp^3.

We note the above discrepancy between the interpretations of the large inward relaxation of the surface Al layer. The argument in terms of bond hybridisation we find less persuasive than the simple electrostatic argument; because (a) ultimately the driving force for any relaxation or reconstruction is the classical electrostatic force on the relaxing ion from the self-consistent charge density (the Hellmann–Feynman theorem) and (b) in the first-principles calculations there is no sign of sp^2 bonding in the charge density distribution, which looks very ionic. There are other seemingly contradictory statements and observations concerning this surface which are worth discussing here. It was noted for example by Godin and LaFemina[9] that the surface Al–O bond length is conserved (to within 10%) as the Al relaxes inwards, which is effected by an associated expansion of the O triangle. Their tight binding model showed this effect, and it is also apparent from the X-ray data referred to above, from which the Al–O bond length is only 4.5% ($\pm 2.5\%$) shorter than the bulk nearest neighbour value. However, it is not the case, as Godin and LaFemina speculated, that the outward lateral relaxation of the O triangle is a *necessary* condition for the large inward relaxation of the Al (by analogy with semiconductor surfaces and their bond-length conserving reconstructions), because the earlier first-principles calculations of large Al relaxation[3,4] did not allow lateral relaxations at all. In the present paper, we report first-principles calculations for this surface including full relaxation of all atomic co-ordinates in the unit cell, and discuss the pattern of relaxation in some detail. The nature of the electrostatic driving force for the Al relaxation has also been described as reducing the electrostatic dipole of the first two atomic layers, in the manner of the attraction between the plates of a capacitor. This seems a satisfactory explanation, as long as one bears in mind that the first *three* atomic layers (Al–O–Al . . .) of the structure form an electrically neutral object in the sense of the ionic model.

Another argument which has sometimes been aired, is that only the single Al terminated surface should be stable because an alternative (1 × 1) surface, such as the one terminated by an O plane, would be charged and therefore have an infinite surface energy. This is clearly only the case within a purely ionic model in which all ions carry their bulk formal charges. However, it is a poor basis for discussing the present surface, especially since it is now known experimentally that there are other rather stable terminations of it. In particular, on heating in UHV, this surface undergoes a series of reconstructions culminating in a $(\sqrt{31} \times \sqrt{31})R \pm 9°$ structure at around 1350 °C; R indicates that the reconstructed layer is rotated with respect to the substrate. This structure has now been characterised[11] and is believed to comprise domains of nearly two Al(111) layers, formed by losing the first two layers of oxygen from the stoichiometric (1 × 1) surface discussed above.

A major aim of the present paper is to describe how the relative energetics of surfaces of different stoichiometry can be calculated theoretically. There is one independent variable needed to compare the surface energies of surfaces of differing stoichiometry (or surface excess of one component), and that is most usefully taken to be the partial pressure of oxygen, since this is what is most directly under the control of the experimentalist. We formulate the theory in Section 3 with this in mind. First, in the following section, we describe briefly our method of calculation of total energies, which provide input to the theory. While we have insufficient computer resources to make calculations for the $(\sqrt{31} \times \sqrt{31})R \pm 9°$ surface, we shall illustrate how this could be done, in principle, by comparing the energetics of five alternative (1 × 1) surfaces of differing stoichiometry. This will also show that previous arguments based on classical electrostatics for the stability of the stoichiometric, Al-terminated surface are of doubtful validity. We find, for example, that the O-terminated surface could in theory be stabilised by a moderately high pressure of oxygen. Our results are described in Section 4 and we conclude in Section 5.

2. Method of calculation

We base our total energy calculations on a supercell with periodic boundary conditions, which enables us to use a basis of plane waves. The supercell has the form of a rhombohedral prism, and in the stoichiometric slab it contains 30 atoms. This slab is exactly the thickness of one bulk unit cell of the corundum structure.

Our surface calculations are made on slabs of this kind repeated periodically in the z direction, with a vacuum space of thickness about equal to that of the slab. This is adequate to isolate the surfaces. The stoichiometric slab has two equivalent surfaces, which are terminated by an Al plane, as in Fig. 1, in which our labelling of the atoms is shown. This termination *defines* the stoichiometric surface, because the slab as a whole is stoichiometric and has two equivalent surfaces. We note as an aside that surfaces which are not stoichiometric are sometimes called polar. Besides the stoichiometric surface, we have studied four non-stoichiometric surfaces, two with an oxygen excess and two with an aluminium excess. By stripping off the surface plane of Al we obtain a surface which is O terminated, with an O excess of $+1.5$ atoms per unit surface cell, or an Al excess of -1 atoms per unit surface cell (see eqn. (5) below). We can then proceed to remove the three surface O atoms one at a time, giving three more surfaces with O excesses of 0.5, -0.5 and -1.5, respectively. The final surface is terminated by two layers of Al. In the actual calculations we ensure that the surfaces on each side of our slab are equivalent (to avoid a dipole moment in the supercell). Thus we actually add a seventh layer of oxygen to create the O-terminated surfaces and remove a layer of oxygen to create the most Al-rich surfaces.

The total energy of the contents of a supercell is minimised with respect to the electronic co-ordinates (the coefficients of each plane wave in each occupied wave function) and the atomic co-ordinates, as described in previous work.[12,13] The ionic potentials are represented by pseudo-potentials of the Troullier–Martens form.[14] All the calculations were made with two k-points in the irreducible (120°) wedge of the Brillouin zone, and with a plane-wave cut-off of 40 Ry. Convergence tests were reported previously.[15]

In order to obtain information about the charge redistribution at these surfaces we calculate the Mulliken populations. These are defined by the formalism of Mayer.[16] In our case we use the pseudo-orbitals $|\varphi_{i\alpha}\rangle$ as a basis for projection, which includes O 2s, O 2p, Al 3s and Al 3p components (labelled α) on each site (labelled i). The "spillage" of each occupied orbital ψ, defined as $1 - \sum_{i\alpha} |\langle \psi | \varphi_{i\alpha} \rangle|^2$, with this atomic-like basis is less than 1.5%.

3. Surface energy and oxygen partial pressure

We have chosen here to discuss the regime we think is more likely to be encountered in practice, in which the Al_2O_3 surface is in equilibrium with Al and O in the vapour phase, rather than the regime in which it is in equilibrium with Al in the solid or liquid phases. Since there are two components, the temperature T and pressure P are two independent variables which can be used to specify the state of the system. To be of use in experimental situations our final formula for the surface energy will be couched in terms of the chemical potential (μ_O) or partial pressure (p_{O_2}) of oxygen rather than the total pressure P.

We adopt the thermodynamic definition of surface excesses due to Gibbs, most clearly described by Cahn.[17] For future application to other oxides, we derive here the general formula appropriate to an oxide of metal M with the stoichiometric composition M_mO_n. The surface energy γ in this case is given by:

$$\gamma(T, P) \cdot A = G_{\text{slab}}(T, P) - N_M \mu_M^v(T, P) - N_O \mu_O(T, P) \qquad (1)$$

in which A is the area of the surface within the supercell, counting both surfaces of the slab. G_{slab} is the Gibbs energy of the contents of the supercell, and N_M and N_O are the total number of atoms of metal and oxygen, respectively, within the supercell. μ_M^v is the chemical potential of the metal, and we use the superscript v to emphasise that at the temperature of interest the metal is in the vapour phase. If this were not the case we would be dealing with an interface of the solid or liquid metal with its oxide, and this would require a different treatment. In a supercell of a few tens of atoms, such as we use for *ab initio* calculations, all the atoms can be assumed to be in the solid phase and approximately half of the supercell is a true vacuum. No serious error is made in eqn. (1) by omitting the statistical occurrence within the supercell of atoms in the vapour phase. Likewise, no significant error is made by not explicitly including the presence of point defects within the solid part of the supercell. It is more problematic that we are ignoring the surface terminations involving a statistical distribution of defects, as in a partial layer of adsorbed oxygen.

The chemical potential of oxygen is given in terms of its partial pressure p_{O_2} by the ideal gas expression:

$$\mu_O = \mu_O^\circ + \tfrac{1}{2} kT \ln p_{O_2} \qquad (2)$$

in which we use superscript $^\circ$ to denote the standard state (STP), and the pressure is in units of atmospheres. We choose to define the chemical potentials per atom rather than per mole. In order to relate $\gamma(T, P)$ to p_{O_2} we wish to insert eqn. (2) into eqn. (1). Two problems arise at this point. First, in *ab initio* calculations the zero of free energy is the energy of separated free ions and electrons at rest; this is not the usual convention, which is the free energy of the elements at STP. With the latter convention, $\mu_O^\circ = 0$, but the convention of our calculations means that if we use eqn. (2) to evaluate eqn. (1) we still have to know the value of μ_O°. Secondly, the free energy of oxygen in the vapour phase is a quantity we wish to avoid calculating explicitly. Because of the paramagnetic nature of oxygen, even the energy of a molecule at rest would require a more accurate treatment than the local density approximation for exchange and correlation, which is adequate for solid phases. We can deal with both these problems by making use of experimental thermodynamic data, as follows.

First, we can use the equilibrium of the two phases to eliminate the chemical potential of the metal vapour in favour of the Gibbs energy per formula unit of the metal oxide, G_{MO}:

$$G_{MO} = m\mu_M^v + n\mu_O. \tag{3}$$

Inserting eqn. (3) into eqn. (1) gives:

$$\gamma(T, P) = \frac{1}{A}(G_{slab}(T, P) - \frac{1}{m} N_M G_{MO}(T, P)) - \Gamma_O \mu_O(T, P) \tag{4}$$

in which we have introduced the surface excess of oxygen with respect to the metal, defined by[17,18]

$$\Gamma_O = \frac{1}{A}\left(N_O - \frac{n}{m} N_M\right). \tag{5}$$

Next we use the definition of the standard Gibbs energy of formation to obtain the oxygen chemical potential in its standard state:

$$G_{MO}^\circ = m\mu_M^\circ + n\mu_O^\circ + \Delta G^\circ. \tag{6}$$

Combining eqn. (6) with eqn. (4) and (2) gives:

$$\gamma(T, P) = \frac{1}{A}\left(G_{slab}(T, P) - \frac{1}{m} N_M G_{MO}(T, P)\right)$$
$$- \Gamma_O\left(\frac{1}{n} G_{MO}^\circ - \frac{m}{n}\mu_M^\circ - \frac{1}{n}\Delta G^\circ\right) - \Gamma_O \frac{1}{2} kT \ln p_{O_2}. \tag{7}$$

Eqn. (7) is the result we were seeking. The first term is calculated at 0 K, and we omit its temperature dependence. In principle, one could perform a molecular dynamics simulation and, by means of thermodynamic integration from 0 K to T, obtain the temperature dependence of this term. Alternatively, at temperatures not too close to the melting point, it could be estimated from a calculation of the phonon frequencies of the slab and of a comparison piece of bulk material, using the quasiharmonic approximation. The accuracy of both approaches is limited by the sample size but we feel that either would be reasonable since the result is an integrated quantity localised near the surface. The first term is also where a large cancellation takes place between the difference of two energies which scale with the number of atoms leaving a superficial quantity. It is therefore advantageous to use an identical cell and k-point mesh for the slab and bulk terms in it.

The quantities G_{MO}° and μ_M° entering the second term are well described by 0 K quantities, which we calculate. It can be verified that correcting them to STP has a negligible effect on the surface energy. Finally ΔG° is taken from experimental thermodynamic data. This leaves the p_{O_2} term as the one that describes the important variation of γ with temperature and oxygen partial pressure. Notice that in the case of a stoichiometric termination, only the first term is non-zero, and this describes the elementary case in which the energy of bulk phase is subtracted from the energy of a slab containing exactly the same number of molecules.

The minimum physically meaningful value of p_{O_2}, which we denote $p_{O_2}^{min}$, is set by the condition that at $p_{O_2} \leqslant p_{O_2}^{min}$ the oxide would spontaneously decompose into metal and oxygen. This would be the case if

$$\mu_M^c(T, P) \leqslant \mu_M^v(T, P) \tag{8}$$

where $\mu_M^c(T, P)$ denotes the chemical potential of the metal in its condensed phase. The result in terms of p_{O_2} is fairly well known, but we briefly rederive it here for completeness. Substituting eqns. (3) and (6) into eqn. (8) and rearranging the inequality gives:

$$\mu_O - \mu_O^\circ \leqslant \frac{1}{n}\Delta G^\circ + \frac{1}{n}(G_{AO}(T, P) - G_{AO}^\circ) - \frac{m}{n}(\mu_M^c(T, P) - \mu_M^\circ). \tag{9}$$

From the specific heat data of the solids we know that the second and third terms are about two orders of magnitude smaller than the first in alumina, and we expect them to be negligible in general. Hence we obtain the relation:

$$\ln p_{O_2}^{min} = \frac{2}{nkT} \Delta G°. \qquad (10)$$

It is worth noting that the above formulae could be applied directly to calculate the energy of oxide/metal interfaces. All that is required is to subtract from eqn. (7) the energy of the same amount of bulk phase of the metal with which the oxide is in contact. In this case G_{slab} would refer to a supercell containing a slab of oxide in contact with metal and there would be no vacuum space.

Note on nomenclature for surface excess

In the remainder of this paper we shall characterise the surface terminations by the natural unit of oxygen surface excess Γ_O, which is atoms per surface unit cell rather than atoms per unit area of the slab. This is equivalent to setting $A = 2$ in the above equations (since there are two surfaces in the supercell). A simple algorithm to determine the surface excess for a given surface which does not rely on having equivalent surfaces on a slab is given by Finnis.[18]

4. Results and discussion

4.1. Surface relaxation

The interlayer relaxations are accompanied by x-y (in-plane) relaxations of the three O atoms per plane in each unit cell. The manner of these in-plane relaxations is illustrated in Fig. 2. The Al atoms do not relax laterally but remain aligned along z to preserve the threefold axes. Since the three O atoms above an empty octahedral site in each plane are symmetry related and lie at the corners of an equilateral triangle, their lateral relaxation can be characterised by two parameters,

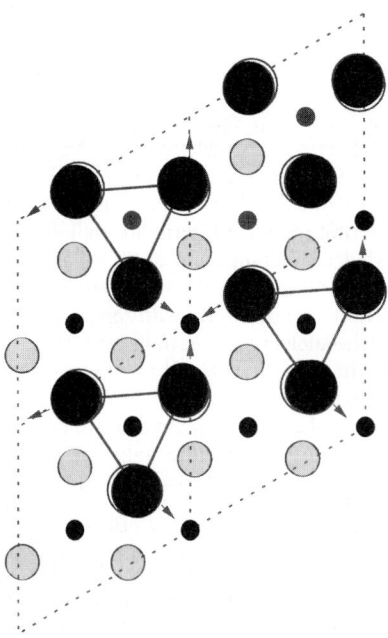

Fig. 2 Plan view of the (0001) surface illustrated in Fig. 1 ($\Gamma_O = 0$), showing the lateral relaxations within the topmost O plane. The relaxed positions are shaded black. The rotation and expansion of the triangle below Al1 is indicated.

which we take to be the rotation of the triangle and the linear expansion or bond length change of the triangle. This is the triangle on which the surface Al atoms labelled Al1 in Fig. 1 are centred. The magnitudes of the relaxations for three surfaces are reported in Table 1.

We find for the $\Gamma_O = 0$ surface, the only one for which a comparison with experiment is possible, qualitative agreement between calculations and experimental measurements. The most marked effect is the well known strong inward relaxation of Al1, which we calculate to be 77%, and which as in the previous published calculations of this quantity, is significantly larger than the experimental value of 51%. The accompanying expansion and rotation of the O1 triangle centred on an axis through Al1 are calculated to be 3.2% and 3.05°, compared to the experimental values of 4.2% and 6.7°. These are quantities for which predictions have not previously been compared directly to experiment. We believe that Al1 as it relaxes is squeezing the O1 triangle open. This triangle rotates so as to allow the O1 atoms to approach Al2 rather than to move towards other O1 atoms in the same plane. There are three symmetry-equivalent O1 atoms which thereby form a more constricting triangle about each Al2 atom, so the interplanar O1–Al2 distance, not surprisingly, increases. This is our interpretation of the 10.6% calculated O1–Al2 interplanar relaxation, which is 16% experimentally. Unfortunately previous calculations give a much smaller or even negative value for this parameter. We tend to discount the tight-binding results in this comparison, because they showed the opposite sign for the third layer (Al2–Al3) relaxation compared to ours and experiment. Nevertheless, there remains a significant discrepancy between theory and experiment and even between the first-principles theories for which we have no satisfactory explanation.

The calculations predict a similar rotation–expansion of the O1 triangle for the $\Gamma_O = 1.5$ (oxygen terminated) surface. In this case, the topmost interlayer spacing contracts by 14.6%, and these relaxations can be consistently explained by the electrostatic pull of the Al2 ions.

4.2. Mulliken charges

The total charges on each ion calculated from the Mulliken population analysis are shown in Table 2. Caution is needed in interpreting these data for two reasons. First, because the local basis is not complete the Mulliken charges do not exactly balance the ionic charges. For example, each formula unit carries an ionic charge of $2 \times 3 + 3 \times 6 = 24$. A spillage of 1% therefore corresponds

Table 1 Calculated relaxations of the O-terminated ($\Gamma_O = 1.5$), Al-terminated ($\Gamma_O = 0$) and double Al-terminated ($\Gamma_O = -1.5$) surfaces[a]

	Γ_O			
	1.5	0	0(exp)	−1.5
O-Rotation	4.12	3.05	6.7	0.60
O-Expansion	4.30	3.20	4.2	0.62
Al1–O1	—	−77.0 (−70.3)	−51	—
O1–Al2	−14.6	+10.6 (+13.9)	16	—
Al2–Al3	+6.8	−34.3 (−38.3)	−29	+13.9
Al3–O2	+12.1	+18.5 (+22.5)	20	+7.9
O2–Al4	−3.6	+1.0 (+9.5)		+7.7
Al4–O5	−14.9	−1.9 (−4.1)		+1.0

[a] The supercells contained 33, 30 and 27 atoms, respectively. Values in parentheses for $\Gamma_O = 0$ were calculated for a thinner slab of only 20 atoms, and are shown in order to give an indication of finite size effects. The third column contains experimental results.[2] The first row shows the rotation of the topmost O triangle, which is measured in degrees, all other relaxations are given in percentages of bulk distances: the second row gives the O–O expansion of the topmost O triangle and subsequent rows give interplanar relaxations. The topmost O triangle contains the atoms labelled O1 except for the double Al termination, for which the topmost O atoms are O2.

Table 2 Mulliken + ionic charges for atoms near the surface[a]

	Γ_O				
	+1.5	+0.5	0	−0.5	−1.5
Al1	—	—	1.72	—	—
O1	−0.58	−0.76	−1.03	−0.93	—
Al2	1.69	1.65	1.54	1.20	0.75
Al3	1.66	1.61	1.57	1.43	0.88
O2	−0.99	−0.99	−0.99	−1.03	−1.01
Al bulk	1.57	1.57	1.57	1.57	1.57
O bulk	−1.00	−1.00	−1.00	−1.00	−1.00

[a] The columns label the five surfaces by their surface excess of O in atoms per unit surface cell. The notation for atomic planes is as in Fig. 1.

to 0.24 "missing" electrons per formula unit. We have not sought to allocate these missing electrons in any way, which would certainly be arbitrary. Secondly, because the local basis itself is not unique, the charge transfers calculated are not unique and a different choice of basis would give different results. However, the *trends* in charge transfer can be interpreted in a meaningful way, which will not be different if the charges themselves are redefined by a different choice of local orbitals. With these caveats in mind we comment on the results in Table 2.

First, considering the $\Gamma_O = 0$ surface, the surface ionic charges are similar to the bulk but the ionicity is slightly enhanced. The charge on Al1, for example, is 1.72 compared to 1.57 in the bulk, an extra charge which is mostly drawn from Al2. There is certainly no sign of increased covalence, which one might have associated with the large relaxation of Al1.

Secondly, on the O terminated surface ($\Gamma_O = 1.5$) the surface O is carrying 0.58 electrons compared to 1.0 in the bulk. As the oxygen excess is reduced to zero the number increases from 0.58 to 1.03, slightly exceeding the bulk value. Correspondingly, as the oxygen excess decreases and becomes negative the topmost Al tends to carry less charge. In the most Al-rich case ($\Gamma_O = -1.5$) the bulk charge on Al of 1.57 is roughly shared between the two surface layers of Al (0.75 and 0.88). These large variations in surface charge highlight the well known difficulty of using an ionic model in all but stoichiometric situations.

4.3. Densities of states

The stoichiometric surface is known from calculations to be insulating with a large band gap like the bulk material, however in the non-stoichiometric surfaces two kinds of surface metallisation can be expected. For Al-rich surfaces the charge on Al is reduced and the extra electrons may occupy the Al 3s and 3p states to give a localised conducting band. Conversely for O-rich surfaces, electrons may be missing from the O 2p states which characterise the top of the valence band in alumina, and this may provide surface localised conducting states of hole character.

The above picture is consistent with the trends in Mulliken charge documented above, and it is confirmed by calculations of the densities of states on the slabs shown in Fig. 3. These show the empty states at the top of the valence band in O-rich surfaces and the metallic band of electrons in Al-rich surfaces. The strong surface localisation of these surface states is illustrated by the way their charge density decays within two or three atomic layers of the surface. An example is plotted in Fig. 4, which depicts the charge density of a HOMO in the $\Gamma_O = 1.5$ slab. The stoichiometric surface in these calculations displays a localised state in the gap just below the conduction band, as was found also with the tight-binding model after relaxing the atomic positions.[9]

4.4. Surface energies

The surface energies are plotted as a function of p_{O_2} in Fig. 5, using eqn. (7). We see that over almost all the range of pressure up to 1 atmosphere the $\Gamma_O = 0$ surface has the lowest free energy. Only in UHV could we hope to see the Al-rich surface. This confirms the experimental picture

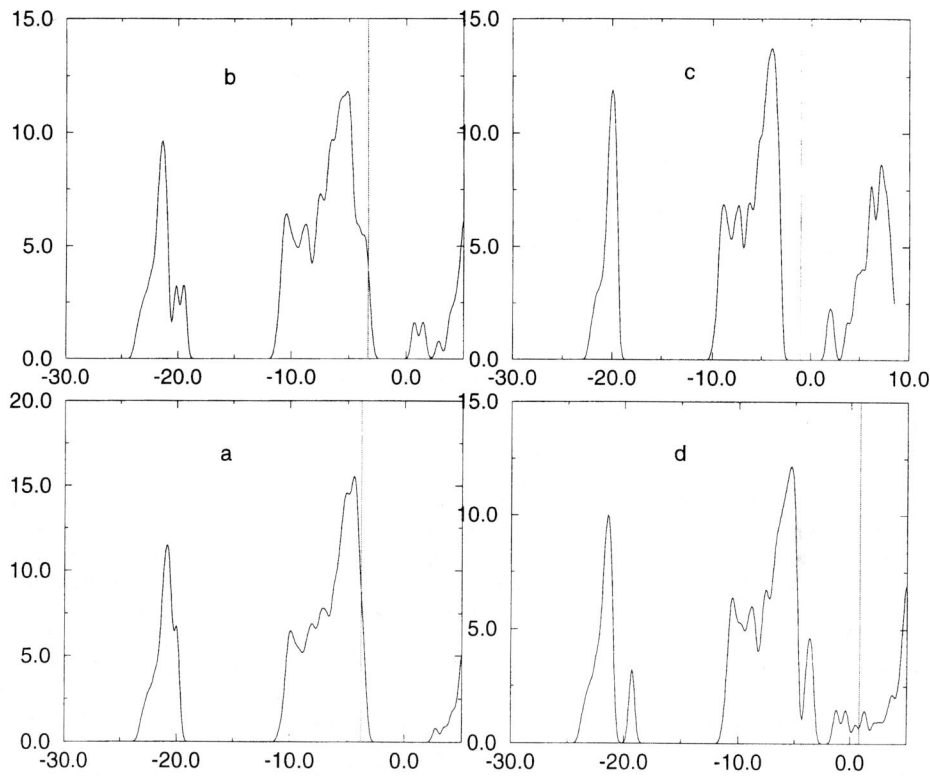

Fig. 3 Total density of states of slabs. (a) $\Gamma_o = 1.5$, (b) $\Gamma_o = 0.5$, (c) $\Gamma_o = 0$, (d) $\Gamma_o = -0.5$.

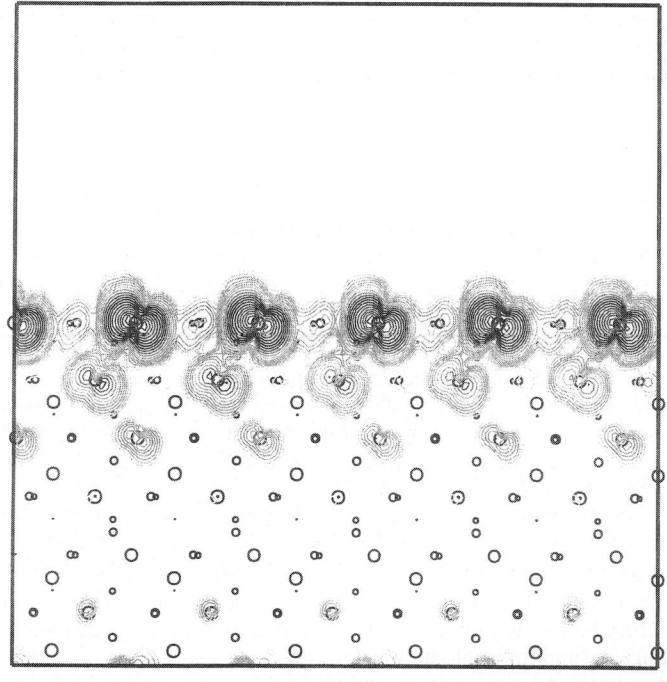

Fig. 4 Charge density of the HOMO for the O-terminated slab ($\Gamma_o = 1.5$).

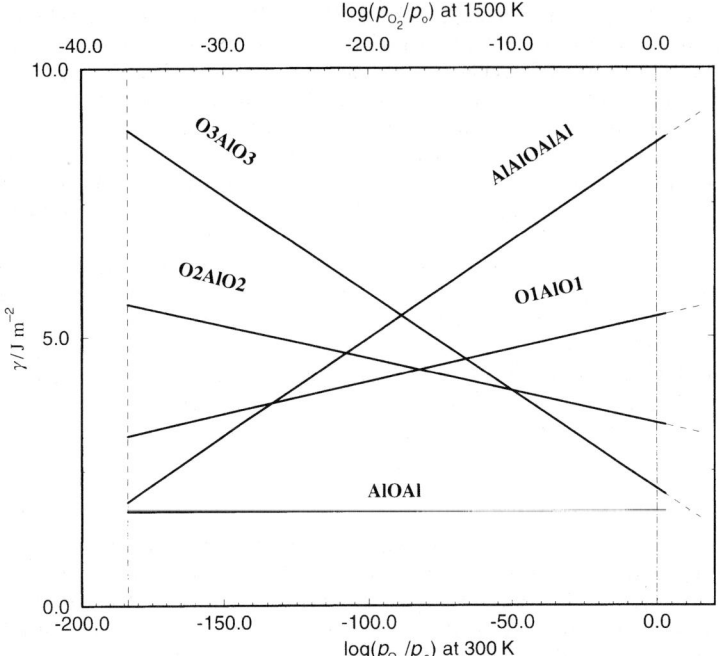

Fig. 5 Calculated surface energies of various 1 × 1 surfaces as a function of oxygen partial pressure; $p_o = 1$ atm. Two different temperatures are shown explicitly by scaling of the horizontal axis according to eqn. (7).

except we have no theoretical data for the observed Al-rich $(\sqrt{31} \times \sqrt{31})R \pm 9°$ surface which has been observed under UHV conditions. It has an oxygen excess $\Gamma_O = -6$ if the postulated structure[1,2,11,19] is correct, so it would have a very steep slope of surface energy vs. p_{O_2} on Fig. 5. Depending on the first term in eqn. (7) it may then cut below the $\Gamma_O = 0$ line on Fig. 5 towards the left of the diagram but at a higher pressure than the $\Gamma_O = -1.5$ line does. Until the calculations are done this is a matter for speculation.

We certainly expect the O-terminated ($\Gamma_O = 1.5$) surface to be stable at oxygen pressures somewhat greater than 1 atm, but we would not claim sufficient accuracy in our calculations, which neglect the temperature dependence of the first terms in eqn. (7), to make a precise prediction.

5. Conclusions

1. We have presented a formalism for calculating the dependence of surface and interfacial energies in oxides on the partial pressure of oxygen, and applied it to compare the energies of five postulated (1 × 1) (0001) surfaces of corundum, differing in their surface stoichiometry. This is essentially a more explicit and somewhat extended version of the formalism used, for example, by Wang et al.,[20] who expressed the surface energy in terms of the chemical potential of oxygen. Some subtleties associated with the definition of the zero of energy have been clarified here, and by means of a thermodynamic cycle it has proved unnecessary to calculate any properties of pure oxygen, which would have been problematic.

2. We find that the observed neutral surface, terminated by Al, is stable up to atmospheric pressure. At some higher pressure, perhaps some tens of atmospheres, depending on the temperature, the oxygen-terminated surface would be more stable (see Fig. 5). An Al-rich surface will probably be stable at sufficiently low oxygen pressure, but the most Al-rich surface considered here (two terminating layers) is only theoretically stable at a pressure just below the very low pressure under which corundum would decompose.

3. The relaxed atomic and electronic structures of the surfaces have been obtained. There is qualitative agreement with experimental X-ray diffraction results for the stable surface; this

includes the sign and approximate magnitude of the first four interlayer relaxations, and also the lateral displacements of the surface oxygen atoms, which can be thought of as a rotation and expansion of the oxygen triangle beneath the surface Al. This is explained in terms of electrostatic forces. Ours and previous results both using the local density approximation (or generalised gradient approximation) and Hartree–Fock consistently overestimate the inward relaxation of the surface Al plane compared to the X-ray data. The discrepancy is unresolved. Regarding the second interlayer relaxation, our result (10.6%) is closer to experiment (16%) than previous calculations, some of which omitted lateral relaxations. We have verified, by further tests, that if we do not allow lateral relaxation of oxygen the second interlayer relaxation is suppressed. However, Verdozzi and co-workers[5] using a Gaussian basis, and Wang et al.[20] using the full-potential Linear Augmented Plane Wave method have predicted smaller second interlayer relaxations than ours, even though they included lateral relaxations; this requires further investigation, although the discrepancy is less than 0.1 Å.

4. Non-stoichiometric surfaces are metallic. In the case of Al-rich surfaces the Fermi energy lies in a band of surface states below and contiguous with the conduction band. Oxygen-rich surfaces are metallic because of holes at the top of the valence band, which is mainly of O 2p character.

Acknowledgements

We thank J. Hütter for technical help with the calculations. This work has been supported by the UK Engineering and Physical Sciences Research Council under grants GR/L08380 and GR/M01753. The Centre for Supercomputing in Ireland is gratefully acknowledged for computer resources. This work has benefited from collaborations within, and has been partially funded by, the Training and Mobility Network on "Electronic Structure Calculation of Materials Properties and Processes for Industry and Basic Sciences" (Contract FMRX-CT98-0178).

References

1   G. Renaud, *Surf. Sci. Rep.*, 1998, **32**, 1.
2   P. Guénard, G. Renaud, A. Barbier and M. Gautier-Soyer, *Surf. Rev. Lett.*, 1998, **5**, 321.
3   I. Manassidis, A. DeVita and M. J. Gillan, *Surf. Sci. Lett.*, 1993, **285**, L517.
4   C. Kruse, M. W. Finnis, V. Y. Milman, M. C. Payne, A. DeVita and M. J. Gillan, *J. Am. Ceram. Soc.*, 1994, **77**, 431.
5   C. Verdozzi, D. R. Jennison, P. A. Schultz and M. P. Sears, *Phys. Rev. Lett.*, 1999, **82**, 799.
6   V. E. Puchin, J. D. Gale, A. L. Shluger. A. A. Kotomin, J. Gunster, M. Brause and V. Kempter, *Surf. Sci.*, 1997, **370**, 190.
7   M. Causà, R. Dovesi, C. Pisani and C. Roetti, *Surf. Sci.*, 1987, **215**, 259.
8   C. Pisani, M. Causà, R. Dovesi and C. Pisani, *Prog. Surf. Sci.*, 1987, **25**, 119.
9   T. J. Godin and J. P. LaFemina, *Phys. Rev. B: Condens. Matter*, 1994, **49**, 7691.
10 W. C. Mackrodt, R. J. Davey, S. W. Black and R. Docherty, *J. Cryst. Growth*, 1987, **80**, 441.
11 G. Renaud, B. Vilette, I. Vilfan and A. Bourret, *Phys. Rev. Lett.*, 1994, **73**, 1825.
12 A. Alavi, J. Kohanoff, M. Parrinello and D. Frenkel, *Phys. Rev. Lett.*, 1994, **73**, 2599.
13 A. Alavi, *Philos. Trans. R. Soc. London Ser. A*, 1998, **356**, 263.
14 N. Troullier and J.-L. Martins, *Phys. Rev. B: Condens. Matter*, 1991, **43**, 1993.
15 I. G. Batyrev, A. Alavi, M. W. Finnis and T. Deutsch, *Phys. Rev. Lett.*, 1999, **82**, 1510.
16 I. Mayer, *Chem. Phys. Lett.*, 1983, **97**, 270.
17 J. W. Cahn, in *Interfacial Segregation*, ed. W. C. Johnson and J. M. Blakely, American Society for Metals, Metals Park, OH, 1977, pp. 3–23.
18 M. W. Finnis, *Phys. Status Solidi A*, 1998, **166**, 397.
19 M. Gautier, G. Renaud, L. P. Van, B. Villette, M. Pollak, N. Thromat, F. Jollet and J.-P. Duraud, *Sci. Alumina*, 1993, **77**, 323.
20 X. G. Wang, W. Weiss, S. K. Shaikhutdinov, M. Ritter, M. Petersen, F. Wagner, R. Schlögl and M. Scheffler, *Phys. Rev. Lett.*, 1998, **81**, 1038.

Paper 9/03278I

Ultrathin alumina film Al-sublattice structure, metal island nucleation at terrace point defects, and how hydroxylation affects wetting

D. R. Jennison* and A. Bogicevic

Surface and Interface Sciences Department 1114, Sandia National Laboratories, Albuquerque, NM 87185-1421, USA. E-mail: drjenni@sandia.gov

Received 7th July 1999

First principles density functional slab calculations have produced the following results: (1) for 5 Å (two O-layer) alumina films on Al(111) and Ru(0001), with larger unit cells than in recent work, the lowest energy stable film was found to have an even mix of tetrahedral (t) and octahedral (o) Al ions arranged in alternating zig-zag rows. This most closely resembles the κ-phase of bulk alumina, where this pattern results in a greater average lateral separation of Al-ions than with pure t or o. A second structure with an even mix was also found, consisting of alternating stripes. These patterns can exist in any of three equivalent directions on close packed substrates. (2) Because of numerical problems associated with the very large relaxations in alumina surfaces, MgO(100) was used to investigate metal island nucleation. Common point defects (vacancies, pairs of vacancies, and water by-products) were placed in supercells and their effects on Pt adatom and ad-dimer binding computed. Unexpectedly, single vacancies were found to *de*stabilize metal dimers, and only the mixed (F_s-V_s) divacancy increases stability. Among the water-by-products, in-surface OH (produced by H^+ reaction with O^{2-}) was uninteresting, but ad-OH was found to both increase adatom binding and significantly stabilize dimers on the surface, suggesting the latter defect nucleates metal islands even at elevated temperatures. We believe these results apply to all highly ionic oxides. (3) Finally, the effect of a substantial coverage of hydroxy on Cu deposition and growth on α-Al_2O_3(0001) was investigated. While Born–Haber calculations show wetting is not thermodynamically preferred on the clean surface, at experimentally relevant ad-OH coverages the strength of the Cu–oxide bond is more than doubled and wetting is strongly favored. This causes a very stable $\approx 1/3$ monolayer (ML) coverage of Cu^{+1}, which then induces layer-by-layer Cu growth, in agreement with recent experiments. Hydroxy coverage can thus control deposited metal morphology across a wide spectrum.

1 Introduction

In this paper, we include for discussion three topics of current interest in metal oxide surface science. Using first principles density functional theory (DFT)[1] calculations, we have investigated: (1) the atomic-scale structure of experimentally relevant ultrathin alumina films, (2) the role of common point defects in metal island nucleation on oxide terraces, and (3) the growth and morphology of metals on oxide surfaces that have high concentrations of a common impurity.

1.1 Ultrathin alumina film structure

Aluminium oxide films have a substantial focus for a variety of reasons. First, they represent structures that can be produced during the oxidation of Al metal, and are thus important for understanding the "barrier layer", which inhibits corrosion. Second, thin films enable the study of adsorption without the charging and hydrogen impurity problems which plague bulk-terminated Al_2O_3. Third, since alumina is an important support material, when metal nanoclusters/crystals are produced by metal deposition, the films serve as model catalysts.[2] Finally, there is growing interest for microelectronics applications.

Considerable experimental and theoretical work on this system has occurred recently, with one main topic being metal adsorption and the islands which result.[3] Obviously, nucleation plays a critical role here (see below), but without knowledge of the atomic scale structure one cannot address the first issue, the basic energetics of adsorption, diffusion, and dimer stability.

Two O-layer 5 Å films are of particular interest because this thickness appears to be self-limited when produced by NiAl(110)[4] or $Ni_3Al(111)$[5] oxidation, and more recently by the deposition and subsequent oxidation of Al on Ru(0001).[6] (Two-layer oxide films have also been recently seen in a completely different system: encapsulated Pt nanocrystals on TiO_2.[7])

High-resolution electron energy-loss spectroscopy (HREELS) evidence from Al_2O_3/NiAl(110) suggests a mixture of octahedral (o) and tetrahedral (t) Al ions is present,[8] which has resulted in the films being called "gamma-like". Recently, transmission electron microscopy (TEM) moiré patterns have indeed indicated that the lattice constant of the film is consistent with a γ-phase,[9] but this result is also consistent with other possible structures. (Of course, film thinness, at two O-layers, prevents a definable hcp *vs.* fcc stacking, which differentiates the α- and γ-phases.) Actual structural details are in fact unknown, but a significant hint has arisen from new experiments by the group of Behm *et al.*[6] on Al deposition and subsequent oxidation on Ru(0001). While islands of 5 Å thick Al_2O_3 were seen at various coverages using scanning tunneling microscopy (STM), low-energy electron diffraction (LEED) evidence for ordering in the Al-sub-lattice was either not or only weakly seen, in spite of annealing well above 300 K.

On the theoretical side, recent calculations on two and three O-layer Al_2O_3 films on Al(111), Mo(110) and Ru(0001) produced three significant findings:[10] (1) the preferred interface between the oxide film and the substrate metal consists in all cases of 1×1 chemisorbed oxygen, (2) on top of which is a nearly coplanar layer of Al and O ions, (3) with the normal bulk preference for octahedral-(o) over tetrahedral (t)-site aluminium ions energetically reversed. The last result is due to the electrostatics and layer separations induced by the interface with the underlying metal. Howeve, this initial work used only small unit cells, and did not allow for the possibility of greater complexity. Here, we report computational results obtained by expanding the unit cell to allow a variety of t/o ratios and structures.

1.2 Metal island nucleation on terraces

Surface topographs from STM or atom force microscopy (AFM) observe metal island nucleation on oxide surfaces not only at line defects (such as antiphase domain boundaries[4,11] and steps,[12] where metal atoms presumably bind more strongly and therefore have a tendency to collect and meet) but also on terraces. For the latter, nucleation could of course occur on a perfect surface, depending on the temperature and density of the adatom lattice gas (which determine the stability of metal ad-dimers, the probability of attaching a third metal atom, *etc.*). In contrast, surface defects might dominate nucleation in experimental conditions.[2,13-16]

Even though it has been speculated that the most common defect in well prepared surfaces, specifically isolated surface oxygen vacancies,[17] may act as a nucleation site,[13,14] this has not been substantiated *via* experiment or theory. Here, we report an investigation of the influence of surface vacancies on Pt island nucleation.[18] For completeness, we also examined how water dissociation products affect nucleation, since there have been several reports that these are common low density contaminants on prepared oxide surfaces.[17,19]

It is very difficult numerically to study these defects in alumina films or with sapphire because of the extremely large surface relaxations.[20,21] Therefore, we have chosen a system with an order-of-magnitude smaller relaxations,[21] MgO(100). From this first study of dimer stability at oxide surface defects, several findings are completely unexpected, but are really quite intuitive in retrospect, and are likely to be general for highly ionic oxides.

1.3 How hydroxylation affects wetting

Cu deposition on oxides has assumed increased recent importance because of microelectronics applications. However, experimental results[22–27] for Cu on alumina have been inconsistent. X-Ray photoelectron spectroscopy (XPS) studies[22] of Cu deposited by thermal evaporation onto bulk truncated α-Al_2O_3(0001) indicated ordered layer-by-layer growth for the first 2–3 atomic layers. The initial Cu ad-layer was observed as oxidized Cu, in the form of Cu(I) ions. Other studies on polycrystalline Al_2O_3 reported layer-by-layer growth[23] and Cu(I) formation at coverages below 0.5 monolayers.[24] In contrast, a study on epitaxial ≈20 Å Al_2O_3 films formed on refractory metal substrates[25] reported the growth of 3-D clusters of metallic Cu, even at sub-monolayer Cu coverages. In particular, XPS and low energy ion scattering (LEIS)[25] indicated Cu cluster formation at the lowest observable coverages at both 300 K and 80 K, with no Cu(I) observed. In addition, X-ray absorption near-edge structure spectroscopy (XANES)[26] measurements carried out on sapphire substrates have reported no evidence of Cu oxidation, and coverage-dependent shifts in Cu core level and LMM peaks have been interpreted in terms of final state screening,[27] rather than ionization of the Cu. Meanwhile, recent ion scattering experiments by Ahn and Rabalais[19] have shown that cut and polished sapphire(0001) surfaces (the basil plane is not a cleavage surface) cannot be made free of hydrogen contamination in the form of hydroxy, even by annealing to 1400 K. Finally, experimental studies of Rh deposited on ultrathin epitaxial Al_2O_3 films[28] suggest that surface hydroxy binds the Rh to the surface as a cation and serve as nucleation sites for Rh clusters. These studies have raised the issue of the role of surface hydroxy groups in producing the apparent disagreements summarized above.

The most recent experimental work, by Kelber et al.,[29] indicates initial layer-by-layer growth of Cu on hydroxylated α-Al_2O_3(0001) at 300 K, and the exclusive presence of Cu(I) during the formation of the first layer. Analysis of X-ray excited Cu(LMM) Auger data indicates that changes in the spectra are due to changes in the initial electronic state of the copper rather than to final state screening effects. The degree of surface hydroxylation is estimated to be high, ≈1/3 ML (here, 1 ML means one adsorbate per surface O ion), on the basis of O 1s XPS, consistent with ref. 19.

In collaboration with Kelber et al.,[29] we have computed the adsorption energy of Cu at 1/3 and 1 ML coverages, both on clean α-Al_2O_3(0001) and on α-Al_2O_3(0001) with 1/3 ML of ad-OH. Born–Haber cycles were then used to study the relative energy of isolated adatoms vs. incorporation into 2-D islands. We find dramatic effects on the binding energies due to hydroxylation and also on the growth mode, suggesting that hydroxylation may explain the discrepancies in the experimental record.

Following a description of the computational details, we present our results and raise some issues for future work and for discussion.

2 Computational method

Our electronic structure calculations were performed using the Vienna *ab initio* simulation package (VASP).[30] This plane-wave based density functional code uses the ultra-soft pseudo-potentials of Vanderbilt,[31] which have good convergence for these systems with a plane wave cut-off of only ≈270 eV. We used either the "standard" local density approximation (LDA)[32] or the PW91 generalized gradient approximation (GGA),[33] as indicated below. The geometric relaxation was done first with a quasi-Newton algorithm using computed interatomic forces. For the alumina systems, where relaxation is large and problematic because of a mix of very hard and soft modes, geometry was refined with a damped dynamics scheme built into VASP. The vacuum gap was in all cases > 18 Å, and k-point sampling was tested to ensure convergence to the quoted level of accuracy.

2.1 Ultrathin alumina film structure

Our slabs had the alumina film on 4–7 layers of Al(111) or Ru(0001). The x–y dimensions of the supercell depended on the structure being studied. Because LDA has shown excellent accuracy in alumina structural predictions[20] and GGA does not improve on same,[34] LDA was used here. Since our study necessitated numerous computations using large supercells, the following tests were performed to ensure accuracy: (1) the relative energies of 2 × 1 supercells (Fig. 1, top) with -o-o-, -t-o-, and -t-t- zig-zag Al rows were computed for a film with seven layers of Ru substrate (bottom four frozen at bulk LDA spacings) and using eight k-points. Errors produced in *relative* energy by reducing the Ru slab to just four layers (bottom two frozen), and/or the number of k-points to two, were found to be <0.1 eV out of energy differences of ≈2 eV per 2 × 1 cell. (2) Because numerical noise (arising from small inaccuracies in force computation) is seen to grow significantly during prolonged geometric relaxation, tests were done to examine the effect of freezing the entire metal substrate (these systems mix strong and soft vibrational modes, and all first-guesses had ions at the ideal positions with respect to the extended metal lattice). It was again found that errors were small, here below 0.05 eV per 2 × 1 cell. These results indicate that the relative energies are determined almost entirely within the oxide film itself which, because the bands are relatively flat, can be described adequately by few k-points.

2.2 Metal island nucleation on terraces

A supercell of five MgO(100) layers with 36 atoms each was used together with GGA.[18] Two types of OH impurity were studied: as a "neutral" species it is produced by adding OH or H to the supercell, while as a "charged" species it occurs when both OH and H are added to the supercell, which naturally charge separate into ad-OH$^-$ and H$^+$, the latter reacting with a surface O^{2-} to produce in-surface OH$^-$. The latter was found to reduce the binding of ad-Pt atoms (due to the reduction in charge compared with the perfect surface) and are thus uninteresting for nucleation, except as a means to concentrate the density of Pt adatoms in defect free regions; we do not discuss them further. Except at the highest coverages, because of its large electron affinity ad-OH would exist as a negative ion rather than as a radical.

2.3 How hydroxylation affects wetting

The sapphire slab had nine O-layers of one Al$_2$O$_3$ unit each, for 45 atoms per unit cell, as was used in a previous study.[20] In order to compare with previously published metal adsorption energies,[20,21] LDA was used for these calculations. The initial Cu positions were at the most favored sites, which at 1/3 ML coverage are atop the deepest lying Al-ion, and at 1 ML coverage

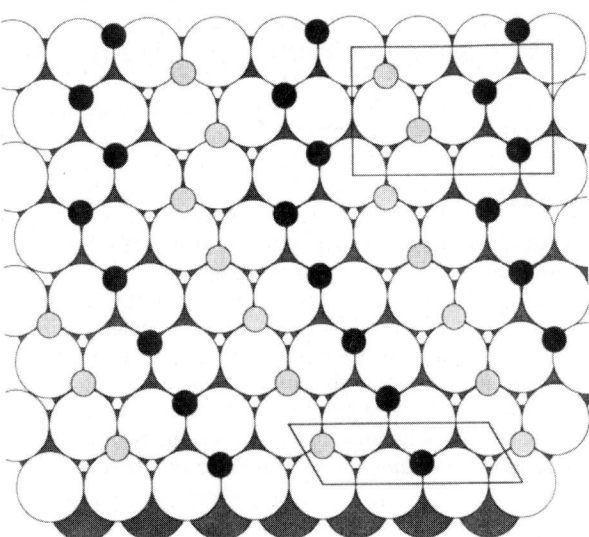

Fig. 1 The preferred structure of the thin film has alternating rows of tetrahedral and octahedral site Al ions in a zig-zag pattern (top) rather than the striped pattern (bottom). The large spheres are O ions, the small spheres Al ions, in either octahedral (black) or tetrahedral (gray) sites.

are atop O. The ad-OH was placed initially at the most favored site also, atop the shallowest Al ion. Relaxations were small laterally, thus preserving these site descriptions for the relaxed geometry.

3 Results and discussion

3.1 Ultrathin alumina film structure

In order to consider only the energetically lowest-lying possibilities, we impose three constraints: (1) we do not allow for non-stoichiometry in the Al/O ratio; (2) we restrict coordination to what is normal, *i.e.*, each surface O has two nearly coplanar[10] Al nearest neighbors; and (3) we do not consider geometries where t and o ions are in sites which are immediately adjacent.

Our analysis indicates it is not possible to produce a localized o-containing "defect" starting with all t ions, or *vice versa*, without violating the above restrictions. However, it is possible to produce a zig-zag row of O ions embedded in an otherwise perfect film of 100% t ions by displacing a row laterally (in the vertical direction in Fig. 1) so as to move all the ions in that row from t to a neighboring o site. Note that the effect of such a movement is to increase locally the average Al–Al interatomic spacing. Thus this electrostatic advantage, maximized by an alternating mixture of t and o rows (*i.e.*, -t-o-t-o-), competes with the t site preference, reported in ref. 10.

Table 1 shows the relative energies of the pure o and t structures compared with the even mix zig-zag structure on both Al(111) and Ru(0001) substrates. We find for both systems that the lateral electrostatic advantage of alternating o and t dominates over the t-site advantage. A second type of -t-o- evenly mixed structure has also been found, consisting of alternating stripes (see Fig. 1, bottom). It is also noted that this structure can mix with the zig-zag one, producing a displacement in the zig-zag axis. However, we find for Ru(0001) that the striped structure is significantly higher in energy than the zig-zag, even though the Al–Al nearest neighbor spacings are the same to several neighbor shells.

It is obvious that it is also possible to have any mix of the two types of rows. For example, the even zig-zag mixture, with -t-o- alternating rows, has a 2×1 unit cell (Fig. 1, top), relative to the primitive cell of three O and two Al ions in the surface plane of the film. Additionally, the 3 : 1 ratio of ion types then has a 4×1 cell (*e.g.*, -t-t-t-o-), but the 2 : 1 ratio results in a 6×1 cell because of the reversal of the phase of the zig-zag after a single -t-t-o- sequence. Thus far, we have been unable to converge these other films geometrically as flat structures. Instead, computationally the energy lowers monotonically as 3-D stripes (three oxygen layers thick) are produced, separated by depleted regions. In other words, the additional degrees of freedom allowed by the larger unit cells permits the geometry relaxation algorithm to find structures, which, while overall energetically preferred, are perhaps not relevant to the (actually metastable) flat structures produced experimentally. Work here is continuing.

If it were not for the effects of relaxation, the relative energies of various mixes could be modeled easily. If short range row–row interactions were to dominate, the cost of converting an o row into a t row or a domain of the perfect 2×1 structure, -o-t-o-t-o-t-, into one with reversed phase on the right, -o-t-t-o-t-o- could be estimated as follows: (1) a reduction in energy of ≈ 0.2 eV per primitive cell for each extra t vs. o row (see Table 1 and Ref. 10; and (2) a penalty of ≈ 1.0 eV

Table 1 Relative energies in eV (per Al_2O_3 unit) *vs.* the Al ion tetrahedral/octahedral site ratio for 5 Å alumina films on Ru and Al substrates (the striped structure on Al(111) has not yet been studied)

	t/o ratio (%), type			
	0, pure o	50, zig-zag	50, stripe	100, pure t
Unit cell	1×1	2×1	3×1	1×1
Ru(0001)	1.2	0.0	0.5	1.0
Al(111)	1.8	0.0	—	1.6

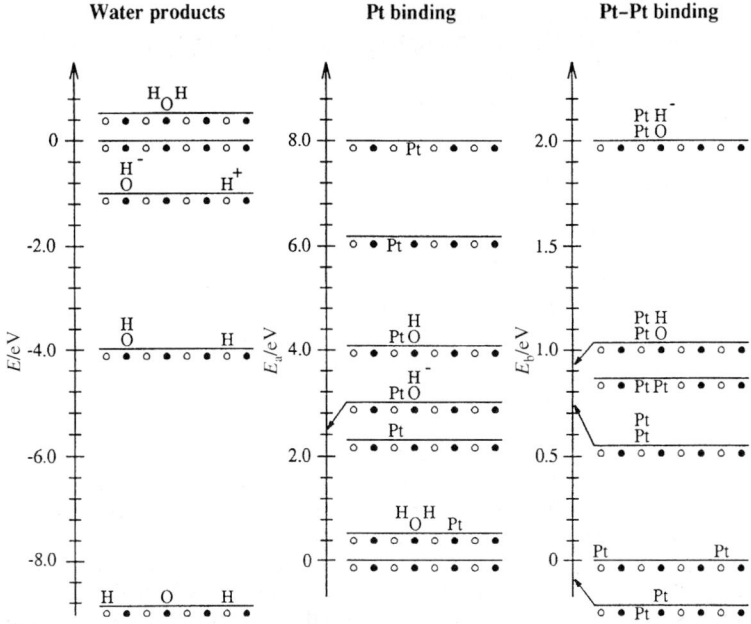

Fig. 2 Pt adsorption, Pt_2 binding energy, and relative stability of water products on MgO(100) and in the presence of surface defects. The open circles under each line denote O ions, the filled circles Mg ions. Negative energies indicate less stable adsorption with respect to (i) isolated gas phase H_2O molecule, (ii) isolated gas phase Pt atom, and (iii) two isolated Pt adatoms, for the three panels, respectively.

per primitive cell (for Ru(0001) substrates, Table 1) for each t-t nearest row–row interaction instead of t-o.

The net cost of phase reversal or other deviations from the 2 × 1 structure would be lowered by relaxation and further-than-nearest row–row interactions. These make it possible that the actual energetic cost may be sufficiently low that real films, where local structure is also influenced by defects and film growth conditions, may display a loss of long-range order in the Al-sublattice by this phase-reversal mechanism. Another possibility for such a loss is presented by the striped structure, in that a small inclusion of stripe between two zig-zag portions results in the zig-zag shifting laterally. Here too, the large relaxations inherent in alumina would reduce the cost of such a defect below that estimated on the basis of the perfect structures (Table 1). Without calculations with very large relaxed supercells, it is difficult to predict the actual energetics.

It seems likely that the 2 × 1 zig-zag dominates in real films (as the lateral Al–Al repulsion is minimized). Domains would then be determined by surface features, such as linear defects, and by film nucleation and growth. While a loss of long range Al-sublattice order would cause an amorphous appearance in scattering experiments such as LEED, the film is still locally dense and its adsorption properties little affected since the surface sites are so similar locally.

Even though our calculations were done using the experimentally relevant substrate of Ru(0001), since the calculations on Al(111) show qualitatively similar results, we suggest these likely apply also to films grown on NiAl[4] and Ni_3Al,[5] as no Ni rises into the film and the film/substrate interface, which drives the film structure, is thus similar to having chemisorbed 1 × 1 oxygen on Al(111).

Of all known aluminium oxide bulk phases, the 2 × 1 zig-zag (Fig. 1, top) most closely resembles the so-called A plane of κ-Al_2O_3[35] (recently structurally determined entirely by DFT and in close agreement with X-ray scattering from chemical vapour deposition (CVD)-grown samples). This phase has -A-B-A-C- bulk stacking of the close-packed near-hexagonal O-layers, and the A plane has an even mixture of o and t, arranged in the alternating zig-zag rows (Fig. 1, top), while the B and C planes have all o ions.[35] If the 5 Å films indeed prefer the A plane structure, one may speculate that this would nucleate κ-like alumina films if grown thicker.

Table 2 The LDA adsorption energy of Cu on a per atom basis (in eV) on clean sapphire(0001) and on hydroxylated sapphire (1/3 ML of ad-OH). The Born–Haber energy ΔE is positive when wetting occurs

	Cu coverage		
	1/3 ML	1 ML	ΔE
Sapphire:	1.8	0.5	−4.5
Sapphire+OH	5.2	1.1	+3.8
Sapphire+O+H[a]	—	1.3	+3.1

[a] With 1 ML of Cu, it was also found exothermic for Cu to dissociate the OH, as shown.

3.2 Metal island nucleation on terraces

In Fig. 2, we find the results of our study. Because water contamination is an issue, we initially studied water adsorption and dissociation, finding it will not dissociate on the perfect surface[15,36] but when it does dissociate (either due to defects or to solvation[36]) it does so as separated ions (Fig. 2, left column—favored structures are always towards the top of Fig. 2). Next, we see the effects of vacancies and water products on Pt adatom binding, noting a general increase in binding due to the presence of hydroxy (Fig. 2, center column). Finally, we note the effects of defects on Pt addimer stability (Fig. 2, right column). Of the vacancies, only the mixed divacancy increases the dimer binding, while both ad-OH species increase the same significantly. Note at moderate ad-OH concentrations, one might expect the charged species to be present, which has the greatest effect on dimer stability.

These results may be simply understood. When a Pt adatom encounters a vacancy of either type, it is drawn to it and becomes ionized by the Madelung potential. It enters the vacancy to the extent allowed by its ionic radius. Because it is an ion, its ability to bind to a second Pt atom is largely destroyed (and is repulsive in the LDA approximation). These results should hold for ionic oxides, all having substantial Madelung potentials. With ad-OH$^-$, however, a Pt adatom and ad-dimer find themselves at a "mini-step", with attractive electrostatic interactions both vertically (to the underlying O^{2-}-ions) and laterally (to the OH$^-$).

3.3 How hydroxylation affects wetting

In Table 2, we see that on perfect sapphire(0001), Cu binds less than half as strongly as on the hydroxylated surface. In particular, the increase in adatom binding (represented by 1/3 ML) not only reverses the preference for wetting given by the Born–Haber cycle, but also pins the Cu adatoms so they are immobile up to very high tempeatures (>1000 K) and are unable to reach and join 3-D islands, presumably nucleated at defects such as steps. This pinning is caused by the large loss of binding energy that would occur were a Cu atom to move laterally away from the adjacent OH, which it sits next to in the relaxed geometry. In the Born–Haber calculation, the tendency to form 2-D islands vs. separated adatoms is given by $\Delta E = E(1 \text{ ML}) + 2E(\text{slab}) - 3E(1/3 \text{ ML})$, where the last three parts refer to the total energy of the slab with 1 ML of Cu, the slab alone, and the slab with 1/3 ML of Cu, respectively. We noted during our study that at 1 ML coverage, it was exothermic for OH to dissociate, placing the H well away from the remaining ad-O, which coordinates locally to two Cu atoms. This observation, however, does not significantly affect the preference for wetting (Table 2).

It thus appears possible not only to increase the dispersion of ad-metal particles by hydroxylation,[2,28] but also to alter the growth mode completely if sufficient OH density is present.[29] It is of course possible that this surfactant effect extends to other impurities as well.

Acknowledgements

We thank Jeff Kelber for extensive discussions on hydroxylated sapphire and for sharing unpublished results. We also thank R. Jürgen Behm for sharing unpublished work and stimulating comments, and Hans-Joachim Freund for valuable discussions. VASP was developed at the Institut für Theoretische Physik of the Technische Universität Wien. Sandia is a multiprogram laboratory operated by Sandia Corporation, a Lockheed Martin Company, for the United States Department of Energy under Contract DE-AC04-94AL85000. This work was partially supported by a Laboratory Directed Research and Development project.

References

1 P. Hohenberg and W. Kohn, *Phys. Rev.*, 1964, **136**, B864; W. Kohn and L. J. Sham, *Phys. Rev.*, 1965, **140**, A1133.
2 G. Ertl and H.-J. Freund, *Phys. Today*, 1999, **52**, 32, and references cited therein.
3 H.-J. Freund, H. Kuhlenbeck and V. Staemmler, *Rep. Prog. Phys.*, 1996, **59**, 283.
4 R. M. Jaeger, H. Kuhlenbeck, H.-J. Freund, M. Wuttig, W. Hoffmann, R. Franchy and H. Ibach, *Surf. Sci.*, 1991, **259**, 235.
5 C. Becker, J. Kandler, H. Raaf, R. Linke, T. Pelster, M. Drager, M. Tanemura and K. Wandelt, *J. Vac. Sci. Technol., A*, 1998, **16**, 1000.
6 R. J. Behm, unpublished work.
7 O. Dulub, W. Hebenstreit, and U. Diebold, unpublished work.
8 J. Libuda, F. Winkelmann, M. Baumer, H.-J. Freund, T. Bertrams, H. Neddermeyer and K. Muller, *Surf. Sci.*, 1994, **318**, 61.
9 S. Nepijko, M. Klimenkov, H. Kuhlenbeck, R. Schlögl and H.-J. Freund, unpublished work.
10 D. R. Jennison, C. Verdozzi, P. A. Schultz and M. P. Sears, *Phys. Rev. B: Condens. Matter*, 1999, **59**, R15605.
11 F. Winkelmann, S. Wohlrab, J. Libuda, M. Baumer, D. Cappus, M. Menges, K. Al-Shamery, H. Kuhlenbeck and H.-J. Freund, *Surf. Sci.*, 1994, **307–309**, 1148.
12 D. A. Chen, M. C. Bartelt, R. Q. Hwang and K. F. McCarty, *Surf. Sci.*, submitted.
13 P. A. Thiel and T. E. Madey, *Surf. Sci. Rep.*, 1987, **7**, 211.
14 G. Haas, A. Mench, H. Brune, J. V. Barth, J. A. Venables and K. Kern, unpublished work.
15 M. J. Stirniman, C. Huang, R. S. Smith, S. A. Joyce and B. D. Kay, *J. Chem. Phys.*, 1996, **105**, 1295.
16 J. Günster, J. Stultz, S. Krischok, D. W. Goodman, P. Stracke and V. Kempter, *J. Vac. Sci. Technol.*, 1999, **A17**, 1657.
17 U. Diebold, J. Lehman, T. Mahmoud, M. Kuhn, G. Leonardelli, W. Hebenstreit, M. Schmid and P. Varga, *Surf. Sci.*, 1998, **411**, 137.
18 A. Bogicevic and D. R. Jennison, *Surf. Sci.*, 1999, **437**, 4741.
19 J. Ahn and J. W. Rabalais, *Surf. Sci.*, 1997, **388**, 121.
20 C. Verdozzi, D. R. Jennison, P. A. Schultz and M. P. Sears, *Phys. Rev. Lett.*, 1999, **82**, 799, and references cited therein.
21 A. Bogicevic and D. R. Jennison, *Phys. Rev. Lett.*, 1999, **82**, 4050.
22 S. Varma, G. S. Chottiner and M. Arbab, *J. Vac. Sci. Technol.*, 1992, **A10**, 2857.
23 J. G. Chen, M. L. Colaianni, W. H. Weinberg and J. T. Yates, Jr., *Surf. Sci.*, 1992, **279**, 223.
24 F. S. Ohuchi, R. H. French and R. V. Kasowski, *J. Appl. Phys.*, 1987, **62**, 2286.
25 Y. Wu, E. Garfunkel and T. E. Madey, *J. Vac. Sci. Technol.*, 1996, **A14**, 1662.
26 S. Gota, M. Gautier, L. Douillard, N. Thromat, J. P. Duraud and P. Le Fevre, *Surf. Sci.*, 1995, **323**, 163; see also M. Gautier, L. Pham Van and J. P. Duraud, *Europhys. Lett.*, 1992, **18**, 175.
27 M. Gautier, J. P. Duraud and L. Pham Van, *Surf. Sci. Lett.*, 1991, **249**, L327.
28 J. Libuda, M. Frank, A. Sandell, S. Andersson, P. A. Bruhwiler, M. Baumer, N. Martensson and H.-J. Freund, *Surf. Sci.*, 1997, **384**, 106.
29 J. A. Kelber, C. Niu, K. Shepherd, D. R. Jennison and A. Bogicevic, *Surf. Sci.*, submitted.
30 G. Kresse and J. Hafner, *Phys. Rev. B: Condens. Matter*, 1993, **47**, 558; 1994, **49**, 14251; 1996, **54**, 11169.
31 D. Vanderbilt, *Phys. Rev. B: Condens. Matter*, 1985, **32**, 8412; 1990, **41**, 7892.
32 J. P. Perdew and A. Zunger, *Phys. Rev. B: Condens. Matter*, 1981, **23**, 5048; D. M. Ceperley and B. J. Alder, *Phys. Rev. Lett.*, 1980, **45**, 566.
33 J. P. Perdew, J. A. Chevary, S. H. Vosko, K. A. Jackson, M. R. Pederson, D. J. Singh and C. Fiolhais, *Phys. Rev. B: Condens. Matter*, 1992, **46**, 6671.
34 Y. Yourdshahyan, U. Engberg, L. Bengtsson, B. I. Lundqvist and B. Hammer, *Phys. Rev. B: Condens. Matter*, 1997, **55**, 8721.
35 Y. Yourdshahyan, C. Ruberto, M. Halvarsson, L. Bengtsson, V. Langer, B. I. Lundqvist, S. Ruppi and U. Rolander, *J Am. Ceram. Soc.*, 1999, **82**, 1365; B. Holm, R. Ahuja, Y. Yourdshahyan, B. Johansson and B. I. Lundqvist, *Phys. Rev. B: Condens. Matter*, 1999, **59**, 12777.
36 See, for example, M. A. Johnson, E. V. Stefanovich, T. N. Truong, J. Gunster and D. W. Goodman, *J. Phys. Chem.*, 1999, **103**, 3391, and references cited therein.

Paper 9/05456A

Electronic properties, structure and adsorption at vanadium oxide: density functional theory studies

K. Hermann,[*a] **M. Witko**[b] **and R. Druzinic**[a]

[a] *Fritz-Haber-Institut der MPG, Faradayweg 4-6, D-14195 Berlin, Germany*
[b] *Institute of Catalysis and Surface Chemistry, ul. Niezapominajek, 30239 Cracow, Poland*

Received 19th April 1999

The local electronic structure at the $V_2O_5(010)$ surface is studied by *ab initio* density functional theory (DFT) methods where embedded clusters as large as $V_{20}O_{62}H_{24}$, representing one or two physical layers of the substrate, are used as models. Results of local binding, charging, and densities of states help to characterize the detailed electronic structure of the surface. In addition, electronic and geometric details of surface oxygen vacancies are studied by $V_2O_5(010)$ surface cluster calculations where oxygen atoms are removed from specific surface sites. A comparison of the data, concerning vacancy energies, charging, and geometric relaxation, shows pronounced variations between different oxygen sites, which gives further insight into possible mechanisms of surface relaxation and reconstruction. Further, cluster calculations of hydrogen adsorption at structurally different surface oxygen sites (leading to surface OH and H_2O) are performed. A comparison of bond strengths of surface OH and H_2O with that of surface oxygen gives valuable information as to which oxygen sites are involved in specific adsorption, desorption, and reaction steps.

1 Introduction

Transition metal oxides are well known for their enormous variety of physical and chemical properties.[1–3] Many of these materials undergo phase transitions with interesting structural, electronic and magnetic behavior.[3] Some exhibit high temperature superconductivity and exciting optical properties or high catalytic activity.[2] Among these, vanadium oxides represent an important class of materials which are widely studied and used in many technological applications.[4,5] In particular, vanadium pentoxide, V_2O_5, or vanadia-based compounds are used as components in various catalysts for mild oxidation, ammoxidation, and dehydrogenation of hydrocarbons and other organic compounds. Further, they are efficient in oxidation of SO_2 to SO_3 and for the removal of NO_x by selective reduction with NH_3.[6,7] Despite the enormous importance of V_2O_5 as a catalyst many microscopic details of its catalytic behavior are still under debate,[4] which makes a detailed study of V_2O_5 surface properties particularly attractive.

The crystal structure of vanadium pentoxide, V_2O_5, is rather complex and can be described in different ways. The orthorhombic crystal has a layer structure in which each physical layer consists of VO_5 sub-units linked by edges and by corners[8–10] with weak inter-layer coupling. The basal (010) and other non-basal planes differ in their bond type and in the degree of coordinative saturation of vanadium and oxygen atoms. This results in rather different behavior with respect to adsorption and catalytically supported reactions.

It is generally accepted that reactions of selective hydrocarbon oxidation at the $V_2O_5(010)$ surface proceed according to a nucleophilic mechanism[6,7,11–13] where an important reaction step involves adsorption and binding of hydrogen at the V_2O_5 surface. In a possible scenario, hydrogen (being abstracted from the organic molecule) adsorbs at an oxygen site forming a surface hydroxy, OH, species which can desorb. Alternatively, the hydroxy group may combine with another hydrogen to form surface H_2O, which desorbs. These processes form oxygen vacancies at the surface which may migrate into the bulk with the equivalent number of metal cations being simultaneously reduced. Gaseous oxygen participates in the oxidation reaction only after adsorption in other parts of the catalyst followed by migration through the lattice to the active site. The key point for understanding these mechanisms is to identify the structurally different surface oxygen sites which take part in the reaction. This issue has been discussed rather controversially in the literature. Some authors assume that terminal vanadyl oxygen (O=V) is removed from the vanadia catalyst surface to form a lattice vacancy[6] while others argue in favor of bridging oxygens (V–O–V or V–O–Me in the case of supported vanadia catalysts) and there are reports suggesting that a mixture of V=O and V–O–V(Me) type oxygens is essential for the selective oxidation process. Previous theoretical cluster studies using semi-empirical[13–20] and *ab initio* methods (Hartree–Fock (HF)[21–23] and density functional theory (DFT)[24,25]) have shown consistently that, after hydrogen adsorption, two- and three-fold coordinated bridging oxygens can be removed more easily from the $V_2O_5(010)$ surface than vanadyl oxygens. This result is confirmed by combined numerical (semi-empirical HF) and experimental (IR) studies[26] on the importance of V_2O_5 surface oxygen for the oxidation of SO_2 into SO_3, which stress the preference of oxygen centers with the highest V coordination. However, the underlying mechanisms have not been verified by experiments on a microscopic basis, for an overview see ref. 6.

In the present theoretical work we examine the local electronic structure at the $V_2O_5(010)$ surface by *ab initio* DFT methods where embedded clusters representing one and two physical layers of the substrate are used as models. In addition, electronic and geometric details of surface oxygen vacancies are studied by cluster calculations where oxygen atoms are removed from specific surface sites. Further, calculations of hydrogen adsorption at structurally different surface oxygen sites (leading to surface OH and H_2O as intermediate reaction products in the hydrocarbon oxidation) are performed using the embedded substrate clusters. In Section 2 we describe briefly the computational details and Section 3 presents the results and a discussion. Finally we summarize our conclusions in Section 4.

2 Theoretical details

Bulk vanadium pentoxide, V_2O_5, forms a layer type orthorhombic lattice structure[9,27,28] (lattice constants $a = 11.519$ Å, $b = 4.373$ Å, $c = 3.564$ Å; V_4O_{10} unit cell) with physical layers extending parallel to the (010) netplane. The physical layers are characterized by periodic arrangements of edge and corner sharing VO_5 pyramids sticking out at both sides of the layer, see Fig. 1. There are three structurally different layer oxygen atoms, terminal (vanadyl) oxygen, O(1), coordinated to one vanadium atom through a short bond ($d_{V-O} = 1.58$ Å) and bridging oxygen, O(2)/O(3), coordinated to two or three vanadium atoms with V–O distances ranging between 1.78 Å and 2.01 Å. This gives rise to five different oxygen centers at the ideal $V_2O_5(010)$ surface (see Fig. 1): terminal (vanadyl) oxygen O(1) located directly above vanadium centers, oxygen O(2), O(2') bridging two vanadyl groups pointing into the bulk and sticking out of the surface, respectively, and oxygen O(3), O(3') connected to three vanadyl groups (one pointing up and two pointing down for O(3), two pointing up and one pointing down for O(3')). The O(2') and O(3') centers (not labeled explicitly in Fig. 1), which are "buried" between vanadyl groups, are of less chemical interest and will not be considered in the following.

In the calculations the local environment at the $V_2O_5(010)$ surface is modeled by clusters $V_{10}O_{31}H_{12}$ (accounting for one physical layer) and $V_{20}O_{62}H_{24}$ ($= 2 \times V_{10}O_{31}H_{12}$, accounting for two physical layers) shown in Fig. 2. All vanadium and oxygen positions are taken from the experimental bulk structure and peripheral oxygen atoms are bond saturated by hydrogen atoms.[25] A full geometry optimization on $V_{10}O_{31}H_{12}$ [29] results in an equilibrium structure of the cluster that deviates only a little from that of the bulk termination (atom shifts by less than 0.18 Å). This justifies the use of $V_{10}O_{31}H_{12}$ with bulk structure as a starting point in the adsorbate and

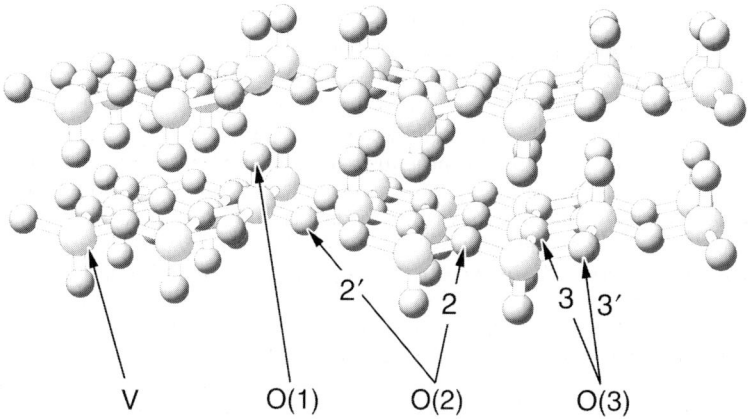

Fig. 1 Crystal structure of orthorhombic V_2O_5 with netplane stacking along (010). Vanadium (oxygen) centers are shown by large (small) balls. Inequivalent oxygen centers, O(1), O(2), O(3), are labeled accordingly. Note that labels O(2) and O(3) point to two centers, O(2), O(2′) and O(3), O(3′), respectively, which are inequivalent at the (010) surface.

oxygen vacancy calculations. Further, comparative cluster studies[30] on the single layer clusters $V_{10}O_{31}H_{12}$ and $V_{16}O_{49}H_{18}$ yield almost identical electronic parameters, which indicates size convergence and shows that $V_{10}O_{31}H_{12}$ can be considered a realistic representation of the extended $V_2O_5(010)$ surface.

Oxygen vacancy formation at different sites, O(1–3), of the $V_2O_5(010)$ surface is considered in calculations on $V_{10}O_{30}H_{12}$ (=$V_{10}O_{31}H_{12}$–O) and $V_{20}O_{61}H_{24}$ (=$V_{20}O_{62}H_{24}$–O) clusters. In a first step, the respective oxygen is removed from the cluster and the electronic structure is evaluated keeping all the atom positions frozen at the bulk geometry. In addition, all cluster atoms in $V_{10}O_{30}H_{12}$ (or $V_{20}O_{61}H_{24}$) except the terminating hydrogen atoms are allowed to rearrange according to the lowest cluster total energy. A comparison with the data of the frozen bulk geometry shows the importance of surface relaxation induced by the vacancy. Further, a comparison of the $V_{10}O_{30}H_{12}$ and $V_{20}O_{61}H_{24}$ data can give information about the influence of vacancy formation on electronic inter-layer coupling. Hydrogen adsorption at different oxygen sites of the $V_2O_5(010)$ surface is modeled by $V_{10}O_{31}H_{12}$ + H clusters where H is approached near the O(1–3) sites, forming a surface OH group, and both the hydrogen and the adsorption site oxygen positions are optimized according to the lowest total energy of the cluster. In addition, two H atoms

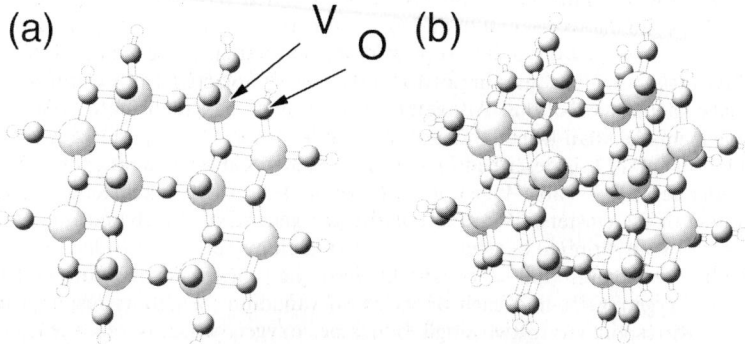

Fig. 2 Geometric structure of the clusters $V_{10}O_{31}H_{12}$ (a, one layer) and $V_{20}O_{62}H_{24}$ (b, two layers). The V (O) atoms are shown as large (small) shaded balls while very small white balls refer to hydrogen atoms used to saturate oxygen atoms at the cluster boundary.

are approached near the O(1–3) sites in a $V_{10}O_{31}H_{12} + 2H$ cluster forming a surface H_2O species which is geometrically optimized analogous to the OH optimization.

The electronic structure of the clusters is determined by *ab initio* density functional theory (DFT) methods where the Kohn–Sham orbitals are represented by linear combinations of atomic orbitals (LCAOs) using extended all-electron basis sets of contracted Gaussians from atom optimizations.[31,32] For the calculations the program package DeMon[33] is applied where electron exchange and correlation is described by the local spin density approximation (LSDA) based on the Vosko–Wilk–Nusair functional[34] as well as by gradient corrected (GGA-II) functionals.[35] In addition to the total energies and equilibrium geometries (based on numerical forces) detailed analyses of the electronic structure in the clusters are performed using Mulliken populations[36] and Mayer bond order indices.[37,38] Further, the dense energetic distribution of the Kohn–Sham valence levels in the clusters allows the definition of a cluster total density of states (DOS), $n_{tot}(\varepsilon)$, and atom projected partial densities of states (PDOS), $n_A(\varepsilon)$, by

$$n_{tot}(\varepsilon) = \sum_k g(\alpha, \varepsilon - \varepsilon_k) \tag{1}$$

$$n_A(\varepsilon) = \sum_k q_k(A) g(\alpha, \varepsilon - \varepsilon_k) \quad \text{with} \quad \sum_A n_A(\varepsilon) = n_{tot}(\varepsilon) \tag{2}$$

where $g(\alpha, \varepsilon - \varepsilon_k)$ denotes a gaussian broadening function of width α centered at cluster level ε_k and the summation goes over all occupied cluster orbitals. Further, $q_k(A)$ gives the population of atom A in cluster orbital φ_k determined by a Mulliken analysis. The computed DOS and PDOS functions can become useful in interpreting experimental photoemission spectra from the V_2O_5 system as will be discussed below.

3 Results and discussion

3.1 Electronic structure of $V_2O_5(010)$

Table 1 lists the geometric and electronic parameters of (a) the $V_2O_5(010)$ surface clusters $V_{10}O_{31}H_{12}$, $V_{20}O_{62}H_{24}$ and of (b) the free (neutral) molecular species, OH and H_2O, where all the values are calculated within the LSDA scheme. Atom charges $q(A)$ are obtained from Mulliken population analyses and bond orders $p(A-B)$ refer to the Mayer bond order indices where the data are given for the V and O atoms closest to the cluster center. In addition, the valence energy width Δ of each cluster is included. An overall comparison of the calculated atom charges and bond orders reveals very close similarity between the one and two layer clusters, which indicates that the electronic inter-layer coupling is rather weak and can be neglected for the electronic surface structure. In agreement with chemical intuition, all the vanadium atoms are positively charged and all the oxygen atoms are negative in the clusters. Vanadium atoms are described by $V^{+1.4}$–$V^{+1.6}$ where the variation reflects the location inside the cluster. Further, the negative oxygen charges scale with coordination, $O^{-0.3}$ for singly coordinated terminal oxygen O(1), $O^{-0.6}$ for doubly coordinated bridging oxygen O(2), and $O^{-0.8}$ for triply coordinated bridging oxygen O(3). This indicates for the $V_2O_5(010)$ surface that bridging oxygen sites are more nucleophilic than terminal vanadyl sites, which becomes important in view of the reactivity of the different sites with respect to surface chemical reactions. Altogether, local charging of the different cluster atoms is found to be much smaller than formal valence charges, V^{+5} and O^{-2}, would suggest. Obviously, inter-atomic binding in V_2O_5 is described by both ionic and sizeable covalent contributions.

The bond order results of Table 1 give a rough estimate of the covalent contributions to the total V–O binding in the clusters. The data confirm the general picture based on simple valence concepts. Bonds between terminal oxygen, O(1), and vanadium yield bond order values close to 2, which suggests double bonds and is consistent with the single coordination of O(1). Bonds between bridging oxygen, O(2), and each of their two vanadium neighbors result in bond order values close to 1, corresponding to two single bonds per oxygen, which is again reasonable based on the coordination of O(2). Finally, V–O bond orders involving bridging atoms, O(3), coordinated to three vanadium neighbors each, give meaningful values of 0.5–0.6 per bond.

The electronic structure of the clusters in the valence region is determined by occupied Kohn–Sham valence orbitals, which are mainly O 2sp type with some V 3d admixture. Their energy

Table 1 Atom charges q from Mulliken analyses and Mayer bond orders p for (a) the $V_{10}O_{31}H_{12}$ and $V_{20}O_{62}H_{24}$ substrate clusters, and (b) the free (neutral) molecular species OH, H_2O obtained by LSDA calculations[a]

(a)	$V_{10}O_{31}H_{12}$	$V_{20}O_{62}H_{24}$
$d_{V-O(1)}$/Å	1.58	1.58
$d_{V-O(2)}$/Å	1.78	1.78
$d_{V-O(3)}$/Å	1.88 (×2), 2.02	1.88 (×2), 2.02
$q(V)$	1.43	1.44, 1.57
$q(O(1))$	−0.29	−0.28
$q(O(2))$	−0.62	−0.63
$q(O(3))$	−0.79	−0.78
$p(O(1)-V)$	2.13	2.16
$p(O(2)-V)$	0.89	0.89
$p(O(3)-V)$	0.55	0.51
Δ/eV	5.40	5.76

(b)	OH	H_2O
d_{O-H}/Å	1.00	0.98
$\angle(H-O-H)/°$	—	105.6
$q(O)$	−0.44	−0.85
$q(H)$	0.44	0.43
$p(O-H)$	0.78	0.80

[a] The two entries of $q(V)$ for $V_{20}O_{62}H_{24}$ correspond to vanadium of the first and second layer respectively. In addition, the valence energy widths Δ of each cluster are given. For further definitions see text.

range corresponds to valence energy widths $\Delta = 5.40$ eV and 5.76 eV, respectively. The width Δ is expected to converge with increasing cluster size towards the total valence band width of the extended $V_2O_5(010)$ surface system. Very recent FP-LAPW (full potential linearised augmented plane wave) band structure calculations[39] yield $\Delta = 5.35$ eV for the V_2O_5 bulk and $\Delta = 5.05$ eV for $V_2O_5(010)$ single layer slabs, which are reasonably close to the cluster results suggesting size convergence for the present clusters as discussed before.

The distribution of the valence levels and their atom character can be described by total DOS and atom projected PDOS curves following the procedure described in Section 2. Fig. 3 shows DOS and PDOS curves for the $V_{10}O_{31}H_{12}$ cluster where the vanadium contributions as well as those from all differently coordinated oxygen atoms, O(1), O(2), O(3), are included and a gaussian level broadening of 0.4 eV (FWHM) is applied. The total DOS in the energy region between −13 and −7 eV shows a multi-peak structure described by mainly O 2sp derived electron states without noticeable energetic separation between O 2s and O 2p and by smaller V 3d contributions. (Note that due to the gaussian broadening the DOSs of Fig. 3 do not exhibit a sharp cut-off at the HOMO energy; −7.07 eV, marked by a thin line in the plot). The additional DOS peaks between −13 and −14 eV reflect split-off energy levels arising from bond saturation of peripheral cluster oxygen atoms by hydrogen terminating atoms. They have to be considered a consequence of the cluster approach and can be neglected for the present purpose.

The PDOS due to vanadium given in Fig. 3 shows moderate variations with larger values near the central part of the valence region. However, its size is overall smaller compared to that of the oxygen derived PDOSs. An integration over the valence region yields populations of 3.6 electrons per V atom, in agreement with the atom charge (1.4) given in Table 1. As a result, the V atoms in the clusters are not fully ionic. The PDOS referring to terminal oxygen O(1) are concentrated near the center of the valence region with smaller contributions above the center and they are described

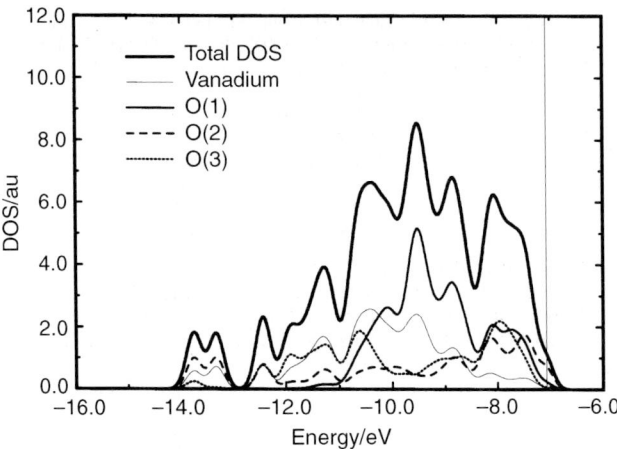

Fig. 3 Total DOS and atom projected PDOS curves for the $V_{10}O_{31}H_{12}$ cluster, see text. The results refer to a gaussian level broadening of 0.4 eV (FWHM) and the HOMO energy is marked by a thin vertical line.

by an overall confined (~3 eV wide) distribution. In contrast, the PDOSs of bridging oxygen O(2, 3) yield a broad distribution covering the full energy range of the total DOS. Obviously, the O(1) derived cluster levels show a dispersion width smaller than that of the bridging O(2,3) species, partly because of the spatial distribution of the different oxygen atoms in the crystal and their effective inter-atomic binding, as discussed elsewhere.[30] The DOS and PDOS curves of Fig. 3 obtained for $V_{10}O_{31}H_{12}$ can be compared with those of the larger clusters, $V_{16}O_{49}H_{18}$ [30] and $V_{20}O_{62}H_{24}$, as well as with results from *ab initio* DFT band structure methods for both bulk V_2O_5 and $V_2O_5(010)$ single layer slabs.[39–42] This comparison shows very good qualitative agreement, which confirms that the different approaches, V_2O_5 surface clusters and the V_2O_5 bulk/slab models, yield basically the same electronic structure for the oxide material and can therefore be applied alternatively to model bulk and surface properties.

Recent angular resolved UV photoemission experiments for the $V_2O_5(010)$ surface[30] yield a spectrum with three peaks, one dominant central and two smaller peripheral, in the O 2sp valence region where the total valence energy width Δ amounts to 5.5 eV, quite close to the cluster results. Further, the variation of the experimental emission intensity with energy is similar to the shape of the calculated total DOS of the V_2O_5 surface clusters. This suggests that the origin of the peaks observed in the photoemission experiment may be identified by a comparison with the calculated PDOSs. As a result, the most prominent central peak in the experimental data is assigned to emission from mainly terminal oxygen, O(1), while the two peripheral peaks at the top and bottom of the valence energy region are characterized as mixtures of vanadium with O(2) and O(3) induced intensity.

3.2 Oxygen vacancies at $V_2O_5(010)$

Table 2 contains results from oxygen vacancy calculation using the $V_{10}O_{31}H_{12}$ cluster, which models one physical layer of the $V_2O_5(010)$ surface. The oxygen vacancy energies, $E_D(O)$, are defined by total energy differences of the corresponding clusters

$$E_D(O) = |E_{tot}(V_{10}O_{31}H_{12}) - E_{tot}(V_{10}O_{30}H_{12}) - E_{tot}(O)| \qquad (3)$$

where in $V_{10}O_{30}H_{12}$ the oxygen has been removed from one of the sites O(1), O(2), O(3) and $E_{tot}(O)$ refers to oxygen in its neutral ground state. In the first step, all the atoms of $V_{10}O_{30}H_{12}$ are kept fixed at their positions from the ideal surface, yielding frozen vacancy energies, $E_D^f(O)$. In the second step, all the cluster atoms of $V_{10}O_{30}H_{12}$, except the terminating hydrogen atoms, are allowed to rearrange according to the lowest cluster total energy, which leads to relaxed values, $E_D^r(O)$. The energetic consequence of relaxation at each oxygen site can be described by a relax-

Table 2 Oxygen vacancy energies, $E_D^{(f,r)}(O)$, with and without surface relaxation at the oxygen sites O(1–3) obtained for the $V_{10}O_{31}H_{12}$ cluster using the LSDA scheme[a]

	$E_D^f(O)$	$E_D^r(O)$	E_{rel}	$q^f(V)$	$q^r(V)$	$\Delta r(V)$
Substrate	—	—	(0.93)	(1.41)	(1.44)	(0.18)
O(1) vacancy	8.13	6.68	1.45	1.32	1.45	0.50
O(2) vacancy	9.36	7.19	2.17	1.15	1.27	0.70
O(3) vacancy	8.52	7.26	1.26	1.17	1.21	0.08

[a] In addition, the table contains values for the relaxation energies E_{rel} as well as for the charges $q^{(f,r)}(V)$ and displacements $\Delta r(V)$ of the central V atom closest to the vacancy. The data of the top row of the table refer to the $V_{10}O_{31}H_{12}$ cluster without a vacancy. For definitions see text. All energies are given in eV and all lengths are in Å.

ation energy

$$E_{rel} = E_D^f(O) - E_D^r(O) \tag{4}$$

Further, Table 2 lists the atom charges $q(V)$ of the central V atom closest to the vacancy (from Mulliken population analyses) where in each case both frozen, $q^f(V)$, and relaxed values, $q^r(V)$, are included. In addition, the values of the displacement $\Delta r(V)$ of the central V atom due to relaxation are shown. For all sites, the oxygen vacancy energies are rather large (7.2–9.4 eV), which suggests that it is quite difficult to remove an oxygen by itself from the $V_2O_5(010)$ surface. Based on the frozen substrate calculations, the $E_D^f(O)$ value is largest, 9.4 eV, for the two-fold bridging site O(2) with that of O(3) and O(1) site being smaller by only 0.8 and 1.3 eV, respectively. Relaxation due to vacancy formation decreases these energies by 1.3–2.2 eV where the effect is again largest for the O(2) site. Thus, the strongest binding of the oxygen to the surface leads to the largest relaxation after its removal. As a result of relaxation, the vacancy energies at the O(2) and O(3) sites are rather close, $E_D^r(O) = 7.2$ and 7.3 eV, while that of the O(1) site is only slightly smaller, 6.7 eV.

When substrate relaxation due to vacancy formation is accounted for, the computed energetic and geometric quantities are influenced by the fact that the $V_{10}O_{31}H_{12}$ cluster without vacancy and with its ideal bulk structure does not correspond to the equilibrium geometry of the cluster. Therefore, a part of the relaxation effect is due to equilibration of the initial $V_{10}O_{31}H_{12}$. However, the corresponding contributions can be neglected for the present purpose. This has been demonstrated by test calculations where all the cluster atoms of $V_{10}O_{31}H_{12}$ except the terminating hydrogen atoms are allowed to rearrange according to the lowest cluster total energy. The corresponding energy lowering, $E_{rel} = 0.9$ eV, as well as atom displacements, $\Delta r(V) = 0.2$ Å, turn out to be rather small and all the atom charges in the relaxed cluster are very similar to those of the cluster in its ideal bulk structure, see the q values in parentheses in Table 2. All energy values of Table 2 are computed applying the LSDA scheme for exchange and correlation. The use of gradient corrected (GGA-II) functionals[35] decreases the $E_D(O)$ values by 0.8–1.1 eV[29] depending on the oxygen site. However, this does not affect the energetic sequence between the different sites and leads to corrections of only 0.2 eV in the relaxation energies E_{rel}. Therefore, the present discussion will be restricted to LSDA results.

As an example of the geometric effect due to substrate relaxation, Fig. 4 shows the geometry of the relaxed $V_{10}O_{30}H_{12}$ cluster for an O(2) vacancy. The main result is that the relaxation effect is found to be locally confined. The strongest relaxation shifts occur for the two vanadium atoms adjacent to the vacancy, which move laterally by 0.63 Å (0.7 Å including a upwards shift) such that the opening of the vacancy is enlarged. However, the lattice topology of the $V_2O_5(010)$ surface is conserved, which suggests that a single O(2) vacancy will not introduce major restructuring of the surface. This result has also been found in calculations of single O(1) and (3) vacancies and even for vacancy pairs.[29]

The atom charges of the central V atoms closest to each oxygen vacancy, see Table 2, are smaller than the corresponding values of the cluster without the vacancy, which suggests chemical

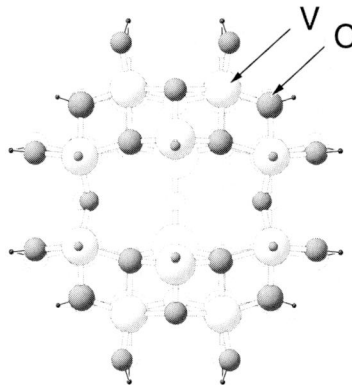

Fig. 4 Relaxed geometry of the $V_{10}O_{31}H_{12}$ cluster with an O(2) vacancy. The cluster atoms are shown by shaded balls where ball radii represent atom charges. Dark (light) shading refers to negative (positive) charge while the radius gives the amount of charge. The white balls behind the relaxed cluster describe the system without vacancy.

reduction of the metal sites. This is due to the fact that the vacancy oxygen is a negatively charged species at the surface. Therefore, when the oxygen is removed, as a neutral species, it leaves a negative excess charge behind that is distributed over the atoms close to the vacancy and compensates some of the positive metal charges. A comparison of the frozen and relaxed cluster values, $q^f(V)$ and $q^r(V)$ of Table 2, shows that the reduction effect decreases by relaxation where, for the O(1) vacancy, the decrease leads to almost no reduction of the respective metal site.

It is interesting to study the consequences of vacancy formation for the electronic inter-layer coupling near the $V_2O_5(010)$ surface. This problem is examined by vacancy calculations using clusters that model two adjacent physical layers, such as $V_{20}O_{62}H_{24}$ sketched in Fig. 2(b). Preliminary results from these calculations[29] show that oxygen vacancies with the substrate cluster $V_{20}O_{61}H_{24}$ frozen at the ideal surface geometry lead to charge distributions very close to those of the one-layer $V_{10}O_{30}H_{12}$ clusters. When atom relaxation is allowed in the two-layer cluster the corresponding relaxation energies E_{rel} amount, for all oxygen vacancies O(1–3), to about twice the values obtained for the single-layer clusters. Further, surface binding and charging as well as geometric consequences of relaxation are somewhat different in the two- and one-layer clusters. In particular, inter layer binding can be affected by relaxation after vacancy formation. As an illustration, Fig. 5 shows the relaxed geometry of the $V_{20}O_{61}H_{24}$ cluster with an O(1) vacancy for views normal and parallel to the surface. Obviously, the vanadium surface atom (labeled by hatching) below the missing O(1) is affected most by relaxation. This atom shifts downwards towards the second layer by 0.94 Å while the second layer oxygen below the vanadium moves upwards by 0.12 Å. As a result, the two atoms interact with each other and form a single V–O bond as can be seen from a bond order analysis.[29] Thus, vacancy formation at the first layer of the $V_2O_5(010)$ surface may increase the electronic coupling with the second layer.

3.3 H adsorption at $V_2O_5(010)$

Table 3 lists the geometric and electronic parameters of the surface cluster $V_{10}O_{31}H_{12}$ + H where the hydrogen species has been added near the three oxygen sites O(1), O(2), O(3). Both the hydrogen and the adsorption site oxygen are optimized in their positions according to the lowest total cluster energy. Atom charges q from Mulliken population analyses and Mayer bond orders p are given for the corresponding oxygen site and for the V atoms closest to the cluster center. The hydrogen binding energy $E_B(H)$ with respect to the clean surface (adsorption energy) is defined by the total energy differences of the corresponding clusters

$$E_B(H) = |E_{tot}(V_{10}O_{31}H_{12} + H) - E_{tot}(V_{10}O_{31}H_{12}) - E_{tot}(H)| \qquad (5)$$

where $E_{tot}(V_{10}O_{31}H_{12} + H)$ refers to the computed equilibrium geometry while the OH desorption energy $E_D(OH)$ (binding with respect to the clean surface with an oxygen vacancy) is defined

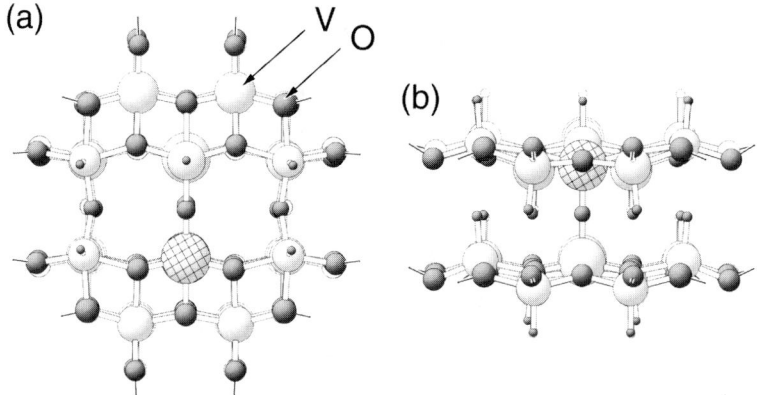

Fig. 5 Relaxed geometry of the $V_{20}O_{61}H_{24}$ cluster with an O(1) vacancy for a view (a) normal and (b) parallel to the surface. Atom shading and radii are defined as in Fig. 4. The white balls behind the relaxed cluster describe the system without vacancy. The central V atom next to the vacancy is emphasized by hatching.

by the total energy differences

$$E_D(OH) = |E_{tot}(V_{10}O_{31}H_{12} + H) - E_{tot}(V_{10}O_{30}H_{12}) - E_{tot}(OH)| \tag{6}$$

where $E_{tot}(V_{10}O_{30}H_{12})$ is computed for the cluster with the appropriate oxygen vacancy. All results shown in Table 3 are calculated within the LSDA scheme, see below.

The results of Table 3 show that hydrogen can stabilize at all oxygen sites with sizable binding energies forming rather stable surface OH groups with equilibrium geometries sketched in Fig. 6. The hydroxy group involving the terminal O(1) site is bent with respect to the surface normal by 73° with the oxygen shifted by 0.37 Å relative to its position without the hydrogen. The hydroxy

Table 3 Surface OH equilibrium geometries, atom charges q, and bond orders p for the $V_{10}O_{31}H_{12}$ + H clusters with H adsorbed at the oxygen sites O(1–3) obtained by LSDA calculations[a]

	O(1) site	O(2) site	O(3) site
d_{V-O}/Å	1.70	1.87	1.97
d_{O-H}/Å	1.01	0.99	0.99
$\Delta r(O)$/Å	0.37	0.27	0.34
ϑ_{inc}/°	73	0	34
$q(V)$	1.46	1.37	1.35
$q(O)$	−0.63	−0.85	−0.96
$q(H)$	0.54	0.54	0.57
$p(O-V)$	1.25	0.51	0.27
$p(O-H)$	0.61	0.66	0.62
$E_B(H)$/eV	3.05	2.76	2.50
$E_D(OH)$/eV	5.96	6.89	5.79

[a] Values d_{V-O}, d_{O-H} refer to equilibrium distances of surface OH while ϑ_{inc} denotes the inclination angle of the OH axis with respect to the surface normal and $\Delta r(O)$ gives the shift of the respective surface oxygen due to adsorption. In addition, binding energies $E_B(H)$ and desorption energies $E_D(OH)$ are given for each site. For further definitions see text.

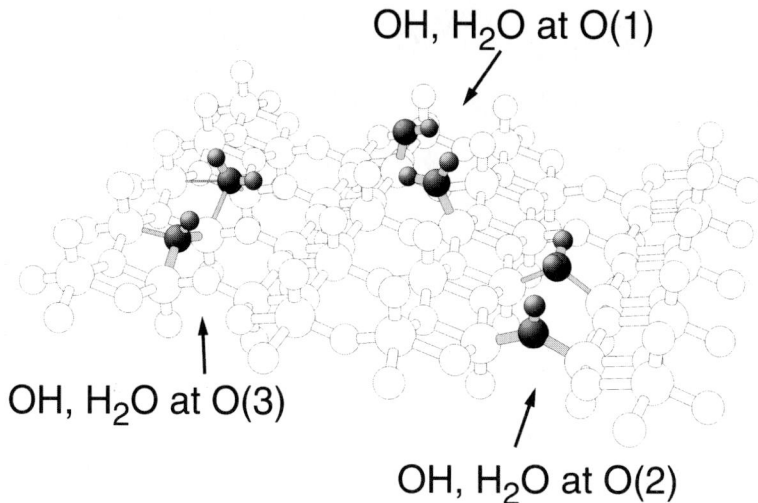

Fig. 6 Equilibrium geometries of surface OH and H_2O at different oxygen sites, O(1), O(2), O(3), obtained from optimizations of $V_{10}O_{31}H_{12}$ + H and $V_{10}O_{31}H_{12}$ + 2H, respectively. The surface species are shown by darker shaded balls while the surface lattice is sketched by light balls.

group formed at the bridging O(2) site points normal to the surface for symmetry reasons, while the oxygen is shifted by 0.27 Å out of the surface. Further, the OH involving the bridging O(3) site is bent by 34° pointing away from the adjacent vanadyl group with the oxygen shifted by 0.34 Å relative to its position before H adsorption. At all sites, the distances between oxygen and its nearest V neighbors at the V_2O_5 surface, d_{V-O}, are enlarged by the adsorption, which suggests a weakening of V–O binding near the adsorption site. This is confirmed by the atom charges and bond orders given in Tables 1 and 3. For all sites the oxygen accumulates negative charge due to adsorbed hydrogen and the surface OH becomes slightly negative, $OH^{-0.1}$–$OH^{-0.4}$, where charging is smallest for the terminal O(1) site. The latter conforms with the result that the positive charge of the vanadium neighbor at the O(1) site remains the same, whereas for the higher coordination sites, O(2, 3), the neighboring V atoms lose positive charge (are reduced) by hydrogen adsorption. The V–O bond weakening near all adsorption sites can be seen by a comparison of Tables 1 and 3 where p(O–V) values are found to decrease due to adsorption. Obviously, the V=O double bond at the O(1) site is reduced to a single bond and the weaker V–O bonds at O(2, 3) are further reduced.

The adsorption energies E_B(H) vary between 2.5 and 3.0 eV depending on the surface oxygen site, which suggests rather stable surface OH groups for all sites. The present data show the strongest adsorptive binding for the O(1) site, followed by the O(2) and O(3) sites. This order is in disagreement with previous results from *ab initio* DFT calculations for V_2O_5 clusters and periodic slab models,[43] which find that E_B(H) is larger for the O(2) than for the O(1) site. This discrepancy is explained by the fact that in the previous study the hydrogen adsorbate and the oxygen site are allowed to relax only perpendicular to the V_2O_5 surface whereas the present results are based on a full OH geometry optimization without constraints. The E_B(H) values listed in Table 3 refer to cluster calculations using the LSDA scheme for exchange and correlation, which is well known to overestimate binding energies. Extended tests[29,43] using gradient corrected (GGA-II) functionals[35] show that the E_B(H) values of Table 3 are decreased by 0.5–0.6 eV[43] due to the improved GGA-II scheme. However, this correction is found to be independent of the adsorption site. Thus, relative binding energies can be estimated using both the LSDA and the GGA-II scheme. Further, equilibrium geometries turn out to be affected only very little by going from the LSDA to the GGA-II scheme. Therefore, we restrict ourselves to a discussion of the LSDA results.

Table 3 also contains results of the OH desorption energy E_D(OH), which is defined as the energy required to remove the OH species from the V_2O_5 surface leaving an oxygen vacancy behind, where the definition in eqn. (6) assumes that OH removal proceeds without an intermediate reaction barrier. Further, in the calculations, the substrate was not allowed to relax due to the oxygen vacancy formation. The desorption energies $E_D(OH)$ vary between 5.8 and 6.9 eV depending on the surface oxygen site, which reveals very strong binding of OH with its $V_2O_5(010)$ surface environment at all sites. The present calculations yield the strongest binding for the O(2) site, followed by the O(1) and O(3) sites. While the computed $E_D(OH)$ values seem to be rather large their actual size is not unreasonable. As discussed in Section 3.2, the removal of oxygen (without pre-adsorbed hydrogen) from the $V_2O_5(010)$ surface is found to require energies (cf. $E_D^f(O)$ values of Table 2) larger than the corresponding $E_D(OH)$ values by 2.2–2.7 eV depending on the oxygen site. These energy differences are obviously due to the hydrogen induced V–O bond weakening at the surface, which makes it easier to remove an OH than an oxygen species.

Table 4 lists the geometric and electronic parameters of the surface cluster $V_{10}O_{31}H_{12} + 2H$ where a second hydrogen species has been added near the OH group formed at the three oxygen sites O(1), O(2), O(3), and the resulting H_2O unit is optimized in its geometry according to the lowest total cluster energy. Here atom charges and bond orders are defined as analogous to the corresponding parameters of Table 3. The $E_B(H)$ values refer to binding of the second hydrogen to the existing surface OH group defined by the total energy differences

$$E_B(H) = |E_{tot}(V_{10}O_{31}H_{12} + 2H) - E_{tot}(V_{10}O_{31}H_{12} + H) - E_{tot}(H)| \quad (7)$$

where $E_{tot}(V_{10}O_{31}H_{12} + 2H)$ and $E_{tot}(V_{10}O_{31}H_{12} + H)$ refer to the computed equilibrium geometries. Further, the H_2O desorption energy $E_D(H_2O)$ (binding with respect to the clean surface with an oxygen vacancy) is defined by the total energy differences

$$E_D(OH) = |E_{tot}(V_{10}O_{31}H_{12} + 2H) - E_{tot}(V_{10}O_{30}H_{12}) - E_{tot}(H_2O)| \quad (8)$$

As mentioned before, all results shown in Table 4 are calculated within the LSDA scheme.

The results of Table 4 show that an additional hydrogen stabilizes at the surface OH groups formed at all the oxygen sites thereby resulting in surface H_2O groups with equilibrium geometries sketched in Fig. 6. The H_2O involving the terminal O(1) site is bent by 33° with the oxygen shifted by 0.75 Å relative to its position at the clean $V_2O_5(010)$ surface. In its most stable geometry, the H_2O formed at the bridging O(2) site points with its molecular axis normal to the

Table 4 Surface H_2O equilibrium geometries, atom charges q and bond orders p for the $V_{10}O_{31}H_{12} + 2H$ clusters with 2H adsorbed at the oxygen sites O(1–3) obtained by LSDA calculations[a]

	O(1) site	O(2) site	O(3) site
d_{V-O}/Å	1.83	2.04	2.27
d_{O-H}/Å	1.08, 0.99	1.00	1.01, 1.00
$\Delta r(O)$/Å	0.75	0.62	1.11
ϑ_{inc}/°	33	0	26
$\angle(H-O-H)$/°	112	113	110
$q(V)$	1.48	1.27	1.26
$q(O)$	−0.88	−0.96	−0.92
$q(H)$	0.55, 0.51	0.59	0.54, 0.56
$p(O-V)$	0.47	0.22	0.19
$p(O-H)$	0.43, 0.71	0.59	0.61, 0.58
$E_B(H)$/eV	2.18	1.36	1.56
$E_D(H_2O)$/eV	2.13	2.23	1.33

[a] Values d_{V-O}, d_{O-H} refer to equilibrium distances of surface H_2O while ϑ_{inc} denotes the inclination angle of the H_2O axis with respect to the surface normal and $\Delta r(O)$ gives the shift of the respective surface oxygen due to adsorption. In addition, binding energies $E_B(H)$ of the second surface hydrogen and desorption energies $E_D(H_2O)$ are given for each site. For further definitions see text.

surface and its molecular plane extends perpendicular to the V–O(2)–V plane, see Fig. 6. Further, the oxygen is shifted by 0.62 Å out of the surface. Finally, the H_2O involving the bridging O(3) site is bent by 26° with its molecular plane perpendicular to the V–O(3)–V plane and the oxygen shifted by 1.11 Å. As for the surface OH, the distances between the oxygen and its nearest V neighbors at the V_2O_5 surface are increased by the adsorption, where the effect is always larger for surface H_2O than for surface OH. This hints at an even more pronounced weakening of V–O binding near the adsorption site. The oxygen becomes more negative due to hydrogen adsorption where the effect is larger for surface H_2O than for surface OH. However, the increased negative oxygen charge is overcompensated by the second (positively charged) hydrogen, which yields a positive surface H_2O species, $H_2O^{+0.2}$, almost independent of the oxygen site. The charge of the vanadium neighbor at the O(1) site remains almost unchanged whereas for the higher coordination sites, O(2, 3), the neighboring V atoms are reduced more strongly by surface H_2O than by surface OH. Further, the V–O bond weakening near all adsorption sites, observed for surface OH, becomes more pronounced for surface H_2O as evidenced by a comparison of the corresponding bond orders p(O–V) of Tables 3 and 4.

The adsorption energies $E_B(H)$ for the second hydrogen vary between 1.4 and 2.2 eV depending on the surface oxygen site, where strongest adsorptive binding is found for the O(1) site, followed by the O(3) and O(2) sites. In all cases, the computed binding energy $E_B(H)$ of the second hydrogen is found to be smaller than the value for the first H species. The H_2O desorption energy $E_D(H_2O)$, listed in Table 4, quantifies the removal of surface H_2O, as analogous to $E_D(OH)$ discussed above. The $E_D(H_2O)$ values vary between 1.3 and 2.2 eV depending on the surface oxygen site, which indicates rather moderate binding of H_2O to its V_2O_5(010) surface environment at all sites. The present calculations yield the strongest binding for the O(2) site, followed by the O(1) and O(3) sites. The $E_D(H_2O)$ values are smaller than the corresponding $E_D(OH)$ values by 3.8–4.7 eV depending on the oxygen site, see Tables 3, 4. This can be explained by the increased V–O bond weakening at the surface for surface H_2O compared to surface OH, which makes the removal of an H_2O species easier that that of OH.

4 Conclusions

Altogether, the present theoretical study provides a clear picture of the electronic structure of the V_2O_5(010) surface and its consequences for oxygen vacancy formation as well as hydrogen adsorption. The electronic parameters calculated for the present cluster models, $V_{10}O_{31}H_{12}$ and $V_{20}O_{62}H_{24}$, are size converged and, therefore, the clusters can be considered realistic models of the extended V_2O_5(010) surface. The theoretical data based on *ab initio* DFT methods confirm the mixed ionic and covalent character of V–O binding and distinguish between the differently coordinated oxygen sites. Both the width of the O 2sp dominated valence band region and its total as well as atom projected DOSs are consistent with angular resolved photoemission (ARUPS) data for freshly cleaved V_2O_5(010) samples.[30] Based on the cluster results, the three-peak structure observed in the experiment can be interpreted as originating from differently coordinated surface oxygen, terminal (vanadyl) O(1) dominating the central peak and bridging O(2,3) characterizing the two peripheral peaks. Thus, the theoretical data suggest that the different O 2sp derived peaks observed in the photoemission experiment may be taken as monitors of differently coordinated surface oxygen and can be used to study details of catalytic reactions at the oxide surface where oxygen participates.

Studies on different oxygen vacancies at the V_2O_5(010) surface yield rather large vacancy formation energies $E_D(O)$, 6.7–7.3 eV depending on the oxygen site, which make it quite difficult to remove oxygen by itself from the surface. Surface relaxation caused by the vacancy contributes 1.3–2.2 eV to $E_D(O)$, which is not negligible. However, the geometric relaxation effect is, for all vacancy sites O(1–3), locally confined, with atom displacements of, at most, 0.7 Å. This leaves the lattice topology of the V_2O_5(010) surface unchanged and suggests that single oxygen vacancies will not introduce major restructuring of the surface. Vacancy formation leads, in all cases, to chemical reduction of the vanadium atoms near the vacancy site, as shown by population analyses. Further, preliminary results from two-layer clusters indicate that surface relaxation due to vacancy formation may affect inter-layer coupling and may lead to additional V–O bonds between the physical layers.

Hydrogen can adsorb at all oxygen sites of the $V_2O_5(010)$ surface forming rather stable surface OH groups. The H adsorption energies $E_B(H)$ vary between 2.5 and 3.0 eV depending on the surface oxygen site. The approach of a second hydrogen to the surface OH group leads to the formation of surface H_2O where the second hydrogen is always bound more weakly ($E_B(H)$ between 1.4 and 2.2 eV) than the first. For both surface OH and H_2O formation the calculations suggest a chemical reduction of the neighboring V metal atoms connected with V–O bond weakening where the effect is larger for H_2O than for OH. The calculations of desorption energies E_D, required to remove surface O, OH, or H_2O, (leaving a surface oxygen vacancy behind) show a general trend, independent of the coordination of the oxygen site: the removal always requires more energy for surface O than for surface OH, and more for surface OH than for surface H_2O. While the desorption energies for surface O are extremely large, those for surface H_2O (1.3–2.2 eV) are within the range of typical chemical reaction energies. This has implications for hydrocarbon oxidation reactions at the $V_2O_5(010)$ surface that involve adsorption and binding of hydrogen as well as oxygen desorption from the surface. Here the present calculations suggest that oxygen removal from the surface occurs preferentially by formation of surface H_2O, which is bound weakly enough to desorb, creating oxygen vacancies at the $V_2O_5(010)$ surface.

Acknowledgement

This work has been supported by Deutsche Forschungsgemeinschaft, SFB 1760, and by Fonds der Chemischen Industrie. Further, support by grant No. 3T09A 14615 of the State Committee for Scientific Research in Poland is acknowledged.

References

1. C. N. R. Rao and B. Raven, *Transition Metal Oxides*, VCH Press, New York, 1995.
2. H. K. Kung, *Stud. Surf. Sci. Catal.*, 1989, **45**, 1.
3. V. E. Henrich and P. A. Cox, *The Surface Science of Metal Oxides*, Cambridge University Press, Cambridge, 1994.
4. B. Grzybowska-Swierkosz, *Appl. Catal. A: Gen.*, 1997, **157**, 409, and references cited therein.
5. E. E. Chain, *Appl. Opt.*, 1991, **30**, 2782, and references cited therein.
6. *Appl. Catal. A: Gen.*, 1997, **157**, 1.
7. *Vanadia Catalysts for Processes of Oxidation of Aromatic Hydrocarbons*, ed. B. Grzybowska-Swierkosz and J. Haber, PWN-Polish Scientific Publishers, Warsaw, 1984.
8. A. Byström, K. A. Wilhelmi and O. Brotzen, *Acta Chem. Scand.*, 1950, **4**, 1119.
9. H. G. Bachman, F. R. Ahmed and W. H. Barnes, *Z. Kristallogr. Kristallgeom. Kristallphys. Kristallchem.*, 1961, **115**, 110.
10. R. W. G. Wyckoff, *Crystal Structures*, Interscience Publishers, John Wiley & Sons, Inc., New York–London–Sydney, 1965.
11. A. Bielanski and J. Haber, *Oxygen in Catalysis*, Marcel Dekker, New York, 1990.
12. A. Bielanski, J. Piwowarczyk and J. Pozniczek, *J. Catal.*, 1988, **113**, 334.
13. J. Haber, M. Witko and R. Tokarz, *Appl. Catal. A: Gen.*, 1997, **157**, 3.
14. M. Witko, R. Tokarz and J. Haber, *J. Mol. Catal.*, 1991, **66**, 205.
15. M. Witko, R. Tokarz and J. Haber, *J. Mol. Catal.*, 1991, **66**, 357.
16. M. Witko, *Catal. Today*, 1996, **32**, 89.
17. M. Witko, R. Tokarz and J. Haber, *Appl. Catal. A: Gen.*, 1997, **157**, 23.
18. R. F. Nalewajski, J. Korchowiec, R. Tokarz, E. Broclawik and M. Witko, *J. Mol. Catal.*, 1992, **77**, 165.
19. R. F. Nalewajski and J. Korchowiec, *J. Mol. Catal.*, 1993, **82**, 383.
20. M. Witko, R. Tokarz and K. Hermann, *Pol. J. Chem.*, 1998, **72**, 1565.
21. M. Witko and K. Hermann, *J. Mol. Catal.*, 1993, **81**, 279.
22. M. Witko and K. Hermann, *Stud. Surf. Sci. Catal.*, 1994, **82**, 94.
23. M. Witko, K. Hermann, and R. Tokarz, *J. Electron Spectrosc. Relat. Phenom.*, 1994, **69**, 89.
24. K. Hermann, A. Michalak and M. Witko, *Catal. Today*, 1996, **32**, 321.
25. A. Michalak, M. Witko and K. Hermann, *Surf. Sci.*, 1997, **375**, 385.
26. R. Ramirez, B. Casal, L. Utrera and E. Ruiz-Hitzky, *J. Phys. Chem.*, 1990, **94**, 8960.
27. L. Kihlborg, *Ark. Kemi*, 1963, **21**, 357.
28. H. Hanke, R. Bunert and H. G. Jetschewitz, *Z. Anorg. Allg. Chem.*, 1975, **109**, 414.
29. R. Druzinic, PhD thesis, Free University Berlin, 1999.
30. K. Hermann, M. Witko, R. Druzinic, A. Chakrabarti, B. Tepper, M. Elsner, A. Gorschlüter, H. Kuhlenbeck and H.-J. Freund, *J. Electron Spectrosc. Relat. Phenom.*, 1999, **98–99**, 245.

31 N. Godbout, D. R. Salahub, J. Andzelm and E. Wimmer, *Can. J. Phys.*, 1992, **70**, 560.
32 *Density Functional Methods in Chemistry*, ed. J. K. Labanowski and J. W. Anzelm, Springer–Verlag, New York, 1991.
33 The DFT-LCGTO program package DeMon was developed by A. St.-Amant and D. Salahub (University of Montreal). Here a modified version with extensions by L. G. M. Pettersson and K. Hermann is used.
34 S. H. Vosko, L. Wilk and M. Nusair, *Can. J. Phys.*, 1980, **58**, 1200.
35 J. P. Perdew, J. A. Chevary, S. H. Vosko, K. A. Jackson, M. R. Pederson, D. J. Singh and C. Fiolhais, *Phys. Rev. B: Condens. Matter*, 1992, **46**, 6671.
36 R. S. Mulliken, *J. Chem. Phys.*, 1955, **23**, 1833; 1841; 2388; 2343.
37 I. Mayer, *Chem. Phys. Lett.*, 1983, **97**, 270.
38 I. Mayer, *THEOCHEM*, 1987, **149**, 81.
39 A. Chakrabarti, K. Hermann, R. Druzinic, M. Witko, M. Petersen and F. Wagner, *Phys. Rev. B: Condens. Matter*, 1999, **59**, 10583.
40 V. Eyert, in *Density Functional Methods: Applications in Chemistry and Materials Science*, ed. M. Springborg, Wiley, Chichester, 1997, and references cited therein.
41 V. Eyert and K.-H. Höck, *Phys. Rev. B: Condens. Matter*, 1998, **57**, 12727.
42 X. Yin, A. Fahmi, A. Endou, R. Miura, I. Gunji, R. Yamauchi, M. Kubo, A. Chatterjee and A. Miyamoto, *Appl. Surf. Sci.*, 1998, **130–132**, 539.
43 K. Hermann, A. Chakrabarti, R. Druzinic and M. Witko, *Phys. Status Solidi*, 1999, **173**, 195.

Paper 9/03109J

The growth of vanadium oxide on alumina and titania single crystal surfaces

Robert J. Madix,* Jürgen Biener, Marcus Bäumer and Andreas Dinger

Departments of Chemical Engineering and Chemistry, Stanford University, Stanford, CA 94305, USA

Received 6th April 1999

Evaporation of vanadium metal onto alumina or titania surfaces at room temperature in an oxygen ambient results in the growth of V_2O_3 overlayers. The results of several complementary methods, including STM, indicate that the oxide grows in clusters 20–30 Å in diameter, eventually covering the surface with a granular thin film at a coverage in excess of one monolayer of vanadium atoms. The similarity of the distribution of vanadium metal and the vanadium oxide on the surfaces observed by STM on a thin, crystalline alumina film suggests that the oxide is formed after the metal nucleates into small clusters. No surface reduction of cations occurs on either the $TiO_2(110)$ or $Al_2O_3(0001)$ surface when the vanadium oxide is formed. On the alumina, the vanadium oxide film appears to be conducting at room temperature, as would be expected for V_2O_3 formation.

1 Introduction

Understanding the interaction of metals and metal oxides with oxide surfaces is important to many technological areas, such as metal/ceramic interfaces, microelectronics, geochemistry and heterogeneous catalysis. Important unresolved questions include the strength of this interaction, the wetting characteristics and the mechanical properties of such interfaces, and the electronic properties and chemical reactivity of the overlayer. At submonolayer coverages metals or metal oxides which disperse themselves over the surface may possess characteristics different from their respective bulk solids, since they would be expected to coordinate strongly with the guest oxide, forming, in a sense, a mixed surface compound.[1] Although a conclusive picture is still missing, it has been shown in several UHV studies that the affinity of the metal with oxygen is a crucial factor in determining the strength of interaction between metal overlayers and the support,[2–4] and the heat of formation of the metal oxide can serve as a guide for assessing the interaction strength.[2,3] However, when the oxygen affinity of both the metal atoms of the deposit and the support is high, the relative heats of formation may not provide a reliable guide to wetting behavior.

In an effort to contribute to a better understanding of such systems, we have investigated the growth and the properties of vanadium and vanadia grown on alumina and titania. As an early transition metal, vanadium is expected to interact strongly with oxide surfaces due to the high heats of formation of the vanadium oxides. For example, titania is reduced upon deposition of vanadium metal which, in turn, is oxidized,[5–7] even though the heat of formation of titania exceeds that of vanadia. In this respect, vanadium behaves similarly to other reactive metals on TiO_2.[2,8,9] Furthermore, vanadium is an interesting metal for study, because of its ability to adopt different oxidation states and thus form different oxides, two of which are either conducting or semiconducting at room temperature.

On alumina, however, the situation is more complicated. In contrast to titania, alumina cannot easily be reduced, so that the question arises as to which way the interaction manifests itself in this case. Several theoretical attempts have been made to elucidate the nature of the bonding between metals and alumina,[1,10–12] but the picture is still unclear.

The deposition of vanadium and vanadia on single crystal surfaces of both titania and alumina was studied in the submonolayer-to-multilayer coverage regime with a combination of UHV methods. Our approach was twofold. On the one hand, we employed a thin alumina film grown on NiAl(110) and a partially bulk-reduced TiO_2(110) single crystal, allowing us to utilize scanning tunnelling microscopy (STM) and electron spectroscopic techniques without charging problems. As described in the literature,[13,14] the alumina film is about 5 Å thick and is oxygen-terminated. Its defect structure is dominated by a network of domain boundaries separating reasonably large rotational and antiphase domains of the oxide overlayer.[13] On the other hand, we employed a Al_2O_3(0001) single crystal.[15] The growth of the vanadium oxide overlayer was studied with LEED, photoemission (XPS) and near edge X-ray absorption fine structure (NEXAFS) spectroscopy. NEXAFS was used to determine the local symmetry and to assist with determination of the oxidation state of the metal cations.

Though XPS is the traditional tool to determine the chemical state of metal cations in metal oxides, based on the correlation between core-level binding energies and the oxidation state, the measurement of reliable binding energies on oxide surfaces can be hampered by sample charging and final state effects. NEXAFS is well suited to investigate metal oxides, since the transition energy and the line-shape can be used as a fingerprint of the chemical state. The line-shape of the L-edges of the 3d transition metal oxides is dominated by multiplet effects, which are sensitive to the local symmetry.[16] The NEXAFS is produced by excitation of electrons into empty or partially empty states above the Fermi level using a tunable X-ray source. Band-structure calculations of the unoccupied density of states of 3d transition metal oxides predict a metal 3d-dominated conduction band and, at higher energies, antibonding states related to oxygen 2p and metal 4sp orbitals.[17] In octahedral coordination the ligand field causes the metal 3d band to split into two sub-bands with t_{2g} and e_g symmetry, respectively. The e_g orbitals ($d_{x^2-y^2}$ and d_{z^2}) point directly towards the oxygen ligands, whereas the t_{2g} orbitals (d_{xy}, d_{xz} and d_{yz}) point in between the oxygen neighbors. Consequently, the e_g band lies higher in energy than the t_{2g} band.

Recently the electronic structure of vanadium deposited on titania has been studied using single crystal surfaces.[5–7,18–21] These studies reveal a strong vanadium/support interaction in the monolayer regime; deposited vanadium reduces titania even at room temperature,[5–7] consistent with the high oxygen affinity of vanadium.[22] The V/TiO_2(110) interface has been characterized by means of diffraction techniques, which indicate an epitaxial growth of vanadium without long-range order.[19,20] The formation of vanadia is observed if the vanadium deposition is performed in an oxygen ambient[5] or if metallic vanadium is exposed to oxygen following the deposition. Using the latter technique, the growth of heteroepitaxial VO_2 layers has been achieved.[21]

2 Experimental

The photoemission experiments were performed at beam line 10-1 of the Stanford Synchrotron Research Laboratory in an UHV chamber with a base pressure below 2×10^{-10} Torr. NEXAFS and XPS spectra were produced with monochromatized synchrotron radiation and monitored with a double-pass cylindrical mirror electron energy analyzer. The spherical grating monochromator (1000 1 mm^{-1}, 50 μm slit) was operated with an energy resolution of 0.2 eV at the Ti L-edge, 0.3 eV at the V L-/O K-edges and 0.5 eV at 650 eV used for photoemission. The resulting overall experimental resolution was approximately 0.3 eV and 1.0 eV for NEXAFS and XPS, respectively. The NEXAFS spectra were recorded either by monitoring the CMA signal in the low energy range dominated by secondary electrons (partial yield, PY) or by setting the CMA on an appropriate Auger transition (Auger yield, AY). The XPS and AY-NEXAFS spectra were normalized by dividing the CMA signal by the photoelectron signal of an Au-covered grid inserted in the optical path (I_0-signal). For the studies on the bulk alumina single crystal a low energy electron gun (4 eV) was used to stabilize the surface potential during the data accumulation.

In order to eliminate any possible residual influence of sample surface charging in the experiments on the bulk alumina crystal, all photoemission spectra were referenced to the Al 2p and O

1s peaks of the clean Al_2O_3 surface, centered at 74.7 eV and 531.6 eV, respectively.[23] In the experiments with $TiO_2(110)$ the photoelectron energy analyzer was repeatedly calibrated using a clean Au sample, assuming the Au $4f_{7/2}$ core-level binding energy to be 84.0 eV.[24] Although the spectrometer calibration was performed repeatedly, it was observed that for the different $TiO_2(110)$ samples used in the experiments, the Ti 2p core-level binding energy varied in energy from 458.5 to 459.3 eV. These values of binding energy are within the range reported previously in the literature;[25] the scatter of the data may arise from the differing electrical conductivities of the samples. In order to eliminate the influence of any charging at the surface on the energy calibration, all the photoemission spectra were referenced to the Ti 2p core line at 458.5 eV. For NEXAFS the photon energy was calibrated by aligning the Ti L-edge and O K-edge of the clean $TiO_2(110)$ sample according to the values reported in the literature.[26] In addition, L-edge NEXAFS spectra from a V foil were recorded to verify the energy calibration.

The $TiO_2(110)$ single crystals were attached to molybdenum disks, which could be heated from the back either by radiation or by electron bombardment. The sample temperature was measured by means of an optical pyrometer and a chromel/alumel thermocouple attached to the sample holder. To introduce n-type semiconducting bulk properties, the TiO_2 sample was bulk-reduced by high temperature annealing (1 h, 1100 K). During this time the crystal color changed from colorless to pale green. Calcium and carbon were removed from the surface by repeated sputter–anneal cycles (500 eV Ar^+, 10 μA). The surface stoichiometry was restored by annealing in oxygen ($p(O_2) = 5 \times 10^{-5}$ Torr, 900 K), and restoration of the stoichiometric surface was judged by the absence of reduced titanium in photoemission spectroscopy and the presence of a sharp (1×1) LEED pattern. Vanadia was deposited at 300 K by evaporating vanadium from a resistively heated vanadium filament (Goodfellows, 0.008 inch diameter, >99.9%) in an oxygen ambient of 2×10^{-6} Torr. The vanadia deposition rate and the coverage were estimated from X-ray spectra following standard procedures,[27,28] making use of the attenuation of the Ti 2p XPS signal caused by the vanadia overlayer. The deposition rate was determined to be approximately 1×10^{-2} ML s^{-1}, assuming that one ML corresponds to approximately 9.4×10^{14} atoms cm^{-2}, the areal density of vanadium atoms in the (0001) plane of V_2O_3.[29]

The α-Al_2O_3(0001) single crystals, 0.8 mm thick, polished wafers, were mounted and heated like the titania crystals. The sample temperature was measured by means of an optical pyrometer, and the error in the temperature measurement is estimated to be less than 100 K. The Al_2O_3 surfaces were cleaned as follows: (1) prolonged degassing (>5 h at 1000 K in UHV) and (2) annealing at 1200 K (1 h, UHV) followed by an oxygen annealing step at 900–1100 K (1 h, 1×10^{-6} Torr O_2). The surface cleanliness was monitored by XPS. The oxygen annealing step was repeated until a sharp (1×1) LEED pattern was observed,[30–33] and surface contamination was reduced below the detection limit. The deposition of vanadium oxide on the Al_2O_3(0001) samples was accomplished by evaporation of metallic vanadium in an oxygen ambient of 1×10^{-7} Torr at 300 K sample temperature. The vanadia deposition rate and the coverage were estimated from X-ray photoelectron spectra following standard procedures [27,28] which make use of the attenuation of the Al 2p and the increase of the V 2p XPS signal intensities as a function of the deposition time. The deposition rate employed was approximately 10^{-3} ML s^{-1}.

STM images were taken in an UHV system equipped with a 'Johnnie Walker' type scanning tunneling microscope purchased from RHK, Tech. The microscope stage allowed heating of the sample by radiation (W filament) and cooling by contact to a liquid nitrogen reservoir. The sample was mounted on a transferable mount, which also contained a trisectional ramp used for the approach of the STM head. This unit could be transferred between the microscope, located in a side chamber, and a sample preparation chamber equipped with LEED and all instruments necessary for the preparation of the surface, including a gas doser and a metal evaporator. A NiCr/Ni thermocouple was used to determine the sample temperature. The sample could be heated in the preparation chamber to at least 1000 °C by electron bombardment, while cooling was possible by directing liquid nitrogen through the manipulator.

The clean NiAl(110) surface was prepared by several cycles of sputtering (Ar^+ ions, 1 keV) and annealing to 1300 K. The ordered Al_2O_3 film was obtained by following procedures previously reported in the literature.[13,14] After dosing about 5000 L O_2 at a sample temperature of 533 K, the crystal was annealed to 1100 K for 3 min. The quality of the oxide was checked *via* its distinctive LEED pattern. Vanadium metal was deposited in the STM system using procedures similar to those employed for the alumina and titania single crystals. The evaporation rate was calibrated by a quartz microbalance. In the STM apparatus the nominal film thickness determined by the thickness monitor was converted into a coverage using the lattice constant of bulk vanadium (3.03 Å [29]). The deposition rates varied between 3×10^{-3} and 10^{-2} ML s^{-1}.

3 Results and discussion

3.1 Photoemission studies of the growth of vanadium oxide on TiO$_2$(110) and Al$_2$O$_3$(0001)

The Ti 2p photoemission spectra for the clean and vanadia-covered TiO$_2$ surface are shown in Fig. 1. The spectrum of the clean surface is identical to those reported in the literature, indicating that the surface is stoichiometric, with the cations in the Ti^{+4} state.[25,34,35] The intensity of the Ti 2p signal decreases with increasing vanadia coverage, until the signal finally vanishes, indicating that the surface is completely covered with a vanadia layer of a thickness larger than the mean free path of the 200 eV photoelectrons. The lineshape of the Ti 2p peaks does not change upon vanadia deposition over the entire range of vanadia coverage examined. Since surface reduction of TiO$_2$(110) due either to sputtering or to submonolayer deposition of vanadium is *easily* detected by XPS, as a shoulder on the low binding energy side of the Ti2p peak,[5] these spectra demonstrates that the vanadia does not reduce the TiO$_2$ surface.

The corresponding O 1s and V 2p regions of the XPS spectra are displayed in Fig. 2. The V 2p signal increases monotonically with vanadium exposure, whereas the intensity of the O 1s signal stays nearly constant, consistent with the deposition of a vanadium oxide. The V 2p peaks are centered at 515.9 eV (V 2p$_{3/2}$) and 523.4 eV (V 2p$_{1/2}$) over the entire range of coverage, lying between values reported for single crystals of V$_2$O$_3$ (515.7/523.3 eV) and VO$_2$ (516.2/523.5 eV).[36] In another investigation of the vanadia/TiO$_2$ system, using a very similar preparation technique, Zhang and Henrich[5] observed the V 2p core-lines at 515.2 and 522.7 eV. For heteroepitaxial VO$_2$ layers on TiO$_2$(110) the V 2p$_{3/2}$ line was reported to be 515.7 and 516.5 eV for submonolayer and

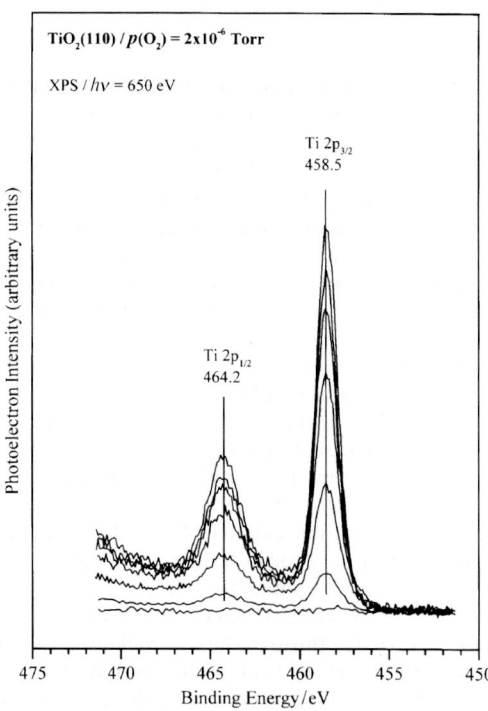

Fig. 1 The Ti 2p$_{1/2}$ and Ti 2p$_{3/2}$ photoemission peaks as a function of vanadium exposure time at room temperature in oxygen. From the top, spectra correspond to exposures of 0, 20, 40, 80, 160, 320 and 640 s. A photon energy of 650 eV was used. Monolayer coverage occurs at exposures near 100 s.

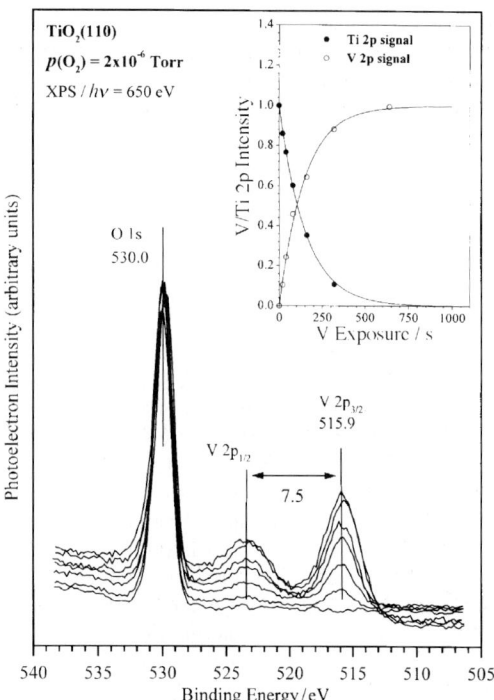

Fig. 2 The O 1s, V $2p_{1/2}$ and V $2p_{3/2}$ photoemission peaks as a function of vanadium exposure time. Exposures and conditions were identical to Fig. 1. The inset shows the dependence of the integrated intensities on vanadium exposure.

multilayer coverages, respectively.[21] The O 1s peak for clean and vanadia-covered TiO_2 is centered at 530.0 eV. For TiO_2, O 1s binding energies in the range of 529.8 to 530.2 have been reported,[21,35] which are similar to the values found for V_2O_3 (530.1 eV), VO_2 (529.9 eV)[36] and VO_2/TiO_2(530.4 eV).[21] Taken together, these XPS results indicate the formation of a vanadia overlayer, but alone they do not allow the determination of the oxidation state of vanadium cations because of the problems involved in determining an absolute binding energy scale, as discussed above.

The inset of Fig. 2 compares the normalized V and Ti 2p photoemission intensities as a function of the vanadium exposure. The exponential decay of the Ti 2p signal intensity and the corresponding increase of the V 2p signal with vanadium exposure are consistent with a simultaneous multilayer (SM) growth mode, characterized by the growth of small particles, which cover the surface at sufficiently high vanadia coverage.[28] However, although the data allow exclusion of a three-dimensional growth mode (Volmer–Weber, Stranski–Krastanov) the number of data points is insufficient to distinguish between a purely layer-by-layer growth (Frank–van der Merve) and a SM growth mode. For reasons discussed below, we favor the simultaneous growth of isolated nuclei of vanadia, which merge into a textured film at multilayer coverage. Consistent with these findings, the initially sharp (1 × 1) LEED pattern of TiO_2(110) nearly vanished following the deposition of a vanadia multilayer, indicating that vanadia covers the surface without any long-range order.

The vanadium oxide formed introduces electronic states in the band gap of the TiO_2(110) surface. The valence band region of the X-ray spectrum for the clean TiO_2(110) surface shows a 6 eV wide O 2p-related valence band centered approximately 5 eV below E_{Fermi}. The band gap is 3 eV wide, typical of the stoichiometric TiO_2.[37] As expected for an n-type semiconductor, the Fermi level is pinned to the bottom of the conduction band. These findings are in accordance with the literature.[5,34] Upon vanadium exposure a new state appears in the band gap, centered approximately 1 eV below E_{Fermi}. Since reduced titania can be excluded as the source of the new feature in the band gap, this state can be assigned to photoemission from the partially occupied V 3d band of the vanadia layer, thus, in agreement with the XPS data, eliminating V_2O_5 (d^0) as the dominant oxide of vanadium present on the surface.

Similarly, the V 2p photoemission signal from the surface of the bulk alumina single crystal, $Al_2O_3(0001)$, increases with vanadium exposure, and the intensity of the O 1s signal decreases rapidly to a constant value, suggestive of the formation of a vanadium oxide (Figs. 3 and 4). After the deposition of multilayer quantities of vanadia the V 2p peaks are centered at 515.5 eV (V $2p_{3/2}$) and 523.2 eV (V $2p_{1/2}$), suggesting the formation of a V_2O_3 overlayer.[36] Furthermore, the V 2p lines exhibit a linewidth of approximately 3.5 eV, indicative of a narrow-band metal like V_2O_3.[36] With increasing vanadium exposure the O 1s peak shifts abruptly from 531.6 eV to 530.1 eV; the latter value is characteristic of vanadium oxides (V_2O_3: 530.1 eV, VO_2: 529.9 eV).[36]

A comparison of the observed binding energy shifts of the O 1s and V $2p_{3/2}$ core-lines as the vanadium coverage increases is shown in Fig. 4. After an exposure time of approximately 2000 s both the O 1s and the V $2p_{3/2}$ peaks shift simultaneously toward lower binding energies by 1.5 eV. As all spectra are referenced to the Al 2p peak, this observation cannot be caused by a shift of the Fermi level within the band gap of Al_2O_3. The shift of the O 1s binding energy can be interpreted in terms of oxygen in different chemical environments, i.e., oxygen bound to aluminum and vanadium, respectively. The shift to lower binding energy occurs with the completion of the second, pure vanadium oxide layer on top of the interface after a vanadium exposure of 2000 s.

It is tempting to interpret the higher value of the vanadium binding energy at low coverage in terms of a higher oxidation state. However, the problems connected with the calibration of an absolute binding energy scale for insulating oxides prevent the unequivocal determination of the oxidation state solely on the basis of the binding energy. For example, for small metallic particles supported on insulating substrates, final state effects influence the observed core hole binding energy of the metal. Depending on the details of the interaction between the metal clusters and the

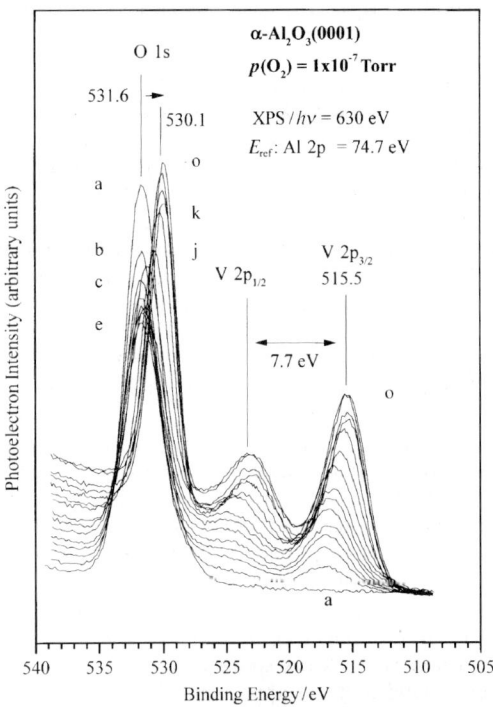

Fig. 3 The O 1s, V $2p_{1/2}$ and V $2p_{3/2}$ photoemission peaks with increasing vanadium exposure times at room temperature in oxygen: (a) clean surface, (b) 250 s, (c) 500 s, (d) 750 s, (e) 1000 s, (f) 1250 s, (g) 1500 s, (h) 1750 s, (i) 2000 s, (j) 2500 s, (k) 3000 s, (l) 3500 s, (m) 4000 s, (n) 6000 s and (o) 9600 s vanadium exposure. Monolayer coverage occurs at exposures near 1000 s.

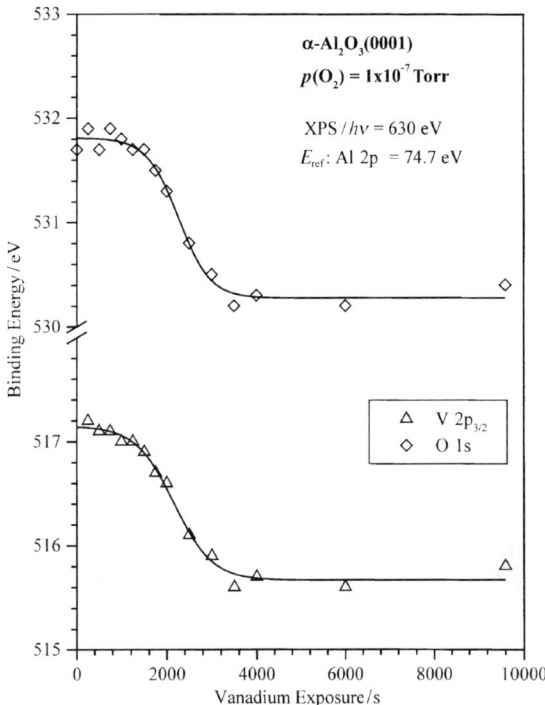

Fig. 4 The dependence of the binding energies of the O 1s and V $2p_{3/2}$ photemission peaks following exposure of $Al_2O_3(0001)$ to vanadium in oxygen at room temperature at varying times.

surface, a shift of the photoemission peaks toward higher binding energies may arise from the Coulomb attraction induced by the positive charge left on the particle as the result of the photoemission process.[38,39] The Coulomb energy scales with $\sim e^2/R$, where R is the cluster radius. Thus, the higher value of the V 2p binding energy at the interface may simply reflect the formation of very small, metal-like clusters with a narrow size distribution during the initial deposition phase, which, in turn, are transformed to V_2O_3 at the completion of a monolayer. This interpretation is consistent with the growth of vanadia on the alumina film observed by STM, as described below.

In order to assess the amount of vanadia deposited and the mode of its growth on the surface, the Al 2p, V 2p and O 1s photoemission spectra were fitted with Gaussian functions. In Fig. 5 the normalized signal intensities are displayed as a function of the deposition time. The Al 2p signal decreases continuously with increasing deposition time, finally reaching the detection limit; this final state indicates a surface completely covered with vanadia. Simultaneously, the V 2p intensity increases to a constant value. The O 1s signal decreases, initially rapidly, and stabilizes after an exposure of 1000 s at about 2/3 of its original intensity. These findings allow us to exclude three-dimensional growth modes, Volmer–Weber and Stranski–Krastanov, respectively. The continous variation of the intensities and absence of break points suggest a simultaneous multilayer (SM) rather than a pure layer-by-layer growth mode (Frank–van der Merve).[28] The attainment of the constant oxygen signal coincides with the deposition of approximately one monolayer of vanadium oxide.

In the course of the study it was observed that the kinetic energy of the Al 2p photoelectrons changed systematically with the vanadia coverage, even though both the analyzer work function and the photon energy were constant during the experiments (Fig. 6). Since the chemical state of aluminum is not expected to change significantly upon vanadia deposition, the shift in the kinetic energy of the Al 2p photoelectrons must originate from surface charging due to an incomplete neutralization of the surface during the experiment. The surface charging decreases rapidly above an exposure time of approximately 1000 s, indicative of the formation of a closed, conducting overlayer, i.e., the deposition of a metallic vanadium oxide. This behavior is consistent with the

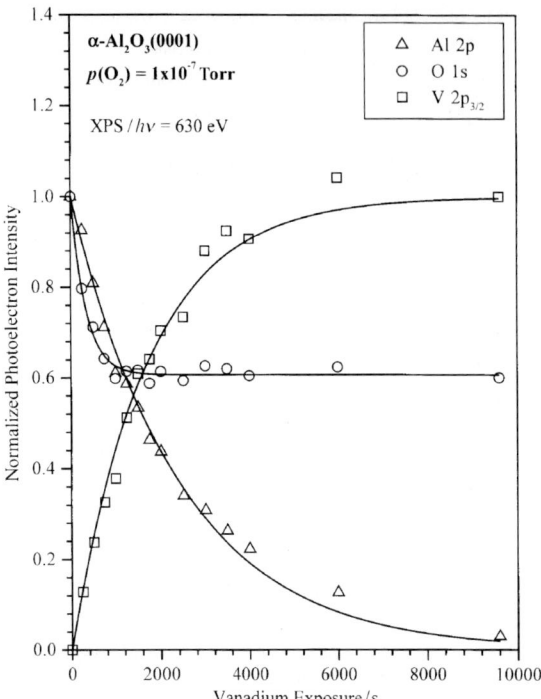

Fig. 5 The evolution of the integrated Al 2p, O 1s and V $2p_{3/2}$ photoelectron intensities following various exposure times of $Al_2O_3(0001)$ to vanadium in oxygen at room temperature. Monolayer coverage is estimated to occur near 1000 s.

shift in the V 2p binding energy, as V_2O_3 is metallic at room temperature,[22] is consistent with the calculated deposition rate, and coincides with the exposure at which the O 1s signal becomes constant.

The lineshape of the Al 2p signal was checked in order to verify the validity of the assumption that the chemical state of aluminum does not change upon the deposition of vanadia. For example, the formation of a reduced alumina surface layer should be reflected by the observation of a new Al 2p photoemission peak shifted ∼3eV to lower binding energies.[40] However, the analysis revealed that the lineshape of the Al 2p peak of the clean and vanadia-covered surface are identical in the *submonolayer* coverage regime. However, for vanadium exposures exceeding 2000 s, the exposure marking completion of a conducting vanadia overlayer, the line width did suddenly decrease from 3.7 eV to approximately 2.7 eV. This decrease is probably a consequence of a more homogeneous surface potential produced by the sudden change in conductivity of the overlayer.

3.2 NEXAFS studies of the growth of vanadia on $TiO_2(110)$ and $Al_2O_3(0001)$

The NEXAFS spectra of the clean $TiO_2(110)$ surface are characteristic of the rutile phase of TiO_2[26] (not shown). In agreement with the information obtained from the photoemission spectra of the Ti 2p region, the only effect of vanadia deposition is an attenuation of the NEXAFS signal intensity. The lineshape remains unchanged, indicating that the vanadia does not disturb the local symmetry of the titanium cations on the TiO_2 surface.

Fig. 7 shows the V L-edge spectra of the vanadia overlayers obtained at progressively higher coverages. The two broad features centered at 517.6 and 524.5 eV are related to excitations from V $2p_{3/2}$ (L_{III}-edge) and V $2p_{1/2}$ (L_{II}-edge) core-levels into empty or partially occupied V 3d orbitals.[26] Both the transition energy [41,42] and the lineshape [43] of the vanadium L-edge are sensitive to the oxidation state of the vanadium cations. The positions of the L-edge features of the deposited

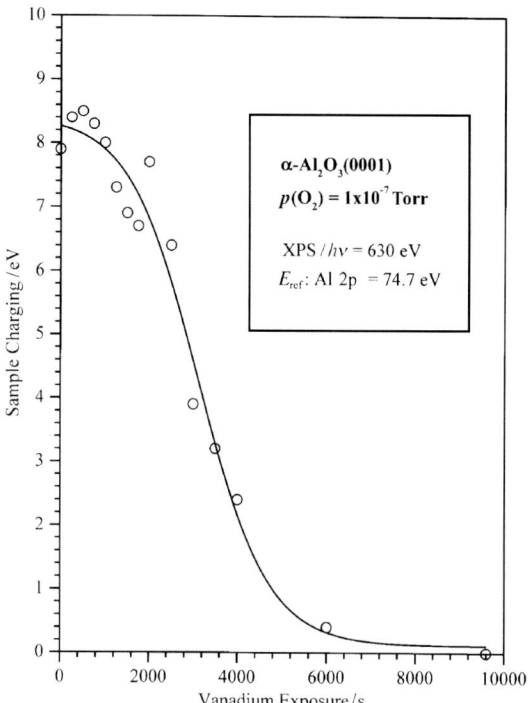

Fig. 6 Relative kinetic energies for Al 2p photoelectrons for Al_2O_3 as a function of the duration of vanadium exposure in oxygen. The energy zero is taken to be the kinetic energy measured for the photoelectrons upon growth of the V_2O_3 multilayer.

vanadia do not shift with the V coverage. The transition energies are in excellent agreement with values expected for V_2O_3 using the calibration curves published by Chen et al.,[42] who reported a linear increase of 0.7 eV per oxidation state in the V L_{III}-edge, ranging from 515.5 eV for metallic V to 519.0 eV for V_2O_5.

In addition to the transition energies, the fine structure of the V L-edge can be used to determine the vanadium oxidation state. The details of the spectral lineshape of the V L-edge can be explained only by atomic multiplet calculations.[44–47] The multiplet structure generally changes with the number of d-electrons. Therefore, the oxidation state can be deduced from the lineshape. The vanadia L-edge fine structure strongly resembles that of single crystal V_2O_3, but it differs from that for VO_2 and V_2O_5.[43] According to calculations from de Groot et al.,[44] the pre-edge feature and a feature between the L_{III}- and L_{II}-edge are characteristic of a $3d^2$ initial-state multiplet. Both features are observed in the experimental spectra of the vanadia overlayer, particularly at the higher coverages, thus indicating the formation of V_2O_3 in the present study.

The O K-edges of the clean TiO_2 surface before and after deposition of a vanadia multilayer are compared in Fig. 8. The O K-edge involves the excitation of an oxygen 1s electron into unoccupied states above the Fermi level with oxygen 2p character, thus providing a picture of the oxygen 2p-projected density of states. In contrast to the L-edges of the transition metals, the oxygen K-edge can be described within the single-particle approximation.[17,48]

In order to extract the positions and the intensities of the Ti and V 3d-related peaks from the clean and vanadia-covered surfaces, respectively, the experimental spectra were fitted with Gaussian functions. The absorption step was approximated with an error function centered at the experimentally determined oxygen 1s binding energy value of 530.0 eV with a FWHM equal to the experimental resolution (0.3 eV).[49] It is obvious from the fitting results that the integrated intensity of the t_{2g} state for V in the vanadium oxide is significantly less that would be expected for a d^0 oxide. As the intensity is related to the number of unoccupied states available to the excited electron, this observation indicates the presence of a partially filled t_{2g} band, in agreement

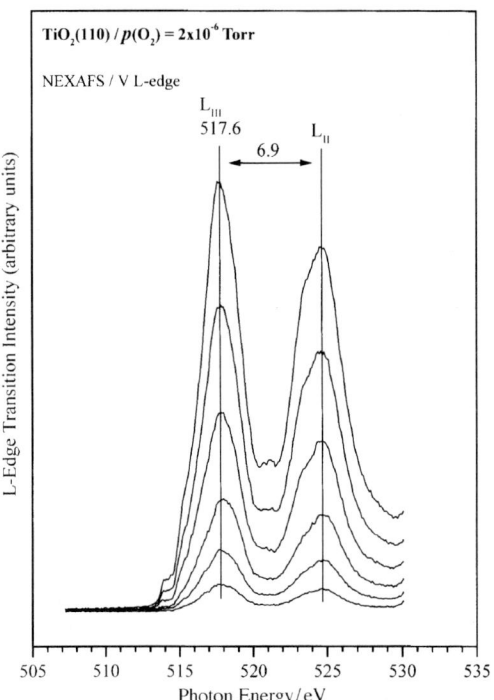

Fig. 7 Transition intensities for V-near-edge absorption fine structure following vanadium deposition at various exposure times in oxygen onto $TiO_2(110)$. Exposure times were 20, 40, 80, 160, 320 and 640 s. Monolayer coverage occurs near a deposition time of 100 s.

with the appearance of states in the band gap observed in the X-ray spectra. The ligand field splitting of the d-orbitals in the vanadium oxide formed on the surface is 2.3 eV, similar to the value of 2.2 eV reported for V_2O_3 and VO_2.[48] Thus, the K-edge data are consistent with formation of V_2O_3. The NEXAFS results provide evidence for the existence of a short-range order of the vanadium cations, even though the sharp (1 × 1) LEED pattern of the clean surface disappeared completely after the deposition of a vanadia monolayer, indicating the loss of long-range order.

Similar results were observed for vanadia deposition on the $Al_2O_3(0001)$ single crystal surface (Fig. 9). The V $2p_{3/2}$ (L_{III}-edge) and V $2p_{1/2}$ (L_{II}-edge) core-level transition energies of 517.5 and 524.4 eV in the multilayer coverage regime are in excellent agreement with the values expected for V_2O_3 using the calibration curve of Chen et al.[26] The fine structure, in particular for V exposures exceeding 1000 s, also indicates the formation of V_2O_3,[43] as discussed above for vanadia formation on $TiO_2(110)$. In the low coverage regime these features are not as pronounced as for the higher exposures, perhaps suggesting a reduced symmetry in the submonolayer regime, however.

3.3 Electron energy loss spectroscopy

Electron energy loss spectra, both in energy loss regimes for vibrational and electronic transitions, are in general agreement with the above conclusions regarding growth of the oxide overlayer. The expected strong Fuchs–Kliewer vibrational excitations are seen at 53 and 94 meV, in accord with previously reported results (Fig. 10).[50,51] For the sake of comparison, spectra were recorded for vanadium metal evaporated onto the surface both with and without oxygen in the background. Based on the attenuation of the titanium Auger features, we estimate monolayer coverage to occur at about 1000 s exposure to vanadium. Conducting adlayers are known to attenuate the Fuchs–Kliewer modes rapidly. However, the presence of these losses, even after 3000 s exposure to vanadium in the oxygen atmosphere, indicates either that a uniform monolayer coverage is not

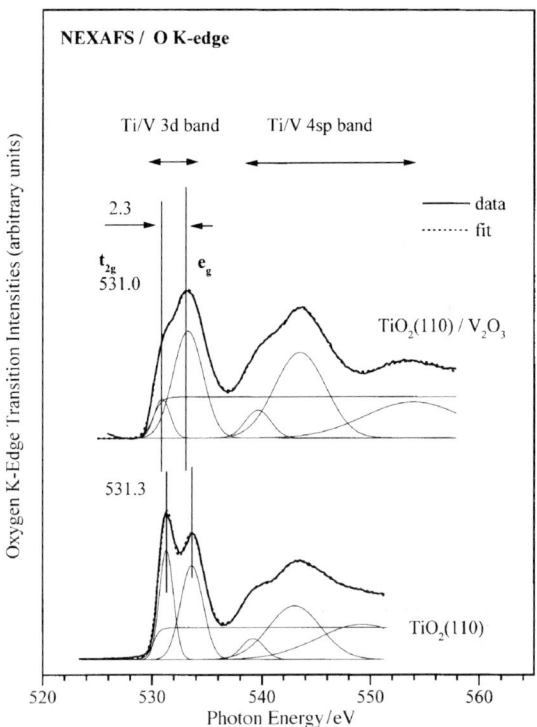

Fig. 8 The oxygen K-edge spectra for clean TiO$_2$(110) and V$_2$O$_3$-covered TiO$_2$(110). The spectra were fitted with Gaussian peaks in order to evaluate the relative transition intensities from the core-levels into the t$_{2g}$ and e$_g$ orbitals, respectively.

attained (assuming the vanadia to be conducting) or that the vanadia film is not sufficiently metallic to screen the Fuchs–Kliewer modes. Furthermore, up to a vanadium exposure of 3000 s there is no detectable shift in the frequency of any of the vibrational losses. Thus, the HREELS results are consistent with a growth mode in which the surface is populated with a patchy distribution of the oxide.

Electronic spectra show clear evidence for an electronic transition accessible within the band gap which arises from the formation of the overlayer of vanadium oxide. This feature, which is due to d–d transitions in the vanadium oxide, grows with the amount of vanadia deposited, and after 1000 s exposure to vanadium in the presence of background oxygen, it appears as a prominent feature at 2.0 eV (Fig. 11). Strong transitions centered at 5.2 and 10.0 eV also appear, corresponding to electronic transitions from O 2p to metal 3d states. The energies for the latter transitions are slightly lower than those observed for the clean surface. Decomposition of the losses observed for the vanadia-covered surface into those for TiO$_2$ and vanadia is difficult, because the relative loss intensities for each of the two oxides are not known. However, the shift of the band centers to lower energies with vanadia deposition is consistent with the NEXAFS results, which show the O 2s to metal 3d transitions at a lower energy for the vanadia overlayer.

3.4 STM and LEED

STM was employed to study the manner in which vanadium deposits grow on TiO$_2$(110)–(1 × 2) and alumina, using the alumina film grown on NiAl(110) as a model for a bulk single crystal of alumina.[52] STM of the thin alumina film reveals smooth terraces interrupted by steps (Fig. 12). Since these results are relevant to the growth of vanadia on these surfaces, they are included here with STM studies of vanadia growth on the alumina thin film. Within a given terrace are large domains of the alumina film, joined by boundaries that are clearly imaged by STM. A series of

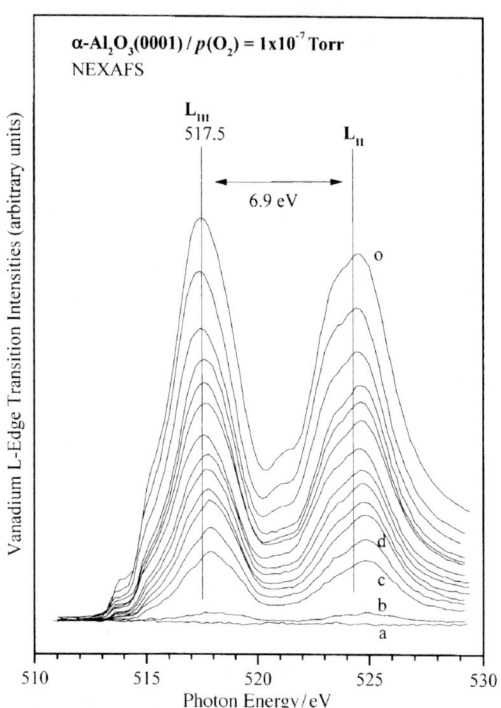

Fig. 9 L-edge spectra for vanadia deposited on Al$_2$O$_3$(0001) at room temperature in oxygen. Spectra a–o are for vanadium exposures given in the caption of Fig. 3.

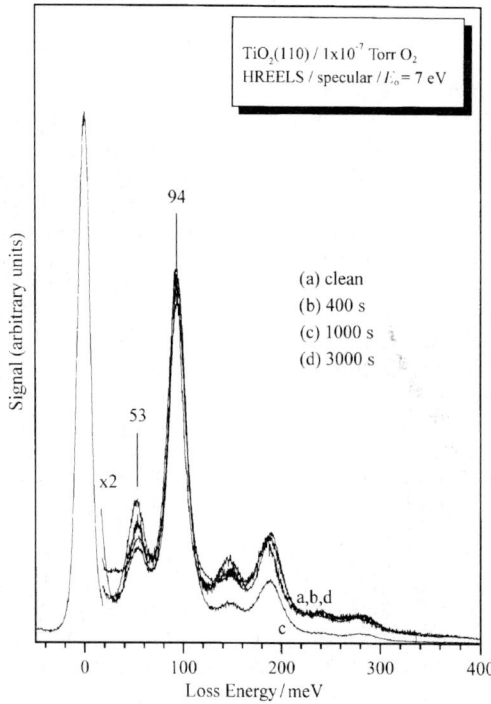

Fig. 10 Electron energy loss vibrational spectra for vanadium deposition times of 0, 400 s, 1000 s and 3000 s in the presence of background oxygen. Monolayer coverage occurs at exposures near 1000 s.

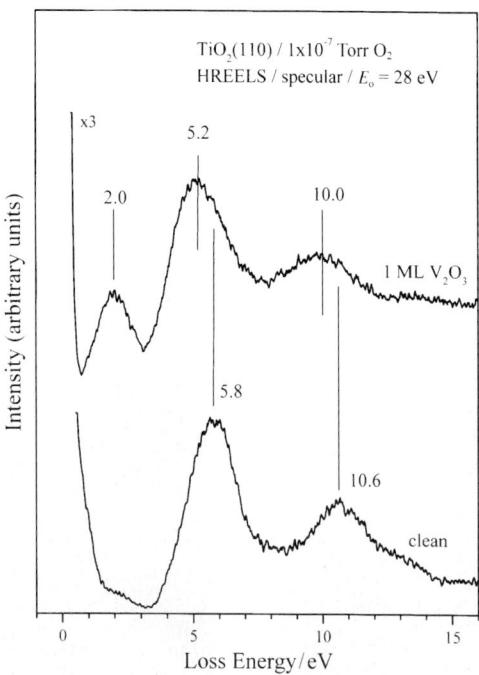

Fig. 11 Electronic energy loss spectra for clean and vanadia covered $TiO_2(110)$.

STM images taken for different exposures of the alumina film to metallic vanadium at room temperature shows that vanadium prefers to nucleate in small clusters on the surface. The particle diameter appears to be nearly identical at all three coverages studied (20–30 Å). The particle density and height, however, depend on the amount of metal deposited, growing larger with exposure. One such image, obtained following an exposure of approximately 0.10 ML is shown on the

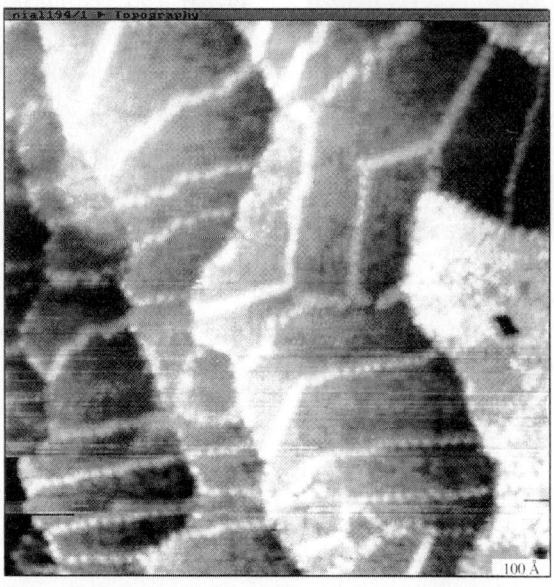

Fig. 12 An STM image of the thin alumina film grown on NiAl(110). The image shows terraces criss-crossed by two-dimensional single-crystal domains of the alumina film as well as steps (constant current image, +3.0 V bias, 0.4 nA).

left-hand side of Fig. 13. These results directly confirm the growth mode to be of the simultaneous multilayer type.

The steps and the domain boundaries are not preferentially decorated by the deposits. Rather the vanadium particles decorate the terraces randomly. Exposure of the film to the same vanadium exposure with a background pressure of 10^{-6} Torr oxygen produces a similar size and distribution of islands on the surface, as shown on the right-hand side of Fig. 13. There is no detectable difference in the size of the islands for the metallic vanadium or the vanadium oxide; however, the corrugation of the oxide particles appears less than that of the metal particles (see the bottom of Fig. 13). Neither is there a preference for the growth of the small oxide islands at defects in the surface structure of the alumina film. The similarity of the islands formed with or without ambient oxygen suggests that metallic vanadium is oxidized after forming small clusters.

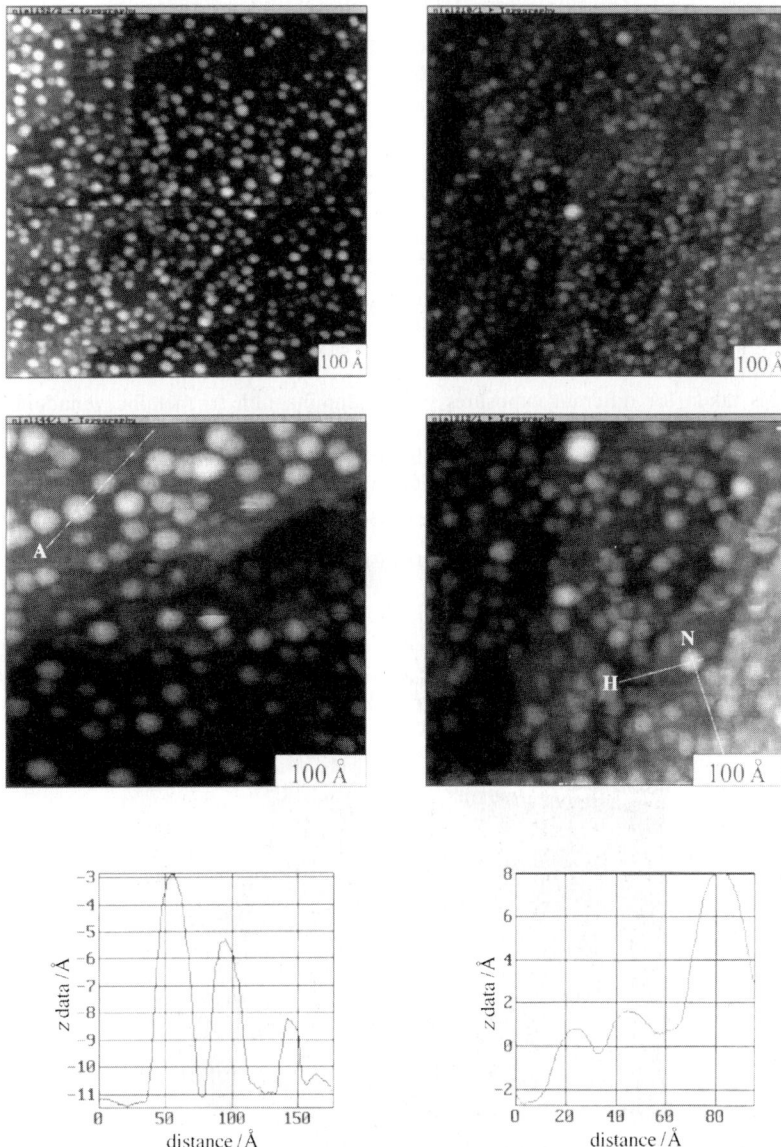

Fig. 13 STM images of vanadium (left-hand side) and vanadia deposited in submonolayer quantities onto the thin alumina film grown on NiAl(110). The particle sizes are similar in both instances.

LEED results (not shown) indicate that the underlying structure of the thin alumina film is not degraded significantly at low coverages of the vanadium or vanadia adlayer. Though the intensities of the diffraction spots in the complex LEED pattern of the alumina thin film diminish with exposure to vanadium (both with and without an oxygen ambient), they are not rapidly extinguished, showing all diffraction spots with significant intensity up to coverages of 0.12 ML. Even at 0.30 ML diffraction features characteristic of the clean surface are readily detectable, though a bright, diffuse background has emerged. Thus the surface structure in the interstitial space between the oxide (or metallic) islands seems to be relatively unperturbed by the presence of the overlayer at appreciable coverages. At a vanadium coverage of 0.5 ML the diffraction spots are no longer visible, however.

STM images indicate a similar pattern in the initial stages of deposition of vanadium on $TiO_2(110)$–(1×2). In Fig. 14 an image of the clean surface, obtained after sputtering at room temperature and annealing near 1100 K, shows a (1×1) terrace, (1×2) overlayer growth and local (2×2) structures, in agreement with results of others.[53–54] Fig. 15 shows the STM images for the $TiO_2(110)$–(1×2) surface, prepared as described above, but after annealing to 1150 K, (a) before and (b) after a vanadium exposure of 0.05 ML. Bright rows parallel to the [001] direction with a periodicity of 13 Å and a corrugation of 1 Å along the [1$\bar{1}$0] axis cover the surface nearly completely. The rows are terminated by small protrusions, 10 Å in diameter and 3 Å high, or by step edges. In agreement with the STM images, LEED indicates the formation of a (1×2) structure. The bright rows in the STM images of the clean $TiO_2(110)$–(1×2) surface have been assigned to the added rows of Ti_2O_3.[55,56] This interpretation of the structure was chosen over a competing model, in which the surface is proposed to be severely reduced, because the (1×2) surface does not react with formic acid. This reaction would be expected for the reduced surface, as even the stoichiometric surface dissociates the acid on exposed titanium cations. The protrusions terminating the added Ti_2O_3 rows indicate the presence of undercoordinated Ti cations.

After the deposition of submonolyer amounts of vanadium, additional protrusions decorating the bright added–rows of the (1×2) phase appeared in the STM images (Fig. 15). The bright features are approximately 1 nm wide and 1–2 nm long, depending somewhat on the tip condition, with a corrugation of 0.1 nm, similar to the structure at the terminus of a (1×2) row. At the lowest coverage the elongated features apparently arise from clusters of vanadium atoms. If these features are caused by single vanadium atoms, the extent to which they perturb the electronic structure must be significantly greater than their ionic radius. However, we must admit the possibility that these circular features arise from the nucleation of several vanadium atoms, since the

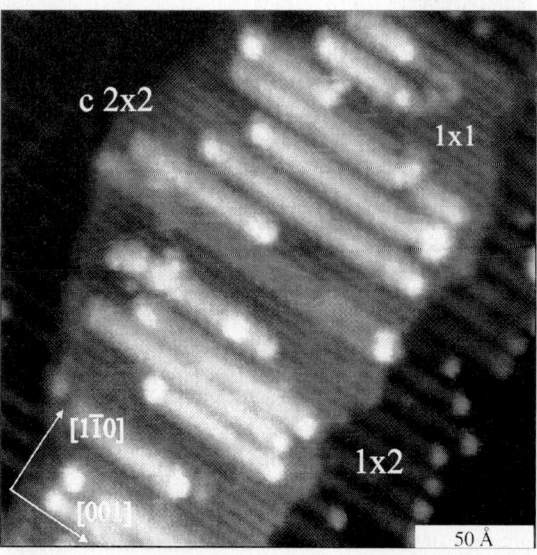

Fig. 14 An STM image of the clean $TiO_2(110)$–(1×2) surface showing strings of the (2×2) structure as well as a shear plane (constant current image, +3.2 V bias, 0.2 nA).

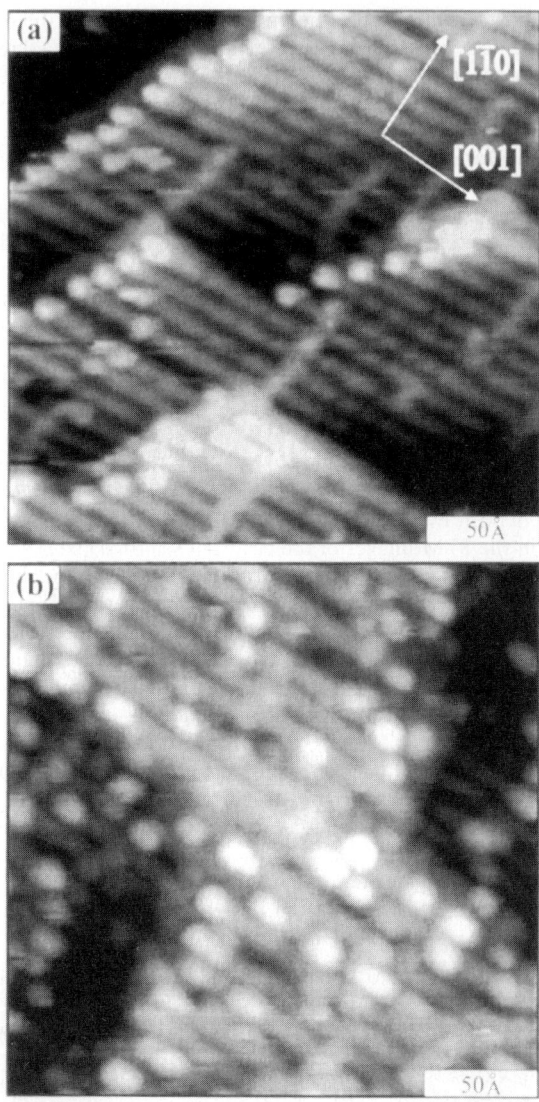

Fig. 15 An STM image of 0.1 ML of vanadium deposited onto TiO_2(110)–(1 × 2) at room temperature. The vanadium atoms appear to form small clusters located preferentially on top of the (1 × 2) added rows. (constant current image, +1.8 V bias, 0.3 nA). (a) Before and (b) after a vanadium exposure of 0.05 ML.

atoms must have sufficient mobility to migrate preferentially to the top of the added rows. The density of these features increases with the vanadium exposure, whereas the size remains nearly constant and similar to the hill-like structures terminating the rows. Due to the high oxygen affinity of vanadium, the most likely adsorption site is on top of the oxygen covered Ti_2O_3 rows, where vanadium could achieve the maximum coordination with oxygen. Thus the additional protrusions on top of the rows are attributed to vanadium. As discussed above, we believe that even submonolayer amounts of vanadium are oxidized to V_2O_3. In view of the similar structures exhibited by titanium and vanadium (Ti_2O_3 and V_2O_3 have the corundum structure[22]), it is not too surprising that the features arising from the deposition of vanadium resemble the 'added row' Ti_2O_3 structures, though it is not yet clear exactly how vanadium is incorporated. The appearance of a linear arrangement of these features along the [001] direction, however, is further indication that they preferentially associate themselves with the underlying (1 × 2) structure.

As the vanadium coverage approaches one ML these bright features coalesce, the surface becomes densely covered with vanadium clusters whose irregular shape seems to reflect the surface symmetry (not shown). On average the clusters have a height of 2.5 Å and a diameter of 20 Å, indicating that 1–2 ML thick *patches* of vanadium and/or vanadia cover the surface. Due to this narrow height distribution the step structure of the underlying surface is still easy to recognize, even after vanadium exposures of one ML. Furthermore, even at this coverage, there is a clear preference for the added vanadium to align along the (1 × 2) rows of the underlying titania surface.

The morphologies adopted by vanadium deposited on the alumina film and the TiO_2(110) single crystal are quite similar. In both cases a small number of vanadium atoms appear to coalesce into small particles dispersed more or less uniformly over the surface. Since vanadium has a high affinity for oxygen, surface migration of the metal atoms is limited. Three dimensional growth of large aggregates is not permitted.

Acknowledgements

The authors gratefully acknowledge the support of the National Science Foundation through NSF CTS-9618807 and through the MRSEC program administered by the Center for Materials Research at Stanford. J.B. and M.B. thank the Deutsche Akademischer Austauschdienst (DAAD) and the Deutsche Forschungsgemeinschaft (DFG), respectively, for research fellowships. The authors are grateful to Dr. Tom Kendelewisc and Prof. Gordon Brown for assistance with the synchrotron experiments and Prof. H. Freund for loan of the NiAl single crystal.

References

1. F. Ernst, *Mater. Sci. Eng., R*, 1995, **14**, 97.
2. U. Diebold, J.-M. Pan and T. E. Madey, *Surf. Sci.*, 1995, **331–333**, 845.
3. T. E. Madey, U. Diebold and J.-M. Pan, *Springer Ser. Surf. Sci.*, 1993, **33**, 147.
4. M. Bäumer, J. Libuda and H.-J. Freund, *NATO ASI Ser., Ser. E*, 1997, **331**, 61.
5. Z. Zhang and V. E. Henrich, *Surf. Sci.*, 1992, **277**, 263.
6. D. Robba, D. M. Ori, P. Sangalli, G. Chiarello, L. E. Depero and F. Parmigiani, *Surf. Sci.*, 1997, **380**, 311.
7. N. Price and R. J. Madix, *J. Electron Spectrosc. Relat. Phenom.*, 1999, **98/99**, 257.
8. C. T. Campbell, *Surf. Sci. Rep.*, 1997, **27**, 1.
9. U. Diebold and N. D. Shinn, *Surf. Sci.*, 1995, **343**, 53.
10. M. W. Finnis, *J. Phys.: Condens. Matter*, 1996, **8**, 5811.
11. G. Pacchioni and N. Rösch, *Surf. Sci.*, 1994, **306**, 169.
12. C. Verdozzi, D. R. Jennison, P. A. Schultz and M. P. Sears, *Phys. Rev. Lett.*, in the press.
13. J. Libuda, F. Winkelmann, M. Bäumer, H.-J. Freund, T. Bertrams, H. Neddermeyer and K. Müller, *Surf. Sci.*, 1994, **318**, 61.
14. R. M. Jaeger, H. Kuhlenbeck, H.-J. Freund, M. Wuttig, W. Hoffmann, R. Franchy and H. Ibach, *Surf. Sci.*, 1991, **259**, 235.
15. J. Biener, M. Bäumer, P. Liu, E. Nelson, T. Kendelewisz, G. Brown and R. J. Madix, *Surf. Sci.*, in the press.
16. G. van der Laan and I. W. Kirkman, *J. Phys. Condens. Matter*, 1992, **4**, 4189.
17. F. M. F. de Groot, J. Faber, J. J. M. Michiels, M. T. Czyżyk, M. Abbate and J. C. Fuggle, *Phys. Rev. B: Condens. Matter*, 1993, **48**, 2074.
18. H. Poelman, K. Devriendt, L. Fiermans, O. Dewaele, G. Heynderickx and G. F. Froment, *Surf. Sci.*, 1997, **377**, 819.
19. M. Sambi, E. Pin, G. Sangiovanni, L. Zaratin, G. Granozzi and F. Parmigiani, *Surf. Sci.*, 1996, **349**, L169.
20. M. Sambi, G. Sangiovanni, G. Granozzi and F. Parmigiani, *Phys. Rev. B: Condens. Matter*, 1996, **54**, 13464.
21. M. Sambi, G. Sangiovanni, G. Granozzi and F. Parmigiani, *Phys. Rev. B: Condens. Matter*, 1997, **55**, 7850.
22. V. E. Henrich and P. A. Cox, in *The Surface Science of Metal Oxides*, Cambridge University Press, Cambridge, 1994.
23. C. D. Wagner, W. M. Riggs, L. E. Davis, J. F. Moulder and G. E. Muilenberg, in *Handbook of X-Ray Photoelectron Spectroscopy*, Perkin-Elmer, Eden Prairie, MN, 1979.
24. S. Hüfner, in *Photoelectron Spectroscopy*, Springer Verlag, Berlin, Heidelberg, New York, 1996, p. 5.
25. J. T. Mayer, U. Diebold, T. E. Madey and E. Garfunkel, *J. Electron Spectrosc. Relat. Phenom.*, 1995, **73**, 1.
26. J. G. Chen, *Surf. Sci. Rep.*, 1997, **30**, 1.
27. M. P. Seah, *Surf. Sci.*, 1972, **32**, 703.
28. C. Argile and G. E. Rhead, *Surf. Sci. Rep.*, 1989, **10**, 277.

29 CRC, *Handbook of Chemistry and Physics*, ed. R. C. Weast, CRC Press, Boca Raton, FL, 76th edn., 1995–1996.
30 T. M. French and G. A. Somorjai, *J. Phys. Chem.*, 1970, **74**, 2489.
31 E. Gillet and B. Ealet, *Surf. Sci.*, 1992, **273**, 427.
32 M. Gautier, G. Renaud, L. P. Van, B. Villette, M. Pollak, N. Thromat, F. Jollet and J.-P. Duraud, *J. Am. Ceram. Soc.*, 1994, **77**, 323.
33 J. Ahn and J. W. Rabalais, *Surf. Sci.*, 1997, **388**, 121.
34 W. Göpel, *Surf. Sci.*, 1984, **139**, 333.
35 L. S. Dake and R. J. Lad, *Surf. Sci.*, 1993, **289**, 297.
36 G. A. Sawatzky and D. Post, *Phys. Rev. B: Condens. Matter*, 1979, **20**, 1546.
37 C. Noguera, in *Physics and Chemistry of Oxide Surfaces*, Cambridge University Press, Cambridge, 1996.
38 G. K. Wertheim, S. B. DiCenzo and S. E. Youngquist, *Phys. Rev. Lett.*, 1983, **51**, 2310.
39 G. K. Wertheim, S. B. DiCenzo and D. N. E. Buchanan, *Phys. Rev. B: Condens. Matter*, 1986, **33**, 5384.
40 S. A. Flodström, C. W. B. Martinsson, R. Z. Bachrach, S. B. M. Hagström and R. S. Bauer, *Phys. Rev. Lett.*, 1978, **40**, 907.
41 C. M. Kim, B. D. DeVries, B. Frühberger and J. G. Chen, *Surf. Sci.*, 1995, **327**, 81.
42 J. G. Chen, C. M. Kim, B. Frühberger, B. D. DeVries and M. S. Touvelle, *Surf. Sci.*, 1994, **321**, 145.
43 M. Abbate, H. Pen, M. T. Czyżyk, F. M. F. de Groot, J. C. Fuggle, Y. J. Ma, C. T. Chen, F. Sette, A. Fujimori, Y. Ueda and K. Kosuge, *J. Electron Spectrosc. Relat. Phenom.*, 1993, **62**, 185.
44 F. M. F. de Groot, J. C. Fuggle, B. T. Thole and G. A. Sawatzky, *Phys. Rev. B: Condens. Matter*, 1990, **42**, 5459.
45 F. M. F. de Groot, *J. Electron Spectrosc. Relat. Phenom.*, 1994, **67**, 529.
46 F. M. F. de Groot, Z. W. Hu, M. F. Lopez, G. Kaindl, F. Guillot and M. Tronc, *J. Chem. Phys.*, 1994, **101**, 6570.
47 F. M. F. de Groot, J. C. Fuggle, B. T. Thole and G. A. Sawatzky, *Phys. Rev. B: Condens. Matter*, 1990, 41, 928.
48 F. M. F. de Groot, M. Grioni, J. C. Fuggle, J. Ghijsen, G. A. Sawatzky and H. Petersen, *Phys. Rev. B: Condens. Matter*, 1989, **40**, 5715.
49 J. Stöhr, in *NEXAFS Spectroscopy*, Springer-Verlag, Berlin, Heidelberg, 1996.
50 G. Rocker, J. A. Schaefer and W. Göpel, *Phys. Rev. B: Condens. Matter*, 1984, **30**, 3704.
51 P. A. Cox, R. G. Egdell, S. Eriksen and W. R. Flavell, *J. Electron Spectrosc. Relat. Phenom.*, 1986, **39**, 117.
52 M. Bäumer, J. Biener and R. J. Madix, *Surf. Sci*, 1999, **432**, 189.
53 M. Sander and T. Engel, *Surf. Sci.*, 1994, **302**, L263.
54 H. Onishi and Y. Iwasawa, *Surf. Sci.*, 1994, **313**, L783.
55 H. Onishi, K. Fukui and Y. Iwasawa, *Bull. Chem. Soc. Jpn.*, 1995, **68**, 2447.
56 K.-O. Ng and D. Vanderbilt, *Phys. Rev. B: Condens. Matter*, 1997, **56**, 10544.

Paper 9/02737H

General Discussion

Prof. Thornton opened the discussion of the Introductory Lecture: A simple explanation for the effect of OH on the growth of Rh on the thin alumina film might be that OH saturates the defects. What models do you have for the effect?

Prof. Freund responded: STM images of Rh deposits on pre-hydroxylated alumina films clearly show that Rh nucleates on the entire surface uniformly. This is not compatible with a preferred site nucleation where the site is formed by hydroxylation of the point and line defects of the clean alumina.

Prof. Madix asked: Prof. Freund, you nicely described that adsorption of molecules onto oxide surfaces may be strongly affected by electrostatic interactions. Is it possible, or is there evidence that there may be stable adsorption states for which there is an activation barrier, largely electrostatic, that could be accessed at higher pressures or higher gas kinetic energy?

Prof. Freund answered: It is quite conceivable that electrostatic interactions between molecules and oxide surfaces establish an activation energy for adsorption which could be overcome by increasing the kinetic energy of the incoming molecule.

Dr Taylor commented: On the subject of the desire for experimental measurements of the dynamics of adsorption at oxide surfaces, for example methane dissociation. It may be worth noting that preliminary results of such experiments are reported by my group in a poster at this meeting. An exposure of approximately 500 L of CD_4, using a seeded supersonic molecular beam ($T \approx 800$ K, $E_{trans} \approx 170$ kJ mol^{-1}), produced no evidence of dissociation at an MgO(100) surface at 300 K. Work continues but these early results suggest that methane dissociation at MgO surfaces does not occur simply *via* an early transition state.

Dr Egdell said: A third strategy for the preparation of ordered oxide surfaces is based on oxide-on-oxide epitaxy, typically involving metal deposition on an ordered oxide substrate in the presence of an oxygen atom source. This technique has been used with particular success by Chambers and co-workers to prepare, for example, ordered (110) surfaces of Nb-doped TiO_2 [1] and RuO_2.[2] Of course impetus to this field comes from the need to prepare oxide thin films with novel superconducting, magnetic or ferroelectric properties.

1　Y. Gao, Y. Liang and S. A. Chambers, *Surf. Sci.*, 1986, **348**, 17.
2　Y. J. Kim, Y. Gao and S. A. Chambers, *Appl. Surf. Sci.*, 1997, **120**, 250.

Prof. Freund added: Dr Egdell is completely right. This is an important field which is not considered in the Introductory Lecture.

Prof. Thomas addressed Prof. Freund: Your reference to hydroxylated oxide surfaces prompts me to recall the classic work of Bradley and Serratosa on single crystal kaolinite. Some 35 years ago they did elegant work using polarised infrared absorption spectroscopy and were able to deduce the precise geometry of the dangling OH bonds at the aluminium rich (and the total absence of OH bonds at the silica rich) surface. My point is simply to draw attention to the convenience and ease of using (plentiful, cheap and readily available) naturally occurring minerals containing OH rich surfaces. It is also relatively easy to introduce substitutional point defects (such as Cr^{3+}, Fe^{2+} or Fe^{3+}, Zn^{2+} or Mg^{2+} ions in place of Al^{3+}) into synthetic variants of these minerals which usually have conveniently large surface areas.

Prof. Freund responded: This comment is particularly useful and the possibility to use hydroxides has been recognized in connection with the preparations of OH-terminated surfaces. Papp and co-workers (refs. 58, 59 in the Introductory Lecture) have used Ni(OH)$_2$ as a template to prepare NiO crystallites with preferential (111) orientation.

Prof. Bowker said: You have nicely described the weak adsorption of CO on oxide surfaces. Perhaps for efficient adsorption, ionic-type reactions lead to stronger reactivity. For example acid–base reactions, which result in OH and ultimately water formation may lead to stronger absorption, and this seems to be the case for formic acid adsorption on TiO$_2$(110) where the reaction may result in ionic species formation.

Prof. Freund responded: Interaction of molecules with perfect oxide surfaces is not necessarily weak, as was demonstrated in the case of CO$_2$ and Cr$_2$O$_3$(0001). The example you mention is another example and is mentioned in refs. 28–31 in the Introductory Lecture. In general, activity is increased by introducing defects.

Prof. Iwasawa asked: As for the optimum feature of CO adsorption quantity against the different Rh particle size you showed, did you assume no restructuring of the Rh particles during CO adsorption?

Prof. Freund replied: We had reported earlier a restructuring of Rh particles under exposure to CO as observed by SPA-LEED (refs. 15, 99 of the Introductory Lecture). The effects were found to be smaller using STM. However, we do observe the formation of Rh(CO)$_2$ by IRAS which is assumed to be involved in restructuring processes (refs. 157, 158 of the Introductory Lecture).

Dr Kantorovich opened the discussion of Prof. Finnis's paper: (1) I understand that you consider the solid/gas interface using DFT. Do you actually include species in the gas phase into the cell in your calculation. Can you explain the trick? (2) Does it actually mean that the gas phase does not interact with the solid one so that there is no effect of the gas on the structure of the surface?

Prof. Finnis responded: (1) No species in the gas phase are included explicitly in the calculation. The trick is that the cell in which calculations are done is relatively small, containing only a few atoms of solid. The probability of finding a gas molecule inside this cell is negligible, so in the expression for the interfacial energy, eqn. (1) of our paper, only molecules in the solid phase need be counted. The gas phase is included purely conceptually in the sense that I related the pressure of the gas to the chemical potential of oxygen, which does enter eqn. (1).
(2) In reality the gas pressure determines the excess (or deficit) of oxygen at the surface such that the gas is in equilibrium with the surface of lowest excess free energy. In calculations by the present formalism, the effect of the gas on the surface is indirect, insofar as the value of its pressure selects a lowest free energy structure of the surface from the set of relaxed structures which we choose to simulate, not from all possible structures. Of course there is no guarantee that we have not missed a lower energy structure! In any case I don't think it is necessary to treat the gas explicitly.

Prof. Catlow said: You estimate the free energy of the oxygen molecule *via* a thermochemical cycle using calculated energies of two solid phases. Could you confirm that this is a more accurate procedure than using a direct calculation on the O$_2$ molecule.

Prof. Finnis replied: Definitely. The local density approximation (LDA), which we use, is not good for the energy of formation of an O$_2$ molecule. Hence we avoid calculating any properties of O or O$_2$. On the other hand the energies of aluminium metal and Al$_2$O$_3$ can be very well calculated with the LDA.

Dr Willock asked: The reaction of an Al terminated (0001) surface with H$_2$O to give a fully hydroxylated surface was found in our work to lead to a much more stable surface than the Al terminated surface (GGA DFT calculation). Could you compare the exchanges of oxygen mediated by H_2O rather than O_2 with your method?

Prof. Finnis answered: In principle, yes. One would have to include the chemical potential of H_2O in eqn. (1) instead of or in addition to that of oxygen. My suggestion would be then to proceed with a calculation for solid ice in order to set the zero of chemical potentials, by analogy with the calculations I have described for solid Al and alumina, and then use thermodynamic tables to integrate up to the chemical potential of water at the required conditions. I believe a satisfactory calculation for ice can be made within the generalised gradient approximation (GGA).

Prof. Vanderbilt asked: Can you clarify what sets the lower limit of oxygen partial pressure in your calculations? Was it the nucleation of particles of aluminium metal, as I would have thought?

Prof. Finnis answered: That's right.

Prof. Catlow said: Your calculations (and others using both electronic structure and static lattice simulation methods) correctly predict (at least qualitatively) the large surface relaxations that have been observed for Al_2O_3. There are, however, quantitative discrepancies between calculated and experimental relaxations. Could you comment on the possible origins of these discrepancies and also on the differences between the results obtained using the electronic structure methods and the earlier shell model/static lattice techniques.

Prof. Finnis replied: Some discrepancies on the theoretical side between different calculations can be due to different thicknesses of slabs, and earlier electronic structure calculations used inadequate bases, or did not relax the oxygen atoms laterally. These problems are under control in the more recent calculations, and we do not understand all the discrepancies that remain. However, it turns out that these relaxations are really rather soft, so that differences between different pseudopotentials or LDA errors could make a difference, e.g. between the variations between 3 and 11% for the second layer relaxation. We have verified that differences of this magnitude can arise. This is seen in the paper by Harrison et al. on TiO_2 at this meeting. We cannot explain the uniformly too large predicted relaxation of the surface compared to experiment, unless the experimental surface is contaminated with OH for example.

Prof. Harrison suggested: Might thermal vibrations be important in the comparison between calculated and experimentally determined geometries?

Prof. Finnis replied: I think so, and your paper has also indicated this. However, for the discrepancy I have just referred to it seems doubtful that thermal vibrations are the answer, because according to classical potential calculations within the quasiharmonic approximation[1] thermal effects do not improve the agreement. We are currently calculating some ab initio vibration frequencies of an alumina slab to investigate this point.

1 N. Allan, private communication.

Dr Gautier-Soyer said: (1) First, I have a question on the driving forces of the relaxations of the Al-terminated α-Al_2O_3 (0001) surface. They have been calculated by classical ionic pair potential methods, as well as ab initio electronic structure calculations. They have also been measured experimentally. However, even though theory and experiment qualitatively agree, there is no consensus about the driving forces of the relaxations. Your interpretation is in favour of an electrostatic effect, while other papers, such as ref. 9, explain them in terms of rehybridization. Using a semiempirical self-consistent Hartree–Fock method carried out on a two unit cell thick slab (collaboration with C. Noguera), we have found that delocalization effects were not negligible, so that their contribution should be an important part in the driving force of the relaxations. Would you like to comment on this point?

(2) Secondly, I would like to make a comment on the covalence/ionicity of the Al–O bond at the surface, compared to the bulk. Again, in the electronic structure papers where atomic charges

are calculated, there is some disagreement: some authors conclude a more covalent bond, others a more ionic one (as in this paper). However, if one thinks in terms of charge transfer per Al–O bond, it seems that the studies are in agreement. According to the recent work of Pojani et al., a simple expression can be used to derive the anion–cation charge transfer Δ from the formal ionic charge and the number of first neighbours: at the surface, $Q_{Als} = 3 - 3\Delta_s$, where Δ_s is the charge transfer between surface aluminium and subsurface oxygen. In the bulk, $Q_{Als} = 3 - 6\Delta$ where Δ is the bulk charge transfer. In this way, $\Delta = 0$ in the purely ionic state and the increase of Δ means a more covalent bond. Taking the charges obtained by different *ab initio* electronic structure calculations (ref. 6 of your paper, ref. 2 given here, as well as your paper), the charge transfer per Al–O bond at the surface is larger than in the bulk. Using the charges of Table 2 (of your paper), one gets for example $\Delta_s = 0.426$ electron at the surface, and $\Delta = 0.245$ electron in the bulk. This means that the Al–O bond is more covalent at the surface than in the bulk. How would you reconcile this point of view with your conclusion that there is no sign of increased covalence at the alumina surface?

1 A. Pojani, F. Finocchi, J. Goniakowski and C. Noguera, *Surf. Sci.*, 1997, **387**, 354.
2 W. C. Mackrodt, *Philos. Trans. R. Soc. London, Ser. A*, 1992, **341**, 301.

Prof. Finnis responded: (1) In the spirit of the Hellmann–Feynman theorem we can describe and explain exactly the relaxation as a purely electrostatic effect, given the exact self-consistent charge density. On the other hand in a tight-binding model, some of the electrostatic energy of which the force is the derivative is hidden in the band energy. The Hellmann–Feynman forces in tight-binding do not appear to be electrostatic, they involve instead the gradients of some hopping integrals, although they are actually modelling forces which are simply electrostatic. It is therefore difficult to tell how much of the forces in tight-binding should be interpreted as the classical electrostatic forces from classical ions. From the *ab initio* point of view, suppose we divide up the force into the contribution from a classical charge distribution (the spherical superimposed ions) and the contribution due to the redistribution of the electronic charge. The fact that the first contribution explains the relaxation suggests to me that the second contribution, which includes rehybridisation effects, is small.

(2) First, let me repeat the caveat, which we all seem to agree about, that absolute numbers for charges and charge transfer in this kind of analysis are not to be interpreted literally, because they are basis dependent. Your question, however, concerns relative values, and these should be meaningful. My answer is that I think the simple formula you have used to measure the ionicity has a particularly arbitrary character at the surface, because the coordination of the surface atoms is different from the bulk. For example, the coordination of the surface aluminiums is 3, whereas their bulk coordination is 6. You have used these numbers in your estimate of charge transfer along the bonds. Consider a hypothetical outcome of a calculation in which all the Al atoms, both bulk and surface, carried the same Mulliken charge, and all the O atoms likewise (*e.g.* $+1.5$ and -1, respectively). We would interpret that by saying that the ionicity at the surface is the same as that in the bulk, just because the same charge X (*e.g.* 1.5) has been transferred from each Al atom. The surface Al atoms might be thought of as having transferred this charge X equally through 3 bonds, $X/3$ through each, whereas the second layer Al atoms might be thought of as having transferred $X/6$ through each of 6 bonds. By this measure, the second layer bonds are twice as covalent. But we could just as well say that the second layer Al transfer their X through only the three bonds to their nearest neighbours, the three O atoms in the layer above them, and none to the three more distant O atoms below them, in which case all the bonds would have the same ionicity. The resulting charges on the O atoms would be the same. It is my impression that either construct is artificial and not particularly helpful for the purpose of defining ionicity, since they both rely on the unphysical concept of charge transferring along a particular bond.

Prof. Harrison addressed Dr Gautier-Soyer: From your analysis of the covalent *vs.* ionic nature of the bond can we understand the reason for the discrepancy between theory and experiment in the surface geometry?

Dr Gautier-Soyer replied: Unfortunately we don't.

Dr Weiss addressed Prof. Finnis: It is very interesting that your calculations predict an oxygen-terminated α-Al$_2$O$_3$(0001) surface to become stable in high oxygen pressure environments. A similar result has been obtained from recent DFT calculations performed for the α-Fe$_2$O$_3$(0001) surface,[1] an oxide with the same corundum crystal structure as Al$_2$O$_3$(0001). These calculations predict the formation of strongly relaxed surfaces exposing outermost close-packed oxygen (111) layers, which in a purely ionic picture are expected to be unstable because of infinite electrostatic surface energies. However, your calculations reveal a strongly changed electron density of states at the surface and a considerably reduced Mullikan ionic charge for the topmost oxygen layer atoms of Al$_2$O$_3$(0001). The bonding at the surface becomes more covalent and less ionic. This represents a mechanism that can stabilise the polar oxygen-terminated Al$_2$O$_3$(0001) surface structure, which also stabilises the polar Fe$_2$O$_3$(0001) surface.

We have recently performed a systematic study of α-Fe$_2$O$_3$(0001) surface structures that are formed in defined oxygen partial pressures ranging over 5 orders of magnitude.[2] Fig. 1 here shows a large scale (a) and an atomic resolution STM image (b) of a surface prepared in 1 mbar oxygen. More than 95% of the entire surface forms an oxygen terminated structure, only small dark patches exposing an Fe-terminated surface can be seen in the large scale image on the left. In oxygen pressures between 10^{-1} and 10^{-5} mbar comparable amounts of iron- and oxygen-terminated surface domains coexist, in 10^{-5} mbar only the iron-terminated surface is formed. Recent dynamical LEED intensity calculations confirm that the two terminations observed in these STM images correspond to oxygen- and iron-terminated surfaces, in analogy to your results for Al$_2$O$_3$(0001). These findings clearly demonstrate that metal oxide surface structures formed in gaseous environments with high pressures can be very different from those formed under UHV conditions. It has important consequences on the surface chemical and catalytic properties of such materials and directly addresses the problem of the 'pressure-material gap' in catalysis research.

1 X.-G. Wang, W. Weiss, Sh. H. Shaikhutdinov, M. Ritter, M. Petersen, F. Wagner, R. Schlögl and M. Scheffler, *Phys. Rev. Lett.*, 1998, **81**, 1038.

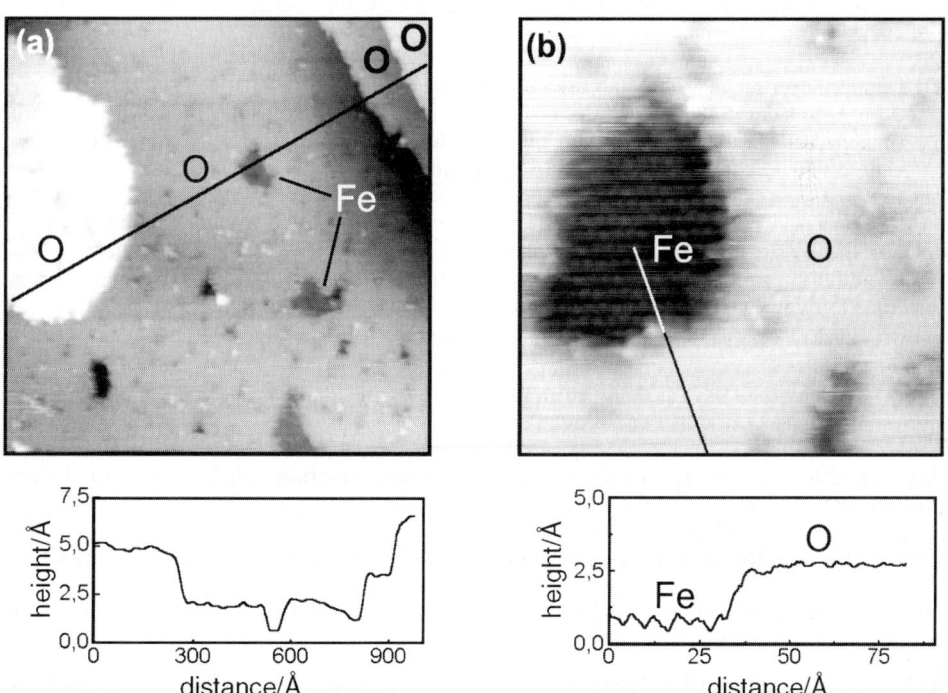

Fig. 1 90 × 90 nm (a) and 15 × 15 nm (b) STM images of an α-Fe$_2$O$_3$(0001) hematite surface, prepared by heating an epitaxially grown hematite film to 1120 K in 1 mbar oxygen for 15 min. Several monoatomic steps can be seen in (a). An oxygen terminated surface structure dominates, small patches of iron-terminated domains cover less than 5% of the entire surface (dark patches).

2 Sh. Shaikhutdinov and W. Weiss, *Surf. Sci.*, 1999, **432**, L627.

Prof. Finnis responded: Thank you for those observations. It would be very interesting to see if you can observe the oxygen terminated surface at around atmospheric oxygen pressure. However, it might only be stable at somewhat higher pressure.

Prof. Vanderbilt commented: I would like to urge caution about the notion that charge transfer can lead to self-compensation of a polar surface. This is misleading. It is necessary to distinguish two cases. One possibility is that the surface can self-compensate by becoming metallic. On the other hand, if the surface remains insulating, *i.e.* the electron chemical potential lies in a gap common to both the bulk and surface, it can be shown that the net excess charge cannot be changed by charge transfer among surface atoms. This follows from a theorem relating this net surface charge to the surface-normal component of the bulk polarisation, which does not change. Thus, if the surface remains insulating, no such self-compensation can occur.

Prof. Finnis said: I completely agree with that. Our calculated polar surfaces are indeed metallic and I don't believe they could conceivably be stable otherwise, for the reasons you describe.

Prof. Jennison addressed Dr Gautier-Soyer: Concerning the degree of covalency at the Al_2O_3 surface: using DFT slab calculations, we noted only small changes in the local DOS compared with the bulk ions, indicating that the high degree of ionicity found in the bulk is preserved at the surface if the surface is fully relaxed.

The very large surface relaxations, which penetrate to the third oxygen layer, make thin slab calculations questionable quantitatively. This places Al_2O_3 in considerable contrast to MgO (with high ionicity but almost negligible relaxations) and TiO_2 (with significant covalency). So I would ask you, how thick was the slab in your calculations?

In fact, we computed the effect of relaxation on the adsorption energy of an isolated Pt atom and found by LDA that it was 2.0 eV for Al_2O_3, 0.4 eV for $TiO_2(110)$, and only 0.2 eV for MgO(100).

Dr Gautier-Soyer replied: The slab used in our calculation was two unit cell thick (about 26 Å), relaxed on both sides. So the slab was made of 12 (–Al–O–Al–) stacking units, which is rather thick. We checked that in the middle of the slab, the Al and O charges were equal to the bulk ones, as derived from the 3D band structure calculation.

Prof. Freund said: On the $Cr_2O_3(0001)$ surface there are no hydroxy groups present as checked with IRAS. Therefore, the presence of OH cannot be used to argue for any disagreement between experiment and theory. We did observe a new phonon mode on $Cr_2O_3(0001)$ and we have strong indications through CO adsorption experiments that this is a surface phonon mode.[1]

1 K. Wolter, PhD Thesis, Fritz-Haber-Institut, Berlin, in preparation.

Prof. Harrison asked Prof. Freund: Can you deduce the symmetry of the vibrational mode at the chromia surface?

Prof. Freund answered: The mode is active in specular reflection indicative of a totally symmetric mode.

Prof. Harrison added: In which case it seems likely to be a vertical motion of the surface Cr ion.

Prof. Freund replied: Isotopic labelling of the oxide indicates that it is a mode involving mainly the motion of Cr ions.

Prof. Thornton addressed Dr Renaud: Perhaps you could comment on the level of agreement between theory and experiment taking into account the experimental error bars.

Dr Renaud responded: This is a question that is also related to previously published material.[1,2] The best way to answer this question is actually to simulate the crystal truncation rod data of the

Al$_2$O$_3$(0001) surface deduced from the different theoretical calculations of the relaxations, and compare them with all the experimental data, with error bars included on the data. I intend to do that in the near future.

1 P. Guénard, G. Renaud, A. Barbier and M. Gautier-Soyer, *Mater. Res. Soc. Symp. Proc.*, 1996, **437**, 15.
2 P. Guénard, G. Renaud, A. Barbier and M. Gautier-Soyer, *Surf. Rev. Lett.*, 1997, **5**, 321.

Prof. Harrison asked: How are the vibrational modes of the Al$_2$O$_3$ surface modelled in the interpretation of your surface X-ray diffraction experiment?

Dr Renaud answered: At the time of the data analysis, no enhanced or anharmonic vibration mode was expected for the top plane Al atom, and hence the thermal vibrations of the top atoms were modelled the standard way by two Debye–Waller factors, one parallel and one perpendicular to the surface. Since the calculations on the TiO$_2$(110) surface suggest that anharmonic vibrations could be present, I will check this possibility for the sapphire surface in the near future.

Prof. Friend opened the discussion of Prof. Jennison's paper: Kinetic factors are important in metal thin film growth, as clearly seen in the work presented by Prof. Freund, in Behm's work that you cited, and also in our work on Co growth on oxidized Mo, metastable structures can form. In your work, you obtain a thermodynamic result. How do you relate your theory to experiment since your results are thermodynamic?

Prof. Jennison responded: I completely agree with your comment on the importance of kinetic considerations, and we are moving towards that with some studies. However, for these mixed metallic and ionic systems, there are no adequate potentials to perform simulations, so we have started with energetic studies to indicate likely structures, the basic nature of the bonding, *etc.* Clearly, with our third topic, which involves water dissociation, hydrogenation, and possible hydrogen evolution with metal deposition, I agree that kinetics is essential to make contact with experiment. I would like to add, that we know we are often interested in metastable structures; for example, in some of our work on the alumina films, we have observed, that with an adequate degree of freedom (a large unit cell) and aggressive geometry relaxation algorithms, 2D films grow into 3D structures, while we were trying to study the 2D metastable structures of experimental relevance.

Dr Venables asked: It is well known that DFT methods, though maybe the best we have available, form an uncontrolled approximation, in that we don't know how accurate they are in any particular case. Can you give us some idea of what accuracy you claim for the results shown in Fig. 2 of your paper. My queries include: what configurations have been taken into account, have long range electrostatic/elastic relaxation been included, particularly in the surface plane, and how has (local) charge compensation been achieved in the case of defect calculations?

I ask this because recent calculations we have undertaken for the notionally similar system Pd/MgO(001) have given a considerably smaller adsorption energy, and the OH-defect is thought to come in pairs linked to a Mg^{2+} vacancy to preserve local charge neutrality.[1] In the latter case we found the Pd to bind at the Mg site with a trapping energy (relative to Pd on the terrace) of 1.15 eV, *i.e.* very much lower than the values given in your Fig. 2. I am left to wonder similarly whether your calculation also overestimates the reduction in metal pair binding from gas phase values, which in our calculation is a relatively small effect.

1 J. A. Venables and J. H. Harding, *J. Crys. Growth*, 1999, in press.

Prof. Jennison answered: Well, I disagree with your first statement, concerning the accuracy for DFT in 'any' particular case, because extensive comparisons have been made for a wide variety of materials, mostly metals and semiconductors, and I think the error bars are usually well known. With the aluminum oxides, we and others have reported that LDA obtains the experimental lattice constant of sapphire, for example, to 0.2%, while GGA does somewhat poorer, being ≈ 0.6% over experiment—this shows electrostatics, ionic radii, and hard-wall repulsion are properly treated. We also know that DFT does an excellent job with metal polarization (see, for example, Jennison *et al.*[1]). I agree with you, that there is more uncertainty for the case of metals interacting with oxide surfaces, as here little is known concerning the energetics from the experimental side. In addition, the rather large differences in LDA *vs.* GGA metal adhesion energies (see our ref. 21) are difficult to understand. We have found that metals, particularly on the right of the periodic table, bind without *any* evidence of covalency (where we would expect LDA to overbind). Rather, metals bind by becoming positively charged by donating electron density to the oxide itself, as may be seen by examining the local DOS changes in the neighboring oxygen ions. The factors here are polarization, ionization, and electrostatics, all well given by DFT. Concerning relaxation around defects on oxides, here again we have electrostatics and ion repulsions, both accurately given because the lattice constants are well given.

Concerning now the results for Pt on MgO, these are *very* accurate DFT calculations. We used slabs of five MgO layers, and the supercell contained 36 atoms per layer. The vacuum gap, which we find needs to be much larger than with calculations involving only metals, was >15 Å. Because this was not a cluster, geometric relaxation and long-range electrostatics are accurately and naturally included (see our ref. 18).

Concerning the importance of relaxations for adsorbates on MgO, I reported that Pt adatom binding on the perfect (100) surface is only lowered by 0.2 eV by relaxation, while the number is more like 0.4 eV for $TiO_2(110)$ and 2 eV (!) for $Al_2O_3(0001)$. However, relaxations are critically important for obtaining the correct binding at surface vacancies. This is because the relaxations determine the height of the ionized metal adatom, and the Madelung potential falls exponentially into the vacuum: Thus the binding energy depends exponentially on the adsorbate height which in turn depends linearly on the outward relaxation at the vacancy.

Concerning your specific queries, the question of configurations (as in 'configuration mixing') does not enter this problem as an obvious DFT failing, as the ionized Pt atoms are similar to s^0d^8, s^0d^9, s^0d^{10}, or s^1d^{10} and are fractionally charged as they should be. Next, as I said above, the long range electrostatic/elastic relaxations have been accurately included. Finally, the (local) charge compensation is not an issue as these supercells are all neutral (*i.e.*, we did not consider the charged defect).

1   D. R. Jennison, P. A. Schultz and M. P. Sears, *Phys. Rev. Lett.*, 1996, **77**, 4828.

Dr Shluger asked: What was the charge of anion vacancies used in your calculations of the Pt adsorption and how does it affect the results? Why does Pt prefer to adsorb in an anion vacancy?

Prof. Jennison responded: The anion vacancy had two trapped electrons, the natural state, as all our calculations used neutral supercells. We cannot compare energies between neutral and charged supercells because the VASP code (our ref. 30) can only compensate a charge cell by the common practice of adding a uniform background, which therefore puts the calculations on an uneven footing. Pt adsorbs at either vacancy type by becoming ionized due to the Madelung potential and entering the vacancy to the maximum extent allowed by its ionic radius and the relaxation around the vacancy. However, it does not prefer the anion vacancy to the cation vacancy (see our Fig. 2), binding several eV more strongly at the latter.

Dr Egdell said: Can your clarify why 'adsorbed hydroxy' and 'in surface hydroxy' differ in their propensity to promote metal nucleation. It would also be interesting to those concerned with photoemission from surface hydroxy species to know details of the location of the localised energy levels associated with the two different surface hydroxy species.

Prof. Jennison answered: Our calculations have shown (see also our ref. 18) that the in-surface hydroxy ion weakens the binding of metals because the binding is mostly electrostatic in nature (see our refs. 20 and 21) and the surface ionic charge is reduced from $2-$ (oxygen ion) to $1-$ (hydroxy ion). However, ad-hydroxy ions strengthen the binding because they sit above the cation sites, allowing the adsorbed metal to sit immediately adjacent and bind both vertically, in the normal way to surface oxygen ions, but also now laterally to the OH^-; *i.e.*, at the binding site, the negative electrostatic potential is deepened. We have not yet published the local DOS of the two species, but appreciate your suggestion for future work.

Prof. Madix asked: What is the origin of the ionization of the metals deposited on the alumina?

Prof. Jennison answered: Electrostatics.

Prof. Madix said: That is very interesting. We have seen formations of V^{3+} on single crystal $Al_2O_3(0001)$ with vanadium coverages well below a monolayer.

Prof. Jennison responded: The electrostatic (Madelung) potential extends further above the surface with alumina than with other oxides for two reasons: (1) the potential falls off exponentially above a surface with strict layer-by-layer neutrality, such as MgO(100), but only as $1/Z$ above a non-strictly-neutral surface, and (2) the dramatically large relaxations of the alumina surface increase the non-neutrality near the adatom site (nearby Al-ion movements are about 1/2 Å downwards, see our ref. 21). These factors serve to substantially increase the ionizing potential at the metal site and it is this potential that can even multiply ionize isolated metal atoms, as we reported for Y and Nb adatoms recently (our ref. 21).

Prof. Thomas said: My question arises from the remark we have just heard from Prof. Madix, namely that when metallic vanadium is deposited on single crystal alumina, individual V^{3+} ions are formed. One can see, from what Prof. Jennison has told us, that the Madelung energy at the surface of the Al_2O_3 accounts for the production of Cu^+ or Pt^+ (in the presence of juxtaposed OH). But why does vanadium opt for V^{3+} (not V^{2+} or lower or V^{4+} or higher)? This presents a nice opportunity for theoreticians.

Prof. Jennison responded: We determined the ionicity of the adatoms by integrating the local DOS of the metal atoms when projected on atomic orbitals. Naturally, we did not obtain integral values nor should we expect to, since DFT is inherently a mean-field theory, and the metal charge is determined by a balance between the ionizing potential, ionic radius, and the ability of neighboring atoms in the oxide to accept electron density. To obtain the approximate charge, we rounded off the integrated LDOS to the nearest integer, which we reported in our ref. 21. This rounding amounted to typically one or two tenths of an electron.

Prof. Campbell said: In contrast to the implications of your calculations that surface OH greatly stabilises Cu on the oxide surface (alumina), our recent microcalorimetric measurements for the heat of adsorption of Cu on MgO(100) showed that surface OH has little effect on the coverage-dependent adsorption energy of Cu.[1] (J. Musgrove, D. Starr, D. Bald, J. Ranney and C. T. Campbell, unpublished results).

1   J. Musgrove, D. Starr, D. Bald, J. Ranney and C. T. Campbell, unpublished results.

Prof. Jennison responded: If, as I expect, your hydroxylated surface contained even mixtures of ad-OH and in-surface OH (produced by water dissociation and the subsequent reaction of H^+ with O^{2-}), I welcome this result. We recently computed the interaction of Cu adatoms with a sapphire(0001) surface saturated with water dissociation products and found it to bind Cu almost identically as the clean surface, for while the ad-species strengthens the binding, the in-surface species weakens it. So I see potential support here for the theoretical results. If, as the calculations suggest is likely, the presence of metal causes hydrogen gas to evolve from the in-surface hydroxy groups, while this event would increase the binding of Cu by several eV per atom, it is nearly neutral energetically (*i.e.*, the heat of desorption of H_2 about equals the increased heat of adsorption of the Cu), so I do not see that this would be seen in microcalorimetry.

Prof. Madey asked: My question concerns the general issue: when is the surface of a thin oxide film a good model for the surface of a bulk crystal?

Experiments on surfaces of bulk Al_2O_3 are generally performed on α-alumina (sapphire samples). Your calculated structure for a two-layer alumina film is more like κ-alumina. What are the conditions (choice of substrate film thickness, T) for which you expect thin sapphire films to be stable? In particular, is it possible to stabilize sapphire in a two-layer film?

On another matter you indicate that water dissociates readily on $Al_2O_3(0001)$ at 300 K but experimental evidence indicates a very low sticking probability. As Prof. Freund indicated in his talk, in order to form hydroxy he "seeds" his alumina surface with a fractional ML of metallic Al, followed by a 20 000 L water dose.

Prof. Jennison answered: The calculations on the two-oxygen-layer ultrathin alumina films clearly suggest that the lateral Al-ion interactions dominate the site preference, thereby producing a 50:50 mix of tetrahedral and octahedral site Al-ion occupations, making the film κ-like. Here, because the problem of structure is essentially two dimensional, the various choices are limited, unless defects are introduced which violate stoichiometry or coordination. I would expect quantitative but not qualitative differences in metal adsorption between this film and its bulk analogue of sapphire(0001), and we have already seen this (to be published). I could only guess that by perhaps six or eight oxygen layers the bulk film energy would be sufficiently large compared with the interfacial energy that, if the surface energy permits,[1] sapphire might form with sufficient annealing.

You raise a very interesting issue: Yes, our results fail to find an energetic impediment for water dissociation on Al_2O_3(0001) at 300 K, nor did Hass et al.,[2] as we would expect since this was also a DFT study. There is also evidence for a lack of sapphire hydroxylation in UHV, as well as for the film studied in Prof. Freund's group, while hydroxylation does evidently occur with a sufficient partial pressure of water, e.g., ref. 3 below. At the present time, we do not know whether this is due just to sticking or also to the reaction dynamics of dissociation, but this is a subject of ongoing work in our group.

1 J. M. McHale, A. Auroux, A. J. Perrotta and A. Navrotsky, *Science*, 1997, **277**, 788.
2 K. C. Hass, W. F. Schneider, A. Curioni and W. Andreoni, *Science*, 1998, **282**, 265.
3 J. W. Elam, C. E. Nelson, M. A. Cameron, M. A. Tolbert and S. M. George, *J. Phys. Chem. B*, 1998, **102**, 7008.

Prof. Freund added: We could not detect water dissociation on a clean alumina film as well as on a well prepared clean single crystal alumina.

Prof. Diebold said: I find the theoretical results on the influence of point defects on nucleation and growth extremely interesting. My comment refers to the experimental data by Jeff Kelber's group.

It is surprising that the Al_2O_3 surface would be hydroxylated with a coverage of $\frac{1}{3}$ monolayer. It is not straightforward to extract the coverage and even the presence of OH groups from non-monochromatized O 1s XPS spectra. In addition, it is difficult to hydroxylate Al_2O_3 under UHV conditions.[1] I think you mentioned that breaks in AES uptake curves indicated a layer-by-layer growth mode. One should note that this procedure is far from reliable, and has led to erroneous results in the past, especially on oxide surfaces.

1 P. Liu, T. Kendelewicz, G. E. Brown, Jr., E. J. Nelson and S. A. Chambers, *Surf. Sci.*, 1998, **417**, 53.

Prof. Jennison replied: I agree that the breaks in the uptake curves do not prove layer-by-layer growth, but are only suggestive (your point is discussed, for example, in ref. 1 below). What does seem clear, however, is the Kelber Auger data (our ref. 29) showing Cu^+ ions at about 1/3 ML—this *cannot* happen unless the Cu adatoms are bound so strongly that they cannot migrate at room temperature to join metallic islands, and we have shown that a surface with a high coverage of ad-OH species is capable of doing just this. This structure is again consistent with the XPS but not proven by it. However, whatever the species seen in the XPS, it is at the surface (as shown by tilting the sample) and is at the correct energy to be ad-OH. Further work is needed here to definitively prove this, and also to answer the question of how this surface came about in the first place, for while this surface was examined in UHV, it apparently was not hydroxylated in UHV.

1 C. Argile and G. E. Rhead, *Surf. Sci. Rep.*, 1989, **10**, 277.

Prof. Iwasawa commented: Recently, we reported the stabilization of Pt^{4+} ions at a MgO surface, replacing Mg^{2+} ions at the top layer of the surface. The Pt^{4+}/MgO sample was very active for catalytic combustion and oxygen-isotope exchange reactions. When the Pt^{4+} ions were reduced to the metallic state, Pt atoms were not stabilized as monomers anymore. Instead, it was suggested that six-atom Pt clusters were produced and attached on the MgO surface. The Pt/MgO sample was selective for catalytic dehydrogenation of propane, butane and isobutane to the corresponding alkenes. The behaviour of Pt atoms is somewhat different from your theoretical consideration.

Prof. Jennison responded: This is very interesting, but I think your surface might be more complex than the model surface we used in our ref. 18. I say this because the Madelung potential at the cation vacancy site is perhaps 20–30 V, and is therefore insufficient to ionize Pt to 4+, which would take perhaps 40 V, since the second ionization potential of Pt is already at 19 eV. Therefore, I suggest that perhaps Pt is not just substitutional for a surface Mg ion. The remainder of your observation, concerning small stable metallic clusters, we have not addressed, but related calculations (our ref. 21) show that just two nearest-neighbour metal atoms are sufficient to reduce metal adatoms to the metallic state.

Prof. Vanderbilt opened the discussion of Prof. Hermann's paper: Previous calculations of Lindan and co-workers have shown the importance of considering spin polarizations for oxygen vacancies in TiO_2. Can you clarify whether you took spin polarization into account for the V_2O_5 oxygen vacancy calculations, and if so, what is the spin structure that you find? Is it a spin triplet? How different would the results have been if spin polarization were neglected?

Prof. Hermann responded: All vacancy calculations were carried out allowing for spin polarisation of the complete cluster system. Here it was found that the energetically favourable cluster states were singlet spin states. We did not determine, however, local projections of spin polarisation in the vicinity of the vacancies, which may well have yielded triplet type character. This is a valuable suggestion for future work.

Dr Noguera asked: For the simulation of oxygen vacancies, did you include basis functions located at the vacancy site, allowing the two electrons left behind by the missing oxygen to be possibly trapped there?

Prof. Hermann answered: We performed the oxygen vacancy calculations without and with additional basis functions at the vacancy site. In the latter case, we used both the original oxygen basis of the system without vacancy and a basis set augmented by diffuse functions. This did not result in localised colour-centre type states where electrons could be trapped. The bridging oxygen vacancies at V_2O_5 surfaces represent large openings where electron localisation is more difficult than in compact systems like MgO, where colour centres have been observed.

Dr Shluger asked: (1) How is the electron density localised in the neutral oxygen vacancy? Does it look more like an E'-centre in SiO_2 or like an F-centre? (2) Did you try asymmetric relaxations where electrons could localise on one V ion? (3) How could long-range lattice polarisation affect these results?

Prof. Hermann answered: (1) As mentioned before, we could not find localised colour-centre type vacancy states in the cluster calculations. Planar sections through the vacancy regions just show a depletion of charge density.
(2) Asymmetric relaxation was a natural result of the optimisation of atom positions near the vacancy and could not be "turned on" in the calculations. The electrons left behind after creating the vacancy did not yield sizeable charge accumulation near selected single V centres.
(3) This could be answered only by large slab calculations, which were not considered in the present study.

Prof. Freund said: Considering the local density of states near the modelled vacancies, does it start to resemble the density of states in VO_2? In particular, do you see the development of a feature near the Fermi energy (E_F) in the density of states?

Prof. Hermann replied: A detailed analysis of changes in the atom projected partial densities of states (PDOS) of the $V_{10}O_{31}H_{12}$ cluster due to the presence of the three different oxygen vacancies, O(1–3), has been performed after and stimulated by the discussion. It is found that vacancy formation results, for all oxygen vacancy types, in additional PDOS contributions near E_F (and located about 2 eV above the O 2sp band region) which are attributed to vanadium centres and are of V 3d type. While this may be interpreted as 'VO$_2$ resemblance' it should be noted that VO$_2$ and V$_2$O$_5$ are substantially different in their bulk crystal structure such that DOS similarities cannot be taken as indicators of structural similarities.

Dr Weiss said: Your calculations reveal a strong covalent bonding character in V_2O_5, with atom charges of +1.5 for V instead of +5 expected in a purely ionic picture. The valence band is formed by O 2sp and V 3d states. I do not understand why the atom projected partial DOS related to the terminal O(1) atoms has the smallest energy width if compared to the O(2) and O(3) atoms, although the V–O(1) distance is the smallest O–V distance occurring in the V_2O_5 lattice. I would expect the valence band width to increase with decreasing distance between neighbouring O and V atoms, because of increasing orbital overlap integrals.

Prof. Hermann replied: The valence bandwidth of the O 2sp dominated region is determined by the lateral O–O interaction rather than by V–O coupling. The bridging oxygen atoms, O(2) and O(3), form a sub-lattice with inter-atomic distances smaller than those of the terminal O(1) sub-lattice. Therefore, the O(2,3) derived valence band dispersion must be larger than that of the O(1) sub-band as discussed in ref. 1. (note that this is ref. 30 of the manuscript).

1 K. Hermann, M. Witko, R. Druzinic, A. Chakrabarti, B. Tepper, M. Elsner, A. Gorschlüter, H. Kuhlenbeck and H.-J. Freund, *J. Electron Spectrosc. Relat. Phenom.*, 1999, **98–99**, 245.

Prof. Kempter asked: In your comparison of the computed V_2O_5 DOS with the intensity curves measured from angular resolved photoemission (ARUPS) on V_2O_5(010) the central peak of the O 2sp band region differs by 1 eV. How can this shift be explained?

Prof. Hermann answered: In the comparison the computed DOS the O 2sp band region has been rigidly shifted in energy such as to coincide with the intensity region of the ARUPS experiment. This is based on the fact that the computed valence bandwidth reproduces the experimental result. However, a more detailed comparison is difficult since there are many factors contributing to differences between theory and experiment. First, a comparison of theoretical DOS with experimental photoemission results ignores the influence of transition matrix elements. Second, the present ARUPS data refer to fixed normal electron emission and peaks may shift as a result of emission angle.

Prof. Catlow said: As is commented on in your paper, the magnitude of the calculated oxygen vacancy formation energies are rather high. Do you have any explanation as to why the techniques used might be overestimating these energies? Also, has it been possible to make any quantitative comparison of your results with values derived from experimental thermochemical data, such as heats of reduction of the oxide?

Prof. Hermann responded: After careful comparison with related literature values we conclude that our calculations do *not* grossly overestimate oxygen vacancy formation energies. The proposed values, corresponding to a transition of oxygen bound at the surface to a free neutral oxygen atom ($O^{-x}_{surface} \rightarrow O^0_{free}$), range between 6.5 and 7.2 eV, depending on the vacancy site. Early experimental vacancy formation enthalpies of bulk V_2O_5 from extrapolations of kinetic data yield 1.3–1.5 eV[1] for a transition of oxygen bound at the surface to a fictitious species $\frac{1}{2} O_2$. This translates to 3.9–4.1 eV for the transition $O^{-x}_{bulk} \rightarrow O^0_{free}$. The experimental atomisation energy of bulk V_2O_5 gives an estimate of average V–O binding energies of 7.92 eV.[1] Finally, the experimental binding energy of the $^4\Sigma^-$ state of the VO dimer amounts to 6.4 ± 0.2 eV[2] (our calculations yield 7.2 eV at the RPBE level). Only a modern experimental determination of oxygen vacancy formation energies at the V_2O_5(010) surface using a well characterised substrate without major imperfections (this may have been a problem with the early extrapolation[1]) can give a definitive answer as to the reliability of the present theoretical data.

1 P. Kofstad, *Nonstoichiometry, Diffusion and Electrical Conductivity in Binary Metal Oxides*, John Wiley, New York, 1972, p. 57 and 180.
2 G. Balducci, G. Gigli and M. Guido, *J. Chem. Phys.*, 1983, **79**, 5616.

Prof. Harrison asked: How accurate are the density functional calculations for the O and OH neutral species used as reference energies in your calculations?

Prof. Hermann answered: In the calculations we have used gradient corrected functionals (GGA-II, RPBE) which are known to yield meaningful total energies for atoms and small molecules such as O ($^3P_2$ reference), OH ($^2\Pi$ reference). Thus, we expect reliable values for adsorption and vacancy formation energies in the present system. This has been confirmed by very recent work on V_2O_5 surface properties[1] where we have applied a revised version of the gradient corrected Perdew–Burke–Ernzerhof functional[2] and where binding energy results similar to the present data are obtained.

1 K. Hermann, M. Witko, R. Druzinic and R. Tokarz, *Top. Catal.*, in press.
2 B. Hammer, L. B. Hansen and J. K. Norskov, *Phys. Rev. B*, 1999, **59**, 7413.

Prof. Finnis commented: In fact I don't think you need to calculate any properties of free oxygen atoms or dimers or pathways but you can use the same trick as we did and formulate the problem in terms of the chemical potential of oxygen.

Prof. Hermann replied: Thank you for the suggestion.

Dr Corà asked: (1) When applying cluster models to simulate solid oxides, the concept of saturating the terminal oxygen with H atoms is mutuated from the field of silica and zeolites. There, the σ O–H bonds replace the σ O–Si bonds that linked the cluster to its crystalline environment. In transition metal oxides such as V_2O_5, in which the M ion has electronic configuration d^0, the frontier crystalline orbitals are those of π symmetry along the M–O direction. In saturating the molecular fragment with Hs, we replace the π M–O bonds with a σ O–H. Can you please comment on which effect this approximation is likely to introduce in the calculated bonding properties of the finite V_2O_5 fragment?

(2) A related question concerns the effect of the finite cluster size and saturation with Hs on the calculated O abstraction energies that you have proposed in the paper. Even though the cluster termination with Hs may reproduce with sufficient accuracy the properties of the perfect V_2O_5 crystal, the finite cluster size limits the possibility to delocalise the effect of perturbations.

In my calculations (described in a paper at this meeting) I have found that an important electronic delocalisation occurs *via* the π M–O crystalline orbitals. In the two-layer cluster, upon reduction of the V ions (which follows the O abstraction) the reduced V ion binds to the O in the sub-surface layer. I would expect that a similar process propagates to the following layers; this would substantially reduce the calculated O abstraction energy.

Prof. Hermann replied: (1) The concept of using hydrogen terminators to simulate cluster embedding is actually much older than silica and zeolite modelling. For example, it has been applied in adsorption studies using silicon surface clusters a long time ago.[1] The statement that vanadium in V_2O_5 has electronic configuration d^0 is incorrect and based on a confusion of the formal chemical valence of the atom with its actual microscopic charge state. The DOS results indicate a $d^{1.5}$ charge state. Extended test calculations have shown that the detailed nature of the bonds being formed between peripheral oxygen and terminator atoms is irrelevant for the chemical behaviour in the cluster centre. This is confirmed by recent full-potential LAPW studies on hydrogen adsorption at $V_2O_5(010)$ using repeated slabs[2] where both binding energies and charge distributions are very close to the present cluster results.

(2) In contrast to the rather compact MgO crystal lattice the V_2O_5 substrate forms an open layer structure where electronic coupling between the layers is found to be weak. Therefore, perturbations, induced by vacancy formation in the first surface layer at $V_2O_5(010)$ and affecting the

second layer, are very unlikely to penetrate into deeper layers. More detailed results require very large cluster and/or slab calculations which exceed the present computational resources.

1 K. Hermann and P. S. Bagus, *Phys. Rev. B*, 1979, **20**, 1603.
2 K. Hermann, A. Chakrabarti, R. Druzinic and M. Witko, *Phys. Stat. Solidi*, 1999, **173**, 195.

Prof. Friend said: Hydrocarbon radicals directly add to oxygen on MoO_3 in calculations and also to oxygen on thin film oxides of Mo in experiments. In our DFT calculations, methyl radicals add to all three types of oxygen. Methyl bound to terminal oxygen is most stable. However, in the experiments there is a kinetic preference for addition to highly coordinated oxygen. Therefore, oxidation of hydrocarbons does not occur *via* insertion of gaseous OH into gas-phase hydrocarbon species.

Prof. Waugh said: Following up on the previous point and on your suggestion that it would be hydroxy species that are responsible for the oxidation of propene, we have looked at the reactivity of propane with vanadium pyrophosphate by temperature programmed desorption and reaction. We have shown that the propene adsorbs as a propyl species and that the hydroxys formed recombine and desorb at lower temperatures than that at which the acrolein desorbs.

Prof. Pettersson addressed Prof. Hermann: I like the proposed mechanism for oxygen removal in terms of successive replacement of the oxygen to crystal bonds by O–H bonds and following easy desorption of water. However, in terms of V_2O_5 as a dehydrogenation catalyst there seems to be a problem with the energetics: the computed hydrogen affinity of the O(1) oxygen, although high at about 70 kcal mol^{-1}, is still far from compensating for the C–H bond strength at 100–110 kcal mol^{-1}. Furthermore, the computed value is at the LDA level and is likely to be somewhat overestimated. Could you speculate on possible interaction mechanisms that could provide the missing energy?

Prof. Hermann responded: Experimental C–H bond strengths are quoted as $D(C–H) = 81$ kcal mol^{-1} in ref. 1 while the computed LSDA values of the hydrogen affinity $E_B(H)$ for the different oxygen sites, O(1–3), at $V_2O_5(010)$ range between 58 and 70 kcal mol^{-1}. Very recent calculations using the RPBE functional[2] yield even smaller $E_B(H)$ values between 42 and 54 kcal mol^{-1}. Therefore, the energy required to split off hydrogen from a free hydrocarbon molecule cannot be fully recovered by the energy gain due to surface O–H bond formation. However, in a hydrocarbon reaction near the vanadium oxide surface the relative energetics will be affected by the hydrocarbon reactant being influenced by the electrostatic potential near the surface as well as by the surface O–H bond being modified due to the presence of the reactant. Both effects may result in effective $E_B(H)$ and $D(C–H)$ values which are different from those mentioned above. Further, corresponding surface reactions may proceed according to complex concerted mechanisms involving different reactants where energy barriers of different intermediate states rather than differences between adsorption and dissociation energies determine the probability of H adsorption at a surface oxygen sites.

1 *Handbook of Chemistry and Physics*, 76th edn., CRC Press, London, 1996, p. 9–52.
2 K. Hermann, M. Witko, R. Druzinic and R. Tokarz, *Top. Catal.*, in press.

Dr Kantorovich said: (1) While doing geometry relaxation, in order to check if there is a Jahn–Teller relaxation, it is important to make sure that the symmetry of the initial configuration would allow you to do that. (2) Local state in the DOS is an indication of the charge localisation in the vacancy created upon the removal of the O atom. The state would pull up from the VB due to lack of the atom there.

Prof. Hermann responded: (1) In all cases, relaxation due to oxygen vacancy formation was evaluated without symmetry constraints. While the O(1) and O(3) vacancies cannot give rise to Jahn–Teller relaxation due to their missing symmetry, the effect was studied for the bridging O(2) vacancy. Here geometry optimisations starting from asymmetric atom arrangements did *not* result in a Jahn–Teller relaxation. (2) So far our DOS results from the cluster levels did not indicate vacancy induced features as you mentioned but this has to be studied in greater detail.

Dr Egdell opened the discussion of Prof. Madix's paper: It is interesting to explore the nature of the surface V-oxide phase in your work in relation to the electronic electron energy loss spectrum in Fig. 11 of your paper. Localised d to d excitations of transition metal ions in octahedral or nearly octahedral environments are both Laporte and parity forbidden and therefore have small oscillator strengths. In HREELS they therefore give fairly weak bands.[1–3] Moreover the V^{3+} ion in an octahedral environment has a $^3T_{1g}$ ground state derived from the ground configuration $t_{2g}{}^2$ with $^3T_{2g}$ and $^3T_{1g}$ excited states derived from the configuration $t_{2g}{}^1 e_g{}^1$. The absorption spectrum of V^{3+} doped as an isolated impurity into Al_2O_3 has absorption peaks at 2.15 eV and 3.12 eV associated with excitation to these states, as well as a third peak at 4.27 eV due to excitation to the $^3A_{2g}$ state associated with the doubly excited configuration $e_g{}^2$ (ref. 4). Thus your observation of a *strong single* new loss peak at 2.0 eV appears inconsistent with an assignment in terms of localised d to d excitations of V^{3+}. An octahedral V^{4+} ion has a $^2T_{2g}$ ground state and a single ^2E_g excited state. However the problem of high band intensity remains.

Bulk V_2O_3 is a metallic oxide at room temperature and the optical properties are therefore dominated by the plasmon mode associated with the 3d conduction electrons. Plasmon losses are strongly allowed and therefore give much stronger loss features in HREELS than corresponding localised d to d transitions.[5] It is tempting to assign your 2.0 eV peak with the conduction electron plasmon of V_2O_3, although the plasmon energy found at room temperature both in thermoreflectance spectra[6] and transmission EELS[7] is about 0.95 eV. Incidentally Goodman and co-workers recently found an HREELS loss peak at 0.9 eV at room temperature for ordered $V_2O_3(0001)$ films grown on $Al_2O_3(0001)$.[8] This showed a blue shift to 1.2 eV on cooling below the metal to insulator transition, in agreement with the thermoreflectance measurements. All in all the simplest assignment of the 2.0 eV loss peak is to some sort of plasmon mode, but the differences from bulk V_2O_3 are telling us that the electronic properties of your vanadia overlayers do differ significantly and interestingly from the simple bulk V_2O_3 phase.

1 P. A Cox and A. A. Williams, *Surf. Sci.*, 1985, **152/153**, 791.
2 A. Freitag, V. Staemmler, D. Cappus, Ca. Ventrice, Ka. Shamery, H. Kuhlenbeck and H.-J. Freund, *Chem. Phys. Lett.*, 1993, **210**, 10.
3 M. Hassel, H. Kuhlenbeck, H.-J. Freund, S. Shi, A. Freitag, V. Staemmler, S. Lutkehoff and M. Neumann, *Chem. Phys. Lett.*, 1995, **240**, 205.
4 M. H. L. Pryce and W. A. Runciman, *Discuss. Faraday Soc.*, 1958, **26**, 34.
5 P. A. Cox, R. G. Egdell, J. B. Goodenough, A. Hamnett and C. C. Naish, *J. Phys. C*, 1983, **16**, 6221.
6 S. Stizza, I. Davoli, R. Bernadini, A. Bianconi and M. Benfatto, *Solid State Commun.*, 1983, **48**, 471.
7 H. Abe, M. Terauchi, M. Tanaka and S. Shin, *Jpn. J. Appl. Phys.*, 1998, **37**, 584.
8 Q. Guo, D. Y. Kim, S. C. Street and D. W. Goodman, *J. Vac. Sci. Technol.*, 1999, **17**, 1887.

Prof. Madix responded: This is a very insightful and interesting suggestion. Taking this interpretation, it is interesting to note that this collective excitation is observed at 2.0 eV from 0.5 to 5 monolayers of the vanadia, suggesting that the metallic character of the vanadia does not change appreciably over this coverage range.

Prof. Freund asked: Do you think the preferred formation of V_2O_3 is structurally driven when formed on Al_2O_3? Could you comment on the growth of vanadium oxides on TiO_2?

Prof. Madix answered: V_2O_3 and Al_2O_3 of course both have the corundum structure, so one could easily imagine the possibility of heteroepitaxial growth. However, the fact that V_2O_3 is also formed on $TiO_2(110)$, which has the rutile structure, suggests that the force behind its formation may not be so simple.

Prof. Kempter asked: Does the vanadia formation take place also by heating the alumina surface in the absence of oxygen? We find that this takes place for Ti/MgO where a TiO epitaxial layer will be formed by heating the Ti-film,[1] even in the absence of oxygen. In this case the oxygen is supplied by the MgO-substrate.

1   T. Suzuki, R. Souda, W. Maus-Friedrichs and V. Kempter, *Phys. Rev. B*, submitted.

Prof. Madix replied: When the vanadium is deposited on the thin alumina film at 800 K, it diffuses to the interface between the NiAl metal and the film. We have not conducted such experiments on the single crystal.

Prof. Diebold said: You used extremely small growth rates for vanadium deposition. Vanadium is a very reactive metal, and will getter residual gases from the UHV chamber with a high probability. Did you experience any problems with contamination of the deposited V films? Is it conceivable that unintentional oxidation of the deposited metal by the residual gas causes some of the observed similarities in the STM results of vanadium metal and vanadium oxide films on TiO_2?

Prof. Madix responded: We have focused in this paper primarily on the growth of vanadium oxide on these surfaces, and the deposition was performed in a background of oxygen. The one comparison I showed between vanadium and vanadium oxide growth is the STM images for the alumina thin films. We have shown by XPS that for very similar vacuum conditions and similar deposition rates we can readily form unoxidized multilayers of vanadium metal. These XPS results show a clear transition from an oxidized form of vanadium to metallic vanadium as the surface coverage of vanadium is increased. We thus believe that oxidation of the vanadium due to residual gases is not a problem in this case.

Prof. Thomas commented: Prof. Madix has given convincing experimental proof that it is V_2O_3, not VO_2, V_2O_5 or any other oxide of vanadium, that is formed under the conditions employed by him. It is intriguing to enquire whether a theoretical analysis, or some computational modeling would have 'predicted' this oxide. The trouble is that such computations cannot readily (if at all) take into consideration the factors that lead to kinetic rather than thermodynamic stability. It could well be that it is simply the case of diffusion of the vanadium and oxygen ions in V_2O_3 that favours its formation.

Prof. Madix replied: This is an interesting question for the theorists to consider.

Dr Carley communicated: In order to provide corroborative evidence for the vanadium oxide overlayer being V_2O_3 have you tried to quantify the vanadium : oxygen XPS intensity ratio? Although in the early stages of growth there will be interference from the oxide substrate, the kinetic energy of the O 1s photoelectrons is *ca.* 100 eV for the photon energies you use, with a correspondingly small mean free path (minimum of the mfp-KE curve). The O 1s intensity from the substrate will thus fall off much more rapidly with overlayer growth than, for example, the Al 2p intensity (Fig. 5 of your paper) and reliable vanadium oxide stoichiometry data should be obtainable for vanadium exposures greater than *ca.* 3000 s (Fig. 5). It is worth noting that if Scofield's cross-section data are used the V 2p cross-section is significantly in error and the V 2s intensity must be employed.

1   J. H. Scofield *J. Electron Spectrosc.*, 1976, **8**, 129.

Prof. Madix communicated in response: This is certainly a helpful suggestion, but, unfortunately, we did not accurately record the V 2s features in these experiments.

Prof. Freund said: The growth of Fe and Co on alumina seem to be very similar, in the case where the metal or the corresponding oxides are grown.

Dr Shluger addressed Prof. Madix: (1) How thick was the alumina film shown in Fig. 12? (2) How thick are the films that can be seen using STM? (3) What is the origin of enhanced contrast at steps on the STM alumina film image?

Prof. Madix replied: (1) Approximately 5 Å. (2) We have not explored the imaging of thicker films. (3) Presumably it arises from differences in the local structure, and hence electronic structure, in the vicinity of step edges.

Prof. Freund said: TEM measurements on thin films can be used to compare the growth mode of metals on thin films and more bulk like materials. In the Introductory Lecture the experimental procedure was explained (see also ref. 132 of the Introductory Lecture). For alumina details of such a comparison are described in ref. 133 of the Introductory Lecture.

Prof. Madey asked Prof. Madix: (1) Your XPS and NEXAFS measurements of V and V_2O_3 growth on $TiO_2(110)$ were performed on the (1×1) surface, whereas the STM measurements you report were carried out on the (1×2) reconstruction. Do you have STM data for V (V_2O_3) on the (1×1) surface? Are there any differences in growth on the two surfaces?

(2) Also, we reported evidence for two-dimensional growth of Cr on $TiO_2(110)$ (1×1) up to ≈ 0.8 ML;[1] are your V (V_2O_3) islands relatively flat, two-dimensional islands, or are they three-dimensional clusters?

1 J.-M. Pan, U. Diebold, L. Zhang and T. E. Madey *Surf. Sci.*, 1993, **295**, 411.

Prof. Madix responded: (1) We do have some images for the growth of V on the (1×1) surface. On this surface exposure to V creates bright circular features, most of which appear centered on the rows of coordinatively unsaturated titanium cations on the undisturbed surface. The underlying (1×1) structure of the (110) surface is disrupted. There is a notable disturbance of the structure in the vicinity of step edges. At higher vanadium exposures the vanadium-induced features increase slightly in size, but the most noticeable increase is in their density.

The general appearance of the STM images is similar on the (1×2) surface at low vanadium exposures. At low coverages the V is centered on the added row features of the (1×2). At higher coverages vanadium forms needle-like growths which orient their long axis along the (001) direction in registry with the underlying (1×2) structure. A preference for a double row spacing of these needles (25 Å) is observed.

(2) The metallic vanadium particles appear to be hemispherical in STM on both the titania and alumina surfaces. STM suggests some flattening of the particles on the alumina surface due to reaction with background oxygen.

Prof. Goodman addressed Prof. Madix: You mentioned that you were able to synthesize an 'anatase-like' phase of TiO_2. Since there is great interest in the community regarding the surface chemistry of anatase TiO_2, could you elaborate on the synthetic procedure and its characteristic structural diagnostics.

Prof. Madix responded: My statement was not meant to be so definitive, but we have observed two different surfaces of $TiO_2(110)$ by NEXAFS after the same nominal treatment to produce the (1×1) stoichiometric surface. These two surfaces show different intensity ratios of the Ti L(III)-edge doublet which correspond closely to the differences in fine structure known for bulk rutile and anatase. This structural effect appears to be confined to the surface.

Prof. Jennison said: Vanadium and other reactive metals are added to brazing compounds, supposedly to cause the metal to wet a ceramic (oxide) surface. When the brazed joints are then cut and examined microscopically, an interfacial compound is often noted which is poorly characterized. My question is then have you seen any indication of reactions which mix the oxide substrate with the vanadium/vanadium oxide you are depositing?

Prof. Madix responded: We have confined most of our work to room temperature, and we have little evidence of the formation of complex oxides at the surface. Our STM results suggest that on the alumina thin film there is some incorporation of vanadium metal into the surface. However, when vanadium metal deposited on either titania or the alumina thin film is heated, it ultimately diffuses away from the surface. The exact nature of the oxides formed in either case is, however, unknown.

Prof. Campbell asked: Did the deposition of V onto the (1×1) surface cause it to transform to the (1×2) structure (which might indicate some reduction of the Ti and formation of a mixed surface V–Ti oxide)?

Prof. Madix answered: No, at a vanadium concentration of 0.03 ML large 'vacancy clusters' appear in the terrace structure of the (1 × 1) surface, and at higher coverages the underlying structure of the $TiO_2(110)$ surface becomes indistinct. The random pattern of coverage by the circular metallic features and their high density may make it difficult for the surface to rearrange into the (1 × 2) structure.

Prof. Bowker asked: Is it possible that the V_2O_3 formation is limited by a low oxygen dissociation probability on the oxide, whereas it is likely to be very high on the initially deposited metal atoms. Is it also possible that if the O_2 pressure were much higher that V_2O_5 would be produced, and did you try such an experiment?

Prof. Madix responded: Yes, we think this conversion is kinetically limited. We tried further oxidation of the vanadia layer in 5×10^{-5} Torr oxygen at 700 K for 100 min and observed no change. Of course, this is not a very high oxygen pressure, and further oxidation might be possible at higher pressures.

Prof. Asscher addressed Prof. Madix: The question/comment was that in order for metallic vanadium clusters to form atoms should diffuse across multiple lattice sites. This suggests that upon initial impact from the gas phase, vanadium atoms cannot be oxidised to the V^{3+} state—it would prevent any surface mobility and thus clusters cannot be formed. It is therefore suggested that initially the vanadium atoms adsorb as neutrals, diffuse to form clusters and only then at a certain point they get oxidised by oxygen molecules which dissociatively adsorb on the metallic cluster.

Prof. Madix answered: The dynamics of the formation of these small particles is not clear, and I am hesitant to speculate on this matter. The deposition rates we employ, when coupled with the size of the vanadium induced features observed in STM, suggest that there are a few vanadium atoms in each of the circular features observed. The coverage calibration relies on calibration by AES, using a characteristic mean free path of the Auger electrons, and has an inherent uncertainty. At very low coverages of vanadium I do find it surprising that metal atoms can hop or migrate the 25 Å on the oxide surface required to 'nucleate'. I think this issue requires further study.

Prof. Freund said: The study of diffusion of metals on oxides is a hot topic. With STM the diffusion of single atoms has not been studied in detail as yet. Field ion microscopy can be used for such a study and activation energies for diffusion can be estimated (ref. 127 of the Introductory Lecture). More detailed studies are in progress.

Dr Yubero communicated: I would like to comment on some of your results related to the characterisation of vanadium oxide on alumina or titania. When you perform the electronic characterisation by XPS you observe (Fig. 3 of your paper) that there is an energy shift to lower binding energies on the V $2p_{3/2}$ peak as the amount of vanadium oxide deposited increases. I agree with you when you say that, in spite of this energy shift, the chemical nature of the vanadium atoms (i.e., V^{3+} species) is always the same. You justify this energy shift by a Coulomb interaction between the positive charge left on the particle as the result of the photoemission process and the photoemitted electron. However, you do not find the same behaviour when you deposit V_2O_3 on titania (Fig. 2 of your paper), which is also an insulator with more than 3 eV band gap.

I would like to point out that this kind of energy shift effect in the first stages of growth of oxides has been extensively studied by the group of González-Elipe et al.[1] In fact, they have developed a theory based on bonding and polarisation effects to predict these energy shifts in the binding energy of the peaks of the deposit as well as in their Auger parameter. Note that variations on the Auger parameter of a given atom are directly proportional to the extraatomic relaxation energy, and that the latter can be related to bonding and polarisation effects. Then, according to their work,[1] if the dielectric properties (i.e., refractive index) of an overlayer of a substrate are similar, no energy shifts are expected. However, when a high refractive index overlayer (e.g., TiO_2) is deposited on a low refractive index substrate (e.g., MgO, SiO_2), the Auger parameter of the overlayer atoms increases as the overlayer thickness increases. Besides, in these particular cases (*i.e.*, TiO_2 on SiO_2 and TiO_2 on MgO), it is observed that the binding energy of the Ti 2p peaks decreases with increasing coverage. On the other hand, when TiO_2 is deposited on a metallic substrate as Ag (even higher refractive index) the observed behaviour for the Auger parameter and TiO_2 binding energies is the opposite. Then, the observed behaviour on the present systems, *i.e.*, V_2O_3 on TiO_2 and V_2O_3 on Al_2O_3, would be explained just by considering that the dielectric properties and in particular the refractive index of V_2O_3 is similar to that of TiO_2 but significatively higher than that of Al_2O_3. Note that polarisation effects of different materials scale with $1/n^2$, with n the refractive index.

1 J. A. Mejías, V. M. Jiménez, G. Lassaletta, A. Fernández, J. P. Espinós and A. R. González-Elipe, *J. Phys. Chem.*, 1996, **100**, 16255.

Prof. Madix communicated in response: This is an interesting suggestion, and it is certainly possible that these polarization effects contribute to the shift in binding energy observed. It is interesting to note that we see larger shifts of the V 2p peak when metallic vanadium is deposited, but both vanadium and vanadia produce shifts toward lower binding energy as the surface coverage is increased. The direction of this shift is compatible with our interpretation that a metal-like oxide is formed, though its refractive index is probably lower. I do not have the data required at this time to make a quantitative evaluation of the relative shifts to be expected, however.

Prof. Møller communicated: I would like to remark that using different preparative conditions (successive cycles of submonolayer vanadium in UHV, each followed by annealing in 10^{-6} mbar oxygen atmosphere at around 150 °C, the exact procedure of Prof. Granozzi and co-workers being reported in ref. 1, and which was later applied in ref. 2), the authors have there been able to grow epitaxial layers of VO_2 as demonstrated from XPD, UPS, ARPEFS and LEED data.

1 M. Sambi, G. Sangiovanni, G. Granozzi and F. Parmigiani, *Phys. Rev. B*, 1997, **55**, 7850.
2 M. Sambi, M. Della Negra, G. Granozzi, Z. S. Li, J. Hoffmann Jørgensen and P. J. Møller, *Appl. Surf. Sci.*, 1999, **142**, 146.

A study of the electronic, magnetic, structural and dynamic properties of low-dimensional NiO on MgO(100) surfaces

William C. Mackrodt,*[a] Claudine Noguera[b] and Neil L. Allan[c]

[a] *School of Chemistry, University of St. Andrews, St. Andrews, Fife, UK KY16 9ST*
[b] *Laboratoire de Physique des Solides, associé au CNRS, Université Paris Sud, 91405 Orsay, France*
[c] *School of Chemistry, University of Bristol, Cantock's Close, Bristol, UK BS8 1TS*

Received 25th May 1999

Recent developments in the growth of ultra-thin epitaxial layers of oxides and the fabrication of a diversity of nanostructures has led to current interest in, and much speculation about, the properties of low dimensional systems. In this paper we report recent calculations for low dimensional NiO on MgO(100) surfaces both from first principles electronic structure calculations and free energy calculations based on surface lattice dynamics. The results include surface structures and dynamics at a range of temperatures and electronic structures of ground, excited, ionised, d → d and charge–transfer excitonic states in different spin alignments.

1. Introduction

Over the past few years, important advances have been made in the fabrication of ultra-thin crystalline oxide layers grown epitaxially on a variety of substrates.[1,2] The methods commonly used are direct oxidation of a metallic surface or deposition of metal atoms followed by controlled oxidation. In this way, films a few Angströms thick may be grown, the crystallinity and perfection of which depend strongly on the ratio of the lattice parameter of the oxide to that of the substrate and details of the preparative conditions such as deposition rate, oxidising temperature, *etc.* These ultra-thin films exhibit several unusual properties which are of interest both from a fundamental point of view[3–5] and with regard to potential technological application. First, as a result of their small thickness, they do not charge when exposed to electron or electromagnetic radiation and can thus be submitted to detailed spectroscopic investigation. They may also exhibit surfaces that are not usually obtained by direct cleavage of the bulk, which is the case for high energy surface orientations such as polar surfaces. The constraint imposed by the substrate may also lead these epitaxial oxide layers to adopt lattice symmetries which differ from those of the thermodynamically stable bulk, and, where structural phase transitions occur, transition temperatures can depend strongly on the thickness of the epitaxial layer.

The system NiO(100)–MgO(100) is prototypical of multilayered oxide films, largely because the lattice mismatch between the two components is less than 1%. It has been shown that MgO grows on NiO(100) epitaxially and in a layer-by-layer mode,[6–8] and that the same is true for NiO on MgO(100).[5,6,9–13] Intermixing occurs only if films are prepared at high temperature. Several types of experiment indicate a strong dependence of the electronic and magnetic properties of these ultra-thin films on their thickness. EELS (electron energy loss spectroscopy) experiments[6,10] have revealed excitations at energies less than the bulk band gap, which have been attributed to d → d transitions. For coverages of NiO on MgO(100) greater than 2.8 monolayers (ML) these excitations are identical to semi-infinite NiO(100), whereas for NiO thicknesses less than 2.8 ML, a new transition at 2.18 eV appears, which has been assigned to surface or interface states and appears to show that the electronic structure of thin NiO films is modified by the presence of the MgO substrate. Other bulk and surface transitions have been reported to shift in energy with coverage.

Two explanations have been suggested to account for these changes. The first is the loss of octahedral symmetry at a surface Ni, associated, not only with the reduced nearest neighbour coordination, but also with a lateral elongation of the NiO lattice and contraction of the interplanar separation. The second is a putative loss of octahedral symmetry at a Ni at the NiO/MgO interface resulting from the difference in covalency between NiO and MgO. With regard to the existence, or otherwise, of NiO/MgO interface states, it is of note that no such states have been detected by EELS[8] for MgO layers grown on NiO. XPS experiments[9,13] have shown that Ni and O core level spectra depend on film thickness, which has been interpreted in terms of the number of Ni second neighbours. Such spectra are also consistent with a negative charge transfer from MgO to NiO at large thicknesses. A detailed Hartree–Fock interpretation of core level shifts in NiO/MgO layered systems in terms of surface lattice relaxation and other effects has been given previously.[14] While magnetic data are generally more difficult to obtain, Alders et al.[5] have recently reported some very elegant linear polarised X-ray absorption experiments, from which they obtain values of the Néel temperature, T_N, of NiO overlayers as a function of thickness and show that even for 20 ML films T_N has not recovered the bulk value.

It is within this context that we have initiated a theoretical study of ultra-thin films and surface properties of NiO. NiO is a paradigm magnetic insulator whose bulk and surface properties have been the subject of extensive experimental and theoretical investigation.[15] It has long been considered as highly ionic and early first principles calculations based on the local spin density approximation (LSD) described it as a Mott–Hubbard system in the AF_2 spin arrangement with a narrow gap spanned by Ni d-states.[16] However, seminal work by Sawatzky et al.[17–19] showed that hole states in Li:NiO were largely of O(p) character, which suggested that the first ionised state of NiO is essentially $d^8\underline{L}$ and the ground state of p → d charge-transfer type. Subsequent first principles calculations confirmed the majority weight of the valence band edge to be O(p), including spin unrestricted *periodic* Hartree–Fock (UHF) calculations.[20–22] The latter have shown that the insulating and (high spin) magnetic properties are the result of large on-site Coulomb and exchange interactions between essentially localised electrons with strong orbital polarisation resulting from the orbital dependence of the one-electron potential. This is determined principally by the non-local exchange interaction which is evaluated exactly within the Hartree–Fock approximation and implemented, again exactly, in the CRYSTAL code.[23] UHF calculations have also provided direct evidence of O(p) holes in Li:NiO and NiO[24,25] and Fe(d) holes in NiO,[25] in agreement with experiment.[26] Thus, despite its approximate nature and inherent limitations, the *periodic* UHF method would seem to be well suited to describing the electronic and magnetic properties of NiO in lower dimensions. Extending a recent UHF study[27] of an unsupported NiO(100) monolayer, which included both ground and d → d and charge-transfer excited states, here we report similar calculations of several NiO/MgO slab configurations, including NiO(100) monolayers on MgO(100) or sandwiched between two MgO(100) monolayers. While there appear to be no other calculations based on *extended/periodic* systems for comparison, there have been high level *cluster* calculations of magnetic interactions[28] and d → d[29–31] and charge-transfer excitations,[31] which provide a guide as to the influence of both electron correlation and non-local/extensive effects.

In view of the growing body of experimental data on the high-temperature properties of ultra-thin films, we also report quasi-harmonic lattice dynamics and free energies calculations of more extended NiO/MgO slab configurations at elevated temperatures using recently developed atomistic lattice methods based on pair potentials.[32] Few calculations have included dynamic effects, such as *temperature*, largely because the full dynamical treatment of surfaces and extended defects, even within the quasi-harmonic approximation, presents severe computational demands if reasonably high precision is required. There are three main simulation techniques available for the calculation of surface lattice properties: Monte Carlo simulation, molecular dynamics and quasi-harmonic lattice dynamics. Of these only the last is capable of giving free energies (as well as derived properties such as the entropy and the heat capacity) *directly* and to high precision. This method is consequently not only the most suitable for structure optimisation as a function of temperature, but in many applications it has also been shown to be a valid approximation up to two-thirds of the bulk melting temperature.[33,34]

Accordingly, the paper is organised as follows. In Section 2 we review briefly the theoretical methods used and in Section 3 present the main body of our results, which we discuss in Section 4.

2. Theoretical methods

2a. Electronic structure calculations

The all-electron *ab initio* LCAO Hartree–Fock method for periodic systems and its computational implementation in the CRYSTAL 95 computer code[23] have been described in detail previously.[35] The calculations reported here use extended Gaussian basis sets and are based on the spin unrestricted (UHF) procedure[36] to describe open-shell electronic configurations. The numerical values of the tolerance parameters involved in the evaluation of the (infinite) Coulomb and exchange series were identical to those used in recent studies:[24,25,27] a detailed account of the effect of these tolerances is discussed elsewhere.[37] The reciprocal space integration utilised the Monkhorst–Pack sampling,[38] with shrinking factors that gave 15–36 k-points in the IBZ, depending on the overall symmetry of the calculation, and the SCF convergence criterion based on differences in the total energy of the unit cell of less than 10^{-6} Ha. *A posteriori* corrections for electron correlation based on three generalised gradient approximations are included in the present study: they are due to Perdew,[39] Perdew and Wang[40] and Perdew, Burke and Ernzerhof[41] and are referred to P, PW and PBE, respectively. As in previous calculations,[20–22,24,25,27] the localised crystal orbitals consisted of 25 atomic orbitals for Ni, 15 for Mg and 14 for O of the type

Ni: 1s(8)2sp(6)3sp(4)4sp(1)5sp(1)3d(4)4d(1)

Mg: 1s(8)2sp(5)3sp(1)4sp(1)

O: 1s(8)2sp(4)3sp(1)4sp(1)

where the numbers in brackets are the numbers of Gaussian functions used to describe the corresponding shell, *e.g.*, 1s, 2sp, 3d, *etc*. The exponents and contraction coefficients were identical to those used for the bulk.[20–22]

In this study we consider various slabs of NiO and MgO, for which we use the notation, (type)$_{nlm}^{N}$, where N refers to the number of layers and nlm to the orientation. Thus, NiO(100) monolayer is (NiO)$_{100}^{1}$, the bi-layer NiO/MgO (NiO/MgO)$_{100}^{2}$ and the three tri-layers we consider here (NiO/MgO/MgO)$_{100}^{3}$, (NiO/MgO/NiO)$_{100}^{3}$ and (MgO/NiO/MgO)$_{100}^{3}$. Furthermore, all our calculations are based on a constant cation–anion distance of 2.1 Å fixed by the MgO substrate, while our axis system equates the xy-plane with (100), so the p_z and d_{z^2} orbitals are perpendicular to the slab planes.

As in a previous study,[27] to investigate the magnetic properties of these slabs we have considered four spin arrangements of a 2×2 surface unit cell shown below,

$$F: \uparrow \overset{\uparrow}{O} \uparrow \quad A_1: \uparrow \overset{\uparrow}{O} \downarrow \quad A_2: \uparrow \overset{\downarrow}{O} \uparrow \quad A_3: \uparrow \overset{\uparrow}{O} \uparrow$$
$$\uparrow \qquad \downarrow \qquad \downarrow \qquad \downarrow$$

where $\uparrow$ and $\downarrow$ represent high spin Ni^{2+} ions and F, A_1, A_2 and A_3 the ferromagnetic, antiferromagnetic, fully-frustrated and ferrimagnetic spin alignments, respectively.[27] In the case of (NiO/MgO/NiO)$_{100}^{3}$ we consider slabs in which two ferromagnetic NiO layers are aligned both ferromagnetically and antiferromagnetically. In this way we have been able to distinguish superexchange coupling within and between NiO planes as a first step towards understanding the recent results reported by Alders *et al.*[9] As described in detail below, we have obtained converged UHF solutions for a number of excited states involving both single and quadruple, *i.e.*, complete, local d → d excitations, which we designate as e_1 and e_4, respectively.[27] Thus, by an 'e$_4$ d$_{xy}$ → d$_{z^2}$ excited state' we mean a state in which all four Ni ions of the unit cell have undergone a d$_{xy}$ → d$_{z^2}$ excitation.

Our treatment of hole states follows that previously used for NiO bulk[24] and (100) monolayer,[27] wherein a renormalisation of the (infinite) inter-cell Coulombic interaction is effected by adding a uniform background charge of opposite sign and equal magnitude to the crystal potential in the plane of the monolayer. As our results indicate, this has no effect on the densities of states, other than a rigid shift in energy of the single particle spectrum. Furthermore, we confine our attention to *differences* in total energy only, as between various electronic and spin states of the hole and small lattice distortions, which, again, are invariant to the uniform background charge.

2b. Lattice simulations

The methods we have used to carry out lattice simulations at finite temperatures based on lattice statics and quasi-harmonic lattice dynamics have been described in full recently.[31,42] For a slab of sufficient thickness to provide what, in effect, is a bulk-like region in the interior and two non-interacting, free surfaces, the Helmholtz free energy of the system, F, at a temperature, T, is minimised with respect to the collection of structural parameters, $\{\mathscr{E}_A\}$, that define the slab. In the quasi-harmonic approximation it is assumed that $F(\{\mathscr{E}_A\}, T)$ can be written as the sum of static and vibrational contributions,

$$F(\{\mathscr{E}_A\}, T) = \Phi_{stat}(\{\mathscr{E}_A\}) + F_{vib}(\{\mathscr{E}_A\}, T)$$

$\Phi_{stat}(\{\mathscr{E}_A\})$ is the potential energy of the static lattice in a given state of strain $\{\mathscr{E}_A\}$ and is evaluated from interatomic pair potentials, while F_{vib} is the sum of harmonic vibrational contributions from all the normal modes. For a periodic structure, the frequencies $v_j(q)$ of modes with wavevector q are obtained by diagonalisation of the dynamical matrix $D(q)$. F_{vib} is given by,

$$F_{vib} = \sum_{q,j} \{\tfrac{1}{2} h v_j(q) + k_B T \ln[1 - \exp(-h v_j(q)/k_B T)]\}$$

in which the first term is the zero-point energy and the associated vibrational entropy, S,

$$S = \sum_{q,j} \frac{(h v_j(q)/T)}{\exp(h v_j(q)/k_B T) - 1} - k_B \ln[1 - \exp(-h v_j(q)/k_B T)]$$

For a macroscopic crystal the sum over q becomes an integral over a cell in reciprocal space, which can be evaluated by taking successively finer uniform grids until convergence is achieved. Since the reciprocal space is two-dimensional the Brillouin zone summation requires a two dimensional mesh of wavevectors.

The minimisation of F, and subsequent thermodynamic manipulation, can, of course, be carried out by brute force from numerical values of F. However, for the slabs of any complexity, such as those considered here, there are large numbers of internal strains, $\mathscr{E}_A$, and it is much more efficient to use analytic expressions for the derivatives of F with respect to strain. The strain derivatives are given by,

$$\left(\frac{\partial F_{vib}}{\partial \mathscr{E}_A}\right)_{\mathscr{E}', T} = \sum_{q,j} \left\{\frac{h}{2v_j(q)}\left(\frac{1}{2} + \frac{1}{\exp(h v_j(q)/k_B T) - 1}\right)\left(\frac{\partial v_j^2(q)}{\partial \mathscr{E}_A}\right)_{\mathscr{E}'}\right\}$$

where the subscript $\mathscr{E}'$ denotes that all the $\mathscr{E}$ are kept constant except for the differentiation variable. In a recently developed code[40] the derivatives, $(\partial v_j^2(q)/\partial \mathscr{E}_A)_{\mathscr{E}'}$, are obtained from analytic expressions for the derivatives, $(\partial D(q)/\partial \mathscr{E}_A)_{\mathscr{E}'}$, by first-order perturbation theory.[42,43] Full details of the particular derivatives of the Parry summation needed for the Coulombic interactions in ionic slabs are given by Taylor et al.;[32] the derivatives required for pairwise short-range potentials are collected together in ref. 42. Since the perturbation is infinitesimal, the procedure is exact. In addition, for thermodynamic properties no special consideration needs to be given to degeneracies in first-order perturbation theory, for the trace of $(\partial D(q)/\partial \mathscr{E}_A)_{\mathscr{E}'}$ is invariant for any complete normal set of eigenvectors of D. To obtain the equilibrium structure a variable metric method is used for the minimisation of F with respect to the $\mathscr{E}_A$. In the initial configuration the *static* energy Hessian, $(\partial^2 \Phi_{stat}/\partial \mathscr{E}_A \partial \mathscr{E}_B)$, which is a good approximation to $(\partial^2 F/\partial \mathscr{E}_A \partial \mathscr{E}_B)$, is calculated from its analytic expression, and its inverse together with the $(\partial F/\partial \mathscr{E}_A)$ are used to obtain an improved configuration. In subsequent iterations the $(\partial F/\partial \mathscr{E}_A)$ are calculated in the new configuration and the inverse Hessian updated by the BFGS formula.[44] An optimisation therefore requires one static Hessian calculation, and a small number of dynamic gradient calculations. In practice it has been found that this is much more efficient than methods involving repeated evaluation of the Hessian, or frequent line minimisations, or methods in which the derivatives are determined numerically.

3. Results

3a. Electronic ground states

We begin with a brief resumé of NiO bulk[20] and (100) monolayer,[27] which, despite their different dimensionality and consequent nearest-neighbour coordination, and different ligand-field splitting, exhibit several common features. Both are predicted to be highly ionic, high spin insulators with lattice constants of ≈ 4.3 Å and ≈ 4.0 Å, respectively. Moreover, for both systems the d^8 ground state configuration is

$$[(d_{xz})^2(d_{yz})^2(d_{xy})^2(d_{z^2})^1(d_{x^2-y^2})^1]$$

with

$$\varepsilon(e_g)_\alpha < \varepsilon(t_{2g})_\alpha < \varepsilon(t_{2g})_\beta < \varepsilon(e_g)_\beta$$

for the bulk and

$$\varepsilon(d_{z^2})_\alpha/\varepsilon(d_{x^2-y^2})_\alpha < \varepsilon(d_{xz})_\alpha/\varepsilon(d_{yz})_\alpha \varepsilon(d_{xy})_\alpha < \varepsilon(d_{xz})_\beta/\varepsilon(d_{yz})_\beta \varepsilon(d_{xy})_\beta$$

for $(\text{NiO})^1_{100}$ where ε are the single particle eigen values and the subscripts α and β refer to spin up (↑) and spin down (↓) electrons, respectively. The local Ni spin magnetic moment is calculated to be $\approx 1.9 \, \mu_B$ for both systems and the stability of low index spin alignments,

$$AF_2 > F > AF_1 \quad \text{(bulk)}$$

and

$$A_1 > A_3 > F > A_2 \quad ((\text{NiO})^1_{100})$$

below the Néel temperature. If, to a first approximation, it is assumed that the differences in energy between the spin alignments for each system separately can be written simply in terms of direct spin–spin (E_d) and indirect superexchange (E_{se}) interactions, UHF total energies lead to values for E_d and E_{se} of -1.6 meV and -8.2 meV, respectively, for NiO bulk at a lattice constant of 4.2 Å, and -1.0 meV and -6.9 meV for $(\text{NiO})^1_{100}$ at the same lattice constant. Another major difference between the two systems is that from an analysis of the unoccupied O(p) density of states (DOS) of the self-trapped hole, the band gap is estimated to be ≈ 4 eV for the bulk[24,25] and ≈ 5 eV for $(\text{NiO})^1_{100}$.[27]

Turning now to the various NiO/MgO(100) slabs, we have obtained converged UHF solutions of the ground electronic states of the F, A_1, A_2 and A_3 spin alignments. As before,[20,27] Mulliken analyses[45] yield effective atomic charges of $\approx 1.9 \, e$, 3d populations of ≈ 8.1 and local spin moments of $\approx 1.9 \, \mu_B$, all of which indicate that the highly ionic, high spin character previously found for the bulk[20] and $(\text{NiO})^1_{100}$[27] is retained in multi-layered NiO/MgO systems. Furthermore, the d^8 ground state configuration, $[(d_{xz})^2(d_{yz})^2(d_{xy})^2(d_{z^2})^1(d_{x^2-y^2})^1]$, of $(\text{NiO})^1_{100}$ is found to remain unchanged in all four spin alignments, the stabilities of which are in the order,

$$A_1 > A_3 > F > A_2$$

as shown in Table 1. As before,[27] the relative energies of A_1, A_2, A_3 and F can be written to a first approximation in terms of direct and superexchange interactions as,

$$E(F) = B + 2E_d$$

$$E(A_1) = B + E_d + 2E_{se}$$

$$E(A_2) = B$$

$$E(A_3) = B + E_d + E_{se}$$

Table 1 Comparison of the energies (meV/Ni) of the F, A_2 and A_3 spin alignments (F and AF_1 in the case of the bulk) relative to A_1 (AF_2) for NiO bulk and various (100) NiO/MgO slabs all at $a_0 = 4.2$ Å

System	A_3	F	$A_2(AF_1)$
Bulk	—	19.7	26.2
$(NiO)_{100}^1$	6.9	12.8	14.8
$(NiO/MgO)_{100}^2$	7.4	13.6	16.0
$(NiO/MgO/MgO)_{100}^3$	7.4	13.5	15.9
$(NiO/MgO/NiO)_{100}^3$	7.4	13.6	16.1
$(MgO/NiO/MgO)_{100}^3$	8.5	15.4	18.6

from which E_d and E_{se} can be obtained. These are given in Table 2, where E_d and E_{se}, corresponding to direct and superexchange interactions *within* a single NiO(100) layer, are seen to increase monotonically with Ni coordination to broadly that of the bulk in $(MgO/NiO/MgO)_{100}^3$. In addition, our results for $(NiO/MgO/NiO)_{100}^3$, in which the two ferromagnetic NiO layers are coupled antiferromagnetically, suggest stronger superexchange between NiO(100) layers than within the layers, at least for the simple tri-layer we have examined.

The densities of occupied states of the A_1 alignment of $(NiO/MgO)_{100}^2$ over the first 12 eV, shown in Figs. 1a and 1c, indicate further close similarities with 2D and 3D NiO. The upper valence band, ≈ 5 eV wide, is essentially O(p) with only a minor contribution from d_{xy} states at lower energies and negligible Ni weight at the upper edge. The local O(p) DOS of the two planes, shown in Fig. 1b, are similar, both in terms of band width and overall profile. The major difference is the weight of states at the upper edge, shown in Fig. 1b, which derives predominantly from the NiO layer. This suggests, though by no means guarantees, that the low energy holes states will be associated with the NiO lattice. The Ni d states occur ≈ 1 eV below the O(p) band and are dispersed over ≈ 4.5 eV in three distinct sub-bands, with

$$\varepsilon(d_{z^2})_\alpha/\varepsilon(d_{x^2-y^2})_\alpha < \varepsilon(d_{xz})_\alpha/\varepsilon(d_{yz})_\alpha \varepsilon(d_{xy})_\alpha < \varepsilon(d_{xz})_\beta/\varepsilon(d_{yz})_\beta \varepsilon(d_{xy})_\beta$$

as in the bulk. This is in marked contrast to the crystal-field splitting,

$$\tilde{\varepsilon}(d_{xz})/\tilde{\varepsilon}(d_{yz}) < \tilde{\varepsilon}(d_{z^2}) < \tilde{\varepsilon}(d_{xy}) < \tilde{\varepsilon}(d_{x^2-y^2})$$

which suggests that, as in the bulk, the single particle spectrum is determined to a large extent by the on-site Coulomb and exchange terms.

3b. d → d excited states

The lowest energy electronic excitations from the ground state in NiO correspond to orbitally forbidden ($\Delta l = 0$) local d → d transitions, or Frenkel excitons, which have been observed in both

Table 2 Comparison of E_d and E_{se} (meV) for NiO bulk and various (100) NiO/MgO slabs all at $a_0 = 4.2$ Å

System	$-E_d$	$-E_{se}$
Bulk	1.6	8.2
$(NiO)_{100}^1$	1.0	6.9
$(NiO/MgO)_{100}^2$	1.2	7.4
$(NiO/MgO/MgO)_{100}^3$	1.2	7.4
$(NiO/MgO/NiO)_{100}^3$ [a]	1.3	7.4
$(NiO/MgO/NiO)_{100}^3$	—	10.5[b]
$(MgO/NiO/MgO)_{100}^3$	1.6	8.5

[a] No superexchange coupling between NiO layers.
[b] Superexchange interaction between two ferromagnetic NiO layers antiferromagnetically aligned.

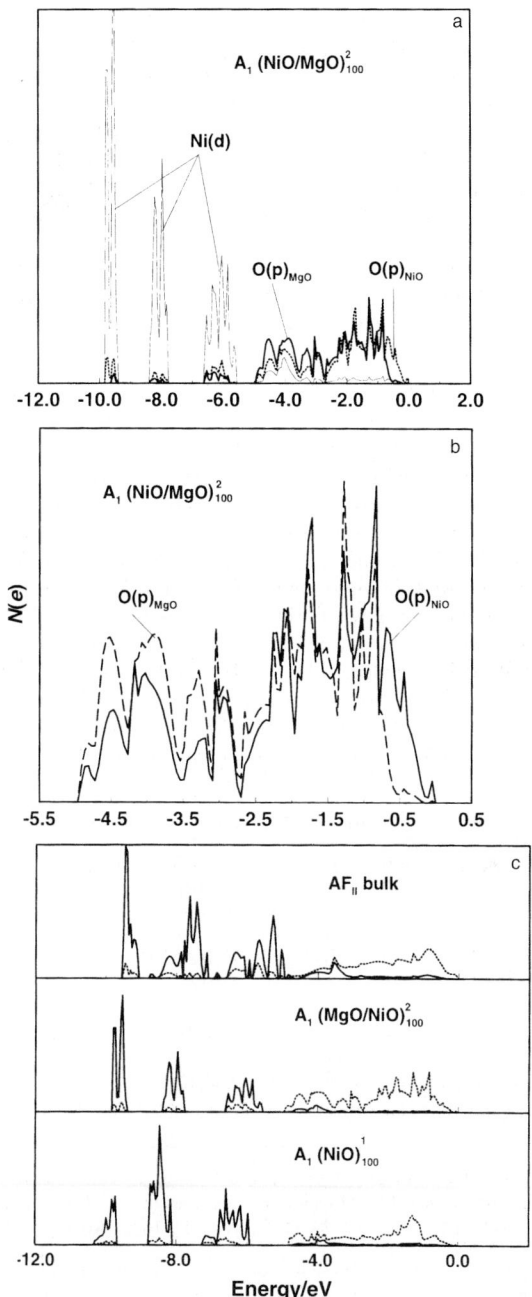

Fig. 1 a, Valence band DOS of $A_1(NiO/MgO)_{100}^2$. b, Comparison of $O(p)_{MgO}$ and $O(p)_{NiO}$ valence band DOS of $A_1(NiO/MgO)_{100}^2$. c, Comparison of the valence band DOS of AF_{II} bulk NiO, $A_1(NiO/MgO)_{100}^2$, and $A_1(NiO)_{100}^1$.

the optical[46,47] and EEL[29,48–51] spectra of the pure material and in EEL spectra of NiO/MgO(100) layered systems.[6,10] We have obtained variationally minimised solutions corresponding to the complete range of one-electron (e_1) excitations of the type, $d_{xy} \to d_{z^2}$, $d_{xy} \to d_{x^2-y^2}$, etc., to the spin-forbidden $d_{x^2-y^2} \to d_{z^2}$ and $d_{z^2} \to d_{x^2-y^2}$ states and the two electron $d_{xy}/d_{yz} \to d_{z^2}/d_{x^2-y^2}$

excited state for F(NiO/MgO)$_{100}^2$. These states remain highly ionic, high spin, insulating with changes in the ionic charges, 3d populations and local spin moments of ⩽1%. The corresponding excitation energies derived from direct total energy differences with respect to the ground state are listed in Table 3, where we include UHF values and those derived from post-Hartree–Fock corrected energies. We have also obtained converged e_4 solutions for a number of these excited states for the F alignment. As found previously for (NiO)$_{100}^1$,[27] the differences between e_1 and e_4 energies are substantially less than 0.1 eV, so that we have not attempted to obtain excitation energies extrapolated to zero concentration. We have calculated excitation energies of the spin-allowed transitions for the A_1 alignment of (NiO/MgO)$_{100}^2$ to confirm that they are essentially independent of the magnetic state of the lattice, and excitation energies for F(MgO/NiO/MgO)$_{100}^3$ to examine the effect of nearest neighbour coordination. These, together with results for F(NiO)$_{100}^1$ are collected in Table 4, from which the excitation energies for (NiO)$_{100}^1$ (b), (NiO/MgO)$_{100}^2$ (c) and (MgO/NiO/MgO)$_{100}^3$ (e) are plotted as a function of Ni coordination in Fig. 2.

3c. First ionised state

The removal of an electron from fully symmetric NiO/MgO multilayers in whatever spin alignment leads to conducting states of essentially $d^8\underline{L}$ character, exactly as suggested by the ground state DOS, where the unpaired electron/hole is delocalised over the O sites, no matter what starting electronic configuration is chosen. As reported previously for NiO bulk[24,25] and (100) monolayer,[27] the removal of this symmetry constraint allows the electronic configuration to relax to non-degenerate *insulating* states of *lower* energy in which the unpaired electron/hole is localised in a p_π orbital at a *single* O site. We have obtained converged solutions for states in which the hole is localised (separately) in both layers of (NiO/MgO)$_{100}^2$, and in different spin configurations. The relative energies of the states are collected in Table 5, in which the nomenclature, XO : S$^+$ and XO : S$^+$(x), correspond to localised holes in the XO layer of (NiO/MgO)$_{100}^2$ where S is the spin configuration of the NiO layer and x the alignment (f–ferromagnetic, a–antiferromagnetic) of the unpaired p_π electron relative to S. As suggested by the valence band DOS of the ground state (Fig.

Table 3 Comparison of UHF, PBE, PW and P91 d → d excitation energies (eV) in F(NiO/MgO)$_{100}^2$

Excitation	UHF	PBE	PW	P91
$xy \rightarrow z^2$	1.98	1.94	1.95	1.95
$xz \rightarrow z^2$	0.83	0.82	0.82	0.83
$xy \rightarrow x^2 - y^2$	0.83	0.83	0.83	0.83
$xz \rightarrow x^2 - y^2$	1.16	1.17	1.17	1.17
$x^2 - y^2 \rightarrow z^2$ [a]	2.97	2.94	2.93	2.88
$z^2 \rightarrow x^2 - y^2$ [a]	4.24	4.14	4.13	4.08
$xz/yz \rightarrow z^2/x^2 - y^2$	1.87	1.85	1.85	1.85

[a] Spin forbidden.

Table 4 Comparison of d → d excitation energies (eV) in (a) F(NiO)$_{100}^1$ (a_0 = 4.0 Å), (b) F(NiO)$_{100}^1$ (a_0 = 4.2 Å), (c) F(NiO/MgO)$_{100}^2$, (d) A_1(NiO/MgO)$_{100}^2$, and (e) F(MgO/NiO/MgO)$_{100}^3$

Excitation	(a)	(b)	(c)	(d)	(e)
$xy \rightarrow z^2$	1.18	1.27	1.98	1.99	2.66
$xz \rightarrow z^2$	—	0.31	0.83	0.84	—
$xy \rightarrow x^2 - y^2$	1.11	0.85	0.83	0.86	0.81
$xz \rightarrow x^2 - y^2$	—	—	1.16	1.20	—
$x^2 - y^2 \rightarrow z^2$ [a]	1.86	2.24	2.97	—	3.68
$z^2 \rightarrow x^2 - y^2$ [a]	—	5.00	4.24	—	3.54
$xz/yz \rightarrow z^2/x^2 - y^2$	1.81	1.57	1.87	1.90	2.14

[a] Spin forbidden.

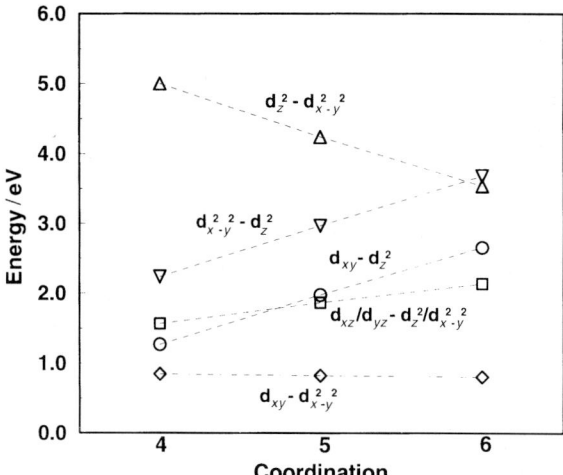

Fig. 2 Comparison of calculated d → d excitation energies in $F(NiO)_{100}^1$, $(NiO/MgO)_{100}^2$ and $(MgO/NiO/MgO)_{100}^3$ as a function of Ni coordination.

1b), the lower energy states correspond to the hole localised in the NiO layer. Mulliken population analyses indicate that in all these states ≈90% of the hole density is localised at a single O site with a moment of ≈0.9 μ_B but with no significant change in local moments at the cation sites. Localisation of a hole at a single O site leads to the creation of a narrow band of unoccupied O(p) states at the top of the valence band, with only minimum changes at the conduction band edge, as shown in Fig. 3 for $A_1^+(NiO/MgO)_{100}^2$, and reported previously for NiO bulk[24,25] and (100) monolayer.[27] This is the exactly equivalent to the changes in the oxygen k-edge spectra obtained originally by Kuiper et al.[17] for $Li_xNi_{1-x}O$, where the energy between the extrinsically controlled unoccupied O(p) states and the conduction band edge approximates the band gap in NiO. We have estimated the band gap, E_g, corresponding to the various holes states given in Table 5 and for the NiO : F$^+$(f) state in $(NiO/MgO/MgO)_{100}^3$ and $(MgO/NiO/MgO)_{100}^3$. The later are given in Table 6 together with that for $(NiO)_{100}^1$, from which the variation of E_g in NiO as a function of Ni coordination can be obtained. This is shown in Fig. 4 where the increase in E_g is seen to be close to linear.

3d. Charge transfer states

In a recent study of NiO(100) monolayer[27] we have calculated the energy of the $p_z \to d_{z^2}$ charge transfer excitonic states as an independent check on estimates of the band gap, E_g, from the unoccupied DOS of the first ionised states. Here we report similar calculations for $(NiO/MgO)_{100}^2$. As before,[27] we have confined our attention to the computationally more convenient F alignment and obtained variationally minimised solutions for the e_4 $p_z \to d_{z^2}$ charge transfer state, (Ni^+O^-),

Table 5 Relative energies, ΔE (eV), and associated gaps E_g (eV) of the first ionised states of $(NiO/MgO)_{100}^2$ and V_{Ni} in $F(NiO/MgO)_{100}^2$

State	ΔE	E_g
NiO : F$^+$(f)	0.0	5.7
NiO : A$_1^+$	0.028	5.9
NiO : F$^+$(a)	0.161	6.0
MgO : F$^+$(f)	0.642	6.0
MgO : F$^+$(a)	0.268	6.0
NiO : F$^+$(f) (V$_{Ni}$)	—	3.6

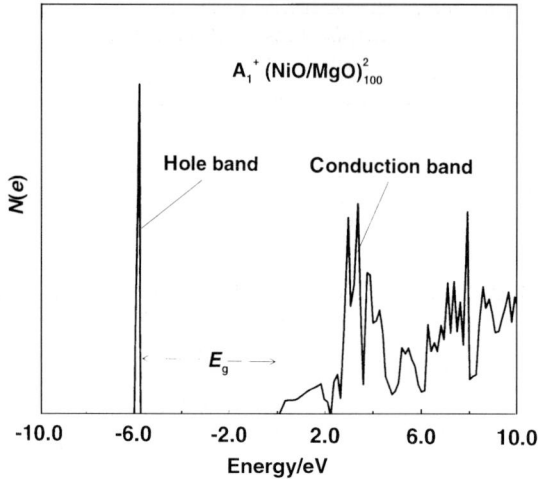

Fig. 3 Empty gap states of $A_1^+(NiO/MgO)_{100}^2$.

in both triplet and singlet spin states, that is to say, configurations of the type,

$$Ni(d^9)_\alpha O(p^5)_\alpha \quad \text{and} \quad Ni(d^9)_\alpha O(p^5)_\beta$$

These can be viewed as fully-condensed, Mott–Wannier (M–W) exciton states, so that in $Ni(d^9)_\alpha O(p^5)_\alpha$ the excited electron is β-spin (which it must be) and the hole β-spin giving a triplet

Table 6 Comparison of E_g (eV) associated with NiO : F$^+$(f) first ionised states in $(NiO)_{100}^1$, $(NiO/MgO)_{100}^2$, $(NiO/MgO/MgO)_{100}^3$ and $(MgO/NiO/MgO)_{100}^3$

System	E_g
$(NiO)_{100}^1$	5.1
$(NiO/MgO)_{100}^2$	5.7
$(NiO/MgO/MgO)_{100}^3$	5.7
$(MgO/NiO/MgO)_{100}^3$	6.6

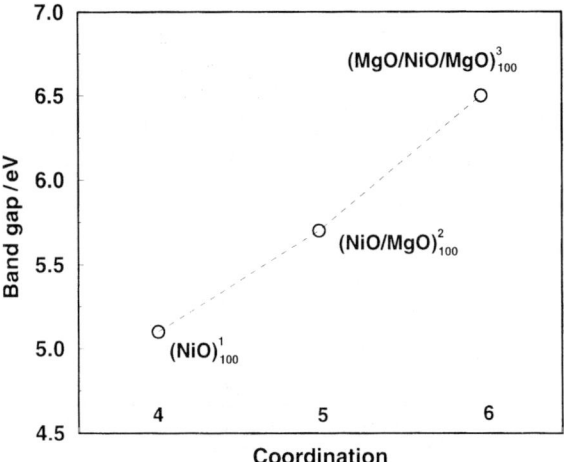

Fig. 4 E_g in $(NiO)_{100}^1$, $(NiO/MgO)_{100}^2$ and $(MgO/NiO/MgO)_{100}^3$ as a function of Ni coordination.

Table 7 Energies (eV/Ni) of the e_4 charge–transfer states of $F(NiO/MgO)^2_{100}$

State	Energy
Triplet $p_\pi \to d_{z^2}$	5.175
Singlet $p_\pi \to d_{z^2}$	5.412
Triplet $p_\pi \to d_{x^2-y^2}$	5.874

exciton. The corresponding formation energies are obtained from *direct energy differences* between the ground and charge transfer states and are *lower bounds* to the optical band gap, thereby providing an alternative estimate of E_g to that from the single-particle eigen values. They are given in Table 7, which shows that the triplet state, as expected, is the lower of the two by ≈ 0.24 eV, which is similar to that found found previously.[27] We have also calculated the first excited configuration of the e_4 triplet excitonic state corresponding to $p_\pi \to d_{x^2-y^2}$ charge transfer, which we find to be ≈ 0.7 eV above the ground state.

3e. Mg substitutional states

In view of the *negative* deviation from ideality of the enthalpy of mixing of (bulk) NiO : MgO,[52,53] low levels of anti-site defects would be expected to occur in ultra-thin films of NiO/MgO, particularly at high temperature, and this is indeed found to be the case. As a prelude to a more extensive investigation of this, we have examined the electronic structure of the $(NiO/MgO)^2_{100}$ bilayer containing 25% Mg_{Ni} substitutional defects in the NiO layer. Mulliken analysis indicates that the substitution of an Mg^{2+} ion for Ni^{2+} in the NiO layer leads to only very minor changes in the electron distribution, with deviations in population of <1% by comparison with the unsubstituted system. Further evidence of the minimal change in electronic structure is contained in the valence band DOS shown in Fig. 5, where it is compared with that of the non-defective $(NiO/MgO)^2_{100}$ bilayer. From this it is evident that while there are small deviations in the DOS, which might well be accounted for by the change in symmetry of the substituted system and consequent modification of the k-space sampling, it remains largely unchanged. This accords well with previous UHF calculations for substitution in the bulk.[21]

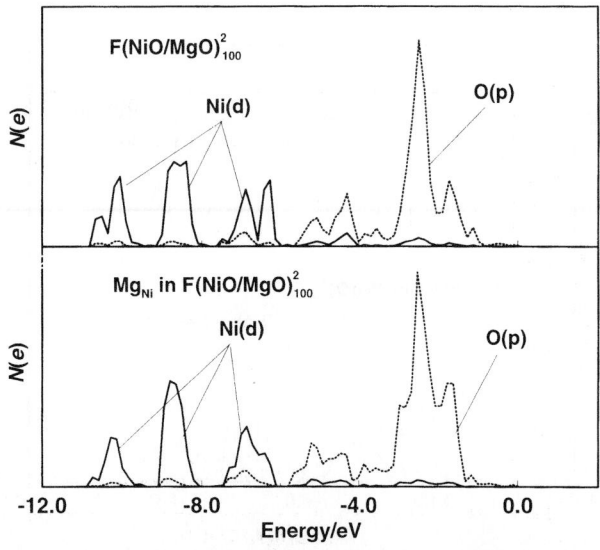

Fig. 5 Comparison of the valence band DOS of $F(NiO/MgO)^2_{100}$ and Mg_{Ni} in $F(NiO/MgO)^2_{100}$.

Table 8 Electron re-distributions (e/atom or e/MgO) associated with $V_{Ni}^{(2-)}$ and $V_{Ni}^{(0)}$ and their differences, δq, in $(NiO/MgO)_{100}^2$

Vacancy	Mg	O_{MgO}	MgO	Ni	O_{NiO}
$V_{Ni}^{(2-)}$	+0.023	+0.044	+0.067	−0.019	−0.048
$V_{Ni}^{(0)}$	+0.020	−0.018	+0.002	+0.031	−0.911; −0.075
δq	−0.003	−0.063	−0.066	+0.050	−0.863; −0.027

3f. Cation vacancy states

The principal oxidative disorder in (bulk) NiO consists of Ni vacancies, V_{Ni}, in different charge states, and (free) holes,[54] with the reasonable likelihood that similar disorder will prevail in NiO/MgO ultra-thin films, principally at high temperature. We have considered two charged states of the Ni vacancy in a $(NiO/MgO)_{100}^2$ bilayer. They are doubly charged, $V_{Ni}^{(2+)}$, which corresponds to the removal of an Ni^{2+} ion, and neutral, $V_{Ni}^{(0)}$, which corresponds to the removal of an Ni^{2+} ion *plus* two electrons. As above, our computational resources have limited us to a concentration of 25% vacancies in the NiO layer. Furthermore, we have not considered any lattice relaxation associated with either defect state. Mulliken analysis of the $V_{Ni}^{(2+)}$ state, which is insulating, indicates that the electron distribution in both layers remains largely unchanged from a non-defective $(NiO/MgO)_{100}^2$ bilayer, with a transfer of $\approx 0.07~e$ from NiO to MgO, as shown in Table 8. The removal of two electrons from this state results in the formation of the neutral, $V_{Ni}^{(0)}$ state, which is also insulating, with the unpaired electrons/holes localised in p_π orbitals at two next-nearest-neighbour O sites adjacent to the vacancy. Once again, Mulliken analyses collected in Table 8 indicate that there is very little re-distribution of electron density between the two layers, and that $\approx 91\%$ of the hole density is localised in the NiO layer with local moments close to 1 μ_B at the two O sites. In both vacancy states there is complete retention of the local Ni moments.

As Figs. 6 and 7 show, despite the minimal re-distribution of electron density within and between the two layers, there are substantial changes in the valence band DOS of both vacancy states. In the doubly charged state the Ni sub-bands retain largely the same profile as in the non-defective state, but are shifted to lower energy relative to the top of the valence band by ≈ 1 eV. There is a more drastic re-arrangement of the oxygen states with a substantial drift of states from the main body of the non-defective O(p) band down to ≈ 10 eV below the valence band

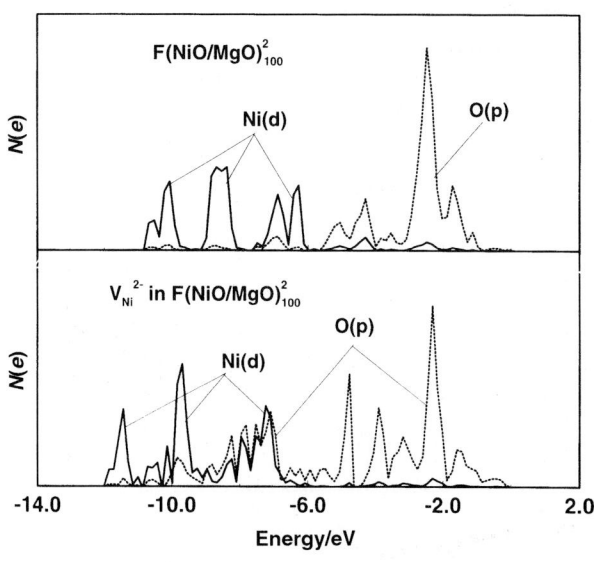

Fig. 6 Comparison of the valence band DOS of $F(NiO/MgO)_{100}^2$ and V_{Ni}^{2-} in $F(NiO/MgO)_{100}^2$.

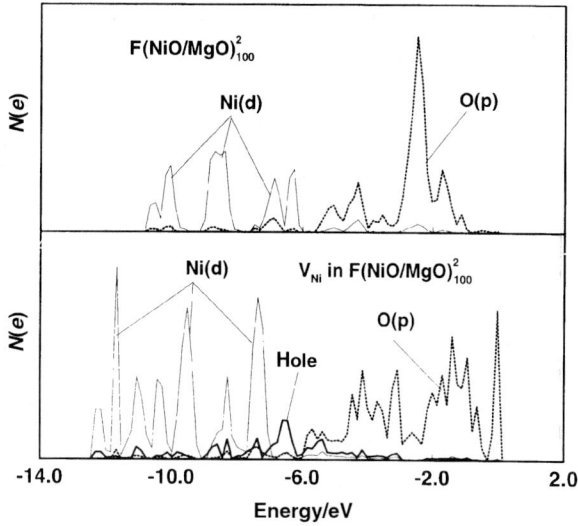

Fig. 7 Comparison of the valence band DOS of $F(NiO/MgO)^2_{100}$ and V_{Ni} in $F(NiO/MgO)^2_{100}$.

upper edge. In the neutral state the re-distribution of states is even more dramatic with the three Ni sub-bands now further split as a result of the reduced symmetry of the neutral vacancy and shifted to lower energy, again by ≈ 1 eV. The profile of the O^{2-} p DOS is changed to a more even distribution of states across the main part of the band width which remains close to 6 eV in width, with the p states associated with the two O^- oxygens, indicated as 'hole' in Fig. 7, shifted to lower energy. As in the case of the free hole, the bound holes of the neutral vacancy give rise to a narrow band of unoccupied states below the conduction band edge. However, as shown in Fig. 8, these states are shifted to higher energy by ≈ 2 eV compared with the free-hole state, leading to a gap of ≈ 3.6 eV between the hole and conduction bands.

3g. {100} Surface free energies

Fig. 9 shows the dynamically relaxed {100} surface free energy of MgO at 700 K as a function of slab thickness, which indicates that approximately ten layers are required for convergence to 0.001

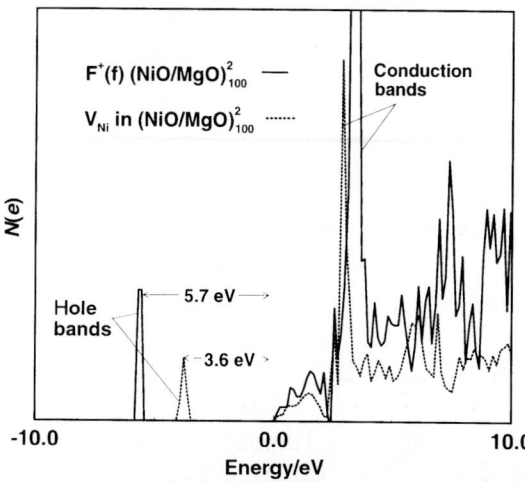

Fig. 8 Comparison of the empty gap states of $F^+(f)(NiO/MgO)^2_{100}$ and V_{Ni} in $F(NiO/MgO)^2_{100}$.

Faraday Discuss., 1999, **114**, 105–127

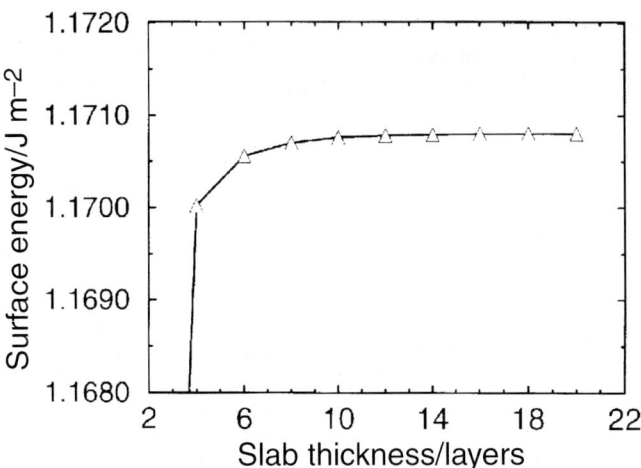

Fig. 9 Variation of the MgO {100} surface free energy at 700 K with slab thickness.

J m^{-2}. This is more than twice the number of layers (4) required to converge the static energy. All the calculations reported here are for slabs comprising twelve layers and 1728 q-vectors used in the Brillouin zone summation.[55] The {100} surface free energy decreases slightly with temperature, by 0.05 J m^{-2} over the temperature range 0–2600 K. Above this temperature, the quasiharmonic approximation breaks down, with the appearance of imaginary frequencies: for the bulk, on the otherhand, imaginary frequencies appear at $\approx$2900 K. Thus the quasiharmonic approximation fails at somewhat lower temperatures for the {100} surface than for the bulk due to the presence of modes with large vibrational amplitudes, which raises the possibility that surface melting occurs at temperatures below that for the bulk.

For NiO our results can be compared directly with those obtained by Mulheran[56] using an Einstein approximation. Fig. 10 shows the calculated temperature dependence of the unrelaxed and fully dynamically relaxed {100} surface energies. Here it is interesting to note that our value of the surface energy in the static limit, 1.18 J m^{-2}, compares favourably with that of 1.23 J m^{-2} from *ab initio* Hartree–Fock calculations.[57] As shown in Fig. 10, both the relaxed and unrelaxed {100} surface energies decrease by $\approx$0.1 J m^{-2} over the temperature range 0–2000 K. The change in the dynamically relaxed surface energy with temperature over this range is only one-third of

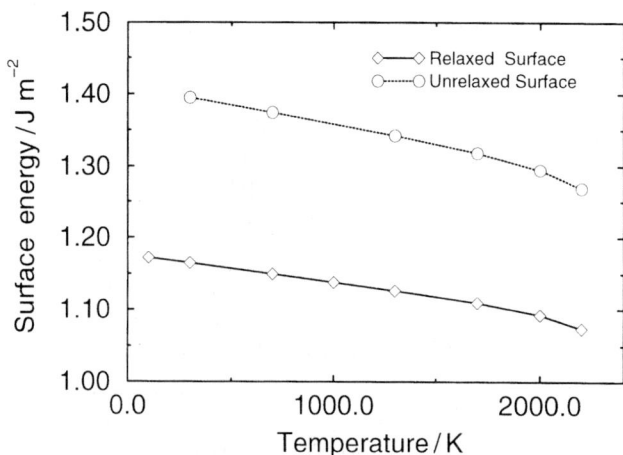

Fig. 10 Calculated temperature variation of the dynamically relaxed and unrelaxed free energy of the {100} surface of NiO.

that reported by Mulheran[56] for the same surface, which indicates of the limitations of the Einstein approximation. There are no experimental data for direct comparison, although the temperature variation appears to be consistent with that noted by Benson and Yun[58] for rocksalt {100} surfaces. The calculated relaxations are also in good agreement with *ab initio* Hartree–Fock calculations.[57] These found a contraction of the first inter-layer spacing of 0.53%, which provides additional support for our use of two-body potentials in free energy simulations. We note, also, that the reported theoretical relaxations are well within the upper bound of $\approx 2\%$ suggested by LEED measurements.[59]

3h. Surface vibrational densities of states

Bulk and surface vibrational densities of states (VDOS) at 300 K for the {100} surface of NiO are shown in Fig. 11, together with the excess VDOS (surface minus bulk) which is responsible for the dynamic contribution to the surface free energies. The form of the bulk VDOS is in good agreement with that reported by Coy *et al.*[60] Once again, there does not appear to be any experimental data for direct comparison. The form of the excess VDOS, with a decrease in intensity at ≈ 6 THz, differs significantly from that presented previously for the {100} surface of MgO, due essentially to the different atomic masses of Ni and Mg.

3i. Free energies of Ni^{2+} segregation

We have calculated heats of segregation, Δh, of Ni^{2+} to the surface of an MgO slab, assuming the formation of ordered, fully relaxed structures. At a given temperature, Δh is obtained from the free energies of fully relaxed twelve-layer slabs with two or four cation surface sites, which may be occupied by Mg or Ni ions, and the free energy of a bulk supercell containing one Ni^{2+} ion, which is an excellent approximation to an isolated impurity ion in the bulk. The top and bottom halves of slabs are the same in any calculation. The heat of segregation is therefore not determined as a continuous function of coverage, but at specific coverages, here 25%, 50%, 75% and 100%. The calculated heats of segregation at a range of temperatures are shown in Fig. 12, where for convenience individual points are connected by straight lines. The negative values of the segregation energy denote the segregation of Ni^{2+} ions from bulk to surface sites. The magnitudes of Δh, ≈ 0.15 eV, reflect the small difference in size between Ni^{2+} and Mg^{2+} ions, and at each temperature the variation of segregation energy with coverage, $\approx 4\%$, is considerably less than that noted previously for larger cations such as Ca^{2+} at the {001} surface of MgO in the static limit.[61] The variation of the segregation energy with temperature over the range studied is somewhat

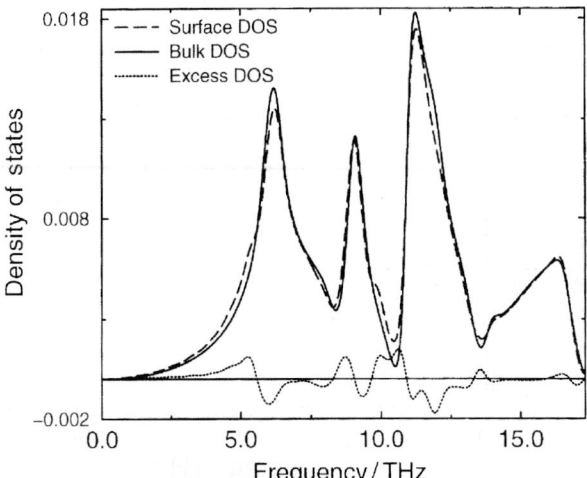

Fig. 11 Calculated bulk, surface and excess (surface minus bulk) vibrational densities of states at 300 K for the {100} surface of NiO.

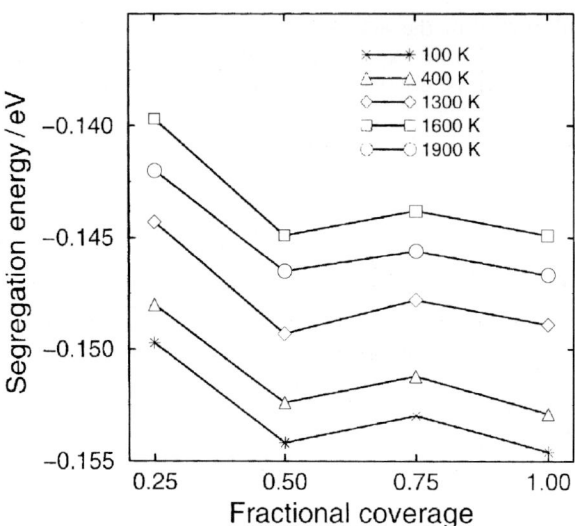

Fig. 12 Calculated heat of segregation, Δh, of Ni^{2+} at the {001} surface of MgO at a range of coverages and temperatures.

greater than the variation with coverage, decreasing, as it does, by $\approx 10\%$ from 100 K to 1900 K. Segregation energies to layers other than the surface layer are found to be negligible. As in the case of the surface energy and lattice relaxation, comparisons with Hartree–Fock calculations are instructive. Differences in energy between the unrelaxed structures $(NiO/MgO/MgO)^3_{100}$ and $(MgO/NiO/MgO)^3_{100}$ range from -0.113 eV, for the A_2 alignment, to -0.103 eV for A_1, which compare with a value of ≈ -0.15 eV deduced for the segregation energy from atomistic simulations.

3j. Vibrational densities of states of defective surfaces

It is straightforward within a lattice dynamics approach to evaluate the vibrational densities of states of defective surfaces at finite temperatures, again, based on fully relaxed structures at the

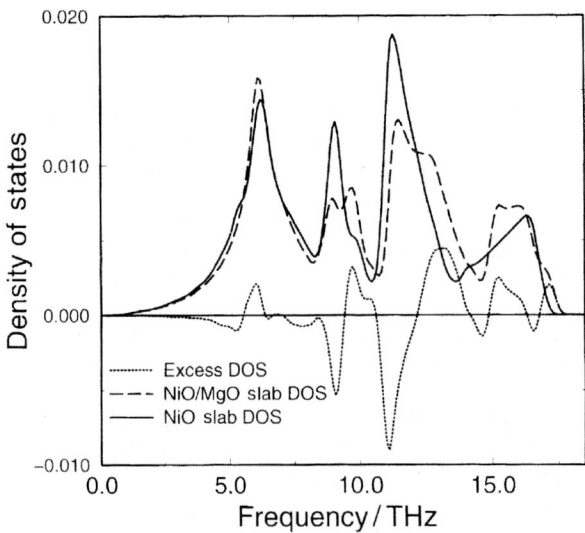

Fig. 13 Vibrational densities of states at 700 K for the {100} surface of NiO with 50% Mg^{2+} coverage. The densities of states of the undefective {100} surface and the excess DOS (defective–undefective) are also plotted.

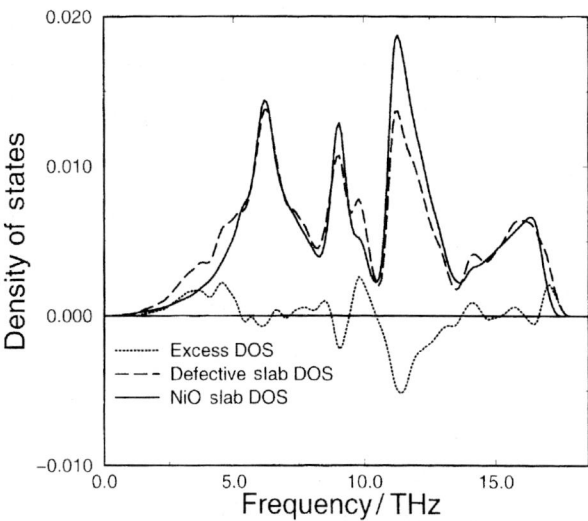

Fig. 14 Vibrational densities of states at 700 K for the {100} surface of NiO with 50% Ni vacancies compensated by oxygen holes. The densities of states of the undefective {100} surface and the excess DOS (defective–undefective) are also plotted.

temperature of interest. Fig. 13 presents two examples, both at 700 K: the first, (Fig. 5) corresponds to a 50% surface coverage of NiO by MgO, the second, (Fig. 6) to a 50% surface coverage by neutral Ni vacancies, $V_{Ni}^{(0)}$, i.e., half the Ni surface sites vacant, with each vacancy charge compensated by two oxygen holes in the surface layer. The difference between the densities of states of non-defective and defective slabs in Fig. 13 indicates the depletion of most low frequency modes (<10 THz) on the introduction of the lighter Mg ions with marked decreases in intensity at ≈9 THz and ≈11 THz. These are accompanied by a pronounced shift of density to higher frequencies. Fig. 14, which admittedly relates to a somewhat unrealistic surface vacancy concentration, shows a pronounced decrease in intensity for modes around 12 THz with increases at <5 THz, ≈10 THz and ≈17 THz.

4. Discussion

4a. Ground state properties

As expected, the ground states of the NiO/MgO multi-layers we have considered here are all ionic, high spin and insulating, with electronic configurations, Mulliken populations and valence band DOS that are very close to those of the bulk[20] and {100} monolayer.[27] While there is some evidence from XPS[5,13] of charge transfer from MgO to NiO for sufficiently thick NiO layers, our results are inconclusive in this respect, as shown in Table 9, due mainly to the limited sizes of slab we have examined. The order of the spin alignments is also identical to that reported for the monolayer. However, there are differences in the direct and superexchange coupling energies within an NiO layer, both of which increase with Ni coordination, as shown in Table 2, leading to values for $(MgO/NiO/MgO)_{100}^3$ that are similar to that for the bulk. We note, in particular, our *decrease* of ≈13% in superexchange coupling from $(MgO/NiO/MgO)_{100}^3$ to $(NiO/MgO)_{100}^3$ and $(NiO/MgO/MgO)_{100}^3$, which compares with a decrease of ≈20% reported by de Graaf et al.[28] based on cluster calculations. The difference in superexchange coupling within (−8.5 meV) and between (−10.5 meV) NiO layers is particularly interesting in relation to the recent report by Alders et al.[5] that somewhere in excess of 20 ML of NiO on {100}MgO are required before the bulk value of the Néel temperature is recovered. As the number of layers increases there is a competition between the A_1 spin alignment of the {100} layered structure and the AF_{II} alignment of the bulk, which the present results suggest might be tilted in favour of the bulk by the interlayer superexchange, since E_{se} for $(MgO/NiO/MgO)_{100}^3$ is calculated to be ≈4% greater than that for the bulk. An important point worth noting here is that our interest here is primarily in the *changes* in the direct and superexchange coupling energies as the local Ni coordination changes and not their absolute values, the reliability of which is the subject of continuing investigation.[62,63]

4b. d → d excitations

Turning now to d → d excitation energies, Table 3 shows that corrections derived from correlation only functionals[39–41] based on Hartree–Fock densities are negligible. We emphasise that this does not indicate that electron correlation is unimportant or that the use of more sophisticated treatments of electron correlation *within the framework of periodic calculations* will come to similar conclusions. A comparison of columns (c) and (d) of Table 4, indicates that the spin alignment also appears to have a negligible effect on d → d excitation energies, which is reasonable, since the differences in energy between the various spin alignments are at least two orders of magnitude less than those between different electron configurations. What Table 4 and Fig. 2 do show quite clearly, however, is that the local Ni coordination can be an important factor, depending on the initial and final states of the excitation. Thus the energy of the $d_{xy} \to d_{x^2-y^2}$ excitation is more or less independent of the local Ni coordination, while that of the two-electron excitation $d_{xz}/d_{yz} \to d_{z^2}/d_{x^2-y^2}$ increases by $\approx 36\%$ from $(NiO)^1_{100}$ to $(MgO/NiO/MgO)^3_{100}$. On the other-hand, the energies of the $d_{xy} \to d_{z^2}$ and spin-forbidden $d_{x^2-y^2} \to d_{z^2}$ excitations increase by $\approx 109\%$ and $\approx 64\%$, respectively, while that of the spin-forbidden $d_{z^2} \to d_{x^2-y^2}$ decreases by $\approx 29\%$. This pattern is readily explained in terms of steric hindrance effects. Since a $d_{xy} \to d_{x^2-y^2}$ excitation involves a transition entirely within a {100} layer, it might reasonably be expected that the presence of planes above and below would influence the energy only to a very minor extent, and this is what UHF calculations predict. On the other hand, the $d_{xy} \to d_{z^2}$ and $d_{x^2-y^2} \to d_{z^2}$ excitations, which increase by 1.39 eV and 1.44 eV, respectively, from 4 to 6 coordination, involve 'out-of-plane' transitions, which might reasonably be expected to be restricted by planes above and below leading to increases in energy, and, again, this is what we find.

While individual points of detail await further clarification, there is general agreement[29,30,51] as to the assignment of the reported optical absorption and EEL spectra below ≈ 3 eV to specific bulk and {100} surface d → d excitations. These can be compared with our values for $(NiO/MgO)^2_{100}$ and $(MgO/NiO/MgO)^2_{100}$, which we take to be representative of the 5-fold coordination of the {100} surface and 6-fold coordination of the bulk. With reference to Table 10, absorptions at ≈ 1.0 eV, ≈ 1.9 eV and ≈ 2.8 eV have been assigned to spin-allowed $^3A_{2g} \to {}^3T_{2g}$ $[(t^6_{2g}e^2_g) \to (t^5_{2g}e^3_g)]$, $^3A_{2g} \to {}^3T_{1g}$ $[(t^6_{2g}e^2_g) \to (t^4_{2g}e^4_g)]$ and $^3A_{2g} \to {}^3T_{1g}$ $[(t^6_{2g}e^2_g) \to (t^5_{2g}e^3_g)]$ excitations in bulk NiO. In our calculations the D_{4h} symmetry of $(MgO/NiO/MgO)^2_{100}$ removes part of the degeneracy of the $^3T_{2g}$ and $^3T_{1g}$ states, from which we obtain energies of 0.81 eV, 2.14 eV and 2.66 eV corresponding to the bulk excitations. We have not obtained a converged solution for the $d_{xz} \to d_{z^2}$ excitation, but, with reference to Fig. 2, note that were this transition to follow the linear relationship between energy and coordination number the energy for the trilayer would be 1.30 eV. These compare with values of 0.79 eV, 1.40 eV and 3.40 eV reported by de Graaf et al.[30] and energies of (0.86–1.04) eV and (1.50–1.81) eV reported by Freitag et al.[29] for the first two excitations, based on cluster calculations which included electron correlation effects.

As shown in Table 10, EEL excitations at ≈ 0.6 eV, ≈ 1.0 eV, ≈ 1.3 eV, ≈ 1.6 eV have been assigned to the {100} surface of NiO,[29,50,51] and that at ≈ 2.2 eV to a possible transition at a {100} NiO/MgO interface.[6,10] These compare with our calculated values for $(NiO/MgO)^2_{100}$ of

Table 9 Comparison of interlayer electron transfer, δq (e/NiO), in various NiO/MgO (100) multi-layers

System	δq
$(NiO/MgO)^2_{100}$	+0.006
$(NiO/MgO/MgO)^3_{100}$	−0.007
$(NiO/MgO/NiO)^3_{100}$	−0.005
$(MgO/NiO/MgO)^3_{100}$	+0.032

Table 10 Comparison of EELS and theoretical d → d transition energies (eV)

Experiment		Theory		
Bulk—				
Transition	Energy/eV	Ref. 29	Ref. 30	Present
$^3A_{2g} \to {}^3T_{2g}$ $(t_{2g}^6 e_g^2) \to (t_{2g}^5 e_g^3)$	1.05,[a] 1.08,[b,c] 1.10,[d] 1.13,[a,e] 1.16,[f] 1.19[g]	0.86–1.04	0.79	0.81
$^3A_{2g} \to {}^3T_{1g}$ $(t_{2g}^6 e_g^2) \to (t_{2g}^5 e_g^4)$	1.79,[a] 1.86,[b] 1.87,[d] 1.95,[a,e]	1.50–1.81	1.40	2.14
$^3A_{2g} \to {}^3T_{1g}$ $(t_{2g}^6 e_g^2) \to (t_{2g}^5 e_g^3)$	2.75,[f] 2.8,[g] ≈ 3[a]	—	3.40	2.66
{100} Surface—				
Transition	Energy/eV	Ref. 29	Ref. 31	Present
$^3B_1 \to {}^3E$	0.57,[d] 0.60[g,h]	0.54–0.65	0.46	0.83
$^3B_1 \to {}^3B_2$	1.0[h]	0.86–1.00	0.83	0.83
$^3B_1 \to {}^3A_2$	1.3[h]	1.11–1.30	1.04	1.16
$^3B_1 \to {}^3E$	1.6,[g] 1.62,[d] 1.63[f]	1.22–1.44	1.17	1.87
$^3B_1 \to {}^3E$	2.18[i]	—	2.55	1.98

[a] Ref. 48. [b] Ref. 47. [c] Optical absorption. [d] Ref. 29. [e] Ref. 46. [f] Ref. 49. [g] Ref. 50. [h] Ref. 51. [i] Refs. 6 and 10.

0.83 eV, 0.83 eV, 1.16 eV, 1.87 eV and 1.98 eV with an average discrepancy of ≈ 0.2 eV. Furthermore, as pointed out by Fromme et al.,[51] the energy of the $d_{xy} \to d_{x^2-y^2}$ excitation would be expected to remain more or less unchanged from 6-fold to 5-fold coordination and that of the $d_{xz} \to d_{z^2}$ excitation to decrease, which, as Table 4 shows, is exactly what we find. Thus, bearing in mind that unrelaxed (NiO/MgO)$_{100}^2$ is only an approximate representation of the non-defective, semi-infinite {100} surface of NiO, the inclusion of which is largely to examine the effects of changing the local Ni coordination, our results would seem to reproduce qualitatively the surface d → d excitation energies. Table 10 also shows that, overall, our results are, at least, comparable to those reported by Freitag et al.[29] and Geleijns et al.,[31] which suggests that for *extended/periodic* systems the neglect of correlation effects beyond Hartree–Fock would seem to be less important than it is for (NiO$_6$)$^{10-}$ and (NiO$_5$)$^{8-}$ clusters.

4c. Hole states and band gaps

In view of its p-type properties, the nature of hole states in NiO has continued to attract considerable attention.[15] Previous UHF studies of the bulk[24,25] and {100} monolayer[27] have provided direct evidence for localised holes with strong O(p) character, in complete agreement with Sawatzky and coworkers.[17–19] They have shown, by analogy with the oxygen k-edge spectra of Li$_x$Ni$_{1-x}$O,[17] that reasonable estimates of the band gap, E_g, in NiO can be obtained from the gap between the polaron band and conduction band edge, as shown in Fig. 3. Here we find that the calculated band gap of the unsupported {100} monolayer, 5.1 eV, which is ≈ 25% greater than that of the bulk, increases in (NiO/MgO)$_{100}^2$ to 5.7–6.0 eV for different hole states. As suggested by the ground state valence band DOS, holes within the NiO layer are more stable than those within MgO by 0.27–0.64 eV, with further differences in energy between different spin configurations of 0.03–0.16 eV for the NiO layer and 0.37 eV for the MgO layer. As in the case of other quantities of interest, we have estimated the variation of E_g with Ni coordination. Fig. 4 shows a near linear increase in E_g from 5.1 eV in (NiO)$_{100}^1$ to 6.6 eV in (MgO/NiO/MgO)$_{100}^3$, where in each case the hole/unpaired electron is in a p$_\pi$ orbital in the NiO plane. This trend indicates that the hole is less stable in (MgO/NiO/MgO)$_{100}^3$ than it is in (NiO)$_{100}^1$, which reflects the change in Madelung potential. Pauli repulsion between the unpaired electron and the neighbouring Mg^{2+} ions might also contribute to this increase.

As in a previous study of the {100} monolayer,[27] we have obtained variationally-converged solutions for $e_4 p_\pi \to d_{z^2}$ charge–transfer states of $F(NiO/MgO)_{100}^2$, from which the corresponding excitation energies are obtained from direct energy differences. The particular significance of the latter within the context of this study is that they are lower bounds to the band gap, and hence provide support, or otherwise, for the values obtained from the single-particle eigenvalues. The formation energy/Ni of the fully condensed triplet excitonic state is ≈ 0.5 to ≈ 0.7 eV less than E_g compared with ≈ 1.7 eV found previously for the monolayer.[27] While there are no experimental values of exciton formation for direct comparison with our values, the well known tail in the optical absorption coefficient of NiO from the onset of absorption at 3.1 eV to the maximum increase in intensity at 4.0 eV[64] might reasonably be explained in terms of bound excitonic states below the conduction band edge. This is supported further by the energy of the first excited excitonic state, ≈ 5.9 eV, which suggests that the first excited state lies very close to the conduction band edge as it does for the F-centre in MgO and CaO.[65]

Structural defects play an important role in controlling the properties of all materials, and here we have considered two classes of defect that might be expected to occur in ultra-thin films of NiO on MgO. They arise from intermixing and non-stoichiometry, both of which are known to be prevalent in the bulk. Despite the fact that intermixing clearly destroys the structural integrity and low-temperature magnetic order of overlayers, it appears to lead to only very minor changes in the electron distribution and densities of states. This in turn suggests that there will be no gross changes in the valence band spectra of layered NiO/MgO systems resulting from interdiffusion. While our results for Ni vacancy states correspond to levels of non-stoichiometry far in excess of that normally encountered and make no allowance for lattice relaxation, there seem to be no obvious reasons why they are inapplicable to lower concentrations. The most significant features of these results are first, that even at a concentration of 25%, both the doubly-charged (V_{Ni}^{2-}) and neutral (V_{Ni}) vacancy states are insulating; second, that in common with the free hole state, there is strong localisation of unoccupied p_π density at two non-adjacent O sites of V_{Ni} with the development of local moments close to 1 μ_B; and third, that in both cases, there is only a minor re-distribution of electrons, or screening, compared with the non-defective lattice, with changes of less than ± 0.05 e in the atomic charges. However, the comparatively minor changes in electron distribution lead to more substantial changes in the valence band densities of states, as shown in Figs. 6 and 7. For both vacancies there is a downward shift of the Ni(d) states relative to the top of the valence band by ≈ 1 eV and, in the case of V_{Ni}, a noticeable broadening and splitting of these states due to a reduction in symmetry. The densities of O(p) states are also broadened, with a significant weight in the region of the Ni(d) levels for V_{Ni}^{2-}. This reduction in energy might, in part, be attributed to a small delocalisation of $O(p_\sigma)$ electrons towards the vacant Ni site which would lower their kinetic energy but leave the Mulliken populations more or less unchanged. By comparison with V_{Ni}^{2-}, the reduced O(p) density of V_{Ni} would be expected to lead to less delocalisation and hence less re-distribution of states to lower energy, other than those specifically associated with the two hole sites which are stabilised by a reduction in the on-site Coulomb repulsion energy, as noted previously for the free hole.[25] In addition to changes in the valance band DOS of V_{Ni}, there is also a significant change to the gap between the polaron band and the conduction band edge which is reduced to ≈ 3.6 eV compared with ≈ 5.7 eV for the free carrier. Since these gaps equate to the E_g of the corresponding holeless states, this indicates that the O(p) states of the doubly charged vacancy state are raised relative to the conduction band edge, exactly as expected from the effective 2- charge of a Ni^{2+} vacancy.

4d. Lattice simulations

The lattice simulations reported here use a recent code[43] based on lattice statics and quasi-harmonic lattice dynamics and designed to minimise the free energy of crystalline slabs which are finite in one direction and infinite in the other two. For simulations involving large unit cells numerical differentiation of the free energy with respect to all the internal coordinates is normally prohibitively expensive, which has led to approaches such as the zero static internal stress approximation (ZSISA) in which the free energy is minimised with respect to the lattice vectors only. In this more recent development the full set of free energy first derivatives is calculated analytically leading to a complete minimisation of the quasiharmonic free energy with respect to all the cell variables. Furthermore, since the sizes of slab used are sufficiently thick to provide a bulk-like region in the interior of the slab and two essentially free surfaces, the simulations reported here represent a radical departure from the two-region strategy used by Tasker[66] and Gay and Rohl,[67] in which the positions (and polarisations) of the ions in the vicinity of the surface *only* are relaxed explicitly by minimising the *internal* energy of the system, while the remainder are constrained to their bulk lattice positions.

Two important points emerge from our surface free energy calculations. The first is the requirement that many more layers are required for convergence than for static simulations. This is due to the contribution to the free energy of long wavelength phonons perpendicular to the slab, particulary at low temperatures. Similar considerations apply to molecular dynamics simulations where the unit cell size prevents the inclusion of small q vibrations. The second is the limitation of the Einstein approximation which, again, severely restricts the range of phonons that contribute to the free energy. Fundamental to all simulations is the quality of the interatomic potentials, particularly where a paucity of experimental data limits the extent to which the validity of the simulations can be examined. Here, support for the potentials used is provided by the close agreement between the static surface energy and lattice relaxation predicted from UHF calculations and lattice simulations. In the absence of data for direct comparison, it is also re-assuring that the temperature dependence of the surface free energy is consistent with that for other rocksalt systems.

With regard to NiO/MgO multilayers, perhaps the most significant of the simulation results concern the free energies of Ni segregation to the {100} surface of MgO, for they appear to confirm the stability of NiO overlayers, even at high temperatures. However, this is clearly a complex issue which involves both thermodynamic and kinetic factors, particulary in view of the reported *negative* deviation from ideality of the enthalpy of mixing of NiO and MgO.[52]

5. Conclusions

The overall conclusion of this paper is that useful information concerning the electronic, magnetic, structural and dynamic properties of NiO monolayers on MgO {100} surfaces can be obtained from a combination first principles unrestricted *periodic* Hartree–Fock calculations and atomistic lattice simulations based on quasi-harmonic lattice dynamics. From an examination of several *semi-infinite* slab configurations, including an unsupported NiO monolayer, an NiO monolayer deposited on MgO {100} layers and an NiO layer sandwiched between two MgO {100} layers, we have been able to obtain the dependence of physical quantities, such as the magnetic exchange coupling constants and the d → d and charge transfer excitation energies, on the Ni coordination number Z, which increases from 4 to 6 in this series. Furthermore, they are all obtained from differences in *total* Hartree–Fock energies and show a remarkably monotonic variation with Z. Further work, currently in progress, compares these results with those of pure bulk NiO and systems with reduced dimensionality.[68] In addition, information of practical importance has been obtained for Ni/Mg substitutional defects and cation vacancies in the NiO layer, both of which might reasonably be expected to occur in ultra-thin films of NiO on MgO as a result of intermixing and non-stoichiometry.

Finally, free energy calculations based on surface lattice dynamics have been used to estimate the surface contribution to the vibrational density of states of NiO {100} and the heat of segregation of Ni^{2+} to the MgO surface as a function of temperature, which appears to confirm the stability of NiO overlayers, even at high temperature. These show that potential-based approaches are capable of providing detailed dynamic and temperature-dependent information, for which, at present, there is little experimental data for comparison. The results reported here also show that the calculation and subsequent minimisation of the free energy *via* quasiharmonic lattice dynamics is sufficiently rapid that an extension to more complex surface structures with more extensive disorder is perfectly feasible.

Acknowledgements

W.C.M. and C.N. wish to thank the British Council and the French Ministere des Affaires Etrangeres for the award of a grant within the Alliance scheme which has facilitated the work reported here.

References

1. H. Kuhlenbeck and H.-J. Freund, in *Growth and Properties of Ultrathin Epitaxial Layers*, ed. D. A. King and D. P. Woodruff, *Chem. Phys. Solid Surf.*, 1977.
2. D. W. Goodman, *J. Vac. Sci. Technol., A*, 1996, **14**, 1526.
3. T. Fujii, D. Alders, F. C. Voogt, T. Hibma, B. T. Thole and G. A. Sawatzky, *Surf. Sci.*, 1996, **366**, 579.
4. R. Hesper, L. H. Tjeng and G. A. Sawatzky, *Europhys. Lett.*, 1997, **40**, 177.
5. D. Alders, L. H. Tjeng, T. Hibma, B. T. Thole, G. A. Sawatzky, C. T. Chen, J. Vogel, M. Sacchi and S. Iacobucci, *Phys. Rev. B: Condens. Matter*, 1998, **57**, 11623.
6. C. Xu, W. S. Oh, Q. Guo and D. W. Goodman, *J. Vac. Sci. Technol., A*, 1996, **14**, 1395.
7. S. Imadduddin and R. J. Lad, *Surf. Sci. Spectra*, 1996, **4**, 194.
8. M. L. Burke and D. W. Goodman, *Surf. Sci.*, 1994, **311**, 17.
9. D. Alders, F. C. Voogt, T. Hibma and G. A. Sawatzky, *Phys. Rev. B: Condens. Matter*, 1996, **54**, 7716.
10. Q. Guo, C. Xu and D. W. Goodman, *Langmuir*, 1998, **14**, 1371.
11. D. M. Lind, S. D. Berry, G. Chern, H. Mathias and L. R. Testardi, *Phys. Rev. B: Condens. Matter.*, 1992, **45**, 1838.
12. S. D. Peacor and T. Hibma, *Surf. Sci.*, 1994, **301**, 11.
13. J. M. Sanz and G. T. Tyuliev, *Surf. Sci.*, 1996, **367**, 196.
14. M. D. Towler, N. M. Harrison and M. I. McCarthy, *Phys. Rev. B: Condens. Matter*, 1995, **52**, 5375.
15. S. Hüfner, *Adv. Phys.*, 1994, **43**, 183.
16. K. Terakura, A. R. Williams, T. Oguchi and J. Kübler, *Phys. Rev. B: Condens. Matter*, 1984, **30**, 4734.
17. P. Kuiper, G. Kruizinga, J. Ghijsen, G. A. Sawatzky and H. Verweij, *Phys. Rev. Lett.*, 1989, **2**, 221.
18. J. van Elp, B. G. Searle, G. A. Sawatzky and M. Sacchi, *Solid State Commun.*, 1991, **80**, 67.
19. J. van Elp, H. Eskes, P. Kuiper and G. A. Sawatzky, *Phys. Rev. B: Condens. Matter*, 1992, **45**, 1612.
20. M. D. Towler, N. L. Allan, N. M. Harrison, V. R. Saunders, W. C. Mackrodt and E. Aprà, *Phys. Rev. B: Condens. Matter*, 1994, **50**, 5041.
21. M. D. Towler, N. L. Allan, N. M. Harrison, V. R. Saunders and W. C. Mackrodt, *J. Phys.: Condens. Matter*, 1995, **7**, 6231.
22. N. M. Harrison, V. R. Saunders, R. Dovesi and W. C. Mackrodt, *Philos. Trans. R. Soc., London, Ser. A*, 1998, **56**, 75.
23. R. Dovesi, V.R. Saunders, C. Roetti, M. Causà, N. M. Harrison, R. Orlando and E. Aprà, *CRYSTAL 95, User Manual*, Università di Torino and CCLRC Daresbury Laboratory, 1995.
24. W. C. Mackrodt, N. M. Harrison, V. R. Saunders, N. L. Allan and M. D. Towler, *Chem. Phys. Lett.*, 1996, **250**, 66.
25. W. C. Mackrodt, *Ber. Bunsen-Ges. Phys. Chem.*, 1997, **101**, 169.
26. C. Springhorn and H. Schmalzried, *Ber. Bunsen-Ges. Phys. Chem.*, 1994, **98**, 746.
27. C. Noguera and W. C. Mackrodt, submitted for publication.
28. C. de Graaf, R. Broer and W. C. Nieuwpoort, *Chem. Phys. Lett.*, 1997, **271**, 372.
29. A. Freitag, V. Staemmler, D. Cappus, C. A. Ventrice, K. Al Shamery, H. Kuhlenbeck and H.-J. Freund, *Chem. Phys. Lett.*, 1993, **210**, 10.
30. C. de Graaf, R. Broer and W. C. Nieuwpoort, *Chem. Phys.*, 1996, **208**, 35.
31. M. Geleijns, C. de Graaf, R. Broer and W. C. Nieuwpoort, *Surf. Sci.*, 1999, **421**, 106.
32. M. B. Taylor, C. E. Sims, G. D. Barrera, N. L. Allan and W. C. Mackrodt, *Phys. Rev. B: Condens. Matter*, 1999, **59**, 6742.
33. D. Fincham, W. C. Mackrodt and P. J. Mitchell, *J. Phys.: Condens. Matter*, 1994, **6**, 393.
34. G. D. Barrera, M. B. Taylor, N. L. Allan, T. H. K. Barron, L. N. Kantorovich and W. C. Mackrodt, *J. Chem. Phys.*, 1997, **107**, 4337.
35. C. Pisani, R. Dovesi and C. Roetti, *Hartree-Fock Ab Initio Treatment of Crystalline Systems*, Springer-Verlag, Berlin, Heidelberg, New York, 1988.
36. J. A. Pople and R. K. Nesbet, *J. Chem. Phys.*, 1954, **22**, 571.
37. R. Dovesi, M. Causà, R. Orlando, C. Roetti and V. R. Saunders, *J. Chem. Phys.*, 1990, **92**, 7402.
38. H. J. Monkhorst and J. D. Pack, *Phys. Rev. B: Condens. Matter*, 1976, **13**, 5188.
39. J. P. Perdew, *Phys. Rev. B: Condens. Matter*, 1986, **33**, 8822; J. P. Perdew, *Phys. Rev. B: Condens. Matter*, 1986, **34**, 7406.
40. J. P. Perdew and Y. Wang, *Phys. Rev. B: Condens. Matter*, 1986, **33**, 8800; J. P. Perdew and Y. Wang, *Phys. Rev. B: Condens. Matter*, 1986, **40**, 3399; J. P. Perdew and Y. Wang, *Phys. Rev. B: Condens. Matter*, 1992, **45**, 13244.
41. J. P. Perdew, K. Burke and M. Ernzerhof, *Phys. Rev. Lett.*, 1996, **77**, 3865; J. P. Perdew, K. Burke and M. Ernzerhof, *Phys. Rev. Lett.*, 1997, **78**, 1396; J. P. Perdew, K. Burke and M. Ernzerhof, *Phys. Rev., Lett.*, 1998, **80**, 891.
42. M. B. Taylor, G. D. Barrera, N. L. Allan and T. H. K. Barron, *Phys. Rev. B: Condens. Matter*, 1997, **56**, 14380.
43. M. B. Taylor, G. D. Barerra, N. L. Allan, T. H. K. Barron and W. C. Mackrodt, *Comput. Phys. Commun.*, 1998, **109**, 135.

44 W. H. Press, S. A. Teukolsky, W. T. Vetterling and B. P. Flannery, *Numerical Recipes in Fortran*, Cambridge University Press, Cambridge, New York, Victoria, 2nd edition, 1992, p. 420.
45 R. S. Mulliken, *J. Chem. Phys.*, 1955, **23**, 1833; R. S. Mulliken, *J. Chem. Phys.*, 1955, **23**, 1841.
46 R. Newman and R. M. Chrenko, *Phys. Rev.*, 1959, **114**, 1507.
47 V. Propach, D. Reinen, H. Drenkhaln and H. Müller Buschbaum, *Z. Naturforsch., B: Anorg. Chem., Org. Chem.*, 1978, **33**, 619.
48 P. A. Cox and A. A. Williams, *Surf. Sci.*, 1985, **152**, 791.
49 S. Hüfner, P. Steiner, F. Reinert, H. Schmitt and P. Sandl, *Z. Phys. B: Condens. Matter*, 1992, **88**, 247.
50 B. Fromme, M. Schmitt, E. Kisker, A. Gorschlüter and H. Merz, *Phys. Rev. B: Condens. Matter*, 1994, **50**, 1874.
51 B. Fromme, M. Möller, Th. Anschütz, C. Bethke and E. Kisker, *Phys. Rev. Lett.*, 1996, **77**, 1548.
52 P. K. Davies and A. Navrotsky, *J. Solid State Chem.*, 1981, **38**, 264.
53 K. D. Heath, W. C. Mackrodt, V. R. Saunders and M. Causà, *J. Mater. Chem.*, 1994, **4**, 825.
54 See, for example, P. A. Cox, *Transition Metal Oxides*, Clarendon Press, Oxford, 1992.
55 It is crucial to use enough q-vectors to achieve convergence. Too small a number of q-vectors leads, in general, to a much smaller decrease in the surface free energy with temperature than is shown by the converged values. A grid of sufficient size must be used in the reciprocal space summation to take adequate account of long-wavelength modes.
56 P. A. Mulheran, *Philos. Mag. A*, 1993, **68**, 799.
57 J. V. Mitchell, *M. Chem. Thesis*, University of St. Andrews, 1995.
58 G. C. Benson and K. S. Yun, *The Solid-Gas Interface*, ed. E. A. Flood, Arnold, London, 1967.
59 M. R. Welton-Cook and M. Prutton, *J. Phys. C: Solid State Phys.*, 1980, **13**, 3993.
60 R. A. Coy, C. W. Tompson and E. Gürmen, *Solid State Commun.*, 1976, **18**, 845.
61 W. C. Mackrodt and P. W. Tasker, *J. Am. Ceram. Soc.*, 1989, **72**, 1576.
62 I. D. Moreira and F. Illas, *Phys. Rev. B: Condens. Matter.*, 1997, **55**, 4129.
63 C. de Graaf, F. Illas, R. Broer and W. C. Niewpoort, *J. Chem. Phys.*, 1997, **106**, 3287.
64 R. J. Powell and W. E. Spicer, *Phys. Rev. B: Condens. Matter*, 1970, **2**, 2182.
65 T. M. Wilson and R. F. Wood, *J. Phys.: Solid State Phys.*, 1976, **7**, 190.
66 P. W. Tasker, in *Computer Simulation of Solids*, ed. C. R. A. Catlow and W. C. Mackrodt, Springer-Verlag, Berlin, Heidelberg, New York, 1982, p. 288.
67 D. H. Gay and A. L. Rohl, *J. Chem. Soc., Faraday Trans.*, 1995, **91**, 925.
68 W. C. Mackrodt and C. Noguera, unpublished results.

Paper 9/04185K

Scanning tunnelling microscopy on the growth and structure of NiO(100) and CoO(100) thin films

Ina Sebastian,[a] Thomas Bertrams,[a] Klaus Meinel[b] and Henning Neddermeyer*[a]

[a] *Martin-Luther-Universität Halle-Wittenberg, Fachbereich Physik, D-06099 Halle, Germany*
[b] *Lettiner Str. 9, D-06120 Halle, Germany*

Received 28th April 1999

We have prepared ordered thin films of NiO and CoO in (100) orientation by evaporating Ni (Co) in an O_2 atmosphere onto Ag(100). The films have been analysed by scanning tunnelling microscopy and low-energy electron diffraction. In the initial stage (coverage up to a few monolayers), growth and structure of the grown films drastically depend on the preparation conditions (in particular, on the temperature of the substrate during deposition and post-annealing). In this case we also observe strong interactions with the substrate. Ag atoms are partially removed from the substrate terraces and form islands or migrate to step edges. No indications for incorporation in the oxide thin films are seen. The oxidic features grow on top of the substrate or in the vacancy islands within the first layer of the substrate left behind by the removed Ag atoms. At low substrate temperatures (near room temperature) an essential part of the oxidic features corresponds to a precursor state rather than to the fully developed (100) oxide film which only develops after post-annealing to higher temperatures (typically around 500 K). I/U characteristics and the sample bias dependency of the contrast of the islands grown have been utilised for identification of whether an oxide reaction had taken place or not. The surfaces of the oxide precursor show a typical defect structure similar to those found on cleaved NiO(100) (M. R. Castell *et al.*, *Phys. Rev. B: Condens. Matter*, 1997, **55**, 7859). This feature shows 'random walk' at room temperature.

I. Introduction

Ordered thin films of oxides are of general importance and not much is known of their formation or of their structural and morphological details. In the case of NiO and CoO, ordered thin oxide films can be grown by oxidation of metal single crystal surfaces (see, for example, ref. 1, where the growth of NiO(100) layers on Ni(100) has been described). The low-energy electron diffraction (LEED) pattern of oxidised Ni(100) surfaces, however, mostly indicates the presence of a rather imperfect NiO(100) film on top of Ni(100),[1] which might result from the large lattice mismatch (of nearly 20%) between Ni and NiO. Although the use of highly oriented Ni(100) as substrate leads to considerable improvement of the LEED pattern from the oxidised surface[2] the large lattice mismatch always remains a problem, with the possible consequence of defect structures in the oxide film.

A different approach to prepare a well ordered NiO(100) film was used by Marre and Neddermeyer[3] who evaporated Ni onto an Ag(100) surface in an O_2 atmosphere. LEED and UV photoelectron spectroscopy have shown that a well ordered and smooth NiO(100) can indeed be obtained in this way. The reason for this more perfect growth of NiO(100) layers by using an

Ag(100) substrate is the smaller lattice mismatch (only 2%) between NiO and Ag. These results were later confirmed by scanning tunnelling microscopy.[4] It has been demonstrated[4] that deposition of Ni onto Ag(100) in an O_2 atmosphere leads to much better NiO(100) films than oxidation of Ni(100), at least for surface orientation, which is accessible by standard Laue techniques.

It is the purpose of the present work to provide some additional details on the growth of NiO(100) films on Ag(100), details that have not been described previously,[4] and to compare the results from the NiO(100) films with those from CoO/Ag(100) prepared and measured in the same way. The lattice mismatch between CoO(100) and Ag(100) is slightly larger (4%) than of NiO(100)/Ag(100) and it might be interesting to look for defect structures associated with the differences of the lattice mismatch. It should be mentioned that Castell et al.[5] have observed, at elevated temperatures (nearly 500 K), atomic defect states on the (100) cleavage planes of bulk NiO. Since similar defect states had already been detected in our previous study of NiO(100) films but had not been described explicitly we have also concentrated on the presence of such defect states for the CoO(100) films. Surprisingly, defect states on the precursor CoO(100) layers, which look similar to those on NiO(100), exhibit a 'random walk' over distances of 1–2 lattice parameters during the observation time (typically 1 min) at room temperature. This implies a particular small activation barrier for surface diffusion of such defects, at least for the oxide precursor.

An additional issue which will be discussed is the possibility of Ag atom removal from the substrate during growth of the oxide layers. In the previous work[4] the large rearrangement of the Ag(100) substrate during post-annealing of the deposited films has been mentioned but without describing it in more detail. For the CoO layers we have observed particularly strong effects for the Ag(100) substrate with direct consequences for the growth mode of the oxide films. Although less drastic, similar effects are present in the NiO/Ag(100) system.

The present work will demonstrate that the growth phenomena and the structural changes of the oxidic films, when changing from a precursor structure to a well ordered (100) oxide layer, are very complicated and are also dependent on the layer thickness. Only for thicker films was a rather uniform appearance of the surfaces detected. In the very thin film limit (1–2 monolayers) the surprising variety of structures and island shapes indicates an unexpectedly large interaction between the growing film and the substrate. We should mention that comparable work has been done for NiO[6] and CoO[7] layers on Au(111).

II. Experimental

The instrumentation used for the preparation and characterisation of the oxidic films has been described in some more details by Berghaus et al.[8] An overview of the apparatus during the measurements of the NiO layers is given in ref. 4. In the meantime we have improved the STM equipment to obtain greater stability and reproducibility for the coarse approach and better positioning of the sample. In addition, a spot profile analysis LEED system (SPALEED) has been added to the system, which allows the measurement of high-resolution LEED patterns of the sample. Such measurements turned out to be important for the analysis of moiré structures that can result from the lattice mismatch of the oxide and the metallic substrate.

The Ag(100) crystals used in the present measurements were cleaned by cycles of ion bombardment (Ar^+) and annealing until LEED showed a satisfactory pattern. In some cases the spots from the substrate were rather broad due to imperfect polishing of the surfaces and the small terrace size. In order to explore the growth mode of NiO and CoO more systematically and to control whether metallic Ni or Co were no longer present within the oxide films, we also studied the condensation of clean Ni and Co layers. It is known from the literature that neither Ni[3] nor Co[9] grows in a layer-by-layer-like growth mode. This has been confirmed in the present experiments which show the growth of three-dimensional clusters of the pure metals in the monolayer regime. The NiO and CoO layers have been grown by deposition of Ni (Co) onto Ag(100) in an O_2 atmosphere of 1×10^{-6} mbar O_2 with the formation of the oxides at a rate of approximately 2/3 (0.2) monolayers (ML) per minute. This means that an O_2 excess is present during the oxide film growth. The influence of sample heating both during and following the deposition was been studied in more detail and showed drastic effects. The thicker films, in particular those obtained after moderate annealing or by deposition on a slightly heated substrate, showed the pattern of the oxides with fourfold symmetry along the surface normal.

The STM data were acquired in the form of constant current topographies (CCTs). To study the electronic effects on oxide formation we systematically varied the sample-bias voltage and its polarity. The stability of the measurements during measuring of the oxidised surfaces was not very good, in general. The state of the tip changed frequently, probably due to transfer of species from the sample to the tip. In some cases the tip changes caused contrast variations of the measured surface structures. By comparison with the results obtained after varying of the sample-bias voltage it turned out that the tip changes could be used for identification of the oxidic structures on the surface. Unfortunately, such tip changes occurred randomly and could not be obtained at will. In order to study electronic effects upon oxide formation more quantitatively we also measured local I/V characteristics. These characteristics showed characteristic differences on the oxidic precursor and the final oxide structures and will be described below for NiO/Ag(100).

III. Results and discussion

During the measurements three effects mainly determined the oxide film growth for both NiO and CoO. Firstly, in the initial stage and for deposition at room (or slightly elevated) temperatures Ag atoms were partially removed from the substrate surface and formed islands with a typical square-like or rectangular shape. The attachment to step edges has not directly been proved but has also to be considered since the surface area of vacancy islands on the substrate was larger than that of the visible Ag islands. Secondly, for room (or slightly elevated) temperatures the condensed material grew in the form of flat islands with a height of one monolayer (ML); this can probably be explained by a kind of oxide precursor. Thirdly, by annealing the oxide precursor a transformation of the islands took place into the (100) oriented oxide film, with a minimum thickness of 2 ML. These observations were made for both materials. Differences exist in the details of the growth process and in the observed microscopic and atomic structure and will subsequently be described for both NiO and CoO.

III.A. NiO(100)

Some results of our experiments on the NiO(100) film growth have already been described in ref. 4 and we will concentrate here on the growth mode of the oxidic precursor, on the defects of the oxide films and on the description of I/U curves measured on the substrate, the precursor and the oxide film.

In Fig. 1 an overview image on the precursor structure is reproduced. As has already been shown and discussed in ref. 4, the CCTs exhibit flat and connected islands on top of the substrate surface with edges mainly along the [110]-like symmetry directions. These directions were determined from atomically resolved STM measurements of the clean Ag(100) surface (not shown here) and from LEED. The growth mode of the features has completely changed in comparison to that of the clean metal. While in the latter case for a coverage of 2/3 ML three-dimensional Ni clusters with a typical lateral extension of 2–3 nm and a height of 0.3–0.5 nm were found, the oxidic precursor grew in the form of two-dimensional islands. These islands show rows of atomically resolved features at a distance of 0.3 nm, which agrees with the Ag lattice parameter. The precursor cannot be explained by an O/Ni adsorption structure which is known to develop either a p(2 × 2) or a c(2 × 2) symmetry.[10] Since neither the observed distances nor the symmetry agrees with either of these adsorption structures we argue that the islands correspond to an oxidic precursor that already contains O ions in sufficient number to form NiO(100) islands, in the approximately correct stoichiometry, after annealing without needing further O_2 exposure. Note that a twofold symmetry has developed on the surfaces of the precursor. In agreement with the fourfold symmetry of the substrate, two orthogonal orientations of the row structures are recognised on the islands.

A more detailed inspection of the CCT reveals that the typical row structure on the surfaces of the islands has also formed within parts of the outermost Ag substrate layer. Its average grey tone level essentially agrees with that of the uncovered part of the Ag substrate. This observation can only be explained if one assumes that the condensation process gives rise to the reactive removal of Ag atoms in the substrate surface layer and that a preferential formation of the precursor is found in the vacancy islands of the substrate. Note that for O/Ni/Ag(100), "empty" vacancy islands are not seen on the surface.

Fig. 1 2/3 ML Ni deposited in an O_2 atmosphere (corresponding to an exposure of 50 L) onto Ag(100) at room temperature. The surface has been measured at $U = -0.5$ V at $I = 2$ nA, the size is 30×30 nm^2. The smooth dark parts of the surface correspond to initial Ag(100) surface. The incorporation of the c(1 × 2) O/Ni precursor structure in the Ag surface is seen on several parts, *e.g.*, at location denoted by "3". The protruding islands have on most parts developed the c(1 × 2) O/Ni precursor structure (*e.g.*, at "2"). The smooth parts on the islands (*e.g.*, at "1" correspond to Ag islands of monatomic height on top of the substrate. They contain the Ag atoms removed from substrate during the deposition. Steps are not available on this part of the surface for the condensation of the migrating Ag atoms.

Annealing the surface produces the final oxide state in the deposit (Fig. 2). The NiO(100) islands correspond to the protruding or depressed structures with rectangular shape in Fig. 2(a) and (b) when measured with larger or smaller sample bias voltage as given in the caption. The surface area covered by these islands is smaller by a factor of two compared to that of the precursor. This means that the minimum thickness of the NiO(100) islands is 2 ML.

Two further points have to be mentioned and discussed. Firstly, the step and terrace structure of the substrate surface after condensation and annealing of NiO is completely different to the clean Ag(100) sample. These large rearrangement effects of the substrate surface have already been described in ref. 4 but can be understood much better now. If one considers the fact that vacancy islands are already formed on the substrate during the initial stage of deposition, then these rearrangements are the consequence of NiO growth on top of the substrate and condensation of the remaining vacancy islands. Note that the NiO islands are always surrounded by Ag, sometimes only in the form of small ridges. The overall morphology of the sample surface is then determined by the position of the NiO islands and by minimising the step length of Ag terraces.

A second point to be mentioned is the observation of a large number of defects on top of both NiO and Ag. These defects have already been mentioned[4] but are represented here again (with improved visibility) in order to compare them with our results on CoO following below.

Since the I/U characteristics are essentially determined by the electronic structure of the sample surface, we measured them on characteristic parts of the surfaces. These measurements frequently gave rise to instabilities in the tunnelling current and could not be used for evaluation of local dI/dU dependences due to the large noise. However, the non-differentiated I/U curves did already show characteristic behaviour on the specific parts of the surface that supported the above assignments of observed features to the metal substrate, Ag islands, oxide precursor and NiO structure.

In Fig. 3(a) the local tunnelling current I is plotted against the sample voltage U while keeping the sample–tip distance constant for smooth parts of the two-dimensional islands (*e.g.*, at location "1") shown in Fig. 1 and for the c(1 × 2) structure (position marked with "2"). The two characteristics are very similar and clearly show the typical behaviour of a metallic state, namely an ohmic dependency at the Fermi level (corresponding to a sample bias $U = 0$ V). For positive values of U the conductivity on top of the smooth parts of the islands (corresponding to Ag) is larger than on the reconstructed parts of the surface. This can probably be explained by the differences in the electronic states of Ag and Ni. The d states of metallic Ni directly cut the Fermi level in contrast

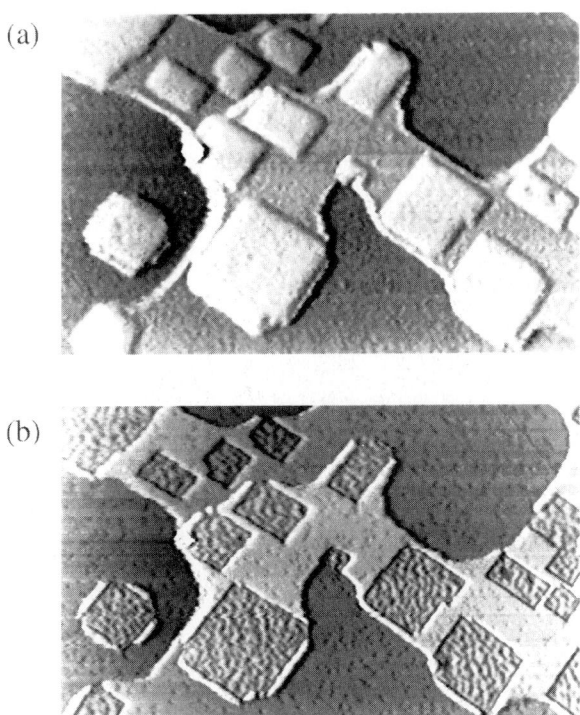

Fig. 2 2/3 ML Ni deposited in an O_2 atmosphere onto Ag(100) at room temperature (the same experiment as shown in Fig. 1) and subsequently annealed at 450 K. (a) $U = 5$ V, $I = 2$ nA, image size $= 50 \times 100$ nm^2. (b) The same part of the surface when measured with small sample bias ($U = 1$ V). The NiO islands now appear as depressions. Note that the surface area covered by the NiO islands is about half of that of the precursor structure in Fig. 1. Individual Ag islands within the NiO double layer structures are no longer detected.

to Ag where only the free-electron sp states may contribute to the tunnelling current. In addition, the presence of the O atoms in the O/Ni structure will also influence the valence and conduction band states. Anyway, the metallic behaviour on the c(1 × 2) ordered parts of the surface indicates that NiO has not yet formed on the surface.

After the O/Ni structure has been transferred into the oxide structure the I/U characteristics (Fig. 3(b)) show drastic differences. For negative and positive sample bias voltages up to about $U = 3$ V the tunnelling current on top of the oxide islands is negligible, while that on the Ag substrate again shows ohmic behaviour. For large positive sample bias voltages (around $U = 4$ V) the tunnelling current on the oxide islands is comparable to that on the substrate. At this voltage the electronic states of NiO are accessible and obviously do contribute to the tunnelling current. As a consequence, the oxide islands are imaged in form of protruding islands.

III.B. CoO(100)

The general growth and formation phenomena observed for the CoO islands were found to be rather similar to that for the NiO films. A closer close inspection of the results revealed a number of differences, which made it worth considering the results in the present work. Some of the results might reflect the differences in the possible oxidation states of Ni and Co, which can lead to the formation of Co_3O_4 and the spinel structure in the latter case.[11] It should be mentioned in this context, however, that for (111) oriented thin films the spinel structure may also develop for NiO.[12] Additional differences might depend on the kinetics of the growth process or on the reactivity during film deposition as will be demonstrated by our experimental results. As in the case of NiO, we consider the results obtained directly after room temperature deposition and after further

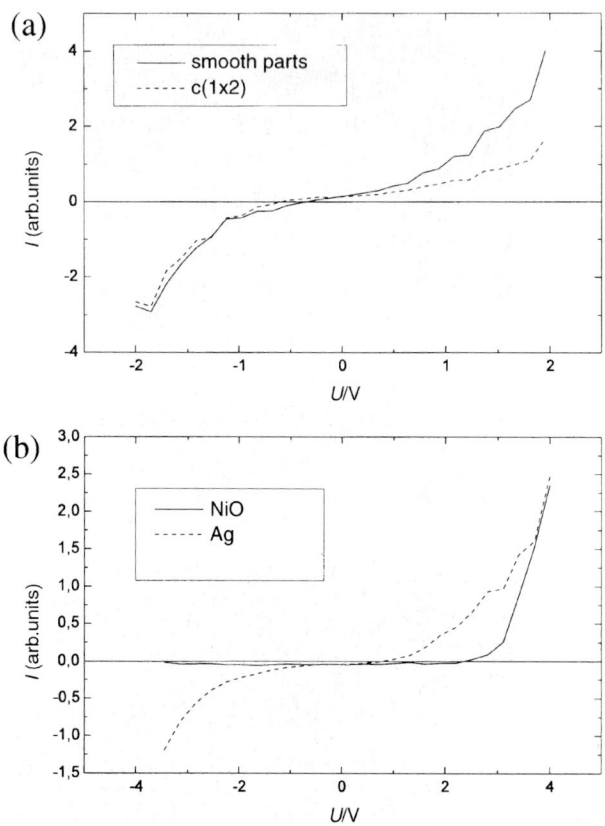

Fig. 3 Local tunnelling current I against the sample bias voltage U on various parts of the samples as reproduced in Figs. 1 and 2. (a) Comparison between Ag islands (*i.e.*, smooth parts of the islands) and the c(1×2) O/Ni precursor. The distance between sample and tip has been set at $U = -2$ V. (b) Comparison between NiO double layer islands and the Ag substrate. The distance between sample and tip has been set at $U = 4$ V.

annealing. Moreover, for the O/Co precursor structure we have observed random walk of the defect states at an atomic level already at room temperature, which might provide evidence for the unstable nature of the CoO precursor. We note that the electronic effects seen for NiO are also present on O/Co/Ag(100). They are even more pronounced than O/Ni/Ag(100) since they are already found in the precursor state, in contrast to the insensitivity of the c(1×2) O/Ni structure against changes of the sample bias. In addition, contrast changes on the O/Co/Ag(100) surfaces due to changes of the tip are also more frequently observed than for O/Ni/Ag(100).

As first results we describe surfaces where a small amount of Co (of the order of 1 ML) was deposited in an O_2 atmosphere onto Ag(100). In Fig. 4(a) the results for deposition at room temperature without further annealing of the sample are shown. Four grey tone levels can clearly be recognised on the individual larger terraces. Note that a step edge of the initial Ag substrate runs from left to right about 1/3 of the image height from the bottom. It might be difficult to recognise the step edge since it no longer follows a straight line but has developed a large number of etched parts and is very rough, therefore. Above the step edge (where the structures appear in a somewhat brighter grey tone level) the four grey tone levels characteristic for one terrace are clearly seen. The darkest features, with a typical lateral extension of 2 nm and mostly of rectangular shape mostly correspond to vacancy islands in the Ag substrate, formed during the deposition process. The Ag atoms removed are mostly condensed in the form of islands with similar lateral extension. They are represented in the brightest grey tone level. The remaining intermediate grey tone levels correspond to the initial substrate level (the darker one) and the brighter ones (whose

Fig. 4 (a) Deposition of 1.2 ML Co in an O_2 atmosphere onto Ag(100). Substrate at room temperature. $U = -2$ V, $I = 0.2$ nA, image size = 100×100 nm^2. (b) Deposition of 1.2 ML Co in an O_2 atmosphere onto Ag(100) at a substrate temperature of 390 K (different experiment). $U = -3$ V, $I = 0.1$ nA, image size = 100×100 nm^2. The nominal thickness of the film should be sufficient in both cases to cover the substrate completely. This is not the case in (a) and (b). We ascribe the observed discrepancy to inaccuracy of the calibration. However, the relative scale of the thickness calibration is the same for all Co deposition experiments.

lateral extension is, by a factor of 2, larger than that of the Ag features) are due to the O/Co precursor. Note that Ag is now found at three atomic levels: the original one, and one deeper and one higher than the initial one. The O/Co precursor which does show distinct electronic effects in its contrast (not reproduced here), essentially occupies one height level on top of the initial substrate. A certain small fraction of the Ag vacancy islands may also contain the O/Co precursor.

If the condensation is performed at a slightly elevated temperature (390 K) the resulting overall surface morphology completely changes (Fig. 4(b)). Ag vacancy islands are no longer found, neither are individual Ag islands seen on top of the substrate terrace due to condensation of the removed Ag islands. The O/Co precursor structure is now seen in two distinct positions (very

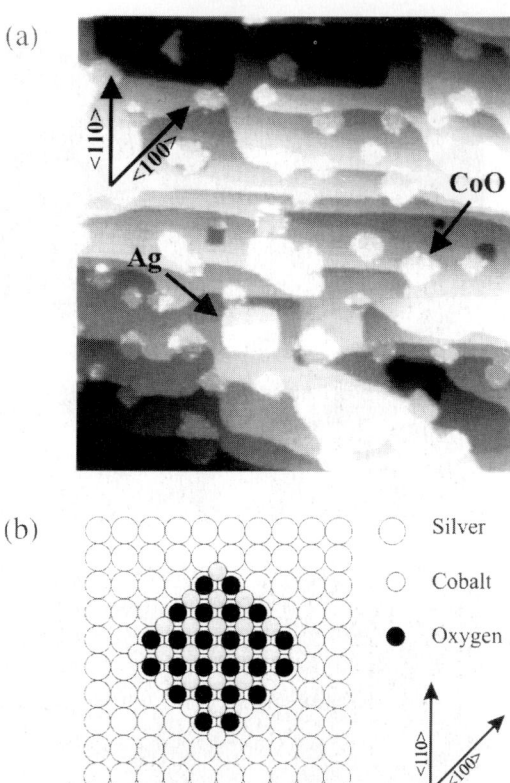

Fig. 5 (a) CoO(100) and Ag islands after deposition of nominally 1.2 ML Co (in O_2) onto Ag(100) at 390 K and subsequent annealing to 470 K. The sample has been measured at $U = -2$ V, $I = 0.3$ nA and a size of 100×100 nm^2. The directions have been determined from atomically resolved measurements. Note that the CoO islands are characterised by a seam of protruding defects on the step edges. (b) Atomic model of CoO(100)/Ag(100) assuming 4-fold hollow sites of the substrate for Co.

similar to O/Ni). One part is condensed in the form of an adlayer (denoted 2 and 3) on the Ag(100) substrate (denoted 1) and the other part is incorporated in the topmost Ag layer of the substrate. Surprisingly and in contrast to O/Ni, the O/Co features develop two kinds of island shape. One is more rounded (*e.g.*, at 2) and the other one is more rectangular (*e.g.*, at 3). The latter islands, whose edges are oriented along [110]-like directions show a row structure of brighter features at a typical distance of a few nm, which means that they are not of atomic origin (it could be a moiré pattern). If we compare the surface area of the O/Co features present on both (a) and (b) we see that it is not too much different, which means that the O/Co precursor state (very similar to O/Ni) is formed by a dense O/Co array with a height of one atomic layer. The differences between 2 and 3 are not yet known precisely but we believe that the more rounded islands correspond to a state that is closer to that of CoO than that of the rectangular islands. For example, the rounded islands sometimes run over the step edges of the substrate without any visible interruption. This behaviour is also found for reacted CoO islands (see below) and can be described by a carpet-like coverage of the surface.

The fact that Ag vacancy islands are no longer seen on the surface (in contrast to the results shown in Fig. 4(a)) is the consequence of the slightly increased substrate temperature during deposition. Obviously, small Ag islands and vacancy structures do not correspond to a thermodynamic equilibrium and are only found for the room temperature experiments due to the kinetic limitations of Ag migration. At slightly elevated temperatures the kinetic limitations can easily be overcome.

If the O/Co islands shown in Fig. 4 are annealed further (470 K was found to be sufficient) they are transferred into the CoO(100) state (Fig. 5(a)). The nominal coverage is the same as in Fig. 4

Fig. 6 (a) Formation of CoO(100) by deposition of 2 ML Co in O_2 atmosphere on Ag(100) at 390 K and subsequent annealing at 470 K. $U = -3.5$ V, $I = 0.1$ nA, the size is 100×100 nm^2. The assignment of the surface either to Ag or CoO is made by the presence of typical defects which could be identified on the two kind of surfaces (in the reproduced image, CoO appears to be somewhat rougher than the Ag metal). (b) CoO(100) layers as prepared by deposition of 10 ML Co in O_2 atmosphere onto Ag(100) at a substrate temperature of 470 K.

(actually, the image was obtained after annealing the layers, as reproduced in Fig. 4(a)). Obviously, the surface is covered by two kind of islands. One kind of island appears with a smooth surface, these islands correspond to Ag formed during the O/Co to CoO transformation process (one is denoted by Ag on the image). The other islands correspond to CoO(100) structures. They exhibit similar electronic effects as described for the NiO(100) islands (not reproduced here). Their shape is essentially quadratic. However, the orientation of the step edges is along the [100]-like directions in contrast to that of the NiO(100) islands. In Fig. 5(b) an atomic model for such islands is shown under the assumption that the Co atoms occupy 4-fold hollow sites of the substrate. It is qualitatively clear that the orientation of the step edges of the observed CoO(100) islands minimises the polar effects of the ionic charges of CoO. It is actually more surprising that the NiO(100)

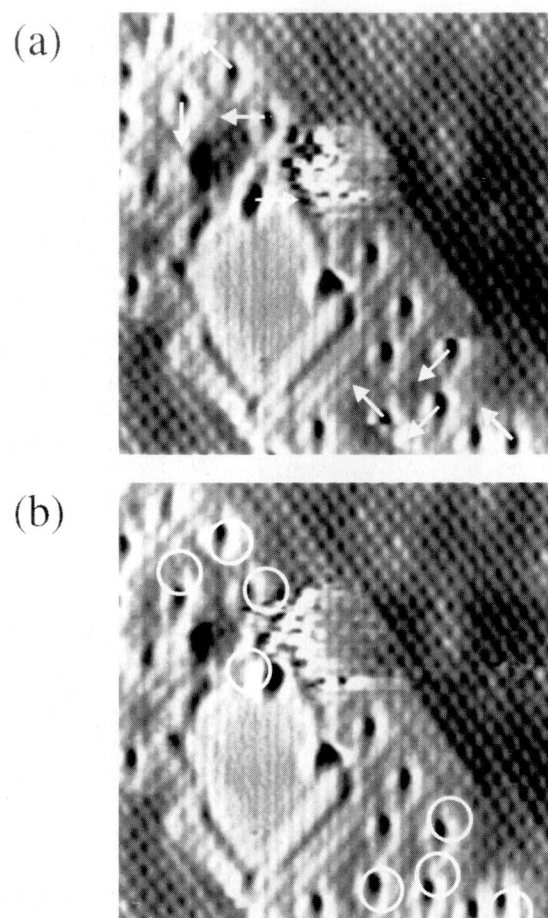

Fig. 7 Random walk to next neighbour and second-next neighbour positions of defects. The images have been obtained on a surface where 2 ML Co has been deposited in O_2 atmosphere onto Ag(100) at 400 K. $U = -2$ V, $I = 0.1$ nA, the image size is 10×10 nm^2. The time difference between measurements of (a) and (b) is approximately 1 min. A movie corresponding to this figure is available as electronic supplementary information. See http://www.rsc.org/suppdata/fd/1999/129

islands are confined by the [110]-like steps. The total surface area covered by the CoO(100) islands is much lower than for the initial film structure. Therefore, the height of the CoO(100) features is at least one double layer.

The carpet-like overgrowth of the substrate by the CoO(100) islands can be seen on some parts of the low-coverage surface displayed in Fig. 5(a). In Fig. 6 the growth mode of CoO(100) can be followed up to Co deposition with a nominal thickness of 10 ML and annealing at 470 K. CoO(100) clearly shows a two-dimensional growth mode, as can be seen in Fig. 6(a). It is evident from the measured image that the step edges always follow [100]-like directions. The observed rounding of the CoO(100) islands may be explained by "freezing-in" of the high-temperature island shape.

As a final example for a grown CoO(100) film the layered structure of 10 ML Co deposited in an O_2 atmosphere onto Ag(100) at 470 K (without further annealing) is reproduced in Fig. 6(b). At this stage, Ag islands are no longer recognised. The growth mode of CoO(100) is nearly of the layer-by-layer type. However, the fact that around 10 uncompleted layers of CoO can be identified within the grown structure strongly indicates a transition to a three-dimensional growth mode.

In the final part of our work on O/Co/Ag(100) we return to the intermediate state of CoO formation as represented in Fig. 4(b). Following the same preparation conditions (deposition at

400 K, no further annealing) we made attempts to improve the resolution in order to understand the differences between the more rounded and the rectangular O/Co precursor structures. It turned out that the atomic defect structures of both kinds of islands were different. While the ordered features seen in Fig. 4(b) on the rectangular islands are extended over a lateral size of typically 5 lattice parameters, the more rounded islands showed atomic-like defect structures (which are not discernible in the overview image). Measuring these structures with better resolution we found that both kind of defects showed random walk at room temperature. Although migration of defects is not an unexpected effect, its observation at room temperature seems to be surprising and further supports our qualitative assignment of the two kind of islands to a precursor state of the oxide. Since the oxide films are readily formed at temperatures of 200 K above room temperature, the mobility of defects and constituents has to increase at a lower temperature and this is apparently the case.

In Fig. 7 two images are reproduced which have been acquired subsequently (on the time scale of 1 min) and where a large number of defects have changed their position. In Fig. 7 the atomically resolved lattice of the O/Co precursors can clearly be identified. The centre part showing no atomic corrugation corresponds to a small Ag island. Attached to this Ag island we find O/Co with a number of local defects (visible as depressions) which correspond to the rounded O/Co islands. The rest of the surface (for example, the parts extending from the left bottom (right top) corner towards the inner part of the measured surface) shows the detailed structure of a rectangular island. Instead of local defects this part shows a kind of modulation on the length scale of several lattice parameters. Both kinds of defect features change their position with time and this can be seen by comparing Fig. 7(a) with Fig. 7(b). In Fig. 7(a) the direction of movement to the resulting positions (as observed in Fig. 7(b)) are indicated by the arrows. The circles in Fig. 7(b) correspond to the initial positions of the defects. A more detailed analysis shows that six defects move to next neighbour and two to second-next-neighbour sites (probably *via* two next neighbour steps). The importance of this observation lies in the fact that the activation barrier for next neighbour hopping obviously is rather small, and that the reason why the transition from the precursor to oxide state takes place at moderate temperatures may be qualitatively understood.

IV. Conclusion

We have demonstrated that by reactive deposition of Ni or Co in an O_2 atmosphere onto Ag(100) either by subsequent annealing (450–500 K) or by condensation of the films on a hot (450–500 K) substrate NiO or CoO films in (100) orientation can be grown. The overall morphology, and the nature of the deposited films and of the substrate drastically depend on the deposition and annealing temperature. For room temperature deposition an O/metal precursor state is identified that appears to have a height of one atomic layer both on top of the substrate and in the vacancy islands of the substrate, which are generated during the deposition process. The oxide film can only be obtained at elevated temperatures and grows with a minimum thickness of a double layer. In the initial stage a two-dimensional growth of the oxide is found which changes to a more three-dimensional growth behaviour for thicker films.

Acknowledgements

This work has been supported by the Deutsche Forschungsgemeinschaft. The initial part of the work (growth of NiO layers) has been financed through the Forschergruppe "Modellkat" at the Ruhr-Universität Bochum.

References

1 M. Bäumer, D. Cappus, H. Kuhlenbeck, H.-J. Freund, G. Wilhelmi, A. Brodde and H. Neddermeyer, *Surf. Sci.* 1991, **253**, 116.
2 D. Cappus, C. Xu, D. Ehrlich, B. Dillmann, C. A. Ventrice, Jr., K. Al Shamery, H. Kuhlenbeck and H.-J. Freund, *Chem. Phys.*, 1993, **177**, 533.
3 K. Marre and H. Neddermeyer, *Surf. Sci.*, 1993, **287/288**, 995.
4 Th. Bertrams and H. Neddermeyer, *J. Vac. Sci. Technol.*, B, 1996, **14**, 1141.

5 M. R. Castell, P. L. Wincott, N. G. Condon, C. Muggelberg, G. Thornton, S. L. Dudarev, A. P. Sutton and G. A. D. Briggs, *Phys. Rev. B: Condens. Matter*, 1997, **55**, 7859.
6 C. A. Ventrice, Jr., H. Hannemann, Th. Bertrams and H. Neddermeyer, *Phys. Rev. B: Condens. Matter*, 1994, **49**, 5773.
7 I. Sebastian, M. Heiler, K. Meinel and H. Neddermeyer, *Appl. Phys. A*, 1998, **66**, S525.
8 Th. Berghaus, A. Brodde, H. Neddermeyer and St. Tosch, *Surf. Sci.*, 1987, **184**, 273.
9 I. Sebastian, M. Heiler and H. Neddermeyer, unpublished results.
10 G. Wilhelmi, A. Brodde, D. Badt, H. Wengelnik and H. Neddermeyer, in *The Structure of Surfaces III*, ed. S. Y. Tong, M. A. Van Hove, X. Xide and K. Takayanagi, Springer, Berlin, 1991, p. 448.
11 P. A. Cox, *Transition Metal Oxides*, Clarendon Press, Oxford, 1995.
12 A. Barbier, personal communication.

Paper 9/03416A

Structure determination of molecular adsorbates on oxide surfaces using scanned-energy mode photoelectron diffraction

M. Polcik,[a] R. Lindsay,†[a] P. Baumgärtel,[a] R. Terborg,[a] O. Schaff,[a] S. Kulkarni,[a] A. M. Bradshaw,[a] R. L. Toomes[b] and D. P. Woodruff*[b]

[a] *Fritz-Haber-Institut der MPG, Faradayweg 4-6, D 14195 Berlin, Germany*
[b] *Physics Department, University of Warwick, Coventry, UK CV4 7AL.*
   E-mail: D.P.Woodruff@warwick.ac.uk

Received 19th March 1999

Using N 1s scanned-energy mode photoelectron diffraction (PhD) combined with full multiple scattering simulations the local adsorption site of NO on NiO(100) has been determined. The molecule bonds through the N atom atop a surface layer N atom, while the N–O axis is tilted away from the surface normal by $59(+31/-17)°$. The special problems presented by adsorbates on compound surfaces, for the direct inversion of PhD data to provide a first-order site determination, are discussed and some alternative schemes tested.

1. Introduction

Despite the considerable growth in the body of quantitative data concerning the structure of surfaces, and adsorbates on them, during the last 20 years or so, the great majority of these data are concerned with metal or semiconductor surfaces. By contrast, there are very few such data for oxide surfaces; indeed, the most recent (1996) edition of the NIST Surface Structure Database[1] contains no entries for adsorbates on oxides (although there are some more recent examples of such results including some on molecular adsorbates[2,3]). One of the reasons for this, of course, is the perceived difficulty in preparing well-characterised oxide surfaces, which has held back surface science studies of oxides in general. Another is the problem of charging for insulating oxides, which makes the use of electron diffraction difficult. It is also not clear to what extent adsorbates on oxides, and especially molecular adsorbates on oxides, commonly involve the long-range ordering required for conventional diffraction studies.

One technique which has proved rather effective in the quantitative determination of adsorbate structures, including molecular adsorbates, on metals and semiconductors, is scanned-energy mode photoelectron diffraction (PhD).[4] In this technique a core level photoemission signal from the adsorbate is measured in a fixed direction as a function of photon (and thus photoelectron) energy. Components of the emitted photoelectron wavefield are elastically scattered by surrounding atoms (notably including backscattering from substrate atoms) and interfere with the directly emitted component of the same wavefield, leading to modulations in the measured intensity as the photoelectron wavelength changes and scattering pathways fall in and out of phase with the

† Present address: Chemistry Department, University of Manchester, Manchester, UK M13 9PL

directly emitted wave. These modulations thus provide local structural information which does not rely on long-range order, and is element specific. Moreover, by exploiting the 'chemical shifts' in the photoelectron binding energy associated with different local environments of atoms of the same element, chemical-state specific structural information is also obtainable.[5,6] This is potentially of especial interest in the case of oxide surfaces to allow full structural studies of oxygen-containing molecular adsorbates (one demonstration of this idea, albeit based on a very small data set, has recently been published[3]). Of course, this method is also only applicable to conducting samples, but even in the case of insulating oxides, oxide surface studies are possible by using thin epitaxial oxide films on conducting metallic substrates, as has proved useful in the application of other forms of electron spectroscopy.

In this paper we present, in detail, our first results of the application of our PhD methodology to determine the structure of a molecular adsorbate on an oxide surface, namely, NO adsorbed on a thin NiO(100) film grown epitaxially on Ni(100). In addition to presenting the detailed analysis of this problem (a brief report of which will appear elsewhere[7]), we also highlight some of the special problems presented by oxide surfaces to this methodology, and briefly review the prospects for further applications in the light of these results.

2. The NiO(100)/NO system and experimental details

The adsorption properties of NO on NiO(100) have been previously characterised rather extensively, notably by Freund and coworkers.[8,9] On the basis of vibrational ('high resolution') electron energy loss spectroscopy (HREELS) and temperature-programmed desorption (TPD) it appears that there is only a single species, NO, on the surface. X-Ray photoelectron spectroscopy, on the other hand, shows two distinct N 1s peaks separated by approximately 5 eV, but it has been proposed that these correspond to excitation to two different final states (strongly screened and weakly screened at lower and higher photoelectron binding energies, respectively) from the same initial state of a single local adsorption geometry. The polarisation-direction dependence of near-edge X-ray absorption fine structure (NEXAFS) indicates that the N–O molecular axis is tilted relative to the surface normal by approximately 45°. Much of this work was conducted on thin films grown epitaxially on Ni(100) to avoid charging problems, and while such films can suffer from a high density of surface defects, it was demonstrated that the NO adsorption behaviour is dominated by ideal (100) terrace sites. The adsorption system has also been studied using total energy cluster calculations; two such independent studies[10,11] both find an energy minimum for a structure involving NO bonding through the N atom atop a surface Ni atom with an N–O molecular axis tilt of approximately 45°. There is, however, no experimental determination of the adsorption site.

Our experiments were performed using NiO(100) films grown *in situ* epitaxially on Ni(100) according to the prescription used in the work of Freund and coworkers.[8,9] The Ni(100) substrate was first prepared by the usual combination of X-ray Laue alignment, spark machining, mechanical and electrochemical polishing and *in situ* cycles of argon ion bombardment and annealing. The cleanness and long-range order of the Ni(100) surface, and of the epitaxial NiO film, were established by soft X-ray photoelectron spectroscopy (including Ni 2p and O 1s spectra) using the incident synchrotron radiation from the HE-TGM (high energy toroidal grating monochromator)[12] on the BESSY electron storage ring in Berlin, and by LEED, respectively. NO exposure was conducted with a sample temperature of approximately 135 K and led to no additional LEED beams, consistent with earlier studies.

N 1s photoemission spectra recorded using the incident synchrotron radiation in the energy range used for the PhD measurements showed that exposure to the photon beam led to a progressive reduction in the emission intensity, attributed to photon-stimulated desorption (although whether this due to direct photoexcitation or involves photoemitted and secondary electrons is unknown); this observation is consistent with previous work using conventional XPS with an Al Kα source.[13] Somewhat surprisingly, however, the N 1s spectra recorded as a function of incident X-ray exposure showed selective attenuation of the higher binding energy state, as shown in Fig. 1. At first sight these spectra suggest that the two well-resolved N 1s peaks are attributable to different species, with only one of them being strongly sensitive to the X-radiation. As we shall show, however, our PhD data show clearly that both components of this initial doublet are associ-

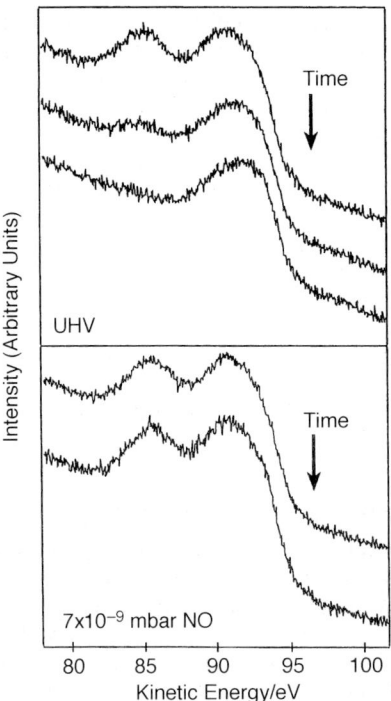

Fig. 1 N 1s photoelectron energy spectra recorded at a nominal photon energy of 500 eV after increasing exposure times; in the upper panel the spectra were recorded under ultra-high vacuum (UHV) conditions, in the lower panel a constant partial pressure of 7×10^{-9} mbar NO was maintained. The (nominal) photoelectron kinetic energy axis is uncalibrated.

ated with the same surface species and that the final single peak is due to an entirely different species, which has, coincidentally, a similar photoelectron binding energy. Moreover, this time-dependence of the N 1s spectra in the X-ray beam could be overcome by working in a partial pressure of NO of approximately 7×10^{-9} mbar, which led to time-independent spectra (see Fig. 1). Our interpretation of these results is that the incident X-rays stimulate two processes, NO desorption and N–O bond scission, the latter leading to oxygen desorption but leaving chemisorbed atomic N on the surface. The N 1s photoelectron binding energy of this atomic species is not distinguishable in our low-resolution experiments from that of the well-screened final state peak of NO. In the presence of a partial pressure of NO, however, any desorbed NO is replaced by new molecular arrival from the ambient gas but, in addition, chemisorbed atomic N is able to react with NO (either from the gas phase or from the coadsorbed species), presumably to form N_2O, which is desorbed from the surface. The results indicate that the NO partial pressure we have used is sufficient to observe a steady-state N 1s spectrum dominated by adsorbed molecular NO. Notice, incidentally, that this state is governed not only by the relative efficiencies of the NO/N reaction and NO photodissociation processes but also by the relative arrival rates of the incident NO molecules and photons. Indeed, detection of the atomic N signal in photoemission requires that a second photon photoionises the adsorbed N atom resulting from photodissociation by a photon (or the associated secondary electrons) that arrived at this same site earlier, before this fragment is removed by chemical reaction.

In order to determine the kinetic energy dependence of the N 1s photoemission signal, which forms the basis of the PhD modulation spectra, short energy range photoelectron energy distribution curves (EDCs) around the N 1s peak were recorded at steps in the photon energy of 2 eV in an energy range corresponding to photoelectron energies of 80–400 eV. These spectra were recorded using a 150° concentric spherical sector analyser (VG Scientific) fitted with three-channeltron parallel detection and mounted at a fixed angle of 60° to the incident radiation, in the horizontal plane, which also is the plane of polarisation of the synchrotron radiation. The angular acceptance of this analyser was set to 5°. Sets of spectra of this kind were recorded at a number of different polar emission angles in the two principal azimuths, [011] and [010], in order to provide a sufficiently large data set to maximise the probability of establishing a unique structural solution.

In order to obtain the PhD modulation spectra from these raw data, each individual EDC was fitted by a sum of two Gaussian peaks plus Gaussian-broadened steps and an experimental background template that accounts for inelastic and secondary electrons. The integrated intensities of these peaks as a function of photoelectron kinetic energy, $I(E)$, were then normalised to produce the PhD modulation function defined as $\chi(E) = ((I(E) - I_0(E))/I_0(E)$ where $I_0(E)$ is the intensity in the absence of diffraction, which is assumed to be described by a smooth spline passing through the measured $I(E)$ data. It is these modulation spectra which form the basis of the quantitative surface structure determination.

3. NiO(100)/NO surface structure determination: multiple scattering simulations

Our standard methodology for surface structure determination through the technique of PhD consists of a two-stage approach.[14] In the first stage, an approximate adsorption site can often be obtained by the use of some form of direct inversion of the experimental data. This stage is intended to play a similar role to Fourier transform methods (such as the Patterson function) in conventional X-ray diffraction, although there are problems associated with this kind of treatment of low energy electron scattering data due to complex scattering factors and the potential importance of multiple scattering. The specific method we have used with considerable success in the past, especially for high symmetry adsorption sites, is the so-called projection method,[15,16] but this approach is formally only applicable to elemental substrates. For this reason we have omitted this stage in the present analysis, although in the following section we discuss this problem in more detail.

The more quantitative stage in PhD analysis involves performing full multiple scattering simulations[17-20] of the data for possible model structures, and adjusting the model to optimise the fit. In the absence of a direct data inversion, this trial-and-error approach included consideration of a more complete set of distinct structural models than would otherwise prove necessary. In all cases, however, assessment of the quality of the agreement between experiment and theoretical simulations is aided by the use of an objective reliability (R)-factor. An appropriate definition has been given previously,[21] but we note here that the R-factor is normalised such that a value of zero corresponds to perfect agreement and a value of unity to experimental and theoretical curves that are wholly uncorrelated. Using this R-factor it is also possible to define a formal test of the precision associated with the structural parameters values obtained in the best-fit structure by defining a variance in the minimum value of the R-factor, R_{min}.[22]

In view of the fact that a PhD modulation spectrum is characteristic of a local emitter geometry, one important first step in the analysis of the NiO(100)/NO data was to demonstrate that the two distinct N 1s photoelectron peaks do, indeed, correspond to a single adsorption site and thus also to a single species. Fig. 2 provides this evidence by comparing the experimental $\chi(E)$ curves obtained for each of the two spectral components at two different emission directions, normal emission and a polar emission angle of 30° in the [011] azimuth. Notice that since the two modulation functions are plotted as a function of photoelectron kinetic energy, E, at any one such energy the two functions were recorded at different photon energies. The spectra for the two states are clearly closely similar. Indeed, if the R-factor normally used to compare experiment and theory is used to compare the two experimental data sets recorded from the two spectral components the values obtained (0.09 and 0.12 for the 0° and 30° spectra, respectively) are comparable to the very best theory–experiment fits we see in PhD. We therefore conclude that the original assignment of these two components was correct, and that the two spectral components do originate from the same local site. To optimise the signal-to-noise quality of the experimental data for the structural analysis, the sum of the two intensities was used in this stage of the analysis. Note, incidentally, that a similar PhD modulation spectrum recorded in normal emission from the single N 1s peak seen when the NO-covered surface was exposed to the synchrotron radiation beam under UHV

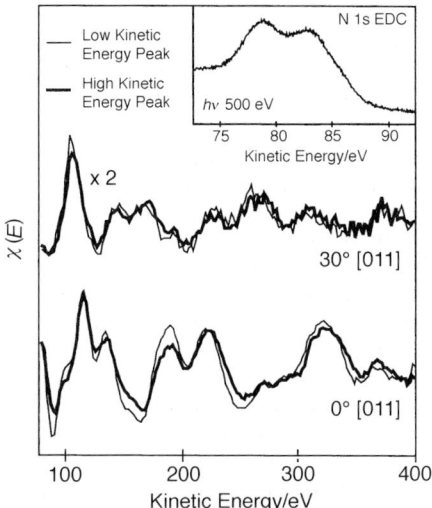

Fig. 2 Comparison of the experimental PhD modulation spectra obtained from the two separate components of the N 1s photoemission spectrum in two different emission directions.

conditions (Fig. 1) did not show these same strong modulations, providing further proof that this peak does correspond to a different surface species.

Although we have not made use of any direct method of data inversion to provide a first estimate of the adsorbate structure, a visual inspection of the raw PhD spectra (Fig. 3) strongly favours an atop adsorption site for the N atom. In particular, we find strong modulations (around ±40%) at normal emission, but the modulation amplitudes fall off rapidly as the emission angle is increased, being less than ±10% at 30° emission in the [001] azimuth. In many previous studies we have shown that PhD typically shows the strongest modulations when a near-neighbour lies directly behind the emitter relative to the detector, placing this neighbour in the favoured 180° scattering geometry (and indeed, this forms the basis of the direct inversion methods[15,16,23–25]). For this reason atop emitter geometries are characterised by strong PhD modulations only close to normal emission.[21,26,27] The strength of these modulations also suggests the near-neighbour backscatterer is Ni rather than O, for which the backscattering cross-section is much smaller.

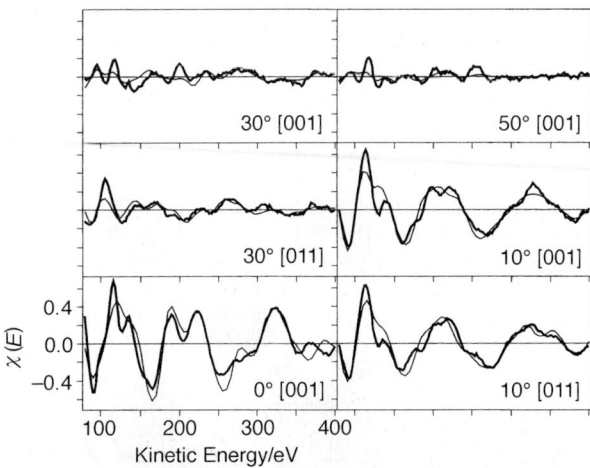

Fig. 3 Comparison of the experimental N 1s PhD modulation spectra (thin lines) for several different emission directions with the results of the best-fit theoretical simulation (bold lines) based on the geometry shown in Fig. 3.

In our search for the optimum structure we have therefore concentrated on the following geometries, all of which involve the N atom being directly above either a Ni or O backscatterer atom: NO bonded through the N atom in sites atop a surface Ni or O atom, and NO bonded through the O atom either atop a surface Ni or O atom, or in bridge or hollow sites. In all cases in these initial searches the N–O axis was assumed to be perpendicular to the surface, but the influence of different N- and O-substrate layer spacings, in increments of 0.05 Å, was explored. Notice that while intramolecular scattering does have some influence on the final PhD spectra, the technique operates under conditions which stress the influence of backscattering, so for the adsorbate atom closest to the surface this is dominated by the substrate scattering and intramolecular scattering influences the results more weakly, mainly through multiple scattering. Of course, if the NO were to be bonded with the O atom down against the surface, intramolecular scattering would include the potentially important N 1s photoelectron backscattering off this O atom. The results of this preliminary search of possible structures strongly favoured the geometry in which the molecule bonds through the N atom which is atop a surface Ni atom; the R-factor for this case was 0.39, whereas the lowest value for any of the other structures was 0.74, vastly greater than this minimum value plus its variance which was found to be 0.07.

Having identified this basic adsorption geometry as favoured, a more detailed optimisation of the associated structural parameters was undertaken with the aid of an adapted Newton–Gauss algorithm to perform automated searches of the multi-parameter space. The specific parameters optimised were the Ni–N and N–O nearest neighbour bond lengths, d_{NiN} and d_{NO}, the tilt angles of these two bonds relative to the surface normal, θ_{NiN} and θ_{NO}, the azimuthal polar angle of these tilts, ϕ, relative to the [011] direction, and the NiO outermost layer spacing, z_{12} (see Fig. 4). Deeper NiO layer spacings were assumed to have the bulk value. In addition, the mean square

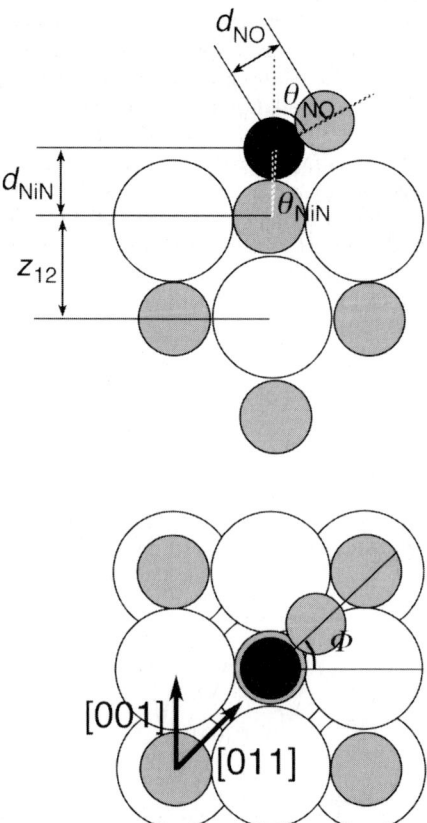

Fig. 4 Schematic diagram of the best-fit geometry for NO adsorbed on NiO(100), showing definitions of the main structural parameters investigated.

Table 1 Best fit structural parameter values for the NiO(100)/NO adsorption geometry (Fig. 3) obtained in this analysis

Parameter	Value[a]
d_{NiN}	1.88 ± 0.02 Å
θ_{NiN}	$3 + 3/-8°$
d_{NO}	$1.12 + */-0.15$ Å
θ_{NO}	$59 + 31/-17°$ *
z_{12}	2.07 ± 0.04 Å
$\langle u_{\parallel}^2 \rangle$	$(1.8 + 4.2/-1.8) \times 10^{-2}$ Å^2
$\langle u_{\perp}^2 \rangle$	$(3.8 \pm 1.9) \times 10^{-3}$ Å^2

[a] For a discussion of the estimated precision (especially in the cases marked *) see the text.

vibrational amplitudes of the N atom parallel and perpendicular to the surface ($\langle u_{\parallel}^2 \rangle$ and $\langle u_{\perp}^2 \rangle$) were optimised to give the lowest R-factor.

The best-fit structure resulting from this optimisation is shown schematically in Fig. 4, while the actual comparison between experiment and the results of the theoretical simulations is included in Fig. 3; the R-factor value of 0.09 is very low, reflecting the excellent fit seen visually. Table 1 lists the structural parameter values associated with this structure, together with estimates of their precision. Notice, in particular, that this best-fit structure clearly does involve a significant tilt of the N–O axis away from the surface normal, consistent with earlier experimental and theoretical determinations of this parameter. As remarked earlier, intramolecular scattering is commonly much less important than substrate scattering in the PhD technique, and in the present case the role of the scattering from the O atom is clearly weak. This leads to the poor precision in the N–O tilt angle (with the near-normal orientation clearly excluded but a significant possibility of very large tilt angles), and also influences the precision of the N–O distance. Indeed, while the optimum value for this distance is 1.12 Å, and values less than 0.97 Å give unacceptable fits, one can increase this bondlength without limit (ultimately therefore removing the O atom completely)

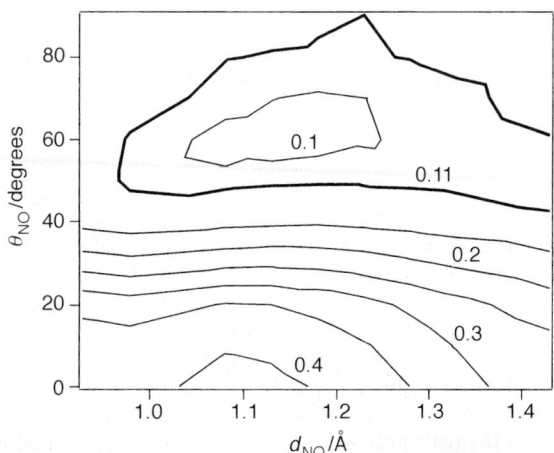

Fig. 5 Contour map showing the variation of the R-factor as a function of the values of the two parameters d_{NO} and θ_{NO}, which define the position of the O atom within the adsorbed NO molecule. The bold contour at a value of 0.11 (corresponding to the sum of the minimum value and its estimated variance) defines the limits of those values estimated to fall within one standard deviation of the best-fit structure.

without increasing the R-factor value above the sum (0.11) of the minimum value (0.09) and its variance (0.02). For this reason we have marked this positive error estimate in Table 1 with an asterisk. Of course, this ambiguity in the N–O bondlength impinges on the error estimate for the N–O bond orientation; if the O atom is removed, this angle is meaningless! Fig. 5 shows a map of the variation of the R-factor as a function of these two parameters, the N–O bondlength and the N–O bond angle, which effectively define the precision with which we can locate the O atom position; the contour corresponding to an R-factor of 0.11 is shown in bold, and as described above all geometries within this contour fall within the estimated limits of our precision. The error for the N–O bond angle quoted in Table 1 is actually a worst-case value for any N–O bondlength of less than 1.43 Å (*i.e.*, falling on the map of Fig. 5); the angular error around the optimum bondlength is significantly smaller ($+22/-8°$). Notice, incidentally, that previous NEXAFS and HREELS studies[8,9] show clearly that the surface species *is* molecular NO, so although the PhD data are relatively insensitive to the presence of the O atom, we do know that this atom *is* present.

One parameter not given in Table 1 is the azimuthal angle, ϕ, of the N–O tilt. In the PhD experiment one necessarily averages over all symmetrically equivalent tilt azimuths, and this is accounted for by a general summing over symmetrically equivalent 'domains', which must be included when any structural parameter lowers the local symmetry below that of the point group of the substrate. The combination of this domain averaging and the insensitivity of the present data to the intramolecular scattering meant that R-factor differences between different azimuthal orientations of the N–O tilt were minimal, and certainly well within the variance, although the actual best-fit structure was one which included additional azimuthal averaging to simulate total disorder of this parameter. The statistics are inadequate to make any clear statement about this aspect of the structure, although we believe that an azimuthally disordered geometry is most probable.

4. Adsorbate structure determination on oxides using PhD: applicability of direct methods

In the previous section we remarked that the present NiO(100)/NO structure determination was performed without using the first stage of our standard methodology, the direct data inversion to provide a first-order structural estimate, because this approach is formally not applicable to a compound substrate. We now consider this problem, and possible ways of circumventing it, in more detail.

The basic objective of the direct data inversion stage is to render more efficient the essentially trial-and-error approach to solving structural problems using time-consuming multiple scattering simulations to search the whole of parameter space. Ideally, one would wish to have a method that provides an accurate three-dimensional 'image' of the atoms surrounding the emitter. Because of the effects of the energy, angle and mass-dependent scattering phase shifts involved in low energy electron scattering from atoms, combined with the complication of multiple scattering, simple Fourier transform methods of data inversion are certainly not capable of yielding quantitatively meaningful results for photoelectron diffraction (or indeed for any other low energy electron scattering method such as LEED). However, the attractiveness of developing a direct inversion method, often referred to as holographic reconstruction, has led to considerable discussion and quite a number of alternative schemes for tackling this problem [see, for example, refs. 28, 29] other than our own.[15,16,24,25] None of these methods claims to be exact, and as such can only provide an approximate structural solution, although regrettably there have been several examples of publications that omit the structural optimisation by quantitative modelling and implicitly claim to have solved structures using only the direct method.[30-32]

The 'projection method', which we have found to be rather successful as a means of providing an approximate 'image' of the real-space structure from PhD data, is based on a relatively minor modification of the Fourier transform approach, although it specifically accounts for the role of the scattering phase shifts (but takes no account of multiple scattering). The basic idea can be best appreciated by considering a simplified formulation of PhD,[4,33] which includes *only* single scattering, and leads to an expression for the modulation function in terms of the electron wavevector,

k as

$$\chi(k) = \sum_j (a_j(\Theta_j, k)/r)\cos(kr_j(1 + \cos \Theta_j) + \delta_j(\Theta_j, k))$$

in which Θ_j is the scattering angle of the jth scatterer at a distance r_j from the emitter. $a_j(\Theta_j, k)$ is an effective scattering amplitude for the jth scatterer, which includes the modulus of the scattering factor and also a polarisation angular term and damping due to inelastic scattering and thermal vibrations, while $\delta_j(\Theta_j, k)$ is the phase of the scattering factor of the jth scatterer. Evidently, in the absence of the extra phase term, $\delta_j(\Theta_j, k)$, a Fourier transform

$$u(r) = \int \chi(k)\exp(ikr)\,dk$$

would yield peaks at the correct scattering pathlength differences, $kr_j(1 + \cos \Theta_j)$ and could therefore be used to invert the data. The presence of the scatterer phase shifts, however, can displace the atomic images by several tenths of an Ångstrom. We note, however, that the essential function of a Fourier transform used in this way is to pick out the periodic components of an oscillatory function, and in the present case we are mainly interested in the *dominant* period, which corresponds to scattering from the nearest neighbour(s). One way of doing this in a way which compensates for the scattering phase shifts is to compute a simple form of the theoretical single scattering modulation function $\chi_{th}(k, r)$ for a scatterer atom at a specific location, r, and to then compute the projection integral of this onto the experimental modulation function $\chi_{ex}(k)$.[15] When the test position corresponds to a real near-neighbour atom position, this integral

$$c(r) = \int \chi_{ex}(k)\chi_{th}(k, r)\,dk$$

should peak. Functions of this type can then be computed for each of a number of experimental modulation functions measured in n different directions, and the results combined according to

$$C(r) = \sum_{i=1}^{n} \exp(Sc_i r)$$

which produces a three-dimensional 'image' of the scatterer positions. The exponential loading is used to pick out only the very strongest features which correspond to the near-180° backscattering near neighbours, r is included to prevent the $(1/r)$ term in $\chi_{th}(k, r)$ causing the function to diverge as $r \to 0$, while the arbitrary scaling factor S can be adjusted to vary the 'contrast' in this image. Evidently this procedure can only be applicable when all the substrate backscatterers are of the same element and have the same $\delta_j(\Theta_j, k)$.

Compound surfaces therefore present a fundamental problem to the correct application of this method. Of course, if the compound comprises elements of very similar atomic number (for example, in GaAs [32]), one may have some success by ignoring this difference, although the resulting image does not then distinguish the different types of scatterer. In the case of oxides, one commonly has elements with very different scattering factors; the scattering factors basically scale with the atomic number although the detailed behaviour is significantly more complex, but oxygen is a rather weak backscatterer while Ni, for example, is relatively strong. The importance of this distinction in the projection method will be discussed further below.

One rather straightforward way in which one might try to gain preliminary information from direct inversion of the data, is first to simply apply the standard projection method assuming first that all the substrate atoms are Ni (*i.e.*, to describe their scattering through the use of Ni scattering phase shifts), and secondly, that they are all O (using O scattering phase shifts). This procedure would give rise to two alternative images of the emitter surroundings; in one of these those features attributable to Ni atoms should be correctly positioned whereas any due to O atoms would be incorrectly represented, while in the other image the roles would be inverted. Of course, the projection method is designed to pick out selectively the strongest backscattering near neighbour, so one possible outcome is that both images will be dominated by the Ni backscatterers, although the true positions of these should be more precisely determined when Ni scatterers are used in the projection integral calculation.

The two panels on the left of Fig. 6 show the results of a test of this idea. The projection method, of course, defines a three-dimensional 'image' in the amplitude of $C(r)$, but it is simplest to show two-dimensional cuts through this function perpendicular or parallel to the surface. In the case of the results presented in Fig. 6, cuts are shown parallel to the surface at a distance below the emitter of 1.9 Å based on inversion of the experimental data using either Ni or O scattering phase shifts and amplitudes; these cuts pass through the maximum intensity for both sets of calculations, and no other features of significant intensity are seen in cuts perpendicular to the surface. The results in both cases show an intense feature directly below the emitter (which is located at (0,0,0)), consistent with the N emitter lying atop a near-neighbour substrate scatterer at a distance of approximately 1.9 Å. The slight splitting of the feature in the Ni scatterer image may indicate slight off-atop positioning, but the significance of this kind of detail in projection maps is marginal. Notice that the use of the O scatterer properties has not significantly displaced the feature, so these maps do not formally distinguish between atop Ni or atop O. Of course, the actual value of $C(r)$ at the peak is much larger in the Ni inversion, because the theoretical modulation amplitudes are larger for Ni scatterers and this leads to larger projection integrals when the correct modulation period and phase is achieved. Potentially, therefore, the projection integral images of Fig. 6 might have been useful. They strongly suggest that the N atom of the adsorbed

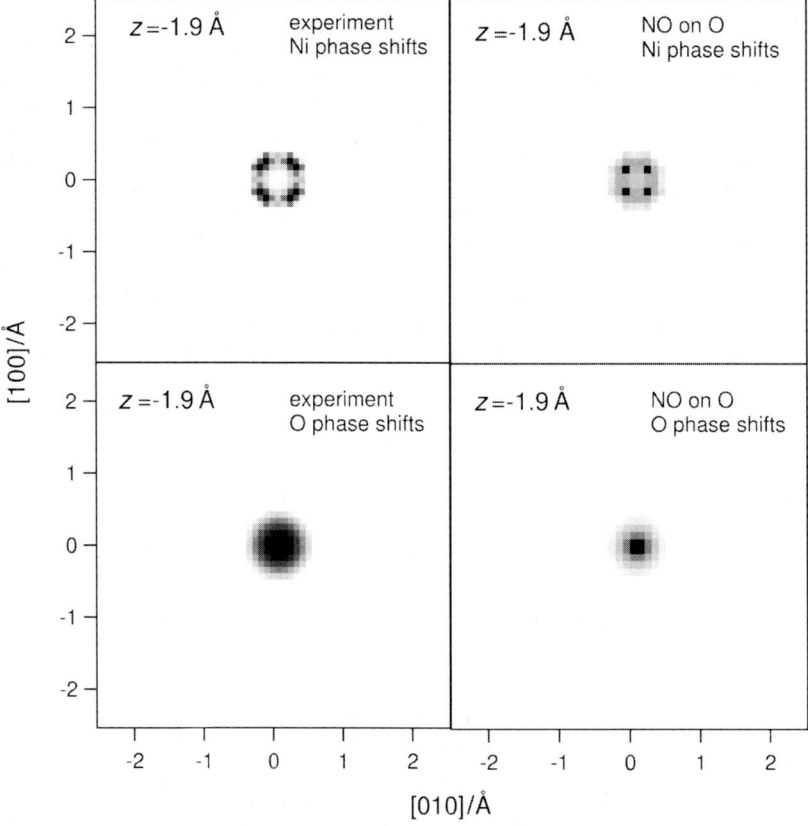

Fig. 6 Standard projection method maps, based on the experimental N 1s PhD data from the system NiO(100)/NO (left-hand panels), and on simulations for NO atop a surface O atom of NiO(100) (right-hand panels) of cuts parallel to the surface but passing through the strongest projection method feature (1.9 Å below the emitter). The inversions have been performed assuming either that all scatterers are Ni atoms or that all scatterers are O atoms.

NO does lie atop a surface layer atom at a distance commensurate with N bonding to the surface, and even if the chemical identity of this atom is unknown, the number of possible structural models is greatly reduced. In reality, of course, we actually drew these conclusions from visual inspection of the raw data.

To explore further the possible utility of this idea, we now extend our consideration to include simulated data for the case of NO adsorbed atop the O atoms of the NiO(100) surface; we might, after all, argue that this test on the experimental data only works well because the nearest neighbour backscatterer to the N emitter in the real surface is a strongly scattering Ni atom. The right-hand panels of Fig. 6 show similar projection method maps applied to these simulated data; in this case too the only strong features found in the projection maps were centred directly below the emitter and the cuts shown in Fig. 6 are again parallel to the surface and centred on these maxima in $C(r)$. Somewhat surprisingly, the dominant features of the two sets of projection maps are very similar; even when the nearest neighbour substrate backscatterer is the more weakly scattering O atom, it is this scatterer which is found in the projection method. Of course, this success in locating (with modest precision) the nearest neighbour substrate atom, even without establishing its chemical identity, would greatly reduce the number of trial structures in a practical application of our structural optimisation. On the other hand, the absence of 'images' of non-nearest neighbour (but more strongly scattering) Ni atoms in the maps obtained from these simulated data is somewhat surprising.

Fig. 7 compares the actual theoretical simulations for the two different adsorption geometries being considered, with the NO molecule placed at the same height above the outermost NiO layer. For the case of NO atop the Ni atom, the theoretical curves are the same as those matching the experimental data; at normal emission a strong short-period modulation is seen associated with the backscattering from the Ni nearest neighbour, although some higher frequency components are also discernible, probably from second layer scattering. By contrast, the equivalent normal emission spectrum from NO atop a surface O atom shows some sign of a similar long period oscillation, but actually appears to be dominated by a period of about half of this value. In NiO(100) a surface O atom lies directly above a Ni atom in the second layer 2.08 Å below, so scattering from this atom (and perhaps others in this layer) is the most probable origin of this contribution; separate calculations performed using only single scattering actually indicate that the shorter period modulations are substantially enhanced by multiple scattering, but a likely event of this kind is forward scattering by the nearest neighbour O atom onto this second layer Ni. Whatever the detailed interpretation of these features is, the very marked difference in these two sets of simulated spectra highlights the value of modelling in distinguishing between different types of structures for a compound surface of this type. Notice, however, that the simulated

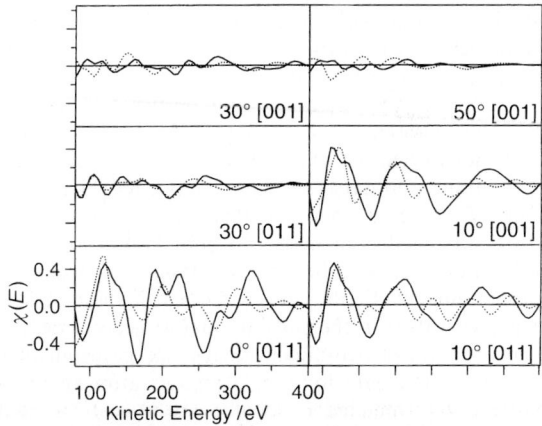

Fig. 7 Comparison of simulated N 1s PhD spectra for adsorption atop Ni (———) and atop O (· · ·) on NiO(100). Apart from the lateral displacement of the emitter all other aspects of the two structure (including their vibrational amplitudes) are identical.

spectra for the NO-atop-O geometry do not show strong modulations at 50° emission in the [001] direction, which should place one of the four top layer Ni nearest neighbour atoms in the favoured 180° scattering geometry. This can largely be attributed to the fact that in conducting these simulations we have used the same large vibrational anisotropy for the N emitter atom as was found for the optimum fit to the experimental data in the atop-Ni site. It is the large vibrational amplitude parallel to the surface which suppresses the importance of scattering from these nearest top layer Ni atoms. If this anisotropy had not been included, the projection method might be expected to distinguish the two atop geometries more clearly, yet our experience is that atop sites are characterised by these large parallel vibrational amplitudes. This detail highlights the difficulty of drawing general conclusions from specific model calculations.

These results do, however, highlight the problem of distinguishing the elemental character of a nearest neighbour backscatterer by the projection method if this scatterer dominates either due to scattering strength or relative vibrational (Debye–Waller) damping. Nevertheless, it is interesting to consider ways in which one might hope to recover information on the location of *both* substrate scatterer species atoms and to distinguish between them. In this regard, the core problem is clearly related to the fact that the stronger scatterer atoms are prone to dominate the actual PhD spectra. However, in the projection method the problem is exacerbated by the fact that the projection integral contains no normalisation to the *intensity* of the theoretical modulations, but simply picks out the common periods and phases of the theoretical and experimental modulations, the intensity of a feature in the transform actually being a product of the modulation intensities of the experimental and theoretical functions at the particular period and phase. This means that if theoretical modulation spectra based on a weak scatterer are inserted into the projection integral, and the experimental data comprise modulations from both weak and strong scatterers, the resulting integral will still peak when the theoretical modulations most closely match the period and phase of the dominant components in the experimental data, which probably arise from the strong scatterers. If it *is* possible to pick out the weak scatterer components, therefore, it is probably important to normalise the projection integral to peak when the amplitude as well as the period and phase of the relevant modulation component match. One way to achieve this is to renormalise the projection integral to

$$c(r) = \frac{\int \chi_{ex}(k) \chi_{th}(k, r) \, dk}{\sqrt{\int \chi_{ex}(k) \chi_{ex}(k) \, dk} \sqrt{\int \chi_{th}(k, r) \chi_{th}(k, r) \, dk}}$$

This normalisation attenuates $c(r)$ when the test modulation spectrum, $\chi_{th}(k, r)$ is too large in amplitude relative to the component of the same periodicity in the experimental modulation function. Of course, this reduces the peak value of the projection integral and potentially greatly lowers the peak intensities, and thus the contrast of the final images, based on $C(r)$, although a modification of the scaling factor S can partially rectify this problem. Nevertheless, it is important to recognise that this change, and indeed its primary objective, shifts away from the original intent of the projection method approach to concentrate only on the strongest features in the PhD data and the location of the near-neighbour scatterers that give rise to them. We may therefore anticipate that some of the spurious artefacts which characterised the early attempts at holographic inversion of photoelectron diffraction data [see, for example, ref. 28] may be encountered. These fears prove well-founded. We have run tests on this modified projection method both on the experimental data and on simulated data based on both NO atop surface Ni atoms and atop surface O atoms. Some features in the resulting images, designed to isolate the Ni and O contributions, do correspond approximately to the true positions of these scatterers, but considerably more spurious features are seen than in the conventional projection method. Fig. 8 provides one example of the results in the form of two-dimensional maps perpendicular to the surface in the [010] azimuth based on simulated data for the N emitter atop an O atom in the outermost NiO(100) surface layer. The map intended to identify the O scatterer is dominated by features centred directly below the emitter but these are split and very spread out. The map based on the Ni scattering similarly shows the dominant features offset as expected for the four-fold coordination relative to these atoms, but here too the locations of these features are not at all sharp or

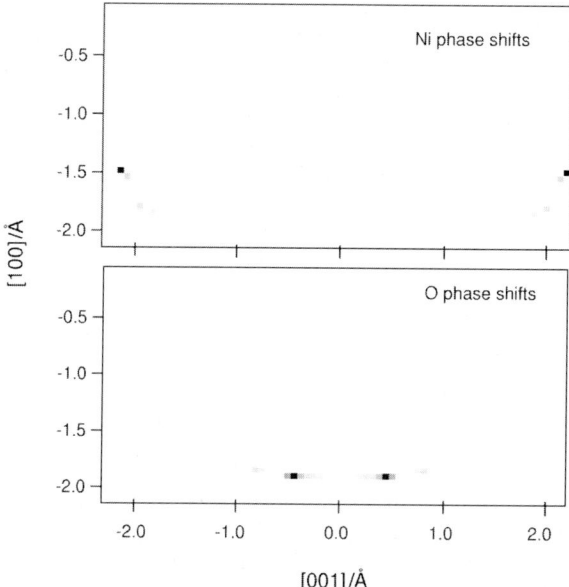

Fig. 8 Modified projection method maps of cuts perpendicular to the NiO(100) surface (with the emitter at (0, 0,0)) based on theoretically simulated data for which the N emitter lies atop a top layer O atom at a layer spacing of 1.88 Å. The upper and lower maps are based on the use of Ni and O scatterers, respectively, in the projection integral. The renormalised projection integral described in the text has been used for these calculations.

well-positioned. These preliminary tests thus suggest that identification of the weaker scatterer locations by this direct inversion approach is far less promising than the rather effective application of the standard projection method to adsorption on elemental surfaces.

5. General discussion and conclusions

Using the scanned-energy mode photoelectron diffraction technique we have determined the local geometry for NO adsorbed on NiO(100); in particular, we find that the molecule bonds through the N atom atop a surface Ni atom, with the N–O axis strongly tilted relative to the surface normal. These findings are consistent with the results of earlier theoretical modelling, and with prior determination of the molecular orientation by NEXAFS, although there was no prior experimental determination of the adsorption site. This is the first complete structure determination of a molecular adsorbate on an oxide surface; an earlier study[3] of the formate species on TiO_2(110) was based on a single low quality PhD spectrum and falls far short of a true experimental adsorbate site determination.

Although this structure determination was conducted exclusively through the use of trial-and-error modelling using multiple scattering calculations, we have also considered the problems and potential of the use of the projection method of direct data inversion as a method of obtaining a first estimate of the structure for subsequent optimisation by full modelling. This approach is formally only applicable to elemental substrates. Attempts to modify the method to locate both the strongly scattering Ni and the more weakly scattering O atoms do appear to achieve this aim, but only at the expense of introducing other spurious features that would render their interpretation ambiguous; this reflects an essential misuse of an approach designed specifically to locate the dominant scatterers. On the other hand, application of the standard projection method to adsorption on NiO does appear to give the approximate location of the nearest neighbour substrate backscatterer atom (but not its elemental identity) and so may be helpful in identifying reasonable starting structures for full multiple scattering analysis. By contrast, at least in the case of NiO in which the two constituent atoms have very different scattering properties, quite simple simulations distinguish the identity of near neighbour scatterers rather readily.

Further studies of other oxide surfaces will be required to establish the generality of the utility of the direct imaging component of the methodology, but it seems clear that the PhD method has considerable potential for the determination of adsorption geometries on these surfaces.

Acknowledgements

This work has been supported by the German Federal Ministry of Education, Science, Research and Technology (contract no. 05 SF8EBA 4), by the Engineering and Physical Science Research Council (UK), and by the European Commission through the Large Scale Facilities component of the HCM programme. The authors thank Volker Fritzsche for providing the multiple scattering computer codes used in this work.

References

1 P. R. Watson, M. A. Van Hove and K. Hermann, *NIST Surface Structure Database, version 2: NIST Standard Reference Database 42*, NIST, Gaithersburg, MD, 1996.
2 D. Ferry, P. N. M. Hoang, J. Suzanne, J.-P. Biberian and M. A. Van Hove, *Phys. Rev. Lett.*, 1997, **78**, 4237.
3 S. A. Chambers, S. Thevuthasian, Y. J. Kim, G. S. Herman, Z. Wang, E. Tober, R. Ynzunza, J. Morais, C. H. F. Peden, K. Ferris and C. S. Fadley, *Chem. Phys. Lett.*, 1997, **267**, 51; S. Thevuthasian, G. S. Herman, Y. J. Kim, S. A. Chambers, C. H. F. Peden, Z. Wang, R. X. Ynzunza, E. D. Tober, J. Morais and C. S. Fadley, *Surf. Sci.*, 1988, **401**, 261.
4 D. P. Woodruff and A. M. Bradshaw, *Rep. Prog. Phys.*, 1994, **57**, 1029.
5 K.-U. Weiss, R. Dippel, K.-M. Schindler, P. Gardner, V. Fritzsche, A. M. Bradshaw, A. L. D. Kilcoyne and D. P. Woodruff, *Phys. Rev. Lett.*, 1992, **69**, 3196.
6 K.-U. Weiss, R. Dippel, K.-M. Schindler, P. Gardner, V. Fritzsche, A. M. Bradshaw, D. P. Woodruff, M. C. Asensio and A. R. González-Elipe, *Phys. Rev. Lett.*, 1993, **71**, 581.
7 R. Lindsay, P. Baumgärtel, R. Terborg, O. Schaff, A. M. Bradshaw and D. P. Woodruff, *Surf. Sci. Lett.*, 1999, **425**, L401.
8 H. Kuhlenbeck, G. Odörfer, R. Jaeger, G. Illing, M. Menges, Th. Mull, H.-J. Freund, M. Pöhlchen, V. Staemmler, S. Witzel, C. Scharfschwerdt, K. Wennemann, T. Liedtke and M. Neumann, *Phys. Rev. B: Condens. Matter*, 1991, **43**, 1969.
9 M. Baümer, D. Cappus, G. Illing, H. Kuhlenbeck and H.-J. Freund, *J. Vac. Sci. Technol. A*, 1992, **10**, 2407.
10 V. Staemmler, in *Adsorption in Ordered Surfaces of Ionic Solids and Thin Films*, ed. H.-J. Freund and E. Umbach, Springer-Verlag, Berlin, 1993, p. 169.
11 L. G. M. Petersson, *Theor. Chim. Acta*, 1994, **87**, 293.
12 E. Dietz, W. Braun, A. M. Bradshaw and R. Johnson, *Nucl. Instrum. Methods Phys. Res., Sect. A*, 1985, **239**, 359.
13 Th. Mull, H. Kuhlenbeck, G. Odörfer, R. Jaeger, C. Xu, B. Baumeister, M. Menges, G. Illing, H.-J. Freund, D. Weide and P. Anderson, in *Desorption Induced by Electronic Transitions, DIET IV*, ed. G. Betz and P. Varga, Springer-Verlag, Berlin, 1990, p. 169.
14 D. P. Woodruff, R. Davis, N. A. Booth, A. M. Bradshaw, C. J. Hirschmugl, K.-M. Schindler, O. Schaff, V. Fernandez, A. Theobald, Ph. Hofmann and V. Fritzsche, *Surf. Sci.*, 1996, **357/358**, 19.
15 Ph. Hofmann and K.-M. Schindler, *Phys. Rev. B: Condens. Matter*, 1993, **47**, 13941.
16 Ph. Hofmann, K.-M. Schindler, S. Bao, A. M. Bradshaw and D. P. Woodruff, *Nature (London)*, 1994, **368**, 131.
17 V. Fritzsche, *Surf. Sci.*, 1989, **213**, 648.
18 V. Fritzsche, *J. Phys.: Condens. Matter*, 1990, **2**, 1413.
19 V. Fritzsche, *Surf. Sci.*, 1992, **265**, 187.
20 V. Fritzsche and J. B. Pendry, *Phys. Rev. B: Condens. Matter*, 1993, **48**, 9054.
21 K.-M. Schindler, V. Fritzache, M. C. Asensio, P. Gardner, D. E. Ricken, A. Robinson, A. M. Bradshaw, D. P. Woodruff, J. C. Conesa and A. R. González-Elipe, *Phys. Rev. B: Condens. Matter*, 1992, **46**, 4836.
22 N. A. Booth, R. Davis, R. Toomes, D. P. Woodruff, C. Hirschmugl, K.-M. Schindler, O. Schaff, V. Fernandez, A. Theobald, Ph. Hofmann, R. Lindsay, T. Gießel, P. Baumgärtel and A. M. Bradshaw, *Surf. Sci.*, 1997, **387**, 152.
23 R. Dippel, D. P. Woodruff, X.-M. Hu, M. C. Asensio, A. W. Robinson, K.-M. Schindler, K.-U. Weiss, P. Gardner and A. M. Bradshaw, *Phys. Rev. Lett.*, 1992, **68**, 1543.
24 V. Fritzsche and D. P. Woodruff, *Phys. Rev. B: Condens. Matter*, 1992, **46**, 16128.

25 K.-M. Schindler, Ph. Hofmann, V. Fritzsche, S. Bao, S. Kulkarni, A. M. Bradshaw and D. P. Woodruff, *Phys. Rev. Lett.*, 1993, **71**, 2054.
26 R. Dippel, K.-U. Weiss, K.-M. Schindler, P. Gardner, V. Fritzsche, A. M. Bradshaw, M. C. Asensio, X.-M. Hu, D. P. Woodruff and A. R. González-Elipe, *Chem. Phys. Lett.*, 1992, **199**, 625.
27 R. Davis, X.-M. Hu, D. P. Woodruff, K.-U. Weiss, R. Dippel, K.-M. Schindler, Ph. Hofmann, V. Fritzsche and A. M. Bradshaw, *Surf. Sci.*, 1994, **307–309**, 632.
28 J. J. Barton, *Phys. Rev. Lett.*, 1988, **61**, 1356.
29 S. Y. Tong, H. Huang and C. M. Wei, *Phys. Rev. B: Condens. Matter*, 1992, **46**, 2452.
30 J. G. Tobin, G. D. Waddill, H. Li and S. Y. Tong, *Phys. Rev. Lett.*, 1993, **70**, 4150.
31 H. Wu, G. J. Lapeyre, H. Huang and S. Y. Tong, *Phys. Rev. Lett.*, 1993, **71**, 251.
32 H. Ascolani, J. Avila, N. Franco and M. C. Asensio, *Phys. Rev. Lett.*, 1997, **78**, 2604.
33 J. J. Barton, S. W. Robey and D. A. Shirley, *Phys. Rev. B: Condens. Matter*, 1986, **34**, 778.

Paper 9/02212K

What can we learn on the structure and morphology of metal oxide/metal interfaces by measurement of X-ray crystal truncation rods *in situ*, during growth

G. Renaud, O. Robach and A. Barbier

CEA-Grenoble, Département de Recherche Fondamentale sur la Matière Condensée / SP2M / IRS, 17, rue des Martyrs, 38054 Grenoble cedex 9, France

Received 6th April 1999

The crystal truncation rods (CTRs) of the substrate's surface were measured during the very first stages of *in situ* deposition of three fcc metals, Ag, Pd and Ni, on the MgO(001) surface. These interfaces are known to form *via* nucleation, growth and coalescence of islands. We show that quantitative analysis of the interferences between the waves scattered by the substrate and the wave scattered by a fraction of the metal film that is long-range correlated *via* the substrate, allows the determination of the adsorption site, the interfacial distance, parameters that are important for theoretical calculation. Some other parameters of the metal/oxide interface are also deduced, in particular, information concerning the morphology. We show that, in the cases of Pd and Ni, the analysis is rather straightforward because most of the signal arises from a few atomic planes that are lattice-strained by the substrate parallel to the interface. Much more complicated is the case of Ag, which is never fully strained by the substrate, whatever the amount deposited, *i.e.*, the island's size. In the three cases, the epitaxial site is shown to be unique, above the oxygen ions of the MgO(001) surface. The evolutions of the interfacial distance during growth are compared. The results are discussed in view of the similarities and differences between the three systems, especially in view of the strongly differing lattice parameter mismatches and the strength of the metal oxide bound at the interface. General trends on the interfacial structure and morphology are deduced.

I. Introduction

Oxide surfaces[1] and metal/oxide[2,3] interfaces are present in numerous technological areas, such as thin films, composite materials, microelectronics, catalysis, and protection against corrosion or industrial glasses. The thermal, mechanical, chemical, magnetic or electrical properties of these materials often depend on the atomic structure of the internal interface they contain. From a theoretical point of view, the properties of metal/oxide interfaces are difficult to predict because the interaction is very complex at the atomic scale.[4-6] The interfacial energy contains several terms[7] that are of the same order of magnitude. Their relative weights are difficult to estimate because of the lack of experimental data. The Ag/MgO(001) and Pd/MgO(001) interfaces have been chosen by numerous theoreticians as prototypical metal/oxide systems[8] because they are relatively simple. They have four-fold symmetry, the epitaxy is cube-on-cube,[9-11] and the contribution of epitaxial strains to the interfacial energy is often neglected. For Ag, this approximation is justified because of the moderate lattice parameter mismatch, -2.98%, between fcc Ag and rocksalt MgO, and, for Pd, despite the larger misfit, −7.64%, because numerous experimental studies show that the first Pd monolayer at the interface is lattice-matched with the substrate. In addition, Ag is a noble metal and hence no chemical reaction takes place at the interface. The Pd/MgO(001) interface is also considered as a model system to study the elementary processes in heterogeneous catalysis,[12] which depend on the exact shape and distribution of the metal clusters that are spread over a ceramic substrate. The Ni/MgO(001) interface has the same symmetry, but with a much larger misfit of 16.4%. It has also been the subject of recent theoretical investigations that show that Ni is expected to strongly interact with MgO(001).[13,14] In addition, the Ni/MgO(001) system has interesting magnetic properties since Ni is a ferromagnetic element. Different studies[15-21] showed that substrate defects and film strain have large effects on the magnetic domain structure.

The MgO(001) surface has been chosen because the relaxation is very small, and thus the surface can be considered as a simple truncation of the bulk.

Two important questions for theoreticians are the determination of the adsorption site among the three possible ones: above O ions of the substrate, above Mg ions, or in between, above the "octahedral site", and the determination of the interfacial distance between the last MgO(001) plane and the first plane of the metal.

However, experimental data are scarce because the insulating character of the substrate hampers most usual experimental surface science methods. To our knowledge, although most calculations estimate the interfacial distance, there are only very few experimental determinations of this parameter. For Ag, a high resolution transmission electron microscopy (HRTEM) study concluded to the co-existence of two alternating epitaxial sites, above Mg and O ions. However, many theoretical calculations favour the O site. For Pd, some recent theoretical results concluded that Pd adsorbs on top of the O at the surface,[13,14,23,24] while others[25,26] concluded to adsorption above Mg ions. A SEELFS (surface electron energy-loss fine-structure spectroscopy) investigation[27] concluded to adsorption above the Mg site. For Ni, the O site is predicted.

In the present study, we show that these structural parameters can be determined by quantitative measurements and analysis of the MgO(001) crystal truncation rods (CTRs). In simple cases, such as for Ni and Pd, where part of the metal is lattice-strained by the substrate, the analysis can be straightforward and a good accuracy obtained. In more complicated cases, such as for Ag, for which the metal is not lattice-strained, but at least partially relaxed, we show that, in general, these parameters can also be deduced, but with a lesser accuracy.

Note that, despite the fact that in most theoretical investigations, the contribution of epitaxial strains to the interfacial energy is neglected, the relaxation of the lattice parameter mismatch is believed to yield an important contribution to the interfacial energy.[28] The state of strain in the metal deposit critically depends on the magnitude of the misfit as well as on the strength of the metal–oxide bonding at the interface. It is thus interesting to compare similar systems, such as Ag/MgO, Pd/MgO and Ni/MgO, with very different misfits and strengths of the interfacial bond.

Grazing incidence X-ray scattering (GIXS)[29] is well suited for characterising the structure and morphology of metal/oxide interfaces during their growth because it is insensitive to the insulating character of the substrate and can be used *in situ*, in ultra-high vacuum (UHV), without perturbing the deposit.

This paper presents GIXS results obtained *in situ*, during the room temperature growth of Ag, Pd and Ni on MgO(001), from the very early stages and up to fairly large thickness.

The results concerning the morphology of the deposit and the accommodation of the lattice parameter mismatch at the metal/MgO interface have been described elsewhere for the Ag/MgO(001),[30,31] Pd/MgO(001)[32,33] and Ni/MgO(001)[34] interfaces. We just recall here that in all three cases, the growth is three-dimensional, of Volmer–Weber type, with nucleation, growth and coalescence of islands, followed by the formation of a continuous film. For Pd, most of the deposit is partially relaxed excepted for the first monolayer, which is partially lattice-matched parallel to the substrate. For Ag, all the metal is partially relaxed and the interfacial plane is not lattice-matched to the substrate. For both Ag and Pd, misfit dislocations enter at the edges of the islands around 5–6 equivalent monolayers of metal deposited, and they next reorder into a square network when the film becomes continuous. For Ni, most of the deposit consists of relaxed Ni with cube on cube epitaxy, but some other orientations, with Ni(110)//MgO(001), are also present in much smaller amounts.

The experimental conditions are first described. Then, the results of the measurements along CTRs of the MgO(001) surface during growth are analysed, in order to deduce the epitaxial and interfacial sites, together with the evolution of several other parameters with the deposited amount. These results are discussed in a last part and a detailed comparison is made between the three interfaces. The differences between these three systems are discussed according to the magnitude of the lattice parameter misfit and the strength of the interfacial bonding.

II. Experiments

For Ag and Ni, the GIXS experiments were performed using the SUV (surfaces in ultra-high vacuum) surface diffraction set-up of the BM32 beamline at ESRF (European Synchrotron Radiation Facility, Grenoble, France).[35] For Pd, the w21v surface diffraction set-up[36] of the ID32 undulator beamline was used. The UHV chambers (base pressure 2×10^{-11} mbar for Ag and Ni; 10^{-10} mbar for Pd), equipped with two Be windows, are mounted on diffractometers that allow simultaneous deposit and the diffraction measurements. The SUV chamber is also equipped with an electronic bombardment furnace, an ion gun, reflection high energy electron diffraction (RHEED) and auger electron spectroscopy (AES) systems. In all cases, the substrate was kept at room temperature and the measurements were performed on cumulative deposits, the growth being interrupted during X-ray scans. Ag was deposited by means of a Knudsen cell, with a deposition rate of 0.73 Å min^{-1}. Pd and Ni were evaporated from electron-beam-heated rods of 99.99% purity. The Pd and Ni deposition rates were 1 Å min^{-1}. In each case, the calibration was performed with a quartz microbalance prior and after the X-ray measurements, and was checked by *in situ* X-ray reflectivity measurements on the last deposit.

The $15 \times 15 \times 0.5$ mm^3 MgO(001) single crystals were supplied oriented $\pm 0.1°$ and both sides polished by Earth Chemical (Japan). The preparation of an MgO(001) surface with a quality adequate for GIXS is difficult and has been described in detail elsewhere.[22] It leads to MgO(001) surfaces that are very flat and of high crystalline quality, free from any impurity, with in-plane domain size larger than 1 μm, an average terrace size of 6000 Å, and an RMS roughness of 2.4 Å.

The X-ray beam energy was set at 18 keV, which allowed measurements of CTRs[37,38] over an extended range of perpendicular momentum transfer, thus yielding a high accuracy on out-of-plane parameters. All measurements were systematically performed with three fixed values of the incident angle α of the X-ray beam with respect to the surface, $\alpha = 0.12° = \alpha_c^{MgO}$, the critical angle for total external reflection of MgO, $\alpha = 2/3\alpha_c^{MgO}$, and $\alpha \approx 2$. Working at $\alpha = 2/3\alpha_c^{MgO}$ was mandatory for small metal thickness, in order to optimise the surface signal over the noise ratio. The sample surface was vertical. The incident X-ray beam size was 0.05 mm (H) × 0.5 mm (V) for Pd, and 0.42 mm (H) × 0.39 mm (V) for Ag and Ni. The opening of the two pairs of detection slits was fixed at 1 mm (H) × 1 mm (V) for the measurements of the MgO Crystal CTRs (corresponding to an angular acceptance of 0.1°) in all three cases.

The Miller indexes ($h\ k\ l$) are expressed in reciprocal lattice units (r.l.u.) of MgO, using the bulk fcc unit cell ($a_{MgO} = 4.2117$ Å). The l index is the component of the momentum transfer perpendicular to the surface.

III. Results and analysis

III.A. Quantitative measurements of the CTRs

The sharp truncation of a surface is known to produce CTRs[37,38] extending perpendicular to the surface, and connecting bulk Bragg peaks. Between Bragg peaks, the intensity variation is very sensitive to the structure of the interface. On the MgO(001) surface, two kinds of non-equivalent CTRs are present: "strong" ones (whose intensity is proportional to the square of the sum of the atomic form factors of O and Mg), with h and k even, and "weak" ones (whose intensity is proportional to the square of the difference of the form factors), with h and k odd. They yield complementary information on the adsorption site.[39]

For the three metals, we have quantitatively measured the (20l) and (11l) MgO CTRs on the bare substrate and during the first stages of metal deposition, between $\theta \approx 0$ and ≈ 12.5 ML where θ is the amount of metal deposited, in equivalent monolayers (ML). For Pd, an additional

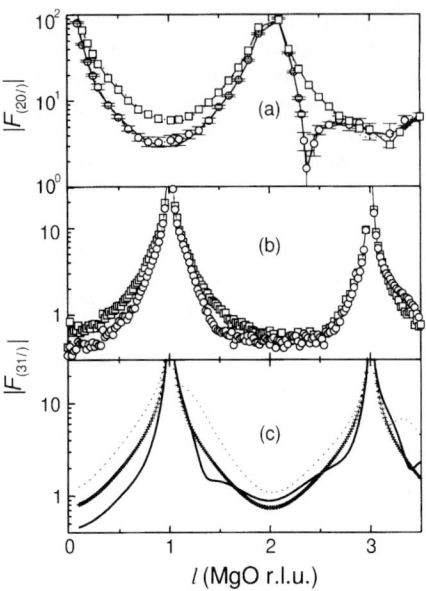

Fig. 1 Modulus of the structure factor of the (20*l*) and (31*l*) CTRs for the bare substrate and for 1 ML of Pd deposited. (a) (20*l*) CTR, obtained by integration and correction of rocking scans measurements, for the bare MgO(001) substrate (open squares with the line showing the best fit); and for $\theta = 1$ ML of Pd deposited (open circles with error bars; the continuous line is the best fit). The best fits yielded the following parameter values: $d_{Pd-MgO} = 2.22 \pm 0.02$ Å, $o_{on-site} = 0.5 \pm 0.1$ ML, $h_{Pd} = 2.7 \pm 0.3$ Å, and $d_{Pd-Pd} = 1.86 \pm 0.03$ Å; (b) *l*-scan measurements of the (31*l*) CTR for the bare substrate (open squares) and after deposition of $\theta = 1$ ML (open circles); (c) calculated (31*l*) CTR, for the bare substrate (open squares), and for 1 ML of Pd deposited, with Pd above either Mg ions (dashed line) or O ions (thick continuous line).

CTR, the (31*l*), was also measured. For the bare substrate and for $\theta = 1$ and 2 ML (and also 10 ML for Ag), the CTRs were measured by rocking the sample around its surface normal for each *l* value. These rocking scans were background subtracted, integrated, normalised to constant monitor counts and corrected for polarisation and Lorentz factors.[40,41] Most other measurements were performed directly in "*l*-scans", *i.e.*, along the CTRs, and corrected with a different Lorentz factor,[41] which is possible because the intrinsic width of the MgO CTRs is always (except close to $l = 0$) much smaller than the width induced by the experimental resolution, and because this width is not modified by the deposition of the metal deposition, at least for $\theta < 10$ ML.

It is important to note that this last observation implies that the possible modifications of the CTRs that may be induced upon metal deposition arise from metal atoms that are correlated *via* the substrate over very long lateral distances, and have the substrate's correlation length (resolution limited). This corresponds to pairs of atoms whose in-plane separation is equal to exact multiples of the MgO lattice vectors, *i.e.*, that have the same internal co-ordinate in the MgO unit cell (in prolonging of the MgO lattice inside the metal). We call this fraction of the deposit "substrate-correlated metal fraction" (SCF or SC metal), or, for simplicity, "on-site" metal. Note that the atoms of the SCF are not necessarily located above the MgO sites and do not necessarily form a continuous crystal.

III.A.1. Pd/MgO(001) interface. Fig. 1 shows the (20*l*) and (31*l*) CTRs, for the bare substrate and for $\theta = 1$ ML of Pd deposited. Pd deposition drastically modifies the shape of the CTRs. On the (20*l*) CTR, it induces a pronounced decrease in the intensity between $l = 0$ and $l = 2$, together with a sharp minimum around $l = 2.35$. Along the (31*l*) CTR, a destructive interference is present both on the low-*l* and high-*l* sides of the (311) Bragg peak, and a positive interference on the right of the (333) peak. These modifications arise from the interference between the waves scattered by the substrate and those scattered along the CTR by that fraction of the Pd film made of Pd atoms correlated *via* the substrate, or "on-site" Pd. By fitting the experimental MgO CTRs with an appropriate model, it is possible to determine the position of the adsorption sites of "on-site" Pd atoms and several structural parameters of the "on-site" fraction.

The (20*l*) and (31*l*) CTRs were simultaneously fitted using a least-squares fitting procedure. The occupancy, $o_{\text{on-site}}(z)$, of the "on-site" Pd plane located at a distance z above the last MgO plane was described by a complementary error function (Fig. 2):

$$o_{\text{on-site}}(z) = o_{\text{on-site}} \frac{1}{2N} \text{erfc}\left(\frac{z}{\sqrt{2}\,h_{\text{Pd}}}\right)$$

where

$$N = \sum_{\substack{\text{Metal plane} \\ \text{nbi}}} \frac{1}{2} \text{erfc}\left(\frac{z_i}{\sqrt{2}\,h_{\text{Pd}}}\right).$$

The fitting parameters are the total amount of "on-site" Pd ($o_{\text{on-site}}$, in ML), the average height h_{Pd} of the "on-site" Pd with respect to the MgO substrate (which includes the root mean squared roughness), the interfacial distance $d_{\text{Pd-MgO}}$, and the mean interplane (002) distance in the Pd film, $d_{\text{Pd-Pd}}$, which was supposed to be independent of the height with respect to the interface. The scale and the substrate roughness (2.4 Å RMS) were first determined by a fit of the clean substrate CTRs (Fig. 1), the substrate roughness being modelled by a Gaussian distribution of terrace heights.[42] The MgO(001) substrate was assumed to be unaffected by the Pd deposit.

The best fits of the (20*l*) CTR for $\theta = 0$ and 1 ML are reported in Fig. 1. Qualitatively, the sign of the interference along this CTR (with h and k even) allows discrimination between epitaxy on top of either oxygen or magnesium on the one hand, or above the octahedral site on the other hand. The strong destructive interference on the right-hand side of the (222) Bragg peaks allows the exclusion of the possibility of the octahedral site. The parameters of the quantitative fit of the (20*l*) CTR for $\theta = 1$ ML were $d_{\text{Pd-MgO}} = 2.22 \pm 0.02$ Å, $o_{\text{on-site}} = 0.5 \pm 0.1$ ML, $h_{\text{Pd}} = 2.7 \pm 0.3$ Å, and $d_{\text{Pd-Pd}} = 1.86 \pm 0.03$ Å. The measurement of a CTR with h and k odd is necessary to distinguish between the two remaining possible epitaxial sites: oxygen or magnesium. Fig. 1(c) shows the simulated (31*l*) CTR, using these parameters, for a Mg site, and for an O site. Clearly *the epitaxial site is above the oxygen atoms of the substrate, and not Mg ones.*

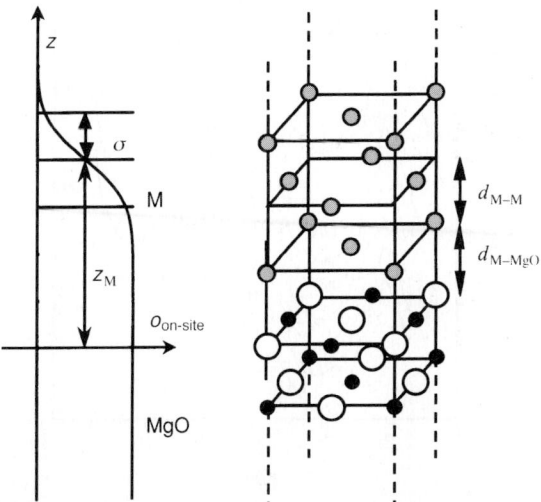

Fig. 2 Schematic drawing of the model used for fitting the CTRs. Right: atomic positions. The metal (= Pd, Ni or Ag) atoms are represented by grey circles, Mg ions by black disks, and oxygen ions by open circles. Left: shape of the profile describing the occupancy of the metal planes as a function of the co-ordinate z perpendicular to the surface. In this figure, the substrate was supposed to be perfectly flat.

Then, for all deposits between $\theta = 0$ and 12.5 ML, the $(20l)$ and $(31l)$ CTRs measured in l-scans were simultaneously fitted over the ranges $l = 1$–3.7 and 0.5–3.7, respectively, assuming the oxygen site. The best fits of the experimental data are reported in Fig. 3, and the corresponding parameters in Fig. 4. For all deposits, the agreement is good, which shows that the chosen model is adequate.

$o_{on\text{-}site}$ [Fig. 4(a)] and h_{Pd} [Fig. 4(b)] are found to first increase quickly with θ, and then slowly reach asymptotes around ≈ 1.3 ML and 6 Å, respectively. d_{Pd-Pd} [Fig. 4(d)] decreases from 1.895 Å for $\theta = 0.2$ ML to 1.79 Å for $\theta = 4$ ML, and then stays nearly constant, with only a very slight decrease to 1.785 Å for $\theta = 12.5$ ML. Finally, d_{Pd-O} [Fig. 4(c)] shows a peculiar behaviour: it first decreases from $\approx 2.23 \pm 0.03$ Å at $\theta = 0.5$ ML to 2.15 ± 0.03 Å for $\theta = 4$ ML, and then increases to reach a steady state value of 2.22 ± 0.03 Å above 10 ML. However, all these d_{Pd-O} values are very close to each other.

The evolution with θ of the structural parameters describing the "on-site" fraction contains a lot of information, in particular, on the nature of this "on-site" fraction.

Let us first look at the average height (or RMS roughness). As shown in Fig. 4(b), below $\theta = 2$ ML, the average height of the "on-site" fraction is equal to the equivalent height of Pd deposited. It then remains much smaller than this equivalent height, reaching only the height of 3 atomic planes for $\theta = 12.5$ ML. This shows that the "on-site" fraction is confined near the interface between Pd and MgO(001). In addition, at the very beginning ($\theta < 0.5$ ML), the "on-site" occupancy [Fig. 4(a)] is nearly equal to the amount deposited: the fraction of Pd that is already relaxed is much smaller than the pseudomorphic fraction. For $\theta = 1$ ML, $o_{on\text{-}site} \approx 0.5$ ML: half of the amount deposited is "on-site" and located at the interface. The "on-site" occupancy next

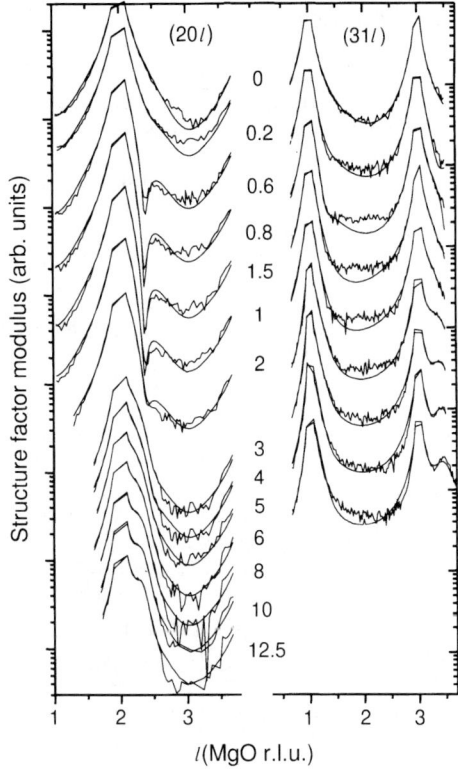

Fig. 3 Comparison between the measured (rough line) and calculated (smooth line) $(20l)$ and $(31l)$ CTRs during the room temperature growth of Pd on MgO(001). The modulus of the structure factor is reported as a function of the out-of-plane momentum transfer. Both CTRs have been simultaneously fitted over a large range of out-of-plane momentum transfer. The amount θ (in ML) of Pd deposited is indicated in the figure. The curves were vertically shifted for clarity.

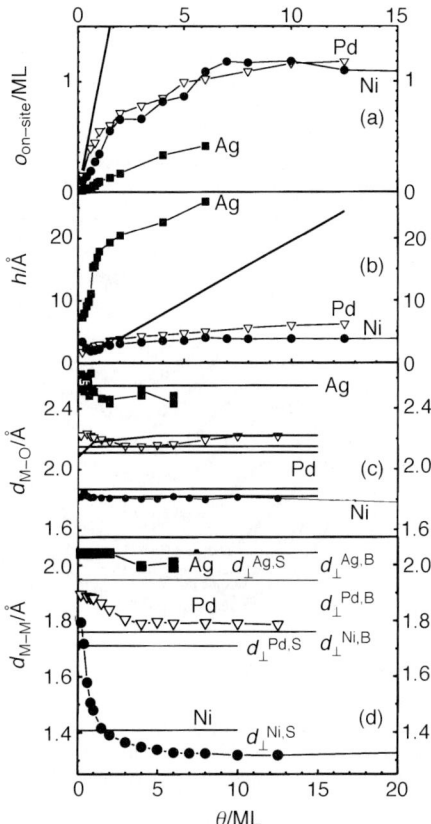

Fig. 4 Evolution with the amount θ (in equivalent ML) of metal deposited of (a) the total amount of "on-site" metal expressed in number of ML for Ag, Ni and Pd compared with the total amount deposited (dashed line). (b) The "on-site" metal thickness for Ag, Ni and Pd. The total equivalent thickness of metal deposited is also represented (solid line). (c) The interplane distance in the metal, perpendicular to the surface, for Ag (filled squares), Ni (filled circles) and Pd (triangles) compared with the distances expected for the bulk materials, and to the distance calculated according to isotropic elasticity for metal strained in-plane to the MgO lattice parameter (horizontal lines). (d) Interfacial distance $d_{\text{M–MgO}}$ deduced from the fits of the CTRs for Ag (filled squares), Ni (filled circles) and Pd (triangles), and average interfacial distance (dashed line).

increases, only much more slowly than θ, saturating at around $o_{\text{on-site}} = 1.2$ ML for large deposited amounts. From these observations, we conclude that, for all deposited amounts, most of the "on-site" Pd is composed of a pseudomorphic Pd layer, i.e., a Pd layer that is lattice-matched with MgO parallel to the surface, and is located either at the interface or in the next two Pd layers. This is confirmed by the fact that the same scale factor is obtained when fitting CTRs of different orders, which would not be the case if the interferences along the CTRs were arising from relaxed Pd, as will be shown below for the case of the Ag/MgO(001) interface. Our conclusions are in agreement with previous HRTEM,[43,44] as well as SEELFS[45,46] measurements, which show that the first Pd layer at the interface is laterally expanded to adopt the lattice parameter of MgO.

The out-of plane Pd–Pd distance [Fig. 4(d)] may be compared to the (002) bulk Pd interplanar distance, $d_{\perp}^{\text{Pd,B}} = 1.945$ Å, and to the value of $d_{\perp}^{\text{Pd,S}} = 1.71$ Å calculated in the framework of linear elasticity for Pd strained in plane to the MgO(001) substrate. This latter is deduced from the lattice parameters according to:

$$\frac{d_{\perp}^{\text{Pd,S}} - d_{\perp}^{\text{Pd,B}}}{d_{\perp}^{\text{Pd,B}}} = -2 \times \frac{C_{12}}{C_{11}} \times \frac{a_{\|}^{\text{Pd,S}} - a_{\|}^{\text{Pd,B}}}{a_{\|}^{\text{Pd,S}}}$$

where C_{11} and C_{12} are the elastic constants of Pd (2.271×10^{11} Pa and 1.761×10^{11} Pa, respectively), $a_{\|}^{\text{Pd,S}}$ is the in-plane lattice parameter of Pd strained by the MgO substrate (i.e., the

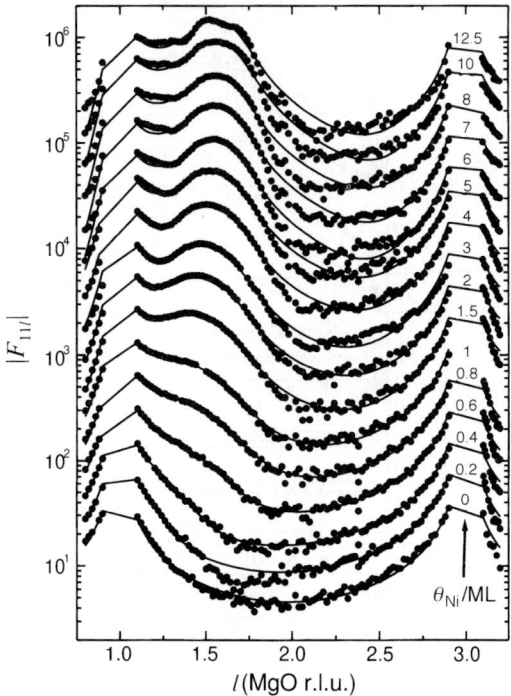

Fig. 5 (11*l*) CTR *vs.* Ni amount deposited. From bottom to top 0, 0.2, 0.4, 0.6, 0.8, 1.0, 1.5, 2.0, 3.0, 4.0, 5.0, 6.0, 7.0, 8.0, 10.0 and 12.5. Comparison for each thickness between experimental data (●) and the best fit (——) obtained with an oxygen epitaxial site and the 4 parameters reported in Fig. 4 for the Ni fraction which sits "on-site".

MgO lattice parameter), and $a_{\parallel}^{Pd,B}$ is the bulk Pd lattice parameter. For all deposits, the interplane distance is intermediate between these two values. At the beginning of deposition it is closer to the bulk value and decreases down to a nearly constant value, closer to that of elastically strained Pd for $\theta > 4$ ML. At the beginning, the "on-site" Pd has a thickness limited to one or two atomic layers, so that the elasticity theory cannot be applied. Since it is covered and surrounded by relaxed Pd[32,33] a value close to the bulk one is not surprising. As the thickness increases, below $\theta = 4$-5 ML, the elasticity theory becomes a better approximation. However, the islands are now composed mainly of relaxed Pd[33] and a small amount of pseudomorphic Pd. This explains the decrease of the interplane distance, although it remains larger than $d_{\perp}^{Pd,S}$.

The fact that both the interfacial distance and the Pd interplane distance show a discontinuity around $\theta = 4$-5 ML can be correlated to the onset of plastic relaxation[33] for this amount of Pd deposited. Above 4-5 ML, dislocations are introduced at the island edges relaxing the misfit, and the "on-site" fraction becomes confined in the regions of "good match" between the Pd and MgO lattices, *i.e.*, laterally located between two misfit dislocations lines. At these locations, the "on-site" Pd is close to truly pseudomorphic, which explains the value of the Pd interplane distance. However, the out-of-plane interplanar distance should no longer be compared to elastic calculation of strained Pd, but rather to a full calculation of the atomic positions between two misfit dislocations, which is beyond the scope of the present study.

Above $\theta = 4$–5 ML, the increase of the interfacial distance would correspond to a weakening of the metal–oxide bond as the interfacial atoms become less isolated, *i.e.*, as they are more and more involved in the metallic bounding of the Pd film. An increase of the interfacial distance with θ is actually what is predicted by *ab initio* calculations[47,48] and is the tendency that is intuitively expected. Why the interfacial distance decreases at the very beginning of deposition is less clear. A tentative explanation would be correlated with a decreasing aspect ratio (height over lateral extension) of the islands with increasing θ. As this aspect ratio would decrease, the influence of

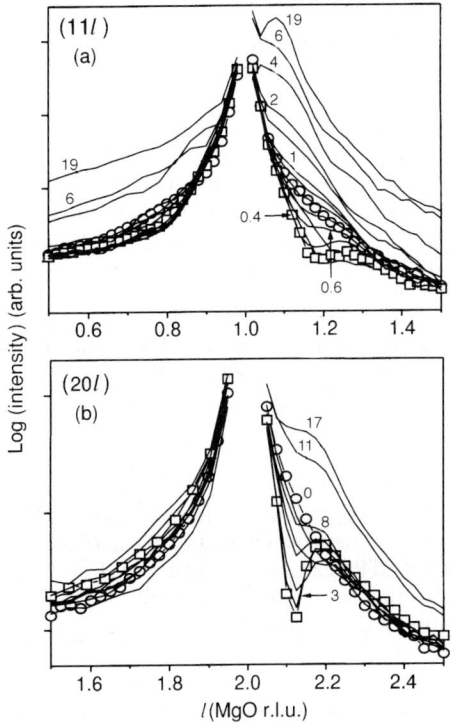

Fig. 6 Logarithm of the measured intensity along the (11*l*) (a) and (20*l*) (b) MgO(001) CTRs, as a function of the out-of-plane co-ordinate *l*, for different amounts of deposited Ag. Incident angle: 0.08° from 0 to 8 ML, 0.12° at 11 ML, 0.15° at 17 ML and 0.22° at 19 ML. (a) (11L) CTR, deposits: 0 ML (open circles), 0.2, 0.3, 0.4 ML (open squares), 0.5, 0.6, 0.7, 0.8, 0.9, 1, 1.5, 2, 4, 6 and 19 ML. (b) (20*l*) CTR, deposits: 0 ML (open circles), 0.2, 0.5, 1, 2, 3 ML (open squares) 4, 6, 8, 11 and 17 ML.

"metallic" bonding *vs.* interfacial bonding would also decrease, resulting in a shorter interfacial distance.

III.A.2. Ni/MgO(001) interface. The (20*l*) and (11*l*) (Fig. 5) CTRs were also measured during the room temperature deposition of Ni on MgO(001).[34] The (20*l*) CTR shows a strong intensity decrease of half an order of magnitude between 0 and 1 ML. A similar intensity decrease is observed between 1 and 125 ML. Such a strong signal decrease is not consistent with the X-ray absorption from one Ni monolayer. It thus arises again from the existence, in the submonolayer regime, of a large part of the Ni atoms sitting on top of the substrate sites.

Fig. 5 shows that, as in the Pd case, that the (11*l*) CTR is well reproduced with the above model of "on-site" Ni. Again, in agreement with the theoretical calculations[13,14,23,24] the evidence of interference along the (11*l*) CTR unambiguously shows that the Ni atoms sit above the oxygen atoms of the last MgO plane. Fig. 5 shows that, within this model, the agreement between the experimental and calculated structure factors along the (11*l*) CTR is good up to 4 ML, and fair up to 12 ML. The values of the four previously described parameters for all thicknesses are reported in Fig. 4. In the very early stages of growth the distance between two successive Ni layers is close to that of bulk Ni (most likely isolated atoms or small clusters) [Fig. 4(d)]; it decreases rapidly with increasing amount of Ni deposited down to a value close to that expected from the elastic theory. The fraction of the Ni layer that sits "on-site" increases first slowly and then more firmly with the amount of Ni deposited and reaches about 1 ML for a total amount of Ni deposited of 10 ML, afterwards it saturates [Fig. 4(a)]. The amount of "on-site" Ni remains always well below the amount of Ni deposited. The average height of the Ni layer passes through a minimum at 0.6 ML and then essentially increases with respect to the amount of Ni deposited [Fig. 4(b)].

On the overall, as concerns the "on-site" fraction, the case of the Ni/MgO(001) interface is very similar to that of the Pd/MgO(001) one, except for the values of the distances, while the structures of the remaining of the metal deposit are markedly different.

III.A.3. Ag/MgO interface. *III.A.3.a. Simplified model.* For Ag, the evolution with θ (Fig. 6) of the intensities along the $(20l)$ and $(11l)$ MgO CTRs looks at first glance very similar to that in the Pd case. However, in-plane grazing incidence small and wide angle scattering[32,33] showed that, from the very beginning of deposition, the Ag grows in the form of islands, with an in-plane lattice parameter close to that of bulk Ag. In other words, there is no pseudomorphic Ag. However, there are residual strains in the Ag islands, which implies that all interfacial Ag atoms are closer to a preferential substrate site, which we therefore call the "adsorption site". When atoms are displaced too much with respect to this site, interfacial dislocations are introduced, yielding "good-match" regions in which the adsorption site can still be defined, and "bad-match" regions in which the Ag atoms do not sit on top of any particular site.

In a first crude approximation, we could use the same model as for the Pd and Ni cases, *i.e.*, to consider that only the Ag atoms that are perfectly "on-site" contribute. The interferences would thus be due either to fully lattice-matched Ag, or to separated "on-site" columns that are located at the centre of the islands during the first three stages of the growth, or exactly halfway between two dislocation lines when the film is continuous. In this last case, the "on-site" fraction does not form a continuous crystal in the directions parallel to the surface, even on the scale of the interatomic spacing, but the MgO CTRs can be modelled as before. Using this model, a qualitative analysis of the sign of the interference observed along the MgO CTRs again shows that the Ag atoms of the first plane sit atop of the oxygen atoms of the last MgO plane.

The parameters of the model (o_{TOTAL}, h_M, d_{Ag-MgO}, and d_{Ag-Ag}), determined by a simultaneous least-squares fit of the $(11l)$ and $(20l)$ CTRs, are represented in Fig. 4 as a function of θ. Fig. 7 shows the comparison between the experimental CTRs and the best fits for selected deposited amounts θ.

The first striking feature is the very small amount of "on-site" Ag [Fig. 4(a)]. It may be surprising that large effects are observed on the MgO(001) CTRs with such a small "on-site" Ag amount. However, the MgO CTRs are extremely sensitive to the presence of Ag because its scattering power is 24 times that of MgO on the "intense CTRs" and 150 times that of MgO on the "weak ones". The second important result is the large values of the height h of the "on-site" fraction.

The very small amount of "on-site" Ag (0.02 ML at $\theta = 0.2$ ML), and also the large values of h (about 3 planes at $\theta = 0.2$ ML) both confirm that the "on-site" Ag does not consist of lattice-matched Ag. If some lattice-matched Ag was present, like in the Pd and Ni cases, it should be confined near the interface, so its thickness should not exceed one or two planes, and the major part of the first Ag plane would be lattice-matched, which would yield, like in the Ni and Pd cases, an amount of "on-site" Ag close to the equivalent amount deposited θ, at least for small θ. The values found for h indicate that Ag is already in the form of islands with a height of several planes at 0.2 ML: there is no stage of two-dimensional growth.

However, although the "on-site" Ag is not lattice-matched, and extends over a significant height, it is not only composed of the central "on-site" columns either. Indeed, the amount of "on-site" Ag obtained with this simple model indicates that, whatever the island size, there is more than one "on-site" Ag column per island.

III.A.3.b. General model: origin of the interference. This observation led us to a more general model, in which, at least for $\theta < 4$ ML, we suppose that all the islands of Ag are *approximately* identical and are "pinned" in the same way on the substrate, so that the whole Ag deposit contributes to the modifications of the MgO CTRs. In other words, we propose a model of partially relaxed islands correlated *via* the substrate. We also come back to the original, more exact notation of "substrate-correlated Ag" (SC Ag) instead of "on-site" Ag, which is misleading. In such a model, where the SC Ag fraction is the whole deposit, all the Ag atoms contribute to the MgO CTRs, but with a weight that decreases with increasing lateral separation from the central "on-site" column. In a small island of relaxed Ag, the central atomic column, which is perfectly "on-site", fully contributes, because all the "on-site" columns are not only above MgO ones, but are, in addition, fully correlated *via* the substrate. As interfacial Ag atoms lie farther away from this

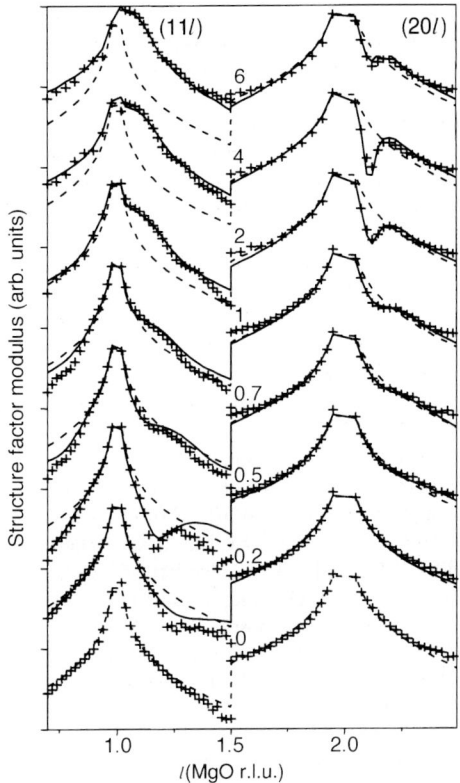

Fig. 7 Comparison between the measured (crosses) and calculated (solid lines) MgO(001) CTRs during the room temperature growth of Ag on MgO(001). The logarithm of the modulus of the structure factor is plotted vs. l. The dashed lines correspond to the clean MgO(001) substrate. The $(11l)$ and $(20l)$ CTRs have been represented on the same l-scale although they are at different h, k values. The curves corresponding to the different amounts of deposited Ag are shifted vertically for clarity.

central column, their contribution to the MgO CTRs decreases for two reasons. The first is that they are more and more displaced from the "on-site" position. The second is that pairs of atoms located in different islands far from the "on-site" column are less likely to be correlated by the substrate, because the exact atomic distribution presumably differs between different islands, especially if they are of different sizes.

Let us first illustrate this model by supposing that all the islands are *exactly* identical and are "pinned" in the same way on the substrate. In this case, the MgO CTRs can be modelled by calculating the intensity scattered by a supercell that comprises a semi-infinite MgO column with a square basis whose lateral size is equal to the inter-island distance, and a Ag island that is either fully relaxed or slightly strained containing at its centre a perfectly "on-site" column. A very simple estimation of the contribution of Ag to the MgO truncation rods allows the prediction of the differences that will appear in the diffracted intensity between this general model and the above simplified one. Let us assume for simplicity that all the Ag has its bulk lattice parameter. The supercell comprises a semi-infinite column of MgO yielding a CTR centred on the parallel momentum transfer value of MgO, and a Ag island yielding a rod centred on the parallel momentum transfer value of Ag. The Ag rod, of full width at half-maximum width $2\pi/d$ (where d is the in-plane island size), yields a contribution below the MgO CTR that decreases as the order of the CTR increases, since the spacing between the MgO and Ag rods increases with h and k while the width of the Ag peak remains constant. We experimentally observed this decrease of the Ag contribution along the CTR as a function of the rod order, which confirms the validity of this last model. By contrast, for lattice-matched or "on-site" Ag, the respective weight of the Ag and MgO contributions to the MgO CTR would not vary with the order of the rod. This last case corresponds to the experimental observation for Ni and Pd deposition, which validates the use of the simplified model in these two cases.

III.A.3.c. Validity of the interfacial parameters. One drawback of this general model is that it contains too many parameters for a quantitative analysis. Even if we were able to calculate the atomic positions in a Ag island given its size and shape, we would have to introduce as free parameters a mean island size, a mean shape, a mean inter-island distance, and the dispersion over all these parameters, in order to estimate the Ag-Ag correlation function and derive the intensity. For the quantitative analysis of the MgO CTRs, the simpler model of "on-site" Ag was therefore used. Of course, this inadequately models the lateral position of the Ag atoms, but it allows deriving the interesting parameters: the height of the SCF, the adsorption site, the inter-plane distance in Ag, and the interfacial distance.

Let us first show that the adsorption site found with this simplified analysis corresponds to the real adsorption site. For this, we have calculated the intensity scattered by a supercell for the extreme case where Ag has its bulk lattice parameter. The lateral size of the island was taken to be equal to 20 Å, the inter-islands distance equal to 66 Å and the interfacial distance was fixed at the experimental steady state value, 2.52 Å [Fig. 4(c)]. Fig. 8 shows the (11*l*) and (20*l*) CTRs calculated with the central atomic column of the symmetric island set either on top of O, or Mg, or the octahedral site. The (11*l*) CTR allows us to distinguish between either the O site on the one hand, or the Mg or octahedral sites on the other hand, while the (20*l*) CTR allows us to distinguish between either the octahedral site on the one hand, or the O or Mg sites on the other hand. The clear destructive interference experimentally observed (Fig. 6) on both sides of the Bragg peaks along both CTRs is consistent only with the O site. The simulation with the O site (Fig. 8)

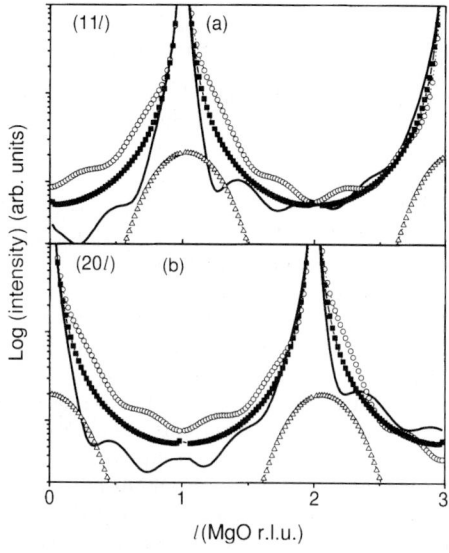

Fig. 8 Calculation of the intensity scattered along the (11*l*) (a) and (20*l*) (b) MgO CTRs, by a supercell composed of a semi-infinite MgO(001) substrate and a small hemispherical island of Ag at its bulk lattice parameter, with a 20 Å diameter. The logarithm of the intensity is plotted *vs. l*. The inter-islands distance (*i.e.*, the lateral size of the supercell) was fixed at 66 Å and the interfacial distance was fixed at the experimental steady-state value, 2.52 Å (Fig. 4). The MgO substrate contribution is shown as black squares and the Ag scattering as open triangles. For the (11*l*) CTR (a), open circles show the CTR intensity for either a Mg or octahedral epitaxial site, and the thick line shows the intensity for an oxygen adsorption site. For the (22*l*) CTR (b), the open circles correspond to the octahedral site, while the thick line shows the intensity for either an O or Mg site.

qualitatively reproduces most of the observed interference at 0.5 ML. For the other sites, the simulated CTRs strongly differ from the experimental ones.

Hence, this very simple simulation shows that, in this system, a qualitative inspection of the sign of the interference along the CTRs using the "on-site" model allows us to determine correctly the adsorption site.

Given this verification, we have to consider that the values deduced for d_{Ag-Ag} [Fig. 4(d)] and d_{Ag-O} [Fig. 4(c)] are average values over the SCF, because these distances are probably non-uniform within a given island, and may vary slightly between islands of different sizes and different strain states. The value of the average height [Fig. 4(b)] is probably representative of the average height of the islands, since it is the central portion of the islands that contribute the most.

The average interplane distance $d_{Ag-Ag} = 2.00 \pm 0.02$ Å is intermediate between the value for bulk Ag ($d^B_{AgAg} = 2.043$ Å) and the value of $d^S_{AgAg} = 1.950$ Å calculated from the linear elasticity theory for Ag strained in-plane to the MgO lattice parameter.

Finally, we do not attach a particular meaning to the total amount of "on-site" Ag [Fig. 4(a)], because this parameter serves to "hide" everything that is not modelled properly, like the real lateral position of the atoms in the islands, and the dispersion on this parameter from one island to the other.

IV. Comparison between the A, Pd and Ni/MgO(001) interfaces

In order to better understand the basic physical phenomena governing the interfacial structure, i.e., the parameters of the interface and the residual deformations within the metal islands, it is interesting to compare similar systems with systematic variations of some physically relevant parameters, among which the lattice parameter misfit and the strength of the interfacial bonding. Indeed, the final state of the islands results from a competition between different mechanisms. The larger the lattice parameter misfit, the sooner the introduction of misfit dislocation or other misfit releasing defects is expected. Also, intuitively, the larger the misfit and the larger the metal stiffness, the less likely the metal is to be strained by the substrate. At the opposite, the stronger the interfacial bonding, the stronger the tendency of the metal to be strained by the substrate, and possibly to form a fully lattice-matched (pseudomorphic) interfacial layer.

According to many recent theoretical calculations, the strength of the metal–oxide bond varies significantly between the three interfaces. For Ag, the binding is weak, of physisorption type, between 0.1 and 0.3 eV atom^{-1}, and mostly of electrostatic origin, while for Pd and Ni, the bond is mostly polar covalent and ranges from 0.65 to 0.81 eV atom^{-1} for Pd, and from 0.88 to 1.24 eV atom^{-1} for Ni.[4,8,13,14,23,24,28,49–56] Comparison of the elastic constants of the three metals also shows that Ag is much softer than Pd and Ni. On the other hand, differences between the surface stress of the three metals can be neglected in a first approximation.[57] Apart from the common three-dimensional growth mode, which is expected by simple thermodynamic arguments, these three interfaces indeed exhibit fairly different characteristics. For Ag and Pd thick films, the lattice parameter misfit is simply relaxed by a network of interfacial misfit dislocations, while for Ni, the relaxation is performed by the growth, in addition to cube-on-cube Ni, of Ni clusters in Ni(110)//MgO(001) epitaxy, with four different in-plane orientational relationships.[34] The absence of ordered misfit dislocations at the Ni/MgO(001) interface is not surprising because the large value of the misfit implies alternative relaxation processes.

The above analysis unambiguously shows that, for Ag, Pd and Ni, the atoms of the first metal monolayer sit on top of oxygen ions of the substrate. The present results are in agreement with all recent theoretical predictions,[4,8,13,14,23,24,28,48–55] and of a recent surface X-ray absorption spectroscopy study in the case of Ag.[58]

We thus believe that the controversy concerning possible epitaxial sites is close to being over: in all cases, the epitaxial site is unique, atop the oxygen ions.

For Pd, whatever the deposit, the interfacial distance lies between 2.15 Å and 2.23 Å, with a steady-state value of 2.22 ± 0.03 Å. This is very close to the values of 2.18 Å for $\theta = 1$ ML and 2.225 Å for $\theta = 2$ ML calculated recently,[47] as well as to the theoretical value of 2.15 Å obtained for a single Pd atom adsorbed on top of the oxygen ions.[13,24] For Ag, the average value of the interfacial distance is $d_{Ag-MgO} = 2.52 \pm 0.1$ Å, which is very close to the most recent theoretical values, of 2.34 Å,[8,23] 2.38 Å,[8,28] 2.47 Å,[48] 2.49 Å,[49,55] 2.50 Å,[56] 2.45 to 2.64 Å,[51] depending on the Ag amount deposited, 2.64 Å,[52] and 2.69[53] as well as to the experimental value of 2.53 Å found both by HRTEM[59] and SEXAFS.[58] For Ni, the interfacial distance remains almost insensitive to the amount of Ni deposited in the validity domain of the analysis. The average distance on top of O sites is found to be 1.82 ± 0.05 Å, which is close to the value of 1.87 Å deduced from recent *ab initio* calculations.[14,23,24]

These results on the epitaxial site and interfacial distance show that the most recent *ab initio* calculations predict these parameters correctly. One important remark is that most published theoretical calculations neglect the lattice parameter mismatch between the metal and the substrate and assume perfectly "on-site" metal atoms. By contrast, in the real situation, the metal is not completely strained to the MgO in-plane lattice parameter. For Ag, in particular, even if most of the interfacial atoms are close to a particular adsorption site, there are only few that are perfectly "on-site". This means that to be accurate, theoretical calculations of the interfacial parameters *should* take into account the lattice parameter mismatch and allow slightly "off-site" Ag atoms.

V. Discussion and conclusions

The differences between the three metals are clear when comparing the parameters of the "on-site" fraction. The significant decrease of the interfacial distance, from 2.52 Å for Ag to 2.22 Å for Pd and 1.88 Å for Ni, directly reflects the increasing strength of the interfacial bond. The most striking differences however lie in the amount and height of the "on-site" fraction, at the beginning of the growth, between $\theta = 0$ and 10 ML. For Pd and Ni, the amount of "on-site" metal saturates around 1.2 ML for large θ, and its average height saturates around 2–3 atomic planes. This shows that, in both cases, a significant fraction of the interfacial atoms is fully lattice-matched laterally with the MgO substrate. The larger misfit for Ni is compensated by its stronger bonding with MgO, yielding a very similar pseudomorphic fraction in both cases. Very different is the case of Ag, with an extremely small amount of "on-site" Ag (less than 0.1 ML for $\theta = 0.5$ ML) extending over a very large height, reaching 15 atomic planes for $\theta = 10$ ML. This, together with the other scattering measurements, could only be interpreted by the absence of lattice-matched Ag at the interface.[30,31] Despite the small lattice parameter misfit and the softness of Ag, which implies a lesser cost of residual deformations as compared to Pd and Ni, all the Ag within the islands has a lattice parameter close to that of bulk Ag, even at the interface. This effect can only be explained by the very weak binding at the Ag/MgO interface.

From all the above comparison, we may conclude that the structure and morphology at these metal/MgO interfaces are mostly influenced by the strength of the bonding at the interface, rather than by the lattice parameter misfit.

In summary, for the growth of Ag, Ni and Pd on a clean and flat MgO(001) surface at room temperature, a new, quantitative analysis of the MgO CTRs allowed the determination of the adsorption site and interfacial distance, as well as a description of the morphology of the deposit for very small deposited amounts. In the three cases, the epitaxial site is shown to be above the O ions of the last MgO(001) plane, and the interfacial distance and its evolution during deposition have been determined. The average values are $d_{Ag-MgO} = 2.52 \pm 0.1$ Å, $d_{Pd-MgO} = 2.22 \pm 0.03$ Å and $d_{Ni-MgO} = 1.82 \pm 0.05$ Å. These results are consistent with the most recent theoretical calculations.

For Pd and Ni, most of the monolayer closest to the interface is pseudomorphic, while most of the Ag is already (partially) relaxed at all stages of the deposition.

Comparison between the three interfaces shows that the main differences between the interfacial characteristics in the three systems mostly arise from the largely differing strength of the metal–oxide bond, rather than from the lattice parameter misfit.

Concerning the technique of surface X-ray diffraction itself, this study shows that measurements of the substrate crystal truncation rods can be used to characterise the deposit even when it does not contain a fully strained part.

This arises because the scattering by the adsorbate always contains a large component with the spatial frequency of the substrate, because the adsorbate is deposited *on* the substrate. Indeed, if we consider for instance a growth mode starting with very small islands made of one or two atoms, all these islands are pinned on a particular site of the substrate, with at least the central atom being "on-site". When more atoms of the adsorbate are deposited, they stick close to these "on-site" atoms, starting to form islands, which may contains different strains. The important point is that the atomic neighbourhood of the adsorbate atoms around the central "on-site" one is approximately the same in all islands, which create a repeating unit in the deposit, the different units being separated by an integer number of substrate lattice parameter parallel to the plane. All these islands thus yield a significant contribution along the CTRs, even if the metal has a lattice parameter that is very different from that of the substrate.

Acknowledgements

We would like to thank A. Bourret, I. K. Robinson, J. Villain, C. Priester, C. Noguera, F. Lançon and T. Deutsch, for valuable help or discussion, and J. Jupille for his participation in one of the experiments. We would like to thank the help of A. Stierle and of the staff of the ID32 and BM32 beamlines during the measurements.

References

1   V. E. Henrich, *Rep. Prog. Phys.*, 1985, **48**, 1481.
2   H. Bialas and K. Heneka, *Vacuum*, 1994, **45**, 79.
3   F. Reniers, M. P. Delplancke, A. Asskali, V. Rooryck and O. Van Sinay, *Appl. Surf. Sci.*, 1996, **92**, 35.
4   M. W. Finnis, A. M. Stoneham and P. W. Tasker, Metal–Ceramic Interfaces, ed. M. Rühle, A. G. Evans, M. F. Ashby and J. P. Hirth, Pergamon, Oxford, 1990, p. 35.
5   V. E. Henrich and P. A. Cox, *The surface science of metal oxides*, Cambridge University Press, Cambridge, 1994.
6   *Acta Metall. Mater.*, 1992, **S40**.
7   C. Noguera, *Physics and Chemistry at Oxide Surfaces*, Cambridge University Press, Cambridge, 1996.
8   J. R. Smith, T. Hong and D. J. Slorovitz, *Phys. Rev. Lett.*, 1994, **72**, 4021.
9   P. Palmberg, T. Rhodin and C. Todd, *Appl. Phys. Lett.*, 1967, **11**, 33.
10 A. K. Green, J. Dancy and E. Bauer, *J. Vac. Sci. Technol.*, 1979, **7**, 159.
11 D. G. Lord and M. Prutton, *Thin Solid Films*, 1974, **21**, 341.
12 D. W. Goodman, *Surf. Rev. Lett.*, 1995, **2**, 9.
13 G. Pacchioni and N. Rösch, *J. Chem. Phys.*, 1996, **104**, 7329.
14 N. Rösch and G. Pacchioni, in *Chemisorption and Reactivity on Supported Clusters and Thin Films*, ed. R. M. Lambert and G. Pacchioni, Kluwer Academic Publishers, Dordrecht, 1997, p. 353.
15 H. Sato, R. S. Toth and R. W. Astrue, *J. Appl. Phys.*, 1962, **33**, 1113.
16 A. A. Hussain, *J. Phys.: Condens. Mater.*, 1989, **1**, 9833.
17 N. I. Kiselev, Yu. l. Man'kov and V. G. Pyn'ko, *Sov. Phys. Solid State (Engl. Transl.)*, 1989, **31**, 685.
18 H. Maruyama, H. Qiu, H. Nakai and M. Hashimoto, *J. Vac. Sci. Technol. A*, 1995, **13**, 2157.
19 H. Qiu, A. Kosuge, H. Maruyama, M. Adamik, G. Safran, P. B. Barna and M. Hashimoto, *Thin Solid Films*, 1994, **241**, 9.
20 H. Bialas and L.-S. Li, *Phys. Status Solidi*, 1977, **42**, 125.
21 M. R. Fitzsimmons, G. S. Smith, R. Pynn, M. A. Nastasi and E. Burkel, *Physica B (Amsterdam)*, 1994, **198**, 169.
22 O. Robach, G. Renaud and A. Barbier, *Surf. Sci.*, 1998, **401**, 227.
23 A. M. Ferrari and G. Pacchioni, *J. Phys. Chem.*, 1996, **100**, 9032.
24 I. Yudanov, G. Pacchioni, K. Neyman and N. Rösch, *J. Phys. Chem. B*, 1997, **101**, 2786.
25 K. Yamamoto, Y. Kasukabe, T. Takeishi and T. Osaka, *J. Vac. Sci. Technol., A*, 1996, **14**, 327.
26 A. Stirling, I. Gunji, A. Endou, Y. Oumi, M. Kubo and A. Miyatomo, *J. Chem. Soc., Faraday Trans.*, 1997, **93**, 1175.
27 C. Goyhenex and C. R. Henry, *J. Electron Spectrosc. Relat. Phenom.*, 1992, **61**, 65.
28 T. Hong, J. R. Smith and D. J. Srolovitz, *Acta Metall. Mater.*, 1995, **43**, 2721.
29 L. K. Robinson and D. J. Tweet, *Rep. Prog. Phys.*, 1992, **55**, 599.
30 O. Robach, G. Renaud, A. Barbier and P. Guénard, *Surf. Rev. Lett.*, 1997, **5**, 359.
31 O. Robach, G. Renaud and A. Barbier, *Phys. Rev. B*, 1999, **60**, 5858.
32 G. Renaud and A. Barbier, *Appl. Surf. Sci.*, 1999, **142**, 14.
33 G. Renaud and A. Barbier, *Surf. Sci.*, 1999, 433.
34 A. Barbier, G. Renaud and O. Robach, *J. Appl. Phys.*, 1998, **84**, 4259.
35 http://www.esrf.fr.
36 G. Renaud, B. Villette and P. Guénard, *Nucl. Instrum. Methods Phys. Res. Sect. B*, 1995, **95**, 422.
37 I. K. Robinson, *Phys. Rev. B: Condens. Matter*, 1986, **33**, 3830.
38 S. R. Andrews, R. A. Cowley, *J. Phys. C: Solid State Phys.*, 1985, **18**, 6247.
39 G. Renaud, *Surf. Sci. Rep.*, 1998, **32**, 1.

40 E. Vlieg, *J. Appl. Crystallogr.*, 1997, **30**, 532.
41 O. Robach, PhD-thesis, University of Grenoble, 1997.
42 J. Harada, *Acta Crystallog. Sect. A*, 1992, **48**, 764.
43 K. Heinemann, T. Osaka, H. Poppa and M. Avalos-Borja, *J. Catal.*, 1983, **83**, 61.
44 S. Giorgo, C. Chapon, C. R. Henry and G. Nihoul, *Philos. Mag. B*, 1993, **67**, 773.
45 C. Goyenex, C. R. Henry and J. Urban, *Philos. Mag. A*, 1994, **69**, 1073.
46 S. Bartuschat and J. Urban, *Philos. Mag. A*, 1997, **76**, 783.
47 J. Goniakowski, *Phys. Rev. B: Condens. Matter*, 1998, **58**, 1189.
48 L. Spiess, *Surf. Rev. Lett.*, 1996, **3**, 1365.
49 J. Purton, S. C. Parker and D. W. Bullet, *J. Phys.: Condens. Matter*, 1997, **9**, 5709.
50 P. Blöchl, G. P. Das, H. F. Fischmeister and U. Schönberger, in *Metal-Ceramic Interfaces*, ed. M. Rühle, A. G. Evans, M. F. Ashby and J. P. Hirth, Pergamon Press, Oxford, 1990, p. 9.
51 E. Heifets, F. Y. Zhukovskii, E. A. Kotomin and M. Causa, *Chem. Phys. Lett.*, 1998, **283**, 395.
52 E. Heifets, E. A. Kotomin and R. Orlando, *J. Phys.: Condens. Matter*, 1996, **8**, 6577.
53 C. Li, R. Wu, A. J. Freeman and C. L. Fu, *Phys. Rev. B: Condens. Matter*, 1993, **48**, 8317.
54 H. X. Liu, H. L. Zhang, H. L. Ren, S. X. Ouyang and R. Z. Yuan, *Ceram. Int.*, 1996, **22**, 79.
55 U. Schönberger, O. K. Andersen and M. Methfessel, *Acta Metall. Mater.*, 1992, **40**, S1.
56 L. Spiess, *Surf. Rev. Lett.*, 1996, **3**, 1365.
57 J. P. Hirth and J. Lothe, *Theory of Dislocations*, Krieger Publishing Co., Malabar, FL, Wiley & Sons, 1992, Appendix 2, p. 838.
58 A. M. Flanck, R. Delaunay, P. Lagarde, M. Pompa and J. Jupille, *Phys. Rev. B: Condens. Matter*, 1996, **53**, 1737R.
59 A. Trampert, E. Ernst, C. P. Flynn, H. E. Fischmeister and M. Rühle, *Acta Metall. Mater.*, 1992, **40**, S227.

Paper 9/02735A

Mg clusters on MgO surfaces: study of the nucleation mechanism with MIES and *ab initio* calculations

L. N. Kantorovich,[a] A. L. Shluger,[a] P. V. Sushko,[ab] J. Günster,[c] P. Stracke,[d] D. W. Goodman[c] and V. Kempter[d]

[a] *Department of Physics and Astronomy, University College London, Gower Street, London, UK WC1E 6BT*
[b] *The Royal Institution of Great Britain, 21 Albemarle Street, London, UK W1X 4BS*
[c] *Department of Chemistry, Texas A&M University, College Station, TX77842-3012, USA*
[d] *Physikalisches Institut der TU Clausthal, Leibnizstraße 4, D-38678 Clausthal-Zellerfeld, Germany*

Received 23rd April 1999

We combined experimental studies using ultraviolet photoelectron spectroscopy (UPS), metastable impact electron spectroscopy (MIES) and temperature programmed desorption (TPD) with *ab initio* calculations of metal adsorption on the perfect MgO surface and at defect sites in order to elucidate the role of surface defects in the initial stages of nucleation and growth of metal clusters at oxide surfaces. MgO films (2 nm thick) grown on Mo and W substrates were used as a prototype system. The MIES and UPS (HeI) spectra were collected *in situ*, and the growth of Mg clusters was observed by monitoring the dynamics of additional MIES peaks during Mg deposition. TPD experiments were made in order to monitor the surface coverage by Mg clusters and to determine the Mg desorption energies. Interpretation of the results was made on the basis of theoretical modelling using density functional theory (DFT) calculations in both periodic and embedded cluster models. The geometric and electronic structures of the surface terrace, F-centre, positively charged anion vacancy, and step edge at the MgO(001) surface were calculated, and their role in adsorption and clustering of Mg atoms on this surface was studied. The absolute position of the top of the surface valence band of MgO with respect to the vacuum was calculated and compared with the MIES results. The MIES spectra were modelled on the basis of surface density of states (SDOS). The calculated SDOS predicted the location of additional peaks in the band gap and their shift as a function of Mg concentration on the surface in agreement with the MIES data. The desorption energies of Mg atoms from small Mg clusters formed at step edges are found to be about 1.3 eV atom^{-1}. Comparison between the theoretical results and the experimental data suggests preferential initial adsorption of Mg atoms at steps and kinks, rather than at charged and neutral vacancies. At larger exposures these Mg atoms serve as the nucleation sites.

1 Introduction

Understanding of the mechanisms of growth and parameters of the geometric and electronic structures of metal clusters and layers on metal oxide surfaces is important for a number of technological applications. In particular, metal addition to oxides leads to an enhanced reactivity *via* electron transfer to a variety of adsorbed molecules leading to the formation of radical anion species.[1,2] The interaction between metal clusters and metal oxide supports plays a key role in catalysis[3,4] and microelectronics.[5] The interaction between metal atoms and oxide surfaces is important for understanding the mechanisms of their segregation,[6,7] diffusion of metal atoms on insulators,[8] formation of metal-induced point defects on oxides,[2,9] and the formation of nanoparticles inside semiconductors and insulators.[10]

Surprisingly, little is known about the structure of metal/oxide interfaces, in particular, the initial stages of the metal adsorption, types of the metal adsorption sites, the nature of bonding to oxides and between the adsorbed metal atoms.[5,11] Although it is clear that the defect sites, such as surface vacancies, step edges and kinks play a significant role, at least at the early stages of metal growth on oxides, the number, distribution and structure of these defects is very difficult to control experimentally. On the other hand, most of the existing theoretical calculations are concerned with metal adsorption on ideal oxide surfaces (see, for example, refs. 12 and 13), and only very few treat metal adsorption on defective oxide surfaces (for a review see ref. 14). In particular, Ferrari and Pacchioni[15] performed cluster Hartree–Fock calculations of MgO surfaces with point defects (neutral and charged anion and cation vacancies), and studied the charge transfer between the Rb atoms and these defective surfaces. While the neutral point defects are not very reactive, the charged anion vacancies can ionize metal atoms provided the electron affinity of the defect is larger than the ionisation potential of the metal atom at the surface. Thus, the interaction of metal atoms with surface point defects can greatly alter their surface diffusion behaviour and, consequently, can be responsible for cluster nucleation in the neighbourhood of defect sites.

In this study we combined several experimental techniques with theoretical modelling in order to elucidate the early stages of Mg cluster formation on the (001) surface of MgO thin films. Metastable impact electron spectroscopy (MIES) using He* (1s2s) projectiles is particularly useful for these purposes because it probes the very top surface layer and is sensitive to quite small concentrations of adsorbed species. It is naturally combined with ultraviolet photoelectron spectroscopy (UPS) using HeI as the light source. Unlike UPS (HeI), MIES is extremely sensitive to features resulting from the charge density of s-electrons specific for adsorption of alkali and alkaline earth atoms. Correlation of the spectroscopic data with the results of the temperature programmed desorption (TPD) experiments is illuminating for the interpretation of the spectral features and understanding of the initial stages of the metal cluster growth.

To understand better the experimental data and to construct a model of metal adsorption, we performed *ab initio* electronic structure calculations using the density functional theory (DFT), and the method of pseudopotentials. The embedded cluster model[16,17] was employed for DFT calculations of the surface ionisation energies, which are compared with the position of the top of the valence band with respect to the vacuum level of the MgO film, determined from the MIES data. Using the periodic DFT calculations we treated the adsorption of up to five Mg atoms on the perfect (001) surface, near an anion vacancy, and at a neutral surface F-center. In order to elucidate the role of extended surface defects in the Mg cluster growth we modelled the adsorption of up to four Mg atoms at a step edge. To facilitate comparison with the experiment, in all these cases we analysed the energetics and the geometric and electronic structures of the metal clusters adsorbed on the surface and near defects, and calculated the surface density of states (SDOS).

What have we learned regarding the adsorption of metal on MgO thin films from this complex study? Both the experimental MIES spectra and the results of our calculations give similar energies for the position of the top of the valence band (about 6.5 eV) of the MgO film with respect to the vacuum level. This is an important parameter for adsorption, photochemistry and interface studies on MgO films. In the context of the present paper, one can compare this value with the 3s electron ionisation energy of the Mg atom (7.6 eV), and use a simple argument in order to predict the nature of the chemical bonding of Mg atoms on the MgO surface. This qualitative prediction was then confirmed using the results of our DFT calculations, which demonstrate the formation of bonding and anti-bonding states due to the interaction of adsorbed Mg atoms with the surface oxygen ions. The electronic states due to Mg adsorption manifest themselves in the MIES spectra, which can be approximately interpreted using the calculated SDOS. The attachment energies of Mg atoms to Mg clusters formed at different surface sites are compared with the TPD data. The best agreement is achieved with the energies at the step edge. These are the largest adsorption energies we found. Combined with the MIES data, this suggests that the initial nucleation of Mg clusters happens at step edges.

The paper is organised as follows. In Section 2 we give a brief account of the experimental techniques used in this study and present the experimental results. The theoretical methods and the results of calculations are described in Section 3. The discussion of the experimental and theoretical results and conclusions are presented in Section 4. A preliminary account of some of our results is presented in ref. 18.

2 Experimental results

2.1 Experimental techniques

The apparatus used in these studies has been described previously.[19,20] Briefly, it is equipped with a cold-cathode gas discharge source, which also serves: (i) for the production of metastable He (^{3}S/^{1}S) ($E^* = 19.8/20.6$ eV excitation energy) with thermal kinetic energy required for MIES, and (ii) as a source for ultraviolet photoelectron spectroscopy UPS(HeI), $E^* = 21.2$ eV. The intensity ratio ^{3}S/^{1}S is found to be 7:1. Additionally, the apparatus is equipped with X-ray photoelectron spectroscopy (XPS), Auger electron spectroscopy (AES) and low energy electron diffraction (LEED). Metastable and photon contributions within the beam were separated by means of a time-of-flight method using a mechanical chopper. The MIES and UPS spectra were acquired with incident photon/metastable beams 45° with respect to the surface normal. The kinetic energy of the electrons emitted in the direction normal to the surface is measured by employing a hemispherical analyzer (Leybold EA10/100) with an energy resolution of 250 meV for MIES/UPS. Collection of each MIES/UPS spectrum requires approximately 140 s.

A second apparatus, described elsewhere,[21] is equipped with a MIES/UPS source of the same type as described above, and a setup for TPD was used to calibrate the Mg coverages. The TPD spectra were collected with a differentially pumped quadrupole mass filter in line-of-sight to the sample while ramping the sample temperature linearly by 3 K s^{-1}. In addition, this second apparatus is equipped with XPS, AES and LEED.

The qualitative interpretation of the results of the MIES experiments is based essentially on a model of refs. 22 and 23, which is shown schematically in Fig. 1. Excited He* atoms approach the surface with thermal velocities. At the distances of about 2.5–4.0 Å, when there is a considerable overlap between the surface, ψ_{nk} (n is a band, k the wavevector), and the He ψ_{1s} wavefunctions, surface electrons tunnel into the 1s He hole states. This may happen from all surface energy levels that are higher than the He 1s level. In the case of MgO, all electronic states of the O 2p VB participate (see Fig. 1). The energy gained in this so-called Auger de-excitation (AD) process is transferred to the electron occupying the He 2s level which is emitted in the *same* process. The kinetic energies, E_{kin}, of the emitted electrons are measured in MIES experiments and their distributions constitute MIES spectra. Conventionally, though, the electron spectra are presented *vs.* the binding energy scale which refers to the Fermi energy, E_F, of metallic substrate, as shown in Fig. 1. Experimentally, the Fermi energy, E_F, is a fixed point on the energy scale, and corresponds to the maximum kinetic energy at which electrons can be measured with MIES and UPS from a metallic substrate. Since substrate and analyser are in electrical contact, E_F appears at the same kinetic energy, irrespective from the substrate work function, *i.e.*, for all Mg exposures. Thus, presenting the spectra with a binding energy scale, with E_F as origin, allows the change of the work function (due to, for example, adsorption or charging) to be determined from the shift of the high-energy cutoff of the spectra. In addition, the absolute value of the work function can be determined from the energetic distance between this cutoff and the point on the energy scale that equals the excitation energy (19.82 eV) of the probe atom. The maximum binding energy, with respect to the vacuum level, E_{vac}, probed by He* equals its excitation energy minus the binding energy of the He* 2s electron.

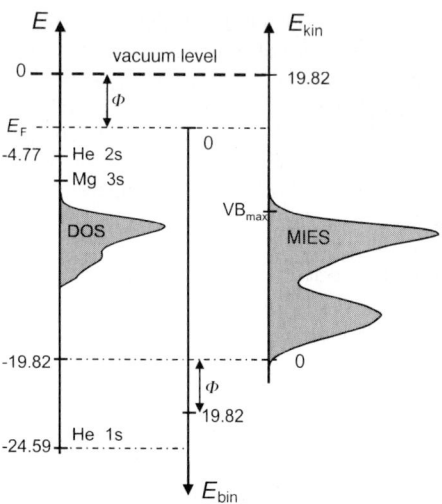

Fig. 1 Energy diagram for a He* probe atom in front of a surface of insulator. Left side: energy levels of the isolated He and Mg atoms and surface density of states in the valence band (VB). Also shown is the position of the Fermi level, E_F, in the insulator band gap; Φ is the work function of the surface. Middle: binding energies, $E_{bin} = E_F - E$, of electrons involved in the Auger de-excitation process are usually presented with respect to this axis, which has its origin at E_F. Right side: schematic of the experimental spectrum of kinetic energies of the electrons emitted in the AD process (E_{kin}). Zero kinetic energy corresponds to a binding energy of 19.82 eV with respect to the vacuum level (or (19.82 − Φ) eV with respect to E_F).

2.2 Electron spectroscopy

MgO layers with an approximate thickness of 2 nm were prepared by evaporation of Mg on Mo(100) and W(110) substrates at room temperature, followed by a subsequent annealing at 800 K in oxygen ambient. The MIES and UPS spectra measured on as-prepared MgO films are very similar to those obtained on MgO single crystals.[24] Fig. 2 shows the MIES and UPS spectra collected during the exposure of the MgO film grown on the Mo(001) surface to Mg atoms at 100 K; no significant changes are found at 300 K. As discussed previously,[24] the MIES spectrum acquired from the MgO film, prior to the Mg exposure (bottom spectrum), reflects the MgO SDOS as seen *via* an AD process. Thus, the spectral feature with binding energies between 4 and 10 eV with respect to E_F is due to the ionisation of the MgO valence band states with O 2p character. The large peak in the MIES spectra located between 10 and 17 eV is, to some extent, affected by secondary and scattered electrons and will not be considered in the following discussion.

One important characteristic of the electronic structure of our MgO film is the position of the top of the valence band with respect to the vacuum level. It depends on the surface preparation and is difficult to determine using conventional methods (see, for example, the discussion in ref. 25). It can be determined using the MIES spectra and the following considerations. The distance (as seen in Fig. 2) between E_F and the top of the valence band is about 4 eV. The distance between E_F and the vacuum level, *i.e.*, the work function Φ, is determined from the high energy cutoff of the spectra. The work function measured for the MgO films grown on the Mo(100) and W(110) substrates is equal to 2.7 eV. Therefore, for the MgO films used in our study we find the top of the valence band at 6.7 ± 0.4 eV with respect to the vacuum level.

As a consequence of the Mg dosing, an additional peak, located at ≃2 eV above the top of the valence band (2 eV binding energy), develops within the band gap. Since the work function of the MgO film (2.7 eV) is considerably smaller than the He 2s binding energy, no unoccupied states are available at the surface into which resonant transfer of the 2s electron can take place, even when taking into account an eventual shift of the 2s level during the He* interaction with the surface. Consequently, the band gap feature arises from the AD process, which involves electrons from the

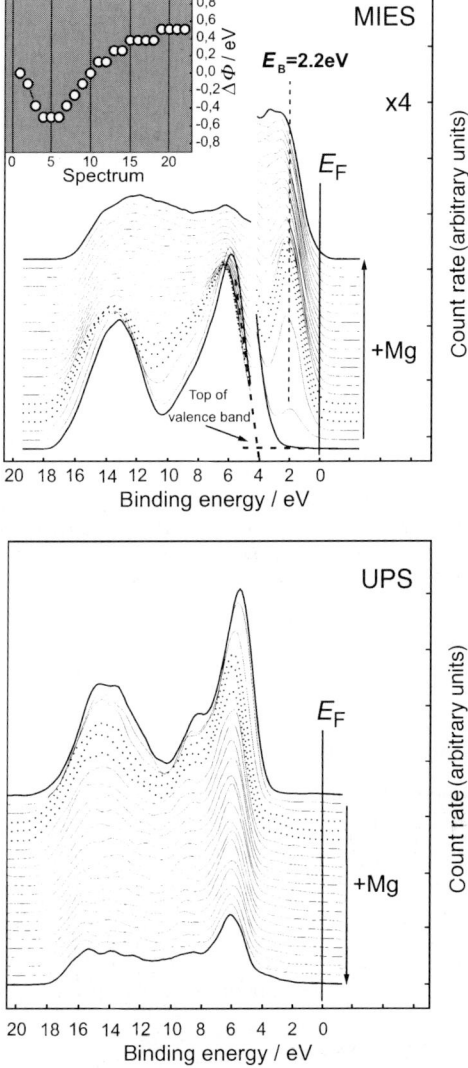

Fig. 2 MIES and UPS spectra acquired from the MgO surface as a function of the Mg exposure. In MIES, the bottom spectrum shows the clean MgO surface; the topmost, the fully covered one; UPS *vice versa*. Inset (top panel): work function change *vs.* exposure time. The dashed spectra were acquired near the work function minimum.

occupied states below the Fermi level due to adsorbed Mg atoms. At larger Mg exposures the MgO valence band emission between 4 and 10 eV weakens considerably. The disappearance of the O 2p structure in the topmost MIES spectrum in Fig. 2 indicates that the entire surface is covered by Mg; the shape of this spectrum is very similar to that for Mg films.

Also shown in Fig. 2 (see inset) is the work function change during the Mg exposure. When Mg atoms are dosed to the oxide, the work function decreases by $\simeq 0.5$ eV while the valence band intensity decreases by $\simeq 10\%$ (dotted MIES spectra). Simultaneously, the top of the valence band shifts towards larger binding energies by approximately the same amount. This coincidental shift of the valence band structures and the high binding energy cutoff indicates a band bending effect rather than a real change of the work function. Such a band bending can be attributed to the creation of additional states on the surface.[5] At larger exposures the work function plateaus at a level typical for metallic Mg films (3.6 eV), which is consistent with a model in which Mg islands grow with lateral bonding similar to bulk Mg.

In order to gain more detailed information about the changes in the low binding energy region during Mg dosage to the MgO surface, MIES spectra were also acquired using a lower Mg evaporation and a higher energy resolution. Fig. 3 summarizes the data obtained at the low evaporation rate. As shown in Fig. 3, both the energy position and the peak width depend (weakly) on the exposure time. An observed feature has a Gaussian shape with 1.8 eV FWHM over the entire studied exposure range. Since this feature appears well separated both from the valence band maximum and E_F, the species responsible for this structure exhibits nonmetallic behavior. The band gap feature can be detected until its intensity falls below a level of 10^{-3} of that of the valence band O 2p emission of the clean MgO surface. Additional measurements show that the band gap feature is stable up to 500 K and has virtually disappeared upon heating to 600 K.

In the valence band region, the UPS measurements (see Fig. 2) provide similar information to MIES, *i.e.*, an attenuation of the MgO substrate intensity at increasing Mg coverages, and a shift of the O 2p structure, which follows, essentially, the work function change of the substrate. On the other hand, due to the fact that UPS probes the average character of several top layers, a significant contribution from the MgO substrate is still noticeable at maximum Mg coverages. In addition, the valence band of the clean MgO surface reveals a two peak structure between 4 and 10 eV, which is discussed in detail ref. 24. However, the most obvious difference is the absence of the Mg-induced band gap feature in UPS; only a small intensity increase in the binding energy range up to 4 eV is observed at high Mg coverages. This can be attributed to the fact that, unlike MIES, UPS probes not just the surface layer but the rather deeper layers, and in addition is very insensitive to metallic s states.[26]

A similar band gap feature was observed in the MIES spectra after dosing with Na atoms; however, it is less stable thermally and disappears between 350 and 400 K upon annealing. Li and Cs additives on the MgO surface also demonstrate similar band gap features.[21]

2.3 Temperature programmed desorption

TPD is an excellent technique for determining surface coverage and studying the interaction between adsorbates themselves and with the surface. Fig. 4 displays TPD and MIES spectra acquired from the Mg-covered MgO surface at the same Mg exposure. Starting at the uppermost MIES spectrum acquired for the clean MgO surface, the Mg exposure increases monotonically towards the bottom spectrum. In the MIES spectrum for the highest Mg coverage (bottom spectrum), the O 2p band has essentially disappeared, indicating complete coverage of the MgO surface by Mg. The relative peak area of the corresponding TPD Mg feature ($m/z = 24$) is 264 times larger than the Mg feature in the uppermost TPD spectrum, which corresponds to the

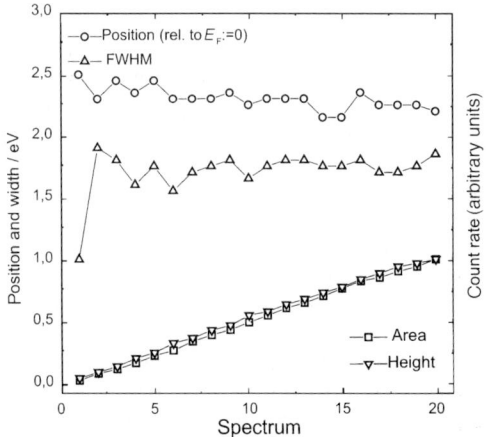

Fig. 3 Intensity, energetic position width (FWHM) of the band gap feature *vs.* time.

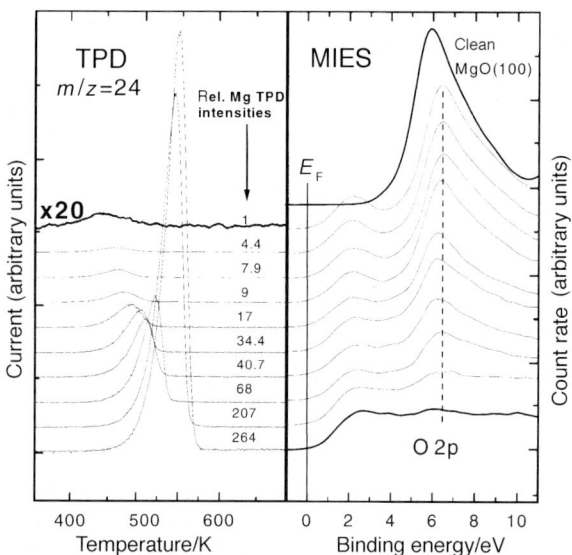

Fig. 4 Comparison between TPD and MIES data. The TPD spectra were taken from the surface characterized with MIES. The topmost MIES spectrum shows the clean MgO surface; the Mg coverage increases monotonically towards the bottom spectrum.

detection threshold of our mass filter. However, even at this very low coverage, the Mg-induced band gap feature appears fully developed in MIES.

Analysis of the Mg TPD peaks reveals an exponential increase in intensity towards higher temperature with increasing coverage (leading edge behaviour), indicating that the desorption follows zeroth-order kinetics. Since this behaviour is typically observed for desorption with scission of adsorbate–adsorbate bonds, this result suggests that even at the lowest Mg concentrations accessible to TPD, the formation of 3D islands occurs. Layer-by-layer growth would lead to a

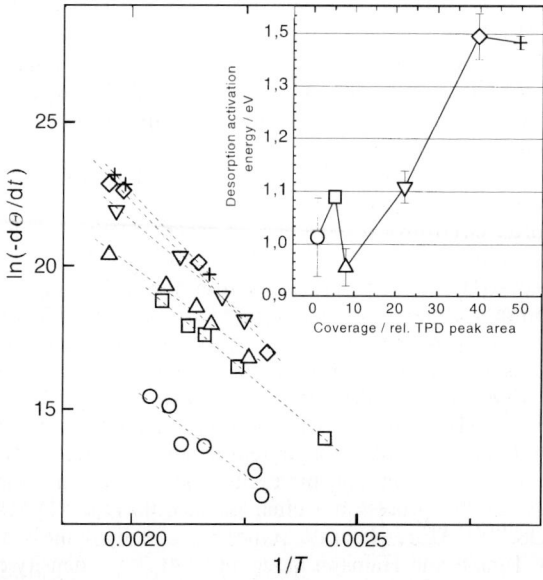

Fig. 5 Arrhenius plot of the desorption rates obtained in the complete analysis[28] of the TPD data in Fig. 4. Inset: calculated desorption activation energy vs. relative Mg coverage.

two-peak structure in TPD (as is actually observed for Na/MgO at 100 to 150 K[27]). The formation of 3D islands is also supported by a non-linear Mg uptake *vs.* exposure time.

The TPD data have been analyzed using the so called *complete analysis*[28] in which an Arrhenius plot of the natural logarithm of the desorption rates yields a straight line for a particular Mg coverage (see Fig. 5). The slope of each line corresponds to the activation desorption energy, E_{des}, for that particular coverage. The inset in Fig. 5 shows that the desorption energy increases initially from about 1.0 eV at very low coverages up to 1.4 eV. This latter value is close to 1.45 eV, the heat of sublimation for bulk Mg.

Even at the lowest exposure accessible for TPD, the band gap feature in MIES is fully developed, and the work function has traversed through its minimum. The band gap feature, as a function of exposure, smoothly transforms into the spectrum characteristics for the full metallic coverage. Our TPD results suggest that this feature is due to Mg clusters which, for sufficiently large exposures, acquire metallic properties. Because the same kind of feature is already present in MIES spectra at much lower exposures, it is reasonable to assume that at these low exposures MIES also detects the presence of clusters at the surface. Thus comparison of Figs. 4 and 5 also supports the formation of Mg clusters with Mg–Mg bond strength similar to that in the Mg bulk metal.

3 Theoretical results

3.1 Theoretical models

Before we go into a detailed description of the theoretical methods and the results of calculations, let us briefly summarise the experimental data. Essentially we are presented with two types of data. The TPD results tell us that the adsorbed Mg clusters have quite large desorption energies of individual Mg atoms, which increase with the coverage. The MIES spectra demonstrate a feature that evolves with the Mg concentration, transforming at large Mg exposures into the spectrum characteristic of full metallic coverage. To construct an atomistic model of adsorption of Mg atoms and further growth of metallic clusters on the MgO film surface, we can also use the STM images[29] of MgO films grown on Mo(001). They demonstrate that these surfaces are very rough and contain 3D MgO islands and a lot of steps within the islands. Such surfaces would normally also have a number of anion and cation vacancies. The anion vacancies can be filled by electrons from the metal substrate forming charged or neutral F-centres. The latter can be ionised directly during the AD process or by trapping electron holes; in addition, holes can localise near cation vacancies (forming the well known stable V-centres). In this study, we assumed that the Mg atom adsorption takes place on terraces, near charged anion vacancies, neutral F-centres, and at step edges. Comparing the calculated Mg atom adsorption energies with the TPD data we can approximately deduce which sites are most favoured. The spectroscopic MIES and UPS data, though, require a much more complex analysis.

The MIES spectra of the MgO films before Mg dosing contain information about their electronic structure. In particular, the largest kinetic energy of emitted electrons corresponds to the AD process, which involves electrons from the highest occupied states localised at the top surface layer, which we call for simplicity "the top of the valence band". The position of the top of the valence band with respect to the vacuum determined from the MIES data is 6.7 ± 0.4 eV. We can calculate the lowest ionisation energy of the surface terrace and compare it with these data. Another prediction that one can deduce from the experimental results is that the top of the valence band and the 3s states of Mg atom have close energies. This suggests formation of partially covalent bonding of Mg atoms to the surface, which can also be checked theoretically.

More detailed analysis of MIES spectra is less straightforward.[30] In the AD process, which is the only one that we took into account in the present paper, a surface electron is transferred into the 1s hole state of the He* 1s2s atom and the excited He 2s electron is ejected. Since only one surface electron is involved in the process, it is often assumed that the AD MIES spectra to a good extent reflect the SDOS.[30,31] More accurate static[30,32] and dynamic[33-35] theoretical models suggest that, like in the Tersoff and Hamann model of STM,[36] the density of states projected on the 1s function of the He* atom and integrated over the incoming trajectory of that atom would better represent the probability of the electron tunneling from the surface to the He*. Our calculations[31,37] have demonstrated that the shape of the experimental MIES spectrum of MgO is different from the bulk and the surface DOS, and is indeed similar to the SDOS projected on the 1s He* function. However, the relative energies of different features in the three DOS and in the experimental spectra remain very similar. Therefore, we believe that the MIES features due to localised band gap states of adsorbed atoms and their relative position with respect to the top of the valence band are likely to be well reproduced by SDOS.

3.2 Theoretical methods

Density functional theory[38] is widely used in surface studies (see, for example, ref. 39). Two models within the DFT method have been used in our work. Due to the importance of the electron correlation for the hole states, we employed the DFT to calculate the surface ionisation energies within an *embedded cluster* model.[16,17] To model the Mg adsorption and MIES spectra, a *periodic model* and the realisation of the DFT in the VASP code[40-43] are more appropriate.

In the embedded cluster calculations, a cluster of up to 50 atoms (quantum cluster) was treated quantum mechanically using DFT. It was embedded into a finite cluster (region I) of $12 \times 12 \times 6$ ions treated in a polarisable ion model. Pair potentials[44] were used to calculate the interactions between these ions, and the shell model[45] to treat the polarisable oxygen ions. The quantum cluster and region I were embedded into an outer region of frozen ions, which makes the total number of ions in the system $20 \times 20 \times 8$. The effective charges on all classical ions were $\pm 2e$ (e is the electron charge). This setup provides the correct values for the Madelung potential and its gradients on the ions in region I and in the quantum cluster. As an example, in Fig. 6 we present the largest quantum cluster $Mg_{37}O_{13}$.

The matrix elements of the electrostatic potential of the rest of the system, including the dipole contributions from the polarized oxygen ions in region I, are included in the Kohn–Sham equations implemented in the modified Gaussian94 code.[46] The B3LYP functional[47] was employed to calculate the electronic structure of quantum clusters. All electrons of oxygen ions and those of the magnesium ions (shown as black in Fig. 6) were described using the 6-31G standard Gaussian basis set.[48] To facilitate calculations of large quantum clusters, the magnesium ions (shown as gray in Fig. 6) were treated using the pseudopotentials of Wadt and Hay[49] and the 1s function described by two contracted Gaussians. In smaller clusters, Mg atoms that had less than two nearest quantum oxygens were treated in the similar way.

Periodic DFT calculations were performed in the slab geometry where an infinite stack of slabs separated by vacuum gap (to suppress the mutual interaction between parallel slabs) is considered in the z-direction. Within each slab, the unit cell is periodically repeated in two dimensions. The

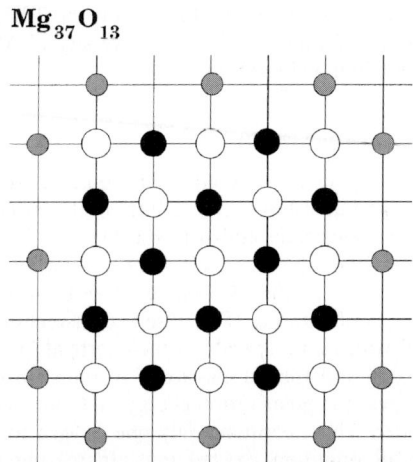

Fig. 6 The top view on the $Mg_{37}O_{13}$ quantum cluster used in the embedded cluster calculations of surface ionisation energy. Open circles represent oxygen ions, black and gray circles are magnesium ions with different basis sets (see text).

Kohn–Sham electronic orbitals ψ_{nk} are expanded into plane waves and the plane wave coefficients are varied to obtain the minimum of the energy for the given ionic geometry. A number of unoccupied states are also included in the variational procedure as it speeds up the calculations and at the same time allows us to consider possible metallisation in the system. The whole system is brought to a mechanical equilibrium by minimising the forces acting on every atom in the unit cell until they become less than 0.1 eV Å^{-1}. Only valence electrons are treated explicitly, which is achieved by using non-local "soft" Vanderbilt pseudopotentials.[50,51] The advantage of using these pseudopotentials instead of the norm-conserving ones[38] is that it is possible to have a relatively small cutoff, E_{cut} = 400 eV, (especially for oxygen), which speeds up the calculations by at least a factor of four. The generalised gradient approximation (GGA)[52] was used in all calculations, which is especially important for surfaces.[39]

The cell sizes, vacuum widths and system geometries for all systems studied in this paper are discussed in detail below. In all calculations the interionic distance of d_0 = 2.122 Å was used to specify the surface unit cells. This was found in ref. 53 to be the equilibrium distance for the bulk MgO using the same GGA functional as in the present study.

The calculations were performed in the following way. First, the singlet ground state of the reference system (perfect surface, step, *etc.*) without adsorbed metal atoms was considered. After, it was relaxed to mechanical equilibrium and Mg atoms were introduced into the calculation. The whole system was relaxed again, except for the atoms in the bottom layer of the slab, which were fixed in the positions corresponding to the perfect system to simulate the crystal bulk. From two to four k-points in the plane (2D) Brillouin zone have been normally used in all such ground-state calculations. This has been shown[54,55] to be sufficient for the cell sizes considered here. The ground state calculations give adsorption energies, relaxed geometries and the electronic densities. The latter were analysed using the general visualisation tool LEV00,[56] which facilitates construction and analysis of arbitrary 3D objects specified on a grid, such as electronic densities and wavefunctions. In addition to the usual contour maps of, for example, electron density, we have also found it very informative to integrate the charge density into spheres of different radii around various positions within the simulating cell, and to compare these results with those obtained for other positions of the same or other similar system (*cf.* ref. 54).

The densities of states were calculated using a method of tetrahedra outlined in ref. 54. Briefly, using the point-group symmetry of the cell, a necessary mesh of k-points was generated for the plane Brillouin zone and the wavefunctions and energies of the surface electrons were recalculated again for all non-equivalent k-points using the VASP code. In most cases, our systems have no symmetry at all and a mesh of 13 k-points corresponding to 750 tetrahedra in the plane Brillouin zone was used. Then, the SDOS was calculated using LEV00. It is smeared by a Gaussian to simulate the effect of phonon broadening[57] at room temperature using a smearing parameter equal to 0.3 eV. All periodic calculations were performed on the T3E parallel supercomputers in the Edinburgh Parallel Computer Center and at the University of Manchester under the Computer Services for Academic Research (CSAR) initiative.

3.3 Calculation of the surface ionisation energy

To calculate the surface ionisation energy, we used several quantum clusters of increasing size, as listed in Table 1. Four of them had quantum oxygens only in the surface plane, and in the case of $Mg_{29}O_{13}$ four additional oxygen ions were added to the $Mg_{25}O_9$ cluster in the second plane to check whether this will have any significant effect. After calculation of the perfect lattice, each cluster was ionised and the difference in the total energy with the ionized state was calculated in two approximations: with, IP(I), and without, IP(0), a self-consistent account of the electronic part of the polarisation in region I, which corresponds to the "vertical" ionisation potential. Only the oxygen polarisation was included and treated classically in the shell model. As the hole delocalisation increases, one would expect the polarisation energy, ΔIP, to decrease.

Except for the smallest cluster, which contains only one oxygen ion, in all the ionised clusters the hole was delocalised by all quantum oxygen ions. Increasing the cluster size we should approach the limit of a completely delocalised band hole state. For the smaller clusters, the hole was distributed almost evenly by all the oxygen ions. However, in the case of the $Mg_{37}O_{13}$ this distribution was more complex, which reflects the fact that as the hole state becomes more delo-

Table 1 The ionisation energies calculated using different quantum clusters, in eV

Cluster	IP(0)[a]	IP(I)[b]	ΔIP[c]	W[d]
Mg_5O_1	9.75	7.64	2.11	0.68
$Mg_{17}O_5$	7.89	6.86	1.03	2.02
$Mg_{25}O_9$	7.29	6.52	0.77	2.26
$Mg_{29}O_{13}$	7.18	6.47	0.71	2.82
$Mg_{57}O_{25}$	7.03	6.52	0.51	3.13

[a] In the calculation of IP(0) the lattice polarisation outside the quantum cluster was not included.
[b] IP(I) are calculated taking into account the classical lattice polarisation.
[c] ΔIP is the difference between the two which reflects the hole localisation and, consequently, the lattice polarisation energy.
[d] W is the valence band width.

calised its eigenvalue approaches the top of the valence band. As a result, several quasi-degenerate hole distributions become possible, which also hampers the convergency of calculations. As one can see in Table 1, this, however, does not significantly affect the calculated energies. The ionisation energies in both approximations first decrease sharply as the cluster size increases and then change slowly. The difference in ionisation energies between the completely localised hole state in the smallest quantum cluster and the most delocalised state in the largest cluster should approximately correspond to half of the valence band width, W, which is also presented in Table 1. This approximately holds in our calculations.

3.4 Adsorption of Mg atoms on the perfect MgO (001) surface and near a surface F-centre

A supercell consisting of three layers of oxygen and Mg atoms (eight surface unit cells in every layer) with the vacuum width between slabs equivalent to three additional layers was used to simulate the perfect MgO(100) surface. Up to four Mg atoms were added to this supercell to model the Mg adsorption. A similar setup was used to model the surface F-centre. The latter was created by removal of one oxygen *atom* from the topmost surface layer in the slab. The whole system was relaxed and its energy was used as the reference energy in the calculations of the corresponding adsorption energies. A detailed account of the surface F-centre calculations with the same method is given in ref. 54. In this paper we focus on the results related to Mg adsorption.

The adsorption energies of one Mg atom on the perfect surface and near the surface F-centre are shown in Table 2. On the regular surface, the most stable position of the Mg atom is above the surface oxygen. The on top F-centre position is much less stable and, although there is a shallow energy minimum corresponding to the adsorption energy shown in Table 2, most of the Mg atoms would probably prefer the nearest oxygen sites to F centres. On top of the F-centre the Mg atom is pulling up part of the electron density from the vacancy. Above the oxygen site the adsorbed Mg atom is about 2.3 Å above the surface plane, with the oxygen ion displaced by about 0.25 Å towards it. No significant relaxation of other surface ions was found. The calculated barrier for

Table 2 Total adsorption energies for all systems studied, in eV

Surface	1 Mg	2 Mg	3 Mg	4 Mg	5 Mg
Perfect surface	0.51	0.96	1.69	2.84	
With F-center	0.21	0.71	2.00	2.51	4.05
With step	1.26	3.05	3.62	4.98	
With anion vacancy	0.51	1.25	2.35	3.64	

diffusion of an Mg atom along the perfect surface is only 0.26 eV, with the barrier point at about 3.2 Å above the centre of the surface unit cell.

The equilibrium distance between Mg atoms in a free Mg_2 molecule, obtained in our calculations, is 3.75 Å, which is much larger than the distance between two nearest oxygen ions and smaller than that between next-nearest ions. Therefore, when more than one Mg atom is added to the system, their lateral interaction does not allow them to occupy the most energetically favourable positions above the oxygen ions. As a result, we find that the potential energy surface in the lateral direction above the surface is very flat with many local minima. Typical geometries for adsorption of four Mg atoms on the perfect surface and near the F-centre are shown in Fig. 7(a), (b).

As one can see, the geometries obtained are similar in both cases. Every Mg atom occupies a surface area containing one surface oxygen. It is positioned above the surface plane in the range 2.2–3.2 Å, depending on the particular arrangement of the Mg atoms and on whether the nearest surface oxygens are displaced significantly towards them from the surface plane. The adsorption energies found for the two systems are summarized in Table 2.

When two Mg atoms are added to the terrace, the adsorption energy per adsorbed Mg atom does not increase more than twice, but is actually a little smaller than the double adsorption energy for one Mg due to the repulsion of Mg atoms. Then it grows slowly reaching 0.7 eV

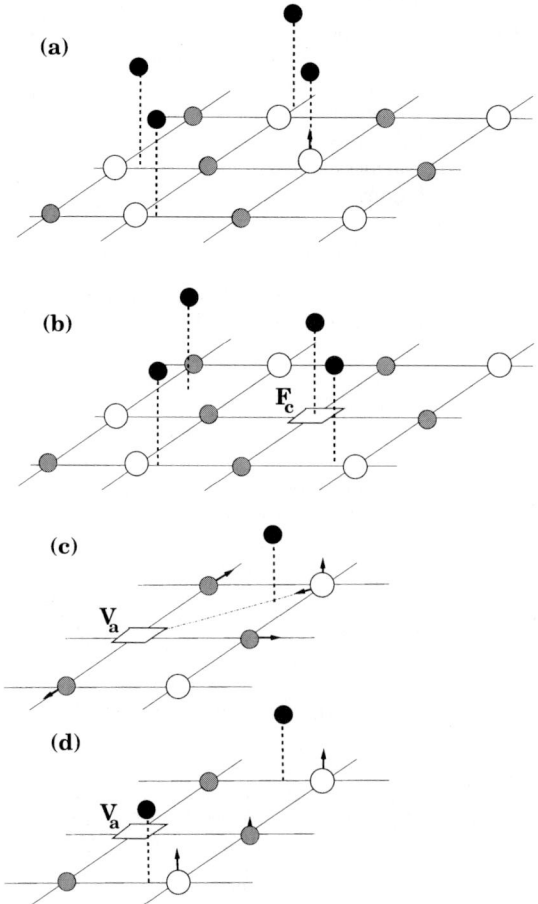

Fig. 7 Typical geometries for the adsorption of four Mg atoms on the terrace (a) and near the F-centre on the flat surface (b); for one (c) and two (d) Mg atoms at the anion vacancy. Oxygens are shown as open balls, surface Mg atoms as shaded smaller balls and adsorbed Mg atoms are shown as black circles.

atom^{-1} when four atoms are added (in this comparison we are using an averaged parameter that corresponds to the dissociation of adsorbed cluster into free atoms). A similar tendency is also observed for the surface with the F-centre. This is due to the creation of mutual bonding between the adsorbed Mg atoms and between them and the surface oxygens.

In order to demonstrate the character of this bonding, we show in Fig. 8 the contour plot of the valence electronic density of the system with one Mg atom adsorbed on the terrace above a surface oxygen. The strong contribution of the electron density from the surrounding oxygen ions into that of the adsorbed Mg atom is clearly visible. To analyse the bonding further, one can integrate the charge density around the adatoms. For one adsorbed Mg atom on the perfect surface, this does not indicate any significant charge transfer to the surface. However, a detailed analysis of the occupied orbitals in this system reveals that the adatom participates in the states at the top of the O 2p valence band (VB) whereas there is a significant contribution of the surface oxygens nearest to the adatom in the charge density localised on the Mg atom. This picture can be further clarified by examining the calculated DOS for this system shown in Fig. 9 (the lowest curve). The last occupied state, n, manifests itself in the DOS as a feature about 1.2 eV from the valence band maximum. The corresponding partial density, $\rho_n(r) = \Sigma_k |\psi_{nk}|^2$, contains contributions from both the adsorbed Mg atom and the surface oxygens underneath. While integrating $\rho_n(r)$ in spheres of different radii, we find that it does not account for the total charge around the adatom, so that part of the density comes from the valence band states. On another hand, there is a considerable localisation of this density on the nearest surface O atoms.

When more than one Mg atom is adsorbed, the electron density is easily shared between them. In all cases we found a rather diffuse density around the adatoms with strong highly localised peaks on the nearest surface oxygens. In the DOS these mixed states manifest themselves as a set of peaks in the gap, which (after smearing) show up as a broad peak around 2 eV above the VB maximum. One can also notice a considerable distortion of the O 2p valence band due to the Mg adsorption, this is seen as a bump at about -1.5 eV.

With three Mg atoms per simulation cell, the partial densities, $\rho_n(r)$, associated with the features in the gap are still well localised. Adsorption of four Mg atoms in our setup corresponds to half of the monolayer. In this case the last occupied state in the system is very diffuse and is spread over most of the simulation cell. This state is mainly due to all four adsorbed Mg atoms and the surface oxygens nearest to them. In the DOS we find that this state has a considerable width (of over 2 eV) and overlaps both with other adatom-related local states at lower energies and with the

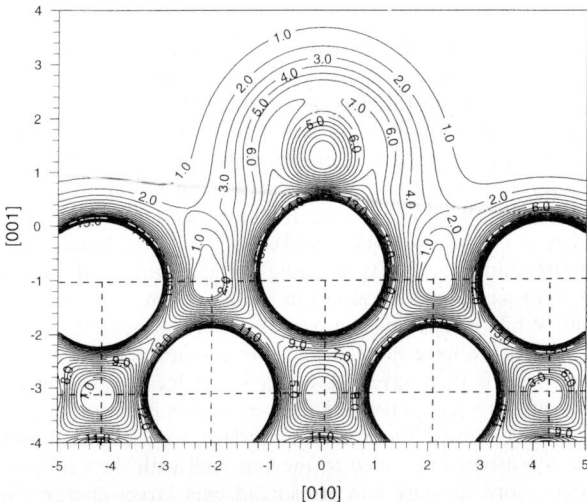

Fig. 8 The contour plot of the valence electronic density of a Mg atom adsorbed above a surface oxygen atom on the terrace (in units of 10^{-2} electron Å^{-3}). To guide the eye, the surface atoms are connected by a dashed line. The cut has been made along [010] axes perpendicular to the surface plane. To avoid high peaks on oxygens, the density has been chopped at 0.2 electron Å^{-3}. Distances are in Å.

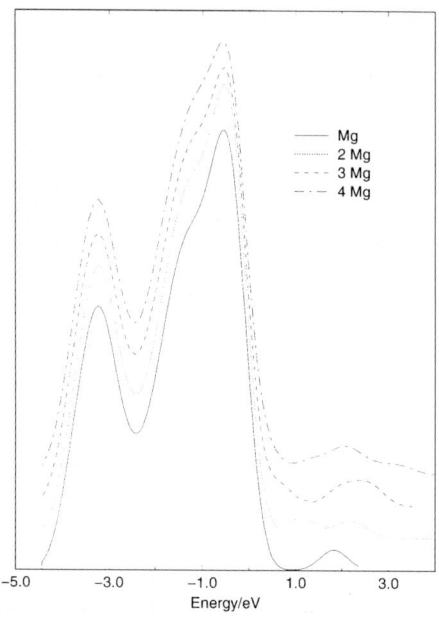

Fig. 9 DOS (arbitrary units) for the perfect surface with up to four adsorbed Mg atoms. The DOS is aligned so that zero energy corresponds to the unsmeared top of the VB.

unoccupied states. Although, strictly speaking, the unoccupied states in the DFT do not have clear physical meaning, we believe, that this behaviour is an indication of the beginning of the system *metallisation*. This is because with four atoms it is already possible to construct a configuration of adsorbed metal atoms in which the distances between the neighboring Mg atoms in the central and adjacent cells are of the same order of magnitude. Then some of the occupied states become very diffuse in the direction of the short distance between the metal atoms, and the system as a whole becomes conductive. This effect is similar to the one in the percolation theory of conductance in disordered systems.

Although some features of the DOS for the system with the F-centre are different, it nevertheless retains the same character. In particular, in the DOS for one adsorbed Mg atom (see Fig. 10, the lowest curve) there are two peaks in the gap: at 0.9 and 2.6 eV above the VB maximum. They correspond to the bonding and antibonding states between the Mg atom and the F-center electrons. These states also contain a significant portion of the density localised on the nearest surface oxygens. Every new atom added to the system generates an additional peak in the gap of the DOS. After smearing (see Fig. 10), all these peaks form a broad feature in the gap approximately 2 eV above the VB maximum. The states that make up the defect band are quite diffuse. They spread over all Mg atoms (some states have bonding, some an antibonding character with respect to the adatoms) and have significant localisation in the anion vacancy and especially on the nearest surface oxygens. While integrating the partial density associated with the states that form the broad feature in the gap, we have not been able to account for all the density, which means that the electrons from the VB themselves have significant localisation on the adsorbed atoms. The VB states therefore are strongly perturbed. This manifests itself in the distortion of the O 2p band, seen in Figs. 9 and 10. Similar to the perfect surface, we found that metallisation starts to form when four or five Mg atoms are added to the supercell with the surface F-centre: the defect states in the gap interact more strongly and are spread over larger energy intervals so that they overlap with unoccupied states.

Thus, we conclude that upon adsorption of Mg atoms, chemical bonding is formed between the Mg atoms, and that at a coverage of roughly half a monolayer the adsorbed layer may become conductive.

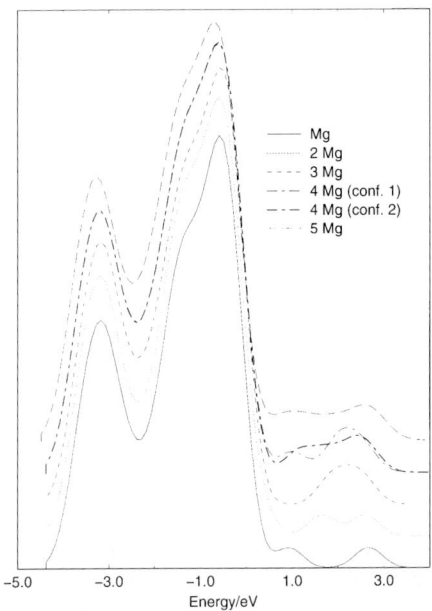

Fig. 10 DOS (arbitrary units) for the perfect surface containing one neutral F-center per simulation cell with up to five adsorbed Mg atoms. The DOS is aligned so that zero energy corresponds to the unsmeared top of the VB.

3.5 Adsorption of Mg atoms near the anion vacancy

Adsorption of Mg atoms near the F-center is similar to that on the perfect surface partly because the F-center bears roughly the same charge (two electrons) as the lattice oxygen O^{2-} ion. A doubly positively charged anion vacancy, V_a, may interact differently with adsorbed metal atoms. However, modelling of charged systems in the periodic model is less straightforward. Formal procedures (see, for example, refs. 58 and 59), which one can use to study charged systems in periodic boundary conditions, are not strictly applicable to surfaces and to slab geometries. In the bulk one can remove the Coulomb interaction energy between the extra charge across the simulation cells by dividing it by the relative permittivity, ε_0. It is not that clear, however, how to model the electronic polarisation in the surface case.

Therefore, in this paper we adopted a different procedure. It is well known that in real systems charged defects tend to be compensated by other defects having the opposite charge. To compensate the anion vacancy at the slab surface we formed a cation vacancy, V_c, at the slab centre. This makes the whole system neutral at the expense of introducing a dipole moment in the cell. The dipole moment is going to be large both along the surface and normal to the surface as we want to separate the two vacancies from each other as much as possible. To check the effect of the dipole moment on the energetics, geometry and the DOS, we have run extensive tests. They have demonstrated that the effect on the energetics and geometries of adsorption is insignificant, and leads only to a shift as a whole of the calculated DOS. Therefore, we used this setup and the 80-atom supercell modelling a five layer slab in further calculations of the Mg adsorption. The cation vacancy was created in the middle layer of the slab, and the oxygen ion was removed in the top layer from the lattice site, which is most separated from the cation vacancy two layers underneath. The distance between the two vacancies is $3d_0 = 6.4$ Å in all calculations. After the geometry relaxation, this system was treated as the reference for further calculations of the Mg atom's adsorption. No significant electron density is localised in the anion vacancy.

One Mg atom is adsorbed between the vacancy and the nearest surface oxygen as shown in Fig. 7(c). The adatom is located 2.4 Å above the surface plane and there is a considerable upward displacement of the nearest oxygen atom. The analysis of the electronic density revealed that two electrons of the Mg atom are strongly pulled towards the vacancy and form a diffuse electronic cloud localised in the area containing both the Mg atom and the anion vacancy. Because of this *direct* charge transfer from the adatom to the surface, the adsorption energy is more than 0.5 eV (see Table 2), which is much greater than in the case of adsorption on the F-centre. In the DOS for this system, shown in Fig. 11 (the lowest curve), one can notice a feature just above the VB top, which is due to this single diffuse state. Because the wavefunction associated with this state penetrates more into the surface than in the other two cases studied above, the extra peak in the gap was found only 0.35 eV above the VB maximum and after smearing appears as a shoulder in Fig. 11.

When one more Mg atom is added to the system, at least two configurations are possible. If the Mg atom finds a surface oxygen within the proximity of the anion vacancy, it shares its electrons with the anion vacancy and the Mg atom already adsorbed nearby. The total adsorption energy (Table 2) is more than doubled. This configuration is depicted in Fig. 7(d). Another possibility is that the second Mg atom adsorbs further away from the vacancy. In this case the adsorption will happen on a terrace above one or two surface oxygens and the adsorption energy increases only up to 0.81 eV, which is less than one would expect from the results for the perfect surface. The corresponding DOS for these two configurations shown in Fig. 11 look very similar after smearing, and one can notice the development of a feature about 1 eV above the VB maximum.

Addition of more Mg atoms leads to the formation of mutual electronic states between them and with the surface, and to a substantial gain in adsorption energy, as seen in Table 2. In the DOS shown in Fig. 11 (two upper curves) one can clearly see the development of a defect band around 1 eV above the VB maximum. It is closer to the VB edge because the defect is positively charged.

3.6 Adsorption of Mg atoms near a monolayer step

Finally, let us turn to the Mg adsorption at a monolayer step. The simulation cell used in these calculations contained 44 lattice sites (for 22 oxygens and 22 Mg atoms) arranged in three layers, as in ref. 54. Each layer, as can be seen in Fig. 12(a), goes like a d_0-high staircase containing a $3d_0$-long and infinitely wide terraces. One Mg atom is adsorbed just in front of the step facing two

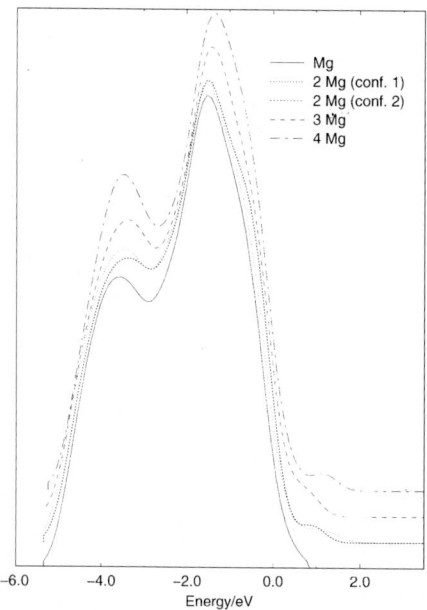

Fig. 11 DOS (arbitrary units) for the surface containing one anion vacancy (compensated by a cation vacancy in the middle of the slab) per simulation cell with up to four adsorbed Mg atoms. The DOS is aligned so that zero energy corresponds to the unsmeared top of the VB.

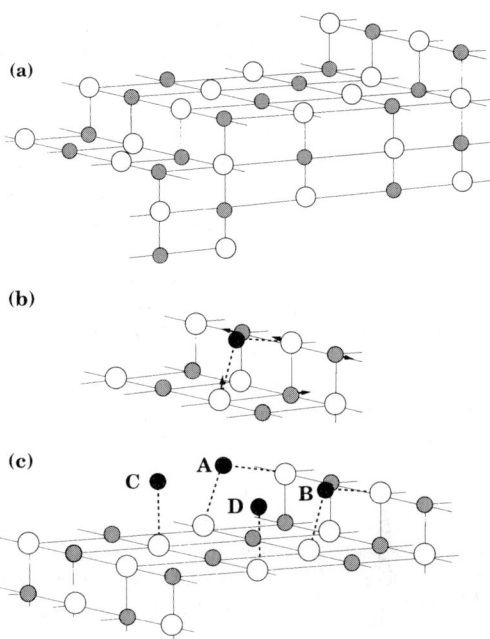

Fig. 12 Three-layer slab system used in the study of the Mg adsorption at the monolayer step (a), the geometry for one (b) and four (c) Mg atoms adsorbed at the step. Notations as in Fig. 7.

surface oxygens, as shown in Fig. 12(b), with the substantial energy gain of 1.26 eV. This is the biggest adsorption energy we obtained for a single Mg atom on the MgO surface. The two oxygens nearest to the adatom slightly displace towards it (by about 0.03 d_0). However, the displacements of the nearest surface Mg ions (up to 0.05 d_0), shown by arrows in Fig. 12(b) are more

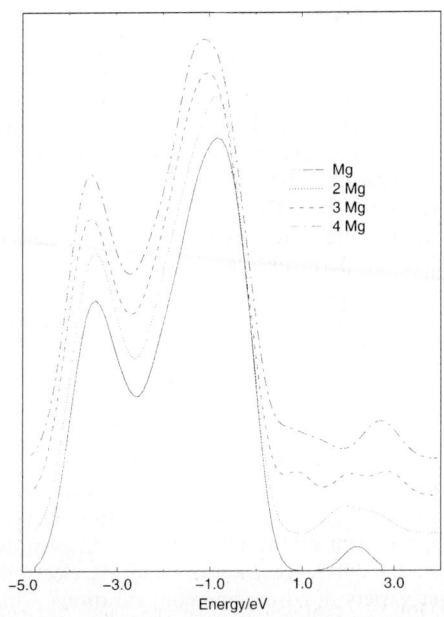

Fig. 13 DOS (arbitrary units) for the monolayer step system with up to four adsorbed Mg atoms. The DOS is aligned so that zero energy corresponds to the unsmeared top of the VB.

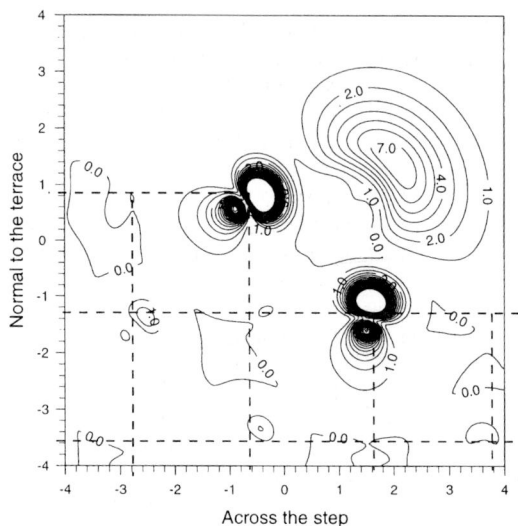

Fig. 14 The contour plot of the partial charge density, $\rho_n(r)$, associated with the defect state in the gap for a single Mg atom adsorbed at the step. The cut has been made perpendicular to the direction of the step through the adsorbed Mg atom and the two nearest surface oxygens. To guide the eye broken lines indicate the surface structure. Other notations are as in Fig. 8.

substantial. This configuration is very stable: the system total energy is, by about 1 eV, higher if the Mg atom is adsorbed on the terrace or just above the center of the unit cell on the step edge. The DOS shown in Fig. 13 (the lowest curve) demonstrates a single peak at about 2 eV above the top of the VB. It is made of the orbitals of the adatom and the two surface oxygens nearest to it. The partial density, $\rho_n(r)$, associated with this state is rather diffuse on the adatom, but forms sharp peaks on the surface oxygens as shown in Fig. 14. The step simulation cell contains two equivalent positions at the step edge (see Fig. 12(b)). It is therefore not surprising that the second Mg atom prefers to stick at this position as well. The adsorption energy presented in Table 2 increases over 1.5 eV atom^{-1}. The band at 3 eV above the VB in the DOS (see the second curve from the bottom in Fig. 13) becomes broader. However, as the third and the fourth Mg atoms are adsorbed, no such positions are available in the simulation cell, and the additional Mg atoms are forced to occupy the terrace sites. The typical geometry for four adsorbed Mg atoms is shown in Fig. 12(c). Nevertheless, the adsorption energy increases substantially (see Table 2). One can also notice that a second defect band in the DOS around 2 eV above the VB maximum is developed due to Mg adsorption on the terrace sites. This band consists of several states equal to the number of adsorbed Mg atoms. Note that the states responsible for the defect features in the gap of the DOS are localised in the direction perpendicular to the step. However, due to a relatively small size of simulation cell along the direction of the step, the states in question are delocalised in this direction. Nevertheless, we have not found any signs of the metallisation at this coverage.

4 Discussion

Let us start with a brief discussion of theoretical results. First, we note a qualitative agreement between our results for the Mg adsorption on the MgO surface terrace and those obtained by Musolino et al.[12] for adsorption of Cu_n ($n = 1, \ldots, 4$) on MgO using a DFT based method. Interestingly, some of the calculated geometries and adsorption energies for clusters, and also the diffusion parameters for one Cu atom on the surface are close, even quantitatively, to our results for the Mg adsorption despite the difference in the Cu and Mg electronic structures. The results,[12] though, demonstrate a richer variety of adsorption configurations, which we did not fully explore in this study. A qualitative agreement also exists with the embedded cluster DFT calculations of M_4 clusters (M = Cu, Ag, Ni, Pd) on MgO by Matveev et al.[13] Both the results of Matveev et al.

and our calculations predict polarisation of adsorbed metal atoms (see Figs. 8 and 14). However, our analysis suggests a more covalent character for chemical bonding of Mg_4 clusters with the MgO surface than suggested for other metals in ref. 13.

We start our comparison with the experimental data from the perfect surface. Although the results of calculations for the surface ionisation energies (see Table 1) are in good agreement with the position of the top of the valence band with respect to the vacuum experimentally determined from the MIES data (6.7 ± 0.4 eV), this agreement is not conclusive. Our results do not show a fast convergency with the quantum cluster size. Although its further increase is not feasible, we believe that the values given in Table 1 are already representative. Thermal fluctuations and the surface roughness broaden significantly the band edge, which is reflected in the UPS spectra (see Fig. 2) and in the experimental error in the determination of the position of the top of the valence band from the MIES spectra. This broadening masks the "electronic" band edge, which can be obtained as a limit in our calculations (see, for example, ref. 60).

Similar to refs. 18, 30, 31, we made the assumption that the MIES spectra due to the AD process reflect to a good approximation the SDOS. This, in fact, does not hold for the states in the lower part of the valence band and those below the valence band. As one can see in Figs. 9–11, the calculated surface DOS has a pronounced maximum at these energies, which is almost completely absent in Fig. 2(a). This is because the wavefunctions of lower energy states decay faster into the vacuum than those of the higher energy states, as discussed in detail elsewhere.[37] Nevertheless we believe that the position of the defect states in the band gap with respect to the top of the valence band can be reproduced more reliably. The comparison of the numerical results and the experimental MIES spectra thus suggests that the band gap feature is due to adsorbed Mg atoms and small Mg clusters. The SDOS calculated for the Mg adsorption on the terrace, near the F-center and at the charged anion vacancy predict a shoulder at about -1.5 eV (see Figs. 9–11). However, despite the fact that the UPS spectrum for the perfect surface is well reproduced by the SDOS, such a shoulder is not seen in the UPS spectra after the Mg exposure, shown in Fig. 2(b). They also do not demonstrate any visible band gap features. We attribute this to the fact that UPS probes mostly deeper surface layers, which are not affected by the Mg adsorption. Another reason is that UPS is very insensitive to metallic s states.[26]

Our theoretical results suggest that individual Mg atoms adsorbed on terraces are fairly mobile at room temperature (the calculated adiabatic barrier for diffusion is only 0.26 eV). The largest adsorption energies were obtained for individual Mg atom adsorption at the step edge. These results indicate that at very low coverages one can expect more Mg atoms to be adsorbed at step edges than at terraces. However, as the Mg exposure increases, the energy gained due to attachment of each atom to existing clusters on the terrace also increases (see Table 2). So, for instance, to desorb one atom from the four atom cluster on the terrace requires 1.15 eV and from the terrace close to the step edge 1.36 eV (see Fig. 12(c)). As one can see in Fig. 5, the desorption activation energies, as derived from the TPD measurements, increase with the Mg exposure from about 1.0 eV at relatively low coverages, to 1.4 eV for almost metallised surface. As discussed in Section 2.3, the TPD data suggest that these energies correspond to desorption of individual Mg atoms from Mg clusters and metallic layer. Although it is tempting to directly compare the theoretical results with the TPD data, this is impossible due to the large variety of both adsorption sites and cluster geometries. Since TPD measures the *smallest desorption energies*, comparison with these data without proper simulation of desorption kinetics can be only qualitative. It suggests that the calculated desorption energies of individual Mg atoms are certainly in the range of desorption energies determined from TPD.

Based on these results, it seems plausible to assume that first the metal atoms occupy all available sites immediately in front of the step edges (decoration of MgO islands or clusters). They form chemical bonds with the nearest surface oxygens accompanied by the considerable lateral interaction between the adsorbed Mg atoms. This assumption is supported by comparison of the dependence of the calculated SDOS on the character of Mg adsorption with the MIES spectra. According to the calculations, when Mg atoms are adsorbed at step edges, this should result in the development of the defect band in the band gap, about 2.5–3 eV above the VB maximum (see Fig. 13). As all such sites are occupied by the adsorbed atoms, the nearby terrace sites also become gradually filled. A band about 2 eV above the top of the VB should develop due to newly adsorbed Mg atoms (see Fig. 9). Since the number of terrace sites is much bigger than that of the edge sites, the peak at 3 eV above the VB should soon become less visible. This implies that the Mg related peak in the MIES spectra should shift to lower energies as the Mg concentration increases. Careful analysis of the experimental spectra demonstrates that this indeed is the case.

It is interesting to note a new structure formed by Mg atoms adsorbed at the step edge (see Fig. 12(c)): the oxygen vacancy created by the three adsorbed atoms A–C–B and the step Mg ion. The electron density plot shown in Fig. 15 clearly shows a considerable localisation of the electron density in the 'pockets' created by the Mg atoms and in the 'vacancy'. The band gap feature corresponding to these states is seen in Fig. 13.

In principle, a band gap feature of similar shape and energetic position could also result from Auger de-excitation of point defects, F-centers in particular. However, no indication of this feature is seen at the clean surface (which certainly is not free of point defects). We also note that, because of the proximity of the metal substrate and also because the position of its Fermi level is several eV above the top of the MgO VB, one can expect that in our experiments all electronic traps, such as anion vacancies and hole centres, will be quickly filled by electrons, tunnelling from the metal. The speed of this should depend, though, on the film thickness. This will not be the case for other types of experiments, *e.g.*, on single crystals.

Our results for Mg/MgO allow for the following, more general qualitative considerations. One of the key parameters responsible for the type and strength of bonding between metal species and oxide surface is the ionisation energy of a metal atom with respect to the valence and conduction bands of the oxide surface. Provided the valence level of the metal atom is in resonance with the occupied valence band states, the bonding will feature covalent and polarisation contributions. This holds in the present case of Mg/MgO and for other metals on MgO with an ionization potential of about 7 eV and larger.

On the other hand, if the metal level is in resonance with unoccupied states of the conduction band, as is the case for alkali atoms adsorbed on TiO_2, a charge transfer from the alkali metal atoms to the oxide can be expected. For the alkali/TiO_2 case this leads to the reduction of the Ti cation (and the appearance of band gap states due to Ti^{3+} 3d formation). Thus, there is ionic chemisorption in this and similar cases. MIES studies on such systems are currently in progress at TU Clausthal.

In cases like Mg/TiO_2, both charge transfer from the metal atom to the cation *and* hybridisation of the Mg 3s and O 2p states might occur *simultaneously*, resulting in mixed covalent *and* ionic bonding. At present we are attempting to verify this prediction in a joint MIES and UPS study on Mg/TiO_2.

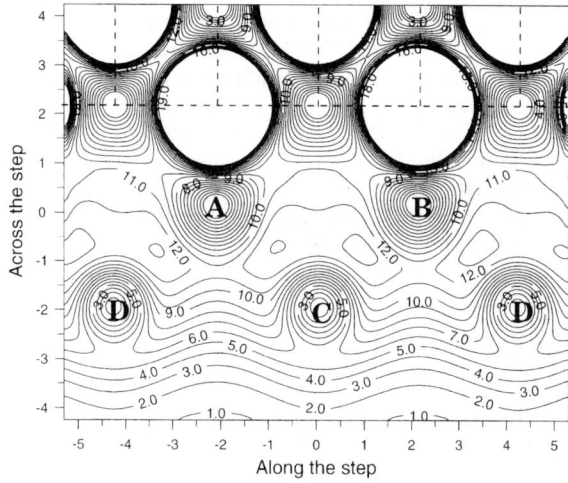

Fig. 15 The contour plot of the valence charge density for the step system with four adsorbed Mg atoms (labelled) as shown in Fig. 12. The cut is made parallel to the terrace in such a way that all four adsorbed atoms are crossed (note that they are not exactly at the same height). Broken lines indicate positions of the atoms at the upper terrace. The adsorbed atoms are in front of this terrace. Other notations as in Fig. 8.

Finally, we note that the presence of extended surface defects, steps in particular, will most likely not change these qualitative considerations. However, as our calculations for Mg/MgO indicate, they may affect the adsorption strength quite appreciably and therefore decide the adsorption kinetics.

Acknowledgements

LNK is supported by EPSRC. PVS would like to acknowledge the support by Kodak. This work has been performed in the framework of a joint collaborative project between University College London and the TU Clausthal, Germany, funded by the British Council (grant ARC 887) and DAAD. We are grateful to A. M. Stoneham and I. V. Abarenkov for useful discussions, and to I. I. Tupitsyn and G. Tsikarishvili for help in calculations, and to A. S. Foster for useful comments on the manuscript. We gratefully acknowledge the allocation of computer time on the Cray T3E provided by the High Performance Computer Initiative through the Materials Chemistry consortium.

References

1 E. Giamello, A. Ferrero, S. Collucia and A. Tecchina, *J. Phys. Chem.*, 1991, **95**, 9385.
2 D. Murphy and E. Giamello, *J. Phys. Chem.*, 1995, **99**, 15172.
3 D. W. Goodman, *Chem. Rev.*, 1995, **95**, 523.
4 C. Xu and D. W. Goodman, *Handbook of Heterogeneous Catalysis*, eds. G. Ertl, H. Knözinger and J. Weitkamp, VCH, Weinheim, 1997, vol. 2, p. 826.
5 C. T. Campbell, *Surf. Sci.*, 1997, **387**, 136.
6 R. Souda, Y. Hwang, T. Aizawa, W. Hayami, K. Oyoshi and S. Hishita, *Surf. Sci.*, 1997, **387**, 136.
7 T. Suzuki, S. Hishita, K. Oyoshi and R. Souda, *Surf. Sci.*, 1997, **391**, L1243.
8 T. Kizuka and N. Tanaka, *Surf. Sci.*, 1997, **386**, 249.
9 E. Giamello and D. Murphy, *Mol. Eng.*, 1994, **4**, 147.
10 Y. Kanzawa, T. Kageyama, S. Takeoka, M. Fujii, S. Hayashi and K. Yamamoto, *Solid State. Commun.*, 1997, **102**, 533.
11 G. K. Wertheim, *Z. Phys. D: At., Mol. Clusters*, 1989, **12**, 319.
12 V. Musolino, A. Selloni and R. Car, *J. Chem. Phys.*, 1998, **108**, 5044.
13 A. V. Matveev, K. M. Neyman, G. Pacchioni and N. Rösch, *Chem. Phys. Lett.*, 1999, **299**, 603.
14 A. M. Stoneham and J. H. Harding, *Proc. 9th World Ceramics Congress (CIMTEC)*, Florence, June 1998, in press.
15 A. M. Ferrari and G. Pacchioni, *J. Phys. Chem.*, 1996, **100**, 9032.
16 A. L. Shluger, P. V. Sushko and L. N. Kantorovich, *Phys. Rev. B: Condens. Matter*, 1999, **59**, 2417.
17 A. L. Shluger, L. N. Kantorovich, A. I. Livsits and M. J. Gillan, *Phys. Rev. B: Condens. Matter*, 1997, **56**, 15332.
18 J. Günster, J. Stultz, S. Krischok, D. W. Goodman, P. Stracke and V. Kempter, *J. Vac. Sci. Technol. A*, 1999, **A17**, 1657.
19 W. Maus-Friedrichs, M. Wehrhahn, S. Dieckhoff and V. Kempter, *Surf. Sci.*, 1990, **237**, 257.
20 W. Maus-Friedrichs, S. Dieckhoff and V. Kempter, *Surf. Sci.*, 1991, **249**, 149.
21 D. Ochs, M. Brause, W. Maus-Friedrichs and V. Kempter, *J. Electron Spectrosc. Relat. Phenom.*, 1998, **88–91**, 757.
22 P. A. Zeijlmans van Emmichoven, P. A. A. F. Wouters and A. Niehaus, *Surf. Sci.*, 1988, **195**, 115.
23 P. Eeken, J. M. Fluit, A. Niehaus and I. Urazgil'din, *Surf. Sci.*, 1992, **273**, 160.
24 D. Ochs, W. Maus-Friedrichs, M. Brause, J. Günster, V. Kempter, V. Puchin, A. Shluger and L. Kantorovich, *Surf. Sci.*, 1996, **365**, 557.
25 P. A. Cox and A. A. Williams, *Surf. Sci. Lett.*, 1986, **175**, L782.
26 *Electron Specroscopy. Theory, Techniques and Applications*, ed. C. R. Brundle and A. D. Baker, Academic Press, London, 1977–79, vol. 1–4.
27 S. Krischok and W. Goodman, to be published.
28 D. A. King, *Surf. Sci.*, 1975, **47**, 384.
29 M. C. Gallaher, M. S. Fyfield, J. P. Cowin and S. A. Joyce, *Surf. Sci.*, 1995, **339**, L909.
30 Y. Harada, S. Masuda and H. Ozaki, *Chem. Rev.*, 1997, **97**, 1897.
31 D. Ochs, W. Maus-Friedrichs, M. Brause, J. Günster, V. Kempter, V. Puchin, A. Shluger, and L. Kantorovich, *Surf. Sci.*, 1996, **365**, 557.
32 H. D. Hagstrum, *Phys. Rev.*, 1954, **96**, 336.
33 Z. L. Mišković and R. K. Janev, *Surf. Sci.*, 1986, **166**, 480.
34 A. T. Amos, K. W. Sulston and S. G. Davison, *Adv. Chem. Phys.*, 1989, **76**, 335.
35 B. L. Burrows, A. T. Amos, Z. L. Mišković and S. G. Davison, *Phys. Rev. B: Condens. Matter*, 1995, **51**, 1409.

36   J. Tersoff and D. R. Hamann, *Phys. Rev. B: Condens. Matter*, 1985, **31**, 805.
37   L. N. Kantorovich, A. L. Shluger, P. V. Sushko and A. M. Stoneham, *Surf. Sci.*, submitted.
38   M. C. Payne, M. P. Teter, D. C. Allan, T. A. Arias and J. D. Joannopoulos, *Rev. Mod. Phys.*, 1992, **64**, 1045.
39   M. J. Gillan, L. N. Kantorovich and P. J. D. Lindan, *Curr. Opin. Solid State Mater. Sci.*, 1996, **1**, 820.
40   G. Kresse and J. Furthmüller, *Comput. Mater. Sci.*, 1996, **6**, 15.
41   G. Kresse and J. Furthmüller, *Phys. Rev. B: Condens. Matter*, 1996, **54**, 11169.
42   G. Kresse, Thesis, Technische Universität, Wien, 1993.
43   G. Kresse and J. Hafner, *Phys. Rev. B: Condens. Matter*, 1993, **47**, RC558.
44   A. L. Shluger, A. L. Rohl, D. H. Gay and R. T. Williams, *J. Phys.: Condens. Matter*, 1994, **6**, 1825.
45   B. G. Dick and A. W. Overhauser, *Phys. Rev.*, 1958, **112**, 603.
46   Gaussian94 (Revision E.1), G. W. Trucks, H. B. Schlegel, P. M. W. Gill, B. G. Johnson, M. A. Robb, J. R. Cheeseman, T. A. Keith, G. A. Petersson, J. A. Montgomery, K. Raghavachari, M. A. Al-Laham, V. G. Zakrzewski, J. V. Ortiz, J. B. Foresman, J. Cioslowski, B. B. Stefanov, A. Nanayakkara, M. Challacombe, C. Y. Peng, P. Y. Ayala, W. Chen, M. W. Wong, J. L. Andres, E. S. Replonge, R. Gomperts, R. L. Martin, D. J. Fox, J. S. Binkley, D. J. Defrees, J. Baker, J. P. Stewart, M. Head-Gordon, C. Gonzalez and J. A. Pople, Gaussian Inc., Pittsburgh, PA, 1995.
47   A. D. Becke, *J. Chem. Phys.*, 1993, **98**, 5648.
48   W. J. Hehre, R. Ditchfield and J. A. Pople, *J. Chem. Phys.*, 1972, **56**, 2257.
49   W. R. Wadt and P. J. Hay, *J. Chem. Phys.*, 1985, **82**, 284.
50   D. Vanderbilt, *Phys. Rev. B:, Condens. Matter*, 1990, **41**, 7892.
51   G. Kresse and J. Hafner, *J. Phys.: Condens. Matter*, 1994, **6**, 8245.
52   J. P. Perdew, J. A. Chevary, S. H. Vosko, K. A. Jackson, M. R. Pederson, D. J. Singh and C. Fiolhais, *Phys. Rev. B: Condens. Matter*, 1992, **46**, 6671.
53   L. N. Kantorovich, M. J. Gillan and J. A. White, *J. Chem. Soc., Faraday Trans.*, 1996, **92**, 2075.
54   L. N. Kantorovich, J. Holender, and M. J. Gillan, *Surf. Sci.*, 1995, **343**, 221.
55   L. N. Kantorovich and M. J. Gillan, *Surf. Sci.*, 1997, **374**, 373.
56   L. N. Kantorovich, unpublished data.
57   G. D. Mahan, *Phys. Rev. B: Condens. Matter*, 1980, **21**, 4791.
58   M. Leslie and M. J. Gillan, *J. Phys. C: Solid State Phys.*, 1985, **18**, 973.
59   G. Makov and M. C. Payne, *Phys. Rev. B: Condens. Matter*, 1995, **51**, 4014.
60   G. K. Wertheim, J. E. Rowe, D. N. E. Buchanan and P. H. Citrin, *Phys. Rev. B: Condens. Matter*, 1995, **51**, 13675.

Paper 9/03241J

A microcalorimetric study of the heat of adsorption of copper on well-defined oxide thin film surfaces: MgO(100), p(2 × 1) oxide on Mo(100) and disordered W oxide

Jeffrey T. Ranney, David E. Starr, Jana E. Musgrove, Dan J. Bald and Charles T. Campbell*

Department of Chemistry, University of Washington, Box 351700, Seattle, WA 98195-1700, USA

Received 1st April 1999

The heats of adsorption as a function of coverage have been determined for copper adsorption onto several well-defined oxide thin film surfaces at room temperature by microcalorimetric measurements. The heats of adsorption are accurately determined as a function of coverage with resolution of 2% of a monolayer. For all three oxide surfaces investigated, MgO(100), a p(2 × 1) molybdenum oxide film on Mo(100) and disordered W oxide, the initial heat of copper adsorption is much lower than the heat of sublimation for Cu (337.4 kJ mol^{-1}). On MgO(100) the initial Cu heat of adsorption in the first 2–4% of a monolayer is 240 kJ mol^{-1} and increases rapidly to the heat of Cu sublimation. Auger spectroscopy shows that Cu grows on MgO(100) as two-dimensional (2-D) islands until ≈0.3 monolayers where it switches to the growth of 3-D islands, at which point the heat of adsorption of Cu reaches ≈92% of its heat of sublimation. The room temperature sticking probability of Cu on MgO was also investigated as a function of coverage and determined to be >0.99. On the ordered p(2 × 1) oxide of molybdenum on Mo(100), the initial Cu heat of adsorption is 287 kJ mol^{-1}. The heat of adsorption then decreases slightly to 278 kJ mol^{-1} in the first 15% of a monolayer, after which it rapidly increases to the heat of sublimation. Similarly, on the disorder W oxide surface the initial heat of Cu adsorption was 280 kJ mol^{-1} at 300 K. These results are compared to Pb adsorption on the same oxide thin films and are discussed in the context of important factors influencing metal island growth.

Introduction

Calorimetric measurements of the heats of adsorption of reactant molecules or metal atoms on surfaces provide important thermodynamic data. Such data greatly enhance both our understanding of the thermodynamic driving forces which influence molecular surface transformations (*i.e.*, catalytic reactions) and of the properties of adsorbed metal films (*i.e.*, oxide-supported metal catalysts). Until fairly recently, detailed studies of the heat of adsorption at low coverages on highly defined surfaces were not possible. However, a new technique of single crystal adsorption calorimetry has been introduced as a powerful surface science tool for measuring the heats of adsorption on well-defined surfaces. This technique was first developed by King and coworkers[1–5] and its applications have been reviewed.[6,7] In this technique, the heat generated by the adsorption of a pulse of gas containing as little as 1–2% of a monolayer (ML) can be measured by monitoring the transient temperature rise caused by the adsorption on an ultrathin single-crystalline sample. Several methods can be used to measure the heat input into the sample as a result of adsorption such as infrared (IR) detection as originally used by King and coworkers,[1-6] or by using pyroelectric LiTaO$_3$ single crystals as heat sensitive substrates.[8] This microcalorimetric technique is particularly useful for non-reversible adsorption processes where temperature programmed desorption (TPD) and equilibrium adsorption isotherm experiments fail to provide adsorption energies. Another difficulty with TPD studies is that the surface structure of the adsorbed species present at the desorption temperature may and often does differ from the structure of the adsorbate of interest present at lower temperatures. This is particularly true in cases for molecules that decompose on the surface during heating and for metal adsorption where thermodynamic equilibrium achieved at high temperatures often provides for vastly different structures than are observed by kinetically limited adsorption and film growth at lower temperatures. In addition, for many systems, the underlying surface is modified during the heating process, either by incorporation of the adsorbate into the surface, such as alloying or mixed oxide formation, or by the decomposition of the surface, which can be observed during TPD of metals on oxide films or of metals on polymer surfaces. King and co-workers[1-6] have used this technique primarily to investigate the adsorption of gaseous molecules (reactants) on single crystal metal (model catalyst) surfaces. We describe here calorimetric measurements of metal atom adsorption energies on well-defined oxide surfaces using a different heat detector than that developed by King and co-workers.[1-6]

Characterization of interfacial systems and the understanding of the interactions between metal films and a variety of substrates used as supports is important for a host of technologies, including catalysis, semiconductor processing, and biotechnology. From a catalytic standpoint, understanding the interactions of metals with underlying support oxides is necessary in order to characterize and predict how the oxide/metal interactions influence the catalyst morphology and catalytic properties. Several reviews of this topic have been published detailing a variety of interesting systems and important effects.[9-14] In the case of metal adsorption on well-defined thin oxide films, very little is known concerning the metal heats of adsorption and most of the existing experimental data is derived from TPD experiments with the aforementioned difficulties. Scanning microelectronic calorimeters have been developed to study metal films and small metal particles.[15-17] These studies provide important information on the heat of fusion for the adsorbed metal islands and investigate melting point changes as a function of particle size for small metal particles. However, these studies are not able to provide the heat of adsorption that is obtained from single crystal adsorption calorimetry. In previous communications, we have presented a modification that extends the microcalorimetric adsorption technique for the study of metal adsorption[18,19] and reported the microcalorimetric heats of adsorption for a variety of metal/metal and metal/oxide systems.[18-24] Systems previously investigated include Pb adsorption on Mo(100), MgO(100) and the p(2 × 1) oxide of Mo(100), all studied at room temperature.[18-23] The observed differences between Pb adsorption on the metal and adsorption on the oxide surfaces follows the expected trend where metals interact strongly with other metals but only weakly with oxide surfaces. In this communication, those heats of Pb adsorption on oxide surfaces are compared to the adsorption energies of Cu on similar surfaces in order to address the factors that influence the growth of metal films on oxide thin film surfaces.

We report here a calorimetric measurement of the coverage-dependent, heat of adsorption of copper onto two different metal oxide single crystal surfaces: a thin film of MgO(100) grown on a Mo(100) crystal and an ordered thin film of Mo oxide produced by mild thermal oxidation of Mo(100). To gain a more complete understanding of the factors that determine the heat of adsorption and play an important role in determining the film growth mode, it is important to compare data from different systems. The results reported here are compared to our previous results for Cu adsorption on an oxidized but disordered W(100) surface,[21] highlighting a trend in adsorption energies for Cu adsorption on oxide thin film surfaces. Lead adsorption has also been studied on these oxide surfaces and shows similar results to the Cu/oxide systems, indicating a trend for metal adsorption on oxides that may, once more systems are tabulated, allow better understanding of catalytic properties from a metal/support interaction standpoint.

Magnesium oxide is one of a large group of oxide materials that is often used as relatively inert supports for active metals in catalytic applications. The Cu/MgO(100) system is a well-studied representative model catalytic system. Numerous studies, both experimental and theoretical, of Cu thin films on oxide materials and on MgO(100) in particular have been previously published.[25–34] The conclusions in the literature for Cu film growth on MgO are inconsistent, although they often show generally similar results. Data for the growth of Cu on MgO(100) have been interpreted to be epitaxial,[30] Volmer–Weber (3-D particle formation, VW)[25] and Stranki-Krastanov (layer-by-layer in the first monolayer with 3-D particles growing on top of the surface wetting layer, SK).[27,28,31] Surfaces of MgO(100) have been studied using both thin films of MgO(100) grown on a suitable refractory metal surface[26] and on cleaved MgO crystals.[25,27–31] LEED, XPS, TPD, EELS, AES and a variety of other spectroscopic techniques have been employed to study Cu on MgO(100) over a range of substrate temperatures and Cu coverages. At high coverage, Cu islands or films grow which have a Cu(100) structure.[30] Below ≈ 180 K, evidence of isolated Cu adatoms on the MgO surface has been presented,[34] while at higher temperatures Cu islands form with the Cu atoms becoming mobile near 200 K and with a diffusion barrier height of around 0.5 eV. The more recent studies indicate that at room temperature Cu islands grow in either VW type 3-D growth mode[25,26] or with SK growth.[27,28] The formation of a Cu_2O species at low Cu coverages has been suggested,[27,28] although this assignment, which is typically based on AES peak shifts, is questionable when considering the importance of final-state effects on these small particles, as discussed below and elsewhere.[9] In theoretical studies, several groups[32,33] estimated the properties of small Cu particles adsorbed on MgO(100). These theoretical studies suggest that on defect-free surfaces the Cu will preferentially bind to surface oxygen atoms, and that the growth of 3-D islands is energetically preferred even for small particles (*i.e.*, VW growth). Additionally, these studies suggest that at low coverages a Cu(100) like overlayer is not expected.[32] Copper adsorption has also been studied on a variety of other oxides, including Al_2O_3, ZnO, TiO_2 and SiO_2,[26,29,35–39] and 3-D Cu particle growth in the sub-monolayer regime is typically reported. Surfaces of MgO(100) have also been studied with respect to several other metals, such as Pd, which are reviewed in detail elsewhere.[10]

Molybdenum finds use in a variety of catalytic applications and molybdenum oxide as a catalyst is active for partial oxidation reactions. The p(2 × 1) structure formed by light oxidation of Mo(100) represents a model molybdenum oxide thin film surface for studying Cu adsorption at room temperature. Oxide surfaces of Mo(100) have been studied previously to understand their structure and chemical/catalytic activity.[40–46] This thin p(2 × 1) oxide is an ≈ 2 atomic layer thick oxide of molybdenum,[40–46] with $\approx 50\%$ oxygen in each of the two top layers of the oxide.[40] Studies using STM support a missing-row reconstruction model of the Mo(100) surface upon oxygen adsorption and annealing.[41,42]

Experimental

The ultra-high-vacuum (UHV) chamber and single crystal microcalorimeter apparatus used to collect the heat of adsorption data and the experimental protocol have been described in detail elsewhere[18–20] and are only briefly described here. The single crystal adsorption microcalorimeter is a modification of the microcalorimetric system first reported by King and coworkers.[1–4] One key difference is the novel, highly sensitive pyroelectric detection method, which utilizes a thin pyroelectric polymer ribbon for sensing the transient temperature change.[47] This detection method does not preclude high temperature treatment of our sample and should also be operable at cryogenic temperatures where IR detection becomes difficult. Detailed analysis of the pulse shape of this detector's response is presented elsewhere.[24] In Fig. 1 a schematic of the calorimetric system is shown illustrating the calorimetric detector and a simplified schematic of the chopped metal atom beam source. The pulsed metal calorimeter train consists of a chopped, collimated beam of metal atoms from an effusive source which is incident onto a thin oxide film grown on a 1 micron thick Mo(100) sample. The thin Mo sample is mounted to a manipulator that allows us to move the sample into the calorimetric position and the various pre-treatment/preparation and electron spectroscopy positions. The pyroelectric detector ribbon used to measure heat input to the sample is gently pressed into contact with the back of the sample and allowed to come to thermal equilibrium. The sensitivity of the ribbon to heat input into the sample is then calibrated by introducing a laser pulse of known energy onto the sample with the same geometric and temporal profile as the metal pulse. The calorimetric measurements are then taken by impinging a

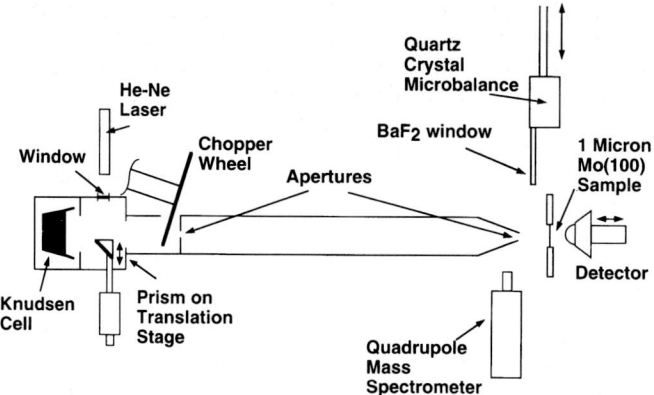

Fig. 1 A schematic of the pulsed metal atom source and the associated apparatus used to obtain the heat of adsorption and sticking probability. The apparatus consists of a high flux Knudsen cell ($T_{source} \approx 1700$ K) and a series of apertures that provide a collimated source of metal atoms that is chopped in 100 ms pulses by a chopper wheel. The microcalorimeter detector is brought into contact with the sample to measure the heat input from the metal atom pulses. The metal flux is measured by a quartz crystal microbalance. The BaF$_2$ window allows separation of the heat input from radiation due to the hot metal source. The laser/prism assembly allows for absolute calibration of the detector signal against a known heat input to the sample. The QMS is used to measure the probability that metal atoms do not stick when they strike the sample.

pulsed metal atom beam onto the surface and measuring the heat input for each metal atom pulse. By first blocking the metal atom flux with the BaF$_2$ window shown in Fig. 1, the contribution to the measured heat attributable to the oven radiation is determined. The absolute metal flux is determined with a quartz crystal microbalance (QCM). We are then able to determine the enthalpy of adsorption at 300 K by subtracting out the oven radiation, correcting for the kinetic energy of the incident atoms to 300 K, and making the appropriate addition to account for changes in pressure times volume. These corrections have been explained in detail elsewhere.[18] The copper charge to the Knudsen-cell effusion metal source for these experiments was copper shot of 99.999% purity (metal basis).

In order to accurately determine the heat of adsorption we need to accurately know the amount of metal in each pulse (metal flux) and the sticking probability of the incident metal atoms as a function of metal coverage, in addition to having an absolute calibration of the detector in contact with the sample to a known heat input. The metal flux is determined by direct measurement of the flux with the QCM and the sticking probability is determined by our line-of-sight modification of the King and Wells method.[48] The chamber is equipped with a UTI 100C quadruple mass spectrometer (QMS), which provides a suitable method to monitor any reflected (non-sticking) metal for calibration of sticking probability in experiments similar to our microcalorimetric experiments. The QMS also allows TPD of the metal films to be performed. The QMS is in a line of sight position at the magic angle (to minimize the effect of any angular distribution[49]) as shown in Fig. 1. In this position we can directly monitor the angle-integrated flux of any reflected metal atoms which do not stick to the sample. It is important for this measurement to have nearly normal incidence of the atom beam, so that non-sticking atoms in both trapping/desorption *and* the quasi-elastic/quasi-specular channels have $\approx \cos^n \theta$ angular distributions, where θ is the polar angle and n is an exponent between one and nine. By comparing the time-integrated signal (per pulse) with the integrated TPD signal of a metal film of known coverage (as determined by the QCM flux × dose time on a surface with unit sticking) we can determine the sticking probability. A temperature correction is also applied to the signal ($\propto 1/T^{1/2}$) to adjust the QMS signal for the differences in velocities between the reflected signal and the TPD signal. In all cases we assume that the metal atoms leave the surface at the surface temperature.[10] The chamber is also equipped with LEED and AES, which are used to characterize the order and cleanliness of the samples as well as to investigate the Cu metal film structure.

The MgO(100) films were grown for these experiments by exposing the Mo(100) sample at 300 K to a flux of magnesium metal ($\approx 4 \times 10^{14}$ atoms cm^{-2} min^{-1}) from a thermal evaporation source under a background pressure of 3×10^{-7} Torr of oxygen similar to the method described elsewhere.[26,50] Subsequently, the MgO films were thermally annealed to ≈ 750 K by electron-beam heating under UHV to order the film. This procedure yielded MgO LEED patterns similar to those reported by Wu and coworkers[50] for MgO(100) films grown on Mo(100). We estimate from AES that the MgO films grown using this procedure are ≈ 40 Å thick. The Mo(100) and p(2 × 1)O–Mo(100) thin films also showed LEED patterns characteristic of well-ordered surfaces. The oxygen concentration on the oxidized molybdenum surface was confirmed by AES.

Results and discussion

Adsorption of Cu on MgO(100)

In Fig. 2, the observed sticking probabilities of Cu on MgO(100) at room temperature as a function of coverage are presented. One monolayer (ML) of Cu is defined here as the packing density of a Cu(111) plane in bulk Cu, 1.77×10^{15} atoms cm^{-2}. The initial sticking coefficient is determined to be ≈ 0.996 and the sticking coefficient slowly increases with coverage towards unit sticking. Although, a significant amount of noise appears in the sticking coefficient as a function of coverage, only sticking probabilities above 0.99 are observed, so this scatter is due to the very tiny reflected signals.

The sticking probabilities that we have observed for copper on the MgO(100) thin films are higher than has been previously reported for this and similar systems.[26,29] For room temperature adsorption of Cu on a MgO(100) thin film very similar to ours, Wu et al.[26] report a value of 0.82 for the initial sticking probability while Zhou et al.[29] report a value of 0.5 for S_o on a cleaved MgO(100) surface, which certainly has a much lower defect density than our thin film. There are several factors that may contribute to the higher sticking probabilities we observe. First, if our MgO film has a larger defect density than that of Zhou et al.,[29] we would expect a greater number of islands to nucleate due to the larger defect density. Since greater island density would effectively lower the average diffusion length between islands for Cu atoms in a mobile precursor state (isolated Cu atoms on MgO), this would result in a higher sticking probability assuming that the sticking probability is governed by competition between desorption and addition to copper islands. This argument for higher sticking with defect density is supported, although rather loosely, by Zhou et al.,[29] who introduced "defects" to the MgO(100) surface by way of small coverages of molybdenum and observed a dramatic increase to unit sticking. Second, and possibly more important, is that the instantaneous metal flux we employ is much higher than the flux used

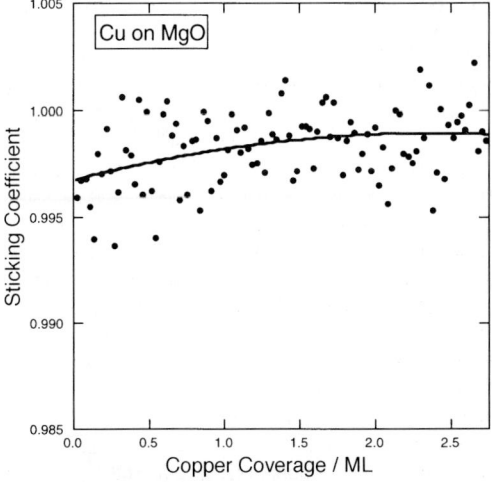

Fig. 2 The measured sticking coefficients of copper as a function of copper coverage on the MgO(100) thin film at room temperature. Each point represents approximately 2.7% of a monolayer of Cu atoms. A best fit curve through the sticking probability is shown as a guide.

in these other experiments. For instance, we routinely operate at a flux during the metal pulse equivalent to one monolayer every 5 s. Wu et al.[26] report a flux equivalent to 1 ML in 160 s, 32 times lower than our flux. Common models for metal island growth[9,51] suggest that the island density is proportional to $(\text{flux}/\text{diffusion})^{1/2-1/3}$. Given that the diffusion constants for Cu should be the same on these similar MgO films, these models suggest an island density 3.1–5.7 times larger in our experiment, neglecting any differences in defect density. One can easily see how this large increase in island density could have a dramatic effect on the sticking probability assuming some effective diffusion length in which a metal atom searches for an island or desorbs. One additional concern is the apparent saturation of the sticking coefficient at 0.82, which has been observed previously upon cooling.[26] This might indicate a small (18%) error in the absolute calibration since one would expect the sticking to increase all the way to unity at lower temperatures. On a similar system, Schaffner et al. estimated a room temperature sticking probability of 0.6 for Ag on MgO(100) thin films.[52] Contrasting our results with those of Schaffner et al. for silver on MgO we see a similar trend. In the Ag/MgO case, the Ag flux was 50 times less than our Cu flux. Also, Ag is expected to have a lower desorption energy, and therefore a lower sticking probability on MgO than Cu, due to Ag's lower sublimation energy. (This trend was observed for Pb adsorption on MgO—see below.) Regardless of the factors leading to the high sticking probability which we observe, this sticking probability allows us to calculate the adsorption energies assuming a unit sticking for copper on the oxide surfaces.

Once we know the sticking probability, we can then determine the copper heat of adsorption as a function of coverage. In Fig. 3 we present the heat of adsorption of a sequence of Cu pulses onto a room temperature MgO(100) thin film. The inset to Fig. 3 shows an enlargement of the experimental heat of adsorption in the low coverage range. The heat of adsorption in the first 2–4% of a monolayer of copper is 240 kJ mol^{-1}, almost 100 kJ mol^{-1} lower than the Cu heat of sublimation of 337.4 kJ mol^{-1}. As the copper coverage increases, the heat of adsorption rapidly increases, reaching 304 kJ mol^{-1} ($\approx 90\%$ of the ΔH of sublimation) by 0.25 ML and 321 kJ mol^{-1} ($\approx 95\%$ of the ΔH of sublimation) by 0.7 ML. The heat of adsorption levels off at $\approx 98\%$ of the Cu heat of sublimation by 2–3 ML and slowly increases thereafter, reaching the heat of sublimation by 4.5

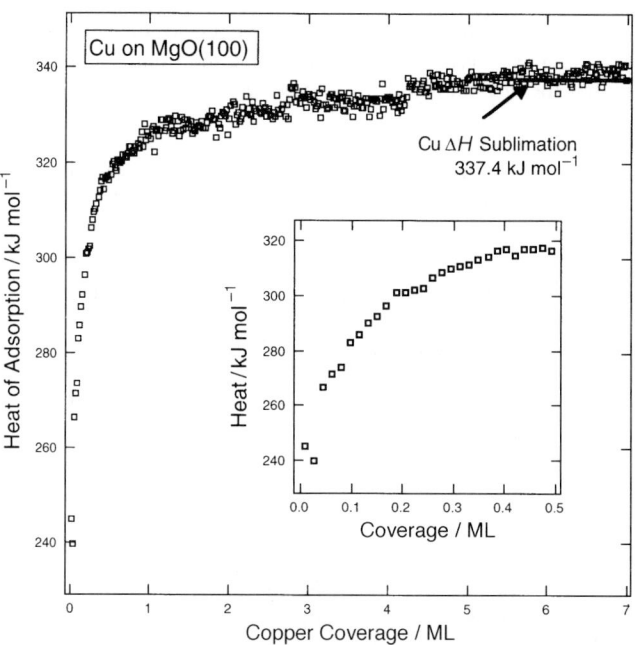

Fig. 3 The calorimetric heat of adsorption (standard enthalpy of adsorption) of Cu on a thin MgO(100) overlayer (grown on the Mo(100) surface) as a function of copper coverage at room temperature. The inset shows a blow up of the low coverage heat of adsorption highlighting the rapid increase in adsorption energy.

ML. Fig. 3 clearly shows that the heat of adsorption of the Cu metal atoms is a strong function of coverage at low coverages and the heat rapidly approaches the expected ΔH of sublimation by a coverage of several ML.

To better understand the growth of Cu on the MgO(100) thin films, we examined the metal film growth as a function of coverage with AES. In Fig. 4 we show representative Cu(MVV) AES spectra of films of different coverages of Cu grown at room temperature on the MgO(100) thin film. The spectra show an increase in the Cu peak to peak intensity with increasing Cu coverage. Upon careful examination of the AES data one also observes a shift in the peak position to higher energy with Cu coverage. A trend line is shown in Fig. 4 to highlight this shift of ≈ 2.5 eV with increasing coverage. Above ≈ 10 ML of Cu the Cu(MVV) transition energy remains constant at 61 eV and is attributed to bulk-like Cu particles.

The AES intensities have been analyzed by comparison to a layer-by-layer model for the growth of Cu films. Fig. 5 compares the experimental Cu AES intensities (data points) to the expected intensity for a layer-by-layer growth model (trend line), both as a function of Cu coverage. In this model, 5.0 Å was used as the escape depth for electrons from Cu in this energy range.[53] Up to ≈ 0.3 ML, the AES data fit the layer-by-layer model reasonably well. Above this coverage a break is observed in the AES data, indicating that the growth has switched from 2-D to 3-D in this coverage range. Since the Cu AES signal falls to about half the value expected for the layer-by-layer growth model by ≈ 3 ML of Cu, these data also show that Cu grows in 3-D islands. The AES signal indicates also that at 3 ML of Cu up to ≈ 30–40% of the MgO surface remains free of Cu, and $\approx 25\%$ is free at 10 ML. This type of transition from 2-D islands to 3-D islands during the growth of metals is often seen on oxides.[9,38] The 2-D islands are kinetically allowed but are not thermodynamically preferred,[9] as also shown by the lower bonding strength of Cu to the oxide relative to the Cu–Cu bonding. Previously published data for the growth of Cu on MgO(100) surfaces have been interpreted to indicate various growth modes including layer-by-layer, VW and SK growth.[25–34] Based on our results and much of this literature, we favor a growth for Cu on MgO(100) at 300 K which is characterized as kinetically controlled 2-D island growth up to a coverage of ≈ 0.3 ML. Above 0.3 ML, the islands thicken into 3-D particles and impinging Cu atoms add on top of these islands faster than they cover the remaining bare oxide surface. This is consistent with both the calorimetric heats, which reach nearly the heat of subli-

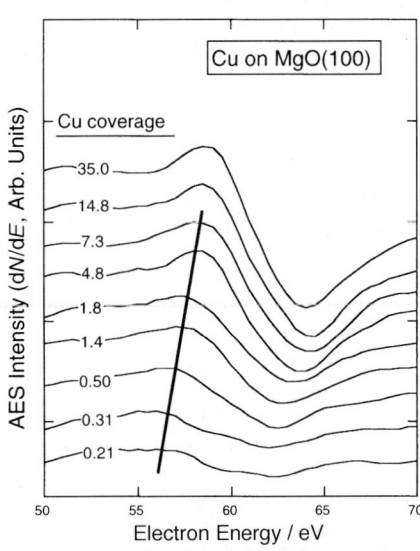

Fig. 4 Representative Cu(MVV) AES spectra of films of different coverages of Cu grown at room temperature on the MgO(100) thin film. A 2.5 eV shift to increasing energy is observed with increasing Cu thickness up to Cu coverages of between 10 ML (not shown) and 14 ML, above which the peak position remains constant. For thicker films the kinetic energy of 61.0 eV is attributed to bulk Cu. The trend line highlights the shift.

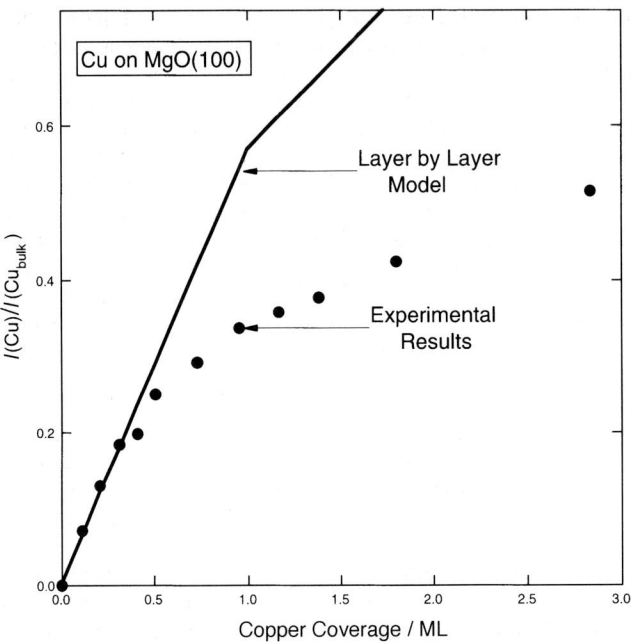

Fig. 5 The Cu AES peak to peak intensities for Cu on MgO(100), as in Fig. 4, plotted *vs.* Cu coverage and compared to the layer-by-layer growth model. The Cu signal shows a deviation from the layer-by-layer model at a Cu coverage of approximately 0.3 ML. Above 0.3 ML the AES data suggest that the growth mode has switched from 2-D island to a 3-D growth mode. The Cu AES intensities are normalized to the intensity from a clean bulk Cu film.

mation by 0.3 ML, as expected for Cu addition onto large 3-D islands. This type of growth could be interpreted as either VW, or as SK up to some coverage less than a complete monolayer, although a non-equilibrium growth model is most appropriate, at least at low copper coverages.

The relatively low initial heat of adsorption observed in Fig. 3 for Cu on the MgO(100) surface indicates that the interactions between copper atoms/islands and the MgO substrate are much weaker than the Cu–Cu interactions. These lower metal–oxide interaction energies clearly indicate that large 3-D Cu particles should be thermodynamically favorable for this system, and any epitaxial or 2-D film growth could only result from a kinetically limited film growth process. This highlights the tremendous advantage in determining heats of adsorption using direct calorimetric measurements, since a TPD approach would be likely to lead to ripening of the Cu islands/particles prior to desorption, resulting in a measurement not truly representative of the non-equilibrium structures formed during metal deposition at room temperature.

At this point we reconsider the shift observed in the Cu AES peak position. In the literature, some reports[27,28] have suggested that at initially low coverages of Cu on MgO(100), Cu is adsorbed as a Cu_2O like species. This interpretation ultimately comes from shifts in the AES peak positions or a shift in the associated Auger parameter. Although we also observe a shift that might be interpreted as Cu_2O formation at low coverages, we feel that the formation of a Cu_2O species is somewhat unlikely. The cause of the shift in AES peak position is most likely a result of final state relaxation effects during the Auger process, which are expected, and indeed cause, these shifts for small, neutral, 2-D and 3-D metal particles.[9,10] Comparing this to a previous study of Cu thin films on ZnO(0001)–O we observed that at the critical coverage of 0.5 ML of Cu (where 2-D island growth switches to a 3-D dominated growth), a shift of 1.8 eV is observed in the position of the Cu AES peaks relative to bulk Cu.[38] This shift is not caused by the formation of Cu_2O on the ZnO substrate, but is the result of final state relaxation effects. In the Cu on MgO(100) experiments reported here, at a coverage of 0.3 ML of Cu we also see a break from the layer-by-layer or 2-D growth mode based on the AES, and we observe a shift of ≈ 2.0 eV from the bulk Cu peak position. This is in excellent agreement with the magnitude of the shift at the 0.5 ML critical coverage of Cu on ZnO, where 3-D growth begins, and strongly suggests that the shift in peak position is due not to Cu_2O formation but instead is due to final state relaxation effects for 2-D charge-neutral Cu islands on the MgO(100) surface as well. This interpretation is in agreement with Zhou and Gustafsson,[25] both in terms of the Cu film growth mode determination and the unlikelihood of Cu_2O formation. In the work of Zhou and Gustafsson[25] on a cleaved MgO(100) surface with a much lower defect density than on our epitaxially growth thin MgO films, the authors found that the Cu islands grew in a VW mode even at coverages as low as 0.14 ML and that at a Cu coverage of 0.99 ML, the islands were on average ≈ 5 layers thick. The higher defect density on our surface may lead to 2-D growth up to the ≈ 0.3 ML which we observe. Theoretical studies find very limited charge transfer even for Cu clusters of just four atoms.[32,33]

From the integral calorimetric heat of adsorption up through multilayer coverage ($\Sigma_n \Delta_{ads}H$), the adhesion energy for the metal on the oxide surfaces (E_{adh}) can be estimated within a simple thermodynamic formula derived elsewhere[9,20]:

$$E_{adh} = -(n\Delta_{sub}H - \Sigma_n \Delta_{ads}H)/A + (1+f)\sigma,$$

where σ is the surface energy of the bulk metal (176 μJ cm^{-2}),[54] A is the area covered by the metal film which contains n atoms, and f is the surface roughness factor of this multilayer film. For the 7 ML Cu film, the integral heat of adsorption is 0.0009378 J or an average of 330.7 kJ mol^{-1}, which is spread over a geometric beam area of 0.138 cm^2. To obtain an accurate value for the adhesion energy between the metal film and the oxide surface, it is necessary to have a good estimation of both the absolute area covered by the multilayer metal film and the roughness factor (f) of the metal film.[20] From the AES data, we estimate that 25% of the MgO surface is not covered by the Cu film at a coverage of 7 ML, therefore we scale the geometric surface area by 0.75. Although we do not measure the surface roughness directly, we can make some reasonable approximations and show that it is near 1.0, so that we can estimate the adhesion energy. The multilayer film from which our adhesion energy is calculated covers $\approx 75\%$ of the surface and is ≈ 7 ML thick on average. The density of islands in such systems is typically[9,10] $\approx 10^{12}$ cm^{-2}, which means that the island centers are separated by 10^{-6} cm from the nearest neighboring island centers. This is a length equivalent to the thickness of ≈ 50 ML. The island diameters are then about the same length, since they must be overlapping if only 25% of the surface is uncovered. As spherical caps, islands with an average height of $\approx 7/\{(50) \times (0.75)\} = 18.5\%$ of their diameter have a surface area that is ≈ 1.14 times their projected area, which provides a crude estimate of their surface roughness factor. A roughness factor even closer to unity would result if we assumed that our Cu islands obtain similar aspect ratios to 3-D Cu mounds grown upon deposition of Cu multilayers on Cu(100).[55] Using 1.14 for the roughness factor, the adhesion energy of Cu on the MgO(100) surface is estimated as ≈ 192 μJ cm^{-2}. This estimation is highly sensitive to the exact surface coverage and the roughness of the film. If, for instance, the island density is actually four times larger, it would give a roughness factor of ≈ 1.5, and an adhesion energy of ≈ 250 μJ cm^{-2}.

Let us consider whether the Cu heats on MgO(100) can be modeled within a nearest-neighbor bond additivity model. For simplicity we will model the Cu particles as growing as fcc(111) platelets, where Cu atoms in the first layer have one Cu–MgO bond downward. When a Cu atom adds to either a large 2-D island or a large Cu island on an existing Cu(111) platelet, it will be assumed that it adds at a kink site, such that it forms three Cu–Cu bonds parallel to the surface. When adding on a large Cu island on an existing Cu(111) platelet, it also forms three Cu–Cu bonds downward to the Cu layer below, for a total adsorption energy equal to the strength of $3 + 3 = 6$ Cu–Cu bonds, and also equal to the bulk sublimation energy of Cu, ≈ 337.4 kJ mol^{-1}. This gives ≈ 56.2 kJ mol^{-1} per Cu–Cu bond. If we assume that at the highest coverage where Cu growth is still following a layer-by-layer model (0.3 ML) that the Cu is adding predominantly to large 2-D Cu islands, then the measured heat of adsorption at this point (310 kJ mol^{-1}) must correspond to the formation of one Cu–MgO downward bond plus three Cu–Cu bonds parallel to the surface. Since the latter contribute 56.2 kJ mol^{-1} each to this energy within this bond-additivity model, the Cu–MgO bond energy must then equal $\approx 310-3(56.2) = 141.4$ kJ mol^{-1}. Within this model, the increase in adsorption energy from a value of $\approx 240-310$ kJ mol^{-1} in the first 0.3 ML (where only 2-D Cu exists according to AES) simply corresponds to the increase in the number of Cu nearest neighbors as the 2-D Cu island size increases. The initial heat of 240 kJ mol^{-1} corresponds then to $(240-141.4)/56.2 = 1.75$ nearest neighbors bonds on average. This value suggests that the initial heat corresponds to the formation of 2-D Cu clusters ranging in size from $\approx 7\text{--}10$ atoms. (Forming Cu dimers gives an average of only 1/2 Cu–Cu bond per Cu atom.) Density functional calculations estimate an average adsorption energy of 206 kJ mol^{-1} for the Cu atoms in a 9-atom 2-D island on Mg(100),[32] and 156 kJ mol^{-1} in a 4-atom cluster.[33]

Within this model, the adhesion energy between Cu and MgO(100) would simply correspond to 141.4 kJ mol^{-1} of Cu–MgO bonds at the interface. Given that the Cu(111) packing density is 1.77×10^{15} Cu atoms cm^{-2}, then this converts to 415 μJ cm^{-2}. This is $\approx 215\%$ larger than the actual observed adhesion energy of 192 μJ cm^{-2}. We attribute this difference to the failure of the bond-additivity concept here. Probably, the heat of adsorption of 310 kJ mol^{-1} measured at 0.3 ML coverage corresponds to much stronger Cu–Cu bonds than the average bulk value of 56.2 kJ mol^{-1}, so that the downward Cu–MgO bond energy is actually somewhat weaker than the value of 141.4 kJ mol^{-1} (estimated assuming pair-wise bond additivity) by $\approx 50\%$, or ≈ 71 kJ mol^{-1}. Density functional calculations[32] indicate that the Cu atoms preferentially reside on top of the oxygen atoms, on the MgO(100) surface with a bond energy of ≈ 84 kJ mol^{-1} which is in reasonable agreement with this value.

Lead adsorption on this same MgO(100) thin film surface has also been studied and is reported elsewhere.[22] The initial heat of adsorption for lead on the room temperature MgO(100) surface is ≈ 100 kJ mol^{-1} and rapidly increases with Pb coverage to the ΔH of sublimation of bulk Pb of 195.2 kJ mol^{-1}. Similar to the Cu film growth on MgO(100), the initial heat of Pb adsorption is ≈ 100 kJ mol^{-1} lower than the heat of sublimation. In this respect, lead should also exhibit a similar equilibrium growth morphology on the MgO(100) surface. Indeed the AES data for Pb on MgO(100) support a VW like growth mode.[22,23] Additionally, the growth of 3-D lead particles at low coverages is supported by the observed sticking probabilities for Pb on the MgO(100) surface. At coverages above 1.0 ML, the sticking probability remains well below unity even though lead sticks to pure lead with unit probability. This suggests large patches of bare MgO still exist.[22,23] Qualitatively the heats of adsorption and growth morphology for Cu seen here are very similar to the results for Pb on the MgO(100) surface.

Adsorption of Cu on the p(2 × 1) oxide of Mo(100)

As a comparison to the heat of adsorption of Cu on MgO(100), we have also investigated the heat of adsorption of Cu on the p(2 × 1)O overlayer on the Mo(100) surface. This overlayer is produced by a saturation dose (≈ 3 L) of O$_2$ at room temperature followed by annealing to 1075 K and is believed to be essentially a bilayer of molybdenum oxide on the underlying Mo(100) structure.[40–46] In Fig. 6 we show the heat of adsorption as a function of coverage for Cu on the p(2 × 1)O–Mo(100) surface at room temperature. The initial heat of adsorption is 286.9 kJ mol^{-1} (average of the first 3 pulses). The heat of adsorption slowly decreases by 3% to 277.8 kJ mol^{-1}, up to a coverage of 0.15 ML. Above 0.15 ML of copper, the adsorption energy increases, reaching 92% of the heat of sublimation of bulk Cu by 1.0 ML and saturating at a value of 332.5 kJ mol^{-1}, within 2% of Cu's heat of sublimation. The nearly constant Cu heat of adsorption for the first 15% of a ML followed by a rapid increase in the heat of adsorption toward the sublimation energy has also been observed for lead adsorption on this same p(2 × 1)O overlayer on Mo(100).[22,23] For lead adsorption, an initial low value of 145 kJ mol^{-1} is observed again and stays relatively constant until ≈ 0.15 ML, above which the heat of adsorption rapidly approaches the heat of lead sublimation of 195.2 kJ mol^{-1}. These similar results suggest that some fundamental aspect of metal adsorption on the p(2 × 1) molybdenum oxide surface must lead to this interesting feature. The similarity between Cu and Pb in this respect is surprising given the differences in covalent bonding usually observed associated with their different electronic configurations (positions in the Periodic Table). The similarity suggests that direct covalent bonding does *not* dominate the metal attraction to the oxide (for the MgO(100), the p(2 × 1)O–Mo(100) and the oxidized W), but instead that the attraction arises mainly from the polarizability of the metal responding the Madelung potential of the oxide surface. Observed trends in measured adhesion and interfacial energies between metals and oxide surfaces (as determined by contact angle methods) have been explained based on a similar physical mechanism for interfacial bonding.[56]

For Pb adsorption on the (2 × 1) molybdenum oxide thin film, AES showed that the growth initially followed a layer-by-layer curve *vs.* coverage. But, at a coverage of ≈ 0.3 ML, the

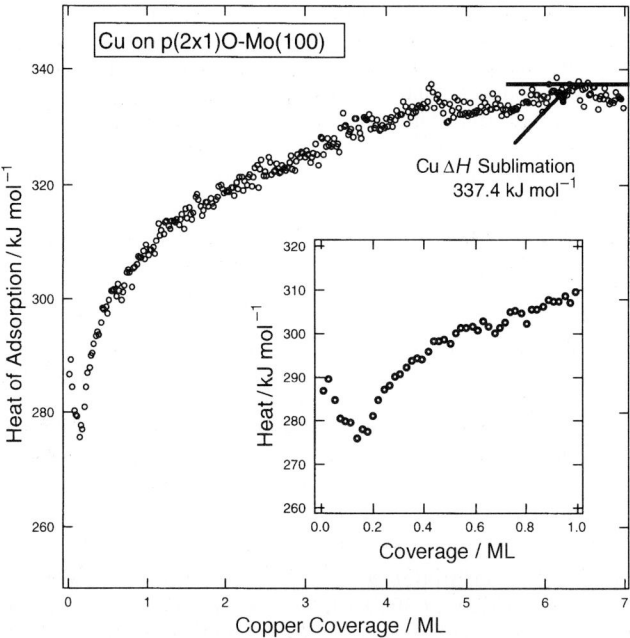

Fig. 6 The calorimetric heat of adsorption of Cu as a function of coverage at room temperature on the Mo(100) single crystal surface pre-covered with an ordered p(2 × 1) overlayer of oxygen. The inset to the figure shows a enlargement of the Cu adsorption in the low coverage regime up to 1.0 ML.

Pb/Mo AES ratio dropped below the layer-by-layer curve, showing that above ≈0.3 ML 3-D islands of Pb predominantly grow.[22] This morphology change was used to explain the abrupt increase in the $\Delta_{ads}H$ of Pb observed at about this same coverage:[22] below ≈0.15 ML, the Pb adatoms bond directly onto the oxide surface (probably forming 2-D islands), and above ≈0.3 ML, they add *on top* of existing Pb islands, thus having larger heats of adsorption, similar to the $\Delta_{sub}H$. We attribute the increase in the $\Delta_{ads}H$ of Cu seen here between 0.2 and 1.0 ML to a very similar phenomenon, the transition from the growth of mainly 2-D to mainly 3-D Cu islands. The growth of Cu on both the Zn- and O-terminated faces of ZnO(0001)[57,58] has been shown to undergo a more abrupt 2-D to 3-D transition in growth, at coverages of ≈0.3 and ≈0.55 ML, respectively. A kinetic model to explain such a transition has been suggested[9,59] which relies on an energetic difference qualitatively similar to that measured here for Cu on this Mo oxide thin film surface and for Pb on the same oxide.

Comparing the results for Cu on the two oxide surfaces and to results to be published elsewhere for Pb on these oxide surfaces[22,23] allows us to develop a better understanding of the important factors in metal growth on oxides. In general, for all four of the aforementioned systems, as well as for metals on hydroxylated MgO(100) thin films,[60] the metal adsorbate interacts much more weakly with the oxide surface than with other metal atoms. This fact is observed through a lower heat of adsorption on the clean surfaces that rapidly approaches the metal heat of sublimation as the coverage increases to a point where the adsorbing metal can interact preferably with large 3-D metal islands, which allow for high coordination of the adsorbing atoms with the underlying substrate.

Very small metal particles often exhibit different catalytic properties than larger metal particles, which behave more like bulk metals in catalytic reactions.[9–14] An example of this interesting phenomena is oxide supported gold particles for both propylene and CO oxidation.[61,62] Certainly numerous effects contribute to the different catalytic properties of smaller metal particles. It has long been thought that smaller particles will be more aggressive in the chemisorption of reactants since the average coordination number of the metal atoms in smaller particles is lower than in larger particles. Stronger adsorption on smaller particles is indeed often observed.[9,11] However, another equally important factor is the role of the underlying oxide support in determining the morphology and chemical properties of the metal particles. Here we have demonstrated that the initial adsorption energy of copper on oxides is much lower than the adsorption energy of the metal onto itself (*i.e.*, the sublimation energy). This much weaker bonding to the surface, for the special case of metal atoms in 2-D islands, compared to metal atoms in the top plane of a 3-D metal island of comparable area, effectively places them at a lower "effective coordination number" as discussed elsewhere.[11] Due to this "effective lower coordination" of metal atoms in 2-D islands, one can expect the metal atoms to be more aggressive in their chemisorption properties (*i.e.*, they bond molecules, fragments, and atoms more strongly). Thus it is easy to understand how the catalytic properties might differ. From this perspective, the metal adsorption energies provide significant insight into how the oxide-metal interactions contribute to the chemisorption and catalytic properties of oxide-supported ultrathin metal particles. This effect of metal–oxide interactions is to be clearly distinguished from an effect of particle size alone. Thus, we predict, based on these energetics, that a 2-D Cu island of the same size on these oxide surfaces will be more aggressive (*i.e.*, less noble) in its chemisorption properties than a 2-D Cu island of the same size and shape supported on another transition metal, or on Cu itself as a 3-D island. Indeed, comparing the TPD characteristics of small probe molecules from 2-D Cu islands on ZnO single crystals[59] with those from 2-D Cu islands on Ru(0001)[63] or bulk-like Cu surfaces indicates this to be the case.

Conclusion

The calorimetric heats of Cu adsorption have been determined as a function of coverage for a series of well-defined oxide surfaces at 300 K. In all cases the Cu bonding to the oxide surfaces is found to be a lower energy interaction than Cu–Cu bonding. On MgO(100) thin films the initial room temperature heat of adsorption for Cu is 240 kJ mol^{-1} and rapidly increases with coverage, eventually approaching the Cu heat of sublimation of 337.4 kJ mol^{-1}. Cu appears to form 2-D islands on the MgO substrate up to a coverage of ≈ 0.3 ML above which 3-D island formation is observed. The coverage dependence of the heat of adsorption suggests that a pair-wise additive nearest neighbor bonding model is insufficient to model the adsorption energies. For Cu adsorption on the ordered p(2 × 1) oxide of Mo(100), the initial room temperature heat of adsorption is 285 kJ mol^{-1} and stays relatively constant up to a coverage of 0.15 ML where it rapidly begins to increase to the sublimation energy. Similar to these results, the heat of Cu adsorption on a disorder oxide thin film on W(100) shows an initial low value of 280 kJ mol^{-1} which increases with coverage as reported elsewhere.[20]

These Cu adsorption data have been compared with calorimetric results for lead adsorption on these same oxide films in order to investigate the effect of the metal on the trends observed for the Cu/oxide systems. In both the MgO(100) and p(2 × 1) oxide of Mo(100) cases, qualitatively similar results are observed for Cu and Pb adsorption. For both Cu and Pb on the MgO(100) thin film, the initial metal adsorption energy is approximately 100 kJ mol^{-1} less than the ΔH of metal sublimation. On the p(2 × 1) oxide, both metals appear to either grow as 2-D islands up to ≈ 0.15 ML or else the metal atoms titrate defect sites on the oxide surface. For both metals the initial heat of adsorption is approximately 55 kJ mol^{-1} less than the ΔH of metal sublimation. This low initial value stays relatively constant (or slightly decreasing) up to metal coverages of 15% of a monolayer. More detailed trends relating growth mode and chemical properties to heats of adsorption both for different metals and different substrates will be developed as the database for these microcalorimetric experiments expands. The weaker interactions between the underlying oxide film and the metal overlayer are significant in controlling the growth of the metal films. These interactions may also lead to modified catalytic properties of thin films and small metal particles when compared to bulk metal like catalysts and contribute to the more aggressive chemisorption properties of thin metal films on oxide supports.

Acknowledgements

The National Science Foundation is acknowledged for financial support of this research project. We would like to thank Dr. Hans Coufal and Professor David King for useful discussions on the calorimetric experimentation and Jacques Chevallier for supplying the thin Mo(100) metal single crystals used in these experiments.

References

1. C. E. Borroni-Bird, N. Al-Sarraf, S. Andersson and D. A. King, *Chem. Phys. Lett.*, 1991, **183**, 516.
2. C. E. Borroni-Bird and D. A. King, *Rev. Sci. Instrum.*, 1991, **62**, 2177.
3. A. Stuck, C. E. Wartnaby, Y. Y. Yeo and D. A. King, *Phys. Rev. Lett.*, 1995, **74**, 578.
4. A. Stuck, C. E. Wartnaby, Y. Y. Yeo, J. T. Stuckless, N. Al-Sarraf and D. A. King, *Surf. Sci.*, 1996, **349**, 229.
5. N. Al-Sarraf, J. T. Stuckless, C. E. Wartnaby and D. A. King, *Surf. Sci.*, 1993, **283**, 427.
6. W. A. Brown, R. Kose and D. A. King, *Chem. Rev.*, 1995, **95**, 797.
7. S. Cerny', *Surf. Sci. Rep.*, 1996, **26**, 1.
8. S. J. Dixon-Warren, M. Kovar, C. E. Wartanaby and D. A. King, *Surf. Sci.*, 1994, **307–9**, 16.
9. C. T. Campbell, *Surf. Sci. Rep.*, 1997, **227**, 1.
10. C. R. Henry, *Surf. Sci. Rep.*, 1998, **31**, 231.
11. C. T. Campbell, *Curr. Opin. Solid State Mater. Sci.*, 1998, **3**, 439.
12. P. L. J. Gunter, J. W. Niemantsverdriet, F. H. Ribero and G. A. Somorjai, *Catal. Rev.-Sci. Eng.*, 1997, **39**, 77.
13. H.-J. Freund, *Angew. Chem. Int. Ed. Engl.*, 1997, **36**, 452.
14. D. R. Rainer and D. W. Goodman, *J. Mol. Catal.*, 1998, **131**, 259
15. S. L. Lai, J. Y. Guo, V. Petrova, G. Ramanath and L. H. Allen, *Phys. Rev. Lett.*, 1996, **77**, 99.
16. S. L. Lai, G. Ramanath, L. H. Allen, P. Infante and Z. Ma, *Appl. Phys. Lett.*, 1995, **67**, 1229.
17. S. L. Lai, J. R. A. Carlsson and L. H. Allen, *Appl. Phys. Lett.*, 1998, **72**, 1098.
18. J. T. Stuckless, N. A. Frei and C. T. Campbell, *Rev. Sci. Instrum.*, 1998, **69**, 2427.
19. J. T. Stuckless, D. E. Starr, D. Bald and C. T. Campbell, *Mater. Res. Soc. Symp. Proc.*, 1997, **440**, 103.
20. J. T. Stuckless, D. E. Starr, D. Bald and C. T. Campbell, *J. Chem. Phys.*, 1997, **107**, 5547.
21. J. T. Stuckless, D. E. Starr, D. Bald and C. T. Campbell, *Phys. Rev. B: Condens. Matter*, 1997, **56**, 13497.
22. D. E. Starr, J. T. Ranney, J. E. Musgrove, D. J. Bald and C. T. Campbell, in preparation.
23. D. E. Starr, J. E. Musgrove, D. J. Bald, J. T. Ranney and C. T. Campbell, in preparation.
24. J. T. Stuckles, N. Frei and C. T. Campbell, *Sens. Actuators B*, in press.
25. J. B. Zhou and T. Gustafsson, *Surf. Sci.*, 1997, **375**, 221.
26. M.-C. Wu, W. S. Oh and D. W. Goodman, *Surf. Sci.*, 1995, **330**, 61.
27. T. Conard, J. Ghijsen, J. M. Vohs, P. A. Thiry, R. Caudano and R. L. Johnson, *Surf. Sci.*, 1992, **265**, 31.
28. T. Conard, J. M. Vohs, P. A. Thiry and R. Caudano, *Interface Anal.*, 1990, **16**, 446.
29. J. B. Zhou, H. C. Lu, T. Gustafsson and E. Garfunkel, *Surf. Sci.*, 1993, **293**, L887.
30. J.-W. He and P. J. Moller, *Surf. Sci.*, 1986, **178**, 934.
31. I. Alstrup and P. J. Moller, *Appl. Surf. Sci.*, 1998, **33/34**, 143.
32. (*a*) V. Musolino, A. Selloni and R. Car, *Surf. Sci.*, 1998, **402–404**, 413; (*b*) V. Musolino, A. Selloni and R. Car, *J. Chem. Phys.*, 1998, **108**, 5044.
33. (*a*) A. V. Matveev, K. M. Neyman and G. Pacchioni, *Chem. Phys. Lett.*, 1999, **299**, 603; (*b*) G. Pacchioni and N. Rösch, *J. Chem. Phys.*, 1996, **104**, 7329.
34. M.-H. Schaffner, F. Patthey, W.-D Schneider and L. G. M. Pettersson, *Surf. Sci.*, 1998, **402–404**, 450.
35. J. B. Zhou, H. C. Lu, T. Gustafsson and E. Garfunkel, *Surf. Sci.*, 1997, **382**, 21.
36. Y. Wu, E. Garfunkel and T. Madey, *J. Vac. Sci. Technol., A*, 1996, **14**, 1662.
37. J.-W. He and P. J. Moller, *Surf. Sci.*, 1987, **180**, 411.
38. K. H. Ernst, A. Ludviksson, R. Zhang, J. Yoshihara and C. T. Campbell, *Phys. Rev. B: Condens. Matter*, 1993, **47**, 13782.
39. U. Diebold, J.-M. Pan and T. E. Madey, *Phys. Rev. B: Condens. Matter*, 1993, **47**, 3868.
40. E. Bauer and H. Poppa, *Surf. Sci.*, 1975, **48**, 31.
41. H. Xu and K. Y. S. Hg, *Surf. Sci.*, 1996, **356**, 19.
42. H. Xu and K. Y. S. Hg, *Surf. Sci.*, 1996, **355**, L305.
43. C. Zhang, M. A. Van Hove and G. A. Somorjai, *Surf. Sci.*, 1985, **149**, 326.
44. S. L. Miles, S. L. Bernasek and J. L. Gland, *J. Phys. Chem.*, 1983, **87**, 1626.
45. R. M. Henry, B. W. Walker and P. C. Stair, *Surf. Sci.*, 1985, **155**, 732.
46. B. W. Walker and P. C. Stair, *Surf. Sci.*, 1981, **103**, 315.
47. H. J. Coufal, R. K. Grygier, D. E. Horne and J. E. Fromm, *J. Vac. Sci. Technol., A*, 1987, **5**, 2875.
48. D. A. King and M. G. Wells, *Surf. Sci.*, 1972, **29**, 454.
49. S. W. Pauls and C. T. Campbell, *Surf. Sci.*, 1990, **226**, 250.
50. M.-C. Wu, J. S. Corneille, C. A. Estrada, J.-W. He and D. W. Goodman, *Chem. Phys. Lett.*, 1991, **182**, 472.
51. J. A. Venables, *Philos. Mag.*, 1973, **27**, 697.
52. M.-H. Schaffner, F. Pattey and W. D. Schneider, *Surf. Sci.*, 1998, **417**, 159.
53. S. Tanuma, C. J. Powell and D. R. Penn, *Surf. Interface Anal.*, 1991, **17**, 911.
54. W. R. Tyson and W. A. Miller, *Surf. Sci.*, 1977, **62**, 267.
55. J.-K. Zuo and J. F. Wendelken, *Phys. Rev. Lett.*, 1997, **78**, 2791.
56. F. Didier and J. Jupille, *Surf. Sci.*, 1994, **314**, 378.
57. J. Yoshihara, J. M. Campbell and C. T. Campbell, *Surf. Sci.*, 1998, **406**, 235.
58. C. T. Campbell and A. Ludviksson, *J. Vac. Sci. Technol., A*, 1994, **12**, 1825.

59 S. C. Parker, A. W. Grant, V. A. Bonzie and C. T. Campbell, *Surf. Sci.*, in press.
60 J. E. Musgrove, D. E. Starr, J. T. Ranney, D. J. Bald and C. T. Campbell, in preparation.
61 T. Hayashi, K. Tanaka and M. Haruta, *J. Catal.*, 1998, **178**, 566.
62 M. Haruta, *Catal. Today*, 1997, **36**, 153.
63 D. W. Goodman and C. H. F. Peden, *J. Chem. Soc., Faraday Trans. 1*, 1987, **83**, 1967.

Paper 9/02649E

Cu atoms and clusters on regular and defect sites of the SiO_2 surface. Electronic structure and properties from first principle calculations

Gianfranco Pacchioni,[a] Nuria Lopez[b] and Francesc Illas[b]

[a] *Dipartimento di Scienza dei Materiali, Università di Milano-Bicocca, Istituto Nazionale di Fisica della Materia, via Cozzi, 53-20125 Milano, Italy.*
E-mail: gianfranco.pacchioni@mater.unimib.it
[b] *Departament de Quimica Fisica, Universitat de Barcelona, Marti i Franques 1, 08028 Barcelona, Spain*

Received 24th March 1999

The interaction of isolated Cu atoms and small Cu clusters, from Cu_2 to Cu_5, with the dehydroxylated surface of silica has been investigated by means of cluster models density functional calculations. The regular, non defective, surface shows very low reactivity towards Cu atoms; the binding is largely due to polarization mechanisms. This implies that impinging Cu atoms will easily diffuse on the surface and re-evaporate unless trapped at a defect site. In fact, strong bonds are formed between Cu atoms and clusters and some typical point defects at the SiO_2 surface. We have analyzed two of these defects, the non-bridging oxygen site (NBO), $\equiv$Si–O·, and the E' center corresponding to a Si singly occupied sp^3 dangling bond, $\equiv$Si·. The Cu clusters interacting with these paramagnetic centers are significantly perturbed by the bonding at the interface, as shown by the different geometrical structures of supported compared to gas-phase clusters. Some observable consequences of the cluster deposition, in particular the appearance of states in the gap of the material, are discussed.

1 Introduction

The metal/oxide interface represents a very important aspect of surface science.[1–3] It is related to a variety of technological applications, from catalysis and gas sensors to microelectronics, from coating and protection of metals to colloidal chemistry and non-linear optics phenomena. The experimental study of oxides using surface science techniques is complicated because oxides are usually insulators and are brittle. This causes experimental difficulties related to sample charging and heating. An approach that has been used in recent years to study oxides and metal/support interactions under controlled conditions is to deposit oxide thin-films on a metal surface.[1–3] Important information about the nature of oxide surfaces comes also from the study of polycrystalline materials.[4] In this case, the high surface area of powders allows the use of techniques like electron paramagnetic resonance (EPR), which would be too difficult to use on single crystals because of the insufficient small number of paramagnetic centers present on the surface.[5,6] Theoretical studies, in particular, if performed with first-principle approaches, provide a useful complement to experiment for the understanding of the microscopic aspects of the metal/oxide interaction.[7–13] The quantum-mechanical description of the interaction of metal atoms or clusters with the surface of oxide materials is still in its infancy and only a few first-principle studies have been performed on this subject so far. This is also because of the lack of accurate structural information. In the last few years, however, several experimental investigations on single crystal or thin film oxide surfaces have been reported, opening the way to a direct comparison of observations with theoretical predictions.[1-3] Another important aspect of the quantum-mechanical simulation of these systems (which severely limits the applicability of parametrized force-fields and semi-empirical approaches) is the great variety of supporting oxides, from very ionic, *e.g.*, MgO, to largely covalent, *e.g.*, SiO_2, or intermediate, *e.g.*, TiO_2 or Al_2O_3. Finally, even when the geometrical and electronic structure of the ideal oxide surface is well characterized, this is not sufficient to fully describe the early stages of the metal deposition since these processes often involve defects, low-coordinated sites and surface irregularities. Since many catalysts consist of small particles on high surface-area powders of SiO_2 or Al_2O_3, an understanding at the microscopic level of the metal/oxide interaction, in particular at the defect sites, is an essential prerequisite for an understanding of the catalytic activity.

Recently, we started a systematic quantum-mechanical study of the interaction of Cu atoms and clusters with regular and defect sites of the dehydroxylated SiO_2 surface using *ab initio* Hartree–Fock (HF) and gradient-corrected density functional theory (DFT) cluster calculations.[14-16] The choice of studying Cu on SiO_2 is due to the existence of experimental data on these systems[17-20] and also to our recent studies on the electronic structure and spectral properties of the point defects in bulk silica.[21-23] We have investigated the interaction of isolated metal atoms with regular and defect sites of SiO_2; then we considered the deposition of small metal clusters. In this paper we briefly review the results of our previous investigations,[14-16] and we present new results on the interaction of Cu clusters on surface defects. The adsorption sites considered are regular bridging oxygens at the dehydroxylated surface, ≡Si–O–Si≡, the non-bridging oxygen site (NBO), ≡Si–O˙, and the E' defect corresponding to a Si singly occupied sp^3 dangling bond, ≡Si˙. Both the E' and NBO defects are paramagnetic and therefore detectable by EPR spectroscopy. They have attracted great interest over the past 20 years because of their role in the degradation of Si/SiO_2 interfaces in microelectronics devices or in the absorption of light in optical fibers.[24] In fact, very similar defects are present in the bulk of amorphous and crystalline silica,[25] on the surface of mechanically activated SiO_2,[26] of SiO_2 thin films,[17] or of α-quartz single crystals surface after Ar^+ bombardment.[27] Proof of the existence of these centers comes from typical EPR spectra and optical absorption bands at 2 eV (NBO) and 5.8 eV (E'). It has been suggested that these defect centers are the primary cause of the interface bond formation.

2 Computational approach

To describe the regular and defect centers of the SiO_2 surface we used cluster models derived from α-quartz. The cluster broken bonds were saturated by H atoms placed 0.98 Å from the O atoms along the O–Si bond directions of the perfect crystal. The H atoms were kept fixed during the geometrical optimization to provide a simple representation of the mechanical embedding of the solid matrix. The optimization was done without any symmetry constraint. The details of the cluster used for each site are given below. Below we also present a detailed analysis of the dependence of the results on cluster size. In general, the adsorption of the metal cluster has little effect on the other geometrical parameters of the SiO_2 substrate model; in particular, the average Si–O–Si angles remain around 145°, *i.e.*, very close to that of α-quartz.[28]

The cluster electronic structure was computed at the spin polarized DFT level using Becke's three parameter hybrid non-local exchange functional[29] combined with the Lee–Yang–Parr[30] gradient-corrected correlation functional (B3LYP). Hartree–Fock calculations were performed for comparative purposes. The basis sets used were: all electron 6-31G* on all Si and O atoms[31] and 3-21G[32] on the terminal H atoms. For the Cu clusters we used an effective core potential, ECP,[33] which explicitly includes in the valence the $3s^2\ 3p^6\ 3d^{10}\ 4s^1$ electrons of Cu. The Cu basis set is [8s5p4d/3s3p2d].[33] For the atomic Cu adsorption a [14s11p6d/8s6p4d] all electron basis set was used.[34] The adhesion energies of the cluster to the silica defects as well as the cohesive energies of the metal clusters were not corrected by the basis set superposition error, BSSE.[35] Only for the isolated Cu atom have we estimated the BSSE, which is of the order of 0.5 eV. Since the bonding of the Cu atoms or clusters to the defect centers is definitely larger than this error, the general conclusions should not be affected. All the calculations were performed with the Gaussian94 program package.[36]

3 Isolated Cu atoms on regular and defect sites of SiO_2

3.1 Regular sites

The interaction of isolated Cu atoms with the regular surface sites of silica has been modeled by the $(HO)_3Si-O-Si(OH)_3$ cluster, Fig. 1. Cu atoms were adsorbed 'on-top' of the two-coordinated bridging oxygen. At the HF level the potential energy curve is purely repulsive; at the DFT-B3LYP level the interaction energies are of the order of ≈ 0.6 eV but after inclusion of the BSSE the binding energies are extremely small, ≈ 0.1 eV, Table 1; the equilibrium Cu–O distance is rather long. A full geometry optimization of the Cu atom position without imposing any constraint leads to an adsorption geometry where the adsorbate interacts with more than one surface oxygen.[15] This is probably what happens in reality: Cu atoms interact, mainly through polarization mechanisms, with two or more surface oxygens. Thus, the regular sites of silica are very unreactive towards Cu atoms and, in general, towards metal atoms. It is worth noting that our calculations indicate for the adsorption of Cu on the O atoms of silica a bond strength even lower than that of the rather unreactive MgO (001) surface.[37] This means that Cu atoms deposited on silica from the gas phase will not be trapped at the regular sites but rather will diffuse on the surface, remain trapped at a defect, or re-evaporate.

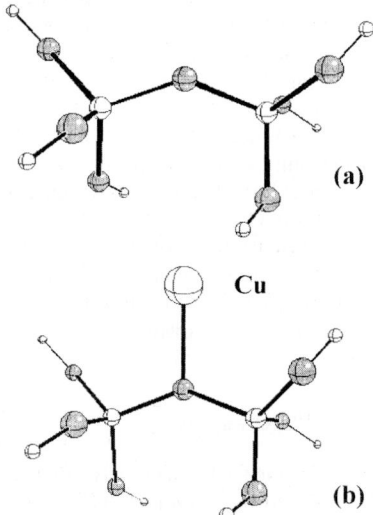

Fig. 1 Cluster model of (a) a bridging oxygen on the silica surface, and (b) a Cu atom interacting with the bridging oxygen.

Table 1 Adsorption properties of Cu on regular sites (bridging oxygen), NBO ($\equiv$Si–O˙) and E' ($\equiv$Si˙) centers on the SiO_2 surface[a]

	Method	Bridging oxygen 2-T model	NBO ($\equiv$Si–O˙) 2-T model	E' ($\equiv$Si˙) 1-T model
$r(SiO_2$–Cu)/Å	HF	Unbound	1.944	2.365
	B3LYP	2.460	1.855	2.238
D_e(BSSE)/eV	HF	Unbound	2.56	1.00
D_e(BSSE)/eV	B3LYP	0.08	3.79	2.32

[a] r = Distance between Cu and bridging oxygen, D_e = dissociation energy corrected by the BSSE (see text).

3.2 Non-bridging oxygens

Non-bridging oxygens at the surface of silica, ≡Si–O˙, probably represent the most important defects for the reactivity of mechanically activated SiO_2. By using metastable impact electron spectroscopy experiments (MIES) these broken bonds have been proposed as the centers where impinging metal atoms are trapped.[38] The ≡Si–O˙ group has been represented by an $(HO)_3$Si–O–Si$(OH)_2$–O˙ cluster, see Fig. 4 in ref. 15.

Cu forms strong covalent bonds with the ≡Si–O˙ groups, having a value of 3.8 eV obtained from B3LYP calculations using a model with two tetrahedra and being BSSE corrected, Table 1, and with an important polarization towards oxygen of the bonding electrons (partial charge transfer). A net residual positive charge forms on adsorbed Cu atoms, as shown by the data of Mulliken population.[15] The fact that the Cu atom donates charge to the substrate favors the appearance of other interactions with the neighboring surface O atoms. The Cu adsorbate binds, mostly electrostatically, with the exposed two-coordinated O atoms of the surface, which then become effectively three-coordinated. This corresponds to the closure of a ring where the Cu atom binds to two surface oxygens; in a sense, the Cu replaces a missing Si atom in the lattice.

3.3 Surface E′ centers

Si sp^3 dangling bonds, ≡Si˙ or E′ centers, are present on the (0001) and (1010) surfaces of α-quartz,[27] on mechanically activated silica[26] as well as on UHV grown thin SiO_2 films.[17] They can be detected by means of EPR, and show a characteristic hyperfine coupling constant of ≈ 470 G with the ^{29}Si nuclide, and an intense electronic transition around 6 eV.[26] The ≡Si˙ surface radical is very reactive towards molecular species; a similar reactivity is also expected towards metal atoms. In fact, on the $[(HO)_3Si–O]_3$Si˙ model of a surface E′ center, Fig. 2(b), Cu forms a rather strong bond of ≈ 1 eV in HF and ≈ 2.3 eV at the B3LYP level (BSSE corrected values, Table 1). The formation of such a strong bond is reflected in the rather short Si–Cu distance, ≈ 2.2 Å, Table 1. The bonding arises from the coupling of the Si sp^3 singly occupied orbital and the metal 4s open shell orbital with formation of a bonding level. This level gives rise to a resonance in the band gap of the material, well above the O 2p band.[15] The character of the Cu–Si bond is largely covalent but the polarization of the bonding electrons is towards the metal, at variance with the NBO center, at least according to the Mulliken charges, Table 1. Unlike the NBO case, the Cu atom does not bind to other O atoms of the SiO_2 surface because the Si–Cu bond does not have the required flexibility for ring closure. Cu passivates the Si dangling bonds without leading to new ring structures.

3.4 Cluster size dependence and QM/MM models

The results described above were obtained with relatively small clusters. It remains to be established to what extent the results obtained on small models can be extended to the real surface. To answer this question we studied the dependence of the results on the cluster size. This was done using clusters of various size but also by employing a hybrid quantum-mechanical/molecular mechanics approach, QM/MM.[14] The method used is the integrated molecular orbital molecular mechanics, IMOMM, method.[39] The analysis was performed for various defects. Here we review the main features of the E′ center. Some of the clusters employed to describe the E′ defect are derived from the structure of α-quartz, in particular $(HO)_3$Si˙ and $[(HO)_3Si–O]_3$Si˙, Fig. 2; these models contain 1 and 4 tetrahedral Si atoms, respectively, and are thereafter denoted for brevity as 1-T and 4-T. We also considered a molecular model of silica, $(H_7Si_8O_{12})$˙ octahydrosilasesquioxane, OSQ, which possess a 'cubic' Si_8O_{12} unit, Fig. 2(c). OSQ and derived compounds, by substitution of terminal H atoms with alkyl or organometallic groups, can be considered as molecular models of SiO_2 surfaces.[40] In the clusters derived from quartz, the position of the H atoms was fixed. No constraint, except for symmetry, was imposed in the optimization of OSQ.

The calculations were performed at the restricted (Open) HF and (spin polarized) DFT-B3LYP levels. Since the emphasis here is on the cluster size dependence the results were not corrected by the BSSE. We described the $(-O)_3$Si unit nearest to Cu as QM, the rest as MM, see Fig. 2. The

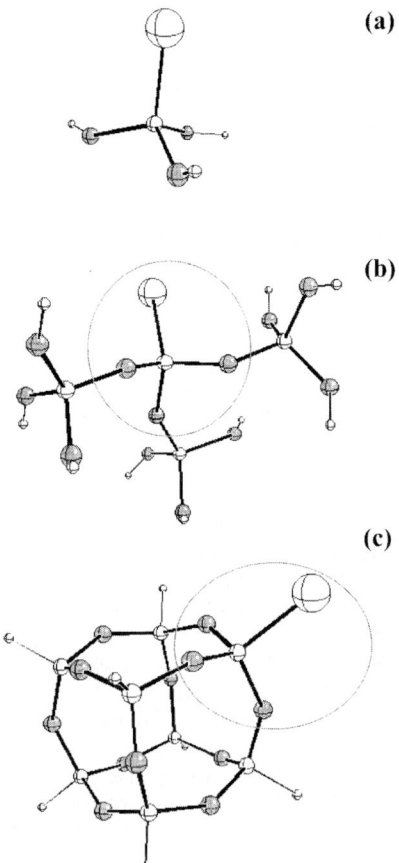

Fig. 2 Models of the E' center, ≡Si˙, (a Si dangling bond) on the silica surface. (a) 1-tetrahedron model; (b) 4-tetrahedra model; (c) octahydrosilasesquioxane model. When a hybrid QM/MM approach has been used the QM part has been enclosed within a grey line.

broken bonds of the QM and MM regions were always filled with H atoms. The IMOMM calculations were performed with a program built from modified versions of the Gaussian92/DFT[41] (QM part) and MM3(92)[42] (MM part) programs. MM calculations used the MM3(92) force field.[43–45] For further details see ref. 14.

In the section 3.3 we have shown that Cu binds to an E' center through direct coupling of the singly occupied Si sp^3 dangling bond with the Cu 4s orbitals. Even at the HF level and using the smallest (HO)$_3$Si˙ QM model the bonding is relatively strong, 1.58 eV (but the overestimate due to the BSSE is about 0.6 eV), with a Si–Cu distance of 2.37 Å, Table 2. The Si–O distances, ≈1.67 Å, are only slightly elongated with respect to the free cluster, 1.65 Å. Modest changes are found in the Si–O–Si angles. At the B3LYP level there is an increase of D_e and a shortening of the Si–Cu bond length, while the other geometrical parameters remain essentially stable. This shows that the bonding is described in a qualitatively correct manner at the HF level. For this reason we have restricted the analysis of the cluster size dependence to HF results. Going from the minimum 1-T cluster to the larger 4-T one, Fig. 2, does not result in significant changes in the adsorption properties, Table 2. In particular, the adsorption energy computed at the HF level goes from 1.58 eV to 1.64 eV by changing the cluster size, Table 2. Also the OSQ model gives bonding properties nearly identical to those of the smaller model; the binding energy, in fact, is 1.63 eV and the Si–Cu distance 2.347 Å, nearly coincident with the HF values obtained with the other clusters, see Table 2. This provides a clear sign of the local nature of the Si–Cu bonding and of the minor effect that this bond has on the surface structure.

Table 2 Adsorption properties of Cu atoms on a E' center, ≡Si·, on the silica surface as a function of cluster size[a]

Model	Method	r(Si–Cu)/Å	r(Si–O)/Å	α(Cu–Si–O)/ degrees	β(Si–O–Si)/ degrees	D_e/eV
(HO)$_3$Si· 1-T	QM HF	2.37	1.67	111	128	1.58
(HO)$_3$Si· 1-T	QM B3LYP	2.24	1.66	110	124	3.08
[(HO)$_3$Si–O]$_3$Si· 4-T	QM HF	2.34	1.66	113	145	1.64
(H$_7$Si$_8$O$_{12}$)· OSQ	QM HF	2.35	1.66	114	154	1.63
[(HO)$_3$Si–O]$_3$Si· 4-T	QM/MM HF	2.36	1.66	113	125	1.98
(H$_7$Si$_8$O$_{12}$)· OSQ	QM/MM HF	2.34	1.65	109	143	1.61

[a] QM = quantum-mechanical treatment; MM = molecular mechanics treatment.

All the results described so far have been obtained with a full QM treatment. To test the validity of the hybrid QM/MM approach we considered the 4-T surface model and the OSQ cluster. In these two clusters the QM part coincides with the minimum 1-T cluster; the rest is treated at the MM level. The results obtained with the QM/MM models are close to the fully QM ones. However, some small changes are found in D_e and in the Si–O–Si angle for the 4-T QM/MM model, Table 2. In fact, D_e is about 0.3 eV higher than in all the other models considered and the Si–O–Si angle is smaller by about 20°. A closer inspection shows that this energy difference is due to the MM part of the cluster.[14] Despite the small inaccuracy in one of the models adopted, the QM/MM approach provides qualitatively similar results at a much lower computational cost.

To summarize, the results show that the bond of Cu with the E' center of the SiO$_2$ surface is local; the results do not change significantly as the size and shape of the cluster is varied. The hybrid QM/MM approach works very well if the definition of the QM part is such that all the important quantum-mechanical interactions between surface and adsorbate are included. When this is done the results are extremely close to those of a full QM treatment. The success of the mixed QM/MM approach is clearly due to the local nature of the Cu/SiO$_2$ bond, specially at defective sites.

4 Cu clusters interacting with SiO$_2$ E' centers

In this section we consider the interaction mode of small Cu clusters, from Cu$_2$ to Cu$_5$, with the E' center, represented by the (HO)$_3$Si–O–(HO)$_2$Si–O–(OH)$_2$Si· cluster, Fig. 3(a). The results for the E' are compared with those on the adsorption of Cu clusters on the NBO sites reported previously.[16] We have seen above that the bonding of a single Cu atom with these two centers is rather strong. It is interesting therefore to examine the modifications occurring in a small cluster as a consequence of bonding with the surface. The Cu clusters were geometrically optimized at the DFT-B3LYP level in the gas-phase and on the oxide support.

4.1 Cluster geometries

At the Cu$_n$/SiO$_2$ interface the clusters are directly bound to the E' and NBO centers with rather different distances, r(Si–Cu) ≈ 2.34 ± 0.06 in E' and r(O–Cu) ≈ 1.95 ± 0.07 Å in NBO, Table 3. However, all the clusters are anchored to the surface through more than one Cu atom, see Figs. 3–5 and Fig. 2–6 in ref. 16. This is a general characteristic; in fact, in addition to the direct O–Cu or Si–Cu covalent bonds, weaker interactions occur between the metal cluster and the bridging oxygens of the surface. In NBO the partial charge transfer from Cu to SiO$_2$ leads to a depletion of electronic charge from the metal cluster. The distance between one of the Cu atoms of the cluster and a bridging oxygen, see r(Cu–O$_2$) in Table 3, is somewhat larger than that of the Cu atom directly interacting with the NBO, see r(Cu–O$_1$) in Table 3. When the Cu clusters interact with a Si dangling bond, however, the distances of the cluster from the bridging oxygens are definitely larger than for the NBO center.[16] In the case of E', the interaction with the bridging oxygens is probably weaker than for NBO and mainly due to the polarization within the metal group induced by the negatively charged O atoms.

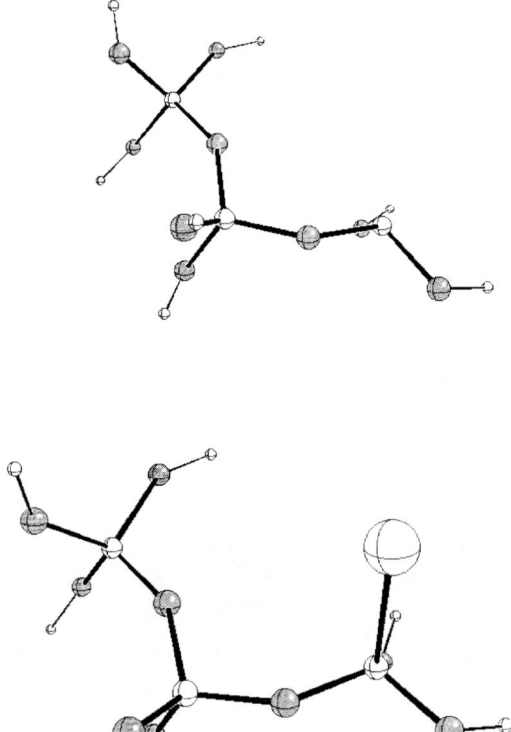

Fig. 3 (Top) (HO)$_3$Si–O–(OH)$_2$Si–O–(OH)$_2$Si˙ model of a Si dangling bond (E′ center) on the silica surface. Small spheres represent the terminal hydrogen atoms, dark spheres oxygen atoms, large spheres silicon atoms. (Bottom) Model of a supported Cu atom (larger sphere) on an E′ center, ≡Si–Cu.

We consider now the geometrical changes within the metal unit induced by the interaction with the substrate. The addition of a second atom to the ≡Si–Cu complex results in the formation of a supported Cu dimer, Fig. 4(a); the ground state of ≡Si–Cu$_2$ is doublet, while gas-phase Cu$_2$ has a $^1\Sigma_g^+$ ground state. The Cu–Cu distance in the supported molecule, 2.34 Å, is about 0.1 Å longer than in gas-phase, 2.26 Å. Notice that the same molecule adsorbed on an NBO center shows an elongation of almost 0.2 Å, consistent with a stronger bond with this surface defect and a more

Table 3 Selected bond distances of NBO (≡Si–O˙) and E′ (≡Si–˙) centers interacting with Cu$_n$ clusters at the Cu/SiO$_2$ interface

	≡Si–O˙	≡Si–O–Cu	≡Si–O–Cu$_2$	≡Si–O–Cu$_3$	≡Si–O–Cu$_4$	≡Si–O–Cu$_5$
r(Si–O$_1$)/Å	1.673	1.614	1.606	1.676	1.622	1.633
r(O$_1$–Cu)/Å^a	—	1.869	1.865	2.001	2.017	1.994
				2.038	2.076	2.004
r(O$_2$–Cu)/Å^b	—	2.090	2.256	3.033	2.116	2.595

	≡Si˙	≡Si–Cu	≡Si–Cu$_2$	≡Si–Cu$_3$	≡Si–Cu$_4$	≡Si–Cu$_5$
r(Si–Cu)/Å^a	—	2.279	2.337	2.329	2.315	2.376
r(O$_2$–Cu)/Å^b		2.205	2.193	2.097		2.217

^{a} Shortest distance(s) of non-bridging oxygen, O$_1$, with the Cu atom(s) of the cluster.
^{b} Shortest distance of a bridging oxygen, O$_2$, from a Cu atom of the cluster.

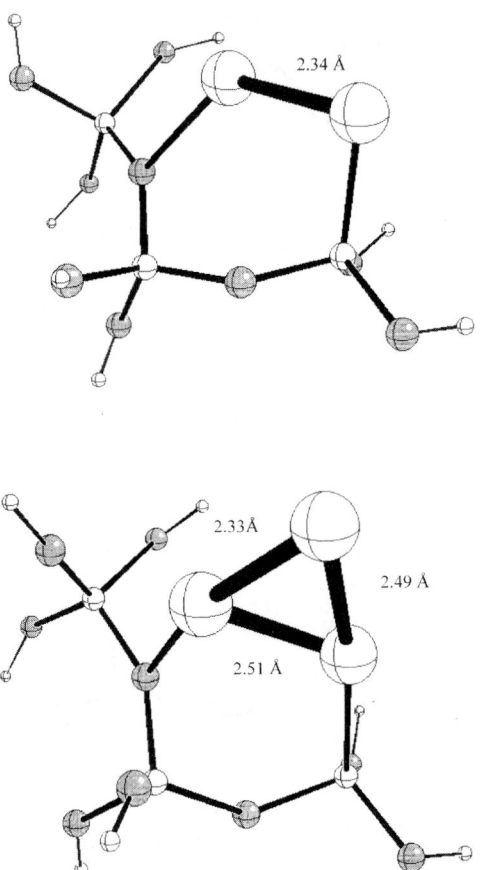

Fig. 4 (Top) Model of a supported Cu_2 cluster (larger spheres) on a E' center, ≡Si–Cu_2. The optimal Cu–Cu distance is given in Å. (Bottom) Model of a supported Cu_3 cluster (larger spheres) on an E' center, ≡Si–Cu_3. The optimal Cu–Cu distances are given in Å. The existence of a second bonding interaction between the cluster and a bridging oxygen of the silica surface has also been represented by a solid line.

pronounced perturbation of the metal dimer. The second Cu atom of the dimer interacts weakly with a bridging oxygen, Fig. 3; the Cu–O distance, 2.20 Å, is very close to that found for the NBO case, Table 3, and the spin is largely localized on the Cu_2 unit, Table 4.

Gas-phase Cu_3 is bent, C_{2v}, with an internal angle of 75.7° and Cu–Cu distances of 2.326 Å while Cu_3^+ is a closed shell equilateral triangle with Cu–Cu distances of 2.394 Å. The addition of a Cu atom (doublet) to the ≡Si–Cu_2 surface complex (doublet) results in the closed shell ≡Si–Cu_3 system, Fig. 4(b). The distances within supported Cu_3 are considerably elongated with respect to the gas-phase unit, but the cluster retains the C_{2v} structure, with two long and one short Cu–Cu distances, Fig. 4(b). On an NBO center, on the contrary, we found a different adsorption mode, with two Cu atoms of Cu_3 interacting with the NBO (see Fig. 4 in ref. 16); on NBO Cu_3 assumes an almost perfect equilateral triangular geometry with internal angles of 60 ± 1° and Cu–Cu distances of about 2.4 Å. Thus, a quite different structure is found for the same cluster interacting with the two defect centers.

Free Cu_4 has a singlet ground state and a planar rhombic structure. The Cu–Cu distances are of 2.455 Å and the short Cu–Cu diagonal of the rhombus is 2.308 Å. When deposited on an E' center of SiO_2, Cu_4 remains nearly planar but one of the Cu–Cu distances is markedly elongated to 2.72 Å, Fig. 5(a). This corresponds to having two Cu atoms of the cluster interacting directly with the surface, one forming a bond with the Si atom and the second with a bridging oxygen atom, Fig. 5(a). Also in this case the structure is different from that of an NBO center where Cu_4

Table 4 Charge q and spin distribution in free and SiO_2 supported Cu_n clusters

	≡Si–O˙	≡Si–O–Cu	≡Si–O–Cu$_2$	≡Si–O–Cu$_3$	≡Si–O–Cu$_4$	≡Si–O–Cu$_5$
q(O)	−0.37	−0.71	−0.70	−0.71	−0.71	−0.71
q(Cu)^a	—	+0.13	+0.11	+0.08	−0.01	−0.05
Spin density	0.93 (O)	—	0.08 (O)	—	0.03 (O)	—
			0.89 (Cu$_2$)		0.96 (Cu$_4$)	
	≡Si˙	≡Si–Cu	≡Si–Cu$_2$	≡Si–Cu$_3$	≡Si–Cu$_4$	≡Si–Cu$_5$
q(Si)	+0.84	+1.06	+1.18	+1.17	+1.17	+1.36
q(Cu)^a	—	−0.24	−0.21	−0.18	−0.11	−0.11
Spin density	0.78 (Si)	—	0.18 (Si)	—	0.11 (Si)	—
			0.73 (Cu$_2$)		0.86 (Cu$_4$)	

^{a} Average values.

assumes a pseudo-tetrahedral shape with Cu–Cu distances scattered in the wide range 2.38–2.61 Å; the structure on the NBO can be better described as that of a bent rhombus (butterfly) (see Fig. 5 in ref. 16). ≡Si–Cu$_4$ has a doublet ground state with the unpaired electron almost entirely delocalized over the four Cu atoms with little spin density on the Si atom of the surface, Table 4.

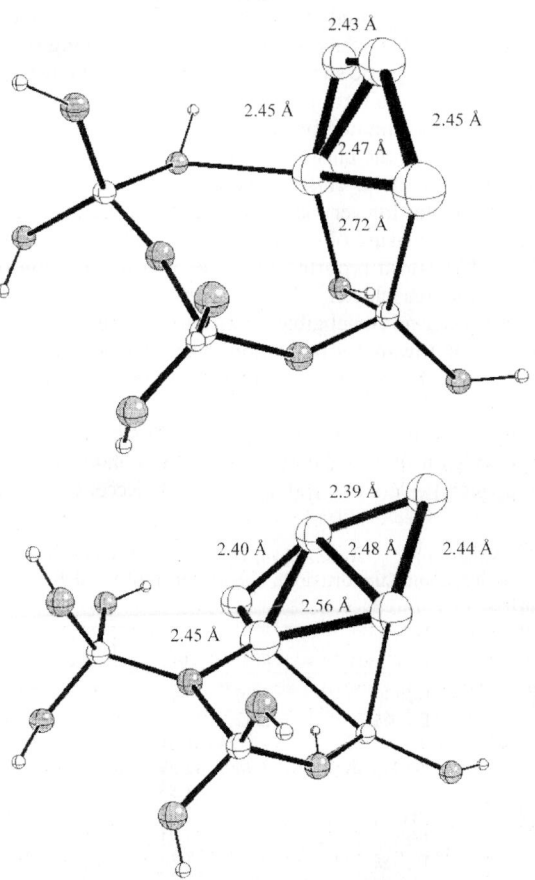

Fig. 5 (Top) Model of a supported Cu$_4$ cluster (larger spheres) on an E' center, ≡Si–Cu$_4$. The optimal Cu–Cu distances are given in Å. (Bottom) Model of a supported Cu$_5$ cluster (larger spheres) on an E' center, ≡Si–Cu$_5$. The optimal Cu–Cu distances are given in Å. The existence of a second bonding interaction with a bridging oxygen of the silica surface has been indicated by a solid line.

The last cluster considered is Cu_5. Free Cu_5 has a planar trapezoidal structure obtained by adding a Cu atom in the plane containing the rhombic Cu_4; the internal angles are close to 60° and the Cu–Cu distances go from 2.404 to 2.475 Å, Fig. 5(b). The structure of supported Cu_5 was obtained starting from that of ≡Si–O–Cu_4 by replacing the NBO oxygen by the fifth Cu. The geometry optimization leads to an almost flat pentamer which resembles that of free Cu_5, Fig. 5(b). This structure is not too different from that found on an NBO center. Cu_5 is bound with two Cu atoms to the surface, with the Si atom of the E' center and with a bridging oxygen, see Fig. 5(b). On average, the Cu–Cu distances of supported Cu_5, 2.45 Å, are practically coincident with those of the free cluster, 2.44 Å, and very similar to those of the ≡Si–O–Cu_4 surface complex (NBO). This suggests that the geometrical distortions within the metal cluster, due to bonding with different substrate defects, disappear quite rapidly as the cluster size increases.

4.2 Adhesion, atomization, and nucleation energies

Small Cu clusters, more polarizable than a single atom, are expected to interact more strongly with the non-defective SiO_2 surface. Still, the role of defects for the diffusion, adhesion and nucleation processes is crucial. All Cu clusters considered, from Cu_2 to Cu_5, interact with the E' center with adhesion energies which go from 1.7 to 3.3 eV, Table 5. These values are always smaller than for the same clusters interacting with NBO, Table 5. The adhesion energies (E_{ad}) were computed for the fully optimized supported clusters with respect to the ground state of the equilibrium gas-phase clusters, $E_{ad} = -[E(Cu_n/SiO_2) - E(Cu_n) - E(SiO_2)]$ (positive values of E_{ad} correspond to bound states). The relatively large oscillations in E_{ad} from cluster to cluster are due to the open shell character of Cu, Cu_3, and Cu_5, which favors the direct coupling of the unpaired electron on the metal with that of the paramagnetic surface center; closed shell clusters, Cu_2 and Cu_4, have to 'open' their configuration in order to form a direct covalent bond. It is also possible that polarization interactions with the two-coordinated bridging O atoms differ from cluster to cluster.

The energy required to atomize the cluster is defined as $D_e/\text{atom} = -\{[E(Cu_n/SiO_2) - nE(Cu) - E(SiO_2)]/n\}$. The D_e/atom in metal clusters increases with cluster size and converges to the cohesive energy of the bulk metal for very large metallic aggregates. For a supported cluster, the atomization energy provides a measure of the additional stability of the cluster due to the bond at the interface. The values of D_e/atom reported in Table 5 show an additional stabilization of the supported, compared with the free, cluster, which is more or less constant for all clusters. Even for a supported pentamer there is a non-negligible contribution from the bond at the interface to the overall stability of the cluster towards atomization. The stronger bonds of the clusters with the NBO centers compared to the E' explain the larger value of D_e/atom, although the effect is not very pronounced, Table 5.

An important quantity determining the mechanism of cluster growth is the nucleation energy, E_{nuc}, defined as the energy gain due to the addition of an isolated Cu atom to a supported Cu_n cluster ($E_{nuc} = -[E(Cu_n/SiO_2) - E(Cu) - E(Cu_{n-1}/SiO_2)]$). Recently, accurate microcalorimetric

Table 5 Adhesion, E_{ad}, atomization, D_e/atom, and nucleation, E_{nuc}, energies of gas-phase and supported Cu clusters on NBO and E' centers at the SiO_2 surface

		Cu	Cu_2	Cu_3	Cu_4	Cu_5
E_{ad}/eV^a	NBO, ≡Si–O'	3.96	3.16	3.46	3.55	4.71
	E', ≡Si'	2.52	1.65	3.32	2.58	3.04
$D_e/\text{atom}/\text{eV}^b$	Free	—	1.01	1.00	1.31	1.41
	NBO, ≡Si–O'	3.96	2.59	2.16	2.19	2.35
	E', ≡Si'	2.52	1.83	2.11	1.95	2.02
E_{nuc}/eV^c	Free	—	2.02	1.00	2.21	1.83
	NBO, ≡Si–O'	—	1.21	1.30	2.30	3.00
	E', ≡Si'	—	1.14	2.67	2.02	2.68

^{a} $E_{ad} = -[E(Cu_n/SiO_2) - E(Cu_n) - E(SiO_2)]$.
^{b} $D_e/\text{atom} = -\{[E(Cu_n/SiO_2) - nE(Cu) - E(SiO_2)]/n\}$.
^{c} $E_{nuc} = -[E(Cu_n/SiO_2) - E(Cu) - E(Cu_{n-1}/SiO_2)]$.

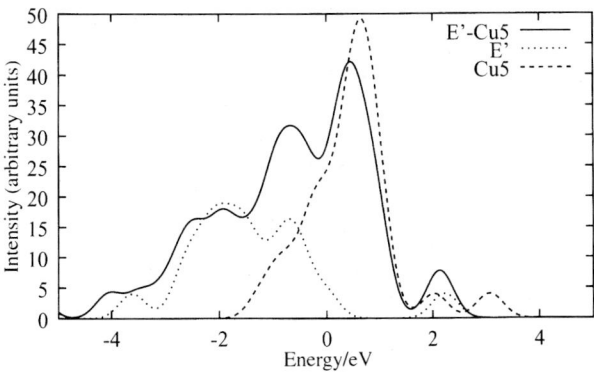

Fig. 6 Valence density of states of a model of the E' center on the silica surface, of a free Cu_5 cluster and of a supported Cu_5 cluster, $\equiv Si-Cu_5$. The zero of the scale has been aligned with the top of the O 2p band.

measurements of the heat of adsorption of a metal atom to a metal cluster supported on an oxide surface have been reported.[46] Therefore, the nucleation energy is a quantity that is becoming available also through experimental studies, even for small aggregates. So far, heats of adsorption of Cu atoms on Cu/MgO have been measured.[46] On oxide surfaces nucleation is believed to occur through diffusion of isolated atoms or eventually dimers; therefore, it is useful to compare the nucleation energy for free and supported clusters. The addition of an extra Cu atom leads to a stabilization that is larger for the supported than for the free Cu clusters, Table 5, with the exception of the dimer (the energy gain for the process $Cu + Cu \rightarrow Cu_2$ is obviously larger than for the $\equiv Si-Cu + Cu \rightarrow \equiv Si-Cu_2$ one because of the closed shell nature of the Si–Cu bond). This is an important conclusion, which shows the role of the substrate in the growth process of a supported particle. In fact, since isolated Cu atoms are weakly bound to the regular SiO_2 surface, they will diffuse with low activation barriers. The diffusion process will stop only at defects or at sites where nucleation has already started. It should be noted that the nucleation energy seems to increase with the cluster size for the supported clusters, going from about 1 eV for the formation of the dimer to about 2.7–3.0 eV for the pentamer. In other words, the energy gain of the process $SiO_2-Cu_n + Cu \rightarrow SiO_2-Cu_{n+1}$ seems to increase for larger n. This is partly because as the cluster becomes larger the cohesive energy increases due to the increase of the coordination number of the atoms in the cluster; another effect, however, is that larger clusters are more polarizable and the interactions with the bridging oxygens also increase. Of course, by further increasing the cluster size the nucleation energy will tend first to that of the corresponding isolated Cu particles and then, for larger crystallites, to the cohesive energy of the bulk metal. In other words, the perturbation induced by the strong bond with the surface defect is rapidly screened by the conduction band electrons of the metal cluster and the electronic modifications induced by the bond with the surface defect disappear for aggregates of a few tens of atoms.

4.3 Valence band and gap states

Recent MIES experiments on metal atoms and clusters deposited on oxide surfaces have shown the appearance of new states in the gap of the oxide material.[38] These states can be attributed to the population of new defects at the surface of the oxide, *e.g.*, F centers in MgO, to the presence of new metal–oxide bonds at the interface, or to features typical of a metal particle (occupied d states, *etc.*). It is therefore of interest to analyze the presence of states in the gap after metal deposition. In principle, cluster calculations are not adequate to determine the gap energies because of the lack of periodic boundary conditions. Nevertheless, it is possible to estimate the size of the gap in a cluster calculation from the HOMO–LUMO separation derived from one-electron orbital energies. It is well known that one-electron energies do not provide a good approximation of an excited state problem like the determination of the optical gap. In particular, Hartree–Fock calculations largely overestimate the gap (by a factor 2 easily), while DFT approaches underestimate it. With a cluster model of the non-defective SiO_2 surface we compute a "HOMO–LUMO" gap of

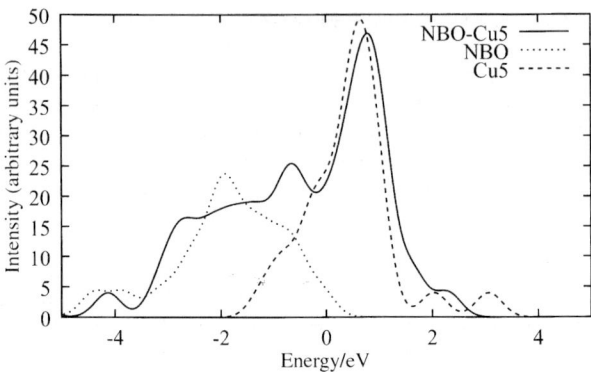

Fig. 7 Valence density of states of a model of the NBO center on the silica surface, of a free Cu_5 cluster and of a supported Cu_5 cluster, $\equiv$Si–O–Cu_5. The zero of the scale has been aligned with the top of the O 2p band.

8.1 eV, underestimated by only 10% with respect to the experimental one, ≈ 9 eV.[47] The computed gap is smaller than in HF but larger than in a pure DFT calculation because of the use of a hybrid DFT approach (B3LYP) where the HF exchange is partially mixed in with the DFT exchange. Kohn–Sham orbital energies can also be used to determine the valence density of states, DOS, by convolution with Gaussian functions of the one-electron energy spectrum. In Figs. 6 and 7 we have reported the DOS of the SiO_2 support, of free Cu_5, and of the $\equiv SiO_2$–Cu_5 surface complex for both the E' and NBO centers. The Cu_5 DOS has been determined for the optimal structure of the gas-phase cluster; the DOS of Cu_5 with the structure adopted on the support is almost identical, at least with the kind of Gaussian broadening used (0.75 eV). The DOS of the E' center shows a small feature in the gap about 2.5 eV above the top of the O 2p valence band, Fig. 6. This state corresponds to the singly occupied Si sp^3 level. On the contrary, no gap states are present in the DOS spectrum of NBO, Fig. 7, since the O dangling bond gives rise to a resonance within the O 2p valence band. The qualitative differences in the O 2p valence DOS of the SiO_2 substrate clusters, Fig. 6 and 7, reflect the slightly different shape of the substrate models. For both interface bonds, $\equiv$Si–Cu_5 (E'), and $\equiv$Si–O–Cu_5 (NBO), a series of new levels of dominantly Cu 3d character appears above the O 2p valence band and extends into the gap. These states are located about 1 eV above the top of the O 2p valence band. The shape and the position of these states is virtually identical in the two cases and similar to those of the unsupported Cu_5 cluster. This shows unambiguously that the gap features are due to the supported metal particle and not to the interface bond. Looking at similar DOS curves for smaller clusters (not shown) we observe a progressive broadening of the peak above the O 2p levels due to metal states. Therefore, it is expected that by growing even larger clusters these features will then appear as broad bands rather than as well resolved peaks.

5 Conclusions

We have performed density functional calculations on the interaction of small Cu clusters with regular and defect sites of the surface of dehydroxylated silica. The non-defective sites, the bridging oxygen atoms, $\equiv$Si–O–Si$\equiv$, are rather unreactive towards adsorbed metal atoms. This is fully consistent with measurements of the sticking coefficient of Cu on SiO_2. Zhou *et al.* found that at 300 K only one third of the initially incident Cu atoms stick to the surface;[19] Xu and Goodman observed that the sticking depends markedly on the temperature, varying from 0.6 at 90 K to 0.1 at 400 K.[17] Both studies agree with the fact that the bonding of Cu with the clean surface is weak and that sticking occurs only at the defect sites. Indeed, strong bonds form between Cu atoms and surface defects. We have considered here two of the dominant point defects at the silica surface: the Si dangling bond (the E' center) and the non-bridging oxygen (an O dangling bond). Both centers are paramagnetic and are possible sites where the nucleation of the cluster begins. We have reported new data on the interaction of small Cu clusters with the E' center and we have compared the characteristic features with those of the same clusters interacting with the NBO centers.

Cu clusters containing from 2 to 5 metal atoms form strong bonds with both E' centers and non-bridging oxygens, although the adhesion energy with these latter centers is higher. The interaction arises in part from the formation of a covalent polar bond between the metal cluster and the paramagnetic center, either E' or NBO, and, in part, from the polarization interaction of the cluster electron density, which leads to direct, although weaker, bonds with a two-coordinated oxygen of silica. As a result, due to the interface bond the fragmentation energy of the supported cluster increases compared to the free, gas-phase counterparts. The effect of the interface bond on the metal–metal distance of the cluster is not large and tends to disappear as the size of the cluster increases. Hence, while the Cu–Cu distances in supported Cu_2 and Cu_3 are larger than in the free clusters, in Cu_4 and, in particular, in Cu_5 the average Cu–Cu distances are close to those of the unsupported clusters. On the other hand, the shape of the supported clusters can differ substantially from that of the gas-phase units. This is true, in particular, for the Cu clusters interacting with the NBO centers, where the interaction is stronger. We expect that by growing larger Cu particles these will assume nearly spherical three-dimensional structures since the metal–metal bonds will dominate over the weak, electrostatic Cu–SiO_2 interactions. In this respect point defects on the SiO_2 surface act as strong anchoring sites for the entire cluster limiting the diffusion process and favoring nucleation. These results are consistent with a Volmer–Weber growth mode of Cu overlayers on silica with formation of 3D particles.[48]

The bonding of Cu clusters with surface defects has some consequences that, in principle, could be observed experimentally. The most important one is that new states appear in the wide gap of SiO_2. These states are located 1–2 eV above the top of the O 2p valence band and extend into the gap. Depending on the cluster size, however, the feature corresponding to the occupied metal states can give rise either to sharp or to broad bands. As the cluster size increases we expect a considerable broadening of the features. Thus, a strong coverage dependence of the width of the states in the gap is expected.

Acknowledgements

N.L. is grateful to the "Direcció General de Recerca Generalitat de Catalunya" for supporting her visit to the University of Milano. G.P. thanks the University of Barcelona for an invited professor position. Financial support from the Spanish "Ministerio de Educación y Ciencia", project CICyT PB95-0847-C02-01, "Acción Integrada Hispano-Italiana, HI1998-0042", Italian INFM (Project PAIS), Italian MURST (Cofin Area 03), "Generalitat de Catalunya" projects 1997SGR00167 is fully acknowledged. Part of the computer time was provided by the "Centre de Supercomputació de Catalunya", CESCA, and "Centre Europeu de Paralel-lisme de Barcelona", CEPBA, through a research grant from the University of Barcelona.

References

1. H. J. Freund, *Angew. Chem.*, 1997, **109**, 444.
2. R. Lambert and G. Pacchioni (ed.), *Chemisorption and Reactivity on Supported Clusters and Thin Films*, NATO ASI Ser., Ser. E, 1997, **331**.
3. C. T. Campbell, *Surf. Sci. Rep.*, 1997, **27**, 1.
4. L. Marchese, S. Coluccia, G. Martra, E. Giamello and A. Zecchina, *Mater. Chem. Phys.*, 1991, **29**, 437.
5. E. Giamello, M. C. Paganini, D. M. Murphy, A. M. Ferrari and G. Pacchioni, *J. Phys. Chem.*, 1997, **101**, 971.
6. J. H. Lunsford, *Catal. Today*, 1990, **6**, 235.
7. G. Pacchioni and N. Rösch, *Surf. Sci.*, 1994, **306**, 169.
8. G. Pacchioni and N. Rösch, *J. Chem. Phys.*, 1996, **104**, 7329.
9. N. Lopez and F. Illas, *J. Phys. Chem. B*, 1998, **102**, 1430.
10. V. Musolino, A. Selloni and R. Car, *J. Chem. Phys.*, 1998, **108**, 5044.
11. C. Li, R. Wu, A. J. Freeman and C. L. Fu, *Phys. Rev. B: Condens. Matter*, 1993, **48**, 8317.
12. Y. Li, D. C. Landgreth and M. R. Pederson, *Phys. Rev. B: Condens. Matter*, 1995, **52**, 6067.
13. I. Yudanov, G. Pacchioni, K. Neyman and N. Rösch, *J. Phys. Chem. B*, 1997, **101**, 2786.
14. N. Lopez, G. Pacchioni, F. Maseras and F. Illas, *Chem. Phys. Lett.*, 1998, **294**, 611.
15. N. Lopez, F. Illas and G. Pacchioni, *J. Am. Chem. Soc.*, 1999, **121**, 813.
16. N. Lopez, F. Illas and G. Pacchioni, *J. Phys. Chem.*, 1999, in press.

17 X. Xu and D. W. Goodman, *Appl. Phys. Lett.*, 1992, **61**, 1799.
18 X. Xu, J. W. He and D. W. Goodman, *Surf. Sci.*, 1993, **284**, 103.
19 J. B. Zhou, H. C. Lu, T. Gustafsson and E. Garfunkel, *Surf. Sci. Lett.*, 1993, **293**, L887.
20 J. B. Zhou, T. Gustafsson and E. Garfunkel, *Surf. Sci.*, 1997, **372**, 21.
21 G. Pacchioni and G. Ieranò, *Phys. Rev. Lett.*, 1997, **79**, 753.
22 G. Pacchioni and G. Ieranò, *Phys. Rev. B*, 1998, **57**, 818.
23 G. Pacchioni, G. Ieranò and A. Marquez, *Phys. Rev. Lett.*, 1998, **81**, 377.
24 *The Physics and Technology of Amorphous SiO_2*, ed. J. Arndt R. Devine and A. Revesz, Plenum, New York, 1988.
25 L. Skuja, *J. Non-Cryst. Solids*, 1998, **239**, 16.
26 V. A. Radsig, *Chem. Phys. Rep.*, 1995, **14**, 1206.
27 F. Bart, M. Gautier, F. Jollet and J. P. Durand, *Surf. Sci.*, 1994, **306**, 342.
28 Y. Le Page, L. D. Calvert and E. J. Gabe, *J. Phys. Chem. Solids*, 1980, **41**, 721.
29 A. D. Becke, *J. Chem. Phys.*, 1993, **98**, 5648.
30 C. Lee, W. Yang and R. G. Parr, *Phys. Rev. B: Condens. Matter*, 1988, **37**, 785.
31 R. Ditchfield, W. J. Here and J. A. Pople, *J. Chem. Phys.*, 1971, **54**, 724.
32 M. S. Gordon, J. S. Binkley, J. A. Pople, W. J. Pietro and W. J. Hehre, *J. Am. Chem. Soc.*, 1982, **104**, 2797.
33 P. J. Hay and W. R. Wadt, *J. Chem. Phys.*, 1985, **82**, 299.
34 A. J. H. Wachters, *J. Chem. Phys.*, 1970, **52**, 1033.
35 S. F. Boys and F. Bernardi, *Mol. Phys.*, 1970, **19**, 553.
36 M. J. Frisch, G. W. Trucks, H. B. Schlegel, P. M. W. Gill, B. G. Johnson, M. A. Robb, J. R. Cheesman, T. A. Keith, G. A. Petersson, J. A. Montgomery, K. Raghavachari, M. A. Al-Laham, V. G. Zakrzewski, J. V. Ortiz, J. B. Foresman, J. Cioslowski, B. B. Stefanov, A. Nanayakkara, M. Challacombe, C. Y. Peng, P. Y. Ayala, W. Chen, M. W. Wong, J. L. Andres, E. S. Reploge, R. Comperts, R. L. Martin, D. J. Fox, J. S. Binkley, D. J. Defrees, J. Baker, J. P. Stewart, M. Head-Gordon, C. Gonzalez and J. A. Pople, *GAUSSIAN 94*, Gaussian Inc., Pittsburgh, PA, 1997.
37 N. Lopez, F. Illas, N. Rösch and G. Pacchioni, *J. Chem. Phys.*, 1999, **110**, 4873.
38 M. Brause, D. Ochs, J. Günster, T. Mayer, B. Braun, V. Puchin, W. Maus-Friedrichs and V. Kempter, *Surf. Sci.*, 1997, **383**, 216.
39 F. Maseras and K. Morokuma, *J. Comput. Chem.*, 1995, **16**, 1170.
40 C. Marcolli and G. Calzaferri, *J. Phys. Chem. B*, 1997, **101**, 4925.
41 M. J. Frisch, G. W. Trucks, H. B. Schlegel, P. M. W. Gill, B. G. Johnson, M. W. Wong, J. B. Foresman, M. A. Robb, M. Head-Gordon, E. Repogle, R. Gomperts, J. L. Andres, K. Raghavachari, J. S. Binkley, C. Gonzalez, R. L. Martin, D. J. Fox, D. J. Defrees, J. Baker, J. P. Stewart and J. A. Pople, *GAUSSIAN 92*/DFT, Gaussian Inc., Pittsburgh, PA, 1993.
42 N. L. Allinger, *MM3(92)*, QCPE, Bloomington, IN, 1992.
43 N. L. Allinger, Y. H. Yuh and J. H. Lii, *J. Am. Chem. Soc.*, 1989, **111**, 8551.
44 J. H. Lii and N. L. Allinger, *J. Am. Chem. Soc.*, 1989, **111**, 8566.
45 J. H. Lii and N. L. Allinger, *J. Am. Chem. Soc.*, 1989, **111**, 8576.
46 J. T. Ranney, D. E. Starr, J. E. Musgrove, D. J. Bold and C. T. Campbell, *Faraday Discuss.*, 1999, **114**, 195.
47 N. F. Mott, *J. Non-Cryst. Solids*, 1980, **40**, 1.
48 X. Xu, S. Vesecky and D. W. Goodman, *Science*, 1992, **258**, 788.

Paper 9/02374G

General Discussion

Prof. Jennison opened the discussion of Dr Mackrodt's paper: When we were working on high-T_c materials, we found that explicit inclusion of correlation, such as through a many band extended Hubbard model which could be solved on a small cluster, was essential to obtain the correct value for the superexchange, as determined for example by neutron scattering.[1]

One important energy in this model is the screened two-hole interaction on the oxygen ion, which we determined from the O(KVV) Auger lineshape to be $U \approx 6$ eV. Since this quantity is likely to be similar for these materials, I might expect antiferromagnetic properties such as the Néel temperature to be poorly given without this added layer of theory. Could you comment on this please?

1 E. B. Stechel and D. R. Jennison, *Phys. Rev. B*, 1988, **38**, 4632.

Dr Mackrodt responded: This is an important point, and I would like to mention three things in this connection. The first is that the inclusion of electron correlation beyond UHF certainly increases the superexchange coupling, as Illas and co-workers have shown from cluster calculations.[1] Whether this is the case in solids to the same extent remains to be established. The second point is that, as with other aspects of magnetism, notably the local moment, the superexchange coupling energy cannot be measured directly, but has to be extracted from experiment on the basis of a model, so that direct comparisons of calculated and 'measured' values need to be viewed with some caution. Finally, there have now been UHF calculations for quite an extensive range of systems, including MnO, NiO, α-Fe_2O_3, α-Cr_2O_3, $KCuF_3$, $CaCuO_2$, Sr_2CuO_3, $SrCuO_2$, $LiCuO_2$, $LiMnO_2$, $LaMnO_3$ and all the known polymorphs of MnS, and without exception these calculations have predicted the observed low temperature magnetic ordering. In addition, UHF calculations for MnO and NiO[2] also yield good agreement with the measured rhombohedral distortion below the Néel temperature due to spin–lattice interaction. Where there are 'measured' values of the moment, in every case other than for Cu ($S = \frac{1}{2}$) systems, for which it is known that zero-point quantum fluctuations renormalise the moment appreciably (in Sr_2CuO_3, for example Kojima *et al.*,[3] have deduced a moment of ≈ 0.06 μ_B compared with the formal value of 1 μ_B), the computed spin moments are in good agreement with the 'measured' values.

1 I. D. Moreira and F. Illas, *Phys. Rev. B*, 1997, **55**, 4129; C. de Graaf, F. Illas, R. Broer and W. C. Nieupoort, *J. Chem. Phys.*, 1997, **106**, 3287.
2 M. D. Towler, N. L. Allan, N. M. Harrison, V. R. Saunders, W. C. Mackrodt and E. Aprà, *Phys. Rev. B*, 1994, **50**, 5041.
3 K. M. Kojima, Y. Fudamoto, M. Larkin, G. M. Luke, J. Merrin, B. Nachumi, Y. J. Uemura, N. Motoyama, H. Eisaki, S. Uchida, K. Yamada, Y. Endoh, S. Hosoya, B. J. Sternlieb and G. Shirane, *Phys. Rev. Lett.*, 1997, **78**, 1787.

Dr Shluger asked: (1) Unrestricted Hartree–Fock (UHF) is known to overemphasise hole localisation in many cases and electron correlation is needed to address the question of hole localisation. Could you please comment on this? (2) Charge transfer transition energies can be affected by lattice repolarisation due to the charge transfer. What effect could lattice polarisation have on your results?

Dr Mackrodt answered: (1) UHF calculations do suggest a substantial degree of localisation which might be reduced by the further inclusion of electron correlation. However, such localisation and the implications of this for the activation energy of hole hopping, for example, are certainly compatible with experimental data, including the recent ^{7}Li NMR measurements for $Li_xNi_{1-x}O$.[1]

(2) There will certainly be some effect due to lattice repolarisation, but separating this effect from the direct interaction of charge transfer excitons, which result from the transition, could be quite difficult.

1   M. Corti, S. Marini, A. Rigamonti and F. Tedoldi, *Phys. Rev. B*, 1997, **56**, 11056.

Prof. Pacchioni asked: How sensitive are direct and superexchange interactions to changes in the lattice parameter due to the epitaxial growth of the thin NiO film?

Dr Mackrodt replied: Superexchange interactions are extremely sensitive to changes in the lattice parameter, so that one might expect substantial differences in the stability of the magnetic order of NiO films grown epitaxially on different substrates such as MgO and CaO.

Prof. Harrison said: The stability of the localised state in Li Ni_7O_8 is due to a competition between the energy gain due to lattice distortion (0.3 eV) and the additional correlation energy of the totally symmetric (delocalised) state. In this case it seems very likely that the UHF solution is correct and a self-trapped hole forms.

Dr Egdell said: Experimentally, Li-doped NiO is not a metallic conductor. Thus the holes are localised rather than itinerant.

Dr Noguera said: As regards the localisation of a hole in NiO, Hartree–Fock (HF) and DFT calculations give contradictory results. While it is true that HF usually overestimates localisation effects, as noted by Dr Shluger, DFT overestimates delocalisation effects because the self interaction of the electrons is not correctly subtracted. It thus seems that standard *ab initio* methods, at present, are not fully satisfactorily in this respect.

Prof. Thornton asked: For low dimensional structures rumpling could be significant. Do you have a feel for how they might affect your results?

Dr Mackrodt replied: There is no experimental evidence of appreciable rumpling of NiO(100), but if there were circumstances in which this did occur, I would expect the magnetism to be affected, since the superexchange interaction is very sensitive to deviations from linearity of Ni(↑)–O–Ni(↓) and to a lesser extent the d → d excitation energies, which is sensitive to the local coordination. However, I would not expect any major changes in the electronic structure, including the d electron configuration, single particle energy levels or high spin insulating behaviour, nor in the nature and stability of hole states.

Prof. Freund asked: From an experimental point of view it is desirable to have predictions on the change in magnetic surface structures as a function of temperature. Are such predictions feasible?

Dr Mackrodt answered: From the differences in energy between different magnetic orderings it should be possible to give crude estimates of the changes in magnetic surface structures with temperature, including that of the disordered state, though the precision would not be high.

Dr Egdell said: The new technique of spin polarised metastable He atom diffraction is now yielding information about magnetic ordering of surfaces. For example the surface ionic layer for NiO(100) shows a (6 × n) magnetic superstructure not found in the bulk.[1] With the related technique of inelastic scattering of metastable spin polarised helium atoms it might be possible to measure surface spin wave dispersions and superexchange parameters.[2]

1   A. Swan, M. Maryowski, W. Franzen, M. Elbatanouny and K. M. Harbui, *Phys. Rev. Lett.*, 1993, **71**, 1250.
2   M. Elbatanouny, C. Murthy, C. R. Willis, S. Kais and V. Staemmler, *Phys. Rev. B*, 1998, **58**, 7391.

Dr Mackrodt responded: These very exciting new developments will certainly have an important impact on our understanding of ultrathin films and provide fresh impetus for further, more extensive calculations of magnetism at the surface and in thin films.

Prof. Campbell asked: Your Fig. 9 shows a surface energy of ≈ 1.2 J m^{-2} for MgO(100). There are experimental values of 0.6–0.9 J m^{-2} for alumina and 0.3–0.6 J m^{-2} for silica.[1] Based on similar wetting behaviour I expect MgO(100) to be close to the value of silica, or ≈ 0.4 J m^{-2}. Can you comment on this difference?

1 C. T. Campbell, *Surf. Sci. Rep.*, 1997, **227**, 1.

Dr Mackrodt answered: I'm not clear as to why you expect the surface energy of MgO(100) to be close to the value of silica, but our value is certainly within the range of experimental values (1.04–1.2 J m^{-2}) reported by Tosi[1] and, as mentioned in our paper, the temperature dependence is consistent with data for rocksalt (100) surfaces (ref. 58 of our paper).

1 M. P. Tosi, *Solid State Phys.*, 1964, **16**, 1.

Prof. Madey said: In Fig. 9 you plot surface energy as a function of slab thickness, and find convergence to a limiting value at ≈ 10 layers. Have you computed other properties of the NiO films (bandgap density of states, magnetic properties) as a function of slab thickness? At which thicknesses do you find convergence?

Dr Mackrodt replied: It is straightforward, and relatively inexpensive, to carry out atomistic simulations as a function of slab thickness, and this we have done for quantities such as the surface phonon densities of states and impurity segregation free energies, for which 10 layers takes us beyond the convergence limit. For quantities computed from first principles electronic structure calculations testing the convergence to the limit is very much more expensive. However, such studies that we have been able to make suggest that for a three layer slab, the central layer is a good approximation to the bulk (see Table 2 of our paper, for example), and I would certainly expect the local electronic properties of the central layer of a five layer slab to be virtually indistinguishable from those of the bulk.

Prof. Diebold opened the discussion of Prof. Neddermeyer's paper: (1) What were the thickest films you could achieve in both the CoO and NiO case, and can you comment on whether the films are n-type or p-type semiconducting? (2) Can you comment on the nature of the defects and the mechanism of motion? Do you think that some of the motion is tip-induced?

Prof. Neddermeyer responded: (1) For the measured samples the maximum coverage was nominally 10 ML for both NiO and CoO. Samples of such thickness could not be measured easily, however, since instabilities of the tunnelling current frequently occurred possibly due to insufficient conductivity. The actual thickness of the films at the measuring position could not be determined and therefore 10 ML should be considered as an upper limit. The doping of the films will depend on the preparation conditions. In the case of NiO islands the I/V curves indicate n-type conductivity (see Fig. 3(b) of our paper), for CoO films the Fermi level was observed in the centre of the band gap (as determined from the dependency of the contrast as a function of the sample bias).

(2) For CoO we observe two kinds of defects (see Fig. 7 of our paper): atomic defect structures (as emphasised by the circles) and more extended irregularities (as displayed on that part of the surface extending to the right upper corner of the images). While the former ones are probably related to vacancies in the surface layer the latter ones might result from small changes of the registry of the CoO films with regard to the Ag(100) substrate. It has to be emphasised that both kinds of feature are preferentially observed on surfaces which have not yet been fully annealed. In our opinion, the changes seen in the images are thermally induced and might be caused by the sample–tip interaction only to a minor extent.

Dr Castell said: As referenced in your paper, we have observed characteristic defects on Li-doped NiO(001) cleavage surfaces. We observe second nearest neighbour brightening around point defects (which we believe are Li dopants) and on top of ⟨001⟩ step edges. Do you observe similar defects on the NiO(001) films you have grown?

Prof. Neddermeyer responded: In the case of the NiO films we observe defects which probably are related to vacancies in the surface layer (see Fig. 2 of our paper). A second nearest neighbour brightening is not seen for these defects. This would be consistent with the special nature of Li dopants in your previous studies which should not be present in our films.

Dr Egdell asked: Could you comment on the fact that the tunnelling spectrum in Fig. 3 of your paper is characteristic of an n-type semiconductor. Bulk NiO is usually considered to be a p-type material.

Prof. Neddermeyer replied: Although the tunnelling spectrum shown in Fig. 3(b) (full line) would be consistent with n-type conductivity in the islands the data might not be simply related to the electronic states of the NiO islands. While the differences on the Ag(100) substrate and NiO islands may qualitatively be understood by the metallic and oxidic nature of the measuring position, respectively, for a detailed analysis of the spectra the nature of the tip has to be known more precisely. We believe that the characteristics shown in Fig. 3 are obtained with a tip which was not metallic on the apex. In this case the rise of the tunnelling current (at both polarities) will be shifted to higher values of the sample bias due to the additional band gap of the tip apex and a lack of electrons at the Fermi level.

Prof. Jennison asked: In your Fig. 4 you show compact islands of oxidized Co on Ag(100). At lower temperatures, one might expect dendritic growth to occur, as Jürgen Behm saw for Al_2O_3 islands on Ru(001).[1] Have you looked at island shapes at lower temperatures?

1 R. J. Behm, unpublished work.

Prof. Neddermeyer replied: We did not perform deposition experiments below room temperature. However, our experiments showed that already deposition at room temperature does not lead to well ordered oxidic films. For deposition at even lower temperature other methods (*e.g.*, pulsed laser deposition) probably have to be applied.

Prof. Thornton asked: When you observe the substrate by tunnelling through the CoO layer, do you have a measure of the oxide film thickness?

Prof. Neddermeyer answered: In our experience, monatomic step heights as measured on three-dimensional oxide islands agree with the expected value (around 0.22 nm). However, measurements of Ag/oxide step heights are influenced by differences of the density of states of Ag and the oxides and consequently show drastic changes with the sample bias. An accurate determination of the thickness of oxide films by the STM measurements alone is therefore difficult.

Dr Shluger said: I would like to comment that in the system which you are studying: STM of oxide film on metal substrates, image forces should be very strong. This can lead to ion instabilities at oxide surface and their jumps on the tip. Your remarks on tip changes during imaging supports this conclusion.

I have two questions. How did the tip–surface interactions in your experiments affect the defect diffusion shown in your video? Can you desorb ions from the surface at large applied voltages?

Prof. Neddermeyer replied: At low measurement speed most of the defects moved to a neighbouring position while scanning the surface. Upon increasing the measuring speed the number of such diffusion jumps per image decreased. This means that the defect movement can at most to a small extent be induced by the sample–tip interaction (whose influence cannot be excluded, however). At higher sample–tip voltages U (*e.g.*, $U > 3$ V and $U < -3$ V) the oxidic films were often destroyed during the measurement. A systematic study of this effect has not yet been performed.

Dr Weiss asked: Fig. 4(a) displays an STM image taken after Co deposition at room temperature, where four different grey levels can be seen. On which basis did you assign these levels to Ag and the O/Co precursor?

Prof. Neddermeyer answered: The most direct evidence for Ag or oxide species as seen in the STM image can be deduced from step heights and contrast changes as a function of sample bias and tip conditions. In addition, in some cases the oxidic layers showed a rougher surface structure as compared with the smooth appearance of the metallic substrate. Note that the Ag/Ag step height is always found in agreement with the expected value (0.2 nm).

Dr Venables said: With this preparation procedure, can you exclude the possibility of Ni and Co metal buried in the Ag substrate? For example, the microstructure of Fig. 4(a) looks remarkably similar to that produced by depositing pure Co on noble metals (Cu or Ag) in the absence of oxygen.

Prof. Neddermeyer responded: Ni and Co atoms may indeed be incorporated in the uppermost atomic layer of the Ag(100) substrate. We have studied the growth behaviour of the clean transition metal on Ag(100) (without the presence of O) in each case. Surfaces as shown in Fig. 4(a) could only be obtained with sufficiently high partial pressure of O. To our knowledge and experience at room temperature clean transition metal/Ag(100) (or Cu(100)) systems show distinct differences. For example, the tendency of three-dimensional growth of the transition metal deposit may be recognised in contrast to experiment with the presence of O described above where a more two-dimensional growth mode of the deposit is found.

Prof. Hayden opened the discussion of Prof. Woodruff's paper: You have suggested that the similarity in the PhD for the two nitrogen transitions is evidence that they derive from the same species. Have you measured the PhD for the nitrogen transition which exists alone (which you ascribe to nitrogen atoms) and compared it with the NO derived levels?

Dr Lindsay responded: We have recorded PhD data from the single N 1s feature at normal emission. We found that there was little or no oscillation, within experimental error. This is in direct contrast to the two normal emission modulation functions extracted from the N 1s doublet feature, which exhibit significant oscillations.

Prof. Thornton said: You could presumably improve the precision of your measurements by using the O 1s signal for PhD. Do you resolve two or three O 1s peaks which would allow you to do this?

Prof. Woodruff responded: We would certainly expect to be able to substantially improve the precision of the location of the O atom within the NO if we had measured the photoelectron diffraction for the O 1s component (or components) associated with the molecule. Unfortunately, in these experiments our spectral resolution was inadequate to make this separation, so no such measurements were recorded. In the future we certainly expect that the availability of undulator radiation at a third-generation synchrotron radiation source such as BESSY II will provide the necessary combination of high flux and spectral resolution to make these measurements possible.

Prof. Freund said: Angle resolved XPS studies, in particular under grazing excidence have revealed indications for the presence of the O 1s ionisation of oxygen of adsorbed NO/NiO. However, due to the substrate oxygen the signal is hard to discern from the background.

Prof. Pettersson asked: Could you amplify your comment that the presumed atomic nitrogen species does not give rise to oscillations in the spectra? In light of the discrepancies between experiment and theory for this system it would be of importance to have as much information as possible also on different adsorbates. It might be that, if you can get atomic nitrogen on the NiO as a decomposition product, this would be easier to treat theoretically and could give insight into the specific problems associated with adsorption on NiO as compared to MgO. Would it be possible for you to make a determination of the structure of the proposed N/NiO product?

Prof. Woodruff responded: It might be possible to obtain structural information on the (presumed) atomic N species using photoelectron diffraction, but we did not make the necessary measurements. In general when we are conducting such a study we measure the PhD spectra in many (typically 10–20 or more) different emission directions. The modulations are typically strongest in directions correcting to 180° back-scattering from a near-neighbour in the substrate, while in directions far from such geometries the modulations can be quite weak. The actual structure analysis is then typically conducted by fitting 5–10 of these spectra, including some of those with the strongest modulations, over a range of emission directions. In the present case our main concern was to establish that this N species was a distinct entity from the adsorbed NO. We recorded the associated PhD spectrum in normal emission and found weak modulations, whereas both the N 1s peaks we attribute to the NO show strong modulations in this direction. This therefore implies that the emitting N atom is in quite a different local site. Indeed, we might infer that it is not atop a surface Ni atom (so normal emission is not a favoured back-scattering direction). Without more data we really cannot say more about this species.

Prof. Joyner said: The explanation involving photodissociation of NO, and its subsequent reactivity that Woodruff proposes seems very reasonable, and in line with the known surface chemistry of NO. It could be relatively simply confirmed by temperature programmed desorption. If his explanation is solid, he should observe desorption of N_2O (probability at a relatively low temperature), and N_2^- at a much higher temperature.

Prof. Woodruff responded: I agree that this would be interesting (although it is not clear that a temperature would exist at which N_2O would be formed as a stable surface species), but no TPD measurements were attempted in our experiments.

Prof. Friend asked: (1) Is it possible that in ambient NO, there are multiple species? It is possible that other techniques, *e.g.* IR would reveal multiple species on the surface; therefore, the two peaks in the absence of NO ambient may not be same as in a vacuum. Is it possible to perform photoelectron diffraction in NO ambient? (2) To what extent will 'floppy' vibrational modes affect your precision?

Prof. Woodruff answered: (1) From the present experiments we can only say that the surface species produced in the stable condition achieved by a low NO ambient pressure appears to be the same as on a freshly prepared surface at UHV, so we believe all the previous characterisation work is valid for our surface. More generally, the ability to conduct the photoelectron diffraction in ambient pressures of reactant gas is limited by three factors: the mean-free-path for the electrons passing from the sample to, and through, the electron energy analyser; the maximum operating pressure for the electron (channeltron) detector; the contamination of the beamline and ultimately the electron storage ring pressure. In practice, I think one could devise ways of working up to about 10^{-6} mbar.

(2) Large amplitude vibrational modes are certainly important in PhD and can limit precision. Vibrations introduce dephasing in the scattering interferences which are treated through a Debye–Waller factor. Because the scattering is basically a local process, the Debye–Waller factors are dependent on the relative movement of emitter and scatterer, and are thus directionally dependent. In fact the present case of a species adsorbed atop provides a good example of a situation in which we believe large amplitude vibrational modes do commonly have a significant influence. Notice that the PhD spectra in the present case show strong modulations at normal emission, but these amplitudes fall off rapidly as the off-normal emission angle is increased. We have seen this behaviour in many atop adsorbates on metal and semiconductor surfaces, and an important factor appears to be the large amplitude frustrated translational mode parallel to the surface which introduces a large Debye–Waller factor for scattering events involving significant components of the scattering path parallel to the surface, thus damping out PhD modulations at off-normal emission angles. Of course, this also leads to lower precision in determining the mean position of the emitter along the directions of these large vibrations. Indeed, we have found a few cases where it is not possible to distinguish between adsorption sites displaced by up to about 0.2 Å off a high-symmetry position, and large amplitude vibrational motions in the same direction.

Prof. Freund said: The system NO/NiO has been looked at with a variety of experimental methods (ref. 85 of the Introductory Lecture). From refs. 111 and 113 of the Introductory Lecture it is clear that the system undergoes photo-desorption and both peaks in the N 1s ionisation spectrum belong to the same species.

Prof. Friend asked: Have there been other types of experiments to confirm that there is a single NO species? The existence of a single type of NO is very unusual because there are many possible types of NO binding.

Prof. Woodruff responded: There has been a great deal of characterisation of this adsorption system performed previously by a variety of methods (notably by Freund and co-workers as cited in the Introductory Lecture). I believe these show rather clearly that there is a single NO adsorbed species under the conditions of our study. As a general statement I should say that it is only sensible to address surface structural problems with PhD after this level of pre-characterisation, and this has been our general policy. In the case of oxide surfaces this certainly will be a limiting factor for us in the near future, especially as we also really need a reasonable starting model for the structure of the clean surface which is not always available. The PhD technique is specialised, and while this gives it real strengths in the area of structure determination, it also means it is blind to other complexities. In addition, if one has multiple species or no clear idea of what the surface species are, this would be a major added complication to the range of models to be tested in a PhD analysis.

Prof. Jennison said: I note that the error bars on the angular position of the NO are very large (could you comment on this?) but I want to point out another source of information on the tilt. In electronically stimulated desorption, the partition of energy into translational *vs.* rotational modes is quite sensitive to the tilt, and considerable theoretical work has recently been done on this system.[1]

Concerning the height disagreement with theory, this molecule is relatively weakly bound and perhaps dispersive forces, not well described by DFT, could play a role in causing DFT to be in error, and while these forces could be included accurately in quantum chemical calculations, here accurate geometric relaxation of the surface cannot be done because the cluster must be small. Perhaps a hybrid approach is indicated.

1 T. Kluner, H.-J. Freund, V. Staemmler and R. Kosloff, *Phys. Rev. Lett.*, 1998, **80**, 5208.

Prof. Woodruff said: As remarked in response to Thornton earlier, our data for the present system comprise only N 1s PhD which means that the first order information is the N site on the surface. In general we would obtain the position of the O atom in the NO in a largely independent fashion from O 1s PhD, but in the present experiment this was not possible because we could not separate this from the O 1s signal from the oxide substrate. The molecular orientation and N–O bond length in the present case can thus only be obtained from the weak intramolecular (mainly multiple) scattering effects, which are really second-order effects. More generally, of course, we do measure PhD from each of the elements within the molecular adsorbate, and obtain much better precision in such parameters.

Prof. Pettersson commented: I should point out that in the theoretical work by Stacmmler *et al.* in ref. 1 the calculations were indeed carried out at the configuration interaction (CI) level so that dynamical correlation was included. I have performed large-scale multireference CI calculations (unpublished work) using different models, including embedded clusters with one or two Ni^{2+} cations, considering rumpling and possible dimer formation, but the results always indicate physisorption rather than a bond formation. Furthermore, in the comparison with CO chemisorption the theoretical results do not indicate any difference in bonding; both show positive vibrational shifts in contradiction with experiment for NO and both show only weak bonding where experiment clearly shows both adsorbates rather strongly bound and with NO substantially more strongly bound than CO.

In light of the recent TDS data from Wichtendahl *et al.*[2] and the present contribution by Prof. Woodruff it seems very clear that we now have some very firm experimental calibration points and that indeed something is missing in the theoretical description. Since the presently determined Ni–N distance is rather shorter than what has been obtained theoretically, earlier (0.2 Å compared to CI calculations by Staemmler et al.[1] I have made a set of calculations (also including correlation) at the proposed geometry; the model was an embedded NiO_5 cluster and the NO was assumed to tilt at 45°. The results are negative, however: for the doublet state the energy was found to increase in going to the shorter distance; introducing rumpling with the Ni relaxing into the lattice had insignificant effects; moving the NO to tilt between two oxygens similarly negligible; allowing an expansion (10%) of the lateral distance between the surface oxygens beneath the oxygen of the NO gives a very small energy lowering; expanding (10%) the oxygens near the nitrogen gives somewhat more but still only in the meV range; finally, investigating the quartet state either transforming as A′ or A″ (C_s symmetry) gave higher energies. It should be pointed out that although NO has a rich chemistry with, *e.g.*, dimer formation with a high electron affinity on many oxide surfaces or dinitrosyl formation, this possibility can be ruled out by the available experimental data. Thus, it would seem that some new idea or understanding of the properties of these surfaces is sorely needed for the theoretical description of the experimental data.

1 H. Kuhlenbeck, G. Odörfer, R. Jaeger, G. Illing, M. Menges, Th. Mull, H.-J. Freund, M. Pöhlchen, V. Staemmler, S. Witzel, C. Scharfschwerdt, K. Wennemann, T. Liedtke and M. Neumann, *Phys. Rev. B*, 1991, **43**, 1969.
2 R. Wichtendahl, M. Rodriguez-Rodrigo, U. Härtel, H. Kuhlenbeck and H.-J. Freund, *Surf. Sci.*, 1999, **423**, 90; R. Wichtendahl, M. Rodriguez-Rodrigo, U. Härtel, H. Kuhlenbeck and H.-J. Freund, *Phys. Status Solidi A*, 1999, **173**, 93.

Prof. Woodruff responded: This discrepancy is clearly potentially important. The one parameter to which our measurements should be most sensitive, and which should therefore yield high precision, is the Ni–N distance. A difference of 0.2 Å is clearly well outside our precision.

As a very general comment (directed at theoreticians generally) it is clear that in comparisons of theoretically computed and experimentally determined structures, a proper understanding of the errors is of paramount importance. Experimental surface crystallographers expend a lot of energy in attempting to quantify their errors, and it would be a big help if one could do the same for theoretical values. Of course, experimentalists usually quote only their *random* errors, and theoreticians may argue that all of their errors are *systematic* and therefore not quantifiable. I can well imagine that this is the issue in the present case. I wonder more generally, however, whether the gradient of the total energy with change in a structural parameter in a theoretical calculation might provide some indication of how reliable different theoretically obtained parameter values might be expected to be. Certainly this should help if one can estimate the precision of the calculated (relative) total energies of the structures.

Prof. Bowker said: I notice that the limits on the tilt angle include the limit of a 90% tilt (normal molecule). This derives from the very flat dependence of the R factor on tilt angle between 40 and 90°. Is this due to error or due to a soft vibrational mode with large amplitude? Would high resolution angle scan photoelectron diffraction, at fixed photon energy give a better idea of the tilt angle?

Prof. Woodruff responded: As no vibrational mode of this type is included in the calculations this cannot account for the observed insensitivity. The reason is probably more to do with the dependence of the scattering cross-section of the O atom with scattering angle. I should stress again, however, that our insensitivity to the NO orientation is not surprising in the present case because we have only measured the N 1s PhD; more surprising is that we have any significant sensitivity at all! I am sure we could obtain a much more reliable orientation with separate O 1s PhD data. However, it is also true that one could try to use angle-scan X-ray photoelectron diffraction (XPD) from the N 1s signal using a conventional laboratory X-ray source. Under these conditions the photoelectron diffraction is dominated by intramolecular forward scattering, and has been used quite successfully to determine some molecular orientations. In the present case, however, the tilt angle away from the surface normal is quite large, so if the tilt is azimuthally random, as we expect, the forward scattering peaks from different molecules would be smeared out over an azimuthal 'ridge'. This could lead to a very weak modulation, perhaps not observable experimentally.

Dr Carley communicated: I have a comment about the assignment of the two N 1s features as arising from a single adsorbed species, since work at Cardiff and elsewhere, studying NO adsorption on different metals as a function of exposure and temperature, suggests that two different molecular species may coexist. These species have been tentaively identified as NO adsorbed in 'linear' and 'bent' configurations (with some theoretical justification) and I wonder how sensitive the experimental technique described in this paper would be in discriminating between such species, especially given the large error quoted for the derived tilt angle for the N–O bond. As the authors comment, the presence of a weakly bound ('linear') and strongly bound ('bent') species would explain the observed photon beam induced changes in a simple way and avoid the need for the complex reaction scheme which they have to invoke.

Prof. Woodruff communicated in response: Because the N 1s PhD data are primarily sensitive to the location of the N atom to the near-neighbour substrate backscattering atoms, it is certainly true that a model which attributed both the N 1s features to NO in atop sites (at the same Ni–N distance) *could* be compatible with our data as shown, in particular, in Fig. 2. One would need to perform further calculations on the level of agreement between the calculations for the different orientations and the PhD spectra from the individual N 1s peaks. On the other hand, the low R factor for the comparison of the experimental PhD spectra from the two N 1s peaks, the low R factor for the experiment–theory comparison for our optimum structure (0.09) and the large value for the R factor when these experimental data (averaged over the two peaks) is compared with the model of a perpendicular N–O bond (0.30–0.40; see Fig. 5 of the paper) suggests that this model will not prove satisfactory. I also believe that previous work on this surface (as opposed to on other, metal, surfaces) by other methods strongly suggests that only a tilted species is present. Moreover, I should stress that our data are certainly not consistent with simple photodesorption of one of these species as implied in the second part of this question; we remark in our paper only that this explanation in terms of two species appears 'at first sight' to present an explanation. However, the PhD measurements show clearly that the species which remains after extended irradiation under UHV conditions is clearly *not* an atop NO species, as discussed in the paper and in the oral discussion. This alternative picture of the basic adsorption states therefore actually makes the situation more, rather than less, complicated.

Prof. Finnis opened the discussion of Dr Renaud's paper: Can you say anything about the misfit dislocations in the Ag/MgO interface? I am recalling the TEM observations suggesting a semicoherent interface.

Dr Renaud responded: Yes, indeed we have studied the interface of fairly thick (≈ 1500 Å) Ag films on the MgO(001) surface by grazing incidence X-ray scattering.[1] We confirm that the interface is semicoherent, with a well ordered array of interfacial misfit dislocations, which yields very nice rods of diffraction from this interfacial superlattice. However, the HRTEM study concluded an erroneous orientation and Burger's vector of the dislocations, along $\langle 100 \rangle$ directions, which was interpreted as arising from an alternate epitaxial site for Ag on MgO, on top of O and on top on Mg. We have unambiguously demonstrated that the dislocations are instead oriented along $\langle 110 \rangle$ directions, with $\frac{1}{2}\langle 110 \rangle$ Burger's vectors, which is what is expected from the O-lattice theory, and corresponds to only one adsorption site: above O ions of the last MgO(001) plane.

1 G. Renaud, P. Guénard and A. Barbier, *Phys. Rev. B*, 1998, **58**, 7310.

Prof. Freund said: You have convincingly shown how GIXS can be used to extract information on the static structure. Do you see routes towards the extraction of dynamical information from the data?

Dr Renaud replied: We have recently performed grazing incidence small angle X-ray scattering (GISAXS) measurements in real time, *in situ*, during the growth of several metals on an MgO(001) surface, and dynamically. This was achieved thanks to a 2D CCD detector, and to fairly low deposition rates, of the order of 0.1 Å min^{-1}. These were the very first dynamical and *in situ* GISAXS experiments, which should provide very nice information on the morphology of the metallic islands during growth, but not, however, on the structure. We intend to try GIXS measurements at wide angles using a 2D detector in the near future. However, a lot has to be done in order to deduce quantitative structural data from such measurements, and this is still an open field.

Prof. Goodman said: This method is obviously the most elegant and preferred approach for addressing metal structure on oxides. However, many of us oftentimes are limited to more available techniques for such structural diagnostics such as TEM. Here, however, one can imagine serious problems with beam damage. With respect to Pd/MgO, Henry has examined this system in great detail with TEM and the results seem to be entirely consistent with your results for Pd/MgO. Is this not the case and if so, it would seem to imply that TEM very well may be generally useful for metals on oxides?

Dr Renaud responded: I would say that it is an elegant method, but not the most elegant one, since every method has its drawbacks and limitations, and GIXS can not provide all the answers by itself. In the case of Pd/MgO, all our results are indeed perfectly consistent with those of Henry, obtained by TEM. This is certainly because Pd is rather strongly bound to the MgO, and because all specimen preparations and TEM studies of Henry were very careful. However, when the metal is less bonded to the substrate, such as in the Ag/MgO(001) case, TEM studies can clearly lead to erroneous results, either because of modifications of the interface during sample preparation, or because the preferred state of the interface in a ≈ 50 Å thick cross-section sample defer from that of the sample before it was thinned (this is the case of the Ag/MgO(001) interface), or because of radiation induced mobility. Another clear advantage of the grazing incidence X-ray scattering technique is that it can be performed *in situ*, during the growth, and that at all stages of the growth, the structure and the morphology can be characterised in great detail in a non-destructive way, as opposed to TEM experiments.

Prof. Jennison said: You remarked that theorists often assume commensurate metal overlayers. I just want to point out that information obtained from such calculations is still useful. For example, while the large lattice mismatch of Cu on TiO_2 causes an incommensurate overlayer, the relatively small mismatch for Ru on Al_2O_3 would likely lead to large areas of commensurability, well described by the calculations, with strain then relieved by periodic misfit dislocations.

Dr Renaud responded: I fully agree with your remark. However, in the Ag/MgO(001) case that I mentioned, the metal–oxide bond was weak as compared to the metal–metal one and to the surface tension effect. In that particular case, it might be important to allow the Ag islands to relax in theoretical calculations, as is experimentally observed at all stages of the growth, despite the relatively small lattice parameter mismatch of 3%.

Prof. Thornton asked: In your data analysis you assume that the MgO(100) substrate is unchanged on metal adsorption. How realistic is this assumption given that the metal–oxygen interaction will be significant?

Dr Renaud answered: The MgO(001) substrate was assumed to be unchanged on metal adsorption mainly because the possible effect of a relaxation or a rumpling of the last layer would in any case be totally negligible as compared to the effect of metal adsorption. We have studied this relaxation and rumpling effect.[1] Clearly, even if the rumpling or relaxation would reach 10%, the effect on the CTRs would still be negligible with respect to metal adsorption, because they are only visible on the 'weak CTRs', whose intensity is proportional to the square of the difference between the atomic scattering factors of O and Mg, *i.e.* $4 \times 4 = 16$ electrons, as compared to the square of the scattering power of Ag for instance, of $47 \times 47 = 2200$!

In addition, all calculations and our experiments show that the MgO(001) surface is extremely stable, with only a very small rumpling ($\approx 1\%$) and relaxation ($\approx -0.6\%$). This is not expected to be strongly modified upon adsorption, especially Ag which is very weakly bound. Moreover, Spiess[2] has calculated the rumpling and relaxation induced upon adsorption of a single Ag atom on top of O. The relaxation was found to be zero, and the rumpling 1.5%, which tends to confirm this hypothesis of very small modifications of the substrate upon adsorption, in the case of MgO(001).

1   O. Robach, G. Renaud and A. Barbier, *Surf. Sci.*, 1998, **401**, 227.
2   L. Spiess, *Surf. Rev. Lett.*, 1996, **3**, 1365.

Prof. Campbell asked: Can you please say what percentage of the Pd is in the first ML at the total coverage of 0.5 ML deposited, and at 1.0 ML deposited (*i.e.* are 2D islands dominating at 0.5 ML Pd). Is it a uniform Pd ML at 1.0 ML of Pd, or is some of the Pd in the second and third layer then?

Dr Renaud answered: The fits of the CTRs (crystal truncation rods) allow determination of this percentage of Pd in each layer, only for that portion of the Pd film that is lattice-matched parallel to the substrate. At 0.5 ML deposited, 90% is in the first ML, 10% in the second. At 1 ML deposited, 50% is in the first ML, $\approx 40\%$ in the second and $\approx 10\%$ in the third. Hence, 2D islands indeed dominate at 0.5 ML Pd, but for larger deposited amounts the second and third layers start to build.

Dr Venables said: You state in Section IV that these metals (Ag, Pd, Ni) are expected to grow on MgO(001) in the island growth mode, yet you observe the first monolayer of Pd to be pseudomorphic. It is of considerable interest to know whether this 2D layer is a result of kinetic rather than thermodynamic mechanisms. Have you confirmed that there is a kinetic barrier, by annealing the (incomplete) first monolayer, and observing that it transforms into 3D islands? This can also be the case I believe for Ag, as first discussed by Jupille and co-workers.[1] In that case, the 2D layer can be suppressed when impurities nucleate 3D islands earlier than otherwise.

1   F. Didier and J. Jupille, *Surf. Sci.*, 1994, **587**, 307; A. M. Flank, R. Delauney, P. Lagarde, M. Pompa and J. Jupille, *Phys. Rev. B*, 1996, **53**, R1737.

Dr Renaud replied: In the case of Ag, we have clearly observed that even a very small increase of the temperature above room temperature induces a fast diffusion, with an associated increase of island size and increase of the height-to-width ratio. We have indeed annealed a 2 ML thick film at 50 °C, which readily resulted in large 3D islands. We have worked under extremely clean conditions, and always found 3D growth, even from the very beginning of deposition. However, again, only the first plane is occupied for a coverage less than 0.2 ML; then the second plane starts to build until ≈ 0.5 ML deposited, and, above 0.5 ML, the third and further planes are also occupied. When we did not anneal under O_2 so that residual C contamination was present, the growth was more 3D, but the most important fact is that a large part of the Ag grew in the (111) instead of the (001) orientation. For a substrate with residual Ca segregation, almost all the Ag is (111) oriented, while on a very clean and flat substrate, all the Ag grows in the (001) orientation.

In the case of Pd, only the first 0.5 ML is flat and pseudomorphic, and this has clearly been shown by Henry to be a kinetic rather than a thermodynamic effect, since growth at higher temperature readily results in 3D islands.

Prof. Diebold asked: How did you prepare the substrate? Do you believe that you don't have any hydroxylated point defects?

Dr Renaud replied: The preparation of the substrates is described in ref. 1. MgO(001) $15 \times 15 \times 0.5$ mm^3 supplied by Earth Chemical (Japan) are first annealed in air at 1500 °C for several hours, which results in single crystals of very high crystalline quality, as shown by high-resolution X-ray measurements as well as X-ray topography. The Bragg peaks are all resolution-limited to a FWHM of $\approx 0.001°$, even near the surface, under grazing incidence conditions, and in most cases, there is only one grain through the whole sample. This also results in extremely flat surfaces, but which are, however, contaminated because of surface segregation of many bulk contaminants (mainly Ca, but also P, Si, K . . .). In order to remove these contaminants while keeping the high crystal and surface quality, we next proceed to *in situ* ion bombardment in the UHV chamber, at the temperature of 1500 °C, which is required to allow a high surface mobility, and thus keep a flat and crystalline surface during the process. The final procedure before measurements or deposition is a 20 min long annealing at $\approx 850\,°C$ under a partial pressure of oxygen $\approx 1 \times 10^{-4}$ Torr. The samples were characterised by X-ray reflectivity, CTRs measurements and STM, all in UHV, without any exposure to air, and found to have an rms roughness of 2.2 Å, which corresponds to a single atomic high step. The average terrace size between steps is 6000 Å, although UHV STM revealed some single atomic-plane deep holes within the terraces a few tens of ångstroms wide.

In the case of the Ag/MgO(001) and MgO(001) clean surface measurements, great care was taken to ensure a very low pressure $\approx 2 \times 10^{-11}$ Torr obtained by several weeks of bake-out followed by very careful degassing, with the result of negligible partial pressure of water and hydrogen, below 10^{-13} Torr, according to our quadrupolar measurements. We also took care that the samples were always maintained in UHV after the ion bombardment preparation, and hence, I am fairly confident that there were no hydroxylated point defects.

1   O. Robach, G. Renaud and A. Barbier, *Surf. Sci.*, 1998, **401**, 227.

Prof. Hayden said: You observe a discontinuity in the interfacial distance and metal interplane distance during Pd growth at 4–5 ML. You ascribe this to relaxation of misfit at the edge of islands. Since the measurement is sensitive to the buried interface, could this relaxation not extend significantly into the structure from the edge of the islands?

Is not sensitivity to the buried interface a significant advantage that your technique brings to understanding such growth?

Dr Renaud responded: Of course, the relaxation of the misfit extends into the structure of the whole island, even far from the edges. Additional metal planes are introduced near the edges of the islands in order to relax the misfit, but this induces a relaxation even at the buried interface, in between two dislocations.

What is very nice with the grazing incidence X-ray scattering technique is that it is sensitive both to the structure and to the morphology, over the whole film thickness, and at all stages of the growth. Hence, in good cases, it can assess the correlation between structural relaxation and change of morphology or change in the interfacial parameters, such as interfacial distance and interplane distance.

Dr Renaud commented: This comment concerns the discussion on the accuracy of the relaxations determined by X-ray crystal truncation rods analysis on the $Al_2O_3(0001)$ surface, published previously.[1]

The first point is that the measurements have been performed three times on two different samples. The first two measurements were done at LURE, on a sample obtained by a 3 h long annealing in air at 1500 °C, in two states: before and after annealing up to 900 °C for 20 min in 10^{-5} Torr of O_2 in the UHV chamber. The third measurement was done on the ID3 beamline at the ESRF, on another sample, also annealed in air for 3 h at 1500 °C, and next annealed at 900 °C for 20 min in 10^{-5} Torr of O_2 in the UHV chamber. The ESRF experimental conditions were defined such that the CTR measurements were very accurate, with a large redundancy of measured structure factors. A total of 9 CTRs were measured at ESRF, as opposed to only 3 at LURE, with a lesser accuracy. Given the uncertainty on the parameters deduced from fits of the LURE measurements, the same interplanar relaxations were deduced from the three different measurements. It has to be mentioned that the CTRs are extremely sensitive to these relaxations. This sensitivity will be discussed in a forthcoming paper.[2]

However, in all three cases, the base pressure was $\approx 10^{-10}$ Torr, with no particular care as regards the residual partial pressure of water or hydrogen. Since it was recently shown that hydrogen always seems to be present on the surface[3] and that it could induce a 'de-relaxation', it would certainly be interesting to perform new measurements, making sure that no water nor hydrogen is adsorbed on the surface.

Another point was the absolute accuracy on the experimental relaxation. It has to be mentioned that the fits were not perfect, which, given the quality and the reproducibility of the measurements, would indicate that the model is not sophisticated enough (we recall that it has only 5 main parameters, 4 out-of-plane relaxations and one in-plane).

As regards comparison with the calculated relaxations, I think the best way is to recalculate the CTRs by using the calculated relaxations, and compare to the experimental data, which I intend to do in the near future.

1 P. Guénard, G. Renaud, A. Barbier and M. Gautier-Soyer, *Mater. Res. Soc. Symp. Proc.*, 1996, **437**, 15; P. Guénard, G. Renaud, A. Barbier and M. Gautier-Soyer, *Surf. Rev. Lett.*, 1997, **5**, 321.
2 G. Renaud, in preparation.
3 J. Ahn and J. W. Rabalais, *Surf. Sci.*, 1997, **388**, 121.

Prof. Flavell opened the discussion of Dr Kantorovich's paper: I would be grateful if you could comment further on the agreement between the experimental MIES spectra and the calculated SDOS. Given that the tunnelling probability of the electron from the surface to the He* is involved, it seems surprising that the agreement between the two is so good.

Dr Kantorovich responded: The full answer to this question is presented in a paper[1] that has recently been submitted for publication. Therefore, we shall give only a very brief answer here.

In the case of the He* projectile and MgO surface the main process which is responsible for the electron emission observed as the MIES spectrum is the Auger de-excitation (AD) process. In this process only one surface electron is involved which tunnels from the surface to fill in an empty He 1s state. Therefore, using very simple intuitive arguments, one can show that the transition rate for this process, $\mathcal{R}(E, R)$ for a given position of the projectile R with respect to the surface and a certain kinetic energy E of the emitted electron, to a good approximation, is proportional to the surface DOS (SDOS) projected on this orbital:

$$\mathcal{R}(E, R) \propto D_{1s}(E - \Delta E, R) \qquad (1)$$

where $\Delta E = E_{He^*} - E_{He}$ is the excitation ($1s^2 \to 1s2s$) energy of the He atom which is 19.82 eV. The projected DOS is defined as

$$D_{1s}(\varepsilon, R) = \sum_k^{occ} |\langle \psi_k | \psi_{1s} \rangle|^2 \, \delta(\varepsilon - \varepsilon_k) \qquad (2)$$

where ψ_k and ε_k are one-particle orbitals and energies of the surface electrons and the summation is carried out over all occupied states, *i.e.* in the case of MgO all electrons from the upper valence band (VB) contribute.

To calculate MIES, one has to account for the fact that transition can actually happen, with a certain probability, at any time along the trajectory of the projectile. For a given trajectory, $R(t)$, of the He* atom the probability to undergo the transition at the time t is given by the product of the probability to survive before this instant (the so-called escape probability) on the incoming part of the trajectory,

$$P_{esc}(R(t)) = \exp\left(-\int_{R(t)}^{\infty} dt' \int dE \mathcal{R}(E, R(t'))\right) \qquad (3)$$

and the probability to undergo the transition during the time interval between t and $t + dt$, which is $\mathcal{R}(E, R(t)) \, dt$. Then, integration with respect to time gives the total probability for the AD process for the given trajectory. Finally, one has to account for all possible trajectories, so that the MIES spectrum is given as

$$P(E) = \left\langle \int_{-\infty}^{0} \mathcal{R}(E, R(t)) P_{esc}(R(t)) \, dt \right\rangle \qquad (4)$$

where $\langle \cdots \rangle$ corresponds to averaging with respect to all possible trajectories and we integrate over the whole incoming part of the trajectory.

The two functions in eqn. (4) (above) have a rather different behaviour: the escape probability changes smoothly from unity at large distances from the surface to zero at close approach, while the transition rate decays exponentially away from the surface. The integrand in eqn. (4) is therefore peaked at some distance from the surface, R_{mp}, called the most probable target distance, which corresponds to the highest probability for the electron emission with energy E in the AD process. Our calculations demonstrate a weak dependence of R_{mp} on energy E. Therefore, the time integral in eqn. (4) appears to be approximately proportional to the SDOS, $\mathcal{R}(E, R_{mp})$, projected on the He 1s orbital centred at the target distance from the surface. Note that target distances depend on the particular trajectory and are in the range of 2–4 Å, above the surface.

The exponential decay into the vacuum of the electronic wavefunctions ψ_k in eqn. (2) is determined by their binding energies, $-\varepsilon_k$. The states with large binding energies (near the bottom of the VB) decay faster and are effectively cut off in the spectrum. Therefore, the largest contribution to the MIES spectra is made by the states in the gap and near the VB top. Since the SDOS and the projected DOS $\mathcal{R}(E, R_{mp})$ have similar structures, the relative position of defect related peaks in the gap with respect to the top of the VB in the MIES spectra can be predicted with good accuracy by SDOS.

This conclusion has been checked in ref. 1 where we directly compared the calculated SDOS and MIES for a peroxide defect, O_2^{2-}, formed at the MgO(001) surface upon adsorption of an O atom which binds to a surface oxygen (see also ref. 55 in our paper). It was found that a defect related feature above the top of the VB present in the SDOS also appears in the MIES at the same position with respect to the top of VB. We think that the agreement between the SDOS and the experimental MIES demonstrated in our paper in the case of Mg adsorption on the MgO(001) surface also supports this conclusion.

1 L. N. Kantorovich, A. L. Shluger, P. V. Sushko and A. M. Stoneham, *Surf. Sci.*, 1999, in press.

Prof. Hermann asked: Did you allow the atoms in the vicinity of the F-centres to relax and how large was the geometric effect. What was the effect on the electronic structure of the F-centre? How does an adsorbing Mg stabilise near the F-centre?

Dr Kantorovich answered: Yes, we did. For all systems studied the complete geometry relaxation was performed as described in detail in Section 3.2. In the case of the F-centre at the MgO(001) surface the relaxation appears to be insignificant (about 1% of the Mg–O distance) due to well localised electronic density in the anion vacancy which is equivalent to two electrons. Note that both valence and the F-centre electrons contribute to the charge in the vacancy as explained in ref. 54 of our paper.

When a Mg atom is added above the F-centre, it pulls out some electronic density from the vacancy and forms bonding and antibonding σ-type states which can be well recognised in the calculated system DOS. The stabilisation energy (see Table 2 of our paper) is only 0.21 eV which is half of that for the perfect surface. We did not calculate, however, the barrier between the two configurations (above the terrace oxygen and the F-centre) which we expect to be of the order of 0.1–0.2 eV.

Prof. Pacchioni said: The adsorption energy of a Mg atom on an anion vacancy is only 0.5 eV according to your calculations (Table 2). An anion vacancy in MgO, F^{2+}, is very electron deficient and based on simple electrostatic arguments I would expect the following process to occur.

$$F^{2+} + Mg \rightarrow F^+ + Mg^+$$

The Mg^+ ion can then become stabilized at a nearby oxygen. Since both Mg^+ and F^+ are characterized by the presence of unpaired electrons this can be modeled only considering a triplet state.

Dr Kantorovich replied: Indeed, simple electrostatics suggest that one may expect significant charge transfer to the anion vacancy from the adatom due to a more favourable Madelung potential in the vacancy. According to our calculations, this does not happen. When a Mg atom is added to the MgO(001) surface with an anion vacancy and the singlet state of the whole system is considered, it indeed loses some electronic charge which is pulled towards the vacancy. However, by integrating the electron density in the vacancy we were able to account for only about 0.5 electron there. Note that the Mg atom is not adsorbed directly above the vacancy but is displaced along the [110] direction towards the nearest oxygen as shown in Fig. 7(c) of our paper.

We also considered this system in the triplet state and found that it lies 4.3 eV higher in energy. In this state the Mg atom is atop the nearest to the vacancy oxygen. We did not study the distribution of the electronic density in the triplet state so that I cannot comment on this.

The fact that the charge transfer from Mg to the vacancy is not very significant as the simple electrostatic argument would suggest is explained by at least two factors. First of all, the large charge transfer would lead to creation of both positively charged F^+ centre at the surface and the Mg^+ adatom nearby and this would contribute with a positive Coulomb interaction energy. Secondly, the electron affinity of the surface vacancy is much smaller as compared to the ionisation potential of the Mg atom which is 7.6 eV. We have calculated the electron affinity of the surface anion vacancy using the embedded cluster method (described in refs. 16 and 17 of our paper) and found that it is only 3.4 eV.

Prof. Joyner asked: What are the relative densities of steps and kinks in the experimental study compared to surface vacancies? I would expect there to be many more steps than vacancies, therefore adatoms have little choice other than to link at steps, almost irrespective of the energetics.

Dr Kantorovich answered: We think that Prof. Joyner is absolutely right and the number of sites available for adsorption at steps and terraces is much larger than the possible number of vacancies at the surface. This is also supported by the STM images of similarly prepared surfaces of MgO films presented in ref. 29 of our paper. They clearly show that these surfaces are very rough and contain 3D MgO islands and a lot of steps within the islands. Our results demonstrate that steps and terrace sites are the most important for the growth of Mg clusters on these surfaces.

Prof. Kempter added: The MgO film preparation procedure used in our experiments leads to a small number of point defects in general, including vacancies. Normally, we start with a Mg-rich MgO film. The film is exposed to molecular oxygen and then to a subsequent annealing procedure. As a consequence, the resulting surface is inert to the exposure of additional oxygen as well as to temperature. This preparation procedure produces valence band spectra, both in MIES and UPS, that are identical to those for MgO(001). In particular, both EELS[1] and MIES (ref. 21 and 24 of our paper) do not display any features in the band gap above the VB maximum of the MgO film that could be attributed to point defects as in particular to F^+ and F centres. Therefore, we expect that the number of extended defects, steps in particular, is indeed much larger than that of point defects. Note that, in principle, the adatoms could link to terrace ions as well. The energetics shows, however, that this is not favoured at the initial stages of the Mg cluster formation.

1   M. C. Wu, C. M. Truong and D. W. Goodman, *Phys. Rev. B: Condens. Matter*, 1992, **46**, 12688.

Dr Noguera said: How do you explain the dramatic change in the Mg–Mg interaction that you obtain between the gas phase state and the adsorbed state? Could your calculations reproduce the weak bonding of the small unsupported clusters (which should be of the van der Waals type, below a critical size)?

Dr Kantorovich replied: We did not perform a comprehensive study of free Mg clusters. However, we did consider a free Mg_2 molecule and indeed found a very small binding energy of 0.15 eV at the interatomic distance of 3.75 Å. There is every reason to believe that the binding of other free Mg clusters will be also weak. I believe, the situation is different at the MgO surface due to covalent bonding between the metal atoms and the surface oxygens.

Prof. Pettersson said: It seems very unlikely that the gas phase Mg dimer bonding should have any influence on the geometry at the surface. Since in the gas phase the bonding is only through van der Waals interactions it is very weak, 429.6 cm^{-1}.[1] This energy is so much smaller than the computed interaction with the substrate that the geometry should be determined only by the bonding to the substrate. The situation could change if the dimer is ionized, but the neutral system represents interaction between two closed-shell systems and is very weak.

1   K. C. Li and W. C. Stwalley, *J. Chem. Phys.*, 1973, **59**, 4423.

Dr Kantorovich responded: I believe, this is not entirely true. The Mg–Mg bond is indeed very weak. However, one has to consider here also the whole potential energy curve of this system, $E_{Mg-Mg}(d)$, because when more than one Mg atom is adsorbed, there is a distribution of distances d between them. We find in our calculations that the energy $E_{Mg-Mg}(d)$ changes weakly for distances $d > d_0$, where $d_0 = 3.75$ Å and is the equilibrium distance between the two metal atoms in a free molecule. However, the repulsion of Mg atoms at closer distances shows much steeper character. The typical distances between adsorbed Mg atoms at the surface (which are indeed primarily determined by the interaction with the surface oxygens) are in the range of 3.1–3.3 Å, *i.e.* they are closer than the gas-phase equilibrium distance d_0. At this distance range the repulsion between two Mg atoms is about 0.1 eV. Therefore, it is not bonding but rather repulsion between Mg atoms which is the second factor affecting the geometry at the surface.

Prof. Asscher opened the discussion of Prof. Campbell's paper: Would it be possible to measure single Cu atom heat of adsorption by avoiding surface diffusion by cooling the sample? Is it not expected that a single atom would be partially ionized and therefore its binding energy should be larger than that of a neutral atom? Maybe even larger than a neutral dimer?

Prof. Campbell responded: There is pretty good evidence from related metal-on-oxide systems (ref. 9 of our paper) that the metal adatoms are cationic when present at such tiny coverages (<2% of a monolayer) and that they remain isolated. However, they adsorb as nearly neutral species as the coverage increases (although still at extremely low coverages), where they are also observed to cluster together (ref. 9 of our paper). This suggests that isolated adatoms are more stable when cationic than when neutral, as you propose. Our first data point for measuring their heats of adsorption averages the heats of adsorption for the first $\approx 2\%$ of a monolayer contained in that first pulse of the metal atom beam. By the end of that pulse, we generally already have clustering of the metal atoms, but the first bit is probably cationic. The low heat for this first pulse suggests that if these cationic adsorbed metal adatoms have a higher heat of adsorption than adatoms in the clusters, it is not dramatically higher.

Prof. Freund said: You have made a very important contribution to our understanding of metal oxide interaction and I expect major insight from the method you presented. Is it conceivable to combine a calorimetric measurement with mass selected exposure of metal clusters to the surface?

Prof. Campbell replied: Thank you. Yes, possibly. I think some groups are now preparing low-energy beams of mass-selected clusters of sufficient intensity to make such measurements feasible. I would guess that nice signals could be obtained if the flux in such a beam were sufficient to deposit $\approx 3 \times 10^{11}$ atoms of metal per pulse, within a pulse width of no more than 0.4 s, and at a repetition period of no shorter than 2 s. With a good contact, our heat detection limit (≈ 3 standard deviations) is about 2×10^{-8} J, so in principle even much smaller fluxes could be measured, but the signal-to-noise ratio in the results would not be so beautiful. Lately we have seen even much, much lower detection limits than that with an aluminium foil as the sample substrate. Detectivity seems to be higher for softer samples, which makes sense since it depends directly on the heat transfer coefficient at the contact between the polymer detector and the sample.

Prof. Kempter said: Our results for Mg/MgO indicate a strong dependence of the heat of adsorption on the presence of extended defects, as in particular steps. I wonder to what extent your results obtained for Cu/MgO (film) are representative for Cu on a MgO(100) single crystal. In fact, your values obtained for the heat of adsorption compare very well with ours for Mg/MgO where MgO is indeed a rough surface.

Prof. Campbell responded: I agree. We probably have a highly defective MgO(100) surface. Since, to my knowledge, there is no way to get a single crystal of MgO(100) that is 1 cm in diameter but thin enough (<6 μm) to use for adsorption calorimetry, we were forced to use a 3.0

nm thick MgO(100) thin film grown on an ultrathin Mo(100) single crystal. Such films have highly defective surfaces and give broad LEED spots indicating average terrace widths of <10 nm.

Prof. Jennison said: You mentioned that the metal Auger spectrum can be shifted several eV in energy if the island sizes are small on an oxide substrate, due to limitations in final state screening. I would like to show some data from the group of Jeff Kelber which illustrate this effect. Note in the lower portion of Fig. 1, for Cu deposition on SiO_2, just the effect you mentioned: the peak in the Auger parameter plot moves continuously with deposition time (as the metal island sizes increase) until it meets the asymptotic position of metallic copper. Now note the contrast with Cu deposited on the heavily hydroxylated sapphire surface in the top portion of the figure: no shift occurs with deposition time, but a second peak of metallic Cu grows, starting when greater than 1/3 ML is deposited. This indicates to us that the first peak is oxidized Cu(I) and is not due to an island size effect.

1 J. A. Kelber *et al.*, *Surf. Sci.*, in press.

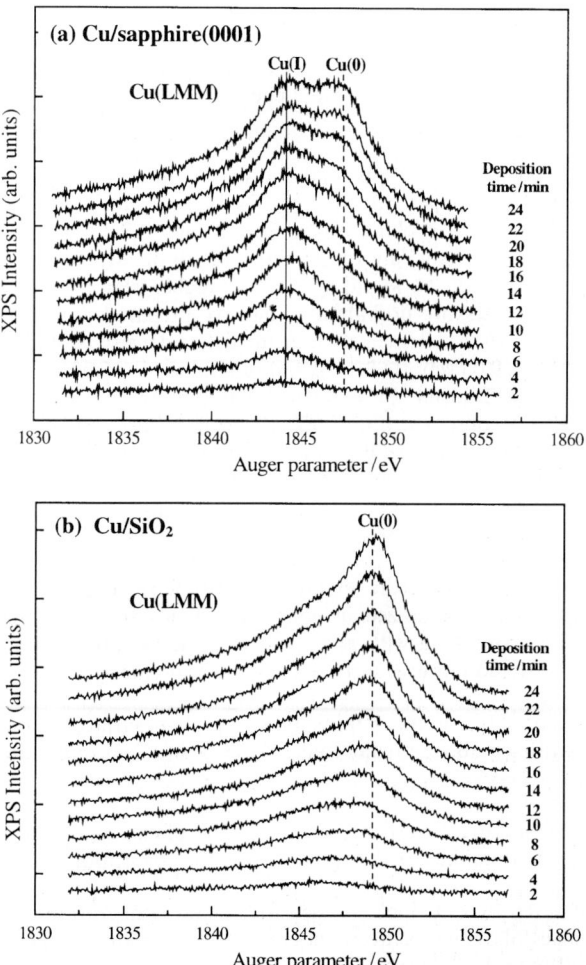

Fig. 1 Cu(LMM) evolution during Cu deposition on (a) sapphire(0001) and (b) SiO_2 with deposition rate at 0.03 ML Cu min^{-1}. Deposition temperature = 300 K. Due to differential charging on sapphire surface, the Auger parameter for Cu(0) on sapphire is different from that on SiO_2. (Reproduced from ref. 1, with permission.)

Prof. Hayden replied: An alternative to a single shifting peak with coverage would be that there is a growth of a second peak at high coverages. Note that the high coverage result spectrum shows a distinct shoulder.

Prof. Campbell also replied to Prof. Jennison: Yes, that is one possible explanation. The result you show from Kelber's group is, however, unlike the result obtained by Madey's group also for Cu on alumina (ref. 36 of our paper), where a continuous shift in the Auger parameter with increasing coverage from 1848.4 to 1851.1 eV was observed and attributed entirely to final state effects. My group also observed a large increase in the Cu Auger parameter up to 1851.6 eV with coverage of Cu on Zn(0001)–O, which was proven by work function and band bending measurements to be due entirely to differences in final state screening as neutral Cu clusters increased in size (ref. 38 of our paper), and not due to the change from Cu(I) to Cu(o). The persistence of an Auger peak due to Cu(I) at high Cu coverages on such a stable oxide is rather unusual, but it could be understood if there were a small amount of oxygen impurity in the background gases, which might oxidise the adsorbed Cu.

Prof. Madey addressed Prof. Campbell: I am curious about your high sticking probability (>0.9 for Cu on MgO(001), which is higher than the values reported by others on cleaved MgO(001) and on MgO films similar to yours. You attribute this to your higher incident flux. Can you rule out the possibility that you have a higher defect density than the other investigators? Do you have adequate signal-to-noise that you could check this by measuring the sticking probability at lower fluxes?

Prof. Campbell responded: Yes, I think that the reason might be that we had higher defect density than in ref. 29 of our paper, which used a very nice bulk crystal surface prepared by cleavage in UHV and reported a sticking probability of only 0.5. Our models for the coverage- and temperature-dependences of sticking probabilities of Pb on MgO(100) (ref. 22 of our paper) which are similar to models in ref. 29, would predict that higher island density and therefore defect density should increase the sticking probability. Since we prepared our MgO(100) following the exact same recipe as in ref. 26, it is less likely that the defect density could explain the differences between the value of 0.82 ± 0.05 reported there and our value of >0.995. It is interesting that this value stays constant at ≈ 0.82 in ref. 26 between ≈ 400 and 100 K, and does not increase upon cooling as expected (and as observed above 400 K, until it saturates at ≈ 0.82 at 400 K). The saturation at 0.82 suggests a possible calibration problem in ref. 26, wherein the Cu TPD intensity corresponding to a unit sticking probability was assigned to a value of 0.82 instead. I assume that this could happen due to differences in mass spectrometer tuning or angular distributions between the calibrant and the real experimental test case.

Prof. Goodman said: It is noteworthy that your calorimetry data at the low coverage limit suggests that the MgO films synthesized in your experiments have a relatively low density of defects, *i.e.* 1%. This is encouraging with respect to using films rather than cleaved or polished oxide surfaces for chemical and physical surface science measurements.

Prof. Campbell replied: Yes, I agree. I think these are good surfaces for most tests. However, our kinetic models mentioned above predict that 1% defect sites can rather dramatically affect the sticking probabilities.

Prof. Finnis asked: Would there be scope with your kind of calorimetry to investigate the heats of any phase transitions in thin films, perhaps induced by the adsorption?

Prof. Campbell answered: Yes, I think our sensitivity and temperature control has become so good that we could also follow temperature-induced phase transitions, and those would be very exciting applications of the calorimeter.

Dr Henderson said: In your data for Pb adsorbed in the clean *vs.* hydroxylated/hydrated Mg(100) surface, the heat of adsorption on the latter is less than that of the former. Is it possible that some of the heat generated by metal adsorption could have gone to desorption of the

OH/H$_2$O layer thus rendering a lower apparent heat of Pb adsorption? Have you looked for H$_2$O desorption or done TPD after the fact to test for this?

Prof Campbell responded: No, we haven't done that but it's a good idea. As with all these adsorption energy measurements we do, it would be far better to have more information about the structure than we do. This is a first-generation instrument, and we are hoping to have more structural characterisation tools on later versions.

Prof. Kirschner asked: If you had a system showing layer-by-layer growth, could you imagine observing oscillations of the heat of adsorption with one monolayer periodicity?

Prof. Campbell replied: Yes, We have actually seen something like that for the case of Pb on Mo(100), which grows layer-by-layer for 2 monolayers. There is a rather sharp change in the heat of adsorption near the completion of both the first and second monolayers (ref. 21 of our paper).

Dr Venables said: You have discussed the possibility that you might measure the adsorption energy for Cu adatoms on MgO at low coverage. May I say, probably in agreement with you, that you will be able to do this in a room temperature experiment, since neutral Cu adatoms are very mobile, with an activation energy expected to be in the range 0.1–0.2 eV atom^{-1} ($\approx$10–20 kJ mol^{-1}). With these low values, island density of a few $\times 10^{12}$ cm^{-2} are formed at a coverage below 10^{-3} ML. In the comparable case of Pd/MgO(001), recent AFM experiments coupled with a rate equation analysis have put an upper limit of 0.2 eV atom^{-1} from the observation of the nucleation density at $T \approx 200$ K.[1] Could you comment on what one might expect if Cu$^+$ adions are present in addition to neutral Cu adatoms, and could you observe any such effects?

1 J. A. Venables, G. Haas, H. Brune and J. H. Harding, *Mater. Res. Soc. Symp.*, 1999, **570**, 51.

Prof. Campbell responded: I agree, the activation energies for diffusion of an isolated metal adatom are very tiny and one needs a very low temperature to prevent them from clustering at 0.02 ML. Like you, I am scared that 100 K may not be cold enough. We can work at smaller pulses now (down to 0.002 ML), albeit with reduced signal-to-noise, so that may help. In principle, we can do these experiments at 20 K or below, although we can not deliver the liquid He needed for that yet.

Prof. Catlow opened the discussion of Prof. Pacchioni's paper: I note that in your simulations you held the positions of H atoms fixed, which I am sure is the correct procedure. Do you have any feeling for how the results are influenced by different choices of geometry of your cluster, or, in other words, different choices of model for the surface structure?

Prof. Pacchioni responded: The general features of the interaction should not change by changing the model of the substrate since the Cu atoms and clusters form relatively strong local bonds with the paramagnetic point defects at the silica surface. However, we observed that as a consequence of a partial charge transfer from the Cu atoms to the non-bridging oxygen atoms at the surface, the Cu atoms and clusters interact also with some neighbouring bridging oxygens with formation of 'ring' structures. The stability of these ring structures will depend on the topology of the surface and on the strain present in the rings. Other models of the surface where more relaxation is possible could result in larger adsorption energies.

Prof. Pettersson said: I like the fact that the coupling to the rest of the crystal matrix has been included and investigated in your models. The restrictions induced by the lattice connectivity on the displacements of atoms in response to the formation of defects or bond dissociation processes are important and should be included in the description. We have recently investigated the dissolution of the siloxane bridge in quartz and find that the barrier is strongly affected by the connectivity of the participating silica units. For instance, in going from a singly connected to doubly and triply connected units the barrier increases by 5 and 10 kcal mol^{-1}, respectively.[1] This is due to the reduced freedom to relax the structure in the transition state. Similar considerations should apply in your study and I am glad that this point is brought up and illuminated in your paper.

1 A. Pelmenschikov, H. Strandh, L. G. M. Pettersson and J. Leszczynski, to be published.

Dr Shluger said: For modelling adsorption from the gas phase, for example, Cu cluster surface centre, the correct presentation of the relative energies of the surface defect and adsorbing atom states is important. However, your cluster model which does not take into account the crystalline potential, does not treat the surface defect states correctly. How could this affect the adsorption energy and character of adsorption?

Prof. Pacchioni responded: Our assumption is that the charge separation which is present within a cluster model of the silica surface includes the most important contributions to the local potential around the defect centre and that the contribution of the long-range potential is not large, in particular if one wants to simulate the surface of amorphous silica. The assumption that the Madelung field is not of major importance in this material is based on the fact that several properties of point defects in silica (optical excitations, hyperfine interactions, defect formation energies, vibrational spectra, *etc.*) have been nicely reproduced using cluster models saturated by H atoms,[1-7] without including the crystalline potential. In this respect we consider SiO_2 a material characterised by covalent polar bonds, certainly not an ionic solid. Of course, since the bond of a Cu atom at the defect sites has a partial charge transfer character, in particular when we consider the non-bridging oxygen site, this can be affected by the Madelung potential of the crystalline phase. In this respect some change in the absolute value of the adsorption energy can be expected when this term is properly taken into account.

1 G. Pacchioni and G. Ieranò, *Phys. Rev. Lett.*, 1997, **79**, 753.
2 G. Pacchioni and G. Ieranò, *Phys. Rev. B*, 1997, **56**, 7304.
3 G. Pacchioni and G. Ieranò, *Phys. Rev. B*, 1998, **57**, 818.
4 G. Pacchioni, G. Ieranò and A. M. Marquez, *Phys. Rev. Lett.*, 1988, **81**, 377.
5 G. Pacchioni and R. Ferrario, *Phys. Rev. B*, 1998, **58**, 6090.
6 G. Pacchioni and M. Vitiello, *Phys. Rev. B*, 1998, **58**, 7745.
7 G. Pacchioni and M. Vitiello, *J. Non-Cryst. Solids*, 1999, **245**, 175.

Prof. Kempter said: You pointed out some observable consequences of the Cu-adsorption to the defects taken into account, in particular the presence of band gap states. In order to observe these states, their intensity relative to that of the valence band emission would be needed. Could you comment on this point?

Prof. Pacchioni responded: Our observation of the appearance of states in the gap when metal atoms are adsorbed on the silica surface is based on the qualitative analysis of the Kohn–Sham orbitals and does not allow us to estimate the relative intensity of these bands.

Prof. Jennison said: I agree with your statement that SiO_2 is best described as having polarised covalent bonding, in contrast with the highly ionic oxides. Aside from theoretical analyses, which I believe all agree in this sense, another piece of evidence comes from Auger O(KVV) spectral data, where the peak position and shape is quite different in SiO_2 from those of the ionic oxides such as MgO.

Concerning your Table 4, I am interested in the ionicity of the single Cu atom attached to the Si–O˙ system. The oxygen gains considerable charge (from -0.37 to -0.71) but the Cu only loses 0.13, which doesn't add up. Could you please clarify the meaning of the reported charges and comment on the ionicity?

Prof. Pacchioni responded: The reason why the charge of the Cu atom and of the non-bridging oxygen do not add up is that we did not report the values of all the atoms present in the model. The Cu atom donates charge to the non-bridging oxygen (0.3 electrons if we consider the change in population of the NBO center), but as a consequence of this charge transfer it interacts also, mostly electrostatically, with a bridging oxygen of the surface, see Fig. 4b in ref. 1. The bridging oxygen donates some charge to the Cu atom which results to be less positively charged than it would be without this additional interaction. However, Mulliken charges are not very reliable and should be considered with some care. More than the net charges computed with the Mulliken scheme, in my opinion what shows the existence of a charge transfer is the shift of the core levels of the oxygen atoms. The occurrence of a partial charge transfer is thus quite evident from the data, although a quantitative estimate of the amount of charge transfer is not easy.

1   N. Lopez, F. Illas and G. Pacchioni, *J. Am. Chem. Soc.*, 1999, **121**, 813.

Prof. Thornton made a general comment: There is a growing body of evidence that 2D clusters of size 7–8 atoms are particularly stable.

Prof. Pacchioni added: The suggested existence of Cu_8 and Pd_8 clusters on TiO_2 raises a question which has been around for a while in the clusters community, *i.e.* the existence of 'magic numbers' corresponding to particularly stable aggregates. Do we have 'magic numbers' also for clusters on oxide surfaces?

Another question is related to the electronic structure of Cu clusters. Since the Cu atom has a ^2S ground state, clusters with an odd number of atoms have an open shell ground state, at variance with clusters with an even number of atoms which can be diamagnetic. This leads to some oscillations in the properties as a function of cluster size (nucleation energy, ionisation potential, *etc.*) In principle, also the heat of adsorption of Cu on MgO should exhibit these oscillations, in the low coverage regime.

Prof. Campbell responded: I don't know if there will be magic cluster sizes, but I agree that there will probably be oscillations in stability with cluster size. Perhaps we could measure their stabilities and see such oscillations if we could get cold enough to control cluster sizes down to such small sizes. This may be possible already with our calorimeter with the proper combination of metal and oxide.

Prof. Freund remarked: With respect to the question whether it is possible to prepare deposited clusters with a small number of metal atoms I would like to mention that deposition of Rh or Ir onto alumina films at 50 K leads to clusters containing 1–7 atoms. After exposure to CO the IRAS are recorded and the analysis of vibrational spectra from isotopically labelled species is in line with the idea of clusters containing 1, 2 and 3 (up to seven metal atoms) (refs. 157 and 158 of the Introductory Lecture).

Prof. Goodman said: With respect to the use of carbonyl precursors in the synthesis of metal and mixed-metal clusters, I should mention our recent work on the synthesis of Ru clusters on TiO_2 using a Ru_3 carbonyl precursor. Although this study is in progress, preliminary results show clearly this carbonyl precursor route to be particularly suitable for the synthesis of highly dispersed metal clusters. In fact, preliminary STM studies show several examples that suggest formation of a significant population of Ru_3 cluster species on the TiO_2 surface. Although, of course, it is impossible to comment on the carbon content of the imaged species, the following STM certainly suggests that ring-opening of the Ru_3 carbonyl could occur to the apparent epitaxial growth of a linear Ru_3 cluster.

Prof. Campbell commented: I just wanted to make a general comment that theory seems to be getting closer to experiments these days on oxide surfaces. For example, the energy of small Cu clusters on MgO(100) that we measured (see our paper presented here) are pretty well reproduced in DFT calculations by Musolino and Car, and by values discussed here by Pacchioni.[1]

1   Musolino and Car, *Surf. Sci.*, 1998, **402–4**, 413.

Oxygen-induced restructuring of rutile $TiO_2(110)$: formation mechanism, atomic models, and influence on surface chemistry

Min Li,[a] Wilhelm Hebenstreit,[a] Ulrike Diebold,*[a] Michael A. Henderson[b] and Dwight R. Jennison[c]

[a] *Department of Physics, Tulane University, New Orleans, LA 70118, USA. E-mail: diebold@mailhost.tes.tulane.edu*
[b] *Pacific Northwest National Laboratory, Richland, WA 99352, USA*
[c] *Sandia National Laboratories, Albuquerque, NM 87185-1413, USA*

Received 5th May 1999

The rutile $TiO_2(110)$ (1 × 1) surface is considered the prototypical 'well-defined' system in the surface science of metal oxides. Its popularity results partly from two experimental advantages: (i) bulk-reduced single crystals do not exhibit charging, and (ii) stoichiometric surfaces, as judged by electron spectroscopies, can be prepared reproducibly by sputtering and annealing in oxygen. We present results that show that this commonly applied preparation procedure may result in a surface structure that is by far more complex than generally anticipated. Flat, (1 × 1)-terminated surfaces are obtained by sputtering and annealing in ultrahigh vacuum. When re-annealed in oxygen at moderate temperatures (470–660 K), irregular networks of partially connected, pseudohexagonal rosettes (6.5 × 6 Å wide), one-unit cell wide strands, and small ($\approx$ tens of Å) (1 × 1) islands appear. This new surface phase is formed through reaction of oxygen gas with interstitial Ti from the reduced bulk. Because it consists of an incomplete, kinetically limited (1 × 1) layer, this phenomenon has been termed 'restructuring'. We report a combined experimental and theoretical study that systematically explores this restructuring process. The influence of several parameters (annealing time, temperature, pressure, sample history, gas) on the surface morphology is investigated using STM. The surface coverage of the added phase as well as the kinetics of the restructuring process are quantified by LEIS and SSIMS measurements in combination with annealing in ^{18}O-enriched gas. Atomic models of the essential structural elements are presented and are shown to be stable with first-principles density functional calculations. The effect of oxygen-induced restructuring on surface chemistry and its importance for TiO_2 and other bulk-reduced oxide materials is briefly discussed.

1 Introduction

The rutile $TiO_2(110)$ surface has evolved as one of the most important model systems for metal oxide surfaces. Titanium dioxide is used in gas sensing, catalysis, and photocatalysis, where surface phenomena play an important role. In 1994, when the surface science of metal oxides was reviewed by Henrich and Cox, $TiO_2(110)$ was already an intensely studied system.[1] Since this time, its popularity has increased steadily, partly because bulk single crystals can be reduced easily (which conveniently prevents charging), and partly because of a desire to perform experiments on a 'well-characterized system'.

The preparation of a clean, atomically flat $TiO_2(110)$ (1×1) surface with a controlled defect density is very important for surface chemistry experiments. Normally, sputtering and annealing in ultrahigh vacuum (UHV) or oxygen at high temperatures are used. Many authors have published preparation recipes, as an example we cite the one given by Pan et al.:[2] "The stoichiometric (or nearly perfect) surface was obtained by sputtering with 500 eV Ar^+, then annealing to 1000 K for 3 min in 2×10^{-6} Torr of O_2, and finally cooling down to room temperature in the same oxygen atmosphere. XPS showed sharp Ti 2p peaks with no indication of reduced Ti states."

Most of the STM studies on $TiO_2(110)$ performed so far[3-12] have focused on UHV annealed surfaces and the nature of the (1×2) reconstruction that evolves at high temperatures. Our STM measurements showed that oxygen-annealed surfaces prepared using Pan et al.'s recipe[2] were considerably rougher than vacuum-annealed samples, and that the appearance of the surface varied greatly when seemingly the same procedure was applied. In order to understand this phenomenon, we re-annealed flat surfaces (prepared by UHV annealing at high temperature) at moderated temperatures in oxygen gas, and found a pronounced morphology change, which we have termed 'restructuring'. A brief report of STM results and conclusions was given previously,[13] and a more complete account (including data not shown here) will be published elsewhere.[14]

In this paper, we apply a combination of STM, LEIS and SSIMS to explore systematically the influence of preparation parameters (temperature, annealing time, oxygen pressure, and reduction state of the crystal) on surface restructuring. Ab initio total-energy density functional calculations are used to test the geometric model for restructured surfaces, and to explore its electronic structure. As the main conclusion, we find that both the surface structure and morphology of a $TiO_2(110)$ surface depend sensitively on the oxidation conditions as well as the history of the crystal. Specifically, we find a new structure (termed 'rosette network') that consists of an incomplete $TiO_2(110)$ layer, where all atoms are in approximate bulk-like positions, but some are missing in a regular fashion. Compared to a (1×1) structure, the rosette networks exhibit quite different bonding geometry, coordination number, undercoordinated sites, and degree of covalency. Hence, $TiO_2(110)$ surfaces prepared by annealing in oxygen may not resemble the flat, (1×1) terminated surfaces that are often assumed in interpretation of surface chemistry experiments. With this work, we would like to provoke thoughts, and invite comments, on how surface chemistry may be affected by the presence of such rosette networks, and to what extent our findings may be transferable to other bulk oxide systems with facile transport of metal interstitials.

2 Experimental and calculation methods

The experiments were performed in two UHV systems described elsewhere.[15,16] Polished TiO_2 single crystals from three different vendors have been used which exhibited a blue color after an initial high-temperature anneal (950 K); details on sample mounting can be found elsewhere.[13-15] Before each experiment, the sample was prepared with sputtering (1000 eV Ar^+, $I_{sample} \approx 8.3$ µA cm^{-2}, 20 min) and annealing to 880 K in UHV for 30 min, which yielded a flat $TiO_2(110)$ (1×1) surface with atomically flat large terraces (up to 500 Å wide) as shown in Fig. 1. The alternating white and dark rows along the [001] direction are located at the positions of 5-fold coordinated Ti atoms and 2-fold bridging O atoms of the (1×1) structure, respectively.[11] Some bright rows typically several tens of nanometers long, are scattered across terraces or connected to step edges. These appear upon UHV annealing of relatively dark crystals.[17] In reference to the features observed upon annealing to higher temperature,[4-7] we call these rows (1×2) strands. Smooth step edges are oriented along the [1$\bar{1}$1] and the [001] direction.[12] Only Ti and O signals were detected by XPS and LEIS, indicative of a clean sample surface after such a treatment, and LEED showed a sharp (1×1) pattern.

Both pure $^{16}O_2$ gas and isotopically enriched $^{18}O_2$ gas ($^{18}O_2 : ^{16}O_2 = 93\% : 7\%$) were employed in oxygen exposure experiments. Gas dosing was performed by backfilling the chamber. To quantify the ^{18}O surface content, we took LEIS spectra[14] which show clearly separated ^{18}O and ^{16}O peaks. These LEIS measurements do not affect the uptake of ^{18}O with carefully controlled parameters (total ion fluence of $\sim 1.6 \times 10^{-15}$ cm^{-2} per measurement and beam energy of

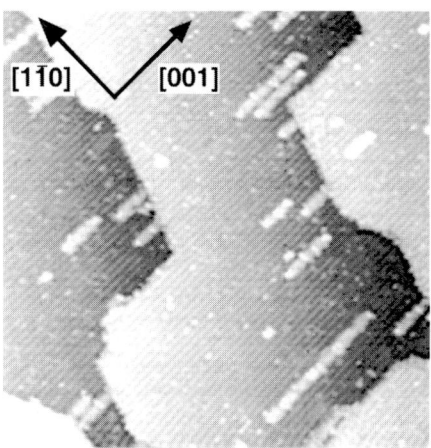

Fig. 1 STM image (500 Å × 500 Å) of TiO$_2$(110), sputtered and annealed in UHV for 30 min at 880 K. The surface shows a regular (1 × 1) termination. The few bright lines are referred to as (1 × 2) strands.

1000 eV). Static SIMS experiments were performed in a separate chamber[16] with a differentially pumped ion gun utilizing a 500 V Ar$^+$ beam with an ion flux in the nA cm^{-2} regime. No ion current was measured at the sample without Ar flowing through the gun, even with a 10^{-6} Torr chamber pressure of O$_2$. This indicates that virtually no O$_2$ entering the gun from the chamber was exposed to the crystal as ions. During a typical experiment, the total Ar$^+$ ion exposure was maintained below about 5% of a monolayer in order to minimize any potential effects due to sputter damage. Secondary ions generated by sputtering were monitored with a quadrupole-based Extrel C50 spectrometer.

The electronic structure calculations used the massively parallel Gaussian-based code QUEST (quantum electronic structure)[18] and density functional theory in the local density approximation (LDA) as described in refs. 14 and 19. Successful tests of computational accuracy included comparisons with the results of Ramamoorthy et al.,[20] who used a plane wave code and different pseudopotentials. Local densities of states (LDOS) are found by the standard projection on the local Gaussian bases. Integrating these to the Fermi level then results in local electron populations having diagonal (same atom) and off-diagonal (interatomic) parts, which provide information on the degrees of ionicity and covalency in local interactions.

3 Results

3.1 Oxygen-induced change of surface structure

In the following, we present LEIS and STM results after re-annealing at various temperatures for a flat, UHV-annealed surface similar to the one displayed in Fig. 1. The following procedure was employed each time: sputtering, UHV annealing, lowering the sample temperature to the specified value, exposing to ^{18}O$_2$ gas (1 × 10^{-6} mbar) for a specified time, and cooling in UHV to room temperature.

^{18}O$_2$ exposure at 500 K for 10 min produces bright features evenly distributed on the (1 × 1) substrate as shown in Fig. 2(a). Most features are assembled into short aggregates (≈40 Å) roughly oriented along the [1$\bar{1}$0] direction. In addition, a few scattered (1 × 1) islands (ca. 40 Å × 30 Å, marked by arrows in Fig. 2(a)) can be seen on top and in between the large (1 × 1) terraces. The step edges of the (1 × 1) terrace become more irregular as compared to a UHV-annealed surface (Fig. 1). After annealing in ^{18}O$_2$ at 520 K (Fig. 2(b)), distinctly different morphological features appear on the large (1 × 1) terraces. Patches of rosette-like networks[13] (labeled R, typically 30 Å wide and elongated along the [1$\bar{1}$0] direction) dominate, see Fig. 4 below. Located in between are small (1 × 1) islands (of typical size 60 Å × 40 Å). In addition, some white clusters are found on top of the (1 × 1) islands. Fig. 2(c) (annealing at 550 K) shows an even rougher

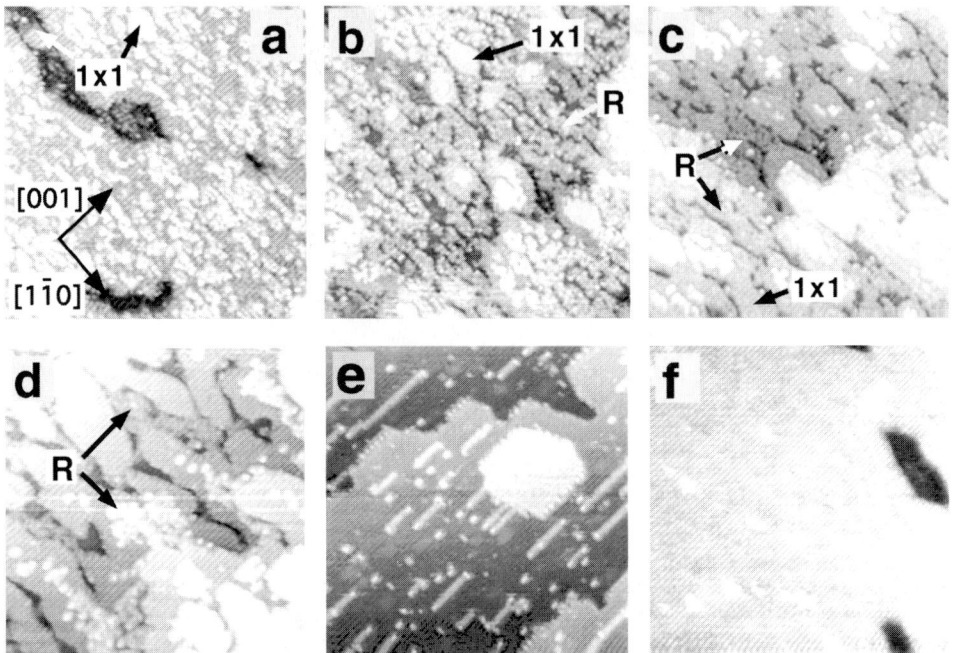

Fig. 2 STM images (500 Å × 500 Å) of a TiO$_2$(110) surface. All surfaces were pretreated by sputtering and annealing in UHV at 880 K for 30 min. ^{18}O$_2$ (1 × 10^{-6} mbar) was dosed at (a) 500 K, (b) 520 K, (c) 550 K, (d) 660 K for 10 min, (e) 710 K for 15 min, and (f) 830 K for 20 min.

surface, consisting of many layers of somewhat larger (1 × 1) islands ($\approx$ 80 Å × 60 Å), partially connected to each other and roughly oriented along the [1$\bar{1}$0] direction. Between and on top of these islands are network patches (R) also elongated along the [1$\bar{1}$0] direction. The flat (1 × 1) substrate (still discernible in Fig. 2(b)) can no longer be identified in this image. After annealing in O$_2$ at 660 K (Fig. 2(d)), the (1 × 1) phase dominates the surface. The (1 × 1) islands are connected to each other to form large (1 × 1) terraces with larger network patches (*ca.* 100 Å × 80 Å) appearing on top of terraces. STM results[14] from a sample annealed in $p_{^{18}O_2} = 1 \times 10^{-6}$ mbar at 660 K for 5, 10, and 20 min gave images very similar to the one displayed in Fig. 2(d) (10 min). (This justifies the comparison between annealing for 10 min (Fig. 2 (a)–(d)) and the somewhat longer annealing times at higher temperatures in Fig. 2(e) and (f)).

Oxygen exposure at 710 K for 15 min leads to a dramatic morphological change (Fig. 2(e)) with much larger (1 × 1) islands and straight step edges (some of which are reconstructed as is observed on UHV annealed surfaces[12]). A number of [001]-oriented bright strands (typically 70 Å long) are distributed uniformly on top of (1 × 1) terrace and extend out from the step edges across the lower terrace. An even higher annealing temperature (Fig. 2(f)) yields large, flat (1 × 1) terraces with a few white clusters and bright strands on top of the bright [001]-oriented rows (the Ti sites) of the substrate. Such strands are also visible in the small scale image (Fig. 4 below).

Annealing in ^{18}O$_2$ leads to incorporation of ^{18}O into the sample surface. LEIS ^{18}O peak areas (after normalization of the total LEIS O signal to 100%) that correspond to the images displayed in Fig. 2 are shown in Fig. 3. The ^{18}O content increases with temperature, up to a maximum value of 75% for Fig. 2(d). Concurrent with the transition to flat, larger (1 × 1) islands and the absence of rosette networks, the ^{18}O content decreases again.

All the surfaces displayed in Fig. 2 exhibit a (1 × 1) LEED pattern. XPS results of an oxygen-annealed surface revealed no difference in the Ti 2p peak position or shape as compared to a surface annealed for 30 min in UHV at 880 K. This may be because of the insufficient sensitivity of our XPS-setup. Previous measurements[2,21] showed a shoulder indicative of Ti^{3+} species (attributed to oxygen vacancies) on the UHV-annealed surfaces.

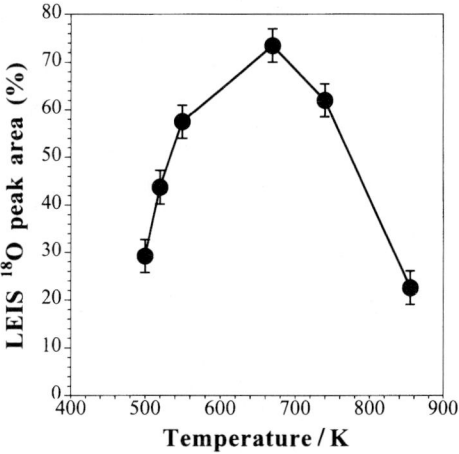

Fig. 3 The surface concentration of ^{18}O (measured with LEIS) on TiO$_2$(110) surfaces prepared as in Figs. 2(a)–(f).

A small scale STM image of both network patches and (1 × 1) islands is shown in Fig. 4. The small isolated (1 × 1) island (*ca.* 60 Å × 30 Å) at the center is partially connected to network patches. The network is atomically resolved as arrays of inter-connected pseudo-hexagonal units (named rosette: R). Usually, one rosette (marked in Fig. 4) is composed of six bright spots with a width equal to the substrate unit cell in the [1$\bar{1}$0] direction (6.5 Å) and twice as long as the substrate unit cell in the [001] direction (2 × 3 Å). Some bigger rosettes composed of more than six bright spots appear occasionally. Incomplete rosettes are incorporated into the edge of the (1 × 1) islands, but rosettes never appear within an island. The rosettes have the same height as the (1 × 1) terraces. The dark centers of the rosettes appear on top of the bright rows (on top of the 5-fold coordinated Ti atoms) of the underlying TiO$_2$ (1 × 1) layer. Structural models of these rosettes are presented below (Figs. 8 and 9).

Some short bright strands (*ca.* 10 Å long) are either connected to or located between network patches and (1 × 1) islands. Usually one of the six bright spots is missing at the connection between a rosette and a strand. In a forthcoming paper,[22] we argue that the strands exhibit the same structure as the double ridges of the TiO$_2$(110) (1 × 2) reconstruction.[7]

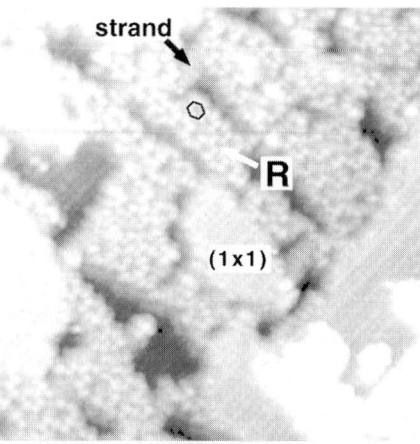

Fig. 4 A small scale STM image (200 Å × 200 Å) of network patches (R), strands, and (1 × 1) islands after annealing in 1 × 10^{-6} mbar ^{18}O$_2$ at 570 K for 25 min.

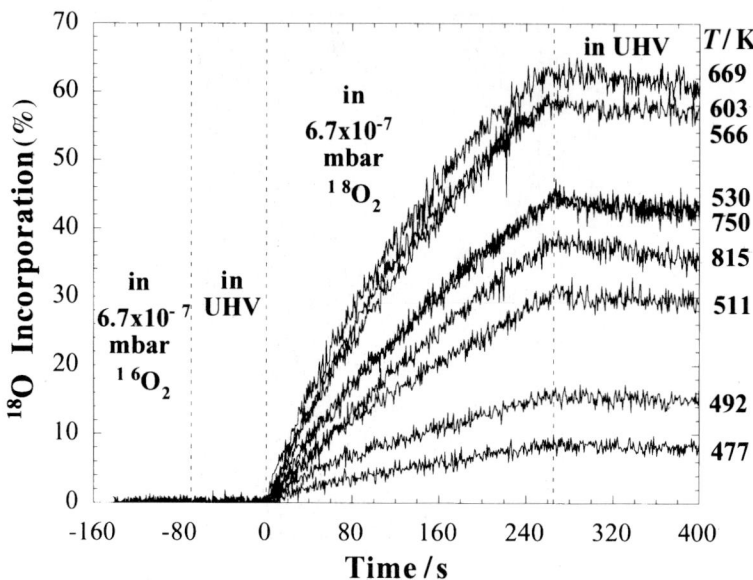

Fig. 5 The ^{18}O surface concentration during annealing in 6.7×10^{-7} mbar ^{18}O$_2$ at various temperatures monitored with SSIMS.

3.2 Kinetics of the restructuring process

A study of the initial ^{18}O incorporation with annealing time was performed using static secondary ion mass spectrometry (SSIMS) in a separate chamber. The sample was prepared by sputtering and annealing in ^{16}O$_2$ (6.7×10^{-7} mbar) at the temperature range 477–815 K. The ^{16}O$_2$ gas was pumped out, the chamber was then backfilled with ^{18}O$_2$ gas (6.7×10^{-7} mbar) which was pumped out after 260 s. During the whole procedure the ^{18}O surface content was monitored (Fig. 5). The ^{18}O uptake occurs more rapidly with increasing annealing temperature, and decreases again above 669 K. This is consistent with the LEIS results displayed in Fig. 3, where a maximum

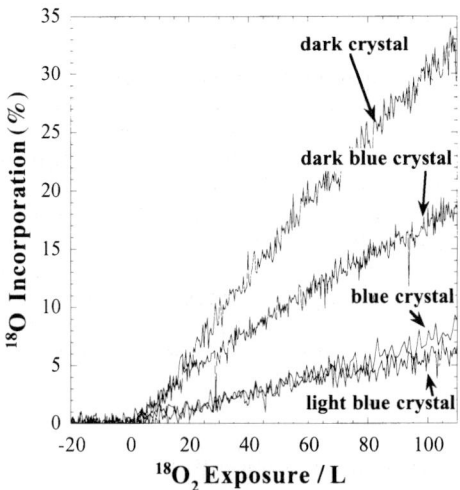

Fig. 6 SSIMS of ^{18}O surface concentration during annealing in ^{18}O$_2$ for rutile crystals with different colors. The color of a crystal is a measure of its bulk defect concentration.

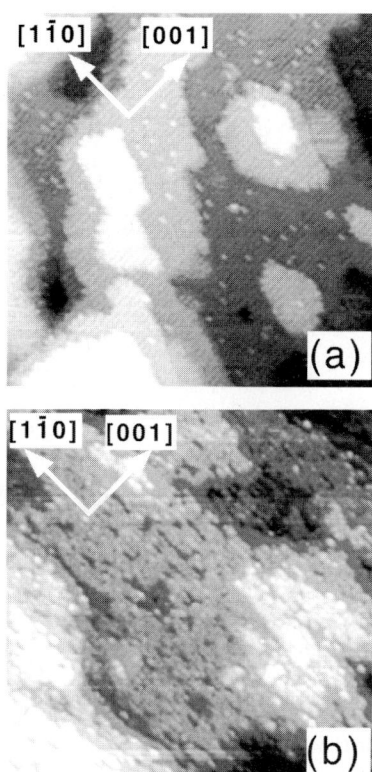

Fig. 7 STM images (500 Å × 500 Å) of (a) a light blue and (b) a dark TiO$_2$(110) sample prepared by sputtering and UHV annealing at 970 K for 20 min followed by annealing in ^{18}O$_2$ (1 × 10^{-6} mbar) at 570 K for 10 min.

of ^{18}O surface content was found after annealing for 10 min at 710 K. From Fig. 5, the ^{18}O uptake rate was determined using a $(1 - \Theta)$ dependence. The rate increases quickly below 566 K, slows down, and decreases above 669 K. From these rates, the ^{18}O activation energy is estimated to be 19 kcal mol^{-1}.

The uptake rate is strongly dependent on crystal 'age' (*i.e.*, reduction state)[17] 'Fresh', as-purchased TiO$_2$ crystals are transparent. With increasing numbers of sputtering/annealing cycles, they change in color from light to dark blue to metallic grayish. This color change is caused by the creation of color centers in the bulk and can be used as a quantitative measure for the degree of bulk reduction. Fig. 6 shows that darker, more reduced samples incorporate ^{18}O at a much faster rate than lighter, more stoichiometric samples. Displayed in Fig. 7 are two TiO$_2$(110) samples with a different degree of bulk reduction. (These samples were mounted next to each other on one sample platen and sputtered and annealed simultaneously to ensure exactly the same treatment. A more detailed and quantitative investigation of the relationship between sample color, type of bulk defects, and surface properties is being published in a forthcoming paper.[17]) A drastically different appearance is visible by STM; annealing in oxygen a light blue sample (top in Fig. 7) shows basically a (1 × 1) surface termination, whereas the much darker sample (Fig. 7, bottom) is quite covered with rosettes. Both samples incorporate ^{18}O, however (Fig. 6).

4 Discussion

4.1 Geometric model for rosette-like network structure

A model for the rosette network has already been presented in a previous paper.[13] The main features are shown in Fig. 8. It consists of an incomplete TiO$_2$ layer, where the O and Ti atoms

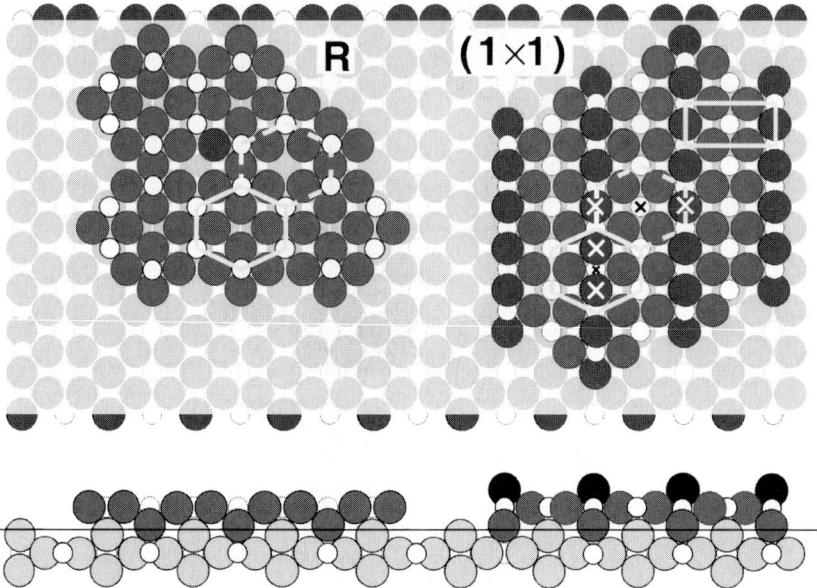

Fig. 8 Atomic model (top and side view) for a restructured surface. A bulk-terminated (1 × 1) island is shown on the right side. The network patch (R) on the left side consists of an incomplete TiO$_2$(110) (1 × 1) layer and contains only atoms at bulk positions. Small white balls are Ti atoms. Shadowed large balls represent oxygen atoms, and darker shading indicates higher z-positions. The rectangle outlines the unit cell of the (1 × 1) structure. The hexagons connect Ti atoms in similar positions on both islands. Atoms missing in the network are marked with large crosses on the (1 × 1) island.

are missing in a regular fashion, and all the remaining atoms are in bulk-like positions. In Fig. 8 two islands are placed onto a (1 × 1) surface, the left one representing a rosette network with missing atoms, and the right one the regular (1 × 1) structure. First consider the (1 × 1) island. Titanium atoms are drawn as small, white balls and oxygen atoms as dark balls. We chose this shading for easier comparison with STM images where Ti sites are generally imaged bright.[11] As is visible in the side view, oxygen atoms at higher locations are shaded darker. For example, the bridging oxygen atoms covering every other Ti row in the regular (1 × 1) structure are shown as black balls. The (1 × 1) structure has a rectangular unit cell as outlined on the far right of Fig. 8. The four 6-fold coordinated Ti atoms on the corners of the unit cell are covered by bridging oxygen atoms, and only the center Ti (5-fold coordinated) is visible by STM.

Outlined (with full lines) on the (1 × 1) island in Fig. 8 is a hexagon connecting four 5-fold-coordinated Ti and two six-fold-coordinated Ti atoms (underneath the bridging oxygens). Suppose the two bridging oxygen atoms marked with crosses as well as the six-fold coordinated Ti atom in the center are missing. Then, one would end up with six Ti atoms arranged in a quasi-hexagon (which is 6.5 Å (one unit cell) wide and 6 Å (two unit cells) high, as observed in the experiment). Exactly these atoms are missing in the 'R' network structure shown in Fig. 4, which consists only of O and Ti atoms in bulk-like positions. So conversely, adding one TiO$_2$ unit into the hexagon (full-lines) of the network structure of Fig. 8 generates the regular (1 × 1) structure. A second kind of hexagon with Ti atoms at the corners is drawn with dashed lines on both islands of Fig. 8. Here, one TiO unit (one 6-fold-coordinated Ti atom at the center and two 'connecting' bridging oxygen atoms, counted as one oxygen atom, at the sides) is missing in the network structure. Hence, the rosette-structure is simply an incomplete TiO$_2$(110) layer with some atoms missing.

4.2 Electronic structure calculations for a rosette on top of TiO$_2$(1 × 1) (110) surface

To test the stability of the proposed rosette structure, we performed LDA calculations using the 258-atom supercell shown in Fig. 9. A single rosette with six Ti atoms and twelve O atoms is

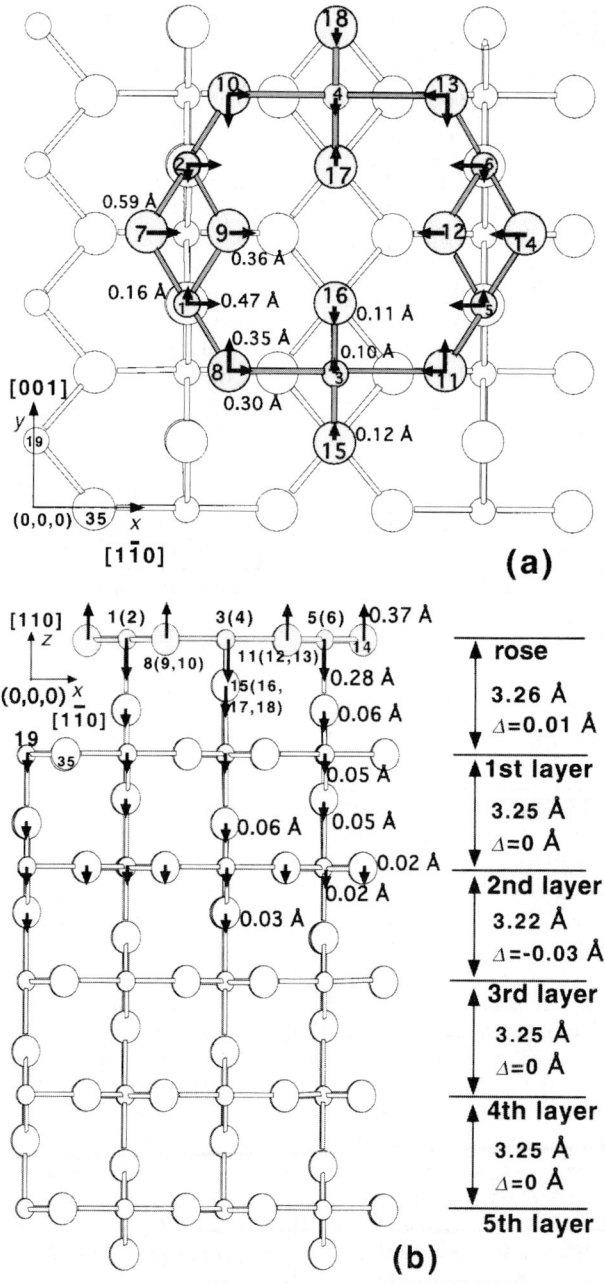

Fig. 9 Supercell used for electronic structure calculations (a) top-view, (b) side-view.

arranged on top of a $TiO_2(1 \times 1)$ (110) substrate. The bottom three TiO_2 layers (3rd–5th layer in Fig. 9(b)) were frozen at the bulk positions and the top two layers of TiO_2 and the rosette were allowed to relax geometrically by minimizing the force of each atom. Relaxations of selected atoms are illustrated in Figs. 9(a) and 9(b).

The rosette shows substantial horizontal relaxations (between 0.1 Å and 0.6 Å, Fig. 9(a)) with a general tendency to collapse towards the center. From the side view (Fig. 9(b)), all Ti atoms (1–6) in the rosette relax downward by 0.3 Å. All inner O atoms (9,12,16,17) sink by 0.3 Å, while 15 and

18 sink slightly (0.1 Å). The other outer O atoms (7,8,10,11,13,14) rise by 0.4 Å. Thus, the rosette is shrinking and buckling to reach its equilibrium position. This results in a shortening of Ti–O bond lengths by 0.1 to 0.2 Å as compared to the bulk. With the rosette on top of the substrate, the relaxation of Ti and O atoms in the first layer is similar to the first-principle calculations of the clean (1 × 1) (110) surface.[23] The existence of rosettes causes small relaxations in the second and third layers.

The LDOS of bulk atoms (O131 and Ti130, not labeled in Fig. 10(b)), first layer atoms (O35 and Ti19) and the DOS of rosette atoms (averaged O, Ti1, and Ti3) are shown in Fig. 10. In each case, the Ti states dominate the unoccupied conduction band and the O states the valence band, similar to other theoretical works on the TiO_2(1 × 1) (110) surface.[23–28] When interpreting our STM images (Fig. 4), we assumed that the Ti atoms in the rosette are also imaged bright. The results displayed in Fig. 10 justify this assumption. In the rosette, the empty Ti 3d states dominate the conduction band and from the theory of Tersoff and Hamann,[29] the tunneling current is found to be proportional to the surface LDOS at the position of the tip.

The valence band width of rosette O atoms is narrower as compared to O atoms on the (1 × 1) surface and in the bulk, and its shape is also different. Such changes should clearly be visible in UPS valence band measurements, especially if performed under conditions where the photon energy is varied to increase sensitivity to Ti 3d derived states.[30] The gap width for both O 2p and

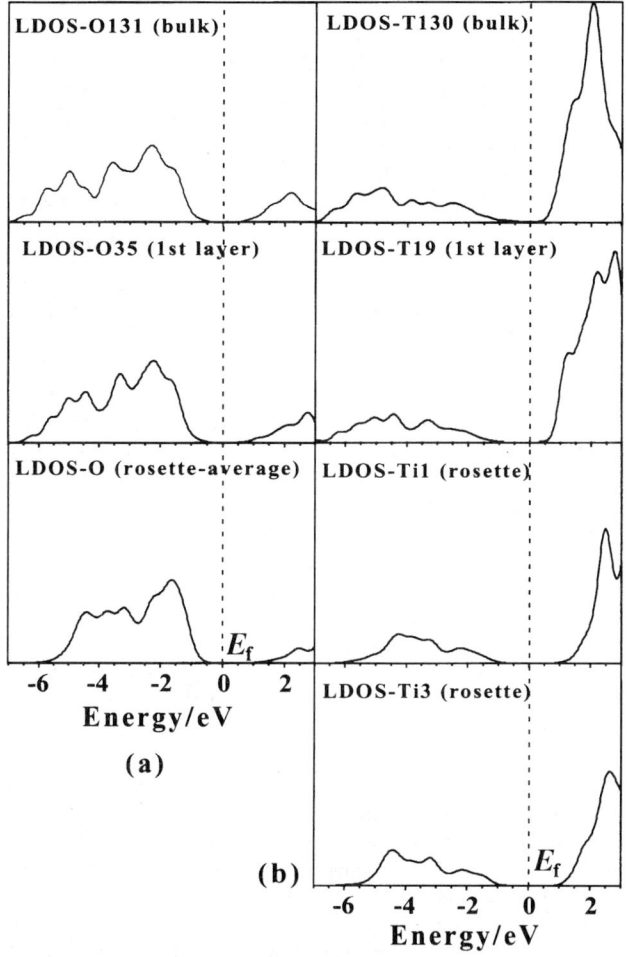

Fig. 10 LDOS of selected atoms within the slab in Fig. 9.

Ti 3d states in the rosette is wider than that in the bulk, as shown in Fig. 10. When analyzing the off-diagonal LDOS of selected Ti–O pairs[14] (not shown here), we find a significant increase in the strength of covalent interaction between Ti and O atoms in the rosette as compared to the bulk. We believe that it is this interaction which broadens the gap, moving the local conduction band minimum higher as seen in Fig. 10.

4.3 Mechanism

Only a few previous accounts of oxygen-induced morphological and structural changes of $TiO_2(110)$ have been given. A brief Letter has been published by this group.[13] Engel and co-workers have observed cross-linked row structures along the $[1\bar{1}0]$ direction after annealing the $TiO_2(110)$ (1 × 2) phase in oxygen (1 × 10^{-7} Torr) at 1000 K followed by heating in UHV.[3,4] Onishi and Iwasawa[9] observed hill-like structures when exposing the $TiO_2(110)$ (1 × 1) surface to O_2 gas ($\approx 1 \times 10^{-7}$ mbar) at 800 K. With time, these features were transformed into added rows and new (1 × 1) terraces. This is consistent with our STM results above 710 K (Fig. 2(e)) where only added strands and (1 × 1) terraces are visible. Onishi and Iwasawa proposed a re-oxidation scheme where Ti^{n+} ($n \leq 3$) interstitial atoms from the reduced bulk diffuse to the surface where they react with O_2 gas to form hills, added rows and new terraces. Consumption of Ti interstitials by reaction with surface oxygen produces a concentration gradient that results in a net diffusion current of these Ti interstitials towards the surface. The same mechanism, i.e., segregation of Ti^{n+}, combined with reaction with gaseous oxygen, is responsible for the formation of rosettes, strands and (1 × 1) islands in our experiments. The rate of surface restructuring is affected by the surface concentration of both reaction partners, Ti interstitials and O_2 molecules. The Ti segregation rate depends on temperature (influencing diffusion to the surface), number of Ti interstitials (reduction state of the crystal), and the chemical potential of oxygen (the oxygen pressure).

The restructuring process can be regarded as the manifestation of reoxidation of a reduced crystal at the atomic scale. It results in the growth of additional TiO_2 layers at the surface with Ti coming from the reduced bulk and oxygen from the gas phase. The kinetic processes and energetics that govern nucleation, growth, and morphology of deposited films are well studied.[31] This case is special; because one constituent of the newly added film comes from the bulk, the kinetics of bulk diffusion must be taken into account as well. (Extensive TiO_2 bulk studies[32–39] revealed titanium interstitial ions as well as oxygen vacancies in a reduced TiO_2 crystal. Bulk diffusion studies[40–48] and a recent SSIMS investigation[49] show that Ti interstitials are the major diffusive species in TiO_2 rutile and not O vacancies.) For heavily reduced crystals, the added features nucleate mainly on terraces, as is visible from their random distribution in low-coverage images (Fig. 2(a)). This suggests that Ti interstitials are driven out in vertical direction from the bulk to the surface, where they react with oxygen. Some step-flow growth occurs also, as evidenced by the relatively rough step edges that develop already at low temperatures and/or gas exposures. Step-flow dominates the growth on less reduced, light crystals (Fig. 7(a)).

The rosette networks are the precursors to the added (1 × 1) islands especially at lower temperatures (<660 K). The transition from rosettes to the (1 × 1) structure is straightforward as discussed above (Section 4.1), one simply needs to add additional atoms to rosettes. This should happen easily upon arrival of new Ti interstitials on the surface, even without additional incorporation of O_2 from the gas phase. At low enough temperatures, the overall island shape of rosette network patches appear to be preserved during the transition to a (1 × 1) island, hence the preferred orientation of the (1 × 1) islands. When the temperature rises above ~700 K (Fig. 2(e)) the (1 × 1) islands assume a more square shape with step edge orientations typical for high-temperature UHV-annealed surfaces.[12]

There are three distinct regions for the rate of ^{18}O uptake, as reported in the context of Fig. 5. Initially, the rate increases with increasing temperature. At higher temperatures, a second process kicks in that first slows and then decreases the reaction rate with annealing temperature. This is also apparent in Fig. 3, where the total ^{18}O uptake for a given exposure time decreases above 700 K. Annealing an ^{18}O-rich surface in UHV decreases the ^{18}O surface content, with a clear break point in the depletion rate around 740 K.[22] The ^{18}O can leave the surface via two routes: exchange with ^{16}O from the bulk, and desorption into the gas phase. The observed changes in surface morphology suggest that the latter process is dominant at high temperatures. Note that the slowdown in ^{18}O uptake under oxygen-rich conditions occurs concurrently with the disappearance of rosettes; upon annealing in ^{18}O$_2$ at 710 K (Fig. 2(e)), only strands are formed on the surface. Similarly, rosettes transform into (reduced Ti$_2$O$_3$) strands when a ^{18}O-restructured surface is annealed in UHV at 690 K.[22] A mere scrambling between surface and bulk oxygen atoms cannot lead to a surface reduction. The assumption that desorption of oxygen from the surface occurs at temperatures above 700 K is also supported by other studies. It has been reported that surface point defects are created when TiO$_2$(110) (1 × 1) surfaces are annealed in UHV at temperatures above 700 K.[2] Xu et al.[10] observed a (1 × 2) reconstructed surface between 700 and 800 K, which reversibly converts to the (1 × 1) surface. They attributed this conversion to O desorption into the gas phase and/or Ti diffusion into the bulk. This is also consistent with our results; at temperatures above 830 K only bulk-terminated (1 × 1) terraces exist (Fig. 2(f)), indicating that the row structure is also a metastable phase that converts into the most stable (1 × 1) terrace with temperature.

5 Conclusion and open questions

The results presented in this paper clearly indicate that both the oxidation conditions and the history of the TiO$_2$(110) sample have a significant bearing on the morphology of the surface. It is somewhat frustrating that even this 'best characterized' of all metal oxide systems is not yet completely understood, and that characterization with spectroscopic and diffraction techniques is not sufficient to reveal the great differences in atomic structure that can form through annealing in oxygen. The STM images displayed in Fig. 7 are a good example. Under exactly the same oxidation conditions, a very light blue crystal exhibits only a (1 × 1) structure, while a dark crystal is covered with rosettes. Hence, it is possible to deliberately prepare either a stoichiometric TiO$_2$(110) (1 × 1) surface or one covered with the metastable structure.

The observed variations in the surface structure with O$_2$ pressure, crystal temperature and bulk defect density are so vast that we suspect chemistry of the TiO$_2$(110) surface should be significantly variant for samples oxidized under different conditions. For example, the issue of whether water is molecularly or dissociatively adsorbed on TiO$_2$(110)[2,16,50–52] may be significantly clouded in the literature because of studies in which the morphology of the surface was unknowingly disordered by the presence of the rosettes and/or strands observed in this study by STM. Another example where restructured surfaces may exhibit quite different chemistry from a (1 × 1) surface is the adsorption of pyridine. Pyridine molecules bind more strongly at 4-fold coordinated Ti atoms at step edges as compared to the 5-fold coordinated Ti on the flat (1 × 1) surface as shown recently by Suzuki et al.[53] Possibly, pyridine interacts strongly with rosette networks, where 4-fold coordinated Ti atoms are prevalent (Fig. 9). Metal overlayer film growth on TiO$_2$(110) may be affected as well,[54] e.g., the roughness, induced by oxygen annealing of dark crystals, may influence nucleation and growth of overlayers. The rosette networks may also provide special adsorption sites for metal atoms, e.g., it is not inconceivable that one could place a single metal atom in the center of the rosette displayed in Fig. 9.

On a larger scale, the results in this study suggest that subsurface (interstitial) Ti is fairly labile in TiO$_2$ rutile, especially as the bulk concentration of these species increases. The bulk of small rutile particles could therefore act as sinks for excess Ti under reductive conditions, with this Ti returning to the surface under oxidative conditions. Such cycling of Ti between the bulk and surface should significantly influence surface properties of small crystalline particles, as suggested by results in this work, but should also affect the bulk electrical and photoabsorptive properties.

Rutile TiO$_2$(110) may not be the only system where such oxygen induced morphology changes occur. A study of the reoxidation mechanisms of other bulk-reduced materials may provide an attractive playing field for surface scientists, where rich and interesting metastable structures may be expected.

6 Acknowledgement

This work was supported in part by NSF-CAREER and DoE-EPSCoR. The SSIMS work was supported by the US Department of Energy, Office of Basic Energy Sciences, Division of Materials Sciences, and was conducted at the William R. Wiley Environmental Molecular Sciences Laboratory, a Department of Energy user facility funded by the Office of Biological and Environmental Research. Pacific Northwest National Laboratory is a multiprogram national laboratory operated for the US Department of Energy by Batelle Memorial Institute under Contract DE-AC06-76-RLO 1830. Sandia is a multiprogram laboratory operated by Sandia Corporation, a Lockheed Martin Company, for the United States Department of Energy under Contract DE-AC04-94AL85000. The electronic structure calculation was partially supported by a Laboratory Directed Research and Development project.

References

1 V. E. Henrich and P. A. Cox, *The Surface Science of Metal Oxides*, Cambridge University Press, Cambridge, 1994.
2 J.-M. Pan, B. L. Maschhoff, U. Diebold and T. E. Madey, *J. Vac. Sci. Technol. A*, 1992, **10**, 2470.
3 A. Szabo and T. Engel, *Surf. Sci.*, 1995, **329**, 241.
4 M. Sander and T. Engel, *Surf. Sci. Lett.*, 1994, **302**, 263.
5 D. Novak, E. Garfunkel and T. Gustafsson, *Phys. Rev. B: Condens. Matter*, 1994, **50**, 5000.
6 P. W. Murry, N. G. Condon and G. Thornton, *Phys. Rev. B: Condens. Matter*, 1995, **51**, 10989.
7 H. Onishi and Y. Iwasawa, *Surf. Sci. Lett.*, 1994, **313**, 783.
8 H. Onishi, K. Fukui and Y. Iwasawa, *Bull. Chem. Soc. Jpn.*, 1995, **68**, 2447.
9 H. Onishi and Y. Iwasawa, *Phys. Rev. Lett.*, 1996, **76**, 791.
10 C. Xu, X. Lai, G. W. Zajac and D. W. Goodman, *Phys. Rev. B: Condens. Matter*, 1997, **56**, 13464.
11 U. Diebold, J. F. Anderson, K.-O. Ng and D. Vanderbilt, *Phys. Rev. Lett.*, 1996, **77**, 1322.
12 U. Diebold, J. Lehman, T. Mahmoud, M. Kuhn, G. Leonardelli, W. Hebenstreit, M. Schmid and P. Varga, *Surf. Sci.*, 1998, **411**, 137.
13 M. Li, W. Hebenstreit and U. Diebold, *Surf. Sci.*, 1998, **L414**, L951.
14 M. Li, L. Groß, W. Hebenstreit, U. Diebold, M. A. Henderson, D. R. Jennison, P. A. Schultz and M. P. Sears, *Surf. Sci.*, 1999, **437**, 173.
15 L. Zhang, M. Kuhn and U. Diebold, *Surf. Sci.*, 1997, **371**, 223.
16 M. A. Henderson, *Surf. Sci.*, 1996, **355**, 151.
17 M. Li, W. Hebenstreit, U. Diebold and M. A. Henderson, *Surf. Sci.*, submitted.
18 M. P. Sears and P. A. Schultz at Sandia National Laboratories, Albuquerque, NM 87185-1111. E-mail: mpsears@sandia.gov
19 C. Verdozzi, D. R. Jennison, P. A. Schultz, M. P. Sears, J. C. Barbour and B. G. Potter, *Phys. Rev. Lett.*, 1998, **80**, 5615; *ibid.*, 1999, **82**, 799.
20 M. Ramamoorthy, D. Vanderbilt and R. D. King-Smith, *Phys. Rev. B: Condens. Matter*, 1994, **49**, 16721.
21 L. Wang, D. R. Baer and M. H. Engelhard, *Surf. Sci.*, 1994, **320**, 295.
22 M. Li, W. Hebenstreit and U. Diebold, *Phys. Rev. B: Condens. Matter*, submitted.
23 K.-O. Ng and D. Vanderbilt, *Phys. Rev. B: Condens. Matter*, 1997, **56**, 10544.
24 O. Gülseren, R. James and D. W. Bullett, *Surf. Sci.*, 1997, **377–379**, 150.
25 S. Munnix and M. Scheits, *Phys. Rev. B: Condens. Matter*, 1984, **30**, 2202.
26 N. Yu and J. W. Halley, *Phys. Rev. B: Condens. Matter*, 1995, **51**, 4768.
27 D. Vogtenhuber, R. Podloucky, A. Neckel, S. G. Steinemann and A. K. Freeman, *Phys. Rev. B: Condens. Matter*, 1994, **49**, 2099.
28 D. Vogtenhuber, R. Podloucky and J. Redinger, *Surf. Sci.*, 1998, **402–404**, 798.
29 J. Tersoff and D. R. Hamann, *Phys. Rev. B: Condens. Matter*, 1985, **31**, 805.
30 Z. Zhang and V. E. Henrich, *Phys. Rev. B: Condens. Matter*, 1991, **43**, 12004.
31 J. A. Venables, *Surf. Sci.*, 1994, **299/300**, 798.
32 M. Aono and R. R. Hasiguti, *Phys. Rev. B: Condens. Matter*, 1993, **48**, 12406.
33 F. Millot, M. G. Blanchin, R. Tetot, J. F. Marucco, B. Poumellec, C. Picard and B. Touzelin, *Prog. Solid State Chem.*, 1987, **17**, 263.
34 P. Kofstad, *J. Less-Common Metals*, 1967, **13**, 635.
35 L. A. Bursill, M. G. Blanchin and D. J. Smith, *Proc. R. Soc. London Ser. A*, 1984, **391**, 373.
36 L. A. Bursill and D. J. Smith, *Nature (London)*, 1984, **309**, 319.
37 J. Sasaki, N. L. Peterson and K. Hoshino, *J. Phys. Chem. Solids*, 1985, **46**, 1267.
38 H. B. Huntington and G. A. Sullivan, *Phys. Rev. Lett.*, 1965, **14**, 177.
39 D. J. Smith, L. A. Bursill and M. G. Blanchin, *Philis. Mag. A*, 1984, **50**, 473.
40 D. J. Derry, D. G. Lees and J. M. Calvert, *J. Phys. Chem. Solids*, 1981, **42**, 57.
41 D. J. Neild, P. J. Wise and D. G. Barnes, *J. Phys. D: Appl. Phys.*, 1972, **5**, 2292.
42 H. Kolem and O. Kanert, *Z. Metallkd.*, 1989, **80**, 227.
43 T. S. Lundy and W. A. Coghlan, *J. Phys. Colloq.*, 1973, 299.
44 K. Hoshino, N. L. Peterson and C. L. Wiley, *J. Phys. Chem. Solids*, 1978, **39**, 457.
45 D. A. Venkatu and L. E. Poteat, *Mater. Sci. Eng.*, 1970, **5**, 258.
46 J. R. Akse and H. B. Whitehurst, *J. Phys. Chem. Solids*, 1978, **39**, 457.

47 M. Arita, M. Hosoya, M. Kobayashi and M. Someno, *J. Am. Ceram. Soc.*, 1979, **62**, 443.
48 M. Someno and M. Kobayashi, *Springer Ser. Chem. Phys.*, 1979, **9**, 222.
49 M. A. Henderson, *Surf. Sci.*, 1999, **419**, 174.
50 L. Wang, D. R. Baer, M. H. Engelhard and A. N. Shultz, *Surf. Sci.*, 1995, **344**, 237.
51 M. A. Henderson, *Langmuir*, 1996, **12**, 5093.
52 R. L. Kurtz, R. Stockbauer and T. E. Madey, *Surf. Sci.*, 1989, **218**, 178.
53 S. Suzuki, Y. Yamaguchi, H. Onishi, K. Fukui, T. Sasaki and Y. Iwasawa, *Catal. Lett.*, 1998, **50**, 117.
54 C. T. Campbell, *Surf. Sci. Rep.*, 1997, **27**, 1.

Paper 9/03598B

The selective adsorption and kinetic behaviour of molecules on $TiO_2(110)$ observed by STM and NC-AFM

Yasuhiro Iwasawa,* Hiroshi Onishi,† Ken-ichi Fukui, Shushi Suzuki and Takehiko Sasaki

Department of Chemistry, Graduate School of Science, The University of Tokyo, Hongo, Bunkyo-ku, Tokyo 113-0033, Japan. Fax: +81-3-5800-6892; E-mail: iwasawa@chem.s.u-tokyo.ac.jp

Received 19th April 1999

In the present paper we report a kinetic aspect of molecules on terraces and steps of a $TiO_2(110)$-(1×1) surface as observed by scanning probe microscopy, which may be relevant to oxide catalysis. When the TiO_2 surface, heated at 400–450 K, was exposed to a formic acid ambient of 1–2×10^{-6} Pa, small particles were formed on terraces. Post-reaction STM observation revealed that the particle formation was strongly suppressed in the vicinity of single-atom height steps. The suppressive effect of the steps ranged 2.4 nm into the terrace. Formate ions (a possible reaction intermediate) were imaged *in-situ* during formic acid exposure at the reaction temperature. The local density of the formates and hence the product distribution, which were very low near the steps, were simulated by a model. The particles produced were suggested to be carbonates, CO_3^{2-} on the bridging oxygen atoms of the $TiO_2(110)$ surface.

Introduction

There are a variety of applications of oxides to industrial materials and processes, *e.g.*, optical applications, electronic applications, catalytic materials (such as photocatalysts, environmental catalysts and so on), electrodes, adsorbing materials, magnetic applications, ceramic applications, coatings, environmental applications, pigments, cosmetics, containers for nuclear materials.[1,2] TiO_2 is a typical oxide and has been widely used in many types of technologies. However, there is little atomic- and molecular-level information on the structure of metal oxide surfaces, which should be uncovered scientifically. The great potential of scanning probe microscopy (SPM) to directly observe surface structures and reactions at a high resolution has been demonstrated on metal oxide surfaces.[1,2] Individual atoms that constitute a sputter-annealed $TiO_2(110)$-(1×1) surface were resolved by scanning tunneling microscopy (STM)[3] and atomic force microscopy (AFM) operated in a frequency modulated non-contact mode (NC-AFM),[4] as shown in Fig. 1. Each Ti atom with five-fold coordination shown in the bright rows of 0.65 nm separation to each other and parallel to the [001] direction was clearly observed at a regular interval of 0.30 nm by STM. The atomically resolved NC-AFM image reproduced the (1×1) unit, 0.65 nm × 0.30 nm, with the alignment of the bridging oxygen atoms. Several dark sites on the bright lines in the NC-AFM image could be vacancies of the bridging oxygen atoms. Further, the molecules and

---

† Present address: Kanagawa Academy of Science and Technology, KSP East-404, Sakado, Takatsu-ku, Kawasaki-shi 213-0012, Japan.

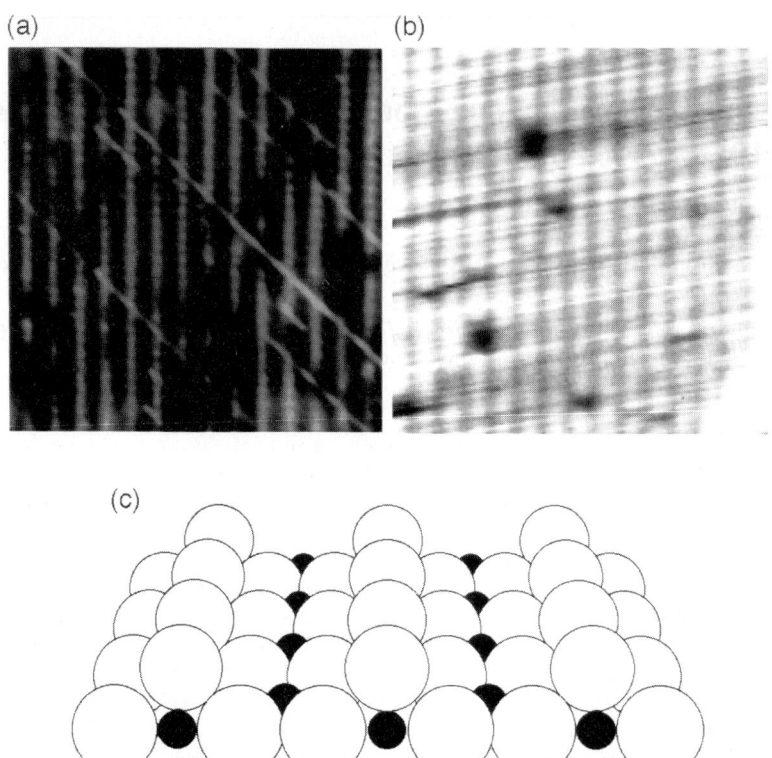

Fig. 1 The atomically resolved images of the (110)-(1 × 1) surface of the rutile TiO_2. (a) A constant current topography by STM.[3] Scan area: 12 × 12 nm^2, sample bias voltage: +1.0 V, tunneling current: 0.3 nA. The diagonal scratches are artifacts. (b) NC-AFM.[4] Scan area = 9.4 × 9.4 nm^2, $V_s = 0$ V, frequency shift = 80 Hz. (c) A stoichiometric truncation of the (110) surface.

reaction intermediates adsorbed on a $TiO_2(110)$ surface can be imaged at temperatures ⩾300 K because their diffusion is fairly restricted over metal oxide surfaces.[5] For example, formate isolatedly adsorbed at room temperature (NC-AFM),[6] formate ions migrating at room temperature (STM)[7,8] and acetate intermediates decomposing at 540 K (STM)[9] have been visualized on $TiO_2(110)$ surfaces. The high-temperature dynamic behavior of oxide surfaces has also been successfully imaged with a $TiO_2(110)$ surface at 800 K.[10]

The most promising ability of SPM expected, in relation to catalysis research, is to discriminate and visualize the reaction events occurring at different sites on a surface. The enhancement of catalytic reactions due to coordinatively unsaturated and atom-rearranged metal centers (like steps) has long been postulated since the 'active center' concept of Tailor.[11] In fact, the site-specific adsorption of pyridine promoted on four-fold coordinated Ti centers at the steps, with, in particular, azimuth orientations, on $TiO_2(110)$ has been visualized.[12] The steps on which the pyridine molecules adsorbed run parallel to the $[1\bar{1}2]$ and $[\bar{1}12]$ directions. Similar activity has been observed on the steps parallel to the $[1\bar{1}3]$, $[1\bar{1}4]$ and $[1\bar{1}5]$ directions. In contrast, the steps of the $[1\bar{1}1]$, $[1\bar{1}0]$ and $[001]$ directions were not active for the adsorption.

In the present paper, we report a long-range suppressive effect of single-atom height steps of a $TiO_2(110)$ surface for a formic acid decomposition reaction that takes place over the terraces far distant from the steps.

Experimental

The experiments were performed in an UHV-compatible scanning tunneling microscope (JEOL-JSTM4500VT) with an electrochemically etched tungsten tip. Constant-current topography was continuously imaged at a rate of 33 s per frame and recorded in video. A polished TiO_2(110) wafer of 6.5 × 1 × 0.25 mm³ (Earth Chemicals Co., Japan) gave a sharp (1 × 1) LEED pattern after repeated cycles of Ar^+ sputtering (3 keV, 0.3 µA) and vacuum annealing at 900 K. The temperature of the wafer was monitored with an IR radiation thermometer. A small deficiency in oxygen concentration, 0.001%, was estimated from the bulk resistivity of the blue crystal, 2 Ω m.[13] Research grade HCOOD gas was purified through trap-thaw cycles.

Results and discussion

When the TiO_2(110) surface heated at 450 K was exposed to a formic acid ambient of 1.3×10^{-6} Pa, small particles were formed on the terraces. Fig. 2 shows the constant-current topography observed on the surface reacted with the atmosphere for 10 min. The exposed surface was cooled to room temperature (RT) in the presence of formic acid vapor and the atmosphere was evacuated. The topography was determined at RT. The diagonal, zig-zag line over the frame was a single-atom height (0.32 nm) step between the (left) upper and the (right) lower terraces. The surface was saturated with adsorbed formate ions. The smallest protrusions presented in a gray scale and arranged in a (2 × 1) order are the formate ions.[14] Many blips larger and brighter than the formate were observed on the terraces. The brighter particles can be assigned to the product of a reaction that took place on the exposed surface, since they were not observed before the exposure at 450 K. Lines of slight contrast parallel to the [001] direction are the added Ti_2O_3 double-strand rows inevitably formed on sputter-annealed (1 × 1) surfaces.[3] They were also observed at the surface before the reaction.

The spatial distribution of the product particles is quite interesting. The local number density depended on the distance from the step. This was obvious on the upper terrace. A few particles appeared in the proximity of the step, whereas they were produced at random on the terrace a long way from the step. The near-step area on the terrace looks a depletion belt and is completely covered with formate ions as a result. This non-uniform distribution of the product particles was a reproducible phenomenon. Depletion belts of similar width were always observed near the steps on the surface (see Fig. 2) and also on similarly reacted TiO_2 surfaces.

To elucidate the non-uniform distribution quantitatively, the local number density of 204 particles observed in a zoomed-out image was plotted as a function of the distance from the step, in a histogram (Fig. 3). The densities observed in the segments nearest and second-nearest to the step

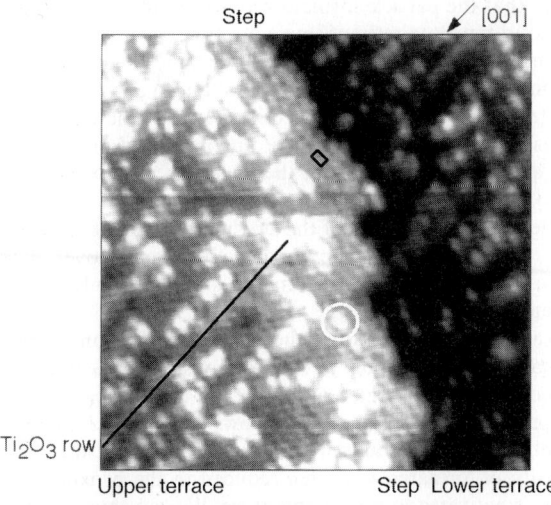

Fig. 2 A post-reaction STM image of the TiO_2(110) surface heated at 450 K and exposed to 1.3×10^{-6} Pa HCOOD gas for 10 min. The constant current topography was determined on the surface cooled to room temperature. Scan area: 25 × 25 nm², sample bias voltage: +2.0 V, tunneling current: 0.10 nA. One product particle is marked by the white circle. One (2 × 1) unit of the formate monolayer is shown with the solid rectangle. A solid line is superimposed on a Ti_2O_3 row for illustration.

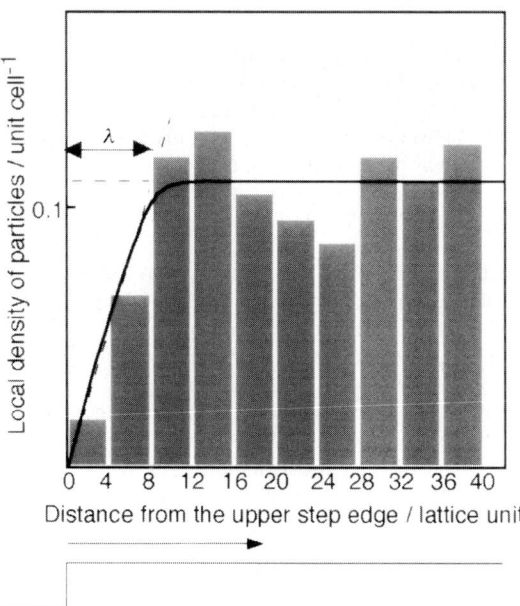

Fig. 3 Local number density of the product (brighter) particles as a function of the distance from the step.

in Fig. 3 were reduced from the value averaged over the third-nearest or farther segments. The degree of the reduction was outside of the random fluctuation predicted from statistics.

The particle was immobile at the reaction temperature (450 K). Fig. 4(a) shows an STM image of a reacted surface with product particles recorded at 650 K in vacuum. At this temperature, the formate ion decomposed unimolecularly,[2,15] leaving no product particles on the surface. Sequential imaging revealed that the migration of the particles was prohibited even at this elevated temperature. When the surface was heated at 700 K, the particles disappeared, probably decomposing, as shown in Fig. 4(b). A product particle stayed at the place where it was formed.

These results indicate that the particle formation reaction, that occurred on the terrace far from the step, was suppressed in nm-proximity of single-atom height steps on $TiO_2(110)$. This may be a novel phenomenon. The inverse effect, step-induced enhancement of a reaction, has long been demonstrated.[16] A typical example of step-induced enhancement has recently been visualized on NO/Ru(001).[17] The nitric oxide molecules were dissociated at the coordinatively unsaturated Ru atoms exposed to steps. The resultant nitrogen atoms migrate from the dissociation sites (steps) into the terraces. The localized distribution of the N atoms was imaged in a post-reaction STM observation.[17]

The long range of the suppressive effect of the single-atom height steps on $TiO_2(110)$ is another interesting feature. Fig. 3 shows that the suppressive effect ranges over eight lattice units, *i.e.*, 2.4 nm. It is quite important to know how an atom-size surface singularity like the step affects the reactivity over nanometer proximity.

In situ STM observation was tried during the particle formation reaction in order to obtain information on the location and transport of formate ions as a possible intermediate of the reaction. A sputter-annealed surface was heated at 420 K and formic acid vapor of 1.6×10^{-6} Pa was dosed in the microscope chamber at $t = 0$. Fig. 5 shows an image of the exposed surface observed at $t = 230$ s. The number of formate ions imaged as small bright dots increased with exposure time. The number density of formates was often reduced in the proximity of the steps. The diffusion of the formate ions is activated at this temperature. We can thus assume that the non-uniform distribution of the product particles results from the non-uniform distribution of the formate intermediates.

The next question is how the step regulates the formate density. The most simple interpretation is the electrostatic effect. The absence of Ti and O ions that would have existed there if the upper

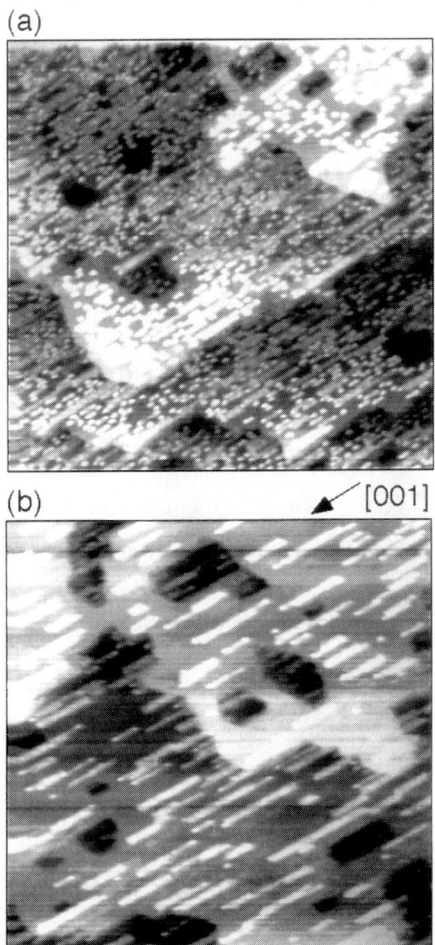

Fig. 4 STM images of the reacted TiO_2 surface observed in vacuum at (a) 650 K and (b) 700 K. Scan area: 100×100 nm^2, sample bias voltage: $+2.6$ V, tunneling current: 0.05 nA.

terrace continued beyond the step, causes a finite modulation of the electrostatic field near the step. Another method of interpretation is a diffusion-derived picture. The formate ion may be consumed very fast at the step site by recombinative desorption or a decomposition reaction, as is postulated in the step-induced enhancement of reaction. The local number density of the formate ion has to reduce near the step, since adsorption-desorption equilibrium and surface diffusion are coupled so as to determine the local population of adsorbed species.

We have simulated the phenomenon by eqn. (1) under the boundary condition of $\theta = 0$ at $x = 0$ (θ particle coverage; x = distance from the step end) assuming a fast desorption from the step. The coverage change of formate ions against the reaction time is given by eqn. (1) including diffusion, adsorption and desorption terms. D is the diffusion constant, k_a is the rate constant for adsorption

$$\frac{\partial \theta}{\partial t} = D \frac{\partial^2 \theta}{\partial x^2} + k_a P(1 - \theta) - k_d \theta \qquad (1)$$

and k_d is the rate constant for desorption. Coverage dependency on the distance was calculated by using the equation. The simulated curve for θ against x is shown in Fig. 3. The simulated curve reproduced the histogram well. In the steady-state, eqn. (2) is obtained and the depletion-zone

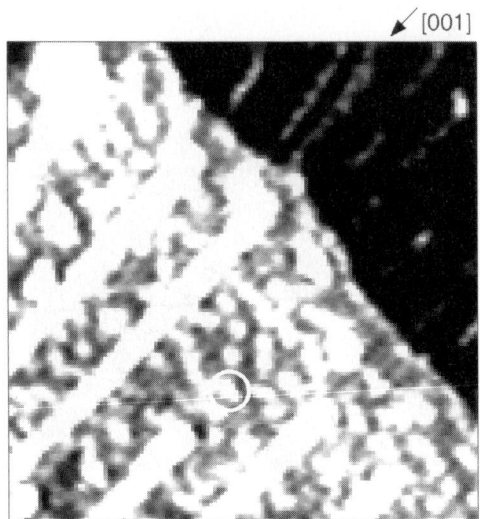

Fig. 5 An *in situ* STM image of the $TiO_2(110)$ surface reacted at 420 K in 1.6×10^{-6} Pa HCOOD gas. The reactant atmosphere was introduced at $t = 0$ s and the topography was observed at the reaction temperature at $t = 230$ s. Scan area: 20×20 nm^2, sample bias voltage: $+1.7$ V, tunneling current: 0.2 nA. One formate ion is marked by the white circle. The seven thick white lines parallel to the [001] direction are Ti_2O_3 rows.

length, λ, is given approximately by eqn. (3), where θ_∞ is $k_a P/(k_a P + k_d)$.

$$\left(\frac{\partial \theta}{\partial x}\right)_{\text{at } x = 0} = \frac{k_a P}{\sqrt{D(k_a P + k_d)}} \quad (2)$$

$$\lambda = \frac{\theta_\infty}{\left(\frac{\partial \theta}{\partial x}\right)_{\text{at } x = 0}} = \frac{\sqrt{D}}{\sqrt{k_a P + k_d}} \quad (3)$$

$$\lambda_{\text{max}} = \sqrt{\frac{D}{k_d}} = \sqrt{\frac{k_h}{2k_d}} L_c \quad (4)$$

When P becomes close to zero, λ becomes maximum (λ_{max}) and λ_{max} is given by eqn. (4). D is $k_h L_c^2/2$, where k_h is a rate constant for hopping migration of formate ions and L_c is the lattice constant of $TiO_2(110)$. When the ratio k_h/k_d is 100, the calculated λ_{max} reproduces the observed depletion-zone length of 2.4 nm.

Finally, it should be considered what the product particle is. Fig. 6 shows a zoomed-in image of a reacted surface. The atom-scale location of the product can be identified by using the coexisting (2×1)-formate structure as a scale. A formate ion in the (2×1) monolayer is adsorbed on two five-fold coordinated Ti atoms in a bridge form with an O-C-O bond angle of 126°.[14,18-20] The product particles were adsorbed at the on-top position of the bridge oxygen atoms. Cross-section analysis revealed that the topographic height of the product was higher by 0.15 nm than the top of the (2×1)-formate monolayer. The formate ion itself exhibits protrusions of 0.14 nm height from the surface in similar tunneling conditions.[14] The accordance in topographic size suggests that the product particle and the formate ion are similar in physical size. In the steady-state reaction of formic acid on $TiO_2(110)$ the main products were H_2 and CO_2 in the temperature range 420-550 K (the activation energy = 15 kJ mol^{-1}).[15] These considerations allowed us to assume that a CO_2 molecule adsorbed at the on-top site of the bridge oxygen atom forming carbonate (CO_3^{2-}) group makes the product particle, as illustrated in Fig. 6(c). The STM image of the particles in Fig. 6 seems elongated, not spheric. A similar elongated feature was observed for carbonate (CO_3^{2-})

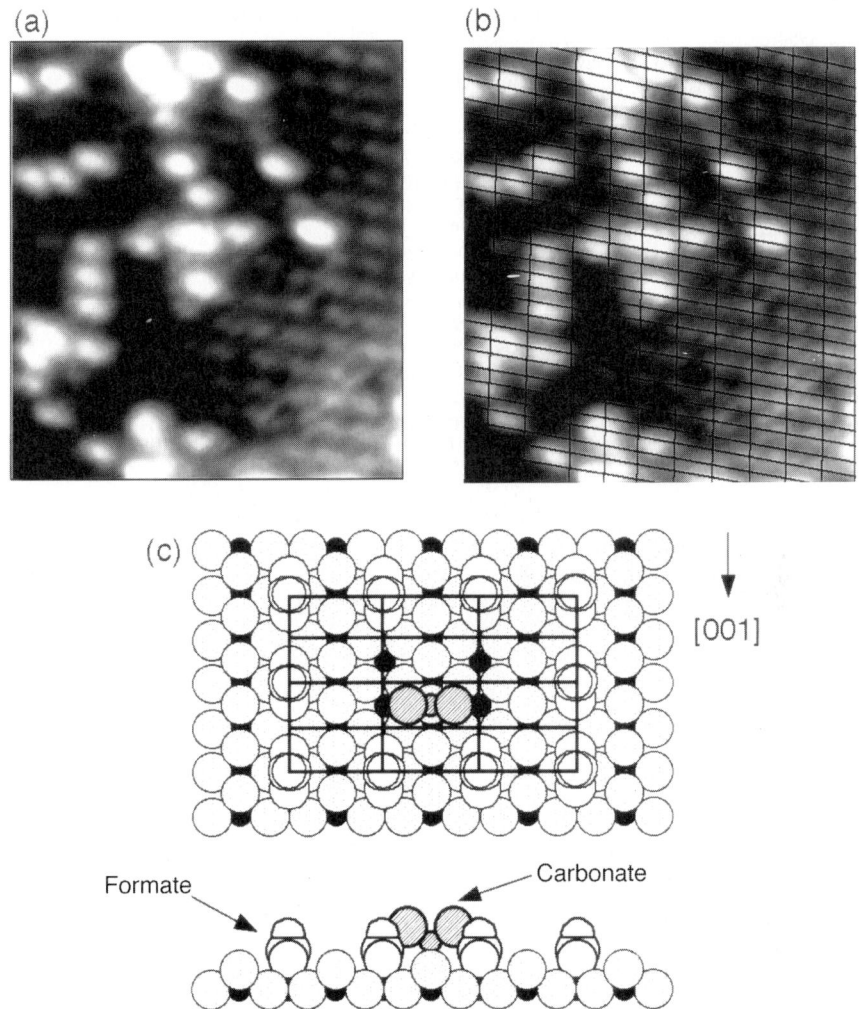

Fig. 6 A zoomed-in STM topography of the reacted TiO_2 surface shown in Fig. 2. (a) Raw image, scan area: 7×8 nm^2, sample bias voltage: $+2.0$ V, tunneling current: 0.1 nA. (b) The (1×1) lattices of the TiO_2 surface, which are scaled on the (2×1) order of the formate monolayer, are imposed with solid lines. (c) A model of carbonate as the product particle. Top and side views are shown.

produced on a Na-deposited $TiO_2(110)$ surface exposed to CO_2.[21] Carbonate formation[22] and CO_2 release[15] have been observed on formate-covered TiO_2 surfaces during heating.

Acknowledgement

This work was supported by Core Research for Evolutional Science and Technology (CREST) of the Japan Science and Technology Corporation (JST).

References

1 Y. Iwasawa, *Stud. Surf. Sci. Catal.*, 1996, **101**, 21.
2 Y. Iwasawa, *Catal. Surveys Jpn.*, 1997, **1**, 3.
3 H. Onishi, K. Fukui and Y. Iwasawa, *Bull. Chem. Soc. Jpn.*, 1995, **68**, 2447.
4 K. Fukui, H. Onishi and Y. Iwasawa, *Phys. Rev. Lett.*, 1997, **79**, 4202.

5. V. E. Henrich and P. A. Cox, *The Surface Science of Metal Oxides*, Cambridge University Press, Cambridge, 1994.
6. K. Fukui, H. Onishi and Y. Iwasawa, *Chem. Phys. Lett.*, 1997, **280**, 296.
7. H. Onishi and Y. Iwasawa, *Langmuir*, 1994, **10**, 4414.
8. H. Onishi, K. Fukui and Y. Iwasawa, *Colloids Surf., A*, 1996, **109**, 335.
9. H. Onishi, Y. Yamaguchi, K. Fukui and Y. Iwasawa, *J. Phys. Chem.*, 1996, **100**, 9582.
10. H. Onishi and Y. Iwasawa, *Phys. Rev. Lett.*, 1996, **76**, 791.
11. H. S. Taylor, *Proc. R. Soc. London Ser. A*, 1925, **108**, 105; *J. Am. Chem. Soc.*, 1931, **53**, 578.
12. S. Suzuki, Y. Yamaguchi, H. Onishi, K. Fukui, T. Sasaki and Y. Iwasawa, *Catal. Lett.*, 1998, **50**, 117.
13. H. Onishi and Y. Iwasawa, *Jpn. J. Appl. Phys.*, 1994, **33**, L1338.
14. H. Onishi and Y. Iwasawa, *Chem. Phys. Lett.*, 1994, **226**, 111.
15. H. Onishi, T. Aruga and Y. Iwasawa, *J. Catal.*, 1994, **146**, 557.
16. G. A. Somorjai, *Introduction to Surface Science and Catalysis*, Wiley, New York, 1994.
17. T. Zambelli, J. Wintterlin, J. Trost and G. Ertl, *Nature (London)*, 1996, **273**, 1688.
18. Q. Guo, I. Cocks and E. M. Williams, *J. Chem. Phys.*, 1997, **106**, 2924.
19. S. Thevuthasan, G. S. Herman, Y. J. Kim, S. A. Chambers, C. H. F. Peden, Z. Wang, R. X. Ynzunza, E. D. Tober, J. Morais and C. S. Fadley, *Surf. Sci.*, 1998, **401**, 261.
20. B. E. Hayden, A. King and M. A. Newton, *J. Phys. Chem. B*, 1999, **103**, 203.
21. H. Onishi, T. Aruga, C. Egawa and Y. Iwasawa, *J. Chem. Soc., Faraday Trans. 1*, 1989, **85**, 2597.
22. K. S. Kim and M. A. Barteau, *J. Catal.*, 1990, **125**, 353.

Paper 9/03106E

Scanning tunnelling microscopy studies of the reactivity of the $TiO_2(110)$ surface: Re-oxidation and the thermal treatment of metal nanoparticles

R. A. Bennett, P. Stone and M. Bowker*

Department of Chemistry, University of Reading, Reading, UK RG6 6AD.
E-mail: M.Bowker@Reading.ac.uk

Received 10th May 1999

We have employed variable temperature scanning tunnelling microscopy (STM) to probe the surface structure of the $TiO_2(110)$ surface with clean, adsorbate and metal covered terminations. The aim of the work is to understand the nature of catalysis on supported metal oxide catalysts for which a good model is an admetal on a single crystal oxide surface. For Pd overlayers, annealing in vacuum shows the formation of metal particles with nanometer sized dimensions, which are comparable to those seen in real catalysts. The clean $TiO_2(110)$ surface has two commonly observed terminations, the (1 × 1) bulk truncation, the (1 × 2) reduced and reconstructed surface. Less commonly, for very reduced crystals, the formation of ordered defects occurs leading to crystallographic shear planes. We have explored all of these surfaces by low energy electron diffraction (LEED) and STM to provide structural information, while we have employed dynamic imaging of the surface in reactive conditions at elevated temperature to assess the chemistry. We find that oxygen rich atmospheres promote a re-growth of the surface that has important consequences for the surface chemistry and morphology. The oxidation and reduction of the support in this system has been shown to modify the reactive properties of the supported metal and we relate our observations to the strong metal support interaction (SMSI).

Introduction

The surface structure of $TiO_2(110)$ has recently been the subject of much interest due to the wide use of titania industrially as a white pigment in paint and cosmetics, as a support for photocatalysts and as a biocompatible interface for medical implants. The behaviour of these disparate functions are critically determined by the surface properties of the oxide, which, as we shall show, are dependent upon bulk defect concentrations. The ability to prepare conducting TiO_2 samples in UHV has led to a wealth of studies on its surface structure and chemistry.[1–16] However, there has been relatively little work exploiting the effect of the variable bulk stoichiometry of reducible oxides on surface structure and reactivity.[9] We have therefore investigated the reactivity of reduced samples at elevated temperatures in UHV conditions. We have identified a novel re-oxidation scheme with strong temperature dependences. Furthermore, we are particularly interested in the use of TiO_2 as a support for (photo)catalysts. To this end we have prepared model catalysts of Pd nanoparticles on $TiO_2(110)$ surfaces and run the same reactions to investigate the influence of the supported particles. We find temperature-dependent enhancements in the rate of re-oxidation around the Pd particles, which we interpret as spillover of oxygen from metal to support. In addition, the nature of the re-oxidation scheme de-activates the Pd particles.

Experimental

STM experiments were performed using an Oxford Instruments variable temperature STM contained within an UHV chamber equipped with additional facilities for Ar^+ ion sputtering, low energy electron diffraction (LEED) and Auger electron spectroscopy (AES), described in detail elsewhere.[17] The chamber was ion pumped to produce a typical base pressure of 1×10^{-10} mbar. $TiO_2(110)$ crystals (PIKEM, UK) were repeatedly sputtered (600 eV, RT) and annealed (1200 K) to initially produce near stoichiometric light blue crystals, which repeatedly produced (1 × 1) LEED patterns after annealing (10 s, 1270 K). In time, continued sputter–anneal[18] cycles further reduced the sample, which then displayed a dark-blue colour, and (1 × 2) reconstructions after annealing. Crystals displaying Magneli phases resulting from crystallographic shear plane formation could be manufactured by prolonged annealing at 1300 K at which point they appeared dark blue/black.[5]

The sample was heated radiatively using a tungsten filament situated close to the rear of the sample to attain temperatures <1000 K while scanning. E-beam heating by biasing the filament was used to reach the extreme temperatures for annealing. Pd was deposited from a home-built source composed of high purity Pd wire tightly wrapped around a resistively heated W wire. Pd was deposited onto the TiO_2 surface at 673 K and subsequently annealed at 773 K for 15 min.

Results and discussion

Surface structures

Early STM studies of TiO_2 reported a variety of interpretations of the images but more recently the dominant contributions to the tunnelling have been established.[8,9] In normal state images the empty states are of 3d character and so Ti^{4+} ions are imaged whereas the bridging O atoms, which are geometrically closer to the tip, appear dark. Fig. 1 shows the (1 × 1) termination, the image contrast is dominated by tunnelling into the 5-fold coordinated Ti sites, which appear as rows in the ⟨001⟩ direction. In the following images the tunnelling parameters are such that the detail imaged within the surface structures is dominated by tunnelling into the Ti 3d states, i.e., positive sample bias.

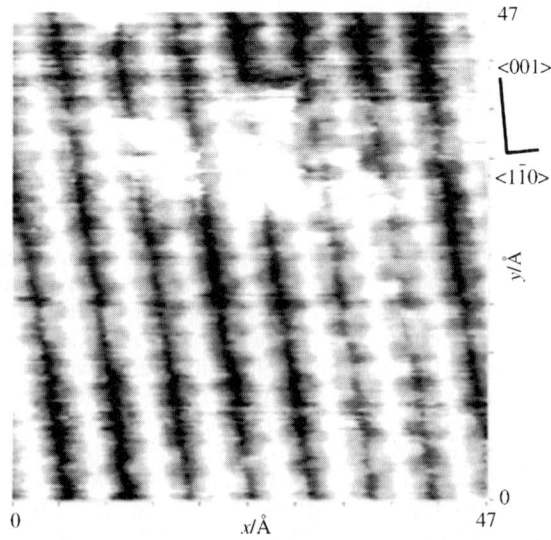

Fig. 1 Small area scan (47 Å square) showing atomic resolution on the (1 × 1) terminated $TiO_2(110)$ surface (1 nA, 1000 mV).

On reduction of the crystal the surface reconstructs to form a (1 × 2) termination that we have recently shown can exist as two distinct reconstructions, which occur for differing levels of reduction. For low levels of reduction added rows of stoichiometry Ti_2O_3 are formed extending from step edges, Fig. 2A, which generate a true (1 × 2) surface reconstruction.[2,3] The rows have a low overall corrugation (≈ 1–1.5 Å), are centred on the bright rows of the (1 × 1) (*i.e.*, the 5-fold co-ordinated Ti) and are relatively unreactive to oxygen exposure. For higher levels of reduction a cross-linked (1 × 2) reconstruction is formed in which the rows running in the $\langle 001 \rangle$ direction are periodically linked roughly ≈ 12 lattice units in the $\langle 001 \rangle$ direction, Fig. 2B.[6,7,11,19] These links most commonly occur as cross-shaped features in the trough between rows, Fig. 2C. The reconstruction is an added row of TiO_2, again centred on the 5-fold co-ordinated Ti of the underlying (1 × 1) surface and displays a larger corrugation in STM, which is close to that expected for a normal step height (≈ 3.1 Å).[19] These surfaces display excellent LEED patterns, Fig. 2D, with extra spots due to the ordered array of cross-links. It is interesting to note that the two (1 × 2) reconstructions have markedly different accommodations of antiphase domain boundaries running in the $\langle 1\bar{1}0 \rangle$ direction, the cross-linked (1 × 2) reconstruction forms antiphase domain walls composed of paired cross-links, Fig. 3, whereas the added Ti_2O_3 type (1 × 2) reconstruction terminates domains with bright points.[20] Both types of reconstruction have similar $\langle 001 \rangle$ directed

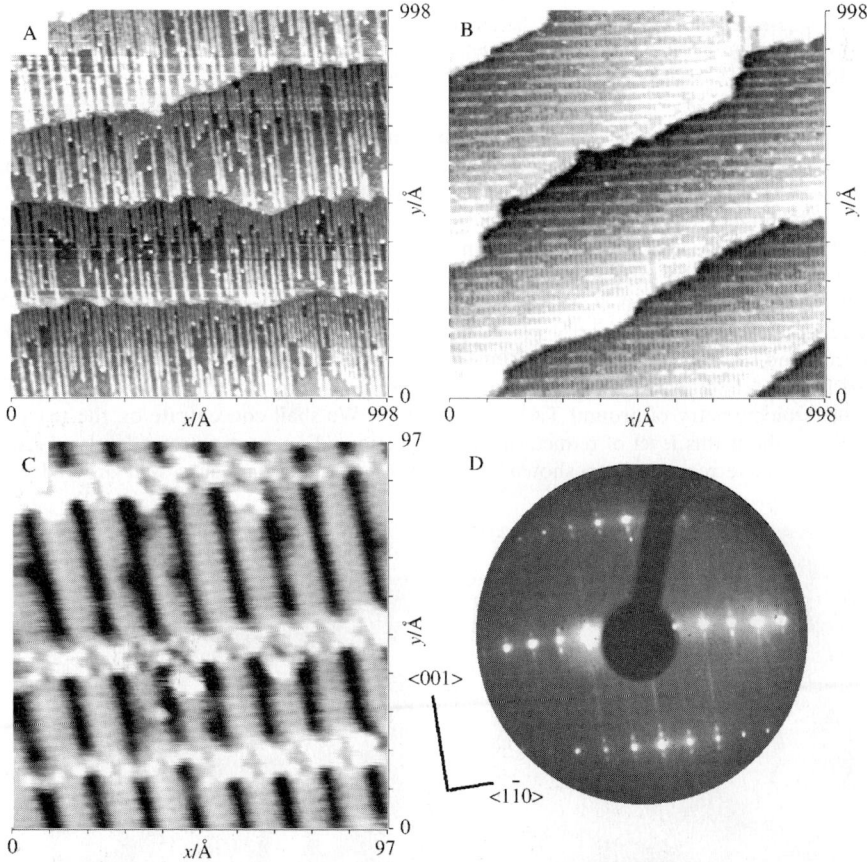

Fig. 2 The two (1 × 2) reconstructions on the reduced TiO_2 surface. A, Added rows of Ti_2O_3 growing out from step edges, which are formed for small departures from stoichiometry (1 nA, 1000 mV). B, Cross-linked added rows of TiO_2 formed on the surface for departures from stoichiometry $\approx TiO_{2-x}$, $x \approx 10^{-4}$. The cross-links form a well ordered array separated by 12 unit cells in the $\langle 001 \rangle$ direction (0.1 nA, 1000 mV). C, Close up of the cross-linking structure showing double rows of Ti within each added row which are periodically cross-linked in the $\langle 1\bar{1}0 \rangle$ direction by bright features in the trough between rows (1 nA, 1000 mV). D, LEED pattern with distinctive 12th order spots and streaking in the $\langle 001 \rangle$ direction.

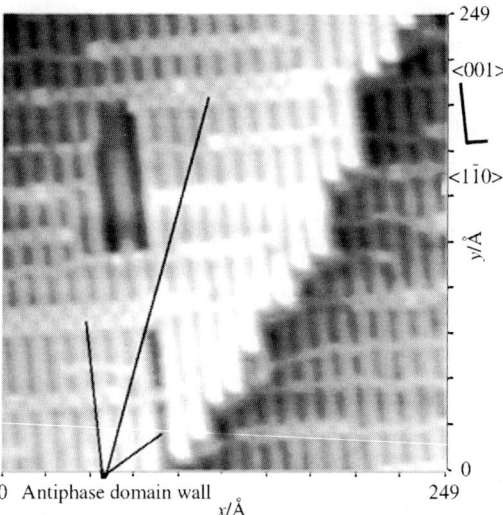

Fig. 3 Antiphase domain boundaries on the cross-linked (1 × 2) surface. Domain walls appear as paired cross-links running in the ⟨1̄10⟩ direction. Domain walls running in the ⟨001⟩ direction have an extra unit cell width separation between added rows (0.1 nA, 1000 mV).

boundaries with a single lattice spacing gap appearing in the ⟨1̄10⟩ direction between the added rows. The cross-linked (1 × 2) surface also has occasional links that bridge these antiphase domain boundaries. This wide bridging is more prevalent when the surface is re-oxidising.

For slightly higher levels of reduction the cross-linked (1 × 2) reconstruction remains but is accompanied by the formation of defects running diagonally across the images, Fig. 4A, B. These are the beginnings of the formation of crystallographic shear (CS) planes, which arise from the ordered clustering of defects within a d^0 metal oxide (TiO_2, V_2O_5, MoO_3 and WO_3) crystal.[1,21,22] The formation of CS planes is also seen in real catalyst TiO_2 supports after reduction[23] and so we believe that using the TiO_2 at this stoichiometry represents a good support material for model catalysts. The formation of CS planes occurs in reduced crystals with departures from stoichiometry of around TiO_{2-x}, $x > 10^{-4}$. We shall concentrate on the reactivity of surfaces with about this level of reduction in later sections. For more extreme levels of reduction the CS planes have recently been shown to terminate at the surface in a well ordered array of

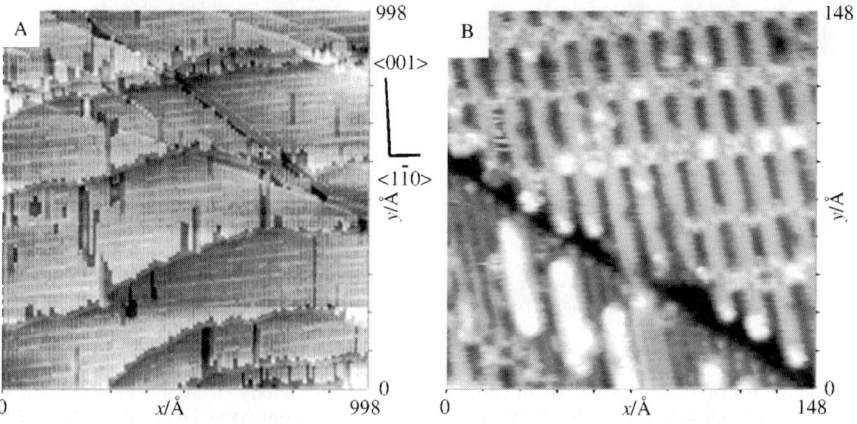

Fig. 4 Crystallographic shear plane pairs, running diagonally across the image, formed on the cross linked surface by slow cooling from 1200 K. Narrow strips of (1 × 1) terminated surface appear between the CS plane pairs. A, Large area image (1000 Å square) showing variety of directions that CS planes may take (0.2 nA, 1000 mV). B, Close up (150 Å square) of (1 × 1) strip between planes (1.5 nA, 1000 mV).

half-height steps[5,24-26] on $TiO_2(110)$. Such ordering of the CS planes produces Magneli phases of variable stoichiometry depending upon separation of the planes and produce complex LEED patterns.[5]

Reduction

Vacuum annealing is well known to reduce TiO_2 surfaces and is commonly applied to produce suitably conductive crystals for STM imaging. In Fig. 5 we show three images of the cross-linked (1 × 2) surface maintained at 1000 K in UHV, taken at time $t = 0$, 1980 s and 2580 s. Towards the top right of all three images are a small group of pairs of CS planes that are used as markers in the images as they remain fixed in position (they are surface terminations of extended bulk defects). The cross-links of the (1 × 2) are not apparent as at this temperature they are mobile on the timescale of scanning. Analysis of the images shows that the step edges move and change shape. Material is predominantly lost from step edges parallel to the $\langle 1\bar{1}0 \rangle$, with movement of $\langle 001 \rangle$ edges arising from removal of the (1 × 2) rows from the $\langle 1\bar{1}0 \rangle$ ends. Some TiO_2 is redistributed and extends from the step edges in the vicinity of the shear planes which may be due to an enhanced stability of steps that incorporate a shear plane displacement, such that the step

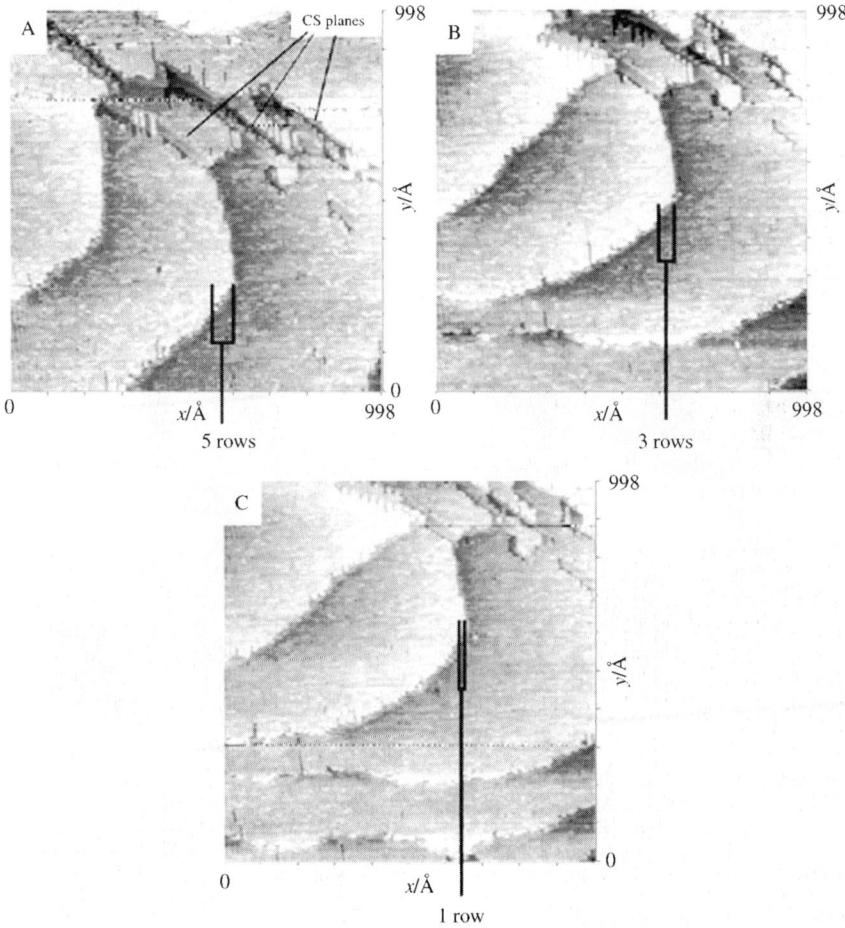

Fig. 5 Sequential images (1000 Å square) taken at 1000 K of the cross-linked (1 × 2) surface reducing by loss of oxygen to the vacuum. A, Time $t = 0$ showing CS plane group and several step edges (0.1 nA, 1500 mV). B, Time $t = 1980$ s showing movement of step edges predominantly in the $\langle 001 \rangle$ direction (0.1 nA, 1500 mV). C, Time $t = 2580$ s showing continued movement of step edges and the 1-dimensional shortening of the added row exposed on the $\langle 001 \rangle$ directed step edge (0.1 nA, 1500 mV).

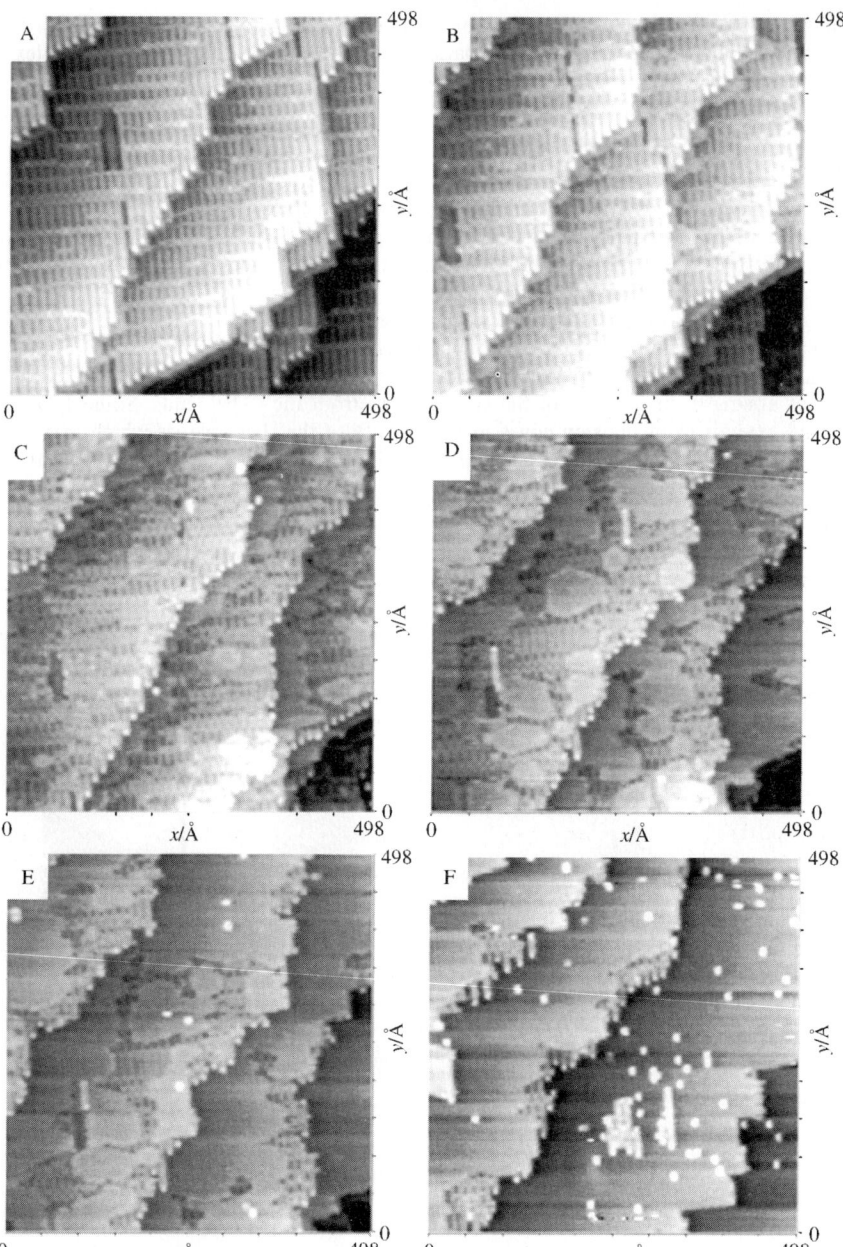

Fig. 6 Sequence of images showing re-oxidation of the cross-linked surface maintained at 673 K in the presence of $\approx 4 \times 10^{-7}$ mbar O_2. A–D, Cross-links increase in number and aggregate to form small islands of (1×1) within the (1×2) terraces. E–F, (1×1) islands coalesce to form extended (1×1) terraces upon which bright points form. These points grow to form rows on the surface which cluster to form a cross-linked (1×2) surface. Tunnelling conditions: A, B and F (0.1 nA, 1000 mV), C and D (0.1 nA, 1500 mV) and E (0.1 nA, 1400 mV).

edges have $\frac{1}{2}$ the normal height. The average rate of retraction of the $\langle 1\bar{1}0 \rangle$ step edges, calculated from these images, is $\approx 0.021 \pm 0.006$ Å s^{-1}.

The net loss of material from the surface implies that oxygen vacancies are not the only defects introduced by vacuum annealing. The production of oxygen vacancies which diffuse into the bulk (and consequently the diffusion of oxygen to the surface) would maintain the surface structure and therefore no loss of material would be seen. However, we have measured a loss of material which implies removal of both O and Ti ions from the surface. Oxygen loss may be accommodated by vacancy formation but as the surface reconstruction is not changed the Ti must dissolve into the bulk and form interstitial Ti ions. Further evidence for the presence of interstitial Ti ions is presented below.

Re-oxidation

Fig. 6 shows a sequence of images of the same area taken at 673 K (the images presented are only a selection taken from the full sequence of over 100 frames) showing the reaction of a cross-linked (1 × 2) surface with a low pressure of oxygen admitted to the chamber (the oxygen pressure varies slightly throughout the sequence but is approximately 4×10^{-7} mbar). Fig. 6A shows the surface prior to reaction with ordered cross-links, antiphase domain boundaries and a small depression of (1 × 1) within the terrace which will act as a marker. The reaction begins with an increase in number of the cross-links between the rows, which also begin to bridge the antiphase domain boundaries directed in the $\langle 001 \rangle$ direction, Fig. 6B. Additionally small areas of (1 × 1) have grown out from the step edges into the lower (1 × 2) terrace. As the cross-links increase in density, islands of (1 × 1) nucleate and grow *within* the (1 × 2), Fig. 6C. Many (1 × 1) islands form on the terraces and grow preferentially in the $\langle 001 \rangle$ direction along the cross-links. These islands coalesce to form larger islands, Fig. 6D, which spread to cover the terraces. At this temperature the (1 × 1) nearly completes a layer before the next stage of the re-oxidation is observed. In this second phase bright points appear *on top* of, and towards the centre of, the (1 × 1) islands, Fig. 6E. The bright points are stable and act as nucleation centres for the formation of bright rows that grow in the $\langle 001 \rangle$ direction, Fig. 6F. The regions in which these rows aggregate form a (1 × 2) surface on top of the (1 × 1) surface. Neighbouring (1 × 2) rows cross-link to form regions with the initial surface termination which spreads across the terrace. The reaction now proceeds in a cyclic fashion with small (1 × 1) islands nucleating and growing within the (1 × 2) terraces. It is possible to grow many layers in this way.

The most striking feature of this re-oxidation reaction is that the surface clearly grows. Step edges move out across the lower terraces, troughs in the (1 × 2) structure fill in to form the (1 × 1) termination, new growth is seen on top of the (1 × 1) terraces in the form of rows which aggregate to form a new (1 × 2) surface. If oxygen vacancies were the dominant defect structure, re-oxidation would simply replace the vacancy with oxygen without growth of the surface. Growth necessitates the incorporation of Ti into the new layer. Thus Ti interstitials, which were distributed throughout the crystal during reduction, are captured by the ambient oxygen as they diffuse to the surface region where they are incorporated into the growing surface.

Nanoparticle reaction

Re-oxidation and reduction cycles are fundamental processes in catalysis, the most familiar being the so called Mars van Krevelen mechanism, whereby oxygen is inserted from an oxide catalyst into a reactant to form the product. The resulting oxygen vacancy is healed by adsorption of oxygen from the gas or adsorbed phases. In the latter case oxygen atoms may be supplied from a second component in the catalyst system and this is termed spillover. Spillover of reactants from one component to another can result in enhanced reactivity due to synergy of action.[27]

We have used nanoscale Pd particles supported on reduced $TiO_2(110)$ as a model catalyst surface and to investigate spillover by the re-oxidation of the support. Fig. 7 shows images of a Pd nanoparticle covered, (1 × 2) reconstructed, $TiO_2(110)$ surface maintained and imaged at 573, 673 and 773 K (each taken from a sequence of over 40 images). The first panel at each temperature shows the particles just before admission of oxygen at $\approx 2 \times 10^{-7}$ mbar. The second panel in the group shows each surface at a later time during the re-oxidation reaction with the final state of the surface being displayed in the last panel.

Starting with the experiment at 573 K, panel A shows three particles on the surface, two at step edges and one in the terrace. After admission of the gas the terrace slowly reacts to form small (1 × 1) islands within the (1 × 2) terraces. These small (1 × 1) islands then promptly develop

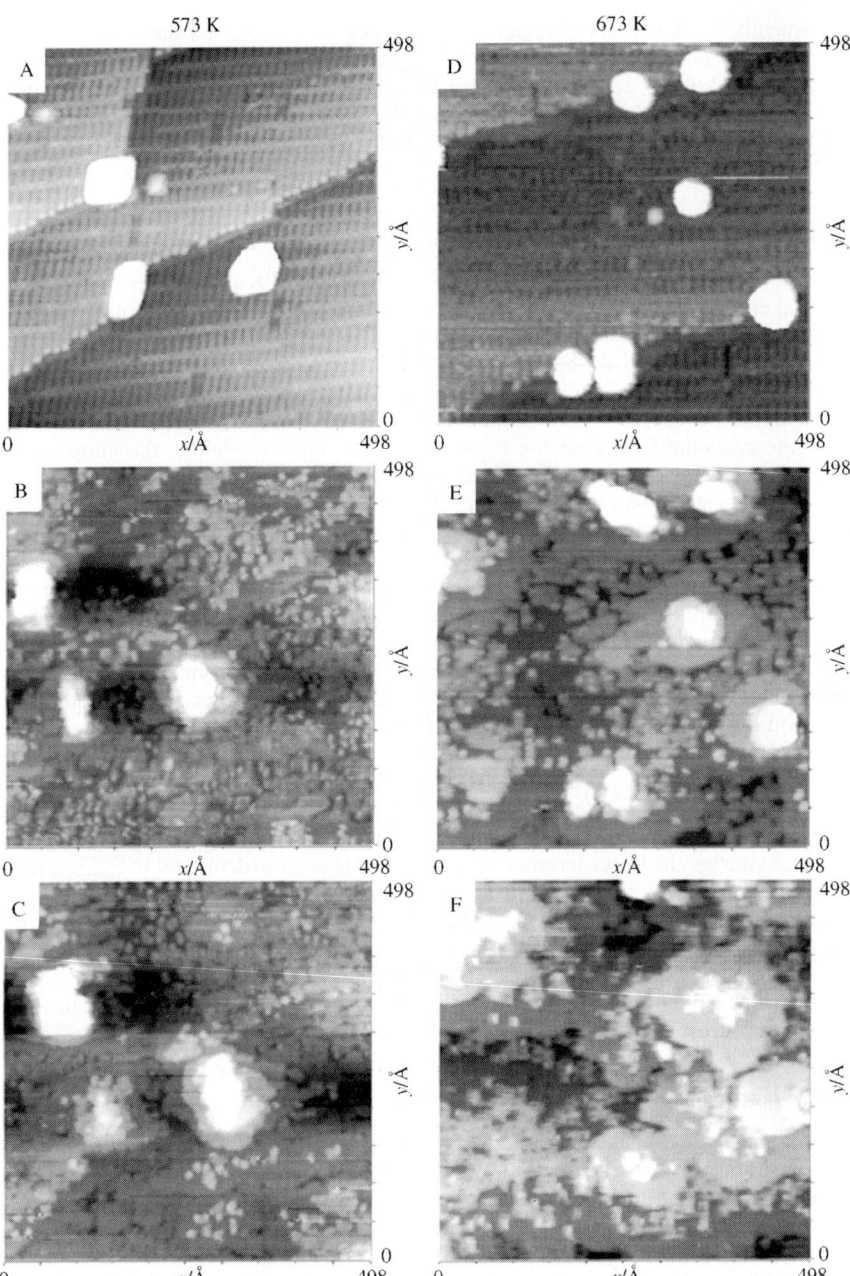

Fig. 7 Three sequences of images taken at 573 K (A–C), 673 K (D–F) and 773 K (G–I) showing the influence of Pd nanoparticles on the re-oxidation of the surface.† At 573 K (A–C) TiO_2 preferentially regrows close to the particles and completely encapsulates them leaving raised features above the terrace, which itself only grows slowly. At 673 K (D–F) particles again become buried but the region of Pd enhanced re-growth spreads out a substantial distance from the particle resulting in raised terraces with (1 × 1) terminations. The surface far from the particles grows slower. At 773 K (G–I) the terrace grows in a layer-by-layer fashion with little preferential growth around the particles. Tunnelling conditions: A (0.1 nA, 1000 mV), B and C (0.1 nA, 2000 mV), D (0.1 nA, 1500 mV), E (0.1 nA, 2000 mV), F (0.1 nA, 1000 mV) and G–I (0.1 nA, 1500 mV).

† An STM movie corresponding to panels D–F is available as electronic supplementary information. See http://www.rsc.org/suppdata/fd/1999/267

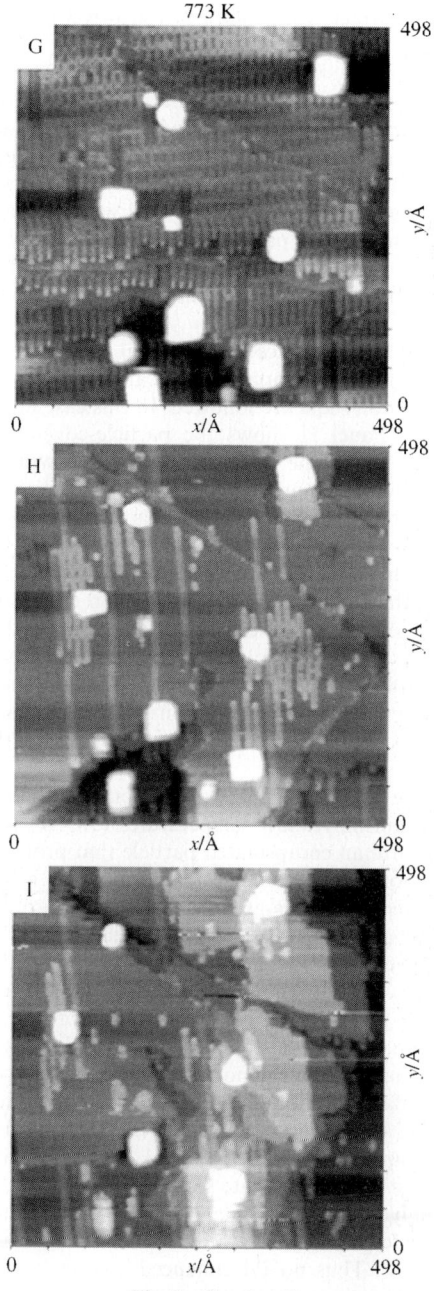

Fig. 7 Continued.

bright points on top of them as another layer begins to grow. This behaviour is identical to that seen for the reduced surface, re-oxidising with no Pd present. By panel B the step edges are obscured by the disordered re-growth. However, the regions closely surrounding the Pd particles appear to have grown faster with the appearance of steps leading up to the Pd particles. The small particle to the lower left is almost buried. In the last panel the same particle has been buried by the growth of the titania while the particle in the centre of the terrace has clearly experienced enhanced growth in the area ≈ 50 Å around the particle. There are ≈ 5 layers of growth up the side on this particle from the terrace, which itself has grown 3 layers since the beginning of the reaction, giving a total of 8 layers growth at the particle.

Panels D–F show images from a similar sequence taken at 673 K. Far from the particles the surface behaves as described above with the sequential growth of (1 × 1) terraces and the formation of bright points on the newly created terraces which aggregate to form a new (1 × 2) layer. However, panel E shows that the reaction leads to a more pronounced peripheral growth around each particle. This extends ≈ 100 Å from each particle until the peripheries merge to produce a (1 × 1) terrace. The new terraces which form and grow outwards from the particles always display the (1 × 1) termination. By panel F the particles have preferentially grown TiO_2 around themselves until they have become buried. In total there has been ≈ 7 layers of growth around the particle and this occurs approximately 16 times as fast as the growth on the non-Pd-covered surface at this temperature.

In the final three panels, G–H, the surface is reacting at 773 K with 2×10^{-7} mbar O_2. Panel G shows many particles supported on a surface, which also displays several CS planes. Again the reaction is started and the reaction sequence followed. In this case little growth is observed preferentially around the particles. Panel H shows the particle-covered surface after the complete growth of the first (1 × 1) layer and the beginnings of the growth of a new (1 × 2) layer. The (1 × 2) rows nucleate and grow in a similar manner to the particle free surface although there seems to be a slight preference for nucleating at the particle. Interestingly the nucleation of the growth at the particle differs at this temperature to the experiments at lower temperature in that (1 × 1) terraces do not form preferentially at the particle. At 773 K the growth is always of the (1 × 2) rows, which suggests that the particles are nucleation centres for growth but do not contribute an excess of spillover oxygen, which can be incorporated to directly form (1 × 1) terraces. In panel I the smaller particles have been covered over by TiO_2 growth (≈ 3 layers). The regions surrounding the particles show little preferential growth but do show the slight preference for nucleating the new layers growth as (1 × 2) rows. Continuing this reaction at higher oxygen pressures buries the nanoparticles on the surface but the growth rate at the particles periphery is the same as on the clean surface.

The observation of three regimes reactivity as a function of temperature for this system allows some discussion of the mechanism. At low temperatures (573 K) the Pd nanoparticles rapidly become covered by TiO_2 to form an encapsulated particle that protrudes from the terrace. At 673 K TiO_2 preferentially grows outwards from the nanoparticle in a layer-by-layer fashion. Eventually the particles are covered by the re-growth. The rate of growth due to the Pd particles presence is ≈ 16 times as fast as that observed for the clean surface and so the surface shows a roughened terrace structure over each particle. At 773 K there is little or no preferential growth of (1 × 1) at the particles. However, the nanoparticles do appear to nucleate the formation of the next (1 × 2) layer which then grows in the same fashion as on the clean surface. The particles again become covered but due to the layer-by-layer growth, which is not preferential around the particles, flat terraces result that display the normal TiO_2 structure. These observations, and the departures from the clean surface reaction, can be rationalised within a scheme which predominantly considers the action of oxygen adsorption and desorption on the Pd particles. At 773 K oxygen desorbs rapidly from single crystal Pd(111) (peak ≈ 800 K[28]) and so with the small oxygen gas phase pressure we expect a low steady state surface coverage of oxygen on the Pd nanoparticles (Pd particles preferentially display low index facets, predominantly (111)[29]). The low steady state coverage on the particle implies that the rate of spillover of oxygen atoms from the particles to the support will also be low. Thus no Pd enhanced growth is observed as the entire flux of oxygen to the surface is directly from the gas phase and the reaction scheme of the clean surface is followed. In contrast, at 673 K, oxygen cannot desorb rapidly from the Pd nanoparticle and a large steady state coverage is produced. However, the oxygen atoms on the Pd can make excursions off the particle and diffuse over the neighbouring TiO_2 surface until they capture a Ti interstitial and become incorporated into the growing layer. This produces an enhancement in growth rate over that of the clean surface as the sticking probability of oxygen on Pd is over an order of magnitude greater than on the reduced TiO_2 surface. Thus the Pd particles supply oxygen atoms to the TiO_2 which allows the growth of the fully oxygen terminated (1 × 1) structure. At 573 K the Pd particles remain a reservoir of oxygen, however, the lower temperature reduces the extent of diffusion off the particle and onto the support. Furthermore, the lower temperature results in a less extended diffusion of oxygen on the TiO_2 and so the growth is restricted to the area in the immediate vicinity of the nanoparticle.

Conclusions

We have shown that non-stoichiometry in TiO_{2-x} samples is accommodated by the dissolution of Ti interstitials into the bulk of the rutile crystal structure. During re-oxidation these ions are extracted and grow on the surface in their normal crystallographic positions. This has wide ranging implications for catalysis and gas sensing since in both areas models of reactivity and sensing ability are based on descriptions of the formation and annihilation of oxygen vacancies. We have shown that titanium interstitials accommodate non-stoichiometry and are key to accurate descriptions of oxidation and reduction at elevated temperature on this surface. The re-growth behaviour has been used to probe spillover of reactive oxygen species from supported metal particles. In this case we show that the reactivity increase due to spillover oxygen is strongly temperature dependent and controlled by the adsorption/desorption from the Pd particle. We note that nanoscale particles supported on this oxide can be buried by re-growth of titania and are thus removed from the surface—a kind of strong metal support interaction (SMSI).

Acknowledgements

PS would like to thank Oxford Instruments PLC for the provision of a CASE award and we thank the EPSRC for support.

References

1 G. S. Rohrer, V. E. Henrich and D. A. Bonnell, *Science*, 1990, **250**, 1239.
2 H. Onishi and Y. Iwasawa, *Phys. Rev. Lett.*, 1996, **76**, 791.
3 H. Onishi and Y. Iwasawa, *Surf. Sci.*, 1994, **313**, L783.
4 D. Novak, E. Garfunkel and T. Gustafsson, *Phys. Rev. B: Condens. Matter*, 1994, **50**, 5000.
5 R. A. Bennett, S. Poulston, P. Stone and M. Bowker, *Phys. Rev. B: Condens. Matter*, 1999, **59**, 10341.
6 M. Sander and T. Engel, *Surf. Sci.*, 1994, **302**, L263.
7 A. Szabo and T. Engel, *Surf. Sci.*, 1995, **329**, 241.
8 U. Diebold, J. F. Anderson, K. O. Ng and D. Vanderbilt, *Phys. Rev. Lett.*, 1996, **77**, 1322.
9 K. O. Ng and D. Vanderbilt, *Phys. Rev. B: Condens. Matter*, 1997, **56**, 10544.
10 P. W. Murray, N. G. Condon and G. Thornton, *Phys. Rev. B: Condens. Matter*, 1995, **51**, 10989.
11 C. L. Pang, S. A. Haycock, H. Raza, P. W. Murray, G. Thornton, O. Gülseren, R. James and D. W. Bullet, *Phys. Rev. B: Condens. Matter*, 1998, **58**, 1586.
12 R. E. Tanner, M. R. Castell and G. A. D. Briggs, *Surf. Sci.*, 1998, **412/413**, 672.
13 M. A. Henderson, *Surf. Sci.*, 1999, **419**, 174.
14 M. A. Henderson, *Surf. Sci.*, 1995, **343**, L1156.
15 W. S. Epling, C. H. F. Peden, M. A. Henderson and U. Diebold, *Surf. Sci.*, 1998, **412/413**, 333.
16 M. Li, W. Hebenstreit and U. Diebold, *Surf. Sci.*, 1998, **414**, L951.
17 M. Bowker, S. Poulston, R. A. Bennett, P. Stone, A. H. Jones, S. Haq and P. Hollins, *J. Mol. Catal. A: Chem.* 1998, **131**, 185.
18 Sputtering while hot (>800 K) greatly increases the rate of reduction of the crystal due to preferential sputtering of surface O atoms and replenishment by diffusion of Ti to the bulk.
19 R. A. Bennett, P. Stone and M. Bowker, *Phys. Rev. Lett.*, 1999, **82**, 3831.
20 R. E. Tanner, M. R. Castell and G. A. D. Briggs, *Surf. Sci.*, 1998, **412/413**, 672.
21 L. A. Bursill and B. G. Hyde, in *Progress in Solid State Chemistry*, ed. H. Reiss and J. O. McCaldin, Pergamon, New York, 1972, vol. 7, p. 177.
22 M. Reece and R. Morrell, *J. Mater Sci.*, 1991, **26**, 5566.
23 S. Bernal, F. J. Botana, J. J. Calvino, C. López, J. A. Pérez-Omil and J. M. Rodríguez-Izquierdo, *J. Chem. Soc., Faraday Trans.*, 1996, **92**, 2799.
24 G. S. Rohrer, V. E. Henrich and D. A. Bonnell, *Surf. Sci.*, 1992, **278**, 146.
25 H. Nörenberg, R. E. Tanner, K. D. Schierbaum, S. Fischer and G. A. D. Briggs, *Surf. Sci.*, 1998, **396**, 52.
26 H. Nörenberg and G. A. D. Briggs, *Surf. Sci.*, 1998, **402–404**, 738.
27 B. Delmon, *Surf. Rev. Letts.*, 1995, **2**, 25.
28 X. Guo, A. Hoffman, J. T. Yates, Jr., *J. Chem. Phys.*, 1989, **90**, 5787.
29 K. Wolter, O. Seiferth, H. Kuhlenbeck, M. Bäumer, H.-J. Freund, *Surf. Sci.*, 1998, **399**, 190.

Paper 9/03731D

Oxygen-induced morphological changes of Ag nanoclusters supported on $TiO_2(110)$

Xiaofeng Lai, Todd P. St.Clair and D. Wayne Goodman*

Department of Chemistry, Texas A&M University, P.O. Box 30012, College Station, TX 77842-3012, USA

Received 8th April 1999

The effect of *in situ* O_2 exposure on $TiO_2(110)$-supported Ag nanoclusters was investigated using X-ray photoelectron spectroscopy (XPS) and scanning tunneling microscopy (STM). An oxygen-induced cluster ripening was observed by STM after $Ag/TiO_2(110)$ was exposed to 10.00 Torr O_2 for 2 h in an elevated-pressure reactor. The Ag clusters exhibit a clear bimodal size distribution after O_2 exposure due to Ostwald ripening: some clusters increase in size while other clusters decrease in size. The cluster density also increased 5–15% after O_2 exposure, indicating redispersion simultaneously occurs with ripening. It is shown that intercluster transport is likely accomplished through the formation of Ag_2O.

1. Introduction

A primary goal of heterogeneous catalysis is to elucidate the relationship between catalyst structure and activity. However, the achievement of this objective is often complicated by the morphological and structural changes that surfaces sometimes undergo at reaction pressures. These adsorbate-induced surface restructurings result from the reorganization of surface atoms near the adsorption site of a chemisorbed atom or molecule. Depending on the strength of the surface–adsorbate bond, the surface may experience different degrees of restructuring, from weak local relaxation to massive transportation of surface atoms. In the case of "real world" catalysts, which often consist of small metal particles supported on oxide substrates, one is not only concerned with changes associated with the substrate surface, but the supported metal clusters as well. Metal nanoclusters contain a relatively small number of atoms, thus the surface atoms can be easily influenced because of their low coordination number, making supported nanoclusters vulnerable to adsorbate-induced morphological changes. If this restructuring of metal clusters results in cluster growth, then active metal surface area has been lost and the catalyst may become deactivated.

The physical and chemical properties of supported metal clusters, and the interactions between these clusters and different substrates, have been widely investigated.[1–9] The scanning tunneling microscope (STM) is particularly suited for studying such interactions because it can directly probe cluster morphology and structure. For example, STM results indicate that exposure of CO pressures between 10^{-3} and 10^{-1} mbar to $Rh/TiO_2(110) - (1 \times 2)$ led to a significant agglomeration of Rh clusters.[10] This phenomenon was attributed to the formation of Rh–CO bonds (185 kJ mol^{-1}) that promote disruption of the weaker Rh–Rh bonds (44.5 kJ mol^{-1}). Similar effects have also been observed for $Ir/TiO_2(110) - (1 \times 2)$.[11]

In a recent study of CO oxidation over model Au catalysts, the well-known structure sensitivity and the stability of Au nanoclusters to CO and O_2 were investigated.[12] STM in conjunction with reaction kinetics measurements revealed that the structure sensitivity was related to a quantum size effect with respect to the thickness of the Au islands: maximum reactivity was observed for Au clusters with diameters of ≈3.5 nm and heights less than 3 atomic layers. It was observed in the course of these CO oxidation experiments that O_2 exposure resulted in sintering of the Au nanoclusters. Fig. 1 shows two STM images of 0.25 monolayer (ML) Au/TiO_2(110) before (top image) and after (bottom image) exposure to 10.00 Torr O_2 for 2 h. The effect of the O_2 exposure was a decrease in cluster density (number of clusters per unit area) and an increase in cluster size, from 2.6 nm × 0.7 nm (diameter × height) to 3.6 nm × 1.4 nm. These results demonstrate the utility of STM for studying supported metal nanoclusters by establishing a correlation between the structural, electronic and reactivity properties of model Au catalysts. Furthermore, given the sintering effect of O_2 on Au nanoclusters, it is expected that other metal nanocluster systems could also be susceptible to O_2-induced morphological changes.

Silver is a metal whose interaction with O_2 environments is of interest primarily because of its use as an industrial catalyst for two important oxidation reactions: ethylene oxidation to ethylene

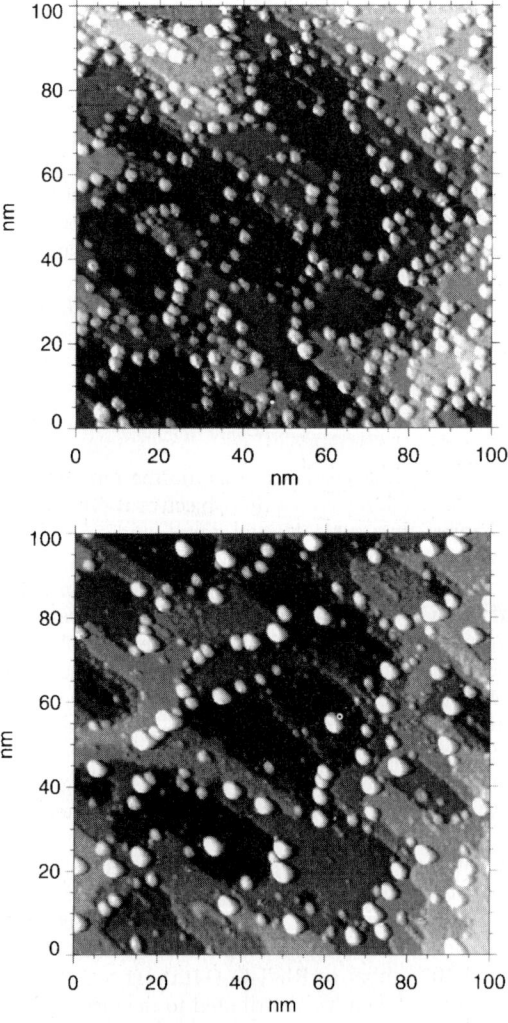

Fig. 1 STM images of Au clusters on TiO_2(110) – (1 × 1) (2.0 V, 2.0 nA). The dosing flux is 0.083 ML min^{-1} and the Au coverage is 0.25 ML. Top: Fresh Au/TiO_2(110) deposited at room temperature and then annealed to 600 K. Bottom: 0.25 ML Au/TiO_2(110) exposed to 10.00 Torr O_2 for 2 h at 300 K. The O_2 exposure results in decreased cluster density and increased cluster size.

epoxide and methanol oxidation to formaldehyde.[13-15] To gain additional insight into the role that O_2 plays in cluster growth, Ag clusters deposited on a $TiO_2(110)$ surface were characterized using XPS and STM before and after O_2 exposures.

2. Experimental

Experimental details have been published elsewhere.[6,7,9,12] Briefly, the experiments were performed in a combined elevated pressure reactor-UHV system with a base pressure of 5×10^{-11} Torr. The system is equipped with a double pass cylindrical mirror analyzer (CMA) for Auger electron spectroscopy (AES) and X-ray photoelectron spectroscopy (XPS), reverse view low energy electron diffraction (LEED) optics, and an Omicron UHV-STM. A $TiO_2(110)$ single crystal (Commercial Crystal Laboratories, Inc.), typically prepared by cycles of Ar ion bombardment and vacuum annealing to 1100 K, was found to be sufficiently clean, flat, and conductive for electron spectroscopies and STM studies. Ag clusters were evaporated onto the $TiO_2(110)$ surface from a source containing high purity Ag wire wrapped around a Ta filament that could be heated resistively. The Ag doser was extensively outgassed prior to use. The experimental Ag flux of 0.125 ML min^{-1} was calibrated with a Re(0001) substrate using AES and STM. One ML Ag coverage corresponds to 1.39×10^{15} atoms cm^{-2}.

A combined elevated pressure reactor-UHV system was used for studies of high-pressure gas exposures and reactions. Following preparation and characterization in the primary UHV chamber, the sample was transferred *in situ* into the elevated-pressure reactor through double-stage differentially pumped Teflon sliding seals. This pumping arrangement provides a convenient way of performing elevated-pressure adsorption and reaction kinetics studies in the range of 1×10^{-10} to 1×10^3 Torr, while maintaining UHV pressures in the main chamber.

3. Results and discussion

The top image in Fig. 2 shows an STM constant current topograph (CCT) image of 2.0 ML Ag deposited on $TiO_2(110)$ in UHV conditions at ambient temperature. Three-dimensional (3D) hemispherical Ag clusters with relatively homogeneous sizes were observed, both on flat terraces and step edges. The Ag clusters have an average size of ≈ 4.8 nm $\times$ 2.6 nm (diameter $\times$ height), corresponding to ≈ 1900 atoms per cluster. In addition to the 3D Ag clusters, bare patches of substrate were also visible, indicating an island growth mode (Volmer–Weber growth mode) for Ag on $TiO_2(110)$, consistent with our previous reports for other transition metals on $TiO_2(110)$.[6,7,9,12] XPS results (not shown) indicated a Ag $3d_{5/2}$ binding energy (E_b) of 368.1 eV, consistent with metallic silver.

After deposition of 2.0 ML Ag on the $TiO_2(110)$ substrate and subsequent investigation by STM, the sample was transferred to the elevated-pressure reactor and exposed to 10.00 Torr O_2 at ambient temperature for 2 h. The sample was then transferred back to the UHV chamber and examined by XPS and STM. XPS results (not shown) indicated no Ag $3d_{5/2}$ E_b shift after O_2 exposure, consistent with other XPS studies of spent Ag catalysts.[14] The bottom image in Fig. 2 presents an STM topograph of O_2-exposed Ag/$TiO_2(110)$ and clearly reveals that the exposure dramatically affected the cluster sizes. A bimodal size distribution of cluster sizes is apparent, with some clusters increasing in size while others decrease. Furthermore, a small increase (5–15%) in cluster density was observed, indicating that redispersion is occurring as well.

Fig. 3 shows the size distributions calculated from the STM images of Ag clusters before and after high-pressure O_2 exposure. Initially, Ag clusters exhibited a unimodal size distribution from 2.0 to 6.5 nm with a maximum diameter ≈ 5.0 nm. However, after O_2 exposure, a bimodal size distribution was observed, with one size domain from 1.0 to 5.0 nm and another from 5.0 to 11 nm. The smaller clusters in the range 1.0–5.0 nm have a higher density and a narrower size distribution, with an average Ag cluster size of ≈ 3.0 nm $\times \approx 1.1$ nm (≈ 260 atoms). The larger clusters, however, have a lower cluster density and a broader size distribution, with an average size of ≈ 6.7 nm $\times \approx 3.1$ nm (≈ 4200 Ag atoms per cluster). It is noteworthy that the total cluster volume before and after O_2 exposure calculated from the STM images agrees to within 10% error. Since Ag oxidation catalysts are typically used in elevated O_2 pressure conditions, an analysis of these induced morphological changes is of general interest to the catalysis community.

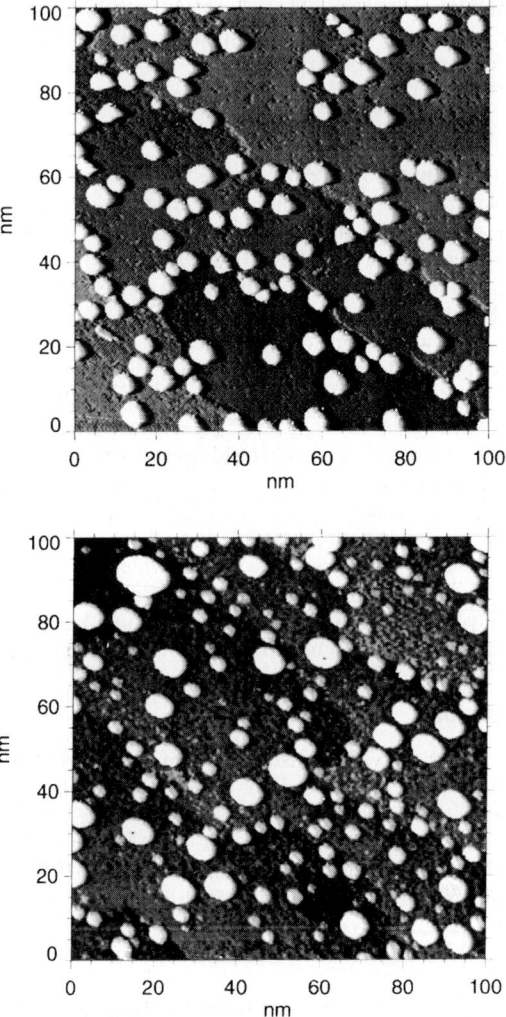

Fig. 2 STM images of Ag clusters on $TiO_2(110) - (1 \times 1)$ (2.0 V, 1.0 nA). The dosing flux is 0.125 ML min^{-1} and the Ag coverage is 2.0 ML. Top: Fresh Ag/TiO$_2$(110) prepared at room temperature. Homogeneous Ag clusters are observed with a uniform size distribution. Bottom: 2.0 ML Ag/TiO$_2$(110) exposed to 10.00 Torr O$_2$ for 2 h at 300 K. The O$_2$ exposure results in a bimodal size distribution: certain clusters increase in size at the expense of other clusters.

In general, cluster growth of supported metal catalysts can proceed by two processes. First, clusters can migrate along the surface until they collide with other clusters, resulting in coalescence. Second, cluster growth can occur by intercluster transport, or Ostwald ripening, which is capillarity driven. In this case, the reduction of the total surface free energy by intercluster transport occurs such that certain clusters grow larger at the expense of other clusters.[4] Thus, in light of the bimodal size distribution observed following O$_2$ exposure, Ostwald ripening is likely the cause of the Ag cluster growth. Regardless of the cause, cluster growth results in catalysts with decreased active surface areas, leading to a decline in catalytic activity.

Intercluster transport of atomic (or molecular) species can occur by either surface diffusion along the substrate or vapor phase transport. Under vacuum or reducing conditions, the transport between Ag clusters can only occur in the form of free metallic Ag atoms, and the driving force should be related to the Ag vapor pressure. However, the Ag vapor pressure depends exponentially on the energy required to break Ag–Ag metal bonds and transfer a Ag atom to the vapor

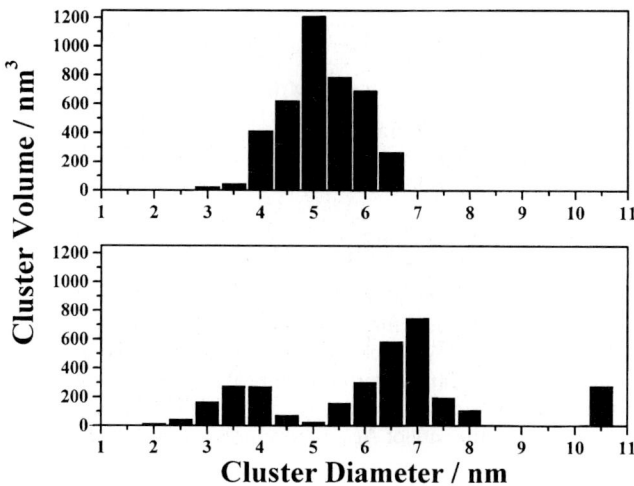

Fig. 3 The size distributions of 2.0 ML Ag/TiO$_2$(110) before and after 10.00 Torr O$_2$ exposure. Top: Fresh Ag clusters with a maximum volume ≈ 5.0 nm cluster diameter. Bottom: Ag clusters after O$_2$ exposure exhibiting a bimodal distribution with average cluster diameters of ≈ 3.0–3.5 nm and 6.5–7.0 nm, respectively.

phase, i.e., the sublimation energy, $\Delta_{subl}H(Ag)$, which is ≈ 285 kJ mol^{-1}. Obviously, such a high energy barrier suggests that intercluster transport by free Ag atoms will be very slow at room temperature. This result is in accordance with our STM observations that Ag clusters are generally stable in UHV conditions.

In an oxidizing environment, the situation is quite different. For example, Wynblatt[16] showed that growth of Pt particles in O$_2$ environments occurred through the formation of volatile PtO$_2$. Platinum oxide has a lower sublimation energy than platinum metal and therefore serves as the mechanism by which intercluster transport occurs. Unfortunately, to the best of our knowledge, no vapor pressure or sublimation energy data is available for silver oxide, rendering it difficult to directly compare such values with Ag metal. However, it will be shown that the formation of Ag$_2$O from Ag particles in 10.00 Torr O$_2$ is at least expected thermodynamically and thus may account for the intercluster transport discussed above.

Thermodynamically, Ag can form silver oxide (Ag$_2$O) by reaction with oxygen at room temperature. This can be illustrated by considering the following simple reaction

$$2Ag(s) + \tfrac{1}{2}O_2(g) \rightarrow Ag_2O(s) \qquad (1)$$

which has a negative standard free energy of formation of Ag$_2$O ($\Delta G_{298} = -11.2$ kJ mol^{-1}) at room temperature. The equilibrium constant K_P for the above reaction can be expressed as

$$K_P = a_P(a_{Ag})^{-2}(p_O)^{-1/2} \qquad (2)$$

where a_P and a_{Ag} are the activities of Ag$_2$O and Ag, respectively (both values are unity), and p_O is the equilibrium partial pressure of oxygen. Substitution of K_p into the standard ΔG equation yields

$$\Delta G_T = \tfrac{1}{2}RT \ln p_{O_2} \qquad (3)$$

where T is the absolute temperature and R is the universal gas constant. Ag oxidation can only occur under the present conditions if the partial pressure of O$_2$ in the cluster environment is higher than the p_{O_2} value in eqn. (4) for T = 298 K. By substituting $\Delta G_{298} = -11.2$ kJ mol^{-1} into eqn. (4), the equilibrium oxygen partial pressure p_{O_2} was calculated to be 1.23×10^{-4} atm (0.094 Torr). Since the partial pressure in the cluster environment (10.00 Torr) is higher than the equilibrium oxygen partial pressure at 300 K, then the oxidation of bulk Ag is thermodynamically allowed.

An additional effect that that must be considered is the influence of the Ag cluster curvature on the free energy. The decrement Δg in free energy due to cluster curvature is given by

$$\Delta g = 2\sigma M/(\rho r) \quad (4)$$

where σ is the surface energy, M is the atomic weight, ρ is the density and r is the cluster curvature radius.[17] At room temperature, taking $\sigma = 1400$ erg cm^{-2} [18] and $\rho = 10.5$ g cm^{-3}, $\Delta G_{298}(r)$, the standard free energy of formation of Ag$_2$O for Ag clusters with curvature radius r (in nm), is given by

$$\Delta G_{298}(r) = \Delta G_{298} - \Delta_g = -11.2 - 28.8/r \text{(kJ mol}^{-1}) \quad (5)$$

For an average cluster diameter of 5.0 nm ($r = 2.5$ nm), the value of $\Delta G_{298}(r)$ is doubled ($\Delta G_{298}(r) = -22.7$ kJ mol^{-1}). As the cluster size decreases further to 3.0 nm ($r = 1.5$ nm), the absolute $|\Delta G_{298}(r)|$ value increases by a factor of 1.7. Therefore, the driving force for oxidation of Ag nanoclusters is increased significantly at room temperature when accounting for cluster curvature effects. While this analysis shows that Ag$_2$O formation from Ag and O$_2$ is possible at room temperature, information about the *rate* of Ag$_2$O formation is not available.

4. Conclusions

In summary, Ag nanoclusters on TiO$_2$(110) are likely to form Ag$_2$O when exposed to 10.00 Torr O$_2$ at room temperature. Ag atoms in the form of volatile oxide molecules can thus be transported between clusters at higher rates than metallic Ag atoms under vacuum or reducing conditions. In principle, this somewhat volatile oxide may diffuse more easily along the surface of the support, or through the vapor phase, from higher-energy sites to lower-energy sites. Consequently, Ostwald ripening is observed when the Ag nanoclusters are exposed to oxygen. Evidence for redispersion was also found as the cluster density increased slightly following O$_2$ exposure. The STM results provide convincing evidence that the TiO$_2$(110)-supported Ag clusters are exceptionally reactive to O$_2$ and reinforce the notion that nanoclusters are particularly susceptible to adsorbate-induced restructurings.

5. Acknowledgements

We acknowledge with pleasure the support of this work by the Department of Energy, Office of Basic Energy Sciences, Division of Chemical Sciences and the Robert A. Welch Foundation.

References

1. D. R. Rainer, C. Xu and D. W. Goodman, *J. Mol. Catal. A: Chem.*, 1997, **119**, 307.
2. C. T. Campbell, *Surf. Sci. Rep.*, 1997, **27**, 1.
3. P. Wynblatt and N. A. Gjostein, in *Progress in Solid State Chemistry*, ed. J. O. McCaldin and G. Somorjai, Pergamon Press, New York, 1975, **9**, p. 21.
4. P. Wynblatt, R. A. Dalla Betta and N. A. Gjostein, in *The Physical Basis for Heterogeneous Catalysis*, ed. E. Drauglis and R. I. Jaffee, Plenum Press, New York, 1975, p. 501.
5. R. Persaud and T. E. Madey, *Chem. Phys. Solid Surf.*, 1997, **8**, 407.
6. C. Xu, X. Lai, G. W. Zajac and D. W. Goodman, *Phys. Rev. B: Condens. Matter*, 1997, **56**, 13464.
7. X. Lai, C. Xu and D. W. Goodman, *J. Vac. Sci. Technol. A*, 1998, **16**, 2562.
8. M. Valden and D. W. Goodman, *Isr. J. Chem.*, 1998, **38**, 285.
9. X. Lai, T. P. St.Clair, M. Valden and D. W. Goodman, *Prog. Surf. Sci.*, 1998, **59**, 25.
10. A. Berkó, G. Menesi and F. Solymosi, *J. Phys. Chem.*, 1996, **100**, 17732.
11. A. Berkó and F. Solymosi, *Surf. Sci.*, 1998, **411**, L900.
12. M. Valden, X. Lai and D. W. Goodman, *Science*, 1998, **281**, 1647.
13. S. Cheng and A. Clearfield, *J. Catal.*, 1985, **94**, 455.
14. V. I. Bukhtiyarov, I. P. Prosvirin, R. I. Kvon, S. N. Goncharova and B. S. Bal'zhinimaev, *J. Chem. Soc., Faraday Trans.*, 1997, **93**, 2323.
15. V. I. Bukhitiyarov, A. I. Boronin, I. P. Prosvirin and V. I. Savchenko, *J. Catal.*, 1994, **150**, 268.
16. P. Wynblatt, *Acta Metall.*, 1976, **24**, 1175.
17. J. D. Verhoeven, in *Fundamentals of Physical Metallurgy*, Wiley, New York, 1975, p. 169.
18. J. D. Verhoeven, in *Fundamentals of Physical Metallurgy*, Wiley, New York, 1975, p. 202.

Paper 9/02795E

First principles simulations of titanium oxide clusters and surfaces

Tristan Albaret, Fabio Finocchi and Claudine Noguera

Laboratoire de Physique des Solides (UMR CNRS 8502) Bât. 510, Université Paris-Sud, 91405 Orsay, France

Received 19th April 1999

Electronic and structural properties of TiO_2 species of various sizes, charges and stoichiometries, ranging from $Ti_nO_m^x$ clusters ($n = 1$–3, $m - n = 0, 1$, $x = -1, 0, +1$) to bulk rutile and its (110) surface, have been obtained by total energy calculation based on the density functional theory (DFT), in the local density and local spin density approximations (LSDA), and complemented by a Bader-type analysis of the total electronic density. Attention has been focused on the electron distribution to better understand how the ionocovalent character of the Ti–O bonding and the screening properties vary as a function of the size of the system, the atomic coordination and the surface orientation.

1 Introduction

TiO_2 is an important material for reasons of both fundamental interest and potential technological applications. For example, TiO_2 is widely used in catalysis, photocatalysis, protective surface coating, etc.[1] The stoichiometric oxide is a non-magnetic insulator with a gap of about 3 eV. The valence and conduction bands have mixed Ti and O character, revealing the mixed iono-covalent character of this oxide.[2] Titania has a wide range of possible oxygen stoichiometries. Eventually, ordered phases such as Ti_2O_3 or TiO are produced, which are no longer insulating. While TiO is metallic,[3] Ti_2O_3 behaves either as a semi-conductor or a metal, as a function of the temperature. This is often assigned to a Mott–Hubbard-type transition, although its exact nature is still controversial.[4,5]

The $TiO_2(110)$ face is the most stable surface of rutile. It has emerged as one of the most important oxide surfaces, which makes it a model system to explore the surface physics and chemistry of transition metal oxides.[1,6] There is evidence that a simple charge-neutral truncation is the correct structure for the stoichiometric (1 × 1) surface, although the amount of surface relaxation remains controversial.[7] Non-stoichiometry, on the other hand, leads to reconstructed surface structures for which several models have been recently debated.[8] This surface has served as the support for metal deposition and systematic trends as a function of the metal electronegativity have been derived.[9]

Unsupported titanium oxide clusters, have been considered both theoretically and experimentally. Aside from the many works devoted to the TiO molecule,[10,11] and from the experimental study of the TiO_2 molecule,[12] positively charged $Ti_nO_{2n-\delta}^+$ ($n = 1$–7, $\delta = 0$–4) clusters have been investigated with mass spectrometry and collision induced dissociation.[13] Recently, anionic TiO_y^- ($y = 1$–3) molecules and $(TiO_2)_n^-$ ($n = 1$–4) clusters have been produced and studied using anion photoelectron spectroscopy.[14] This has permitted the determination of the electron affinities and excitation gaps of these clusters as a function of their size. From a theoretical point of view, TiO_y molecules ($y = 1–2$) have been studied using various methods.[15] Ab initio Hartree–Fock calculations have been performed on neutral and cationic $Ti_nO_{2n-\delta}$ clusters ($n = 1–3$, $\delta = 0–1$), in order to determine the equilibrium geometries, ionization potentials and vibrational spectra.[16] By using an ab initio DFT-LSDA approach we have obtained preliminary results on the geometry and electronic properties of the lowest energy stoichiometric Ti_2O_4 and Ti_3O_6 isomers.[17]

An open question related to both systems (free clusters and surfaces) is the extent of charge redistribution when they are either ionised or non-stoichiometric. To deflect free clusters into a mass spectrometer, for example, a preliminary ionization of the clusters has to be performed, so that it is actually the abundance of positively charged species that is directly probed. Similarly, anion spectroscopy is performed on negatively charged species. Whether the excess charge in free clusters is delocalised on all atoms or is mainly trapped on specific sites has been addressed for charged and/or non-stoichiometric NaCl,[18–20] NaF[21,22] and Li_2O[23] clusters, which are very ionic compounds with large gaps. However, the general answer to this question is related to the screening properties of the systems under consideration.

On surfaces, usually no direct charge injection is performed, but, in the presence of oxygen vacancies or metal deposits, charge redistributions take place, the amount and spatial extent of which are directly related to the electronic structure of the oxide. For example, TiO_2 has recently been examined in relation to alkali deposition. The adsorption characteristics and the charge transfer from K to $TiO_2(100)$[24] and from Na, K and Cs atoms to $TiO_2(110)$[25–29] have been studied. From a theoretical point of view, both K[30] and Na[31] adsorption on $TiO_2(110)$ have been considered.

The aim of the present paper is to analyse the charge redistribution effects in free clusters and surfaces of titanium dioxide, to better understand how the ionocovalent character of Ti–O bonding and screening properties of the oxide vary as a function of the size of the system, the atomic coordination number and the surface orientation. Section 2 gives the details of the computational method that we use, which is based on the DFT. Some emphasis will be given to the choice of the pseudopotentials that account for the interaction between valence and core electrons. We will also describe the charge analysis, which is performed subsequently to investigate the electron or hole redistributions. Sections 3 presents our results for the neutral, cationic and anionic $(TiO_2)_n$ lowest energy isomers ($n = 1–3$), and the addition of one oxygen atom on the neutral clusters. We also give results on the geometry and charge analysis of bulk TiO_2 and the clean (1×1) $TiO_2(110)$ surface. Sections 4, 5 and 6 are devoted to a discussion of our results. They focus on the modifications of the degree of covalency of the Ti–O bonds as a function of the different environments, on the localization of the excess charges (holes or electrons) and on screening effects, respectively.

2 Computational method

The electronic structure calculations have been performed within the density functional theory, by using both local density approximation (LDA) and local spin density approximation (LSDA) for the exchange and correlation energy. In particular, LSDA is used for clusters with an odd number of electrons, and for some even-numbered clusters that could present a non spin–paired ground state. We have thus checked the stability of the calculated electronic structure with respect to spontaneous spin polarization.

The Kohn–Sham orbitals are expanded in a plane–wave basis set, and soft, norm-conserving pseudopotentials are used to describe the interaction between the ionic cores (consisting of 1s oxygen atomic states, 1s, 2s and 2p titanium states), and the valence electrons.

The interaction between core and valence electrons is accounted for by norm-conserving pseudopotentials in the Kleinman–Bylander form[32] including s, p and d components for oxygen and titanium. The reference state is $1s^2 2s^2 2p^6$ for O and Ar $3d^4$ for Ti. A local reference potential is chosen (d component for oxygen, s for titanium). We followed the prescriptions given by Troullier and Martins[33] in the generation of pseudopotentials. The core radii chosen for O (Ti) are 1.38 (1.30) a_0, 1.60 (1.40) a_0 and 1.38 (1.80) a_0 for the s, p, and d components, respectively (a_0 = Bohr radius). The O pseudopotential was generated for the neutral atom, and extensively tested in previous studies.[23,34]

The inclusion of the titanium 3s and 3p states in the valence was found necessary to obtain a good agreement with the experiments and advanced calculations performed on TiO.[15] This appeared to be especially important to ensure the transferability of the pseudopotential in systems having sites of various coordination numbers. For bulk rutile titanium oxide, it is well known that many pseudopotential types give results in good accordance with respect to the experiments, and usually the differences with the latter are not discussed in terms of the pseudopotential characteristics but rather in terms of the approximation used for the exchange-correlation energy. The LDA, for example, is known to underestimate interatomic distances by 1% or 2%, and the GGA (generalized gradient approximation) to overestimate them. To perform a more stringent test than bulk rutile, we chose the TiO molecule, which is an open shell system of very small size for which experimental data are available.[11] The ground state electronic configuration is $^3\Delta$ and the equilibrium distance and vibrational frequency are equal to 3.06 a_0 and 1000 cm^{-1}, respectively. We have built a first pseudopotential (I) including Ti 3s, 3p, 4s and 3d states in the valence and a second one (II) only including Ti 4s and 3d states in the valence. Both pseudopotentials yield the right symmetry ground state and the discrepancies with experimental values are -3% $(+6\%)$ for the interatomic distance and $+9\%$ $(+0.1\%)$ for the vibrational frequency, for the I(II) pseudopotentials. II is seen to give abnormally high equilibrium distances.

These differences between the two pseudopotentials show the active contribution of Ti 3s and 3p electrons in the Ti–O bonding. More precisely, we have checked that, using pseudopotential I, the Ti 3p orbital, which points in the bond direction, slightly hybridizes with an O 2p orbital, resulting in a ppσ bonding orbital. Evidence for this hybridization is also supported by the energy splitting between the Ti 3p, σ and π states, which decreases from 2.90 eV to 0.45 eV when increasing the interatomic distance from 2.60 a_0 to 3.20 a_0. As a consequence of the interaction between Ti 3p and O 2p orbitals and the enhanced electron density along the bond, the ordering and relative splitting of the electronic states depends sensitively on the inclusion of the Ti 3p states. Ti 3s electrons, on the other hand, do not form hybrids but are polarized toward oxygen and it was not possible to take account of this polarization without treating these states explicitly. In addition, their removal may cause technical problems like the appearance of ghosts states.[35] Our results using pseudopotential I compare well with previous CI (configuration interaction) calculations,[15] regarding the symmetry and the shape of the electronic wavefunction. As a consequence, in the following, we adopt the pseudopotential I. The logarithmic derivatives show an optimal transferability, over a wide energy range, well above the atomic vacuum level. Although the number of electron states and the energy cut-off needed to get convergence with respect to the plane-wave basis set are high, as a result of the inclusion of the 3s and the 3p Ti electrons in the valence band, they participate in the bonding, and can modify the computed structural parameters, especially for low-coordinated atoms, as it is the case for clusters and surfaces.

A supercell geometry, with a face-centered cubic cell of lattice parameter equal to 35 a_0, was used for the cluster calculation, and integration in the Brillouin zone was performed using the Γ point. Convergence on the total energy as a function of the cut-off energy was checked: a precision of 0.1 eV on total energy is reached for $E_{cut} = 60$ Ry, and of 0.1 eV on total energy difference between isomers at $E_{cut} = 40$ Ry. The calculations were performed with $E_{cut} = 60$ Ry for TiO$_2$ and (TiO$_2$)$_2$, and with $E_{cut} = 40$ Ry for larger clusters.

The charged clusters are simulated by introducing a compensating uniform background, which takes into account the finite contribution of the background–background interaction to the long range part of the electrostatic energy.[36] This method allows a faster convergence of the total energy as a function of the size of the unit cell, as confirmed by extensive tests on charged molecules. We checked that the size of the unit cell is sufficient to ensure that spurious interactions between neighboring periodic images do not bias the computed energy differences between the various isomers and the calculated interatomic distances.

In order to find out the ground state configurations, we started from structures suggested in ref. 13 by simple pair-potential simulations. The geometries were then relaxed until the atomic forces did not exceed 0.01 eV Å^{-1}, and the calculated electronic structures thus refer to the stable configurations optimized in a fully self-consistent way.

The (110) unreconstructed face of rutile is simulated by a slab containing three O–(Ti$_2$O$_2$)–O units, consisting of 18 atoms in total (see Fig. 4 below) in the two-dimensional unit cell, plus a vacuum whose width (≈ 5.5 Å) is enough to prevent spurious interactions between periodic images.

Two $\vec{k}$ points in the irreducible two-dimensional Brillouin zone are used for the charge integration. A full geometry optimization of all TiO$_2$ layers is performed in order to determine the equilibrium geometry of each surface configuration. The search for the energy minimum is stopped when atomic forces are less than 0.01 eV Å^{-1}.

Our charge analysis follows the scheme proposed by Bader,[37] who introduced the concept of atoms-in-molecules or solids. According to Bader, an atom is seen as an open system that can be described by a Schrödinger equation. Consequently, its volume must be enclosed by a surface through which no electron flux passes. The mathematical condition which defines the partitioning of space into atomic basins is thus:

$$\vec{\nabla}\rho(\vec{r}) \cdot \vec{n} = 0 \quad (2.1)$$

in which $\rho(\vec{r})$ is the total electron density, and $\vec{n}$ the normal to the surface at $\vec{r}$. Usually, and more specifically in ionic systems, each atomic basin includes one nucleus. In the course of our study, we got no evidence that this condition breaks. Bader analysis includes a topological discussion of the electronic density, in terms of critical points, that we will not use in the present work. We will only focus on the determination of the atomic basins and the calculation of the integral of the total electronic density within each basin.

From a numerical point of view, in order to obtain the total electronic density $\rho(\vec{r})$, we sum the valence density, issued from our DFT calculation, and the core level density, obtained from the pseudopotential generation code. $\rho(\vec{r})$ is written on a regular grid, and is integrated in each atomic basin, as follows. Starting from a grid point close to a given nucleus j, the point $\vec{R}_0^{(j)}$ of the mesh corresponding to the density maximum is determined. Then a simple algorithm determines the mesh points $\vec{R}_i^{(j)}$ on which $\rho(\vec{R}_i^{(j)})$ decreases monotonically. The space is therefore partitioned in domains $\mathscr{D}^{(j)}$ consisting of all paths on which the sign of $\vec{\nabla}\rho(\vec{R}_i^{(j)}) \cdot \vec{l}_i$ is constant, where $\vec{l}_i$ is the elementary displacement along the paths. The electron density is then summed up within each domain, and the contributions from the border regions are shared between the neighbouring basins. By using this method on a cubic grid with a parameter of 0.15 a_0, we obtain the charge integral with a precision better than 0.1 electron per basin.

3 Charge distribution in titanium oxide: results

3.1 Neutral (TiO$_2$)$_n$ clusters ($n = 1$–3)

We have previously described[17] the geometry and energetics of the low energy isomers issued from the geometry optimization for $n = 2$ and 3. They are represented in Fig. 1. For (TiO$_2$)$_2$, the two configurations have C_{2v} and C_{3v} symmetry, respectively. The C_{2v} isomer is non-planar, as a result of the bent configuration of the TiO$_2$ trimer. For (TiO$_2$)$_3$, we have found two low energy isomers of C_s and S_4 symmetry.

Integration of the electron density in the regions $\mathscr{D}^{(j)}$ associated with each atom j, yields the Bader charges borne by j in the clusters. We find titanium charges in the range [+1.5, +2] and the oxygen charge in the range [−0.7, −1.1]. These values are far from the formal charges +4 and −2 assigned, in the fully ionic limit, to Ti and O in titanium oxides of TiO$_2$ stoichiometry. This is an indication that the Ti–O bonds have a large part of covalent character, in agreement with resonant photoemission results.[2]

To quantify the relative part of ionic and covalent characters, we do not directly compare the charge values, which are the complex result of O–Ti electron transfer through orbital hybridization and local environment factors, such as coordination numbers. In order to discriminate the different effects, we analyze the charges in the following way. Starting from the purely ionic limit, in which Ti and O have +4 and −2 charges, we take into account the effect of Ti–O orbital hybridization between first neighbour atoms, by introducing parameters Δ, which describe electron transfers per bond. For example, in the TiO$_2$ molecule, by symmetry, a single Δ is required and the atom-in-molecule charges are equal to:

$$Q_{Ti} = +4 - 2\Delta \qquad Q_O = -2 + \Delta \quad (3.1)$$

According to this picture, each O^{2-} atom donates Δ electrons back to the Ti^{4+}. In this simple case, the orbital hybridization is included in Δ, and the coefficient in front of Δ in eqn. (3.1) is the

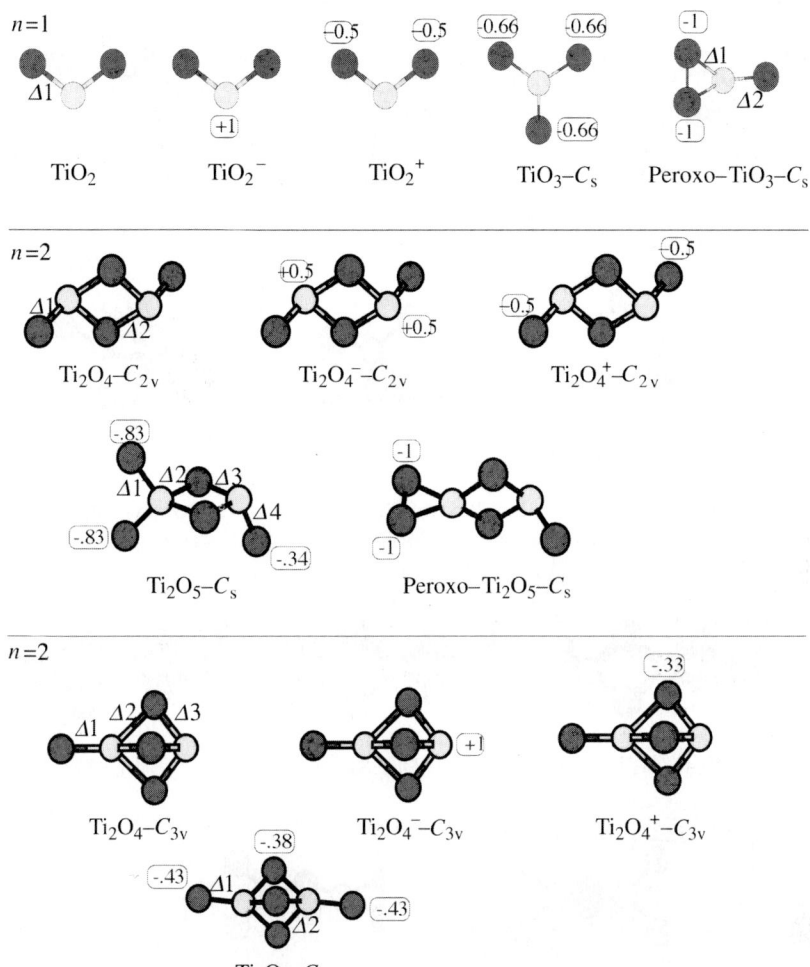

Fig. 1 Low energy isomers of $(TiO_2)_n$ clusters ($n = 1-3$). Ti atoms are drawn in pale grey and O atoms in dark grey, in all ball and stick representations. The inequivalent electron transfers per sites are labelled according to their values in Table 1. In charged and non-stoichiometric clusters, are indicated the values of the excess electron density (positive sign) or excess hole density (negative sign) on the atoms on which the electron or hole(s) are localized (see text). For the sake of clearness, we have indicated the excess electron density on the three bridging oxygens in $Ti_2O_4^+$-C_{3v}^+ and Ti_2O_5-C, only once, although these oxygens are equivalent.

atom coordination number. We have shown[38,39] that, in a tight-binding approach, Δ is a monotonic function of the ratio $\beta/(\varepsilon_C - \varepsilon_A)$ between the resonance integral β associated to the orbital hybridization and the difference in energy between the cation and anion atomic orbitals under consideration. Δ vanishes in the ionic limit (either $\beta = 0$ or $\varepsilon_C - \varepsilon_A \to \infty$). Its magnitude thus characterizes the strength of covalent bonding. When inequivalent bonds are present in a system, distinct parameters Δ_j are required, and the charge of an atom-in-molecule can be obtained from the values of the electron transfers from or to its Z first neighbours.

For example, in Ti_2O_4-C_{3v}, there exist three inequivalent electron transfers Δ, that we label Δ_1, Δ_2 and Δ_3 (see Fig. 1). The atoms bear charges equal to:

$$Q_{Ti} = 4 - \Delta_1 - 3\Delta_2 \qquad Q_{Ti} = 4 - 3\Delta_3$$
$$Q_O = -2 + \Delta_1 \qquad Q_O = -2 + \Delta_2 + \Delta_3 \qquad (3.2)$$

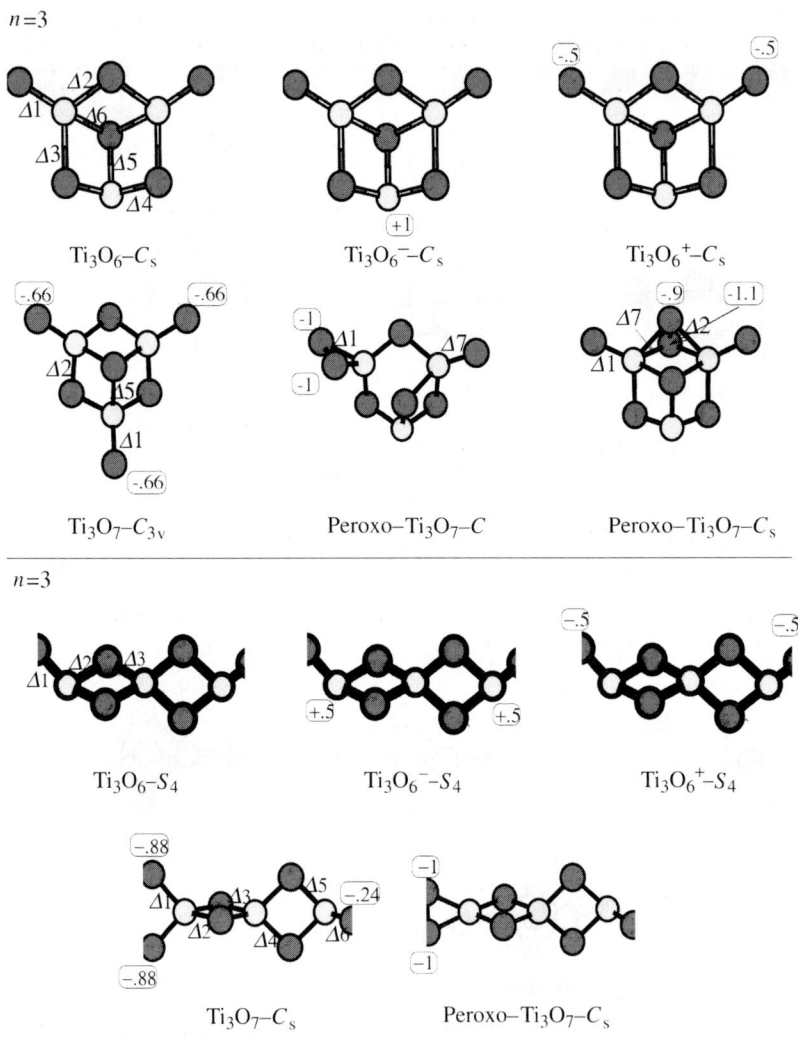

Fig. 1.—Continued

for 4-fold and 3-fold coordinated Ti, respectively (first line), and for 1-fold and 2-fold coordinated oxygens, respectively (second line). More generally, we have used the relations:

$$Q_{Ti} = 4 - \sum_j \Delta_j \qquad Q_O = -2 + \sum_j \Delta_j \qquad (3.3)$$

with the sum index running over bonds in which the atom is involved. This has allowed us to derive the values of the Δ parameters in the five neutral $(TiO_2)_n$ clusters ($n = 1-3$), that are collected in Table 1 together with the computed corresponding bond lengths.

The Δ_i parameters, and consequently the degree of covalency of the Ti–O bonds, vary in large proportions, ranging from 0.2 to 1.26 electrons. The same is true for the interatomic distances which vary from 3.0 to 3.9 a_0. Inequivalent local environments of the atoms thus induce strong changes in the mixed ionocovalent character of the Ti–O bonds. This is akin to rather covalent systems, such as TiO_2, and would not have been found in very ionic clusters.

3.2 Anionic $(TiO_2)_n^-$ and cationic $(TiO_2)_n^+$ clusters ($n = 1-3$)

The addition or removal of an electron does not modify the cluster conformations deeply. Bond lengths and bond angles vary slightly, while the electronic structure changes dramatically.

Table 1 Electron transfers per bonds Δ and interatomic distances (in atomic units) in neutral, anionic, cationic $(TiO_2)_n$ and in non-stoichiometric Ti_nO_{2n+1} clusters ($n = 1$–3), in bulk TiO_2 rutile and at the (110) surface. The charges borne by the atoms may be derived from the Δ_j, using eqns. (3.3) for neutral stoichiometric clusters, eqn. (3.4) for anionic clusters and eqn. (3.5) for cationic or non-stoichiometric clusters.

$n = 1$	TiO_2	TiO_2^-	TiO_2^+	TiO_3-C_s	Peroxo-TiO_3-C_s	
$\Delta_1(d_1)$	1.24 (3.076)	.94 (3.134)	1.16 (3.083)	0.78 (3.25)	0.54 (3.380)	
$\Delta_2(d_2)$					1.23 (3.106)	
$n = 2$	Ti_2O_4-C_{2v}	$Ti_2O_4^-$	$Ti_2O_4^+$	Ti_2O_5-C_s	Peroxo-Ti_2O_6-C_s	
$\Delta_1(d_1)$	1.24 (3.045)	1.03 (3.101)	0.98 (3.07)	0.66 (3.30)	0.48 (3.385)	
$\Delta_2(d_2)$	0.52 (3.426)	0.46 (3.461)	0.61 (3.413)	0.46 (3.51)	0.55 (3.404)	
$\Delta_3(d_3)$				0.68 (3.35)	0.52 (3.416)	
$\Delta_4(d_4)$				1.0 (3.16)	1.22 (3.098)	
$n = 2$	Ti_2O_4-C_{3v}	$Ti_2O_4^-$	$Ti_2O_4^+$	Ti_2O_5-C		
$\Delta_1(d_1)$	1.07 (3.075)	1.02 (3.133)	1.32 (3.00)	0.85 (3.247)		
$\Delta_2(d_2)$	0.29 (3.692)	0.37 (3.666)	0.22 (3.795)	0.44 (3.505)		
$\Delta_3(d_3)$	0.90 (3.307)	0.65 (3.398)	0.87 (3.306)			
$n = 3$	Ti_3O_6-C_s	$Ti_3O_6^-$	$Ti_3O_6^+$	Ti_3O_7-C_{3v}	Peroxo-Ti_3O_7-C	Peroxo-Ti_3O_7-C_s
$\Delta_1(d_1)$	1.26 (3.11)	1.18 (3.145)	1.0 (3.175)	0.76 (3.251)	0.54 (3.38)	1.26 (3.092)
$\Delta_2(d_2)$	0.44 (3.434)	0.40 (3.456)	0.54 (3.416)	0.57 (3.424)		0.21 (3.712)
$\Delta_3(d_3)$	0.31 (3.639)	0.51 (3.541)	0.40 (3.552)			0.34 (3.610)
$\Delta_4(d_4)$	0.86 (3.264)	0.53 (3.330)	0.84 (3.273)			0.84 (3.27)
$\Delta_5(d_5)$	0.63 (3.324)	0.42 (3.401)	0.54 (3.373)	0.33 (3.615)		0.67 (3.313)
$\Delta_6(d_6)$	0.2 (3.932)	0.22 (3.843)	0.25 (3.808)			0.15 (4.041)
$\Delta_7(d_7)$						0.17 (3.844)
$n = 3$	Ti_3O_6-S_4	$Ti_3O_6^-$	$Ti_3O_6^+$	Ti_3O_7-C_s	Peroxo-Ti_3O_7-C_s	
$\Delta_1(d_1)$	1.24 (3.107)	1.06 (3.144)	0.92 (3.145)	0.62 (3.31)	0.48 (3.382)	
$\Delta_2(d_2)$	0.50 (3.42)	0.42 (3.45)	0.65 (3.37)	0.46 (3.493)	0.53 (3.41)	
$\Delta_3(d_3)$	0.53 (3.405)	0.55 (3.405)	0.52 (3.40)	0.60 (3.39)	0.53 (3.421)	
$\Delta_4(d_4)$				0.53 (3.416)	0.57 (3.386)	
$\Delta_5(d_5)$				0.55 (3.402)	0.50 (3.419)	
$\Delta_6(d_6)$				1.1 (3.115)	1.24 (3.095)	
$n = \infty$	Bulk	(110) Surface				
$\Delta_1(d_1)$	0.35 (3.668)	0.54 (3.399)				
$\Delta_2(d_2)$		0.17 (3.826)				
$\Delta_{21}(d_{21})$		0.42 (3.616)				
$\Delta_3(d_3)$		0.28 (3.888)				
$\Delta_{31}(d_{31})$		0.47 (3.405)				
$\Delta_4(d_4)$		0.43 (3.585)				
$\Delta_{41}(d_{41})$		0.29 (3.669)				
$\Delta_5(d_5)$		0.29 (3.737)				
$\Delta_{51}(d_{51})$		0.46 (3.577)				

In neutral stoichiometric $(TiO_2)_n$ clusters, the filled electronic states are mainly oxygen-derived, with a non-negligible contribution from titanium orbitals. For the sake of conciseness, we refer to them as valence band states, which is formally correct only for extended systems. Symmetrically, in the neutral clusters, the lowest empty states (conduction band states, in the following) have mainly Ti character. In anionic clusters, this state (HOMO) becomes half-filled. Direct visualization of the HOMO shows the localization of the additional electron. A typical picture, relevant for $Ti_3O_6^-$-S_4 is shown in Fig. 2, where we can see that the probability of finding the electron is equally shared on the two titaniums equivalent by symmetry. More generally, we have found that the HOMO wavefunction is highly localized on one or a few titaniums, with a d-orbital-like shape, consistently with a Ti non-bonding character for the HOMO. When inequivalent titaniums are present, the electron is generally localized on the titanium(s) with the lowest coordination number (i.e., the three-fold coordinated Ti in Ti_2O_4-C_{3v} and Ti_3O_6-C_s, and the two three-fold coordinated Ti in Ti_3O_6-S_4). The distribution of the added electron N_i^e can be unambiguously determined for all the anionic clusters, and is given in Fig. 1.

Due to the perturbation brought in by the negative charge, all electron transfers along the Ti–O bonds are modified, and we have calculated them, from the following expression:

$$Q_{Ti} = 4 - N^e - \sum_i \Delta_i \qquad Q_O = -2 + \sum_j \Delta_j \qquad (3.4)$$

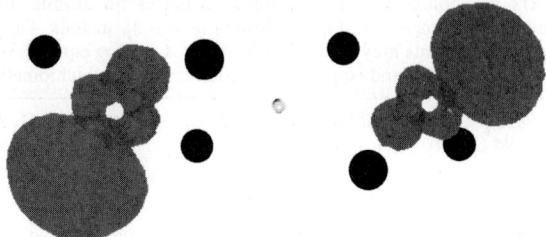

Fig. 2 Highest occupied orbital in $Ti_3O_6^- - S_4^-$. Ti and O atoms are represented as small light grey atoms and in black, respectively.

with N^e the excess electron number on the Ti site under consideration. The resulting values of the electron transfers per bonds are written in Table 1. As a general trend, Δ decreases in the vicinity of the perturbed sites, and the bonds are elongated.

In cationic clusters, the removal of an electron is associated with the creation of a hole in the valence band. Due to electron–electron interactions, the valence band states are strongly rearranged and the hole wave function cannot be simply expanded in one or few wavefunctions of the neutral cluster. However, the hole is fully spin polarized, so that the polarization density $\zeta(\vec{r}) = \rho_\uparrow(\vec{r}) - \rho_\downarrow(\vec{r})$, gives a good hint of the hole localization. Interestingly, the sign of $\zeta(\vec{r})$ agrees almost everywhere with that of the net spin, showing that the hole creation affects the total electronic distribution nearly independently of the spin. As a general trend (see Fig. 1), we find that the hole is shared between the two terminal oxygens (one-fold coordinated oxygens) in all clusters, except Ti_2O_4-C_{3v}. We define positive parameters N_i^h, equal to the hole densities on atom i. We have written their values in Fig. 1, with a '−' sign to emphasize the decrease in electron numbers.

Due to the perturbation that results from the positive charge of the cluster, all electron transfers along the Ti–O bonds are modified. We calculate them, through:

$$Q_{Ti} = 4 - \sum_i \Delta_i \qquad Q_O = -2 + N^h + \sum_j \Delta_j \qquad (3.5)$$

where N^h is the hole concentration on the O site under consideration. The charge transfers Δ are collected in Table 1. As a general trend, Δ decreases in the vicinity of perturbed oxygen sites and the corresponding bond lengths increase.

3.3 Non-stoichiometric Ti_nO_{2n+1} clusters

Starting from the five stoichiometric clusters described above, we have considered the addition of one oxygen atom. In each case, several distinct isomers are formed. We have represented in Fig. 1 those of lowest energy, with their space groups. In most cases, the additional oxygen binds to a

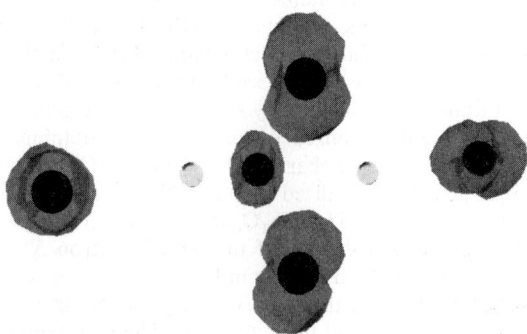

Fig. 3 Surfaces of equal spin polarisation in Ti_2O_5-C. Ti and O atoms are represented as small light grey atoms and in black, respectively.

single titanium atom, the only exception being the peroxide cluster Ti_3O_7-C_s. The conformation of the clusters is not deeply modified by the oxygen addition, but, as we will see below, important modifications of bond lengths take place.

The electronic structure of the non-stoichiometric clusters is characterized by a valence band with two holes. As in the case of cationic stoichiometric clusters, we have determined the localization of these holes, by studying the spin density distributions. The case of Ti_2O_5-C is represented in Fig. 3. In all cases, the hole localization is very similar to what it is in cationic stoichiometric clusters, i.e., essentially on the terminal oxygens. Ti_2O_5-C is slightly more complex, due to a sharing of the hole density on the five oxygens of the clusters. We have indicated in Fig. 1 the hole distribution on the different atoms. In most cases, and especially in complex cases such as Ti_2O_5-C, the estimation of the N_i^h results from a procedure in two steps, with a first rough determination followed by a refinement as commented in Section 4. As for the cationic clusters, we have derived the modified values of the charge transfer per bonds, using eqns. (3.5), and we have collected them in Table 1.

Actually, two families of non-stoichiometric clusters may be discriminated, according to whether an O–O bond forms or not. When the latter case occurs, an energy stabilization takes place and the isomers with such O–O bond have the lowest energy. These O–O bonds are known in the condensed phase or in oxidized metallic clusters.[40,41] When the O_2 group bears a -1 charge, it is named superoxide anion and when its charge is -2, peroxide anion. The O–O bond lengths are different in the two cases, d being of the order of 1.34 Å and 1.49 Å for superoxides and peroxides, respectively. For the five isomers which present such an O_2 group, we find that $1.483 < d/Å < 1.498$, which unambiguously points towards the existence of an O_2^{2-} peroxide group. The hole distribution reported in Fig. 1 supports this conclusion, since the two holes are entirely localized on the O_2 group. In Section 5, the charge distribution in these clusters will be further discussed.

3.4 TiO_2 bulk and the (110) surface

Bulk TiO_2 rutile is tetragonal and may be characterized by lattice constants a and c and by an internal parameter u, which governs the location of the oxygen atoms. We have computed the structural properties of rutile, by allowing a full atomic relaxation. The calculations are performed at a 60 Ry cut-off, and are fully converged with respect to the Brillouin zone sampling. The results are compared to the experimental data and to previous calculations in Table 2. One can note that: (i) our results well agree with the experimental data,[42] from which they differ by a slight underestimate of less than 1% of the lattice parameters a and c. This is usual when using the LDA. (ii) Our results agree almost perfectly with those obtained by Ramamoorthy and coworkers,[43] who also treated the semicore Ti 3s and 3p electrons self-consistently. On the other hand, the lattice parameters obtained by Glassford and Chelikowsky,[44] who kept the semi-core Ti electrons frozen, are overestimated. We attribute this disagreement mainly to the different treatment of the 3s and the 3p Ti electrons.

As regards the (110) surface, there is a longstanding controversy about its surface conformation. On one hand, grazing X–ray diffraction experiments[7] give a large inward relaxation for the bridging O, which was then seen to be consistent with shadowing effects in the diffraction of O^+ ions.[45] On the other hand, most of the theoretical calculations agree with the experimental positions for the Ti ions, while those computed for the oxygens show quantitative discrepancies. For these

Table 2 Equilibrium lattice parameters of bulk rutile

	Experiment Ref. 42	Theory Ref. 44	Theory Ref. 43	This work
a/Å	4.5936	4.653	4.567	4.556
c/Å	2.9587	2.965	2.932	2.930
u	0.3048	0.305	0.307	0.306

Table 3 Ionic displacements (in Å) of the surface atoms at the (110) surface of rutile, with respect to the bulk-truncated surface

	Ref. 7		This work				Ref. 43	
	Experimental		Theoretical lattice parameter		Experimental lattice parameter		Theoretical	
	[110]	[1̄10]	[110]	[1̄10]	[110]	[1̄10]	[110]	[1̄10]
6-fold Ti	$+0.12 \pm 0.05$	0	$+0.12$	0	$+0.11$	0	$+0.13$	0
5-fold Ti	-0.16 ± 0.05	0	-0.15	0	-0.15	0	-0.17	0
Bridging O	-0.27 ± 0.08	0	-0.09	0	-0.12	0	-0.06	0
In plane O	$+0.05 \pm 0.05$	0.16 ± 0.05	$+0.14$	$+0.05$	$+0.12$	$+0.06$	$+0.13$	$+0.04$

reasons, we carried out a careful relaxation of our slab, by using both theoretical and experimental lattice constants a and c, in order to assess whether the theoretical results could be biased by the various computational ingredients. The atomic positions were relaxed until the atomic forces do not exceed $\approx 5 \times 10^{-3}$ eV Å^{-1}. Our final results are in line with previous LDA calculations. In Table 3, our computed ionic displacements are collected and compared to those obtained by Ramamoorthy and coworkers,[43] and to those inferred by X-ray surface diffraction.[7] One can note that a substantial agreement exists between the calculations. Our calculation performed at the experimental lattice parameters is slightly closer to the experimental result, though the gap between the latter and our numerical values is still significant.

For both geometries, bulk and surface, we have applied the same charge analysis as in clusters to determine the electron numbers for each atom-in-molecule. The electron transfers per bond, Δ_i, for each class i of equivalent bonds, are obtained using the same formalism as in the neutral stoichiometric clusters (eqn. (3.3)). In the bulk, the Ti atoms have six neighbouring oxygens placed at two inequivalent distances, so that only the average of Δ can be determined. As regards the (110) surface, Fig. 4 indicates the labels used to specify the inequivalent bonds. We have reported the values deduced for the Δ_i in the lower part of Table 1. An enhancement of covalency is evidenced on some bonds, and more specifically on the bond which links the bridging oxygens to the underlying titaniums.

To summarize, in the various systems that we have considered, the charge fluctuations are rather weak and do not exceed 0.3 electrons. The only exception is Ti_2O_4-C_{3v} in which the charge difference between the two inequivalent titaniums is 0.75 electrons. On average, the Ti charge is close to $+1.7$ and the oxygen charge is close to -0.85. The rough constancy of these values results from a partial cancellation between the large variations of the electron transfers per bonds,

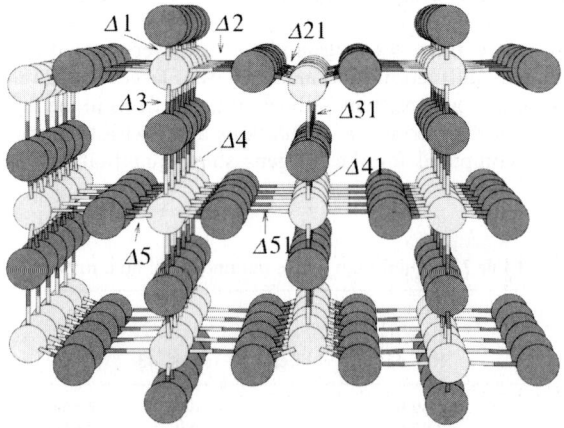

Fig. 4 Electron transfers per bonds on the (110) surface of TiO_2. Ti and O atoms are represented in light grey and dark grey, respectively.

and the large variations in the coordination numbers of the atoms.

4 Iono covalent character of the Ti–O bonds

The electron transfer per bond Δ varies over a wide range, as a function of the type of the system (clusters, (110) surface, bulk), of the net charge, of the stoichiometry and of the specific conformation. This shows that the Ti–O bond in some cases has a large part of covalent character (for example, in TiO_2), while in some other cases, it is much more ionic (for example, one of the central bonds in Ti_3O_6-C_s). In this section we point out the factors that drive these strong variations of Δ. Morever, we show that there is a relation between the electron transfers per bond and the bond lengths, valid for all the systems under consideration, which connects a theoretical concept, the mixed ionocovalent character of a bond, quantified by the non-measurable quantities Δ_i, with the experimentally accessible structural parameters d_i.

4.1 Driving factors underlying electron transfers

As shown in refs. 38 and 39, in a tight binding picture, the electron transfer per bond Δ is a monotonic function of the ratio $\beta/(\varepsilon_C - \varepsilon_A)$ between the resonance integral β associated with the orbital hybridization and the energy difference between the cation and anion atomic orbitals under consideration. β is strongly dependent upon the interatomic distances. The atom effective levels ε_C and ε_A are renormalized, with respect to their values in the neutral atoms, by Coulomb and exchange interactions due to the surrounding electrons.

Coulomb effects show up in the Madelung potential exerted by the neighbouring charges on a given atom. Since in most compounds, anions are surrounded by cations and *vice-versa*, the Madelung potential is positive on an anion and negative on a cation, which shifts their effective levels towards lower and higher energies, respectively. For a given pair of Ti and O atoms, connected by the ith bond, the difference $\delta V_{Mad}^{(i)}$ between the Madelung potential acting on Ti and O renormalizes the difference $\varepsilon_C - \varepsilon_A$. Compact environments are expected to induce large Madelung potentials, *i.e.* large $\varepsilon_C - \varepsilon_A$ values, and thus small electron transfers per bond Δ. We have checked this idea in the neutral stoichiometric clusters and shown the correlation between Δ_i and $\delta V_{Mad}^{(i)}$ in Fig. 5. Actually, the Madelung potential results from a self-consistent relation between the charges and the total potentials. In order to point out the role of the *morphology* on the renormalization of the atomic levels, we have computed the $\delta V_{Mad}^{(i)}$ for the actual geometry, by using a constant charge q_{Ti} for all cations and q_O for all anions, related by $q_{Ti} + 2q_O = 0$.[46] The current value for q_{Ti} has no importance in the following discussion apart from an overall scaling in Fig. 5.

We have also added the values calculated in bulk rutile, which result from Madelung potential data given in ref. 47 correctly renormalized by the q_{Ti} charge value, and the values that we have calculated for the (110) surface. Considering the large range of morphologies of the systems under consideration, from the TiO_2 molecule, in which Ti is 2-fold coordinated, to the bulk in which the local symmetry is nearly octahedral, Fig. 5 clearly demonstrates that our computed $\delta V_{Mad}^{(i)}$, and thus the cluster morphologies, are the relevant parameters that drive the values Δ_i of the electron transfers per bond.

4.2 Correlation between electron transfers and bond lengths

In addition, the Madelung potential also depends upon the bond lengths d, and so do the resonance integrals β, which enter the expression for Δ in a simple tight binding approach. Again, the relationship between the Δ and the d values is self-consistent. Strong hybridizations (associated with large Δ) are expected to yield high bond energies, and thus small interatomic distances.

To check this correlation, we report in Fig. 6, all the Δ and d values collected in Table 1, for the neutral or charged stoichiometric clusters as well as for the non-stoichiometric ones, the bulk and the surface.

The trend is a rapid decrease of Δ as a function of d. If we consider the ratio $\beta/(\varepsilon_C - \varepsilon_A)$, which drives the value of Δ, both its numerator and its denominator decrease when the bond lengths are

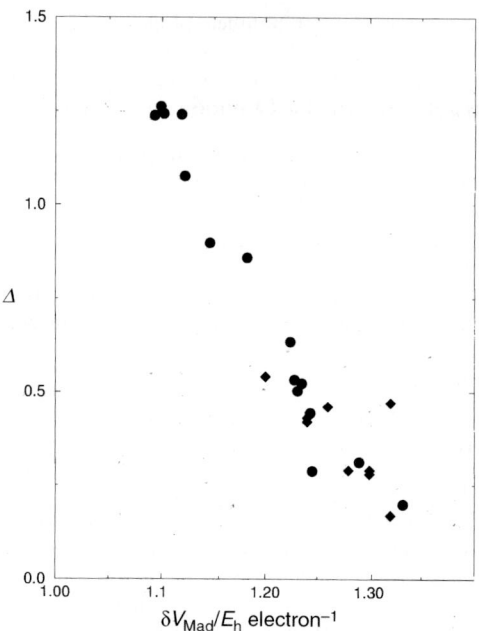

Fig. 5 Electron transfer per bond Δ as a function of the differences δV_{Mad} in the Madelung potentials acting on the Ti and O in the neutral stoichiometric clusters (filled circles), in bulk rutile TiO_2 (plus), and on the TiO_2(110) surface (diamonds).

elongated. We can conclude that the variations of the resonance integrals β are stronger since, as a whole, Δ follows the trend given by β.

In addition, the dispersion of the points in Fig. 6 is very small, considering the large differences among the systems that were investigated. We consider that this excellent correlation between Δ

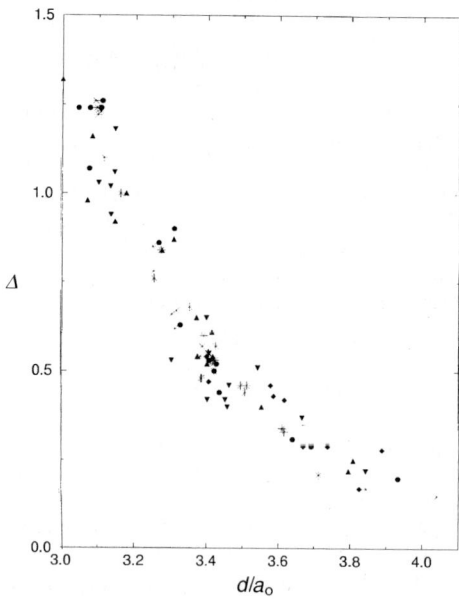

Fig. 6 Electron transfer per bond Δ as a function of the Ti–O bond length d in neutral (filled circles), anionic (down triangles) and cationic (up triangles) stoichiometric clusters, in non-stoichiometric clusters (stars), in bulk rutile TiO_2 (plus) and at the (110) surface (diamonds).

and d, whatever the morphology and the size of the system, is a strong indication that the Δ parameters, although they are not measurable quantities, are very relevant quantities to discuss the nature of the anion–cation bonding, especially in these systems, which, at the same time, are highly inhomogeneous and present a mixed ionocovalent character.

As a matter of fact, we have used this correlation in several instances. In Ti_3O_6-C_s clusters, in which an undetermination exists in the linear system of equations which gives the Δ_i values, by using Fig. 6 we have assumed a reasonable value of Δ on one bond (usually the longest one) and deduced all other electron transfers per bonds from the equations. We have also used Fig. 6 to refine the values for the excess hole distribution N_i^h, in complex cases. This happened for Ti_2O_5-C, where the holes are shared between two types of oxygens (Fig. 3), in Ti_2O_5-C_s, where the same occurs, although to a lesser extent, and to estimate the difference of hole concentrations in non-stoichiometric clusters with O_2^{2-} groups. However, among the 92 values reported in Table 1 and Fig. 6, 85 are unambiguously determined.

To summarize, the charge analysis in the titanium oxide finite and infinite systems has allowed us to study the bonding characteristics as a function of the size, the local environment and the atom coordination number. We have found that the degree of covalency of the Ti–O bonds, quantified by the values of the electron transfers per bond Δ_i, strongly varies as the local environment of the atoms changes, and is driven by the value of the Madelung potentials acting on the atoms. Electron transfers per bond and bond lengths are monotonic functions of each other, through a "universal" relationship valid from the molecule to the bulk, whatever the charge state and the stoichiometry. This conclusion supports the use of this universal curve as a predictive tool for other isomers or geometries.

5 Excess charge localization in Ti_nO_{2n} and Ti_nO_{2n+1} clusters

The localization of excess charges in free clusters has been considered in alkali-halides NaCl[18–20] and NaF[21,22] and Li_2O[23] systems. In most cases, these excess charges arose from an excess of metal atoms. The authors found that the excess electrons are very delocalized in the clusters and that excess metal atoms induce a kind of metallization of the clusters. However, for some special geometries and stoichiometries, namely cuboid Na_nF_{n-1}[22] with a single anion vacancy, the electrons remain trapped at the vacancy site. This also happens when a neutral oxygen vacancy is created in the bulk or at the surface of MgO.[34,48]

The excess electron in anionic clusters has a quasi-atomic d-type wave function (see Fig. 2) which is highly localized on titaniums and presents a non-bonding Ti–O character, as exemplified in Fig. 2. Such Ti–O non-bonding states of Ti character are usually found to form the bottom of the conduction band in defective systems. When inequivalent titaniums are present, the excess electron rests on the titanium(s) with the lowest coordination number, which have the weakest Madelung potential.

Symmetrically, in cationic or non-stoichiometric clusters, the hole wave function has a quasi-atomic p-type shape and is localized on oxygens. It is a Ti–O non-bonding wave function, of O character, as expected at the top of the valence band in systems where the oxygen atoms have lost some of their cation neighbours. In most clusters, when inequivalent oxygens are present, the hole rests on the oxygen(s) with the lowest coordination number, or more precisely, on the oxygen(s) which experience the smallest Madelung potential.

$Ti_2O_4^+$-C_{3v} and Ti_2O_5-C seem not to obey this rule. As proved by a direct visualization, the HOMO consists of O–O hybridized 2p orbitals, with a non-bonding Ti–O character. This is not due to the existence of especially short O–O bonds, but rather to the presence of three atoms which take part in the hybridization. The O–O antibonding state is thus higher in energy than the non-bonding 2p orbital of the terminal oxygen. The hole is therefore shared between the bridging oxygens, with an additional weight on the terminal oxygens, in the case of Ti_2O_5-C. This effect is visible in the clusters with the C_{3v} geometry and not in the other clusters, because their two Ti are bridged by three oxygens.

In the non-stoichiometric clusters with an O_2 group, it is found that the two holes are completely shared between the two oxygens of the O_2 group, with no noticeable weight on other oxygen atoms. This supports the prediction of the existence of a peroxide O_2^{2-} group. We will see in the next section that the actual charge on the two oxygens is reduced from the -2 value by screening effects. The short-hand notation O_2^{2-} has thus to be understood, in this framework, as indicative of the hole localization. The comparison between non-stoichiometric clusters of similar conformation which either present or do not present a peroxide group shows that the proximity of the two oxygens in the O_2 group enhances the hole localization on the group. This is especially clear on the Ti_2O_5 clusters issued from Ti_2O_4-C_{2v}. We assign this effect to the enhanced electrostatic interaction between the two oxygens of the O_2 group that raises their effective levels, and to the strong hybridization effects, which raise the O–O antibonding states.

6 Screening effects

When atoms, molecules or solids are submitted to an electrostatic perturbation, the ground state electron distribution changes to screen the perturbation, lowering the energy of the system. The fruitful framework to understand and describe these effects is the theory of the relative permittivity. In the homogeneous electron gas, the basic model for simple metals, the perturbing potential is totally screened at large distance. The characteristic screening length is proportional to the inverse of the Thomas–Fermi wave vector. In semiconductors or insulators, the existence of a gap in the spectrum of electronic excitations and the strong inhomogeneities of the electron density make the description of screening more involved. It is known that the degree of screening is incomplete and related to the value of the optical relative permittivity, ε_∞. In addition, local field effects are important, which are a manifestation of the discrete atomic structure of the material.[49] For example, screening of a test charge located close to an anion is different from what takes place close to a cation. Also, in small finite systems, such as molecules or clusters, there exists an effect called anti-screening.[50] In the case of a positive test charge, for example, the electron cloud moves towards the site of the perturbation, so that there are regions, especially those far from the perturbation, where the perturbing potential can be reinforced.

6.1 Strength and characteristics of screening effects

The charge analysis performed in the titanium oxide clusters that we have considered allows us to get some insight into the strength and characteristics of screening effects. In analogy with the theory of electrostatic screening, for charged stoichiometric clusters, we write the electron number on a given atom i:

$$N_i = N_i^0 + N_i^{\text{ext}} + N_i^{\text{scr}} \qquad (6.1)$$

as the sum of the electron number N_i^0 in the neutral cluster (the "unperturbed charge"), plus a modification which includes the charge perturbation N_i^{ext}, and the contribution from screening N_i^{scr}. The charge perturbation is related to the excess number of electron N_i^e or holes N_i^h that we have determined previously. By using eqns. (3.4), (3.5) and (6.1), one can relate the screening charge to the variations $\Delta_j - \Delta_j^0$ of the electron transfers induced by the perturbation (Δ_j in the charged clusters, and Δ_j^0 in the neutral ones), as follows:

$$\delta N_i^{\text{scr}} = \pm \sum_j (\Delta_j - \Delta_j^0) \qquad (6.2)$$

The sum runs over the bonds that link atom i to its neighbours, and the $\pm$ sign refers to cations (+) and anions (-). The screening charge thus appears as due to the modification of the electron transfers per bonds. Moreover, the sign of $\Delta_j - \Delta_j^0$ indicates the direction of displacements of electrons to screen the perturbation. A positive (negative) value is associated to a displacement from (to) the oxygens toward (from) the titaniums.

The largest variations $\Delta_j - \Delta_j^0$ are of the order of 0.3 electrons. These large values are consistent with the breakdown of Koopman's theorem as discussed in ref. 17. The general trend (see Fig. 7) is a displacement of electrons away from a perturbed site, which bears an excess of electrons, and toward a site with an appreciable hole density, with a magnitude which decreases as a function of the distance from the perturbation.

In non-stoichiometric clusters, the description of screening effects is slightly more involved. The oxygen addition is considered as the perturbation, together with the formation of new Ti–O bonds. One thus has to compare the final non-stoichiometric cluster to an initial state, which

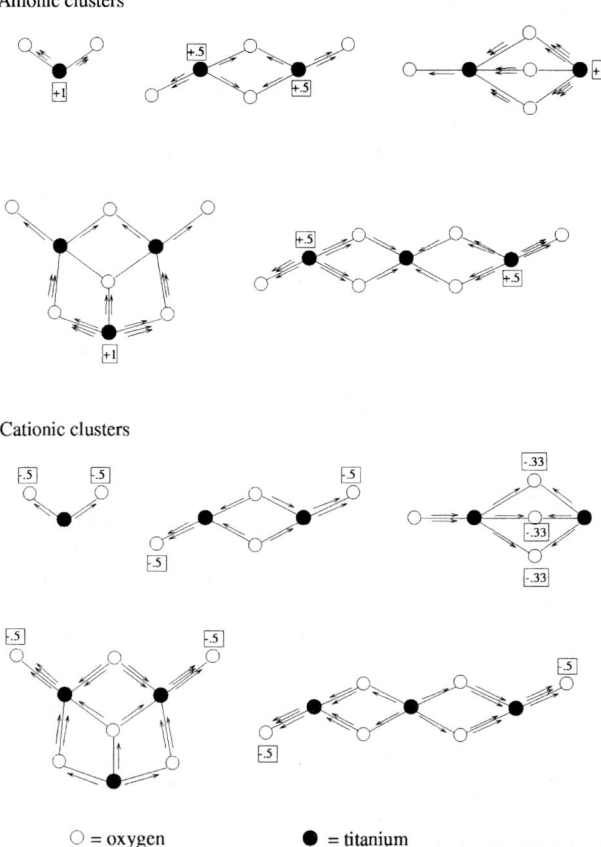

Fig. 7 Screening effects in charged $(TiO_2)_n^{\pm}$ clusters ($n = 1-3$). Top panel: addition of an electron to form anionic clusters; lower panel: subtraction of an electron to form cationic clusters. The arrows indicate the electron displacements to screen the perturbation. One, two or three arrows are drawn as a rough indication of the magnitude of the electron transfers.

consists of a neutral stoichiometric cluster plus an oxygen atom far away. At variance with charged stoichiometric clusters, the perturbing charge N_i^{ext} is thus $+(2 - N^h)$, and along the newly formed bond(s) $\Delta_j^0 = 0$. Screening effects in non-stoichiometric clusters present the same characteristics as in charged stoichiometric clusters.

6.2 Mechanism of screening

The screening mechanism[6] can be understood on the basis of the model already used in Section 4 to rationalize the magnitude of the electron transfers in terms of the energy differences between the effective levels $\varepsilon_C - \varepsilon_A$ of the neighbouring anion and cation. Similarly, in charged or non-stoichiometric clusters, we can understand the modifications $\Delta_j - \Delta_j^0$ of the electron transfers, through the modifications of $\varepsilon_C - \varepsilon_A$ induced by the perturbation. When $\varepsilon_C - \varepsilon_A$ increases, the bond is more ionic and electrons are backdonated by Ti to O. When $\varepsilon_C - \varepsilon_A$ decreases, the bond gets more covalent and electrons are transferred from O to Ti.

An excess (depletion) of electrons raises (lowers) all the energy levels, with a magnitude which is a decreasing function of the distance from the perturbed site i_0. As depicted schematically in Fig. 8, whatever the sign of the perturbing charge, the bonds j_1 involving i_0 become more ionic (Δ_{j_1} decreases). Further away, the energy level differences (ΔE_j in Fig. 8) vary with alternating signs, which results in oscillating variations $\Delta_j - \Delta_j^0$ of the electron transfers, associated with oscillating variations of bond lengths (Fig. 6).

Electron addition

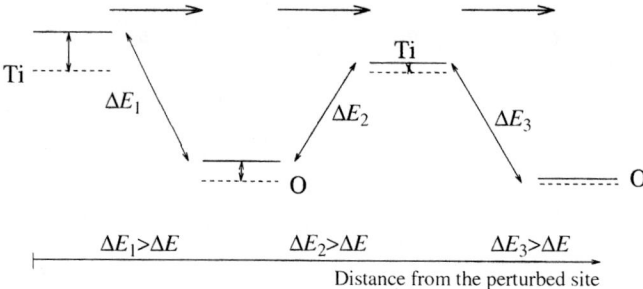

Electron subtraction

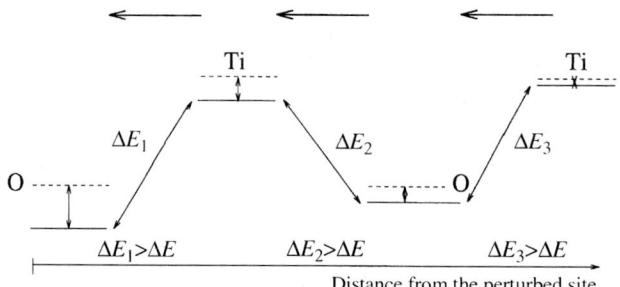

Fig. 8 Modification of the positions of the effective atomic levels (dashed lines: positions in the neutral cluster, plain line: positions in the charged clusters) as a function of the distance from the site where the perturbation is localised. Top panel: addition of an electron on a Ti; lower panel: subtraction of an electron to an O. The arrows indicate the direction of the electron displacements to screen the perturbation.

These oscillating variations of $\Delta_j - \Delta_j^0$ allow a cooperative displacement of the electrons to screen the perturbation. This mechanism becomes less and less effective as the difference $\varepsilon_C - \varepsilon_A$ increases, i.e. as the ionicity grows. In the limit of perfect ionic bonding ($\varepsilon_C - \varepsilon_A \to \infty$), the charges coincide with their formal values and no screening takes place. In TiO$_2$ species that we have

Table 4 Ratios $X = |N_i^{scr}/N_i^{ext}|$ between the screening charge and the charge perturbation on the site on which it is localised, in Ti$_n$O$_m^x$ clusters[a]

$n = 1$	TiO$_2^-$ 0.60	TiO$_2^+$ 0.17	TiO$_3$-C_s 0.70 0.59	Peroxo-TiO$_3$-C_s 0.70 0.54	
$n = 2$	Ti$_2$O$_4^-$-C_{2v} 0.68	Ti$_2$O$_4^+$-C_{2v} 0.53	Ti$_2$O$_5$-C_s 0.70, 0.70 0.56	Peroxo-Ti$_2$O$_5$-C_s 0.76 0.48	
$n = 2$	Ti$_2$O$_4^-$-C_{3v} 0.75	Ti$_2$O$_4^+$-C_{3v} 0.27	Ti$_2$O$_5$-C 0.79, 0.52 0.54		
$n = 3$	Ti$_3$O$_6^-$-C_s 0.88	Ti$_3$O$_6^+$-C_s 0.52	Ti$_3$O$_7$-C_{3v} 0.76 0.57	Peroxo-Ti$_3$O$_7$-C 0.64 0.48	Peroxo-Ti$_3$O$_7$-C_s 0.49 0.40
$n = 3$	Ti$_3$O$_6^-$-S_4 0.71	Ti$_3$O$_6^+$-S_4 0.64	Ti$_3$O$_7$-C_s 0.67, 0.79 0.58	Peroxo-Ti$_3$O$_7$-C_s 0.76 0.48	

[a] The perturbed sites are Ti when $x = -1$ and O when $x = 1$ or $m - n = 1$. In non-stoichiometric clusters, screening on oxygens belonging to the stoichiometric clusters is written on the first line (in the order from left to right in Fig. 1) and screening on the added oxygen is written on the second line.

considered, the ionic charges are far from their formal values, as can be expected from a compound with a narrow gap, and screening effects are important.

6.3 Local field effects

In Table 4, we report the values of the ratios $X = |N_i^{scr}/N_i^{ext}|$ between the screening charge and the charge perturbation on the site i on which it is localized. This ratio is an indication of the strength of screening effects on inequivalent perturbed sites, for the various clusters under study. It is independent of the magnitude of the perturbation, in a linear response theory framework.

The results given in Table 4 raise several issues.

1. X is in the range 0.17–0.88 and shows large variations as a function of the nature of the perturbed site (type, coordination number, *etc*). This is the indication of a strongly inhomogeneous screening, consistently with the cluster morphologies. It reveals the importance of local field effects.

2. For cationic stoichiometric clusters of increasing sizes (*i.e.* TiO_2^+, $Ti_2O_4^+$-C_{2v} and $Ti_3O_6^+$-S_4, in the order), the hole is shared between the two terminal oxygens. The bigger the distance between them, the larger X. As shown in the Appendix, this result is due to a kind of destructive interference effect between the hole perturbing potentials on the atomic effective levels.

3. As far as holes are concerned, screening is much more efficient in neutral non-stoichiometric clusters than in cationic stoichiometric ones. This may be rationalized by Madelung field arguments, which are explained in the Appendix. In most non-stoichiometric clusters, X is close to 0.75 on the regular oxygen sites, which is indicative of strong screening.

7 Conclusions

Electronic and structural properties of TiO_2 species of various sizes, charges and stoichiometries, including $Ti_nO_m^x$ clusters ($n = 1$–3, $m - n = 0,1$, $x = -1, 0, +1$), bulk rutile and its (110) surface, have been obtained by total energy calculation based on the density functional theory, in the local density and local spin density approximations. Using a Bader-type analysis of the total electronic density, we have focused our attention on the electron distribution, to better understand how the ionocovalent character of the Ti–O bonding and screening properties vary as a function of the size of the system, the atomic coordination numbers and the surface orientation.

We have found that the mixed ionocovalent character of the Ti–O bonds varies dramatically according to the local environment of the atoms, and can be quantified by a parameter Δ, which represents an electron transfer per bond due to orbital hybridization. We found a unique correlation between the values of Δ, as issued from the charge analysis, and the bond lengths, for all the systems considered here. Moreover, we have rationalized the variations in Δ from bond to bond through arguments based on the values of the Madelung potentials on the different atoms.

On charged clusters, we have found that the excess charge is localized in a quasi-atomic state (d-type orbital on Ti and p-type on O) on the atom(s) with the lowest Madelung potential. Non-stoichiometric neutral clusters with an excess O have two holes in the valence band. As regards the hole localization, a striking analogy is found with cationic stoichiometric clusters. It is worth emphasizing the existence of low energy isomers with a short O–O bond. The bond lengths, which are close to 1.49 Å whatever the cluster size, and our charge analysis point towards the presence of peroxide O_2^{2-} groups.

We described screening effects around localized electrostatic perturbations in charged or non-stoichiometric clusters and proposed a mechanism for electron redistribution based on the behaviour of the Madelung potential. It predicts an increase of ionicity around the perturbed site, while further away the bonds become alternately more covalent and more ionic. This allows an electron displacement to take place away from a negatively charged perturbed site and towards a positively charged site. This mechanism for screening in materials with a gap in the excitation spectrum is markedly different from screening in simple metals, which takes place naturally through the very delocalized character of the electron wave functions in metals. We have found that, in TiO_2 species, screening processes are very efficient, as can be expected from a compound with a relatively narrow gap, and are very dependent upon the chemical nature and the location of the perturbed site in the cluster, thus exemplifying the importance of local field effects.

Finally, it should be stressed that these small TiO_2 clusters represent ideal model systems to study the ionocovalent character of the anion–cation bonding and the screening properties in insulating compound systems. They provide a wide range of conformations associated to very different local environments of the atoms. In addition, the non-negligible part of covalent character in the Ti–O bonding in bulk rutile is responsible for the strong responses to electrostatic perturbations and for the large electron redistributions. We have seen in this work that the understanding on covalency gained in the clusters extends to bulk and clean surfaces. Current work is under progress to show that the same is true as regards surface modifications by adsorption or non-stoichiometry.

8 Acknowledgements

Calculations were performed on the Cray C98 at the IDRIS computational centre in Orsay (projects 984089 and 994089). We acknowledge fruitful discussions with C. Lecomte and M. Souhassou.

9 Appendix

In this Appendix, we give a theoretical background to local field effects in anionic and non-stoichiometric clusters. Local field effects are characterized by the ratio $X = |N_i^{scr}/N_i^{ext}|$ between the screening charge and the charge perturbation on the site on which it is localized.

We first consider cationic clusters and restrict ourselves to the analysis of local screening around a one-fold coordinated oxygen.

$$X = \left| \frac{\Delta - \Delta_0}{N^{ext}} \right| \tag{9.1}$$

The variation $\Delta - \Delta_0$ of the electron transfer results from the effect of the perturbation on the energy difference $\varepsilon_C - \varepsilon_A$, A being the oxygen and C the neighbouring Ti. It has been said that Δ is a monotonic increasing function of the ratio $\beta/(\varepsilon_C - \varepsilon_A)$, i.e., a monotonic decreasing function F of $\varepsilon_{CA} = \varepsilon_C - \varepsilon_A$. In a linear response theory framework, the variations of Δ are obtained by differentiation of this function:

$$\Delta - \Delta_0 \approx F' \times (\varepsilon_{CA} - \varepsilon_{CA}^0) \tag{9.2}$$

and:

$$X \approx \left| \frac{F' \times (\varepsilon_{CA} - \varepsilon_{CA}^0)}{N^{ext}} \right| \tag{9.3}$$

We consider the electrostatic contribution of N^{ext} to $(\varepsilon_{CA} - \varepsilon_{CA}^0)$. It is equal to the difference between the Hartree intra-atomic term $U_O N^{ext}$ on the O and an interatomic one N^{ext}/d on the Ti, if no other holes are present in the vicinity:

$$X \propto U_O - \frac{1}{d} \tag{9.4}$$

If another atom bearing a non-zero hole density is present in the neighbourhood, such as in TiO_2^+, $Ti_2O_4^+$-C_{2v} and $Ti_3O_6^+$-S_4, its contribution has to be taken into account, via a potential $V_1 N^{ext}$ on the oxygen and $V_2 N^{ext}$ on the titanium.

$$X \propto U_O - \frac{1}{d} - \delta V \tag{9.5}$$

In the geometries under consideration, $\delta V = V_2 - V_1 > 0$ and its value grows as the two holes get closer. Smaller values of X result. This explains the evolution of X ($X = 0.17, 0.53$ and 0.64) in the series TiO_2^+, $Ti_2O_4^+$-C_{2v} and $Ti_3O_6^+$-S_4. It can be said that the two holes interact destructively on screening.

We then consider the difference in screening of holes in positively charged clusters and non-stoichiometric ones. Actually, the values of X are very different in the two cases, despite the fact that in many instances, the same oxygen atoms are involved. Quite systematically, screening of the non-stoichiometry is more efficient than screening of a positive charge. We assign this difference to the fact that no global positive charge exists in the non-stoichiometric clusters, which remain neutral, as shown in the following way.

We again restrict ourselves to the analysis of local screening around a one-fold coordinated oxygen, and use eqn. (9.3) to estimate X. The electrostatic contribution of N^{ext} to $(\varepsilon_{CA} - \varepsilon_{CA}^0)$ is of the order of $U_O N^{ext}$ as regards the oxygen levels. The titanium levels, on the other hand, experience a different Madelung contribution according to whether the cluster is cationic or non-stoichiometric. In the former case, the Madelung contribution is of the order of N^{ext}/d, as shown above. In the second case, it amounts to $N^{ext}W$, with $W < 1/d$ due to the influence of a surrounding negative charge, which balances N^{ext} and assures the neutrality of the cluster. As a result, in non-stoichiometric clusters:

$$X \propto U_O - W \tag{9.6}$$

is larger than in cationic clusters.

References

1. V. E. Henrich and P. A. Cox, *The Surface Science of Metal Oxides*, Cambridge University Press, Cambridge, 1994.
2. Z. Zhang, S. P. Jeng and V. E. Henrich, *Phys. Rev. B: Condens. Matter*, 1991, **43**, 12004; J. Nerlov, G. Qingfeng and P. J. Møller, *Surf. Sci.*, 1996, **348**, 28.
3. S. R. Barman and D. D. Sarna, *Phys. Rev. B: Condens. Matter*, 1994, **49**, 16141.
4. L. F. Mattheiss, *J. Phys: Condens. Matter*, 1996, **8**, 5987 and references therein.
5. M. Catti, G. Sandrone and R. Dovesi, *Phys. Rev. B: Condens. Matter*, 1997, **55**, 16122.
6. C. Noguera, *Physics and Chemistry at Oxide Surfaces*, Cambridge University Press, Cambridge, 1996; *Physique et Chimie des Surfaces d'Oxydes*, Eyrolles, 1995, collection Alea, Paris.
7. G. Charlton, P. B. Howes, C. L. Nicklin, P. Steadman, J. S. G. Taylor, C. A. Muryn, S. P. Harte, J. Mercer, R. MacGrath, D. Norman, T. S. Turner and G. Thornton, *Phys. Rev. Lett.*, 1997, **78**, 495 and references therein.
8. M. Li, W. Henbenstreit and U. Diebold, *Surf. Sci.*, 1998, **414**, L951; C. L. Pang, S. A. Haycock, H. Raza, P. W. Murray, G. Thornton, O. Gulseren, R. James and D. W. Bullet, *Phys. Rev. B: Condens. Matter*, 1998, **58**, 1586; H. Onishi and Y. Iwasawa, *Surf. Sci.*, 1994, **313**, L783.
9. U. Diebold, J. M. Pan and T. E. Madey, *Surf. Sci.*, 1995, **331–333**, 845.
10. M. Barnes, A. J. Mercer and G. F. Metha, *J. Mol. Spectrosc.*, 1997, **181**, 180; L. A. Kaledin, J. E. McCord and M. C. Heaven, *J. Mol. Spectrosc.*, 1995, **173**, 499.
11. A. J. Merer, *Annu. Rev. Phys. Chem.*, 1989, **40**, 407.
12. N. S. McIntyre, K. R. Thompson and W. Weltner, Jr., *J. Phys. Chem.*, 1971, **75**, 3243.
13. W. Yu and R. B. Freas, *J. Am. Chem. Soc.*, 1990, **112**, 7126.
14. H. Wu and L. S. Wang, *J. Chem. Phys.*, 1997, **107**, 8221.
15. C. W. Bauschlicher, P. S. Bagus and C. J. Nelin, *Chem. Phys. Lett.*, 1983, **101**, 229; R. Bergström, S. Lunell and L. A. Eriksson, *Int. J. Quant. Chem.*, 1996, **59**, 427; M. V. Ramana and D. H. Phillips, *J. Chem. Phys.*, 1988, **88**, 2637.
16. A. Hagfeldt, R. Bergström, H. O. G. Siegbahn and S. Lunell, *J. Phys. Chem.*, 1993, **97**, 12725.
17. T. Albaret, F. Finocchi and C. Noguera, *Appl. Surf. Sci.*, 1999, **144–145**, 672.
18. U. Landman, D. Scharf and J. Jortner, *Phys. Rev. Lett.*, 1985, **54**, 1860.
19. S. Pollack, C. R. C. Wang and M. M. Kapper, *Z. Phys. D*, 1989, **12**, 241.
20. R. N. Barnett, H. P. Cheng, H. Hakkinen and U. Landman, *J. Phys. Chem.*, 1995, **99**, 7731.
21. G. Rajagopal, R. N. Barnett and U. Landman. *Phys. Rev. Lett.*, 1991, **67**, 727.
22. E. C. Honea, M. L. Homer, P. Labastie and R. L. Whetten, *Phys. Rev. Lett.*, 1989, **63**, 394.
23. F. Finocchi and C. Noguera, *Phys. Rev. B: Condens. Matter*, 1998, **57**, 14646; F. Finocchi and C. Noguera, *Eur. Phys. J. D*, 1999, **9**, in press.
24. K. Prabhakaran, D. Purdie, R. Casanova, C. A. Muryn, P. J. Hardman, P. L. Wincott and G. Thornton, *Phys. Rev. B: Condens. Matter*, 1992, **45**, 6969.
25. R. Souda, W. Hayami, T. Aizawa and Y. Ishizawa, *Surf. Sci.*, 1993, **285**, 265.
26. R. Heise and R. Courths, *Surf. Sci.*, 1995, **333**, 1460.
27. J. Nerlov, Q. F. Ge and P. J. Møller, *Surf. Sci.*, 1996, **348**, 28.
28. J. Nerlov, S. V. Christensen, S. Weichel, E. H. Pedersen and P. J. Møller, *Surf. Sci.*, 1997, **371**, 321.
29. H. Onishi and Y. Iwasawa, *Catal. Lett.*, 1996, **38**, 89; *Surf. Sci.*, 1997, **371**, 321.
30. P. J. D. Lindan, J. Muscat, S. Bates, N. M. Harrison and M. Gillan, *Faraday Discuss.*, 1997, **106**, 135.

31 M. A. San Miguel, C. J. Calzado and J. F. Sanz, *Int. J. Quant. Chem.*, 1998, **70**, 351.
32 L. Kleinman and D. M. Bylander, *Phys. Rev. Lett.*, 1982, **48**, 1425.
33 N. Troullier and J. L. Martins, *Phys. Rev. B: Condens. Matter*, 1991, **43**, 1993.
34 F. Finocchi, J. Goniakowski and C. Noguera, *Phys. Rev. B: Condens. Matter*, 1999, **59**, 5178.
35 X. Gonze, R. Stumpf and M. Scheffler, *Phys. Rev. B: Condens. Matter*, 1991, **44**, 8503.
36 M. Leslie and M. Gillan, *J. Phys. C.*, 1985, **18**, 973.
37 R. F. W. Bader, *Chem. Rev.*, 1991, **91**, 983; R. F. W. Bader, in *Atoms in Molecules-A Quantum Theory*, Oxford University Press, Oxford, 1990.
38 C. Noguera, A. Pojani, F. Finocchi and J. Goniakowski, in *Chemisorption and reactivity on supported clusters and thin films: Towards an understanding of microscopic processes in catalysis*, ed. R. M. Lambert and G. Pacchioni, Kluwer, ASI series E: Applied Sciences, 1997, **331**, 455.
39 A. Pojani, F. Finocchi and C. Noguera, to be published.
40 F. A. Cotton and G. Wilkinson, in *Advanced Inorganic Chemistry*, Wiley, New York, 5th edn., 1980.
41 S. D. Elliott and R. Ahlrichs, *J. Chem. Phys.*, 1998, **109**, 4267.
42 S. C. Abrahams and J. L. Bernstein, *J. Chem. Phys.*, 1971, **55**, 3206.
43 M. Ramamoorthy, R. D. King-Smith and D. Vanderbilt, *Phys. Rev. B: Condens. Matter*, 1994, **49**, 7709.
44 K. M. Glassford and J. R. Chelikowsky, *Phys. Rev. B: Condens. Matter*, 1992, **46**, 1284.
45 B. Hird and R. A. Armstrong, *Surf. Sci.*, 1997, **385**, L1023.
46 Given the small fluctuations of the actual charges around a mean value, our approach may be thought of as a perturbation calculation on the effective atomic levels.
47 J. Q. Broughton and P. S. Bagus, *J. Electron Spectrosc. Relat. Phenom.*, 1980, **20**, 261.
48 G. Pacchioni, A. M. Ferrari and G. Ierano, *Faraday Discuss.*, 1997, **106**, 155.
49 M. S. Hybertsen and S. G. Louie, *Phys. Rev. B: Condens. Matter*, 1987, **35**, 5585.
50 G. Onida, L. Reining, R. W. Godby, R. Del Sole and W. Andreoni, *Phys. Rev. Lett.*, 1995, **75**, 818.

Paper 9/03066B

The influence of soft vibrational modes on our understanding of oxide surface structure

N. M. Harrison,*[†][a] X.-G. Wang,[b] J. Muscat[‡][a] and M. Scheffler[b]

[a] *CCLRC Daresbury Laboratory, Daresbury, Warrington, UK WA4 4AD*
[b] *Fritz-Haber-Institut der Max-Planck Gesellschaft, Faradayweg 4-6, D-14195 Berlin-Dahlem, Germany*

Received 5th August 1999

We examine the reasons for the poor quantitative agreement between the structures predicted from the minimum energy configuration of first principles calculations and those deduced from surface X-ray diffraction experiments for the structure properties of the $TiO_2(110)$ surface. In order to confine all numerical approximations very large scale all-electron first principles calculations are used. We find a very soft, anisotropic and anharmonic surface rigid-unit vibrational mode which involves displacements of the surface ions of approximately 0.15 Å for thermal vibrations corresponding to room temperature. It is concluded that in order to perform an accurate comparison between theory and experiment for this and perhaps other oxide surfaces it will be necessary to take account of such anisotropic vibrations in models used to interpret experimental data. In addition the contribution of the vibrational entropy to the surface free energy is likely to be significant and must be taken into account when computing surface energies and structures.

1 Introduction

Titanium dioxide (TiO_2) is an interesting and industrially important material which has been studied extensively in recent years. This interest is, in part, due to the existing applications of TiO_2 as a white pigment and as a catalyst support[1,2] but also due to the many new applications currently under investigation. Recent examples include self cleaning paint coatings,[3] catalytically active paving stones,[4] solar cells[5] and water disinfection.[6] TiO_2 is also of interest as a model transition metal oxide. It is readily reduced in the bulk and at the surfaces resulting in the occupation of Ti d orbitals which had a profound effect on the physical and electronic structure.[7–9] However, the bulk structure of TiO_2 is simple relative to many oxides and thus a variety of empirical and first principles theories can readily be used to compute its physical and chemical properties. It is therefore an excellent model system displaying many of the properties of more complex oxides which can be studied relatively easily using a variety of experimental and theoretical techniques.

Many of the important and useful properties of TiO_2 depend on the physical and electronic structure of its surfaces. The structure of the most stable (110) surface has attracted enormous

---

† Also at Department of Chemistry, Imperial College London, London, UK SW7 2AY.
‡ Present address: CSIRO Minerals, Box 312, Clayton South, Victoria 3169, Australia.

interest in recent years. Experimental studies have included low energy electron diffraction (LEED),[10] scanning tunneling microscopy (STM),[11–13,7] surface X-ray diffraction (SXRD)[14] and ion scattering.[15] There have also been a large number of first principles theoretical studies.[16–20] Early contributions included a periodic Hartree–Fock study[16] within a linear contribution of atomic orbitals (LCAO) formalism and a density functional theory study (DET) which used a full potential linear augmented plane wave (FP-LAPW) method.[17] In each case the model systems studied were rather small and the surfaces only partially relaxed. More recently plane-wave (PW) pseudopotential calculations have been used based on both the local density (LDA) and generalised gradient (GGA) approximations to DFT.[18–20] These calculations included extensive relaxations of the surface structure and were based on larger structural models. Despite these extensive efforts the agreement between computed and measured structures is semi-quantitative at best. The extensive experience of calculations on bulk oxides which has been built up in recent years leads one to expect that DFT and HF calculations will reproduce experimental bond lengths to somewhat better than 0.1 Å. At the (110) surface the inward relaxation of the bridging oxygen ion determined by SXRD is -0.27 Å while most calculations find a relaxation of less than -0.1 Å. The current article is concerned with a detailed examination of this discrepancy.

Transition metal oxides represent a significant challenge to first principles calculations. The localised nature of the oxygen 2p and in particular the titanium 3d states makes the PW pseudopotential method particularly demanding. The expansion of localised orbitals in plane waves requires large kinetic energy cut-offs to converge the total energy. In addition the separation of the sp and d electron eigenvalues in the periodic system is dependent on the choice of the atomic reference state from which the pseudopotential is constructed. Great care must be taken to understand the effect of these approximations on computed material properties. Recently, Hamann[21] had discussed in detail calculations for bulk TiO_2 and concluded that computed geometries and energies varied significantly with the choice of the local component in the pseudopotential.

With these difficulties in mind we have chosen to use two complementary all electron techniques to study the TiO_2 (110) surface. Firstly the FP-LAPW method[22] employs a basis set consisting of plane waves which, (inside atom-centered, non-overlapping spheres) are matched continuously in value and slope to an expansion in terms of spherical harmonics (here up to $l_{max}^{wf} = 10$) and numerical solutions of the radial Schrödinger equation. This basis set has maximum flexibility and ensures the high accuracy of the calculations. Secondly, the LCAO method which employs a basis set of atom centred Gaussian functions for which a hierarchy of basis sets approaching complete convergence has recently been developed and tested for TiO_2 surfaces.[24,25] By confining all numerical approximations we provide definitive DFT-GGA results for this surface. Using the energy surface obtained we are able to examine the comparison of theory and experiment and thus resolve this long standing problem.

The next section contains details of the computational methods used, the results are then presented and discussed and our conclusions are summarised in the final section.

2 Methodology

In this section we give details of the structural model used to describe the (110) surface and of the FP-LAPW and LCAO methods used to perform our calculations.

2.1 Structural model

The (110) surface is modelled as a slab periodic in [001] and [$\bar{1}$10] directions but finite in the [110] direction (as shown in Fig. 1). As essential requirement for quantitative studies is that computed properties are fully converged with respect to the thickness of the slab. A systematic series of tests in which the structures of slabs of varying thickness were fully relaxed revealed that for a slab containing 21 atomic layers (i.e., 7 O–Ti_2O_2–O layers) the surface energy was converged to better than 0.1 J m^{-2} (6 eV Å^{-2}) and geometric displacements to better than 0.02 Å.[23–25]

2.2 FP-LAPW

In the FP-LAPW calculations a supercell periodic in 3 dimensions was used in which the slab geometry described above was repeated in the [110] direction with slabs separated by a large vacuum region of 9 Å to ensure that there were no significant interactions between the slabs. The

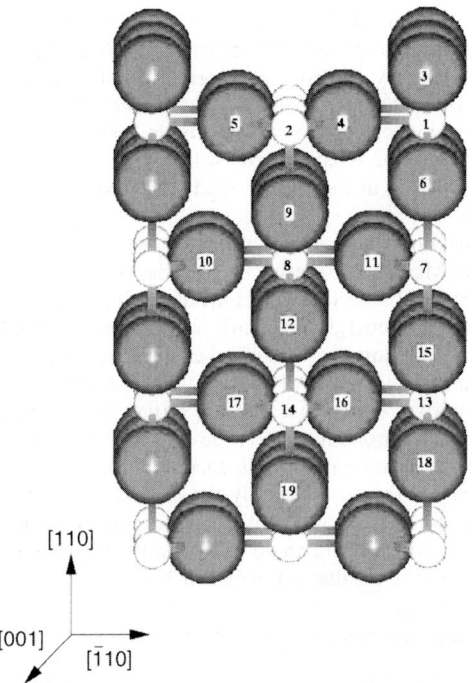

Fig. 1 A section through a (110) surface viewed in the [001] direction. The surface is based on a 21 layer slab with a mirror plane through the centre of the slab. All symmetry inequivalent ions in the top half of the slab are labelled.

size of this structural model is significantly larger than that used in previous studies of surface structures within the FP-LAPW method. These calculations have been made feasible by recent developments and improvements of the method.[26]

In order to minimise the number of k-points required to converge the surface energy care was taken to ensure a systematic cancellation of errors between the calculations on the slab and bulk crystal. This is achieved by describing the bulk crystal with a unit cell corresponding as closely as possible to that used to describe the surface and using identical computational parameters in both sets of calculations. A bulk unit cell with six times the volume of the primitive cell was used. With this arrangement we found that a uniform k-point mesh with three points in the irreducible part of the Brillouin zone was adequate. A kinetic energy cut-off for the plane-wave basis of $E_{max}^{wf} = 22$ Ry was used. This is a rather high value for such huge systems. However, because of the large surface relaxations we had to use rather small muffin-tin spheres ($R_{Ti}^{MT} = 0.90$ Å, $R_{O}^{MT} = 0.80$ Å) and therefore a large value for E_{max}^{wf} was mandatory to ensure good numerical accuracy. The electron density and potential are expanded in lattice harmonics up to $l_{max}^{pot} = 6$ inside the spheres, and the wavefunctions are expanded in angular momenta up to $l_{max}^{wf} = 10$. The electron density and potential in the interstitial region are expanded in plane waves up to 144 Ry. The core states are treated fully relativistically. The Ti 3s, 3p and O 2s, which are represented by local orbitals, as well as the valence states (Ti 3d, 4s, O 2p) are treated scalar-relativistically.

The relaxations of all atoms in the slab were considered, and the surface structure was determined by relaxing the entire system to equilibrium. All the atoms were relaxed according to the force directions and total energy minimization until all atom forces for a geometry fall below a certain limit. The process of the structure optimization has been described in ref. 27.

2.3 LCAO

The LCAO calculations were performed with the CRYSTAL program.[28] In contrast to the FP-LAPW calculations the slab geometry was modelled as periodic in two dimensions and finite in the third removing the need to define a vacuum gap.

The main approximation in the LCAO formalism is the choice of the local basis set used to expand the Bloch orbitals of the crystal. The basis set is made up of atom centered Gaussian functions with s, p or d symmetry. A systematic hierachy of basis sets was developed in a recent study of the TiO_2 (100) surface.[24] In this study it was shown that sets employing two basis functions to describe the valence electrons (so called, double valence—DV) can predict surface ionic relaxations to an accuracy of 0.02 Å compared to the basis set limit. Tests for the (110) surface confirm these conclusions and so in the current study a DV basis set has been used the details of which are given elsewhere.[24,25,29]

The total energy of the bulk crystal and surface were explicitly converged with respect to k-point sampling. A Pack–Monkhorst mesh[28,30] of order 4 which yields 10 k-points in the irreducible Brillouin zone of a (110) slab and 36 in that of the bulk crystal were used. This procedure of converging the bulk and slab energies explicitly with respect to k-space sampling removes the reliance on a systematic cancellation of errors when computing surface properties.

CRYSTAL computes matrix elements of the Coulomb, exchange and correlation matrix elements by direct summation over the infinite periodic lattice. Very efficient computational schemes for truncating the lattice summations have been developed.[31] The accuracy of the summation is based on overlap criteria for the atomic orbitals. Details of the control of these criteria have been described elsewhere.[32–34,24] In the current study the criteria were chosen to achieve an accuracy in the relative energies of the surface and bulk structures of the order of 1 meV per cell.[35] The surface relaxations were performed using an adapted conjugate gradient minimisation algorithm[36] to a tolerance of 0.01 Å in atomic positions and 10^{-5} eV in the total energy.

2.4 The Exchange-correlation functional

The main results of this article have been computed using the GGA functionals recently introduced by Perdew, Burke and Ernzerhof (PBE).[37] In addition a number of alternative treatments of the electron exchange and correlation interactions have been used in order to establish the sensitivity of key results. Within the LCAO formalism Hartree–Fock (i.e., non-local exchange with no treatment of correlation) and LDA calculations were also performed.[38,39,25] Within the FP-LAPW formalism the GGA functional proposed by Perdew and Wang (PWGGA)[40] was also used.

We find that the structural properties of the bulk crystal and the surface are very insensitive to the choice of functional. The PBE and PWGGA approaches agree to well within the numerical tolerances and differences between HF, LDA and GGA are confined to less than 0.02 Å in any surface or bulk displacement.[25] It is likely that differences between previous calculations which have been assigned to differing treatments of exchange and correlation are in fact due to incomplete convergence of the calculations.[16–20]

3 Results and discussion

The atomistic structure of the TiO_2 (110) surface is depicted in Fig. 1 in which labels are assigned to the atoms in the surface region. The relaxations of the top few layers computed here and in a number of recent studies are compared to those deduced from surface X-ray diffraction experiments in Table 1.

At first sight the most notable feature of this data is that the agreement between theory and experiment is poor. This is particularly true for the position of the bridging oxygen ion ($O_{(3)}$) for which the computed relaxation is never more than 0.16 Å while the experiment finds -0.27 Å. On closer examination it is the discepancy between the various theoretical approaches which gives most cause for concern. This is especially true for the current study in which, as stated above, great care has been taken to control the effects of all numerical tolerances on two different all-electron approaches. Nevertheless the relaxation of the bridging oxygen ion is computed to be -0.02 or -0.16 Å and the relaxation of the six-fold coordinated Ti-ion directly "beneath" it ($Ti_{(1)}$) to be 0.23 or 0.08 Å in the LCAO and FP-LAPW methods, respectively. These variations are significantly larger than the numerical errors that one would expect. It is however evident that the Ti–O separation in the [110] direction is more consistent (this is reported in the final row of Table 1). In the current study this is 1.04 Å and 1.03 Å in the FP-LAPW and LCAO calculations,

Table 1 The displacements (in Å, from bulk terminated positions) of the ions at the (110) (1 × 1) surface computed within the FP-LAPW and LCAO formalisms compared to previous plane wave calculations and those deduced from surface X-ray diffraction data[a,b]

Label	FP-LAPW-PBE [110]	FP-LAPW-PBE [$\bar{1}$10]	LCAO-PBE [110]	LCAO-PBE [$\bar{1}$10]	PW-LDA[18] [110]	PW-LDA[18] [$\bar{1}$10]	PW-GGA[20] [110]	PW-GGA[20] [$\bar{1}$10]	SXRD[14] [110]		SXRD[14] [$\bar{1}$10]
$Ti_{(1)}$	0.08	0.00	0.23	0.00	0.13	—	0.23	0.00	0.12	±0.05	0.00
$Ti_{(2)}$	−0.23	0.00	−0.17	0.00	−0.17	—	−0.11	0.00	−0.16	±0.05	0.00
$O_{(3)}$	−0.16	0.00	−0.02	0.00	−0.06	—	−0.02	0.00	−0.27	±0.08	0.00
$O_{(4)}$	0.09	−0.06	0.13	−0.05	0.13	−0.04	0.18	−0.05	0.05	±0.05	−0.16
$O_{(5)}$	0.09	0.06	0.13	0.05	0.13	0.04	0.18	0.05	0.05	±0.05	0.16
$O_{(6)}$	−0.09	0.00	0.02	0.00	−0.07	—	0.03	—	0.05	±0.08	0.00
$Ti_{(7)}$	0.07	0.00	0.14	0.00	0.06	—	0.12	—	0.07	±0.04	0.00
$Ti_{(8)}$	−0.13	0.00	−0.10	0.00	−0.08	—	−0.06	—	−0.09	±0.04	0.00
$O_{(9)}$	−0.05	0.00	0.00	0.00	0.02	—	0.03	—	0.00	±0.08	0.00
$O_{(10)}$	−0.04	0.03	0.03	0.03	−0.03	0.05	0.00	0.02	0.02	±0.06	0.07
$O_{(11)}$	−0.04	−0.03	0.03	−0.03	−0.03	−0.05	0.00	−0.02	0.02	±0.06	−0.07
$O_{(12)}$	−0.04	0.00	−0.01	0.00	−0.01	—	0.03	—	−0.09	±0.08	0.00
$Ti_{(13)}$	0.02	0.00	0.05	0.00	—	—	—	—	—	—	—
$Ti_{(14)}$	−0.08	0.00	−0.06	0.00	—	—	—	—	—	—	—
$O_{(15)}$	−0.07	0.00	0.01	0.00	—	—	—	—	−0.12	±0.07	0.00
$O_{(16)}$	−0.03	0.02	0.01	−0.02	—	—	—	—	—	—	—
$O_{(17)}$	−0.03	−0.02	0.01	0.02	—	—	—	—	—	—	—
$O_{(18)}$	−0.02	−0.02	0.01	0.00	—	—	—	—	—	—	—
$O_{(19)}$	0.02	−0.02	−0.01	0.00	—	—	—	—	—	—	—
O–Ti[a]	1.04		1.03		1.09		1.03		0.89±0.13		

[a] The last row gives the $O_{(3)}$–$Ti_{(1)}$ separation in the [110] direction. [b] The labels and directions refer to Fig. 1.

respectively, while 0.89 ± 0.13 Å is deduced from the SXRD experiment and ion scattering measurements.[15] This observation leads one to the hypothesis that the energy surface with respect to vertical displacement of the bridging oxygen is rather flat and that the displacement involves a cooperative motion of the surface ions which, at the least, involves the six-fold coordinated surface Ti-ion ($Ti_{(1)}$).

In order to explore this possibility a number of relaxations have been performed in which the position of the bridging oxygen ion has been constrained but the surrounding surface ions fully relaxed. For reasons of efficiency these calculations were performed on a relatively small system containing 9 atomic layers. The minimum energy structure of this system is somewhat displaced from that of larger systems but it displays all of the features necessary to examine the qualitative structure of the energy surface. The resultant energy surfaces computed within both the FP-LAPW and LCAO formalisms are displayed in Fig. 2. The LCAO calculations on this smaller system were performed with a very large basis set (the TVAEd basis set reported in ref. 24 and 25). In order to give some feeling for the energy scale a line representing a typical room temperature thermal energy per degree of freedom ($k_B T = 0.025$ eV) has been drawn on the figure. The energy surface is sufficiently flat for thermal vibrations to leave the minimum undetermined to about 0.15 Å. From this it is clear that the discrepancy between different theoretical approaches is due to the difficulty in finding an absolute energy minimum in this very flat energy surface.

We may associate the flat energy surface with a highly anisotropic and anharmonic surface vibrational mode. The nature of the mode is easily seen from an animation of the atomic positions. The displacements explored at thermal energies approximately corresponding to room temperature are displayed in Fig. 3. During this vibration the $O_{(3)}$–$Ti_{(1)}$ separation along [110] remains very close to 1.03 Å and the separation of $Ti_{(1)}$ and $O_{(6)}$ is also nearly constant. The displacements of $Ti_{(2)}$, $O_{(4)}$ and $O_{(5)}$ are very small. Thus, to the first approximation, we may understand the vibration as a "rigid unit mode" of the square planar TiO_4 unit containing $Ti_{(1)}$ and its four nearest neighbours ($O_{(3)}$, $O_{(6)}$) and their periodic images (this is depicted in the lower left panel of Fig. 3).

An immediate consequence of the rigid unit mode is that the structure of the TiO_2(110) surface apparent in the experimental probes applied at finite temperature does not correspond to the minimum energy configuration computed within a total energy calculation. In order to make such a comparison further treatment of the effect of surface vibrational modes on our interpretation of experimental data must be explored. In the case of SXRD and LEED experiments this necessitates the modelling of an anharmonic thermal vibration which is also highly anisotropic. The current practice is to fit the diffraction rods within an harmonic and often isotropic Debye–Waller model which is inadequate for the current case. A quantitative interpretation of STM images will require a treatment of the tip–surface interaction as this is likely to result in significant distortions of the surface structure.

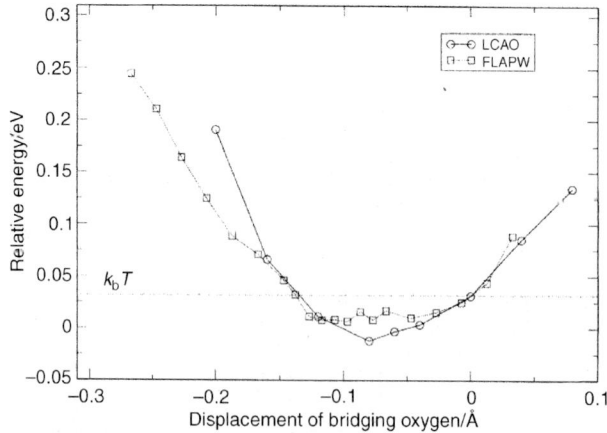

Fig. 2 The relative energy of the (110) surface computed within the LCAO and FP-LAPW formalisms for various fixed positions of the bridging oxygen ion relative to the unrelaxed bulk terminated position.

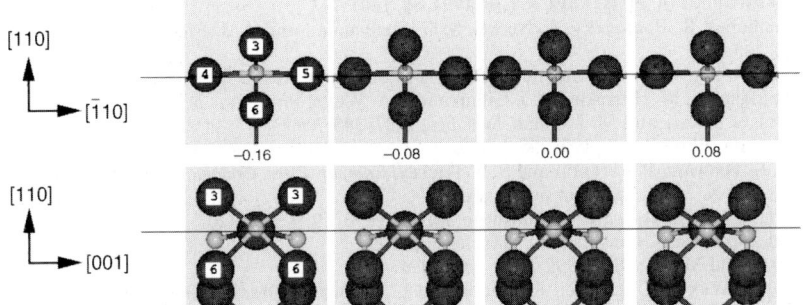

Fig. 3 The approximate room temperature thermal motions of the atoms within the soft rigid unit mode of the (110) surface.§ The atom labels correspond to those in Fig. 1. The oxygen atoms comprising the rigid unit are labelled in the lower left panel.

In addition the free energy associated with soft surface modes cannot be neglected. This has been demonstrated in recent first principles free energy calculations on the Ag(111) surface. In this system the minimum energy structure corresponds to a contraction of the outer layer spacing of -1.0% while the free energy minimum at 1150 K yields an expansion of 6.3%—a shift in the interlayer spacing of 0.16 Å.[41] We expect a significantly larger effect at the TiO_2(110) surface due to the presence of a soft, anharmonic, surface vibrational mode.

4 Conclusion

There is poor quantitative agreement between the structures predicted from the minimum energy configuration of first principles calculations and those deduced from X-ray diffraction experiments for the TiO_2(110) surface. We find that a very soft and anharmonic surface rigid-unit vibrational mode involves displacements of the surface ions of approximately 0.15 Å for thermal vibrations corresponding to room temperature. In order to perform an accurate comparison between theory and experiment for this and perhaps other surfaces it will be necessary to take account of such anisotropic vibrations in models used to interpret experimental data. In addition the contribution of the vibrational entropy to the surface free energy is likely to be significant and must be taken into account when computing surfaces energies and structures.

References

1 V. E. Henrich and A. F. Cox, *The Surface Science of Metal Oxides*, Cambridge University Press, Cambridge, 1993.
2 A. Fujishima and K. Honda, *Nature (London)*, 1972, **238**, 37.
3 *New Scientist*, 1998, 8.
4 *New Scientist*, 1997, 15.
5 *New Scientist*, 1998, 11.
6 I. M. Butterfield, P. A. Christensen, A. Hamnett, K. E. Shaw, G. M. Walker, S. A. Walker and C. R. Howarth, *J. App. Electrochem.*, 1997, **27**, 385.
7 R. A. Bennett, P. Stone, N. J. Price and M. Bowker, *Phys. Rev. Lett.*, 1999, **82**, 3831.
8 J. Muscat, N. M. Harrison and G. Thornton, *Phys. Rev. B*, 1999, **59**, 15457.
9 A. T. Paxton and L. Thien-Nga, *Phys. Rev. B*, 1998, **57**, 1579.
10 V. E. Henrich and R. L. Kurtz, *Phys. Rev. B*, 1981, **23**, 6280.
11 H. Onishi and Y. Iwasawa, *Surf. Sci.*, 1994, **313**, L783.
12 P. W. Murray, N. G. Condon and G. Thornton, *Phys. Rev. B*, 1995, **51**, 10989.
13 U. Diebold, J. F. Anderson, K. O. Ng and D. Vanderbilt, *Phys. Rev. Lett.*, 1996, **77**, 1322.
14 G. Charlton, P. B. Howes, C. L. Nicklin, P. Steadman, J. S. G. Taylor, C. A. Muryn, S. P. Harte, J. Mercer, R. McGrath, D. Norman, T. S. Turner and G. Thornton, *Phys. Rev. Lett.*, 1997, **78**, 495.
15 B. Hird and R. A. Armstrong, *Surf. Sci. Lett.*, 1997, **385**, L1023.

§ Two movies of these motions are available as electronic supplementary information. See http://www.rsc.org/suppdata/fd/1999/305

16. P. Reinhardt and B. A. Hess, *Phys. Rev. B*, 1994, **50**, 12015.
17. D. Vogtenhuber, R. Podloucky, A. Neckel, S. G. Steinemann and A. J. Freeman, *Phys. Rev. B*, 1994, **49**, 2099.
18. M. Ramamoorthy, D. Vanderbilt and R. D. King-Smith, *Phys. Rev. B*, 1994, **49**, 16721.
19. P. J. D. Lindan, N. M. Harrison, M. J. Gillan and J. A. White, *Phys. Rev. B*, 1997, **55**, 15919.
20. S. P. Bates, G. Kresse and M. J. Gillan, *Surf. Sci.*, 1997, **385**, 386.
21. D. R. Hamann, *Phys. Rev. B*, 1997, **56**, 14979.
22. P. Blaha, K. Schwarz, P. Sorantin and S. B. Trickey, *Comput. Phys. Commun.*, 1990, **59**, 399.
23. J. Muscat and N. M. Harrison, in preparation.
24. J. Muscat, N. M. Harrison and G. Thornton, *Phys. Rev. B*, 1999, **59**, 2320.
25. J. Muscat, PhD Thesis, University of Manchester, 1999.
26. M. Petersen and M. Scheffler, 1999, to be published.
27. B. Kohler, S. Wilke, M. Scheffler, R. Kouba and C. Ambrosch-Draxl, *Comput. Phys. Commun.*, 1996, **94**, 31.
28. R. Dovesi, V. R. Saunders, C. Roetti, M. Causà, N. M. Harrison, R. Orlando and E. Aprà, *CRYSTAL95 User's Manual*, University of Turin, Turin, 1996.
29. http://www.dl.ac.uk/TCS/Software/CRYSTAL/. The CRYSTAL Basis set library, 1998.
30. J. D. Pack and H. J. Monkhorst, *Phys. Rev. B*, 1977, **16**, 1748.
31. C. Pisani, R. Dovesi and C. Roetti, *Hartree–Fock Ab Initio Treatment of Crystalline Systems*, Springer-Verlag, Berlin, 1988, vol. 48.
32. R. Orlando, R. Dovesi, C. Roetti and V. R. Saunders. *J. Phys.: Condens. Matter*, 1990, **2**, 7769.
33. R. Dovesi, M. Causa, R. Orlando, C. Roetti and V. R. Saunders, *J. Chem. Phys.*, 1990, **92**, 7402.
34. *Quantum Mechanical Ab Initio Calculation of the Properties of Crystalline Materials*, ed. C. Pisani, Springer-Verlag, Berlin, 1996, vol. 67.
35. Explicitly the ITOL parameters were set to 10^{-6}, 10^{-6}, 10^{-6}, 10^{-6} and 10^{-12}.
36. C. Zhu, R. H. Byrd, P. Lu and J. Nocedal, *L-BFGS-B—Fortran Subroutines for Large Scale Bound Constrained Optimisation*, Dept of Elec. Eng. and Comp. Sci, Northwestern University, Illinois, 1994.
37. J. P. Perdew, K. Burke and M. Ernzerhof, *ACS Symposium-Series*, 1996, **629**, 453.
38. P. A. M. Dirac, *Proc. Cambridge Philos. Soc.*, 1930, **26**, 376.
39. J. P. Perdew and A. Zunger, *Phys. Rev. B*, 1981, **23**, 5048.
40. J. P. Perdew, J. A. Chevary, S. H. Vosko, K. A. Jackson, M. R. Pederson, D. J. Singh and C. Fiolhais, *Phys. Rev. B*, 1992, **46**, 6671.
41. J. Xie, S. Gironcoli, S. Baroni and M. Scheffler, *Phys. Rev. B*, 1999, **59**, 970.

Paper 9/06386B

The chemistry of methanol on the TiO$_2$(110) surface: the influence of vacancies and coadsorbed species

Michael A. Henderson,[*a] **Sary Otero-Tapia**[b] **and Miguel E. Castro**[b]

[a] *Interfacial and Processing Sciences Group, Environmental Molecular Sciences Laboratory, Pacific Northwest National Laboratory, Richland, WA 99352, USA. E-mail: ma.henderson@pnl.gov*

[b] *Department of Chemistry, University of Puerto Rico, Mayuagues, Puerto Rico 00680*

Received 16th March 1999

The chemistry of methanol was explored on the vacuum annealed TiO$_2$(110) surface, with and without the presence of coadsorbed water and oxygen, using temperature programmed desorption (TPD), high resolution electron energy loss spectroscopy (HREELS), static secondary ion mass spectrometry (SSIMS) and low energy electron diffraction (LEED). The vacuum annealed TiO$_2$(110) surface possessed about 8% oxygen vacancy sites, as determined with H$_2$O TPD. Although evidence is presented for CH$_3$OH dissociation to methoxy groups on the vacuum annealed TiO$_2$(110) surface using SSIMS and HREELS, particularly at vacancy sites, the majority of the adlayer was molecularly adsorbed, evolving in TPD at 295 K. Although no evidence of irreversible decomposition was found in the TPD, dissociative CH$_3$OH adsorption at 135 K on the vacuum annealed TiO$_2$(110) surface led to recombinative desorption states at 350 and 480 K corresponding to methoxys adsorbed at non-vacancy and vacancy sites, respectively. Coadsorbed water had little or no influence on the chemistry of CH$_3$OH on the vacuum annealed TiO$_2$(110) surface, however new channels of chemistry were observed when CH$_3$OH was adsorbed on the surface after O$_2$ adsorption at various temperatures. In particular, O$_2$ exposure at 300 K resulted in O adatoms (*via* dissociation at vacancies) that led to increased levels of CH$_3$O–H bond cleavage. The higher surface coverage of methoxy then resulted in a disproportionation reaction to form CH$_3$OH and H$_2$CO above 600 K. In contrast, low temperature exposure of the vacuum annealed TiO$_2$(110) surface to O$_2$ resulted in low temperature state of O$_2$ (presumably an O$_2^-$ species) that oxidized CH$_3$OH to H$_2$CO by C–H bond cleavage. These results provide incentive to consider alternative thermal and photochemical oxidation mechanisms that involve the interaction of organics and oxygen at surface defect sites.

1 Introduction

The role of thermal reactions is often ignored in the studies aimed at understanding the photochemical oxidation of organics on TiO$_2$ surfaces, perhaps because the importance of electron-hole pair processes is easily over-conceptualized. However, key thermal processes may proceed, be in conjunction with or follow photoexcitation events. In a similar sense, thermal processes occurring at photo-generated surface sites may also be important. Focusing on this last issue, a number of groups have shown that photon irradiation of TiO$_2$ surfaces (with photon energies in excess of the bandgap) results in the creation of electronic surface defects.[1–4] Although their exact nature has not been determined, many groups have suggested that these surface defects are oxygen vacancies. If so, photoinduced surface oxygen vacancies may exhibit the same kinds of rich chemistry for activation of adsorbates that is known to occur on thermally induced surface oxygen vacancies.[5]

In this work, the chemistry of methanol was studied on the vacuum annealed surface of $TiO_2(110)$ both with and without coadsorbed water and oxygen. Dissociative methanol chemistry forming methoxy groups occurs at both vacancy and non-vacancy sites, and is enhanced by the presence of oxygen adatoms. Methanol is also oxidized to formaldehyde *via* two processes. The first is at high temperature (>600 K), and is associated with disproportionation of methoxy groups on the fully oxidized surface. The second, and more interesting, results from a low temperature (<250 K) oxidation of C–H bonds by coadsorbed O_2 molecules adsorbed in the vicinity of oxygen vacancies. These results suggest that oxygen molecules adsorbed at vacancy sites, generated by photoabsorption processes, may promote catalytic and/or stoichiometric oxidation reactions of organics. This is a previously unexplored mechanism for the photooxidation of organics on TiO_2.

2 Experimental

This study was conducted in an ultrahigh-vacuum (UHV) vacuum chamber equipped with a quadrupole mass spectrometer (QMS) for temperature programmed desorption (TPD) and static secondary ion mass spectrometry (SSIMS) studies, a high resolution electron energy loss spectrometry (HREELS) and a low energy electron diffraction (LEED) apparatus. Procedures for use of these techniques in studying TiO_2 single crystals are described in greater detail elsewhere,[6-8] but the essential experimental details are discussed below. HREELS experiments were performed in the specular scattering direction with a primary beam energy of 8–9 eV. Multiple phonon losses resulting from two or more losses associated with the oxide's intense optical phonon modes were removed using Fourier deconvolution.[6,9-11] All HREELS spectra were obtained at 110 K. SSIMS data were collected with a focused and unrastered 500 eV primary Ar^+ beam. The current density in the beam was maintained in the static sputtering domain, and the total surface damage done during any SSIMS experiment did not exceed 5% of a monolayer. All temperature programmed studies were conducted with a heating rate of 2 K s^{-1}.

The $TiO_2(110)$ crystal used in this study was bulk reduced to a blue color by annealing in UHV at 850 K. It was cleaned by Ar^+ ion sputtering followed by annealing at 850 K in UHV. Methanol (Aldrich, ACS HPLC grade, 99.9 + %) and water (Aldrich, ACS HPLC grade) were purified by several freeze–pump–thaw cycles using liquid nitrogen. Research grade oxygen and isotopically labeled oxygen were used without additional purification. The gas handling system was conditioned to each molecule prior to use. Gases were dosed either by backfilling (oxygen) or through a 0.25 in OD directional doser with a calibrated pinhole aperture (methanol and water). Additional details on directional dosing can be found elsewhere.[8]

3 Results and discussion

Annealing $TiO_2(110)$ in UHV results in the formation of surface oxygen vacancy sites at a surface concentration of about 5–10%. These vacancy sites are detectable with a variety of electron spectroscopies,[5,12-15] with scanning tunneling microscopy[16,17] and with temperature programmed desorption.[7,15,18,19] The surface referred to throughout this paper as the 'vacuum annealed' surface is prepared by annealing at 850 K for 10 min in UHV resulting in about 8% (or 0.4×10^{14} sites cm^{-2}) oxygen vacancies (as gauged by water TPD). Results for CH_3OH on the clean vacuum annealed $TiO_2(110)$ surface are presented first, followed by the coadsorption studies of CH_3OH with water and oxygen.

3.1 Methanol on the vacuum annealed $TiO_2(110)$ surface

3.1.1 TPD results. Fig. 1 shows a series of CH_3OH ($m/z = 31$) TPD spectra as a function of CH_3OH exposure on the vacuum annealed $TiO_2(110)$ surface at 135 K. Each spectrum in Fig. 1 was obtained without subjecting the surface to any additional treatment other than the initial 850 K anneal and previous CH_3OH TPD experiments. The surface was carbon-free after CH_3OH

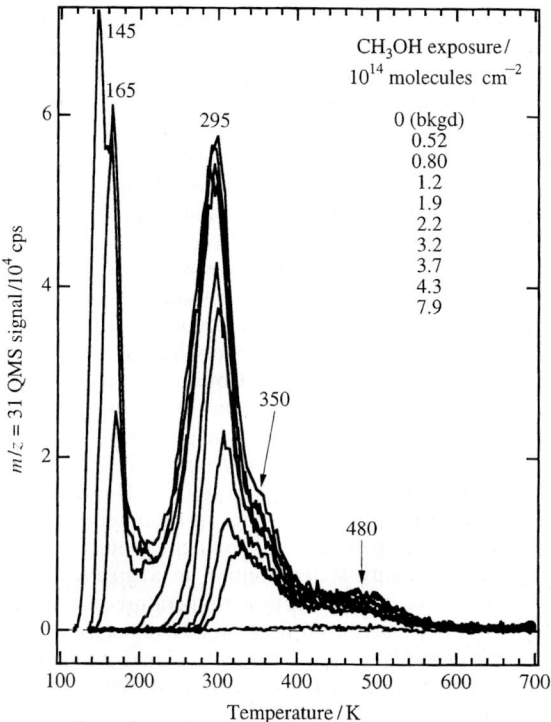

Fig. 1 CH$_3$OH ($m/z = 31$) TPD spectra from various exposures of CH$_3$OH on the vacuum annealed TiO$_2$(110) surface at 135 K. The lowest exposure spectrum corresponds to background adsorption of CH$_3$OH.

TPD to 700 K as gauged by AES (data not shown). CH$_3$OH was the only carbon-containing product detected in TPD, although a small amount of H$_2$O desorption was also detected, presumably from background adsorption. (The role of preadsorbed H$_2$O on CH$_3$OH chemistry will be discussed below.) This is in contrast to CH$_3$OH TPD behavior on the TiO$_2$(001) surface, which shows various decomposition products including formaldehyde and methane.[20,21]

For any given CH$_3$OH exposure, the TPD spectrum of CH$_3$OH was reproducible for consecutive TPD experiments indicating that the vacuum annealed TiO$_2$(110) surface was not altered by CH$_3$OH TPD to 700 K. The lowest exposure spectrum in Fig. 1 corresponds to accumulation of background CH$_3$OH during a 20 min period in UHV. This spectrum shows a broad CH$_3$OH TPD state centered between 430 and 500 K with a TPD peak area of approximately 2% of the saturated monolayer (see below). The lowest exposure dosed on the crystal, 5.2×10^{13} molecules cm^{-2}, gave two TPD states at about 350 and 480 K. Both peaks slightly increased in intensity with increasing CH$_3$OH exposure, but a third state appearing at about 295 K dominated for exposures above 1×10^{14} molecules cm^{-2}. The 350 K TPD state became unresolved as the 295 K TPD state increased with increasing CH$_3$OH exposure. Assignments for these TPD features will be discussed below. Monolayer saturation was reached after a CH$_3$OH exposure of about 3.2×10^{14} molecules cm^{-2}. At higher CH$_3$OH exposures, two additional CH$_3$OH TPD features developed at about 165 and 150 K. The 165 K CH$_3$OH peak appeared first for exposures above 3.2×10^{14} molecules cm^{-2}, and saturated after an exposure of about 4.5×10^{14} molecules cm^{-2}. This peak is assigned to desorption of methanol molecules hydrogen-bonded to bridging oxygen sites based on similar behavior observed for water.[6,18] The 150 K CH$_3$OH TPD peak, observed for adsorption temperatures below 130 K, is due to multilayer CH$_3$OH desorption based on comparisons with TPD studies of methanol on metal surfaces.[22,23] (Notice that the adsorption temperature for the 7.9×10^{14} molecules cm^{-2} exposure in Fig. 1 is at the leading edge of the multilayer TPD peak. The total CH$_3$OH TPD peak area of this spectrum suggests that a multilayer can not be sustained on the surface at an adsorption temperature of 120 K.)

Fig. 2 shows the plot of the total CH_3OH TPD peak area (circles) and the TPD peak area above 200 K (triangles) vs. the CH_3OH exposure on the vacuum annealed $TiO_2(110)$ surface at 135 K. The total peak area data (circles) form a straight line that intersects the x-axis at the origin. Since only CH_3OH was observed as a TPD product, the linear yield implies that the sticking probability of CH_3OH on the vacuum annealed $TiO_2(110)$ surface at 135 K is constant with exposure, and is likely unity. The CH_3OH exposure corresponding to monolayer saturation can be approximated from the data in Fig. 2. A delineation temperature of 200 K was used to differentiate TPD peak area associated with the physisorbed and chemisorbed (or 'monolayer') states of CH_3OH. The CH_3OH peak area data above 200 K (triangles) as a function of CH_3OH exposure intersects with the total peak area data (circles) at an exposure of about 3.2×10^{14} molecules cm^{-2}, implying a monolayer saturation coverage of the same value (assuming the CH_3OH sticking probability at 135 K is unity in this exposure regime (as suggested by the total TPD peak area in Fig. 2). Total monolayer coverages on rutile powders have been estimated at 4.8×10^{14} molecules cm^{-2}[24] and 4.3×10^{14} molecules cm^{-2}.[25] Considering the fact that the (110) surface is more open compared to the other major faces found on rutile powders,[26] the 3.2×10^{14} molecules cm^{-2} value obtained in this study is consistent with the powder results. Previous estimates of the saturation CH_3OH coverage for adsorption *at room temperature* based on XPS yielded values of 3×10^{14} molecules cm^{-2}[27] and 2.6×10^{14} molecules cm^{-2}.[28] However, the TPD results in Fig. 1 suggest that the rate of CH_3OH desorption is too high at room temperature to sustain a coverage of greater than about 1×10^{14} molecules cm^{-2}. One might presume that CH_3OH adsorption is activated into a higher binding energy state resulting in a greater coverage in that state for adsorption at 300 K than at 135 K (the adsorption temperature of the data in Fig. 1). Although such an activation barrier has been predicted for CH_3OH dissociative adsorption on $TiO_2(110)$ using *ab initio* cluster calculations,[29] but not with molecular dynamics simulations,[30] results by Kim and Barteau show no enhanced CH_3OH adsorption at 300 K *vs.* 200 K for the $TiO_2(001)$ surface.[20] Based on arguments presented below, if activated CH_3OH dissociative adsorption were to occur on the $TiO_2(110)$ surface at room temperature, it would be into the 350 K TPD state whose leading edge is below room temperature. The discrepancy between the coverage estimate in this study and those based on XPS probably lie in the difficulty of obtaining absolute coverages with XPS.

The total surface cation site density (vacancy and non-vacancy related) on the vacuum annealed $TiO_2(110)$ surface is approximately equal to 5.6×10^{14} molecules cm^{-2}, with 0.4×10^{14} molecules cm^{-2} associated with the vacancies (see above). If one assumes that each available vacancy sites binds one CH_3OH molecule (an assumption discussed below), then the coverage of non-vacancy related CH_3OH species at saturation is about 2.8×10^{14} molecules cm^{-2}. However, it is by no coincidence that the saturation CH_3OH TPD peak area in the 480 K state corresponds to about

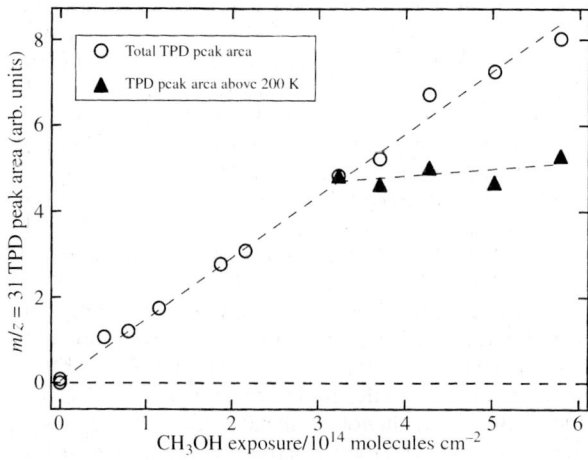

Fig. 2 Total CH_3OH (m/z = 31) TPD peak area (circles) and CH_3OH (m/z = 31) TPD peak area above 200 K (triangles) *vs.* CH_3OH exposure.

0.4×10^{14} molecules cm^{-2}, with the remaining peak area (corresponding to about 2.8×10^{14} molecules cm^{-2}) in the 295 and 350 K TPD states. The former value is equivalent to the vacancy population as determined by water TPD (see below). Although the 480 K CH$_3$OH TPD peak appeared to be related to oxygen vacancies, the identity of the species bound in the vacancy (molecular or dissociative) is not discernable from the TPD data. Also, the identity of the surface species associated with the 295 and 350 K CH$_3$OH TPD states is not known. The next sections focus on identifying the surface species associated with these TPD features.

3.1.2 HREELS results. HREELS analysis was attempted in order to identify the adsorbed states of methanol associated with the various TPD states in Fig. 1. Fig. 3 shows Fourier deconvoluted HREELS data for a multilayer exposure of CH$_3$OH on the vacuum annealed TiO$_2$(110) surface at 110 K, followed by heating to various temperatures. The '×1' spectra (Fig. 3a) are the raw and Fourier deconvoluted data from the clean surface, the latter of which is also shown as a '×50' spectrum (Fig. 3b). The deconvolution process removed the majority of the intensity above 1000 cm^{-1} associated with the multiple phonon loss processes.[6,9–11] The multiple phonon remnants at 1520 and 1255 cm^{-1} shown in the '×50' clean surface spectrum (Fig. 3b) are the result of combinations of dipole and non-dipole scattering processes which are not removed by the Fourier deconvolution process.[6,10] After adsorption of multilayer CH$_3$OH at 110 K, new losses are present at 3245, 2930, 1480, 1150, 1025, 730, 445 and 350 cm^{-1} (Fig. 3c). Part of the loss intensity at 730 cm^{-1} and all of the loss intensity at 445 cm^{-1} can be assigned to the phonon modes of the oxide, which are strong enough to be detected through the methanol multilayer. The 730 cm^{-1} loss can also be assigned to the ρ(OH) of CH$_3$OH ice.[22,23,31] The losses at 3245, 2930, 1480, 1150, 1025, and 350 cm^{-1} can be assigned to the ν(OH), ν(CH$_3$), δ(CH$_3$), ρ(CH$_3$), ν(CO) and τ(CH$_3$OH) modes, respectively, of CH$_3$OH ice.[22,23,31] Heating to 200 K desorbed the multilayer and second

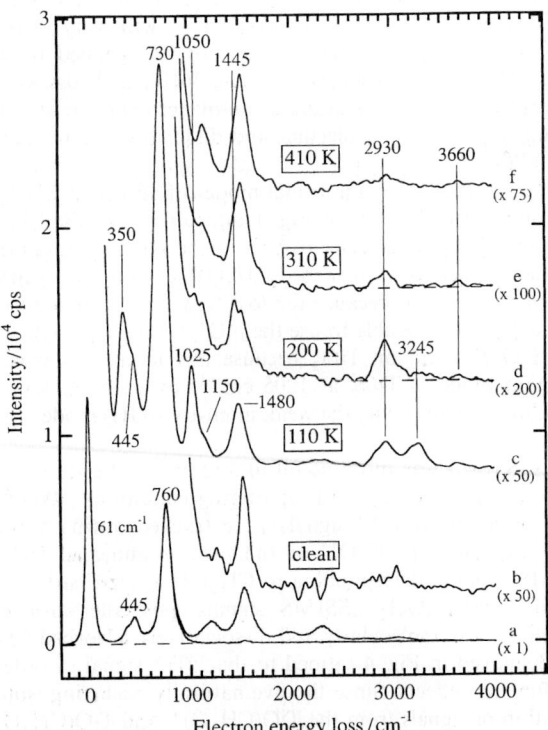

Fig. 3 Fourier deconvoluted HREELS spectra from the clean vacuum annealed TiO$_2$(110) surface (a and b), and after a multilayer CH$_3$OH exposure at 110 K (c), followed by heating to 200 (d), 310 (e) and 410 K (f). The heated surfaces were recooled to below 110 K before commencing data collection. The '×1' spectra are for both the raw and deconvoluted data.

layer (see Fig. 1) resulting in the disappearance of the 730 and 350 cm^{-1} losses (Fig. 3d), and a resurgence of the intense phonon modes (data off scale). After heating to 200 K, the multiple phonon remnants partially obscured losses associated with the ν(OH), δ(CH$_3$) and ν(CO) modes, with the last two shifting to 1445 and 1050 cm^{-1}. The 3245 cm^{-1} loss suggests that some molecular CH$_3$OH is still present on the surface after heating to 200 K in the form of hydrogen-bonded species, either to bridging oxygen atoms or to other methanol-related species. However, there was also loss intensity at 3660 cm^{-1} suggestive of a surface hydroxy group, possibly indicating that some CH$_3$OH molecules dissociated to methoxy groups. CH$_3$OH dissociation has been detected on TiO$_2$(110),[28] TiO$_2$(100)[28] and TiO$_2$(001) surfaces,[20] as well as on rutile and anatase powders.[24,25,32–40] Several theoretical studies predict CH$_3$OH dissociation on TiO$_2$(110).[29,30,41,42] (Dissociation of CH$_3$OH has also been detected on other single crystal oxide surfaces.[43–45]) The 3660 cm^{-1} loss is similar to the ν(OH) loss observed after dissociation of formic acid to formate and a bridging OH group.[46,47] Although the 3660 cm^{-1} loss could conceivably be due to the ν(OH) mode of an isolated CH$_3$OH molecule (the gas phase value for this mode is at 3680 cm^{-1},[48] it is most likely due to a surface hydroxy group. Based on data for H$_2$O on TiO$_2$(110),[6] the ν(OH) mode of the isolated CH$_3$OH molecule should be red-shifted from its gas phase value by at least 100 cm^{-1} due to chemisorption on the surface.

Heating to 310 K desorbed the majority of the 295 K TPD state (see Fig. 1), leaving weak losses at 3680 and 2945 cm^{-1} (Fig. 3e). Loss intensity was still present at about 1445 and 1050 cm^{-1}, but was not clearly resolved. Heating to 410 K left only the 480 K CH$_3$OH state in TPD and weak losses at 3660 and 2960 cm^{-1} (Fig. 3f). The surface coverage of CH$_3$OH species at this point is about 0.08 ML (see above).

The HREELS data in Fig. 3 suggests that some CH$_3$OH molecules were dissociatively adsorbed on the vacuum annealed TiO$_2$(110) surface after desorbing the multilayer, but whether any dissociation occurred upon adsorption could not be determined from multilayer exposure data. The HREELS spectrum from a submonolayer coverage of CH$_3$OH on the vacuum annealed TiO$_2$(110) surface at 110 K gave losses at 2970, 1450 and 1375 cm^{-1}, with very weak features at 3660 and 3120 cm^{-1} (data not shown). The 3120 cm^{-1} loss can be assigned to the ν(OH) mode of a hydrogen-bonded CH$_3$OH molecule, whereas the weak 3660 cm^{-1} loss was presumably due to a hydroxy group that was observed after heating a multilayer exposure to 200 K (Fig. 3d). Therefore, HREELS data suggest that both molecular and dissociative forms of CH$_3$OH were present upon adsorption at 110 K.

Similarities in the frequencies of the vibrational modes of adsorbed CH$_3$OH and CH$_3$O makes assignments of the various TPD features in Fig. 1 difficult using HREELS. Based on HREELS studies of CH$_3$OH adsorbed on metal surfaces,[22,23,31] the most reliable means of differentiating between molecularly and dissociatively adsorbed CH$_3$OH is with the disappearance of the ν(OH) and ρ(OH) modes upon dissociation because the loss frequencies related to the methyl group do not significantly change. It is not possible to use the ρ(OH) mode of CH$_3$OH to reliably follow the dissociation of the CH$_3$O–H bond on TiO$_2$ because this mode was partially obscured by the multiple phonon remnant mode of TiO$_2$ at 1205 cm^{-1}. In a similar sense, the cleavage of the CH$_3$O–H bond is difficult to follow using the weak intensity ν(OH) mode.

3.1.3 SSIMS results.
SSIMS was more useful in addressing the state of adsorbed CH$_3$OH in the various TPD states of Fig. 1. Using static sputtering conditions (500 eV Ar$^+$, $\leqslant$2 nA cm^{-2}), CH$_3$O- and CH$_3$OH-containing SSIMS signals were tracked during temperature programmed heating of a multilayer exposure of CH$_3$OH on the vacuum annealed TiO$_2$(110) surface (Fig. 4). Although several SSIMS signals corresponding to CH$_3$OH-related surface species were detected, the TiO(CH$_3$O)$^+$ and TiO(CH$_3$OH)$^+$ SSIMS signals gave the strongest signals, and best described the surface changes in molecularly and dissociatively adsorbed CH$_3$OH during heating. The SSIMS data are displayed in Fig. 4 ratioed to the ^{48}Ti$^+$ signal in order to remove ion yield changes due to work function effects. Since the five naturally occurring isotopes of Ti ($m/z = 46$–50) result in a convolution of signals from the TiO(CH$_3$O)$^+$ and TiO(CH$_3$OH)$^+$ SSIMS ions, the individual ^{48}TiO(CH$_3$O)$^+$ and ^{48}TiO(CH$_3$OH)$^+$ SSIMS signals were deconvoluted from the other ion contributions at $m/z = 95$ and 96, respectively, using matrix algebra and the known natural abundances of Ti isotopes. (The 'negative' signals in the 48TiO(CH$_3$OH)$^+$ deconvoluted data resulted from the deconvolution process and not from a vertical displacement of the spec-

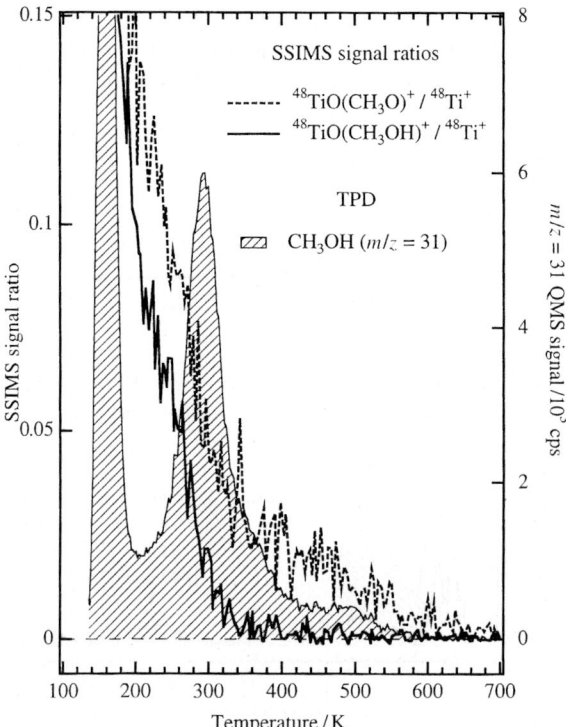

Fig. 4 TPD ($m/z = 31$) and temperature programmed SSIMS (^{48}Ti(CH$_3$O)$^+$ and ^{48}(Ti(CH$_3$OH)$^+$) data from multilayer CH$_3$OH exposures at 135 K on the vacuum annealed TiO$_2$(110) surface. The SSIMS data are presented as ratios with respect to the ^{48}Ti$^+$ signal. The Ar$^+$ ion energy and current for the SSIMS measurement were 500 eV and $\leqslant 2$ nA cm^{-2}, respectively.

trum in Fig. 4.) The ^{48}TiO(CH$_3$O)$^+$ and ^{48}TiO(CH$_3$OH)$^+$ SSIMS signals possessed significant intensity at 135 K arising from multilayer CH$_3$OH on vacuum annealed TiO$_2$(110) (both signals are off-scale in Fig. 4). As the surface was heated, the ^{48}TiO(CH$_3$OH)$^+$ SSIMS signal attenuated to near zero intensity during desorption of the 295 K CH$_3$OH TPD state. In contrast, the ^{48}TiO(CH$_3$O)$^+$ SSIMS signal, which also attenuated during the 295 K desorption process, retained significant intensity above this temperature and attenuated to zero signal above 500 K. These data suggest that the 290 K TPD state is associated with desorption of molecularly adsorbed CH$_3$OH, while the 480 K TPD state is due to recombinative desorption of dissociatively adsorbed CH$_3$OH, presumably at vacancy sites based on the TPD results discussed above. The identity of the 350 K CH$_3$OH TPD peak is not discernable from this data, and will be discussed below. The behavior of the ^{48}TiO(CH$_3$O)$^+$ and ^{48}TiO(CH$_3$OH)$^+$ SSIMS signals matches that for the ^{48}TiO(OH)$^+$ and ^{48}TiO(H$_2$O)$^+$ SSIMS signals for H$_2$O adsorbed on the vacuum annealed TiO$_2$(110) surface.[15] The ^{48}TiO(H$_2$O)$^+$ SSIMS signal attenuated to zero during desorption of molecularly adsorbed H$_2$O at 270 K, while the ^{48}TiO(OH)$^+$ SSIMS signal attenuated to zero during desorption of the 540 K H$_2$O TPD peak, ascribed to recombinative desorption of OH groups at vacancies (see below). Assignment of the 295 and 480 K CH$_3$OH TPD states to molecularly adsorbed CH$_3$OH at non-vacancy sites and dissociatively adsorbed CH$_3$OH at vacancy sites, respectively, is therefore consistent with the known chemistry of H$_2$O on the vacuum annealed TiO$_2$(110) surface.[6,7,14,18,49] H$_2$O adsorbs primarily in its molecular form on the clean, vacancy-free TiO$_2$(110) surface, but dissociation occurs at oxygen vacancy sites and possibly also at step sites.[7]

3.1.4 Oxygen isotope studies. Although both CH$_3$OH and H$_2$O dissociate at oxygen vacancy sites, results using oxygen-18 reveal a significant difference between the surface chemistries of the two molecules. Fig. 5 shows TPD spectra for multilayer exposures of H$_2$^{16}O and CH$_3$^{16}OH

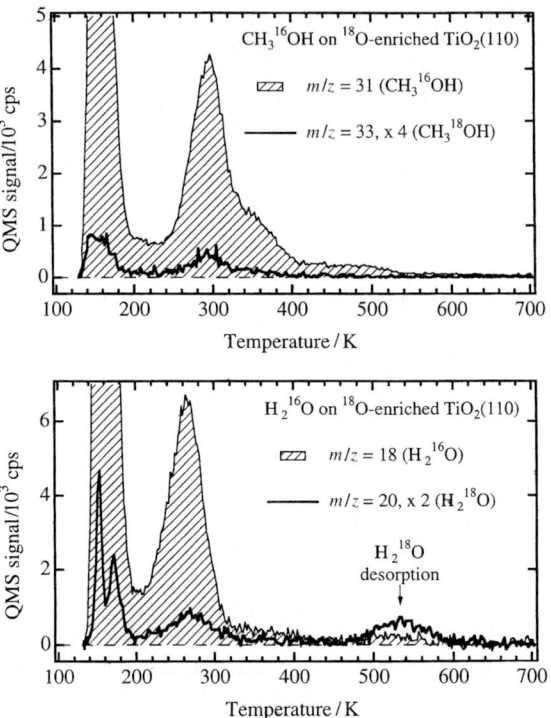

Fig. 5 Exchange of lattice oxygen with adsorbed $CH_3^{16}OH$ (upper panel) and $H_2^{16}O$ (lower panel) on the ^{18}O-enriched vacuum annealed $TiO_2(110)$ surface. Upper panel: TPD of $CH_3^{16}OH$ ($m/z = 31$; filled spectrum) and $CH_3^{18}OH$ ($m/z = 33$; solid line) from a multilayer exposure of $CH_3^{16}OH$ on the ^{18}O-enriched vacuum annealed $TiO_2(110)$ surface at 130 K. Lower panel: TPD of $H_2^{16}O$ ($m/z = 18$; filled spectrum) and $H_2^{18}O$ ($m/z = 20$; solid line) from a multilayer exposure of $H_2^{16}O$ on the ^{18}O-enriched vacuum annealed $TiO_2(110)$ surface at 130 K. See text for details on preparation of the ^{18}O-enriched surface.

adsorbed at 135 K on the ^{18}O-enriched vacuum annealed $TiO_2(110)$ surface. (The ^{18}O-enriched surface was prepared by oxidation in 5×10^{-6} Torr $^{18}O_2$ at 750 K, followed by annealing in UHV at 850 K to smooth the surface and regenerate vacancy sites. SSIMS indicated that the surface was essentially $Ti^{18}O_2$ after this treatment.) H_2O dissociation at oxygen vacancy sites is evidenced by preferential desorption of $H_2^{18}O$ ($m/z = 20$) in the 540 K TPD state (lower panel of Fig. 5). The 540 K H_2O TPD state was assigned to recombinative water desorption at vacancy sites based on these isotope experiments,[7] and on the quantitative similarities between the relative TPD peak area in this state and the coverage of vacancies typically observed on annealed $TiO_2(110)$ by STM.[16,17] The 270 K H_2O TPD peak, which HREELS[6] and SSIMS[15] have identified as desorption of molecularly adsorbed water, does not scramble oxygen atoms with the lattice. This result not only indicates that the 540 K water TPD state is due to dissociative recombination of hydroxy groups, but also that the hydrogen atoms, residing on the rows of bridging oxygen sites, are highly mobile along the oxygen rows. (The presence of $H_2^{18}O$ in the water desorption states below 300 K occurred with roughly the same $m/z = 18$ to $m/z = 20$ ratio indicating that it resulted from exchange processes in the QMS. The QMS was unavoidably exposed to $^{18}O_2$ during the surface ^{18}O-enrichment process.) In contrast, TPD of a multilayer exposure of $CH_3^{16}OH$ on the ^{18}O-enriched vacuum annealed $TiO_2(110)$ surface does not result in preferential $CH_3^{18}OH$ evolution from the surface in the 480 K TPD state (upper panel of Fig. 5). Although one might expect that the deposited alcohol hydrogen atom should be as mobile as that from water dissociation, the methyl group on the vacancy-bound methoxy did not diffuse presumably because the barrier to C–O bond cleavage was too high. Therefore, the methoxy group remained fixed in its vacancy site, while the alcohol hydrogen atom was free to diffuse along its bridging oxygen row. The similarity in the temperature of recombination of two bridging hydroxy groups with that of one hydroxy and a methoxy suggests the energetics for these two processes are similar.

3.1.5 LEED results and monolayer CH_3OH structural model.

Results to this point suggest that the monolayer saturation coverage of CH_3OH on the vacuum annealed $TiO_2(110)$ surface is 0.62 ML (or 3.2×10^{14} molecules cm^{-2}), of which 0.08 ML (0.4×10^{14} molecules cm^{-2}) is associated with oxygen vacancy sites. No irreversible decomposition in TPD was detected. The majority of the CH_3OH monolayer desorbed below 380 K, and that associated with vacancies desorbed above this temperature. Adsorption at vacancy sites was dissociative, whereas the majority of adlayer was molecularly adsorbed. LEED analysis was conducted to determine if an ordered structure of CH_3OH existed on the vacuum annealed $TiO_2(110)$ surface. Fig. 6A–C shows LEED patterns from the clean (A) and methanol-covered (B and C) vacuum annealed $TiO_2(110)$ surface. A saturated CH_3OH monolayer was prepared by briefly heating a multilayer CH_3OH exposure to 197 K.[50] In the absence of CH_3OH, LEED from the clean vacuum annealed $TiO_2(110)$ surface exhibited a (1 × 1) pattern with sharp spots (Fig. 6A). In the presence of a saturated CH_3OH monolayer, the main substrate spots were still sharp, but third-order streaks were observed running along the [1$\bar{1}$0] direction (Fig. 6B and C). The greatest intensity in the streaks appeared to be at the (1/3,1/2) and/or (2/3,1/2) positions (the first vector coordinate corresponds to the [001] direction). However, the CH_3OH adlayer responsible for these streaks was extremely sensitive to electron exposure, because the LEED pattern faded in a matter of seconds in the LEED beam of microamp 53 eV electrons. This is not surprising based on the very high-cross sections ($>10^{-16}$ cm^2) measured for 100 eV electron induced decomposition of CH_3OH adsorbed on the vacuum annealed $TiO_2(110)$ surface.[50] For this reason, the LEED features in the images of Fig. 6B and C may not perfectly depict the order exhibited in the saturated CH_3OH monolayer. No ordered patterns, other than the (1 × 1) were observed after heating the CH_3OH monolayer to temperatures higher than 200 K.

The third-order streaks indicate a higher degree of order along the [001] direction, but poor order along the [1$\bar{1}$0] direction. This ordering will be referred to as a (3 × n) surface structure, which translates into a surface coverage of either 1/3rd or 2/3rd of a monolayer. The latter value is close to the monolayer saturation coverage of 0.62 ML from TPD, although only about 0.54 ML was associated with non-vacancy sites. A (3 × n) structure, coupled with 2/3rd ML surface coverage, implies that two methanol molecules/fragments reside in every three $TiO_2(110)$ unit cells along the [001] direction. A schematic model explaining both the (3 × n) surface structure and the less than 2/3rd ML coverage is shown in Fig. 6D. The model shows the $TiO_2(110)$ surface with rows of bridging oxygen atoms running right–left. The white and gray atoms are the in-plane and bridging oxygen atoms, respectively, and the smaller black atoms are the titaniums. The right side of the figure shows a clean $TiO_2(110)$ surface with isolated oxygen vacancies, as characterized by STM.[16,17] The left side of the figure shows a hypothetical model for the saturated CH_3OH monolayer. The first step in constructing this model was to fill the vacancy sites (which were placed in the model at a surface coverage of about 0.08 ML) with methoxy groups, as suggested by the TPD, HREELS and SSIMS discussed above. These vacancy-bound methoxy groups are depicted in the model as large gray circles with the label 'D_v' for 'dissociated at a vacancy site'. Assuming the methoxy groups at vacancy sites stand normal to the surface, both adjacent five-coordinate cation sites (one on either side of the vacancy along the [1$\bar{1}$0] dircction) should be blocked because the van der Waals radius of a methyl group (2.0 Å[51]) precludes occupation of either site. This assumption, in itself, goes a long way in explaining why the saturation coverage estimated by TPD is less than 2/3rd ML. If each vacancy-bound methoxy group blocks two non-vacancy cation adsorption sites, then complete occupation of the remaining non-vacancy sites with a (3 × n) surface structure would give a saturation coverage of: 0.67 ML (3 × n) surface coverage) + 0.08 ML (vacancy-bound methoxys) – 0.16 ML (sites blocked by vacancy-bound methoxys) = 0.59 ML. This value assumes, of course, that all of the vacancies are in registry with each other so as not to disrupt the (3 × n) order of the adlayer. This is probably not the case, so the saturation surface coverage should be less than 0.59 ML, in agreement with the 0.54 ML value obtained with TPD.

As mentioned above, a (3 × n) surface structure implies that two methanol-related species exist per every 3 unit cells along the [001] direction. There are two possible surface arrangements of

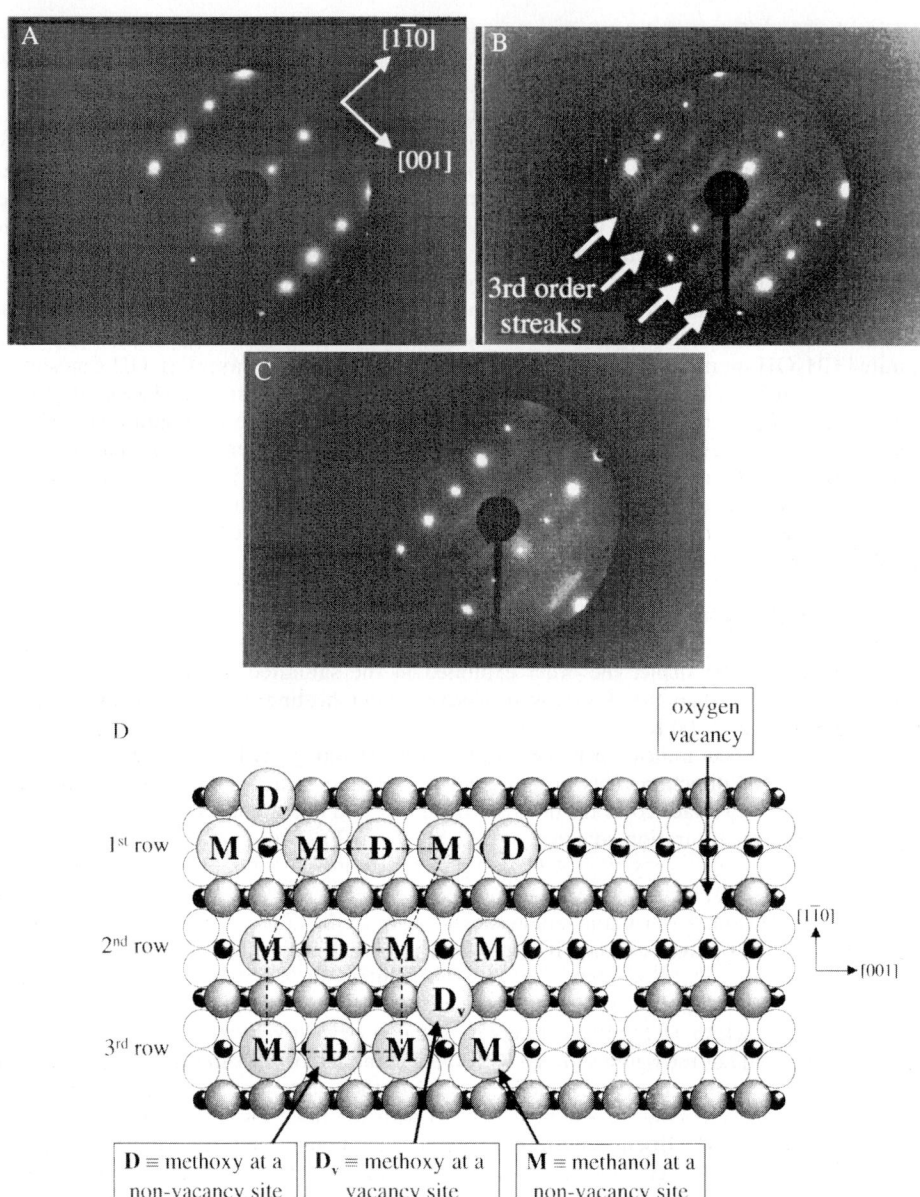

Fig. 6 LEED patterns from: A, the clean TiO$_2$(110) with 8% vacancies; B, a monolayer of CH$_3$OH on the vacuum annealed TiO$_2$(110) surface with 8% vacancies; and C, same as B but in off-normal geometry. D, A schematic model for the CH$_3$OH–covered vacuum annealed TiO$_2$(110) surface.

these methanol-related species along the [001] direction consistent with the steric properties of methanol/methoxy. The first is two atop (on five-coordinate cation sites) and one bridged (between two five-coordinate cation sites) species, and the second is two bridged and one atop species. The 2 atop to 1 bridged option is probably the correct structure for the following reasons. The majority of methanol desorption below 380 K is from molecules that are molecularly adsorbed based on the SSIMS and TPD results. From a structural and acid–base perspective, it seems more likely that molecular methanol would adsorb in an atop position, whereas methoxy could adsorb in either atop or bridged positions. Bridging methoxy ligands are not uncommon in the inorganic literature, and an example of this exists for a binuclear titanium complex.[52] Bridged bonded methoxy species have also been observed on ZrO_2 powders,[53,54] but have not been proposed on TiO_2 rutile powders. No examples of bridged bonded molecular CH_3OH ligands was found in the literature. The identity of the 350 K TPD peak has not been addressed to this point. It seems most reasonable to assign this peak to dissociatively adsorbed methanol bridging two five-coordinate cation sites for the following reasons. First, the structural model discussed above is most consistent with at least 1/3 of the non-vacancy bound species being methoxys. A rough estimate of the TPD peak area in the 350 K peak vs. that in the 295 K peak is consistent with this assumption.[50] Second, the behavior of the 350 K and 480 K CH_3OH TPD peaks under electron bombardment were nearly identical (both are converted in formaldehyde), whereas the 295 K CH_3OH TPD peak behaved very differently (no products left on the surface).[50] Third, the similarities and differences between the TPD spectra of H_2O[6] and CH_3OH suggest that the 350 K CH_3OH TPD peak was due to recombinative desorption. Water, which does not dissociate at non-vacancy sites,[6,7,49] desorbs molecularly at 270 K (see below), whereas the peak assigned to molecular methanol desorption is at 295 K (Fig. 1). Water does not exhibit a prominent peak on the high temperature side of its 270 K peak, whereas methanol has the 350 K shoulder on its 295 K TPD feature. Additionally, both molecules exhibit recombinative desorption from vacancy sites in the temperature window between 450 and 550 K.

Finally, although there was obvious order in the LEED along the [001] direction for the saturated CH_3OH monolayer on the vacuum annealed TiO_2(110) surface, there was no strong preferential ordering along the [1$\bar{1}$0] direction (across the rows of bridging oxygens), as evidenced by the streaks (Fig. 6B and C). The streaking can be explained by the model in Fig. 6D. The presence of the methoxys in the vacancies dictated how the methanol molecules and methoxy groups filled in the non-vacancy sites. Mis-registry between vacancies on the same, the adjacent or even the next-over rows of bridging oxygen atoms (as shown in Fig. 6D) would result in mis-registry in the adlayer. Fig. 6D illustrates one possible example of this mis-registry with different adsorbate unit cells between the 1st–2nd rows of cation sites vs. the 2nd–3rd rows. A variety of such mis-registries would result in a surface structure that gave a streaked LEED pattern along the [1$\bar{1}$0] direction.

3.2 Coadsorption studies

Coadsorption studies of methanol with water and oxygen were undertaken in an effort to better understand how these species might influence the chemistry of methanol on TiO_2 surfaces under aqueous and/or aerated conditions, which are typical of most photochemical studies. Water is perhaps the most pervasive coadsorbate on oxide surfaces. Molecular oxygen is of obvious importance in many thermal and photochemical oxidation processes.

3.2.1 Methanol coadsorbed with water.
In order to determine what effect, if any, H_2O coadsorption has on the surface properties of CH_3OH, experiments were conducted with pre-adsorbed H_2O on the vacuum annealed TiO_2(110) surface. The upper and lower portions of Fig. 7 show H_2O ($m/z = 18$) and CH_3OH ($m/z = 31$) TPD spectra, respectively, for a multilayer exposure of CH_3OH at 130 K to approximately 2 ML of adsorbed H_2O on the vacuum annealed TiO_2(110) surface. For comparative purposes, TPD spectra for H_2O and CH_3OH without coadsorption are also shown (as dashed lines). Without subsequent exposure to CH_3OH, TPD of 2 ML of H_2O on the vacuum annealed TiO_2(110) surface gave the dashed line H_2O TPD spectrum shown in the upper panel of Fig. 7. As discussed above and in previous works,[6,7,18,49] the 270 K TPD peak is from desorption of molecularly bound water at five-coordinate Ti^{4+} sites and the 520 K TPD state is due to recombinative desorption of hydroxy groups formed by water dissociation at oxygen vacancy sites. The sharp H_2O TPD feature at 175 K (off scale) is due to desorption of water molecules physisorbed to bridging oxygen sites.[6,7,18,49] The TPD of CH_3OH by itself (dashed line in the lower panel of Fig. 7) is discussed in previous sections.

TPD of a multilayer exposure (6×10^{14} molecules cm^{-2}) of CH_3OH on 2 ML of H_2O yielded the solid line spectra of CH_3OH and H_2O spectra shown in the lower and upper panels of Fig. 7, respectively. The H_2O TPD spectrum will be discussed first. Although some chemisorbed H_2O remained on the surface, as evidenced by the 245 and 520 K TPD states, the majority of the monolayer H_2O was displaced, along with much of the second layer H_2O, to the multilayer, as

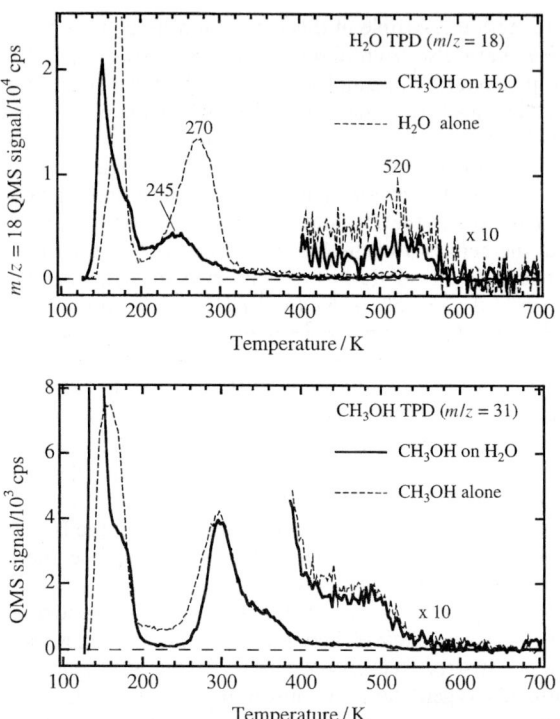

Fig. 7 The effect of preadsorbed water on the desorption properties of CH_3OH on the vacuum annealed $TiO_2(110)$ surface. Upper panel: H_2O ($m/z = 18$) TPD spectra from approximately 2 ML of H_2O alone (dashed line) and from a multilayer CH_3OH exposure adsorbed on 2 ML H_2O (solid line). Lower panel: CH_3OH ($m/z = 31$) TPD spectra from a multilayer exposure of CH_3OH alone (dashed line) and from a multilayer CH_3OH exposure adsorbed on 2 ML H_2O (solid line). All exposures were carried out at a sample temperature below 130 K.

evidenced by the 155 K TPD state. This displacement process probably occurred on adsorption at 135 K. Some H_2O molecules likely were displaced into vacuum during CH_3OH exposure at 135 K since the total H_2O TPD peak area decreased by about 34% after dosing CH_3OH. There was also a 50% reduction in the peak area of the 520 K H_2O TPD state, ascribed to recombinative desorption of bridging OH groups. Since no new H_2O TPD states were formed above 300 K as a result of coadsorption with CH_3OH, it appears that CH_3OH adsorption not only displaced molecular water, but also resulted in the removal of bridging hydroxy groups. This has also been observed for CH_3OH adsorption on hydroxylated TiO_2 rutile and anatase powders.[24,25,35,40] The most likely explanation for this is that a reaction between bridging OH groups and CH_3OH molecules formed H_2O molecules and CH_3O groups, with the latter displacing the former at the vacancy sites (see below). An alternative explanation is that H_2O does not dissociate at oxygen vacancy sites upon adsorption at 135 K, and is molecularly displaced from these sites by CH_3OH. This does not, however, appear to be the explanation since preheating the water layer to 400 K prior to adsorption of CH_3OH at 135 K also resulted in displacement of roughly half of the bridging hydroxys (data not shown).

Inspection of the CH_3OH ($m/z = 31$) TPD traces in the lower panel of Fig. 7 with (solid line) and without (dashed line) preadsorbed H_2O shows virtually no difference above 300 K. In particular, the amount of vacancy-related CH_3OH desorption in the 480 K TPD peak was approximately the same irrespective of the presence of H_2O. Based on these data, CH_3OH was not significantly influenced by the presence of preadsorbed H_2O in either its molecular or dissociative forms. This conclusion is consistent with somewhat higher binding energy of molecular CH_3OH compared to molecular H_2O, as reflected the 25 K higher desorption CH_3OH TPD peak temperature, but is inconsistent with the apparent greater stability of vacancy-bound OH vs. vacancy-bound CH_3O (recombinative desorption of the former occurred above 500 K, while the latter occurred below 500 K). The major influence of H_2O on CH_3OH is restricted to the more weakly bound CH_3OH molecules that desorb below room temperature. These results suggest that under conditions where water is present, adsorbed organics such as methanol may not be displaced from TiO_2 surfaces.

3.2.2 Methanol coadsorbed with oxygen.

As mentioned above, studying the coadsorption of organics and molecular oxygen on TiO_2 surfaces is important for developing a deeper understanding of processes such as photochemical oxidation. However, the chemistry of O_2, by itself, on the $TiO_2(110)$ surface is extremely complex, and varies significantly depending upon the level of bulk reduction and the temperature of the surface.[55-58] These variations, discussed briefly below, may have significant impact on the chemistry of coadsorbed organics such as methanol. Oxygen interactions with the $TiO_2(110)$ surface can result in: (1) a fully oxidized, well-ordered surface; (2) 'over-oxidation'; (3) dissociative O_2 species; (4) molecular O_2 species.

In general, most single crystal TiO_2 researchers use oxygen in an effort to generate ideal and fully stoichiometric surfaces. In particular, O_2 is often used to oxidize vacancies resulting from high temperature annealing.[5] However, it is becoming evident that the formation of a fully oxidized, well-ordered $TiO_2(110)$ surface is difficult to obtain (at least with O_2) once the bulk defect density of interstitial titanium cations (these are the dominant defects in vacuum reduced TiO_2; see references cited elsewhere[59]) and the surface density of vacancy sites are high. As the concentrations of bulk and surface reduction increase, the interaction of oxygen with the $TiO_2(110)$ surface increasingly becomes characteristic of the processes in categories (2)–(4) above. The 'over oxidation' phenomenon results from bulk Ti interstitials drawn to and stabilized at the surface by O_2 oxidation, resulting in extremely rough surfaces as gauged by STM.[57,58] This process occurs above a temperature of about 500 K. Below 500 K, O_2 adsorption fills vacancies, but at a 1:1 ratio leaving oxygen adatoms on the surface.[55] These adatoms readily dissociate O–H and N–H bonds in coadsorbed water and ammonia. The oxygen adatoms can be removed by heating, but apparently not without regeneration of vacancy sites. O_2 adsorption at temperatures below 200 K results in a variety of molecularly bound O_2 species, some of which dissociatively fill vacancies on heating to form adatoms.[56] In this section, the chemistry of CH_3OH on $TiO_2(110)$ with these three conditions is explored using TPD and SSIMS.

Fig. 8 shows $m/z = 29$ (dashed lines) and 31 (solid lines) TPD spectra from a multilayer CH_3OH exposure on the vacuum annealed $TiO_2(110)$ surface (A) and on the surface exposed to 40 L O_2 at 715 (B), 300 (C) and 150 (D) K. Only TPD data above 180 K in each spectrum are shown in order to accentuate events occurring in the monolayer. The $m/z = 29$ signals result from the HCO^+ QMS cracking fragment and the $m/z = 31$ signals result from the CH_2OH^+ QMS cracking fragment (see references cited previously[50]). Both ions are major QMS cracking fragments for CH_3OH, whereas the HCO^+ ion is also observed for H_2CO electron-impact ionization. The $m/z = 29$ and 31 signals nearly perfectly track each other over the entire temperature range for the TPD experiment of CH_3OH adsorbed on the vacuum annealed $TiO_2(110)$ surface (Fig. 8A). This is consistent with the premise that CH_3OH is the only TPD product from CH_3OH exposure to the vacuum annealed $TiO_2(110)$ surface.

Exposure of 40 L O_2 at 715 K to the $TiO_2(110)$ surface has been shown to generate a rough surface composed of hexagonal rosettes, strands and small (1×1) islands of TiO_2.[57,58] TPD of CH_3OH exposed to this surface resulted in new peaks at 210 and 625 K (Fig. 8B). The 480 K TPD state ascribed to recombinative desorption of methoxy groups at vacancy sites is nearly absent, as one would expect for an oxidized surface. The TPD peak areas of the 295 and 350 K states were slightly less than those observed on the vacuum annealed surface (Fig. 8A). Because of the complicated nature of this surface (see STM images in refs. 57 and 58), it is difficult to interpret these changes at the site-specific level. However, comparison of the $m/z = 29$ and 31 signals indicates that the 210 K state was due to desorption of weakly bound CH_3OH, but the 625 K state was from concomitant desorption of CH_3OH and a second organic species. Based on a comparison of the $m/z = 29$ and 30 QMS signals (the latter is not shown), the second organic species is identified as H_2CO. The relative intensities of the $m/z = 29$ and 31 signals in the 625 K TPD peak indicates that roughly half of the $m/z = 29$ signal was due to H_2CO and half was due to CH_3OH. Assuming the ionization cross sections for HCO^+ formation from CH_3OH and H_2CO are approximately

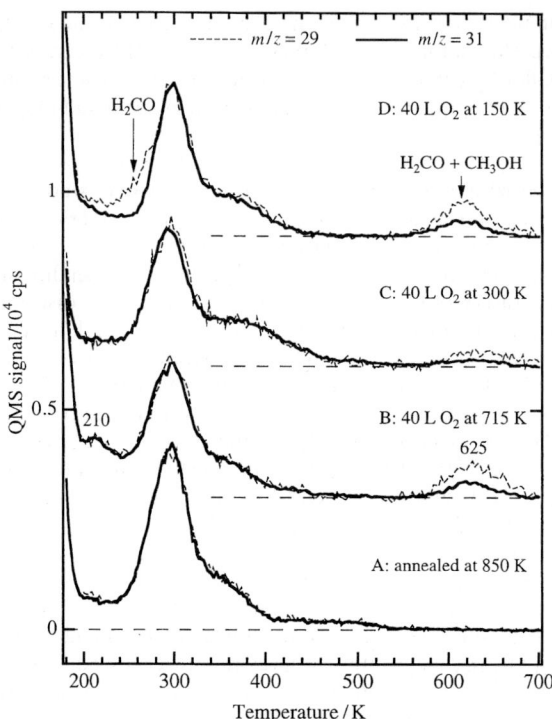

Fig. 8 The effect of O_2 adsorption temperature on the TPD properties of CH_3OH from the vacuum annealed $TiO_2(110)$ surface. TPD spectra for the $m/z = 29$ (dashed lines) and 31 (solid lines) QMS signals from a multilayer CH_3OH exposure at 135 K for: A, the vacuum annealed $TiO_2(110)$ surface; B, the vacuum annealed $TiO_2(110)$ surface exposed to 40 L O_2 at 715 K; C, the vacuum annealed $TiO_2(110)$ surface exposed to 40 L O_2 at 300 K; and D, the vacuum annealed $TiO_2(110)$ surface exposed to 40 L O_2 at 150 K.

equivalent, these data suggest that a disproportionation reaction between two methoxy groups was the source of the concomitant desorption of CH_3OH and H_2CO at 625 K. A similar process at about the same temperature was observed by Gamble and coworkers[60] for ethanol on $TiO_2(110)$, except that the TPD products were ethanol and ethylene. The amount of CH_3OH, and hence the amount of H_2CO, desorbed in the 625 K TPD peak was about 0.04 ML. It is unclear why these methoxy groups remained on the oxidized surface to above 600 K, whereas those formed on the reduced surface (at oxygen vacancies) were unstable above 500 K. Gamble et al.[60] have suggested that the ethoxy disproportionation process occurred at vacancy sites formed at lower temperature by abstraction of lattice oxygen by protons to make water. The protons originated from cleavage of the alcohol's OH group. A water desorption peak comprised about 0.12 ML was detected at 302 K in the TPD experiment of Fig. 8B (data not shown). Abstraction of lattice oxygen by deposited protons has also been observed for formic acid decomposition on $TiO_2(110)$.[46] It is interesting, however, that the vacancies created by thermal annealing (Fig. 8A) influence CH_3OH differently from the (proposed) vacancies formed by water abstraction on the oxygen-induced roughened surface (Fig. 8B). A possible explanation for this is that the vacancies formed by thermal annealing possess Ti^{3+} cations, whereas those formed by abstraction of lattice oxygen by protons to form water possess Ti^{3+} cations. Vacancies possessing Ti^{3+} sites may not be able to support the disproportionation of 2 methoxy groups to formaldehyde and methanol. It is also interesting that little or no water formation occurred on the vacuum annealed $TiO_2(110)$ surface from the protons of those CH_3OH molecules that were dissociatively adsorbed.

Results in Fig. 8C show that 300 K exposure of O_2 on the vacuum annealed $TiO_2(110)$ surface yielded slightly different CH_3OH chemistry than that obtained after higher temperature O_2 exposure. As Epling et al.[55] have shown, O_2 exposure at 300 K to vacancies on $TiO_2(110)$ resulted in vacancy filling and oxygen adatom formation at five-coordinate Ti^{4+} sites. These oxygen adatoms readily dissociated coadsorbed H_2O or NH_3. Results in Fig. 8C using CH_3OH concur with these findings that oxygen adatoms from O_2 adsorption at 300 K facilitated CH_3O-H bond dissociation. This is evidenced by the increase in the proposed recombinative CH_3OH desorption at 350 K relative to the TPD spectrum on the vacuum annealed $TiO_2(110)$ surface (Fig. 8A). The alcohol proton did not end up on the bridging oxygen sites but formed a terminal OH groups with the oxygen adatom. A small fraction of these terminal OH groups combined to form water at 300 K (data not shown), presumably stranding methoxy groups that eventually disproportionate to CH_3OH and H_2CO above 600 K. The extent of this reaction is less than that observed on the 715 K O_2 exposed surface (Fig. 8B).

SSIMS results in Fig. 9 provides evidence for the presence of methoxy groups at temperatures above 550 K on the vacuum annealed $TiO_2(110)$ surface that was pre-exposed to 40 L O_2 at 300 K (the same experimental conditions as the TPD experiment of Fig. 8C). As was shown in Fig. 4, the $^{48}TiO(CH_3O)^+$ and $^{48}TiO(CH_3OH)^+$ SSIMS signals, ratioed to the $^{48}Ti^+$ signal, are superimposed on the CH_3OH ($m/z = 31$) TPD spectrum from Fig. 8C. The behavior of the $^{48}TiO(CH_3OH)^+$ SSIMS signal resembles that shown in Fig. 4 in that this signal decreased as the multilayer and second layer desorbed (below 180 K) and attenuated to zero during the desorption of the 295 K CH_3OH TPD state. The $^{48}TiO(CH_3O)^+$ SSIMS signal also decreased as the surface was heated through these TPD states, however this signal did not attenuate to zero until above 650 K. As discussed above in regards to the results in Fig. 4, the $^{48}TiO(CH_3O)^+$ SSIMS signal in Fig. 9 can be assigned to methoxy groups that are retained on the surface until they disproportionate above 600 K to methanol and formaldehyde.

Returning to Fig. 8, exposure of the vacuum annealed $TiO_2(110)$ surface to 40 L O_2 at 150 K has an additional effect on CH_3OH chemistry (Fig. 8D). Although the 625 K disproportionation TPD states were similar in intensity to those observed on the 715 K O_2 exposed surface (Fig. 8b),

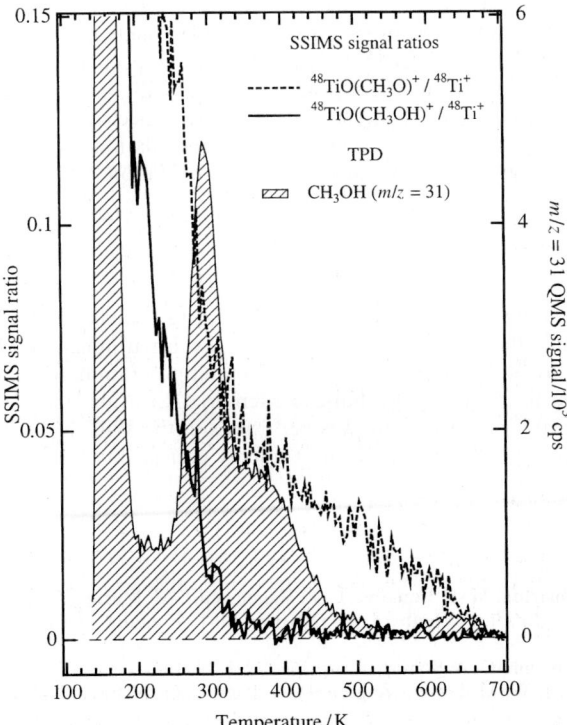

Fig. 9 TPD ($m/z = 31$) and temperature programmed SSIMS ($^{48}TiO(CH_3O)^+$ and $^{48}TiO(CH_3OH)^+$) data from a multilayer CH_3OH exposure adsorbed at 135 K on the vacuum annealed $TiO_2(110)$ surface exposed to 40 L O_2 at 300 K. The SSIMS data are presented as ratios with respect to the $^{48}Ti^+$ signal. The Ar^+ ion energy and current for the SSIMS measurement were 500 eV and $\leqslant 2$ nA cm^{-2}, respectively.

O_2 exposure at 150 K resulted in a new $m/z = 29$ TPD state at about 250 K due to H_2CO desorption. The 250 K H_2CO TPD peak temperature resembles desorption-limited evolution of H_2CO observed by Lu and coworkers[19] suggesting that a low temperature reaction between an O_2-related species and either CH_3OH or CH_3O produces adsorbed H_2CO. The major difference between O_2 exposure at 300 and 150 K is that O_2^- species are formed at the lower temperature, as evidenced by a 2.8 eV loss ELS.[56] Although the exact nature of these O_2^- species is not fully understood, detailed TPD experiments indicate they were formed both at vacancy sites and at cation sites adjacent to the vacancies (along the [1$\bar{1}$0] direction). The most obvious explanation is that an O_2^- species abstracts a hydrogen atom from the methyl group of a methoxy species to form H_2CO. However, the TPD peak areas of both the molecular (295 K) and dissociative (350 K) CH_3OH TPD peaks are diminished relative to the 300 K O_2 exposed surface spectrum (Fig. 8C), suggesting that either (or both) of these species could have been oxidized. Note also that the amount of the 625 K disproportionation reaction has increased, which may have contributed to the diminished 295 and 350 K TPD peaks. Surface analyses with HREELS and SSIMS were unfortunately not useful in further characterizing the evolution of H_2CO from low temperature oxidation of methanol/methoxy, presumably because their respective signals were low or indistinguishable from those originating from the rest of the adlayer.

4 Conclusions

Methanol chemistry on the vacuum annealed TiO_2(110) surface is very similar to that of water in that the majority of the adlayer is molecularly adsorbed and that dissociation occurs at oxygen vacancy sites. However, methanol appears to dissociate at non-defect sites to an appreciable extent, in contrast to the absence of water dissociation at non-defect sites. In coadsorption studies, methanol is preferentially adsorbed and displaces preadsorbed water into the multilayer.

Small, but noteworthy, changes occur in methanol chemistry when coadsorption with molecular oxygen. In particular, thermal reactions between coadsorbed methanol and molecularly or dissociatively adsorbed O_2 leads to C–H or O–H bond cleavage, respectively. Although a direct correlation between thermally generated defects and photochemically generated defects does not yet exist in the literature, these results suggest that alternative mechanisms other than traditional electron-hole pair redox processes may occur during photooxidation. In particular, photogenerated oxygen vacancies may be important sites for thermal oxidation processes during photooxidation of organics on TiO_2.

Acknowledgements

This work was funded by the U.S. Department of Energy Office of Basic Energy Sciences, Materials Sciences, and the U.S. Department of Energy Environmental Management Science Program. Pacific Northwest National Laboratory is a multiprogram national laboratory operated for the U.S. Department of Energy by Battelle Memorial Institute under Contract DE-AC06-76RLO 1830. The research reported here was performed in the William R. Wiley Environmental Molecular Science Laboratory, a Department of Energy user facility funded by the Office of Biological and Environmental Research.

References

1. T. Le Mercier, M. Quarton, M.-F. Fontaine, C. F. Hague and J.-M. Mariot, *J. Appl. Phys.*, 1994, **76**, 3341.
2. T. Le Mercier, J.-M. Mariot, F. Goubard, M. Quarton, M.-F. Fontaine and C. F. Hague, *J. Phys. Chem. Solids*, 1997, **58**, 679.
3. W. J. Lo, Y. W. Chung and G. A. Somorjai, *Surf. Sci.*, 1978, **71**, 199.
4. A. N. Shultz, W. Jang, W. M. I. Hetherington, D. R. Baer, L.-Q. Wang and M. H. Engelhard, *Surf. Sci.*, 1995, **339**, 114.
5. V. E. Henrich and P. A. Cox, in *The Surface Science of Metal Oxides*, Cambridge University Press, Cambridge, 1994.
6. M. A. Henderson, *Surf. Sci.*, 1996, **355**, 151.
7. M. A. Henderson, *Langmuir*, 1996, **12**, 5093.
8. M. A. Henderson, *Surf. Sci.*, 1994, **319**, 315.

9 P. A. Cox, W. R. Flavell, A. A. Williams and R. G. Egdell, *Surf. Sci.*, 1985, **152/153**, 784.
10 K. W. Wulser and M. A. Langell, *Phys. Rev. B.*, 1993, **48**, 9006.
11 W. T. Petrie and J. M. Vohs, *Surf. Sci.*, 1991, **259**, L750.
12 W. Göpel, J. A. Anderson, D. Frankel, M. Jaehnig, K. Phillips, J. A. Schaefer and G. Rocker, *Surf. Sci.*, 1984, **139**, 333.
13 G. Rocker, J. A. Schaefer and W. Göpel, *Phys. Rev. B.*, 1984, **30**, 3704.
14 R. L. Kurtz, R. Stockbauer, T. E. Madey, E. Román and J. L. De Segovia, *Surf. Sci.*, 1989, **218**, 178.
15 M. A. Henderson, *Surf. Sci.*, 1998, **400**, 203.
16 U. Diebold, J. F. Anderson, K.-O. Ng and D. Vanderbilt, *Phys. Rev. Lett.*, 1996, **77**, 1322.
17 U. Diebold, J. Lehman, T. Mahmoud, M. Kuhn, G. Leonardelli, W. Hebenstreit, M. Schmid and P. Varga, *Surf. Sci.*, 1998, **411**, 137.
18 M. B. Hugenschmidt, L. Gamble and C. T. Campbell, *Surf. Sci.*, 1994, **302**, 329.
19 G. Lu, A. Linsebigler and J. T. Yates, Jr., *J. Phys. Chem.*, 1994, **98**, 11733.
20 K. S. Kim and M. A. Barteau, *Surf. Sci.*, 1989, **223**, 13.
21 E. Román, F. J. Bustillo and J. L. de Segovia, *Vacuum*, 1990, **41**, 40.
22 J. Hrbek, R. A. DePaola and F. M. Hoffmann, *J. Chem. Phys.*, 1984, **81**, 2818.
23 J. E. Parmeter, X. Jiang and D. W. Goodman, *Surf. Sci.*, 1990, **240**, 85.
24 Y. Suda, T. Morimoto and M. Nagao, *Langmuir*, 1987, **3**, 99.
25 I. Carrizosa, G. Munuera and S. Castanar, *J. Catal.*, 1977, **49**, 265.
26 P. Jones and J. A. Hockey, *Trans. Faraday Soc.*, 1971, **67**, 2669.
27 H. Onishi, T. Aruga, C. Egawa and Y. Iwasawa, *Surf. Sci.*, 1988, **193**, 33.
28 L.-Q. Wang, K. F. Ferris, J. P. Winokur, A. N. Shultz, D. R. Baer and M. H. Engelhard, *J. Vac. Sci. Technol., A*, 1998, **16**, 3034.
29 K. F. Ferris and L.-Q. Wang, *J. Vac. Sci. Technol., A*, 1998, **16**, 956.
30 S. P. Bates, M. J. Gillan and G. Kresse, *J. Phys. Chem. B*, 1998, **102**, 2017.
31 B. A. Sexton, *Surf. Sci.*, 1979, **88**, 299.
32 N. Aas, T. J. Pringle and M. Bowker, *J. Chem. Soc., Faraday Trans.*, 1994, **90**, 1015.
33 G. Busca, P. Forzatti, J. C. Lavalley and E. Tronconi, *Stud. Surf. Sci. Catal.*, 1985, **20**, 15.
34 R. P. Groff and W. H. Manogue, *J. Catal.*, 1984, **87**, 461.
35 G. A. M. Hussein, N. Sheppard, M. I. Zaki and R. B. Fahim, *J. Chem. Soc., Faraday Trans. 1*, 1991, **87**, 2655.
36 K. S. Kim, M. A. Barteau and W. E. Farneth, *Langmuir*, 1988, **4**, 533.
37 V. S. Lusvardi, M. A. Barteau and W. E. Farneth, *J. Catal.*, 1995, **153**, 41.
38 G. Ramis, G. Busca and V. Lorenzelli, *J. Chem. Soc., Faraday Trans. 1*, 1987, **83**, 1591.
39 P. F. Rossi and G. Busca, *Colloid Surf.*, 1985, **16**, 95.
40 E. A. Taylor and G. L. Griffin, *J. Phys. Chem.*, 1988, **92**, 477.
41 S. P. Bates, G. Kresse and M. J. Gillan, *Surf. Sci.*, 1998, **409**, 336.
42 A. Markovits, J. Ahdjoudj and C. Minot, *Mol. Eng.*, 1997, **7**, 245.
43 H. E. Sanders, P. Gardner and D. A. King, *Chem. Phys. Lett.*, 1994, **231**, 481.
44 K. W. Wulser and M. A. Langell *J. Electron Spectrosc. Relat. Phenom.*, 1992, **59**, 223.
45 M.-C. Wu, C. A. Estrada, J. S. Corneille and D. W. Goodman, *J. Chem. Phys.*, 1992, **96**, 3892.
46 M. A. Henderson, *J. Phys. Chem. B*, 1997, **101**, 221.
47 S. A. Chambers, M. A. Henderson, Y. J. Kim and S. Thevuthasan, *Surf. Rev. Lett.*, 1998, **5**, 381.
48 M. Falk and E. Whalley, *J. Chem. Phys.*, 1961, **34**, 1554.
49 D. Brinkley, M. Dietrich, T. Engel, P. Farrall, G. Gantner, A. Schafer and A. Szuchmacher, *Surf. Sci.*, 1998, **395**, 292.
50 M. A. Henderson, S. Otero-Tapia and M. E. Castro *Surf. Sci.*, 1998, **412/413**, 252.
51 *CRC Handbook of Chemistry and Physics*, ed. R. C. Weast, CRC Press, Cleveland, OH, 5th edn., 1977.
52 Z. A. Starikova, A. I. Yanovsky, N. M. Kotova, M. I. Yanovskaya, N. Y. Turova and D. Benlian, *Polyhedron*, 1997, **16**, 4347.
53 F. Ouyang, J. N. Kondo, K.-I. Maruya and K. Domen, *J. Phys. Chem. B*, 1997, **101**, 4867.
54 M. Bensitel, V. Moravek, J. Lamotte, O. Sauer and J. C. Lavalley, *Spectrochim. Acta, Part A*, 1987, **43**, 1487.
55 W. S. Epling, C. H. F. Peden, M. A. Henderson and U. Diebold, *Surf. Sci.*, 1998, **412/413**, 333.
56 M. A. Henderson, W. S. Epling, C. L. Perkins, C. H. F. Peden and U. Diebold, *J. Phys. Chem. B*, 1999, **103**, 5328.
57 M. Li, W. Hebenstreit and U. Diebold, *Surf. Sci.*, 1998, **414**, L951.
58 M. Li, W. Hebenstreit, L. Gross, U. Diebold, M. A. Henderson, D. R. Jennison, P. A. Schultz and M. P. Sears, *Surf. Sci.*, 1999, **437**, 173.
59 M. A. Henderson, *Surf. Sci.*, 1999, 419, 174.
60 L. Gamble, L. S. Jung and C. T. Campbell, *Surf. Sci.*, 1996, **348**, 1.

Paper 9/02070E

General Discussion

Dr Onishi opened the discussion of Prof. Diebold's paper: Your model for the rosette contains six titanium and twelve oxygen atoms. Do you have experimental evidence for the stoichiometry (Ti : O = 1 : 2)? In the model all the titanium atoms are four-fold coordinated. What stabilises the coordinatively unsaturated (four-fold) atoms in the rosette?

Prof. Diebold responded: The surfaces were prepared by annealing in oxygen which, as many previous studies have shown, results in stoichiometric TiO_2. Our own XPS measurements are consistent with this conclusion, *i.e.*, they do not show a shoulder in the Ti 2p spectrum that would be indicative of lower oxidation states and sub-stoichiometry. The rosettes are metastable, and can be considered a kinetically limited phase. Upon heating to elevated temperatures, they convert into (1 × 1) islands and (1 × 2) strands.[1]

1 M. Li, W. Hebenstreit and U. Diebold, *Phys. Rev. B*, in press.

Prof. Vanderbilt asked: In your DFT calculations, did you calculate the difference between the total energy of the rosette structure and that of the 1 × 1 surface?

Prof. Jennison responded: We did not, because it was not clear how to compare easily the total energies. We only showed that the rosette structure was metastable and reasonable.

Prof. Thornton said: The three bright spot motif appears to be the same as those seen in the local 2 × 2 strands observed by Sander and Engel,[1] ourselves[2] and shown earlier in this meeting by Prof. Bowker. Prof. Bowker presented a model where such a motif was described in terms of our simple added row model.[3] Could you comment on the possibility that a similar model would describe your structures.

1 M. Sander and T. Engel, *Surf. Sci.*, 1994, **302**, L263.
2 P. W. Murray, N. G. Condon and G. Thornton, *Phys. Rev. B*, 1995, **51**, 10989.
3 C. L. Pang, S. A. Haycock, H. Raza, P. W. Murray, G. Thornton, O. Gülseven, R. James and P. W. Bullett, *Phys. Rev. B*, 1998, **58**, 1586.

Prof. Diebold responded: I believe that ours and Bowker's work complement each other, and the proposed structures have many similarities. In a recent publication,[1] Bowker's group suggests a model for 'single links' (the three bright spot motif) and 'cross links' that resembles the rosettes. As shown in Fig. 8 of our paper, there are two different types of rosettes with their centres either above a five-fold coordinated Ti atom (marked with a full line in Fig. 8) or a bridging oxygen atom (marked with a dashed line). Bowker's cross-link consists essentially of two half-rosettes of the dashed kind, situated next to each other. Our explanation for the bright strands in our images (commonly called '2 × 1' strands, see Figs. 2 and 4) is different from the explanation of the ones in Bowker's work. As discussed in a forthcoming publication,[2] the appearance of the strands and adjoining rosettes cannot be reconciled with the added model for the (2 × 1) structure you had suggested. Ti_2O_3 strands (originally proposed by Onishi and Iwasawa) form under our experimental conditions. One of the main reason for the observed differences may be the different reduction states of our crystals as opposed to Bowker's.

1 R. A. Bennett, P. Stone, N. J. Price and M. Bowker, *Phys. Rev. Lett.*, 1999, **82**(19), 3831.
2 M. Li, W. Hebenstreit and U. Diebold, *Phys. Rev. B*, submitted.

Prof. Iwasawa asked: (1) The metastable rosette species has the same composition as stable TiO_2. Is there any reason why the stable TiO_2 islands are not produced from the beginning? It seems that the coordinatively unsaturated rosette species has a higher surface energy and its formation may be unfavourable kinetically and thermodynamically.

(2) Did you observe the precursor species for the rosette structure?

Prof. Diebold answered: (1) Rosettes are produced simultaneously with small (1 × 1) islands, *e.g.*, see Fig. 4 in our paper. While rosettes probably do have a higher surface energy, kinetics plays an important role in their formation. Both, rosettes and added (1 × 1) islands are formed in a 'building block' fashion, by adding TiO_2 unit after TiO_2 unit. Rosettes are structures where this formation process was stopped when some of these units were still missing.

(2) Fig. 2a in our paper shows some sort of 'precursor' structure, *i.e.*, small clusters distributed over flat terraces. Unfortunately, the resolution in our STM images was not good enough to allow conclusions about the structure of these small clusters.

Dr Lindsay opened the discussion of Prof. Iwasawa's paper: My comment concerns your proposed geometry (Fig. 6 of your paper) for the surface carbonate species, which you suggest gives rise to the bright elongated features in your STM images. Previously (ref. 21), you have observed an almost identical feature following CO_2 adsorption on a Na-deposited TiO_2(110) surface. This was also assigned to a surface carbonate species with its molecular plane aligned along the [1$\bar{1}$0] azimuth. However, C K-edge NEXAFS data measured as a function of photon incidence angle, which we have recorded from the latter system,[1] demonstrate that although the species is indeed carbonate its molecular plane is twisted $58 \pm 5°$ out of the [1$\bar{1}$0] azimuth with a tilt of $46 \pm 5°$ away from the surface normal. One possibility for this discrepancy is that there is more than one carbonate geometry, and that you have only observed one of them in STM whereas we probe all with NEXAFS. It should be noted that we did investigate the effect of including a second 'defect' related species, oriented such that the molecular plane was in the [1$\bar{1}$0] azimuth, and found that we obtained good fits to the data with as much as 20% concentration of the minority species. Therefore it is not inconceivable that you are imaging only one of two species in this case. If this is not true and in fact both NEXAFS and STM are probing the same single species then the feasibility of extracting adsorbate geometries from STM images must be questioned.

1 A. Gutiérrez-Sosa, J. F. Walsh, R. Lindsay, P. L. Wincott and G. Thornton, *Surf. Sci.*, 1999, **435**, 538.

Prof. Iwasawa responded: Thank you for your comments on the orientation of adsorbed carbonates on a Na-deposited TiO_2(110) characterized by NEXAFS. We have reported the STM images of carbonate species in two different ordered structures on a Na-deposited TiO_2(110), where the origin of strong basicity of the surface of Na/TiO_2(110) has been discussed in relation to the ordered structures (ref. 21). I am interested in the NEXAFS data for the carbonates which indicate a twisting of *ca.* 58° from the [1$\bar{1}$0] azimuth and a tilt of *ca.* 46° from the surface normal. In the present paper we have also found the formation of carbonate species under the steady state reaction of formic acid on a TiO_2(110) surface without Na deposition. When precisely looking at the STM image of the carbonates, lots of twisting images out of the [1$\bar{1}$0] azimuth are observed (Fig. 2 of our paper). The O–C–O plane of the rest of the images seems to be directed to the azimuth. The twisting observed with the carbonates on the TiO_2(110) by STM is not so large as that observed with the carbonates on the Na/TiO_2(110) by NEXAFS. The difference may be due to a Na effect. It is to be noted that NEXAFS data are averaged for all the carbonate species, while STM images are given to individual species in a smaller area. Unfortunately, we were not able to image defect sites of the surface under the tip conditions used for imaging the surface carbonates. We can not discuss relative arrangements of the species against the defects. The STM image can not provide information on tilting of the carbonates from the surface normal either.

Dr Henderson said: This question somewhat depends upon one's perspective regarding the relative ability of formic acid/formate to act as an oxidising *vs.* a reducing agent. Given that your reaction temperatures fall close to the onset temperature at which interstitial Ti^{3+} cations are able to diffuse from the bulk to the surface of reduced TiO_2(110), could it be possible that the white spots you observe on the bridging oxygen rows are actually TiO_2-subunits formed by formic acid/formate oxidation of these Ti^{3+} interstitials and not due to carbonate species, as you propose?

Prof. Iwasawa responded: The $TiO_2(110)$ surface with 0.001% oxygen-defect employed in this study did not give such bright contrasts that were characterised as elongated features when the surface was heated simply in a vacuum at 400–450 K. The elongated STM images appeared after and under the steady-state reaction of formic acid at 400–450 K. We have not observed the surface in this temperature range in the presence of oxygen though the surface heated at 800 K in the presence of low-pressure oxygen changed dynamically to make Ti_2O_3 added rows and TiO_x suboxide clusters (ref. 10 of our paper). In general, formic acid has a strong acidity and formate anion has a basic character. In the steady-state reaction conditions at the relatively low temperature the dehydrogenation of formic acid to hydrogen and carbon dioxide occurred selectively in agreement with the basic character of surface formate anions, which means no oxidation behaviour of the surface (also Ti^{3+} interstitials) by the formates. At the higher temperature the dehydration of formic acid to water and carbon monoxide preferentially proceeded where no redox process is involved. The product carbonates were adsorbed at the on-top position of the bridge oxygen atoms as suggested from the STM image. Cross section analysis revealed that the topographic height of the product was higher by 0.15 nm than the top of the (2 × 1)-formate monolayer. The formate ion itself exhibited protrusions of 0.14 nm height from the surface in similar tunneling conditions (ref. 14). The accordance in topographic size suggests that the product particle and the formate ion are similar in their physical size. These data indicate that the species produced during the steady-state dehydrogenation of formic acid at 400–450 K are carbonates.

Prof. Diebold said: I bring two pieces of evidence concerning Mike Henderson's idea that the bright spots are actually TiO_x species instead of CO_3^{2-}. The first one supports Henderson's, and the second one Iwasawa's proposal.

(1) Fig. 2(a) in our paper presented at this meeting shows white TiO_x clusters formed upon annealing at 500 K in oxygen. These clusters, as well as the rosettes formed at somewhat higher temperatures, are always located *ca.* 20 Å away from the step edge. We believe that diffusion mechanisms cause this behavior.[1] The clusters are formed by migration of Ti interstitials from the reduced bulk to the surface where they are oxidized by gaseous oxygen (or, perhaps, formate). A Ti interstitial that has migrated into the second-from-top (110) layer into a region close to a step edge has a higher probability for moving out onto the terrace than hopping up into the top layer. If such an interstitial attaches right at the step edge, the denuded zone will move laterally in parallel with the flow of the step edge during growth.

(2) We have recently performed a study of chlorine adsorption on $TiO_2(110)$[2] where we also see a denuded zone next to the step edge with STM. The adsorption has been performed at room temperature, and there is no doubt that bright spots in STM images are Cl atoms in this case.[3] Cl is probably adsorbed in a negative charge state, and, similarly to CO_3^{2-}, might be repelled from step edges.

1 M. Li, W. Hebenstreit, L. Groß, M. A. Henderson and D. R. Jennison, *Surf. Sci.*, 1999, **437**, 173.
2 W. Hebenstreit, E. L. D. Hebenstreit and U. Diebold, unpublished work.
3 U. Diebold, W. Hebenstreit, G. Leonardelli, M. Schmid and P. Varga, *Phys. Rev. Lett.*, 1998, **81**(2), 405.

Prof. Iwasawa responded: (1) As described in reply to Dr Henderson, the reactivity of oxygen is entirely different from those of formic acid/formate anion. We have succeeded in *in situ* imaging both formate anions as the reaction intermediate and carbonate species as the product particles in the formic acid dehydrogenation at 420 K (Fig. 5 of our paper). The *in situ* STM image indicates that the carbonate's location on the terrace more than 2 nm away from the step edges is the onset distribution of the formate intermediates. The topographic analysis is likely to be compatiable with carbonate molecules on the bridging oxygen atoms rather than with larger TiO_x suboxide clusters. The flow of the step edge was not observed during the catalytic dehydrogenation of formic acid at 420 K, while the denuded zone did not move laterally either. When the surface with the carbonate species was heated to 650 K in vacuum, the white carbonate particles in the STM image did not change (Fig. 4). At 700 K the particles disappeared by decomposition, where the (1 × 1) terraces and the Ti_2O_3 added rows were observed.

(2) The similar feature of denuded zone observed with Cl adsorption on $TiO_2(110)$ to that reported in the present paper for carbonate species is interesting. The carbonates are located on-top of the bridging oxygen atoms but the origin of the carbonate formation is the formate

intermediate anions which are located on the Ti atoms of five-fold coordination. Assuming that the Cl anions are also located on the Ti atoms, the characteristic denuded zones for both the anion species may be derived from the similar mechanism involving surface diffusion.

Prof. Campbell asked: Can you comment on the possibility that CO production, seen in TPD from adsorbed formate on TiO_2, should lead to surface "OH" as well. Two such surface OHs would then give the H_2O product plus an oxygen atom. This oxygen could possibility serve to oxidize the surface as Dr Henderson just proposed, at least during TPD.

Prof. Iwasawa responded: TPD data for adsorbed formate on $TiO_2(110)$ showed similar Arrhenius parameters such as preexponential factor and activation energy to those obtained in the steady-state decomposition of formic acid to water and carbon monoxide at the higher temperatures than the present temperatures for the dehydrogenation, indicating probably a unimolecular decomposition mechanism. In this case adsorbed formate leads to surface OH. This OH should react preferably with OH (proton OH formed on the surface bridging oxygen atoms by formic acid dissociation, recovering the bridging oxygen atoms). The data presented here are all those for the catalytic dehydrogenation of formic acid which selectively occurs on $TiO_2(110)$ at 420–550 K. We have reported that the catalytic dehydrogenation proceeds by a bimolecular mechanism between formate and formic acid.

Prof. Madey opened the discussion of Prof. Bowker's paper: You referred to encapsulation of Pd by heating in O_2 as an 'SMSI-like' effect. The SMSI (strong metal–support interaction) effect in catalysis is widely reported for metal particles supported on TiO_2 and heated in reducing atmospheres. In this case, the metal particles are encapsulated by a TiO_x species. We (and others) have also seen evidence for encapsulation of Pt, Fe and Pd on $TiO_2(110)$ after heating in vacuum. Have you ever seen encapsulation of Pd upon heating in vacuum? Can you compare your oxygen-induced encapsulation effects with SMSI encapsulation? Do you have a suggested mechanism for SMSI?

Prof. Bowker responded: We do not have good evidence for encapsulation merely by heating in vacuum, although we cannot rule out the possibility of a thin skin on the Pd particles. Indeed, a possible indication for such encapsulation is that there is an induction period during which the spillover effect is not seen. Also the particle apparently narrows during this first phase, perhaps due to particle de-encapsulation, prior to activation. We have shown this in a recent publication.[1]

The comparison with traditional SMSI is not straightforward, since, as you say, such effects are normally seen in reducing conditions. Further, significant growth of the oxide (ca. 7 monolayers) occurs around the particle in the work we reported here. The necessary interstitial Ti ions, although very dilute in the bulk, are plentiful in terms of monolayer equivalents on a single crystal. However, on a small catalyst particle of, say 5 nm diameter there will be few such interstitials at the same low level of bulk reduction. It is therefore likely to be a quite different process. Furthermore our 'encapsulation' doesn't appear to occur as a monolayer covering, but, so to speak, as scaffolding erected around the particle.

1 R. A. Bennett, P. Stone and M. Bowker, *Catal. Lett.*, 1999, **59**, 99.

Prof. Diebold commented: As was pointed out before, what is generally called the 'SMSI state' is reached by heating a group VIII metal on TiO_2 under reducing conditions. When a similar system, $Pt/TiO_2(110)$, is heated in UHV, it behaves very differently from the oxygen-exposed Pd clusters studied in this work. Pt clusters do not get buried completely, but are covered with a two monolayer thick film of reduced TiO_x. The geometry of this overlayer is reported in a forthcoming paper.[1]

1 O. Dulub, W. Hebenstreit and U. Diebold, submitted.

Prof. Madix said: We see exactly the same cross-linking structures as you simply by heating *in vacuo* to 1150 K. Can you relate the formation of these structures to the mechanism you propose in an oxidising atmosphere?

Prof. Bowker replied: Our cross-linked structures are formed after a period of sputtering and annealing of the sample. We believe this is characteristic of a sample which is more reduced than one which presents the (1 × 1) structure, or even the simple (1 × 2) of Iwasawa. This is, we believe, a sample with interstitial Ti^{3+} ions in the bulk. These ions then are in a state of flux between the bulk and surface and can be trapped at the surface in an oxygen atmosphere, hence growing new layers on the surface.

Dr Henderson said: During your STM movies of the 're-oxidation' process, bright dots appear after (1 × 1) islands are formed. These bright dots appear suddenly in the middle of the islands. Is it your opinion that these dots are Ti^{3+} cations or TiO_x subunits and what is your interpretation of these appearance in the middle *vs.* on the edges of the islands?

Prof. Bowker replied: I believe that they are oxidised Ti units and from the size of these features it may be a dimer which is the nucleus for growth. It's possible it is a Ti_2O_3 unit.

Prof. Joyner asked: I have a general question regarding the extent to which STM images can be interpreted. Can you be sure that the metal particle is really encapsulated, rather than simply disappearing in the STM image? Is there independent, non-STM image, evidence supporting your conclusion?

Prof. Bowker responded: The point is well made. We are sure that (1 × 1) titania grows where the Pd particle was, since we can resolve the structure with the correct lattice parameter. Further we can still see the Pd feature until the final layer of titania is added. Further, we collided with one of the particles (one out of 30 in the scan area) in the widescan image, and for some reason this has modified that particle such that it doesn't show spillover or encapsulation during oxidation (it possibly has some tungsten or tungsten oxide on it). If other particles were 'disappearing' from the image we would not expect to see this.

Prof. Diebold addressed Prof Joyner: In the case of reducing conditions (the 'classic' SMSI treatment), low-energy ion scattering and XPS measurements clearly show encapsulation of clusters. For both Pt and Pd clusters, the metal signal completely disappears in low-energy ion scattering spectra, and recovers after a slight sputter.

Prof. Friend addressed Prof. Bowker: (1) Could you comment on what characteristics of the metal particle are important in the spillover? While you observe spillover from Pd? Prof. Goodman did not from Ag. (2) In addition, can you generalise your arguments to predict which other oxides might exhibit this behaviour?

Prof. Bowker responded: (1) Yes, and in particular in relation to Ag, it is necessary that the dissociation probability is significantly higher on the metal than on the oxide. From our measurements on the (1 × 2) TiO_2 surface alone we estimate the dissociation probability to be *ca.* 10^{-3}, whereas on Pd it is around 0.4.[1] Hence the oxygen dissociates much faster on the Pd and spills over by diffusion onto the support. In the case of Ag, the dissociation probability on reactive single crystal planes is *ca.* 10^{-3} at best;[2] on Ag particles it is *ca.* 10^{-6},[3] but will probably be strongly particle size dependent. Thus, adsorption is not favoured on the Ag. It could even be the case that reverse spillover of oxygen from the support to the metal may occur, although this is unlikely due to the higher binding on the support.

(2) Yes, this should be seen for many reducible oxides which form interstitial cations after reduction. The spillover which we have reported is, of course, a form of oxygen storage, a phenomenon of importance in automobile exhaust catalysis (especially noted with ceria).

1 I. Z. Jones, R. A. Bennett and M. Bowker, *Surf. Sci.*, 1999, **439**, 235.
2 M. Bowker, M. Barteau and R. J. Madix, *Surf. Sci.*, 1980, **92**, 528.
3 M. Dean and M. Bowker, *Appl. Surf. Sci.*, 1988, **35**, 27.

Prof. Goodman commented: It is important to note that the rosette structures or generally the overgrowth of TiO_2 is a very relevant subject with respect to catalytic applications. For example, similar temperature/pressure treatments of catalysts, as seen to induce TiO_2 overgrowth in the STM experiments, are known to lead to metal, *e.g.* Au, re-dispersion on TiO_2. This re-dispersion is critical to the employment of catalysts in practical applications, however, little is known regarding this critical re-dispersion step. These beautiful STM data illustrate the power of a direct imaging method for detailing phenomena that otherwise are very elusive with the usual array of surface science methods.

Dr Mitchell opened the discussion of Prof. Goodman's paper: My comment relates to the ripening of Au clusters on $TiO_2(110)$ (1 × 1). The two limiting possibilities for cluster growth are (i) the Ostwald ripening mechanism discussed by Prof. Goodman, and (ii) the migration and coalescence of entire clusters. There is some evidence from 'before' and 'after' studies that in addition to Ostwald ripening, migration of clusters does occur in metal/oxide systems.[1,2] Recently, we have directly observed cluster motion using elevated temperature STM to follow individual Au clusters on $TiO_2(110)$ (1 × 1). During a sequence of STM images measured over a period of 6 h at 750 K, we see some small clusters (diameter 2–3 nm) move over distances of several nm. An example of this behaviour is shown in Fig. 1.

1 R. A. Dixon and R. G. Egdell, *J. Chem. Soc., Faraday Trans.*, 1998, **94**(9), 1329.
2 P. Stone, S. Poulston, R. A. Bennett and M. Bowker, *Chem. Commun.*, 1998, **13**, 1369.

Prof. Goodman responded: This is a very interesting and impressive result. Of course, this is an alternative mechanism by which cluster ripening can occur. In fact, recent dynamical calculations by Landmann and co-workers indeed suggest that for relatively small clusters, in the range of those shown in your figure, cluster mobility could be quite substantial. Our measurement, which employed a room temperature microscope, preclude us from any observations while heating, however, a new microscope recently acquired does have this capability. I should add, however, that no appreciable ripening of the clusters upon heating was observed in the absence of an oxygen background gas. This shows, in any case, that an oxygen environment is critical to the sintering of silver clusters in our experiments.

Prof. Campbell said: Very nice! I just wanted to support your statement that the sintering of the metal particles is highly important in practical catalysis. We've been studying the sintering kinetics of $Au/TiO_2(110)$ using LEIS and find that the full sintering can only be observed by ramping the catalysts up through a 700 K change in temperature, reflecting a huge range of important time scales. We think this is due to the huge change in stability with metal particle size, such as seen by calorimetry in our paper here.

Prof. Goodman responded: I agree completely with your assessment.

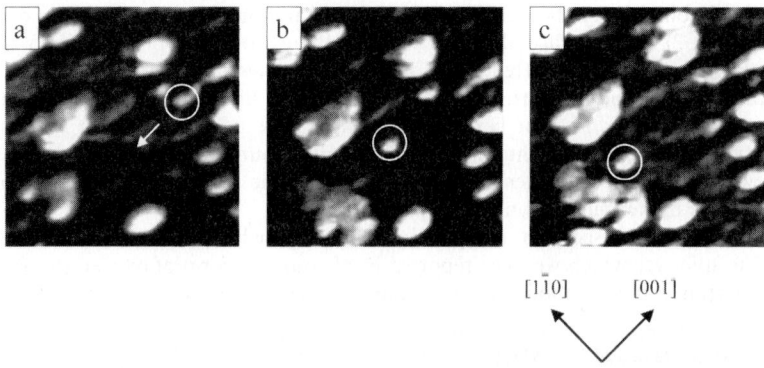

Fig. 1 Three STM images (+1.9 V, 0.2 nA) of the same 170 × 170 Å area of $TiO_2(110)$ (1 × 1), obtained at intervals of approximately 1 h, showing a Au cluster migrating across the surface in the [001] direction. Changes in the morphology of some of the larger clusters is also evident. The images are taken from a 6 h sequence during which the sample temperature was maintained at 750 K.

Prof. Thornton said: With high pressures of O_2 in your reaction cell, presumably there will be a significant partial pressure of CO in the cell. Could it be that CO causes the redispersion of Ag?

Prof. Goodman responded: In any experiments involving exposure to elevated pressures of a gas, *e.g.* Torr, one can never discount the possibility of monolayer quantities of impurity gases being present, CO included. We take special care to purify all gases prior to use, with very special care taken with those introduced at elevated pressures. Frequently IRAS can be used *in situ* to verify the absence of CO, however, no such experiments have been carried out with the Ag surfaces described. Similar experiments with Au on TiO_2 have utilised IRAS and no evidence for low levels of CO were apparent.

Prof. Iwasawa said: The ripening of Ag particles promoted by O_2 was monitored by STM, in which the surface and bulk were oxidised. Similarly, ripening of Au particles in the presence of O_2 was observed by STM, in which the surface and bulk of Au particles may not be oxidised. The features are similar, but the states are different. Could you make comments on this matter?

Prof. Goodman replied: Indeed the sintering behaviour of Ag in these experiments are in many respects very similar to our previously published data on Au on TiO_2. As the paper here details, the growth mode of Au on TiO_2 is quite different from that of Ag, therefore the influence of O_2 on the ripening could be substantially different in Au compared to Ag. However, it is tantalizing to suggest that oxygen could perhaps stabilize the migration of Au in similar fashion as we suggest for Ag. Au oxides would certainly be less stable than for Ag, but, on the other hand, their stabilities might be sufficient for them to serve as a mobilization intermediate.

Prof. Bowker said: It was very nice to see such direct evidence for Ostwald ripening as your bimodal distribution of particle sizes. Regarding the diffusion which causes this ripening it is a wonder to me, in terms of our knowledge of metal surface behaviour, that small particles exist at all under these circumstances, (and clearly they do). For reaction of oxygen with Ag and Cu crystal surfaces it is known that added structures incorporating new supersurface metal atoms, are formed. It is thought that diffusion of metal atoms away from step edges occurs, even at room temperature. Such surfaces probably have a higher metal atom coordination than small metal particles, so I wonder if you have any view about why sintering of such particles doesn't occur much faster? Could it be due to trap sites on the oxide?

Prof. Goodman responded: Although we do not have direct evidence that surface defects serve to stabilize the dispersion of small metal clusters, much direct evidence along with theory supports this view. In my opinion there is little question that defects are critical to the nucleation and stabilization of supported metal cluster, especially those of Au. Future work with our heated-atmosphere microscope should be illuminating with regard to this important aspect of cluster sintering.

Prof. Thornton opened the discussion of Dr Noguera's paper: Did you calculate the relative band gap in the bulk and at the surface? We see a decrease from 4 eV (photoemission)[1] to 3 eV at the surface (STS)[2] for $TiO_2(100)$.

1 P. J. Hardman, G. N. Raikar, C. A. Muryn, G. van der Laan, P. L. Wincott, G. Thornton, D. W. Bullett and P. A. D. M. A. Dale, *Phys. Rev. B*, 1994, **49**, 7170.
2 P. W. Murray, F. M. Leibste, H. J. Fisher, C. F. J. Flipse, C. A. Muryn and G. Thornton, *Phys. Rev. B*, (*Rapid Commun.*), 1992, **46**, 12877.

Dr Noguera responded: We have not yet calculated the gap of TiO_2, neither in the bulk nor at the surface. In previous studies,[1] using a different method, we found a systematic gap narrowing at the surfaces of various oxides. The key effect is the reduction of the Madelung potential on the surface sites, compared to bulk sites, which shifts the anion and cation levels, respectively, towards higher and lower energies. While the gap narrowing is usually relatively strong if a rigid cut of the lattice is assumed, surface relaxation smooths the effects.[2] Work is currently under progress to obtain gap values in the present systems.

1   J. Goniakowski and C. Noguera, *Surf. Sci.*, 1994, **319**, 68; J. Goniakowski and C. Noguera, *Surf. Sci.*, 1994, **319**, 81; J. Goniakowski and C. Noguera, *Surf. Sci.*, 1995, **323**, 129.
2   C. Noguera, *Physics and Chemistry at Oxide Surfaces*, Cambridge University Press, Cambridge, 1996.

Prof. Vanderbilt said: In our earlier work,[1] we did calculate the band gap for a five-layer (110)-terminated slab with stoichiometric (1 × 1) TiO_2 surfaces. We found a band-gap reduction of 0.2 eV for this case, but such a small effect should not be considered significant in view of the approximations involved. Thus, it is best to say that we did not observe any substantial band-gap reduction at the surface.

1   R. Ramamoorthy, R. D. King-Smith and D. Vanderbilt, *Phys. Rev. B*, 1994, **49**, 7709.

Prof. Kempter said: It was stated that the band-gap width of TiO_2 at the surface is larger than in the bulk. In principle, MIES determines the band gap width at the surface. Therefore, it would be interesting to have an estimate for the difference between the surface and bulk band-gap width.

Prof. Vanderbilt responded: As commented previously, our calculations[1] did not indicate any significant band-gap change for the stoichiometric (110) surface.

1   R. Ramamoorthy, R. D. King-Smith and D. Vanderbilt, *Phys. Rev. B*, 1994, **49**, 7709.

Dr Noguera commented: I fully agree, and, as I said before, gap calculations for the surface and comparison with bulk results are in progress in my group. However, the band-gap in an insulator is a property related to the excited states of the systems. This remains a challenge from a computational point of view. Starting from standard DFT or HF methods, it is necessary to introduce self energy corrections, in one way or another, which is very demanding when the latter are computed from first principles. By no way is this an established procedure for the complex systems we are all looking at. In addition, at the surface, our present calculation, as many others before, seems to underestimate the relaxation of the bridging oxygens, with respect to available experimental data. If this is the case, the gap variation between bulk and surface could be incorrectly predicted.

Dr Corà commented: When comparing the band-gap of surface and bulk systems, we have to distinguish between two physically different surfaces: either layers of material of finite thickness, for instance, an oxide mono- or bi-layer deposited on a support, or the surfaces of a single crystal large enough to have bulk behaviour in the centre. Slab calculations, which represent the surface with a layer of finite thickness in vacuum, can only refer to the second case. The surface should in this case be represented by a slab thick enough to reproduce the bulk properties in its centre. If the calculations are properly converged, the band-gap should be the same in bulk and slab calculations, except for the eventual presence of localised surface states in the slab. The effect of surfaces on the energy levels is linked to the under-coordination of the surface ions, which alters the balance between covalence and ionicity in the bonding of atoms at the surface. In the case of oxides such as TiO_2 in which the band-gap is of the type O 2p → M d, the lower coordination reduces the Madelung potential at the surface and hence the band-gap. The importance of covalence at the surface is inversely proportional to the energy separation between the O 2p and the M d levels (this parameter is called $\varepsilon_C - \varepsilon_A$ in Noguera's paper and $\Delta\alpha$ in our paper). The increased mixing of AOs in the crystalline orbitals caused by the lower ($\varepsilon_C - \varepsilon_A$), *e.g.* the increased covalence, stabilises the O 2p based level and destabilises the level based on the M d, and hence increases their energy separation. Surface states will therefore be inside or outside the main band-gap according to which effect prevails. From my experience, in binary early transition metal oxides the increased covalence prevails over the lower Madelung field, and surface states are located towards the middle of both the valence and conduction bands of the material.

To investigate the above feature, I have calculated the density of states (DOS) for the ⟨001⟩ and ⟨102⟩ surfaces of WO_3, and compared them with the DOS of bulk WO_3. In each case, I employed the energy-minimised geometry. The structure, labelling of the ions, and electronic distribution in the two surfaces examined are described in ref. 1. The calculated DOSs are shown in Fig. 2. In the two surface DOSs, the dashed line highlights (magnified) the contribution to the DOS of the ions directly exposed on the surface: the five coordinate $W_{\{5s\}}$ and the terminal $O_{\{1s\}}$

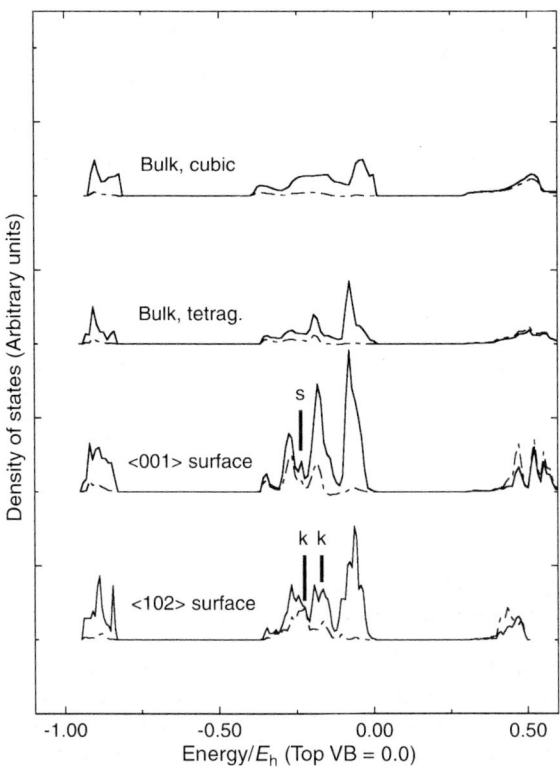

Fig. 2 Calculated density of states for bulk (cubic and tetragonal antiferroelectric phases), ⟨001⟩ and ⟨102⟩ surfaces of WO_3; the energy scale is shifted to have $E = 0$ at the top of the valence band for each system.

for the ⟨001⟩ surface; the four coordinate W_k and the one-coordinate $O_{\{2k\}}$ ions at the kinks of the ⟨102⟩ face (the labelling is the same as in ref. 1). Although the Hartree–Fock Hamiltonian employed in the calculations overestimates band-gaps compared to experiment, the error is systematic, and affects bulk and surfaces in the same way. The calculated DOS for bulk and surfaces are therefore directly comparable. It is clear in Fig. 2 that no surface state is present in either of the surfaces examined. The contribution of the ions directly exposed at the surface (dashed line) appears at the centre and bottom of the valence band, i.e. in energy regions far from the band-gap. The electronic states associated with the ⟨001⟩ surface and the kink of the ⟨102⟩ face are labelled with 's' and 'k'.

The behaviour can be rationalised with the diagram of Fig. 3: the W–O bonding is partitioned for convenience into ionic and covalent contributions, as suggested by the description given in our paper at this meeting. In bulk WO_3, the Madelung (ionic) field causes a splitting $\Delta\alpha_0$ between the relevant O 2p and W 5d AOs (the situation is labelled as i in Fig. 3). W–O covalence overlaps with the splitting of levels due to ionicity (the situation is labelled as i + c in Fig. 3), and increases the separation of the (O 2p)–(W 5d) levels; covalence scales as $1/\Delta\alpha_0$. The energy of the O 2p and W

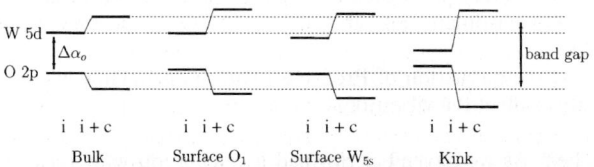

Fig. 3 Schematic representation of the O 2p and W 5d energy levels in the different WO_3 systems examined. The splitting of levels is due to the combined effect of ionicity (i) and covalence (c) in the W–O interaction.

5d levels in the bulk material determines the band-gap. The bulk system is represented in the leftmost diagram of Fig. 3.

On the $\langle 001 \rangle$ surface, in a purely ionic description, the energy levels on the terminal O_1 ion (second diagram in Fig. 3) are shifted to higher energy; they would therefore be in the band-gap. However, this feature causes $\Delta\alpha$ to decrease, and enhances the covalent interactions involving O_1. The energy splitting of the O 2p and W 5d levels due to covalence is consequently higher than in the bulk, and stabilises the O_1 2p levels. Similarly, for the five-coordinate $W_{\{5s\}}$ ions exposed at the $\langle 001 \rangle$ surface (third diagram in Fig. 3), in the ionic-only description the 5d levels would be at lower energy and inside the band gap, before considering the enhanced W–O covalence. At the kink, the ionic-only levels of both W_k and $O_{\{2k\}}$ ions would enter the band-gap; however, the $O_{\{2k\}}$ 2p and W_k 5d levels hybridise much more effectively due to their lower energy separation $\Delta\alpha$.

The absence of surface states in the band-gap, in the calculated DOS of Fig. 2, means that in WO_3 the covalent effect prevails over the change in the Coulomb field. The ionic levels inside the band-gap give rise to a more effective covalent interaction, and are expelled from the band-gap energy region. The latter finding is of general validity for WO_3: the stronger the perturbation created by the surface, the higher the electronic rearrangement at the surface.

In materials where the energetic importance of M–O covalence is less pronounced, the ionic shift of the energy levels may prevail, thus causing the appearance of surface states in the main band-gap.

1 C. R. A. Catlow, L. Ackermann, R. G. Bell, F. Corà, D. H. Gay, M. A. Nygren, J. C. Pereira, G. Sastre, B. Slater and P. E. Sinclair, *Faraday Discuss.*, 1997, **106**, 1.
2 F. Corà and C. R. A. Catlow, *Faraday Discuss.*, 1999, **114**, 421.

Dr Noguera responded: In previous studies related to simple oxides, but also TiO_2, using a semi-empirical Hartree–Fock method, we found that the first effect generally prevails. However, the balance is delicate. From a fundamental point of view, the competition between the two effects is controlled by the behaviour of the dielectric constant. One should thus check that the method used correctly reproduces the screening properties of the material.

Prof. Finnis asked: You highlighted the use of Bader analysis for calculating charge transfer. This looks very attractive as it is the only tool we have which is completely independent of the basis functions. Do you see it as a method which can tell us more than a Mulliken analysis?

Dr Noguera replied: As you say, the Bader analysis is completely independent of the choice of the basis set, which is not the case for the Mulliken analysis. In addition, it does not involve the arbitrariness of equally sharing the bond charge between the two protagonist atoms. When we started this study, we had no *a priori* idea of its validity. We simply wished to try and see what the method would yield, in these systems with strongly diversified local environments and charges. The results seem quite encouraging, since all of them can be rationalised through reasonable physical arguments.

Dr Shluger said: Does comparison between the cluster and bulk crystal data provide any evidence for the origin of the soft modes in the bulk of TiO_2 which are responsible for abnormally large dielectric constant of TiO_2, and possibly for the soft surface mode discussed by Prof. Harrison?

Dr Noguera answered: We have made no attempt to extract vibration frequencies, from dynamical runs. We have simply performed a geometry optimisation to find the equilibrium configuration of the most stable isomers, bulk and surface of TiO_2. This remains an open question.

Prof. Diebold opened the discussion of Prof. Harrison's paper: How will an oxygen vacancy in the bridging oxygen rows affect the vibrational mode?

Prof. Harrison replied: As we haven't performed a calculation with vacancies I cannot say for certain. We have, however, studied the effects of reduction on the electronic structure of titania surfaces so perhaps an educated guess is in order. I think that the excess electrons would localise on the Ti sites directly beneath the bridging oxygen row (*i.e.*, Ti_1 in Fig. 1 of our paper) and that this would result in a rather local distortion of the structure. So, if the density of vacancies were not high I think that the rigid unit mode would be largely unaffected. The shift in the frequency of vibration near the vacancy is harder to estimate—perhaps it is an opportunity for some further calculations. More detail about the effects of reduction on the geometry and electronic structure of titania surfaces, is given in refs. 1–3 below.

1   J. Muscat, N. M. Harrison and G. Thornton, *Phys. Rev. B*, 1999, **59**(23), 15457.
2   P. J. D. Lindan, N. M. Harrison, M. J. Gillan and J. A. White, *Phys. Rev. B*, 1977, **55**(23), 15919.
3   W. C. Mackrodt, E. A. Simson and N. M. Harrison, *Surf. Sci.*, 1977, **484**, 192.

Prof. Jennison asked: You said you did not know the frequency of the soft mode because the effective mass was difficult to determine, but could you give us a rough estimate based on the number and types of atoms which dominate the motion?

Prof. Harrison answered: Yes—I think roughly 2 THz which puts it at the bottom of the bulk vibrational density of states and indicates that it will make a significant contribution to the relative free energy of the surface.

Prof. Flavell said: Following up from Prof. Jennison's comment, the vibration looks as though it should be HREELS active (normal dipole moment). Would its energy be sufficiently large to be resolvable? (Say a few meV?)

Prof. Harrison responded: It seems to be at the lower limit of the energy scale resolvable in HREELS. I do not think that this surface will be unique in having a soft vibrational mode—perhaps we can find one which is not quite so soft and will show up in an HREELS experiment.

Dr Henderson replied to Prof. Flavell: There are the issues of resolution and of dynamic dipole intensity. To my knowledge, a soft phonon mode has not been observed by HREELS of $TiO_2(110)$, but this may be due to limitations in resolution or sensitivity (or both).

Dr Willock addressed Prof. Harrison: To generate your soft mode energy profile you displace surface oxygen and then fully relax the rest of the system. Have you tried holding the rest of the system fixed to give an upper bound on the vibrational frequency to compare with the lower bound generated by your method?

Prof. Harrison replied: No we haven't tried this although it sounds like an interesting idea. Our intention is to compute the dynamical matrix and generate the vibrational modes directly—when time allows.

Dr Noguera asked: Have you estimated quantitatively—for example, through the static displacement of the atomic positions due to anharmonic effects—how the presence of this soft mode at the surface may yield a better agreement between theoretical and experimental relaxations?

Prof. Harrison answered: I think that the discrepancy we are seeing with the geometry deduced from the experimental data is more due to the way the experiment is interpreted than to a correction for anharmonic effects; it will be interesting to study these when the larger contributions have been taken into account. The surface X-ray diffraction is interpreted using a vibrational model (the Debye–Waller model) in which each atom vibrates independently—clearly this is not adequate to account for the rigid unit vibrations present at the surface. The effect of this on the geometry deduced is not immediately apparent but one would not be surprised if it were of order 0.2 Å. Our approach will be to seek a better model with which to interpret the experimental data including previously unpublished LEED data.

Prof. Thornton said: We did assume a Debye–Waller model when analysing the SXRD data.[1] What do you think would happen if we cooled to 25 K?

1 G. Charlton, P. B. Howes, C. L. Nicklin, P. Steadman, J. S. G. Taylor, C. A. Muryn, S. P. Harte, J. Mercer, R. McGrath, D. Norman, T. S. Turner and G. Thornton, *Phys. Rev. Lett.*, 1997, **78**, 495.

Prof. Harrison replied: It is hard to be sure but I would expect a change in the measured diffraction rods and thus changes in the deduced displacements. We could use the computed energy surface to form a thermal average and try to predict the changes!

Dr Shluger said: The results regarding the softness of the oxygen motion perpendicular to the $TiO_2(110)$ surface you have presented are perhaps not surprising. 'Soft' potentials for ionic displacements normal to the surface are characteristic for many materials. Our modelling of non-contact atomic force microscopy demonstrates that they are partly responsible for the contrast formation in the AFM images. I want to stress that although the presented results may not necessarily explain the existing discrepancies regarding the relaxation of the $TiO_2(110)$ surface, the issue of vibrations of surface ions normal to the surface has a much broader importance for the understanding of the structure and properties of other surfaces and interfaces.

Prof. Harrison responded: Perhaps I can comment that I have every expectation that the mode we observe will explain the discrepancy between theory and experiment for this surface. I would also like to emphasise that the experiments performed to determine the structure of oxide surface at present do not take such modes into account and that we hope that the current work will help to provide a more complete framework within which such experiments can be better understood.

Prof. Hermann asked: (1) Is the amount of total energy data that you obtained enough to carry out an anharmonic analysis? (2) VO_2 forms a crystal structure which is extremely close to that of TiO_2. In view of this result, would you expect VO_2 to exhibit the same type of soft mode?

Prof. Harrison answered: (1) Yes. (2) I think that there is every possibility that similar modes will be present at many other oxide surfaces including VO_2.

Dr Renaud said: In your article, you mention experimental results from SXRD, LEED, STM and ion scattering experiments. However, you only compare your calculation with the SXRD results, in particular as concerns the abnormal displacement of the O_3 atom. Would not LEED and ion scattering be very precise in determining this displacement? What are the LEED and MEIS results?

Prof. Harrison replied: The MEIS data are sensitive to the inter-layer spacing at the surface—in particular the O_3–Ti_1 separation in the [110] direction. This spacing is reported in the last row of Table 1 and we see that for the spacing the geometry computed and that deduced from SXRD are in reasonable agreement. Of course as spacing is preserved in the rigid unit mode the MEIS data are not particularly sensitive to the mode. The LEED data, collected by Geoff Thornton and Steve Tear, are currently unpublished. We hope to analyse this data in the near future.

Prof. Thornton added: In the LEED analysis we may have a problem with phase shifts which we are currently trying to resolve.

Dr Egdell said: One problem in attempting to observe dipole active soft phonon modes by HREELS lies in overlap with the tail of the elastic peak to low energy and the tail of the Fuchs–Kliewer phonon modes to high energy. The latter would typically be around 20 times stronger than a mode localised in the surface layer and have an intrinsic linewidth that will remain significant even when the experimental resolution becomes very good. The intensity of Fuchs–Kliewer modes is strongly attenuated in metallic oxides,[1] where the conduction electrons screen out coupling to the long range dipolar fields associated with the Fuchs–Kliewer modes.

The best prospects for observing a soft mode at a rutile (110) surface would therefore be to study a metallic rutile oxide such as RuO_2 with the best possible experimental resolution.

1 P. A. Cox, M. D. Hill, F. Pepkinskii and R. G. Egdell, *Surf. Sci.*, 1984, **141**, 13.

Prof Freund commented: It is possible to improve the resolution of EELS instrumentation, as realised in the latest design based on ideas by Ibach.[1] We have observed a surface phonon near 20 meV excitation energy well below the onset of the Fuchs–Kliewer modes in $Cr_2O_3(0001)$.

1 H. Ibach, *Electron energy loss spectrometers: the technology of high performance*, Springer series in optical sciences, Springer, Berlin, 1991, vol. 63.

Prof. Jennison communicated: Another system where a large finite temperature effect may occur is $Al_2O_3(0001)$, where the surface Al ions relax typically ≈ 0.5 Å, even in the presence of relatively weakly bound adsorbates, suggesting a soft mode. Here we have disagreement between thick-slab DFT results[1] and the X-ray scattering experiments of Renaud *et al.*[2] However, another possible contributing factor is hydrogen contamination, as we have computed the sapphire surface with adsorbed hydrogen and relaxations for the surface Al ions are in the direction (outward) to improve agreement with experiment. Could we have a comment on the possibility of hydrogen contamination during the X-ray experiments?

1 C. Verdozzi, D. R. Jennison *et al.*, *Phys. Rev. Lett.*, 1999.
2 P. Guénard, G. Renaud, A. Barbier and M. Gautier-Soyer, *Surf. Rev. Lett.*, 1997, **5**, 321.

Dr Renaud communicated in response to Prof. Jennison: From the experimental conditions during the X-ray experiments, the possibility of hydrogen contamination can not be ruled out, because the base pressure was of the order of 1×10^{-10} Torr, and no specific care was taken to keep the H_2O or H_2 partial pressure to negligible levels. Since we now know that hydrogen adsorption may strongly modify the surface relaxation, I think we should redo the X-ray measurements with a base pressure in the low 10^{-11} Torr, and negligible partial pressures of hydrogen and water.

Prof. Friend opened the discussion of Dr Henderson's paper: I think your studies of vacancies are very important and that there are parallels with other systems. We have shown that methoxy irreversibly decomposes on vacancies in high coordination sites. First of all, have you specifically looked for other products, *e.g.* methyl radicals or C–C coupling products? Secondly, have you observed any oxygen exchange, in particular with O_2^-? It is possible that the formation of formaldehyde is not simply a H transfer from methoxy to O_2^-.

Dr Henderson responded: Regarding the first point, the only detectable desorption product from TPD of methanol adsorbed on the vacuum annealed $TiO_2(110)$ surface (*i.e.*, the surface with about 8% vacancies) is methanol. The uptake curve (methanol TPD peak area *vs.* exposure) indicates that all adsorbed states of methanol, even those bound at vacancies, are recovered in TPD as methanol. In short, vacancies are not oxidized by methanol (under UHV conditions), and detectable quantities of C_2 species or of methyl radicals are not formed. On the oxygen predosed surfaces, methanol is still the dominant TPD product, although formaldehyde appears to be a secondary product.

On the second point, there is no evidence for oxygen isotope scrambling between methanol (or its decomposition product formaldehyde) and predosed oxygen or the lattice oxygen. For example, coadsorption of $CH_3^{16}OH$ with $^{18}O_2$ yields only $CH_3^{16}OH$ and $H_2C^{16}O$ in TPD. The same is the case if the surface is preoxidized in $^{18}O_2$. Therefore, simple hydrogen abstraction appears to be the correct mechanism for formaldehyde formation.

Prof. Bowker asked: I would like to ask if you are sure that the minor product you see is formaldehyde and not ethane. The reason I say this is that in work we did on $SrTiO_3$ (admittedly a different surface) we saw coupling of methanol to produce ethane, which also has major mass 29 and 30 peaks.

Did you check mass 27? This is an important check for ethane *vs.* formaldehyde.

Further in experiments we carried out and published several years[1] ago on TiO_2 (rutile) powder, we found methane as a major product, *i.e.* the methanol oxygens were dumped into lattice vacancies. Did you check masses 15 and 16 for evidence of methane production?

1 T. Pringle, N. Aas and M. Bowker, *J. Chem. Soc., Faraday Trans.*, 1994, **90**, 1015.

Dr Henderson responded: In rechecking our data files, we definitely *did not* monitor mass 27 in any of our TPD experiments. To be honest, we never thought of ethane as a potential candidate for the product at 260 K, which was obviously a mistake on our part. So the issue of ethane *vs.* formaldehyde must be addressed on other grounds. Although, we routinely monitored mass 15 (from which there is no evidence for methyl radical or ethane production), mass 15 is not a significant cracking fragment of ethane. The ratio of mass 29 to mass 30 is also not a good indicator since this ratio is about the same for cracking of both molecules. The mass spectrometry cracking ratio of mass 29 to mass 28 ratio is about 3 : 1 for formaldehyde but is about 1 : 5 for ethane. Our data shows a ratio of slightly less than 3 : 1 suggesting that the 260 K peak is due to formaldehyde. There is also indirect evidence that supports formaldehyde. First, the 260 K peak temperature for the product is the same as that for desorption-limited evolution of formaldehyde from clean $TiO_2(110)$.[1] Second, the 260 K product is only observed when the O_2^- species is present in the vacancies, and not on the vacuum annealed surface where vacancies are unoccupied. The presence of vacancies provides a mechanism for ethane formation, but such does not occur. On the other hand, there is no obvious mechanism for methyl radical or ethane production from the reaction of O_2^- with CH_3OH or CH_3O, but there is if formaldehyde is the product. So at this point we favour formaldehyde as the product.

Regarding your last point, one needs to be careful in comparing chemical processes on powders *vs.* single crystals, as Kim and Barteau have pointed out.[2] They note that although formaldehyde is observed from formic acid decomposition on single crystals, it has not been observed in any study of formic acid on powdered TiO_2. This is because formaldehyde itself has a finite probability of irreversibly decomposing on TiO_2. As Kim and Barteau have stated, 'The net result of readsorption of the formaldehyde produced by formate coupling is likely to recycle this product to extinction before it can exit the powder sample and be detected by the mass spectrometer.' Therefore, low probability reaction channels on single crystals (where only one surface encounter is probed) become detectable on powders (where multiple surface encounters are unavoidable) if the end product of that channel is unreactive. This is likely the case for formaldehyde decomposition to methane on TiO_2.

1 G. Lu, A. Linsebigler and J. T. Yates, Jr., *J. Phys. Chem.*, 1994, **98**, 11733.
2 K. S. Kim and M. A. Barteau, *Langmuir*, 1990, **6**, 1485.

Prof. Friend said: Following up on Prof. Bowker's point, it is important to specifically monitor low masses. We observe methyl radical evolution from methoxy decomposition on thin films of molybdenum oxide; therefore, these types of processes are possible on single crystals, not just powders.

In our work, co-adsorbed NO, which is somewhat analogues to your O_2^-, opens a C–C coupling pathway from CH_3O decomposition on oxidised Mo. For this reason, I would like to see data for $CH_3^{18}OH$ reaction in order to test for possible carbon coupling. These experiments would unequivocally determine if your product identified by you as H_2CO, actually contains oxygen.

Dr Henderson responded: As mentioned in response to both your previous question and Prof. Bowker's question, we can exclude the possibility of either methyl radical or ethane formation on the vacuum annealed surface (where vacancies are present) since no irreversible decomposition of methanol occurs. Experiments with $CH_3^{18}OH$ would perhaps, as you say, be unequivocal, however, these were not done. The existing data favours formaldehyde. Also, there is no reasonable mechanism for the formation of methyl radicals from the reaction of O_2^- with methanol or methoxy, whereas abstraction of hydrogen from a C–H bond to form formaldehyde is sensible. This is consistent with what is known in the literature about the chemistry of alcohols, aldehydes and carboxylic acids on TiO_2 surfaces. Your observations on molybdenum oxide are quite interesting. However, it should come as no surprise that the surface chemistry of TiO_2 is quite different from that of the various molybdenum oxides. The latter are rich in both acid–base and redox chemistry, whereas the former has weak redox capabilities and is known primarily for its strong Lewis acidity.

Prof. Madix said: The temperature at which the methoxy decomposes to formaldehyde and methanol by disproportionation is very close to that at which formate yields products on the same surface. The disproportionation is most likely the result of activated C–H bond cleavage in methoxy to yield formaldehyde and hydrogen atoms, the latter combining with methoxy to form CH_3OH. Thus, both for methoxy and formate decomposition C–H bond cleavage appears rate-limiting and to have a similar activation energy.

Now since we are led to believe that formate prefers to coordinate to titanium cations in the exposed rows of titanium, whereas the low amount of formaldehyde formed at 600 K from methanol is suggestive of a reaction at defect sites, it is interesting that the activation energies appear so similar. Do you have an understanding of this similarity?

Dr Henderson responded: In the case of formate decomposition, it is not clear what the rate limiting steps are, much less the factors that dictate CO_2 vs. CO production. Irrespective of this, the peak temperature for formate decomposition is at about 550 K, whereas the disproportionation reaction is at about 625 K. The methoxys that disproportionate are not necessarily at oxygen vacancies (remember that those methoxys on the vacuum annealed surface recombine with their alcohol protons to evolve methanol at 480 K), but are probably at the five-coordinate Ti^{4+} sites. The disproportionation channel is the result of a high methoxy coverage resulting from coadsorption with oxygen. The oxygen, in the form of adatoms, facilitates CH_3O–H bond cleavage and removes a significant coverage of the alcohol protons from the surface as water (at about 300 K) preventing recombination to methanol. In the absence of its alcohol proton, the methoxy is stable on the surface until it decomposes at about 625 K. The mechanisms by which these species (formate and methoxy) unimolecularly decompose must be complex. We are a long way from understanding the potential energy surfaces associated with these kinds of decomposition processes.

Prof. Waugh asked: My question relates to Bob Madix's comment on the nature of the surface intermediate particularly with respect to the effect of dosing oxygen at 150 K. The formaldehyde observed at 260 K must have been formed at or below that temperature. My question relates to the nature of the intermediate responsible for the co-desorption of H_2CO and CH_3OH at 625 K.

Dr Henderson responded: The low temperature formaldehyde is probably formed below 260 K since this temperature corresponds to the peak for desorption-limited evolution of formaldehyde from $TiO_2(110)$.[1] The intermediate for the high temperature formaldehyde is methoxy at five-coordinate Ti^{4+} sites formed by O–H bond cleavage of methanol by oxygen adatoms.

1 G. Lu, A. Linsebigler and J. T. Yates, Jr., *J. Phys. Chem.*, 1994, **98**, 11733.

Dr Carley said: You have some interesting data on the effect of pre-adsorbed oxygen species on the reaction of methanol. Have you tried reacting methanol–oxygen mixtures with the surface, *i.e.* dynamic coadsorption? We have performed such studies for some years and find that they reveal surface chemistry not expected or predictable from conventional sequential adsorption experiments, and by varying the mixture composition one can select different reaction pathways. In your study the peroxo induced reaction would probably be observed at 300 K rather than 150 K due to trapping of an O_2^- species by simultaneous coadsorbed methanol.

Dr Henderson replied: This is a good suggestion. However, there are some complications to it. In independent studies,[1] we have observed that the O_2^- species is only formed by adsorption below 200 K. Above this temperature, O_2 dissociatively fills vacancies in a 1:1 ratio leaving an oxygen adatom on the Ti^{4+} sites. These adatoms are great at breaking O–H bonds but do not do the same chemistry as the O_2^- species.

1 M. A. Henderson, W. S. Epling, C. L. Perkins, C. H. F. Peden and U. Diebold, *J. Phys. Chem. B*, 1999, **103**, 5328.

Prof. Thornton said: Recently we recorded low temperature STM data which points to a model for water adsorption which does not necessarily involve O defects. The data show features on the bright (Ti) rows at 150 K, which transform to features on the dark rows by 280 K. At 270 K clusters are observed.

Dr Henderson responded: This is a very interesting result, which shows we have not fully plumbed the depths of water on $TiO_2(110)$. We also have evidence, although indirect, that water clusters around OH groups that are formed from the reaction of water with oxygen adatoms on $TiO_2(110)$.[1] These clusters appear to fall apart in TPD at about 330 K.

1 W. S. Epling, C. H. F. Peden, M. A. Henderson and U. Diebold, *Surf. Sci.*, 1998, **412–413**, 333.

Dr Lindsay asked: This question concerns the $(3 \times n)$ LEED pattern you observe following methanol adsorption. The streaking in the (110) direction is ascribed to the presence of methoxys in oxygen vacancies. Can one alter the degree of streaking by changing the defect concentration?

Dr Henderson replied: This is a good idea, which unfortunately we have not tried. Our experience has been that vacancies on $TiO_2(110)$ are not filled 'cleanly' in UHV by O_2 treatment; that is, vacancies are oxidized but oxygen adatoms are left. Upon heating, it appears that the adatoms recombine with bridging oxygens to reform vacancies and evolve O_2. If one were to adsorb methanol on the vacancy-free surface one might expect a LEED pattern with spots instead of streaks, although streaks could still form if there were no registry between the adsorbate rows.

Prof. Madix said: I would like to open the general discussion with a question directed to Prof. Iwasawa and a more general comment. In my laboratory at Stanford Qing-gen Wang has with some difficulty reproduced the steady state catalytic reaction of formic acid on $TiO_2(110)$, first reported by Prof. Iwasawa's group. The product distributions and temperature dependence of the decomposition we measure agree with their results to within 20%. However, in order to obtain reproducible results we must condition the $TiO_2(110)$ surface at 800 K for at least an hour in the beam of formic acid. Given what we have heard about the complexity of structures on TiO_2, it is fairly certain that we do not understand the nature of the $TiO_2(110)$ surface after this conditioning and the role of surface structure on determining reactivity and selectivity. Prof. Iwasawa, do you have any additional information regarding this matter?

More generally, another aspect of studies of surface reactivity on $TiO_2(110)$ as practised by most of us here is that we used bulk reduced samples. This disequilibrium promotes a variety of processes leading to many transient structures that have been reported here. My concern is that transport of material to and from the partially reduced interior of the crystal may affect reactivity at the surface that would not occur were the oxide truly stoichiometric throughout. Thus we may be led to conclusions in our studies that do not pertain to the reactivity of the pure oxide. Are there any comments on this issue?

Prof. Iwasawa responded: Formic acid does not adsorb on $TiO_2(110)$ at 800 K. At 800 K formic acid decomposes to water and carbon monoxide as major products, where the impinging rate is rate-determining for the surface reaction. Thus the surface pretreated with formic acid beam at 800 K would not be so different from that annealed in vacuum. It is, however, to be noted that the surface depends on the history of the crystal.

Our $TiO_2(110)$ with 0.001% oxygen-defects may be close to neutral. Regarding TiO_2 powder catalysts, it is known that the pretreatments in the appropriate manners for each purpose are necessary to obtain the optimum activity and selectivity; for example, alkene isomerization which is controlled by surface acidity/basicity that is regulated by annealing temperature. The surface acidity/basicity is related to surface defects of Ti and O atoms. At least from catalysis view-points not only stoichiometric oxide but also non-stoichiometric oxide are of great interest, though their surfaces should be scientifically characterized on an atomic/molecular scale.

Prof. Bowker also replied to Prof. Madix: Yes. We should have a better general characterisation of the bulk reduction state of our samples; if possible we should all measure the bulk conductivity *in situ*. I am quite sure that (unlike for metals) the surface reactivity is strongly dictated by the bulk structure, even though levels of defects may be apparently low. Nevertheless these have a very strong effect on conductivity, and as we know, electrons are crucial to surface reactivity and catalysis.

Regarding the relationship to powder catalysts, it has been reported that such samples can also be non-stoichiometric. Thus Baker *et al.*[1] report Ti_4O_7 formation during the reduction of Pt/TiO_2

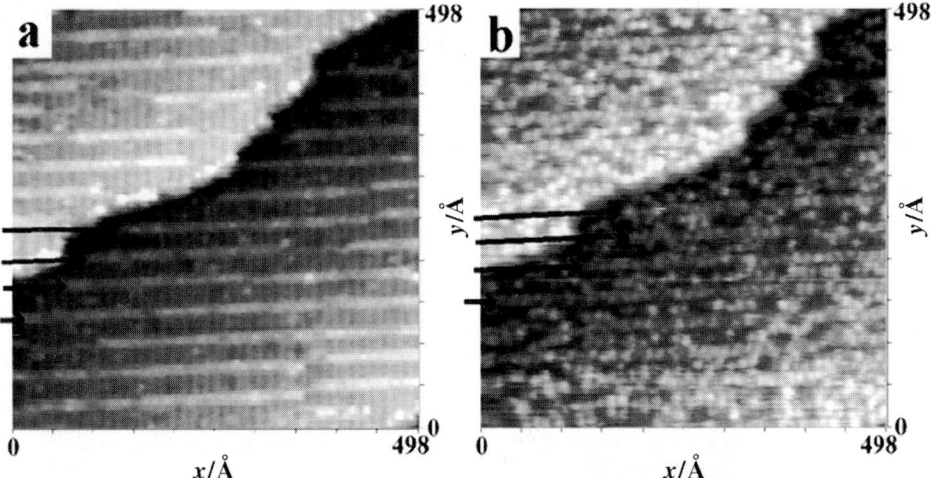

Fig. 4 STM images of the same area of the cross-linked (1 × 2) reconstructed surface before (a) and after exposure to ~240 L formic acid (b). The bright points in (b) predominantly occupy the cross-linked sites which can be seen most clearly in the rows indicated by the arrows. Images taken at 0.1 nA, 1000 mV sample bias and room temperature.

and Bernal et al. find the formation of magneli phases (shear planes) upon reduction of Rh/TiO$_2$ catalysts.[2] So it may be the case that the surfaces we are studying here may be closer in behaviour to some catalytic samples, and indeed may be less reduced than in the SMSI situation.

1 R. Baker, E. Prestridge and R. Garten, *J. Catal.*, 1979, **59**, 293.
2 S. Bernal, F. Botana, J. Calvino, C. Lopez, J. Perez-Omil and J. Rodriguez-Izquierdo, *J. Chem. Soc., Faraday Trans.*, 1996, **92**, 2799.

Prof. Thornton said: On the question of studying the surfaces of stoichiometric TiO$_2$ samples, or other insulating materials, there are now techniques available that do not rely on the substrates being conducting. One is non-contact atomic force microscopy, which has been used to image TiO$_2$ surfaces at atomic resolution. Thus far, however, these measurements have been of reduced crystals.[1,2]

1 K. Fukui, H. Onishi and Y. Iwasawa, *Phys. Rev. Lett.*, 1997, **79**, 4202.
2 H. Raza, C. L. Pang, S. A. Haycock and G. Thornton, *Appl. Surf. Sci.*, 1999, **140**, 271.

Prof. Diebold said: I believe that the different adsorbed structures are relevant to more applied problems, because unless one operates under very oxidising conditions, one will always form bulk objects which will give rise to what is observed on bulk reduced single crystalline surfaces.

Prof. Goodman commented: It is indeed a challenge to establish a chemical connection between those oxide surfaces used by surface scientists and the corresponding 'real-world' surfaces, *e.g.* catalysts. For metals, this connection was made by relating kinetic measurements at realistic conditions of pressure and temperature on the model surfaces with similar measurements on the realistic systems, normalized to the exposed surface area. These measurements are tedious and difficult, however, absolutely essential to establish the chemical validity of the model systems. These parallels need to be established for an array of oxide model systems in similar fashion.

Prof. Campbell said: Regarding the prior discussion as to whether formic acid is reducing or oxidising, consider the equilibrium:

$$H_2 + CO + 2TiO_2(s) \rightleftharpoons H_2 + CO + Ti_2O_3(s) + \tfrac{1}{2}O_2 \rightleftharpoons$$
$$H_2 + CO_2 + Ti_2O_3(s) \rightleftharpoons HCOOH + Ti_2O_3(s)$$

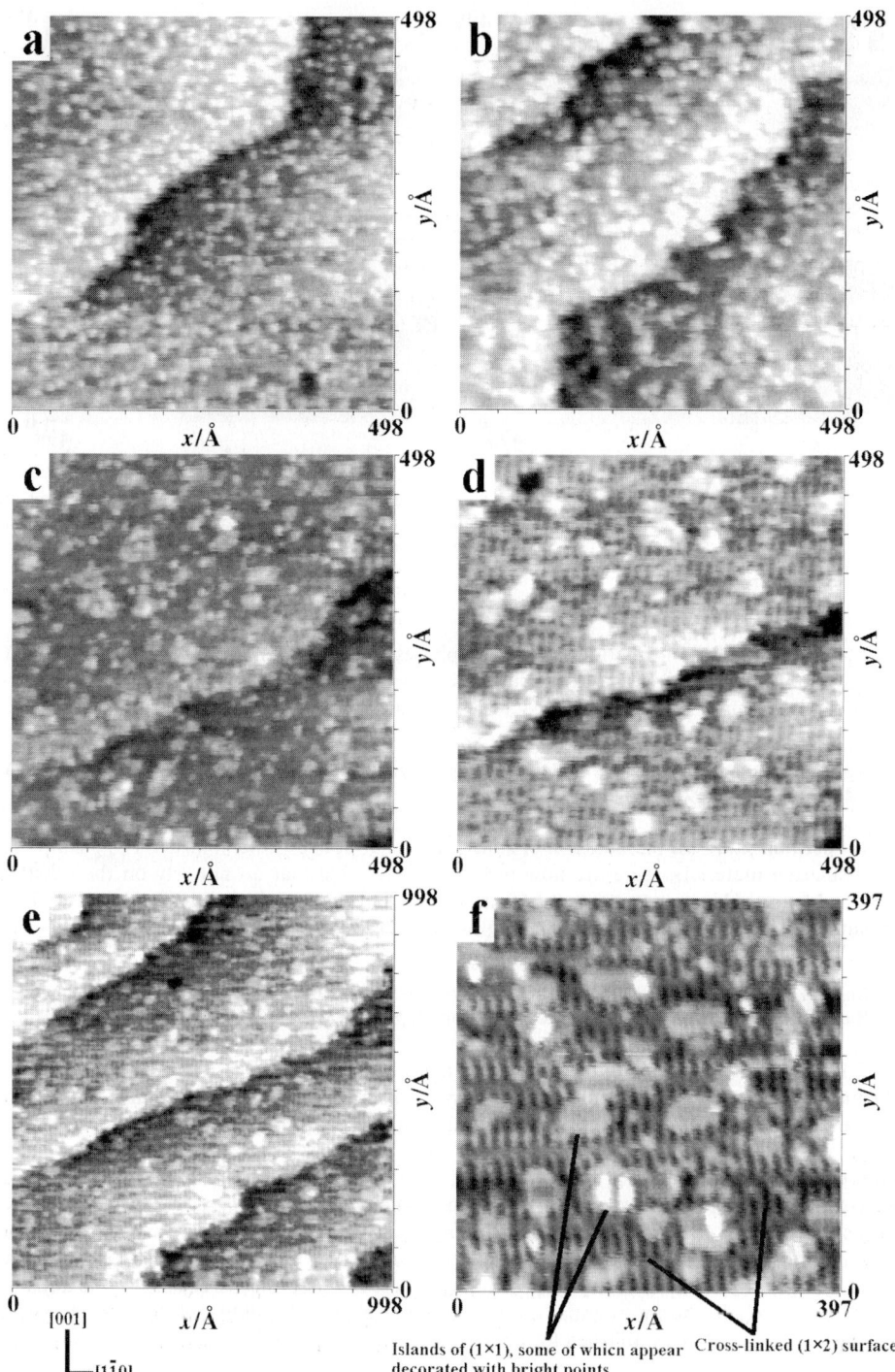

Fig. 5 STM images of the surface after exposure to 500 L formic acid during a temperature ramp of 2 K min^{-1}. The images were recorded at (a) ∼390 K, 52 mins, (b) ∼420 K, 66 mins, (c) 460 K, 82 mins, (d) 470 K, 90 mins, (e) 480 K, 97 mins and finally after stabilising the temperature (f) 570 K, 180 mins. With increasing time and temperature the number of individual bright formate features declines while small islands form within the (1 × 2) reconstructed terraces. The islands show a (1 × 1) termination at 570 K. All images 0.1 nA, 1000 mV.

The $\Delta G°$ for this net reaction (and each sub-reaction) can be determined from thermodynamic data, found easily in the *Handbook of Chemistry and Physics*. Therefore, the equilibrium constant for this net reaction ($K_{eq} = e^{-\Delta G^0/RT}$) is known. Its value gives $K_{eq} = P_{HCOOH}/(P_{H_2})(P_{CO})$, the equilibrium pressure ratio at that temperature. For pressure ratios above this equilibrium ratio (which is for pressure units in bar), then formic acid will oxidise Ti_2O_3 to $2TiO_2$. For pressure ratios below this equilibrium ratio, then H_2 and CO will reduce $2TiO_2$ to Ti_2O_3, uninhibited by the weak oxidising power of the low formic acid pressure.

My main point is that one can make quantitative statements about this, and not just speculate.

Dr Onishi said: When a $TiO_2(110)$ (1 × 1) surface maintained at 700 K was exposed to formic acid at ambient pressure (1 × 10^{-6} Pa), we observed that (1 × 1) islands shrank with time. This shows that formic acid acts as a reducing agent against the TiO_2 surface.

Dr Bennett commented: We have adsorbed formic acid on the non-stoichiometric $TiO_2(110)$ surface which displays the cross-linked (1 × 2) reconstruction.[1] Fig. 4 shows images taken at room temperature before and after exposure to 240 L of formic acid. The resulting adsorbed formate appears to be found preferentially on the cross-links of the reconstruction. On heating this surface in vacuum the formate is removed from the surface and small islands of (1 × 1) termination remain. This is shown in the temperature programmed STM images of Fig. 5 in which a temperature ramp of 2 K min^{-1} has been applied to the surface. Fig. 5a ($\approx$390 K) shows a high density of bright points in the image which we believe are formate moieties. By 420 K these bright features appear to have clustered together producing a lower density of isolated bright features. Continuing the temperature ramp through Figs. 5c and d at $\approx$460 and $\sim$470 K, respectively, still further reduces the population of isolated bright features and allows the underlying (1 × 2) reconstruction to reappear. Crucially, however, small islands, have appeared within the terraces, often with bright features on top of the islands. Fig. 5e ($\approx$480 K) shows a larger area (1000 Å^2) to highlight the extent of the growth of the islands embedded in the (1 × 2) terraces. The final panel, Fig. 5f, shows the surface with the temperature stabilised at 570 K where the cross-linked (1 × 2) reconstruction is apparent along with small islands displaying the (1 × 1) termination. This shows that during decomposition formate can insert oxygen into the surface which then recombines with interstitial Ti^{n+} ions in the surface region to re-form TiO_2 and generate the (1 × 1) termination from the (1 × 2).[1] This re-oxidation scheme is a similar process to that which occurs for oxygen adsorption at elevated temperature.[2-4]

1 R. A. Bennett, P. Stone, R. Smith and M. Bowker, *Surf. Sci.*, in press
2 R. A. Bennett, P. Stone and M. Bowker, *Phys. Rev. Lett.*, 1999, **82**, 3831.
3 P. Stone, R. A. Bennett and M. Bowker, *New J. Phys.*, 1999, **1**, 8, (www.njp.org).
4 R. A. Bennett, P. Stone and M. Bowker, *Faraday Discuss.* 1999, **114**, 267.

Prof. Thomas commented: As well as STM (or AFM) there are other powerful scanning probe methods, and the one I wish to commend to this community is scanning transmission electron microscopy (STEM) with high-angle annular dark field imaging (HAADF). The experimental set-up is as shown below (Fig. 6), the specimen being held in vacuum. Bright field (BF) and dark field (DF) images are produced in any electron microscope, depending upon whether one forms the image by Fourier transforming (using an electromagnetic lens) the forward (Bragg) scattered or the (Bragg) diffracted beams, respectively. With HAADF, however, scattering is recorded at much larger angles, so that one is witnessing not Bragg but Rutherford scattering. Such scattering occurs so close to the nucleus that is intensity is proportional to Z^2 (Z being the atomic number of the nucleus). This therefore means that a Z-contrast map of the specimen is recorded on the image plane. Moreover, since Rutherford scattering (unlike Bragg scattering) is an incoherent process, no abnormalities, (such as image contrasts) arises in the image plane. The technique is at its best when material of high Z value is supported by low Z material such as silica. We have recently shown such images of Pd_6Ru_6 clusters encapsulated in mesoporous silica. But the technique was developed in the Cavendish Laboratory by Prof. Howie and his colleagues, especially Dr M. M. J. Treacy,[2,3] who has put it to very good use in imaging small clusters (1–5 atoms) of Pt in zeolite-L.

STEM offers two further advantages over STM and AFM. First, it has chemical sensitivity, in that, by electron-stimulated X-ray emission or electron-energy loss spectroscopy, it can identify

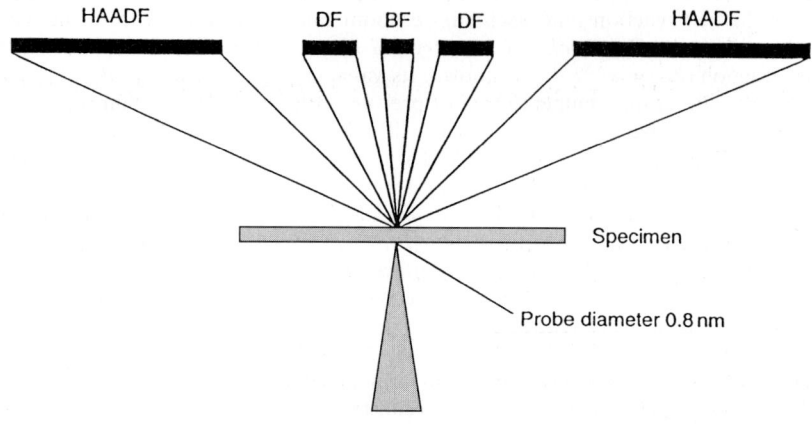

Fig. 6 Schematic diagram illustrating the geometry of the detectors in the STEM. Typical detector collection angles are: BF 0–10 mrad, DF 10–50 mrad and HAADF 80–200 mrad. (Reproduced with permission from ref. 1.)

the elements present in the specimen. Second, it yields micro diffraction patterns, which therefore means that the crystallographic phase of (or phases within the) sample may also be determined.

1 D. Ozkaya, W. Zhou, J. M. Thomas, P. A. Midgeley, V. J. Keast and S. Hermans, *Catal. Lett.*, 1999, **60**, 113.
2 A. Howie, *Microscopy*, 1979, **177**, 11.
3 M. M. J. Treacy and S. B. Rice, *Microscopy*, 1989, **156**, 211.

Prof. Thornton commented: Regarding the subject of our understanding of STM images—at least for clean oxide surfaces, there are now a number of examples where rather good agreement has been achieved between theory and experiment (see for example refs. 1 and 2).

1 C. L. Pang, S. A. Haycock, H. Raza, P. W. Murray, G. Thornton, O Gülseren, R. James and D. W. Bullett, *Phys. Rev. B*, 1998, **58**, 1586.
2 M. R. Castell, S. L. Dudarev, G. A. D. Briggs and A. P. Sutton, *Phys. Rev. B*, 1999, **59**, 7342.

Metal oxides: O^{2-} chemistry and dynamical effects on oxide reactivity

Luciano Triguero,[a] Stefano de Carolis,[a] Micael Baudin,[b] Mark Wójcik,[b] Kersti Hermansson,[b] Martin A. Nygren[a] and Lars G.M. Pettersson*[a]

[a] *FYSIKUM, Stockholm University, Box 6730, S-113 85 Stockholm, Sweden*
[b] *Inorganic Chemistry, The Ångström Laboratory, Uppsala University, Box 538, S-751 21 Uppsala, Sweden*

Received 22nd June 1999

Doping of CeO_2 with calcium introduces defects and oxygen vacancies and leads to a strong increase of the catalytic activity. Desulfurization of SO_2 with CO involves oxygen abstraction from the lattice and charge-transfer (CT) excitation; this reaction runs at a 70 °C lower temperature on the doped substrate. The doping reduces the CT energy cost and the oxygen binding energy, but oxygen abstraction by CO is still not favorable for the lattice at 0 K. The charge state of the ion depends on the Madelung potential, which depends on the lattice structure. Introducing changes in temperature is found to generate vibrations of sufficiently large amplitudes that oxygen anions and cerium cations sometimes can be found at positions where they are sufficiently destabilized so as to be reactive. As the CT energies and oxygen binding energies depend on the instantaneous positions of the ions, active sites appear and disappear at the surface dynamically. The activity of the catalyst substrate is a dynamical quantity that depends on the amplitudes of thermal motion of the surface ions.

I. Introduction

CeO_2 plays an important role as a support and for oxygen storage in many materials with a catalytic function; most notably in three-way car emission catalysts. The activity may be modified through doping with cations with a different formal charge from that of Ce(IV) ions of the perfect lattice, and, for instance, doping with Ca leads to a material with a 70 °C lower light-off temperature for the desulfurization reaction $SO_2 + 2CO \rightarrow S + 2CO_2$.[1] Doping with calcium furthermore enhances the oxidation of methane over CeO_2, but here the effect is rather small.[2] In order to maintain charge neutrality in the crystal each dipositive dopant ion replaces a cerium and oxygen ion pair; thus both cation defects and oxygen vacancies are introduced into the lattice as a consequence of the doping in this case.

In order to understand the effects of doping at an atomic level one needs to consider both the effects on the geometrical structure of the material and, in addition, the effects on the electronic properties. We have recently presented a study of Ca-doped CeO_2 where both these effects were considered:[3] geometrical distortions through static and molecular dynamics (MD) simulations employing interatomic potentials at varying temperatures combined with quantum chemical studies of the electronic structure, in particular charge-transfer (CT) energies. It was found, in agreement with earlier work,[4,5] that doping leads to a substantially reduced energy cost to transfer an electron from the O^{2-} anion to the cerium; this CT excitation is a prerequisite for the formation of covalent bonds involving the lattice oxygens and a reactive adsorbate. However, even though the CT energy cost is reduced by a factor of two,[3] it is still computed to be positive when working with cluster models based on the optimized (or experimental) crystal structure; thus the reactions are still found to be endothermic and additional contributions to the enhanced reactivity must be sought.

In oxide chemistry the stabilizing crystal potential plays a very important role in determining the chemical properties of the material. The doubly negative oxygen anion is unstable in vacuum and would auto-ionize to O^- and an electron. In a strong stabilizing potential, as in a crystal with a small lattice parameter, such as MgO, however, the electrons become strongly bound leading to a closed-shell structure more like that of the rare-gas atom, neon. By changing the lattice parameter, as in the sequence MgO (4.2112 Å), CaO (4.8105 Å), SrO (5.1602 Å) and BaO (5.523 Å),[6] one modifies the electron donating properties of the oxygen anions; this affects the bonding of electronegative species to the surface and the activity of the anions in this respect may be called an "O^{2-} chemistry". One example is the bonding of CO_2 and SO_2 to MgO and CaO [7] and another is the bonding of an oxygen atom to an anion at either the MgO or CaO surfaces leading to a peroxo (O_2^{2-}) complex at the surface: at a regular site of the MgO surface this is exothermic by 0.4 eV while on CaO a much higher binding energy, 1.7 eV, is found.[8] This leads to the conclusion that oxygen abstraction from N_2O (energy cost ≈ 1.7 eV) by binding to an O^{2-} at the surface should be possible at regular, fully coordinated anion sites for CaO, but not for MgO; this difference is due to the difference in stabilization of the ions in the two lattices. Note that a recent paper by Snis and Miettinen[9] has a higher value for the adsorption on MgO, but the difference between the two surfaces is still maintained.

In order to have reactions involving charge donation from the oxygen anions these must thus be sufficiently destabilized. One way, which has been mentioned above, is through changing the material to an oxide with a larger lattice constant, another is to modify the structure by doping the material with ions of different formal charge. In the present work we will investigate yet another possibility to generate a reactive surface for an oxide charge transfer based catalyst, which involves thermal motion and connected displacements of the ions from their equilibrium (most stable) lattice positions; through heating the material, ions are forced to vibrate out of the stabilizing potential. While undergoing large-amplitude motions they will be electronically destabilized and, for a short time, be reactive as they move away from the equilibrium positions. Thus, in this picture the surface can be seen, from a catalysis point of view, as a dynamic entity where reactive sites are created and disappear as a result of thermal motion. Furthermore, this implies that the reactivity of the average or crystallographic surface structure may have very little relevance for the reactivity of the material, but that the overall reactivity is rather given as the probability-weighted activity of the different distorted structures that occur as a result of the thermal motion.

In the present work we will present results from a quantum chemical study of the reactivity of Ca-doped CeO_2 based on computed charge-transfer energies and the abstraction of lattice oxygen by CO in the first step of desulfurization of SO_2. We use embedded cluster models to describe specific structures obtained as snapshots from the MD simulations of refs. 3 and 10 where the entire MD simulation box encompassing a 2-D periodic slab model of the (110) surface is fed directly into the quantum chemical embedding program. The average structure is found to be electronically too stable to show any appreciable reactivity as measured by the computed CT energies and hydrogen affinity; only by considering instantaneous structures showing large deviations from the average have we been able to find sufficiently low CT energies to allow this energy cost to be balanced by the gain from bond formation. Similarly, the oxygen binding energy to the lattice is, viewed as an average over time, too high to be compensated by the bond formation of CO to form CO_2; only for specific distortions does this become energetically possible.

In the following two sections we give a brief presentation of the theoretical techniques. This is followed by sections giving our results, discussion and conclusions.

II. Quantum chemical calculations

Since the (111) face of CeO_2 shows the highest stability and furthermore has been shown to be inefficient for oxygen exchange,[11] we have selected to study only the (110) surface, which is the second most stable surface; this facet may thus be expected to occur on the nanocrystals used in the experiments. The quantum chemical calculations are based on embedded cluster models of ions at the (110) surface of 12.5% Ca-doped CeO_2. It has been shown[12] that relatively small and also non-stoichiometric clusters, if properly embedded, can provide a good model of the crystal. In the description of strongly ionic systems, embedding of the cluster into a lattice of unscreened point-charges will lead to an artificial polarization of the anions in the cluster.[12–15] This may be avoided by a proper embedding scheme using effective core potentials (ECP) based on the frozen ionic charge distribution[16–18] to describe the nearest neighbors to the cluster. In the present work we have used the *ab initio* core model potential (AIMP) embedding approach[16,17] together with a representation of the crystal potential of the lattice based on structures obtained from individual MD coordinate dumps. The crystal potential was included through evaluation of all integrals over the explicit Ewald[19–21] sums over the infinite crystal based on the simulation cell, always using the nominal ionic charge for the ions. The embedding scheme is based on total-ion model potentials, which include an approximate description of all quantum mechanical and electrostatic interactions with the surrounding crystal without including additional basis functions. External ionic wavefunctions, suitable for generating the embedding model potentials, were obtained from self-consistent embedded ion (SCEI) calculations.[16,17] Briefly, self-consistent field (SCF) wavefunctions appropriate for the embedded Ce^{4+} and O^{2-} ions are found and then used to generate the corresponding total-ion model potentials.[3] These give an approximate description of the short-range Coulomb (incomplete screening), exchange and orthogonality interactions, together with the major relativistic effects (Darwin and mass-velocity potentials), obtained within the Cowan–Griffin (CG) approximation.[22] All integrals involving embedding were generated through the ECPAIMP program.[23]

In the different embedded cluster calculations, ions within a certain distance (always larger than 13 Å) from any cluster ion are represented by total-ion model potentials. In all the cluster calculations, the basis set for the oxygen atoms is a (9s5p) primitive basis taken from ref. 24, augmented with one s and one p function (these extra functions have been taken from ref. 13) and two d polarization functions (exponents $\alpha_1 = 0.5657$, $\alpha_2 = 0.2828$), so that the final contraction is (5s4p2d).

For the cerium atom, an AIMP [25–28] representation was used.[3] This is based on a [Kr]-core AIMP (CG-quasirelativistic[28,29]) and a (13s10p6d6f) valence basis set optimized for the ^3H state of the cerium neutral atom, corresponding to the electronic configuration ($[Xe]4f^25d^06s^2$), the SCF ground state. This valence basis set, that is used to describe the 4d4f5s5p6s valence of the cerium atoms, was finally contracted to (5s4p4d2f).

III. MD simulations

The structures used in the quantum chemical modelling were obtained from specific dumps from the MD simulations reported in ref. 3; here only a very brief summary will be given. The $CeO_2(110)$ surface system in the MD simulations was described as a slab, 12 atomic layers thick and periodic in two dimensions, with faces perpendicular to the (110) and ($\bar{1}\bar{1}0$) directions. The doped slab system contained 1/8 Ca^{2+} ions and was then charge balanced by oxygen vacancies. The MD box volume was $16 \times 15 \times 23$ $Å^3$ and contained altogether 432 ions. All ions were allowed to move in the MD simulations, which were run for 1.0 ps with temperature scaling invoked every 10th step to equilibrate the slab, followed by production runs for up to 25 ps for different temperatures using a time-step of 0.1875 fs.

The interatomic potential parameters employed were those used by Sayle *et al.*[30] in an earlier static modeling investigation of CeO_2; the non-Coulombic cation–cation interactions were set to zero and the shell model[31] was used to describe the polarization energy. This description has been found[3] to give good agreement (within 0.01 Å for distances) with experimental diffraction data for bulk CeO_2 and reproduces $g(r)$ and the experimentally observed lattice expansion (2%) upon 12.5% doping with Ca^{2+}.[10] The simulations were performed with a constant-stress, constant-temperature molecular dynamics program for systems periodic in three- and two-dimensions, with dynamically variable lattice vectors (lengths and angle). Details about the MD program have been given elsewhere.[32]

IV. Results

The pure CeO_2 bulk lattice has the fluorite structure, in which each Ce^{4+} cation is surrounded by eight equivalent O^{2-} ions forming the corners of a cube, and with each O^{2-} coordinated to four Ce^{4+}. The experimental lattice parameter is 5.41 Å leading to a Ce–O distance in the bulk of 2.343 Å.[6] The (110)-terminated CeO_2 surface layer consists of six-coordinated Ce^{4+} ions and three-coordinated O^{2-} ions, and all planes parallel to the surface are charge neutral. The experimental level of doping was 10%, which was approximated in the simulations by replacing one cerium and oxygen pair with a calcium in every second unit cell; i.e., 1/8 of the cations were dopant, which gives a 12.5% level of doping. The effects of this on CT energies and oxygen binding energies were investigated based on the simulated structures. The structure of the relaxed surface unit cell after doping is shown in Fig. 1; two different, five or six-fold, coordinations around the cerium cations are found.

The chemistry of cerium compounds is considered to be largely determined by the balance between the two oxidation states Ce(III) and Ce(IV), where the Ce(III) state has an additional 4f electron. Thus we will begin by presenting our computed results for the in-crystal CT energies.

IV.A. Charge-transfer energies

In order to compute, using quantum chemical techniques, the CT energies as a function of the time-steps in the MD simulations, we must build embedded cluster models based on combinations of ions; the minimum model is a Ce–O pair of ions. With a simulation box that in the surface region contains 12 unique Ce ions and 21 oxygens, this becomes a formidable task if every combination needs to be investigated for all available dumps. Thus, a simplified technique to predict the pair-wise CT energies based on individual or average single-ion data is valuable.

For a Ce^{4+} ion to be stable in the bulk CeO_2 its electron affinity must be smaller than the O^{2-} ionisation potential. As the gas phase electron affinity $Ce^{4+} \rightarrow Ce^{3+}$ is computed, at the MCPF (Modified Coupled Pair Functional) level of theory,[33] to be 35.3 eV, the Madelung potential must be of the same order of magnitude in order to stabilize the ion. If this is not so, a change of charge state would be energetically favorable. Similarly, oxygen cannot exist in a charge state of minus two without a stabilizing external potential, as O^{2-} automatically decomposes in the gas phase.

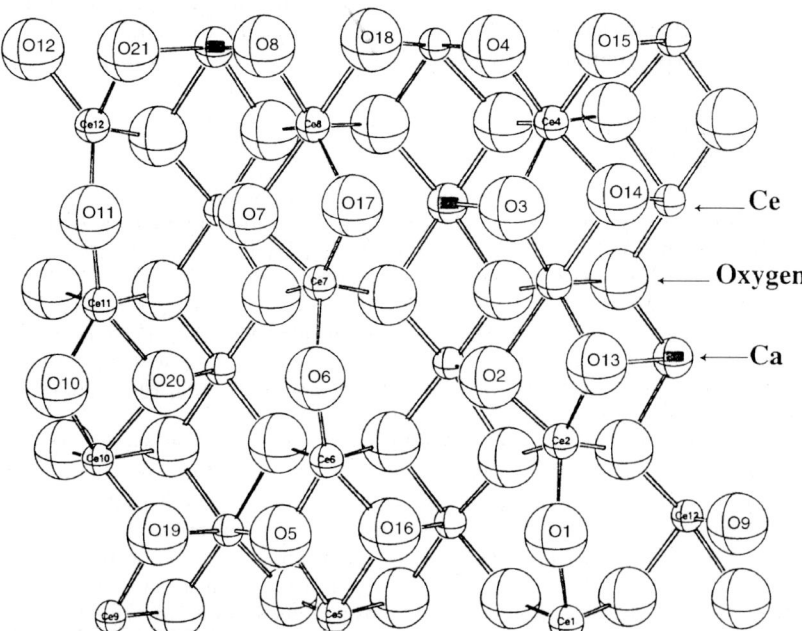

Fig. 1 Structure of surface layer of Ca-doped CeO_2. Investigated surface ions indicated through numbering.

Table 1 Computed in-crystal Ce^{4+} ion electron affinities (E_{ea}) and contributions from Madelung potential (V_{Mad}) and Pauli repulsion[a]

Ion	T = 300 K			T = 700 K		
	$e \cdot V_{Mad}$	Pauli	E_{ea}	$e \cdot V_{Mad}$	Pauli	E_{ea}
1	−33.78	3.76	−1.77	−36.38	3.94	−4.49
2	−34.45	3.63	−2.33	−35.81	3.65	−3.82
3	−35.76	3.55	−3.65	−36.56	3.67	−4.54
4	−36.30	3.82	−4.43	−37.10	3.64	−4.99
5	−35.71	3.60	−3.56	−36.61	3.36	−4.22
6	−33.98	3.78	−1.96	−35.62	3.75	−3.63
7	−34.09	3.36	−1.75	−36.34	3.63	−4.24
8	−36.63	3.48	−4.51	−37.83	3.52	−5.68
9	−35.30	3.48	−3.10	−37.23	3.64	−5.19
10	−35.10	3.58	−3.03	−36.00	3.51	−3.70
11	−34.68	3.55	−2.55	−35.96	3.72	−3.91
12	−35.67	3.60	−3.48	−37.88	4.07	−6.14
Average	−35.12	3.60	−2.99	−36.61	3.73	−3.81
σ	0.89	0.18	0.82	0.72	0.19	0.82

[a] All energies in eV. A negative E_{ea} indicates the Ce^{4+} state is most stable.

If a charge-transfer reaction occurs between two infinitely separated ions the charge-transfer energy will simply be the ionisation potential of the oxygen anion minus the electron affinity of the cerium cation. In a purely ionic crystal the cation electron affinity is negative, *i.e.*, it does not accept additional electrons. It is an important point that the energy of the Ce^{3+} ion in CeO_2 is lower than the $2s \rightarrow 3p$ excited state of the oxygen anion such that the CT state can exist. In ionic materials where this criterion is not fulfilled, such as strongly stabilized rocksalt oxides, charge-transfer reactions between oxygen anions and cations are too high in energy to be of relevance.

With a finite distance separating the formed Ce^{3+} and O^- pair their interaction energy must be included in the charge-transfer energy. Neglecting all electronic contributions, polarization of the ions and at short distances covalent contributions, the interaction energy becomes the change in Coulomb energy between a Ce^{3+} and an O^- infinitely far apart and at a finite distance, R_{Ce-O}, in the crystal. Denoting the oxygen anion in-crystal ionisation potential $E_{i,O}^{CeO_2}$ and the cerium cation in-crystal electron affinity $E_{ea,Ce}^{CeO_2}$ the charge-transfer energy E_{CT} can thus be estimated from

$$E_{CT} = E_{i,O}^{CeO_2} - E_{ea,Ce}^{CeO_2} - 1/R_{Ce-O}.$$

Neglecting all electronic contributions to the interaction energy this gives an upper bound to the CT energy for the selected combination of ions, which can be used to sift through all possible Ce–O pairs and identify interesting candidates. This reduces the workload of finding candidate Ce–O pairs with low CT energy from CeO embedded cluster calculations for all pairs to one set each of single-ion Ce^{4+} and O^{2-} embedded cluster calculations. These linear searches over the cerium and oxygen sites can easily be automated.

Before leaving the subject of simplified ways of calculating charge-transfer energies it is appropriate to discuss whether the charge-transfer energy can be estimated from only the Madelung potential at the involved sites or not. To examine this question we start by decomposing the electron affinity and ionisation potential into an internal and an external part

$$E_{i,O}^{CeO_2} = E_{i,O}^{int} - V_O$$

$$E_{ea,Ce}^{CeO_2} = E_{ea,Ce}^{int} + V_{Ce},$$

where V_O and V_{Ce} are the Madelung potentials at the oxygen anion and cerium cation sites, respectively. If these internal parts are reasonably constant it is sufficient to calculate the Madelung potential in order to estimate the charge-transfer energies.

For the cerium cations the internal part can be estimated as the gas-phase electron affinity plus the change in Pauli repulsion against the surroundings. with a change in charge state from four to three. The Pauli repulsion contribution is estimated from the difference between calculating the embedded electron affinity using only a point-charge array and including an embedding AIMP description of the surrounding ions. This is possible as the electron density of the cation does not polarize appreciably towards the now unscreened negative nearest-neighbor point-charges. With values of 3.8 eV for the average Pauli repulsion and 35.3 eV for the gas-phase electron affinity, the internal part of the cerium cation electron affinity becomes 31.5 eV. From this value and the Madelung potential it is possible to estimate the cerium cation electron affinity within a few tenths of an electron volt, see Table 1.

Turning to the oxygen anions their substantial polarizability introduces large errors if they are embedded in naked point-charges.[12,13] For this reason it is not possible to decompose the oxygen anion ionisation potential into a gas-phase part and a Pauli-repulsion part. The total internal stress of the oxygen anion can, of course, be calculated as the difference between the computed ionisation potential and the Madelung potential, see Table 2. The Madelung potential corresponds to the stabilisation of a unit charge at the site and would thus provide an idealised estimate of the ionisation potential. The computed E_i values are about 16 eV smaller than this, which shows the magnitude of this internal repulsion in the anion. However, with a standard deviation of ± 0.4 eV for the oxygen anion total repulsion useful estimates of the charge-transfer energies can still be obtained. These are upper bounds due to effects of incomplete charge transfer and polarization effects not accounted for in the single-ion estimates (Table 3).

IV.B. Desulfurization reaction

The activity of doped and undoped CeO_2 towards desulfurization of SO_2 has been investigated by Palmqvist and coworkers.[1] The reaction takes place under near-stoichiometric conditions with

Table 2 Computed in-crystal O^{2-} ionisation potentials (E_i) and contribution from Madelung potential (V_{Mad})[a]

Ion	$-e \cdot V_{Mad}$ (T = 300 K)	Repul.	E_i	$-e \cdot V_{Mad}$ (T = 700 K)	Repul.	E_i
1	−25.94	15.47	10.46	−24.79	15.77	9.01
2	−24.23	16.11	8.11	−23.33	16.22	7.10
3	−23.48	16.58	6.90	−22.07	16.29	5.77
4	−25.34	16.81	8.53	−23.71	16.82	6.89
5	−25.32	16.59	8.73	−23.14	15.96	7.17
6	−26.60	15.59	11.00	−24.58	15.62	8.95
7	−22.94	15.63	7.31	−23.69	15.97	7.71
8	−22.49	16.34	6.14	−22.05	16.33	5.72
9	−25.96	15.97	9.98	−23.72	15.98	7.73
10	−24.29	16.39	7.90	−23.13	16.66	6.46
11	−25.28	15.63	9.64	−24.74	15.83	8.90
12	−25.20	17.04	8.16	−22.54	17.14	5.39
13	−24.41	16.41	8.00	−23.32	15.96	7.35
14	−24.19	16.25	7.93	−23.55	16.40	7.15
15	−25.55	16.60	8.94	−23.30	16.77	6.53
16	−25.39	15.92	9.47	−24.61	16.57	8.03
17	−23.86	16.38	7.47	−21.63	16.43	5.20
18	−23.61	16.38	7.23	−22.70	16.84	5.85
19	−24.76	16.24	8.52	−24.59	16.63	7.96
20	−24.29	16.06	8.22	−23.22	16.07	7.14
21	−22.53	16.46	6.06	−21.25	17.17	4.07
Average	−24.56	16.24	8.16	−23.32	16.35	6.96
σ	1.11	0.41	1.36	0.99	0.43	1.36

[a] The internal repulsion plus Pauli repulsion is obtained as the difference. All energies in eV.

Table 3 Predicted and computed CT energies for selected embedded ion pairs (numbered as in Fig. 1) from the different dumps[a]

	Ion pair		E_{ea}(emb.)	E_i(emb.)	R(Ce–O)	CT(pred.)	CT(SCF)
	C	O					
$T = 300$ K							
Dump 1	7	17	−1.75	7.47	3.85	2.2	1.6
	7	17	−1.75	7.31	4.38	2.8	2.1
Dump 2	7	17	−3.64	7.09	3.85	3.6	2.5
	5	8	−2.05	6.44	3.99	1.7	1.3
Dump 3	7	17	−1.81	8.15	4.18	3.4	2.2
	9	21	−1.33	7.26	4.00	1.8	1.3
Dump 4	7	17	−1.02	9.30	3.95	3.4	2.2
	2	13	−1.82	8.28	3.95	3.1	2.2
$T = 700$ K							
Dump 1	7	17	−2.52	8.25	3.90	3.8	2.1
	5	8	−0.83	7.83	4.00	1.9	1.4
Dump 2	7	17	−2.22	9.63	3.90	4.9	3.4
	6	5	−1.20	8.78	3.91	3.0	2.1

[a] All energies are in eV. Dynamical correlation increased all CT(SCF) energies by 1.2 eV. The dumps are arbitrarily numbered.

CO according to:

$$SO_2 + 2CO \rightarrow S + 2CO_2$$

where elemental sulfur and CO_2 are produced. For the undoped CeO_2 the T_{50} (temperature for 50% conversion) for CO is 490 °C while for SO_2 it is 505 °C. Since CO reacts before oxygen begins to be removed from the SO_2, the reaction is believed to proceed as an initial abstraction of a lattice oxygen by a CO leading to CO_2 and an oxygen vacancy associated with two electrons. In a second step the vacancy is filled by abstraction of an oxygen from the SO_2. Since the reaction between CO and O to form CO_2 is exothermic by 5.4 eV, this represents the maximum binding energy of the surface oxygen in order for the abstraction reaction to be energetically favorable. Furthermore, since the binding energy between the SO initial product and an oxygen atom is about 5.1 eV, there is a fairly narrow window within which the binding energy of a reactive surface oxygen must lie. For abstraction of the second oxygen, i.e., the reaction

$$SO + vacancy \rightarrow S + O_{surf}$$

an even higher oxygen affinity of the vacancy is required. In the present discussion we will focus only on the first abstraction step.

Doping with Ca has a large effect on the reactivity, both in terms of the respective T_{50} values and on the difference between the two species. For the CO oxidation, T_{50} now becomes 430 °C and for the SO_2 reduction, 435 °C; the doping thus lowers the value for CO by 60 °C and by even more for SO_2.

As a first step towards an understanding of these reactions we will simply consider the oxygen abstraction from the lattice from an energetic point of view. Thus, in the present case we will not consider the full reaction path to form the CO_2, i.e., not search for transition states, but focus on whether or not the reaction is energetically favorable. We note, in passing, that for the reaction between an atomic oxygen, adsorbed on a surface O^{2-} anion in MgO(001), and CO the computed barrier was found to be quite low, of the order of 0.3 eV.[8]

The abstraction of the lattice oxygen creates an oxygen defect and two electrons. These two electrons can either be left at the defect as a surface F_s center, one of them transferred to a neighboring cerium resulting in a Ce^{3+} and a charged F_s^+ center or, finally, both of them transferred to neighboring cerium ions. In modelling the abstraction using the embedded cluster approach, all three possibilities will be accounted for through a minimal model using one oxygen and the two nearest cations; when the oxygen is removed the basis set description of the oxygen is left in the vacancy to enable a description of the F_s center. The results from these calculations are given in Table 4.

For the 10 K case the vibrational motion of the ions is very small and it is sufficient to consider only one dump; the surface is fully relaxed, but static. This can be considered as representative of the normal modelling approach where either the experimental crystal structure, appropriately cut to generate the desired face or, as in the present case, a fully relaxed model of the surface, is used to define the model geometry. As is seen from the computed oxygen binding energy (5.8 eV, MCPF-level) the conclusion is that, in spite of the doping, the energy cost to extract an oxygen exceeds the gain from attaching it to the CO and no "active site" is thus found on this surface. It should be noted that the doping has led to a substantial decrease in the oxygen binding energy, which with the same approach as above is computed to be 7.6 eV at the SCF-level, which, with the about 2 eV contribution from correlation, becomes an estimated 9.6 eV. Even with this reduction through the doping the binding energy is too high and normally this would lead to the conclusion that defects of different kind should be involved. However, the reaction proceeds with 50% conversion rate at 700 K (430 °C), which could be due to kinetic energy requirements to overcome a barrier for the abstraction or to a change in the energetics as a result of thermal motions in the substrate. We will, in this work, continue to investigate the latter possibility.

In structures sampled from the simulation at 300 K we find larger variations in the computed oxygen binding energies. Following one set (Ce7–O17–Ce8, with notation defined in Fig. 1) in the different dumps we find a variation in the oxygen binding energy between 6.1 and 5.4 eV; Fig. 2. This is the same set of ions as was used in the 10 K model and which there gave a value of 5.8 eV. We may view this as the 10 K model (or crystal structure) resulting in the average value, while thermal motions induce a variation around this. The lowest value, 5.4 eV, found in the present set of dumps is on the border of being isothermal with the CO_2 plus oxygen vacancy. For the present investigation it was not practical to extend the quantum chemical sampling beyond the five well-separated MD dumps reported here. We can thus not assert that the value of 5.4 eV represents a lower limit to the surface oxygen binding energy at this temperature. In comparison with the 10 K result we do, however, see a dependence between the thermal distortion of the lattice and how strongly the lattice oxygen is bound at the surface. The mean square displacements for the surface

Table 4 Computed oxygen SCF- and MCPF-level binding energies (eV) and resulting total Mulliken charges in the oxygen vacancy and on the two cerium ions for the different dumps (numbered as in Table 3)[a]

	Ions			$E_{b,O}$					
	Ce	O	Ce	SCF	MCPF	$q(O_{V_{ac}})$	$q(Ce)$	$q(Ce)$	4f
T = 10 K									
Dump 1	7	17	8	3.5	5.8	−0.49	+3.21	+3.28	−1.98
T = 300 K									
Dump 1	7	17	8	2.9	5.4	−0.48	+3.21	+3.26	−1.97
	7	7	8	5.2	7.7	−1.14	+3.10	+4.04	−1.05
Dump 2	7	17	8	3.6	6.1	−0.49	+3.21	+3.27	−1.98
	5	8	8	3.1	5.6	−0.42	+3.16	+3.27	−1.96
Dump 3	7	17	8	3.5	6.0	−0.46	+3.19	+3.27	−1.98
	9	21	10	4.8	7.3	−1.16	+3.17	+3.99	−1.03
Dump 4	7	17	8	2.9	5.4	−0.51	+3.22	+3.28	−1.98
	2	13	4	3.2	5.7	−0.43	+3.19	+3.25	−1.98
Dump 5	7	17	8	3.6	6.1	−0.54	+3.26	+3.28	−2.00
	12	21	9	3.2	5.7	−0.38	+3.18	+3.20	−1.99
T = 700 K									
Dump 1	7	17	8	3.4	5.9	−0.41	+3.24	+3.18	−2.00
	5	8	8	2.7	5.2	−0.40	+3.22	+3.18	−1.99
Dump 2	7	17	8	3.9	6.4	−0.59	+3.28	+3.30	−1.99
	6	5	5	4.1	6.5	−1.24	+3.19	+4.05	−1.09

[a] A 4f population of two electrons indicates that both cerium ions are in the Ce(III) state.

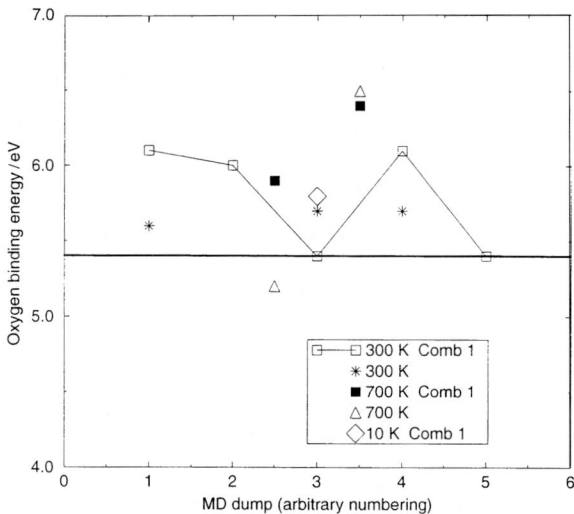

Fig. 2 Computed oxygen binding energies (eV) for different embedded Ce–O–Ce combinations at the different temperatures. The combination of Ce7–O17–Ce8 is indicated for each dump at 300 K. The CO + O → CO_2 reaction energy (≈ 5.4 eV) indicated by solid line.

ions is of the order of 0.006 Å^2 giving a root mean square displacement of about 0.08;[3] however, instantaneous individual ion displacements of up to 0.3 Å are observed in the dumps from the 300 K simulation.

Turning now to the results from the higher temperature simulation at 700 K we find an even larger spread in computed oxygen binding energies, from 6.5 down to 5.2 eV. The latter value is actually below the 5.4 eV approximately required for the reaction to become energetically favorable; this is, furthermore, a temperature at which the reaction runs in the Ca-doped case. At this temperature only two dumps are included in the analysis so there are no statistics on how often this situation occurs or whether the distribution of oxygen binding energies can be correlated with the conversion rate. This would constitute an interesting topic for further study. The obtained results are summarized graphically in Fig. 3, which illustrates the computed oxygen binding energies based on dumps at given temperatures. The highest computed values from Table 4 have been

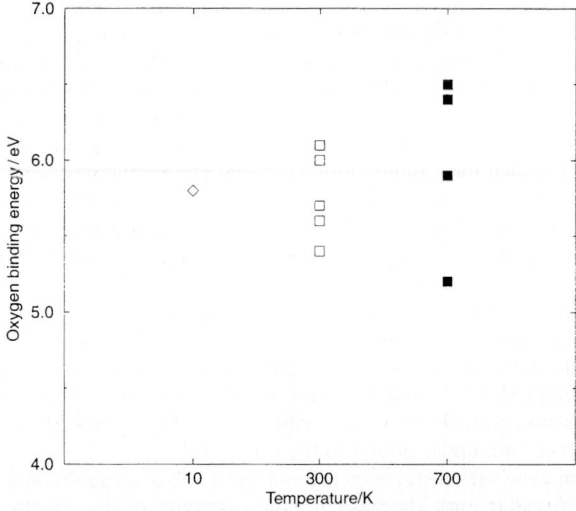

Fig. 3 Statistics of spread in computed oxygen binding energies (eV) as function of temperature.

excluded from the figure since they will not contribute to the rate of oxygen abstraction from the surface.

The computed oxygen binding energies represent upper bounds since they are obtained within a frozen lattice approximation. By this we mean that the structure from the MD dump at a given time-step has been kept and not reoptimised after the oxygen atom was removed. Full geometry optimizations within the embedded cluster model are beyond our present capabilities. We have performed some test calculations to check the reliability of the obtained structures and have found only very minor differences between the MD structure and a (partial) reoptimization within a CeO_6 cluster model. Furthermore, the effects of reoptimization on the charge-transfer states were investigated within that model, again with only very minor effects. Thus the results seem stable, but the possibility of relaxations around the vacancy introduces some uncertainty.

In most cases the electronic structure around the oxygen vacancy corresponds to a complete charge transfer involving two ceria. Only in three of the cases shown in Table 4 do we find formation of one Ce(III) and an F_s^+ center. This seems to occur for the higher binding energies, which we take to indicate that this is not the preferred path for oxygen abstraction from the lattice.

V. Discussion

The present results have some strong and important implications on the way that oxide-based catalysts can be viewed and also provide some simplified connections between simulations of oxide materials and their chemical activity. It is clear that the concept of O^{2-} chemistry is a powerful one giving a useful entrance to the understanding of charge-donating properties and of charge-transfer excitations of the oxides.

The stability of the ionic charge states in the lattice is due to the crystal potential, which must be large enough to compensate for the removal of the electrons from the cation, *e.g.*, in the case of CeO_2, four electrons from the cerium. A typical value in the present case is around 140 eV total stabilization of the Ce^{4+} even in the Ca-doped system. The electronic structure associated with the doubly negative oxygen anion has a strong internal stress that is overcome by the potential; the higher the potential the larger internal repulsion can be accommodated such that the size of the anion is directly related to the magnitude of the Madelung stabilization. Thus, given the same crystal structure the largest internal stress can be compensated for in the crystal with the smallest lattice parameters; *i.e.*, O^{2-} in MgO is a much harder ion than O^{2-} in BaO.

In order to activate the metal oxide for charge donation from the anions or internal charge-transfer, the stabilizing crystal potential needs to be reduced. Doping CeO_2 with calcium introduces both defects (cations with lower charge) and oxygen vacancies, which directly reduces the CT energy by about 50%,[3] but the internal redistribution of charge is still associated with a substantial energy cost. The resulting computed oxygen abstraction energy, which depends directly on the ability of the vacancy or surrounding cations to accommodate the two extra electrons, is still well above that gained by adding the oxygen to CO to form CO_2. Interestingly the thermal motion, as modelled using the MD simulations, introduces a variation in the electrostatic stabilization of the ions over the unit cell. The ions in the material vibrate around the most stable lattice positions, which for a purely ionic material will be the positions determined through the electrostatics; sometimes it seems that the distortions are of a sufficiently large amplitude to strongly destabilize the ions and to make the charge transfer more accessible. It should be noted, however, that in the cases investigated so far we have never observed the CT state to be the lowest in energy.

As has been demonstrated in the present work the CT energies may be estimated reasonably well from single-ion quantities determined from embedded ion models. This avoids having to investigate all possible pairs for the study of charge transfer in all the dumps. Since the cation electron affinity can be predicted to within about 0.2 eV based only on the Madelung energy and an average Pauli repulsion contribution, this facilitates the search through the MD dumps for ion combinations within a certain upper bound to the CT energy.

The thermal motion generates distortions around the average structure and it is these distorted, instantaneous structures that form the basis for the reactivity of the system; we find no specific active site in the average (or 0 K) structure, but active sites are created and disappear as a function of thermal motion in the substrate. Thus, part of the effect of temperature in oxide-based catalysts is likely to provide the required destabilisation of the substrate. From the point of view of theoretical modelling of the reactivity of the system, we find that the ions in the average structure are too stable to be reactive; thus, a model based on the optimum or crystallographic structure cannot describe the CT processes that occur at the temperatures where the catalyst is active; for this we need to compute the reactivity of the instantaneous structures and then form the average, rather than compute the reactivity of the average structure. This requires more calculations, but takes into account the effects on the CT energies that have been found in the present work.

For charge donation from the anion to a surface-adsorbed electronegative species it is sufficient that only the anion becomes destabilised. For the charge transfer both anion and cation need to move away sufficiently from the lattice positions for this to become accessible. From the present work, one conclusion is that simulated thermal motion in the temperature range of the experiment is sufficient to generate these distortions. Thus we must consider two effects of temperature for the reactivity: the kinetic energy distribution of the adsorbate, such that a sufficiently large fraction of the reactants has the required energy to overcome the relevant barriers, but also the effects on the substrate where the surface ions need to be activated through thermal motion. A conclusion is that the lowest CT energies will be found for the largest amplitude deviations and that these should be monitored in a theoretical search for active sites on a charge-transfer-based oxide catalyst. Doping with calcium may have the effect, at a given temperature, to give larger amplitude vibrations through the introduction of defects and vacancies in addition to the static effects on the crystal potential. This opens the perspective to analyse theoretically possible dopants using the present combination of MD and quantum chemical embedding calculations simply through the effects on the maximum amplitude of the correlated anion–cation thermal motion; a larger amplitude at a given temperature holds the promise of a lower temperature for the reaction, *i.e.*, a more efficient catalyst.

VI. Acknowledgements

This work has been financially supported by the Materials Consortium of Clusters and Ultrafine Particles, and the Swedish Research Council for Natural Sciences (NFR).

References

1　A. E. C. Palmqvist, M. F. M. Zwinkels, Y. Zhang, S. G. Järås and M. Muhammed. *Nanostruct. Mater.*, 1997, **8**, 801.
2　A. E. C. Palmqvist, E. M. Johansson, S. G. Järås and M. Muhammed, *Catal. Lett.*, 1998, **56**, 69.
3　S. de Carolis, J.-L. Pascual, L. G. M. Pettersson, M. Baudin, M. Wójcik, K. Hermansson, A. E. C. Palmqvist and M. Muhammed, *J. Phys. Chem.*, 1999, **103**, 7627.
4　G. Balducci, J. Kašpar, P. Fornasiero, M. Graziani and M. Saiful Islam, *J. Phys. Chem. B*, 1997, **101**, 1750.
5　G. Balducci, J. Kašpar, P. Fornasiero, M. Graziani and M. Saiful Islam, *J. Phys. Chem. B*, 1998, **102**, 557.
6　R. W. G. Wyckoff, *Cryst. Struct.*, Wiley, New York, 1963.
7　G. Pacchioni, J. M. Ricart and F. Illas, *J. Am. Chem. Soc.*, 1994, **116**, 10152.
8　M. A. Nygren and L. G. M. Pettersson, *Chem. Phys. Lett.*, 1994, **230**, 456.
9　A. Snis and H. Miettinen, *J. Phys. Chem. B*, 1998, **102**, 2555.
10　A. E. C. Palmqvist, P. Berastegui, S. Eriksson, A. C. Hannon, L. R. Furenlid, M. Baudin, M. Wójcik and K. Hermansson, unpublished work.
11　H. Cordatos, T. Bunluesin, J. M. Vohs and R. J. Gorte. *J. Phys. Chem.*, 1996, **100**, 785.
12　J. L. Pascual and L. G. M. Pettersson, *Chem. Phys. Lett.*, 1997, **270**, 351.
13　M. A. Nygren, L. G. M. Pettersson, Z. Barandiarán and L. Seijo, *J. Chem. Phys.*, 1994, **100**, 2010.
14　M. Pöhlchen and V. Staemmler, *J. Chem. Phys.*, 1993, **97**, 2583.
15　A. M. Ferrari and G. Pacchioni, *Int. J. Quant. Chem.*, 1994, **58**, 241.
16　Z. Barandiarán and L. Seijo, *J. Chem. Phys.*, 1988, **89**, 5739.
17　Z. Barandiarán and L. Seijo, in *Studies in Physical and Theoretical Chemistry: Vol. 77(B), Computational Chemistry: Structure, Interactions and Reactivity*, ed. S. Fraga, Elsevier, Amsterdam, 1992, p. 435.
18　J. A. Mejías and J. Fernández Sanz, *J. Chem. Phys.*, 1995, **102**, 327.
19　P. P. Ewald, *Ann. Phys.*, 1921, **64**, 253.
20　D. E. Parry, *Surf. Sci.*, 1975, **49**, 433.
21　D. E. Parry, *Surf. Sci.*, 1976, **54**, 195.
22　R. D. Cowan and D. C. Griffin, *J. Opt. Soc. Am.*, 1976, **66**, 1010.

23 ECPAIMP is an integral program for ECP and AIMP calculations written by L. G. M. Pettersson, L. Seijo and M. A. Nygren.
24 T. H. Dunning, Jr., *J. Chem. Phys.*, 1970, **53**, 2823.
25 S. Huzinaga, Z. Barandiarán, L. Seijo and M. Klobukowsky, *J. Chem. Phys.*, 1987, **86**, 2132.
26 L. Seijo, Z. Barandiarán and S. Huzinaga, *J. Chem. Phys.*, 1989, **91**, 7011.
27 Z. Barandiarán, L. Seijo and S. Huzinaga, *J. Chem. Phys.*, 1991, **94**, 3762.
28 Z. Barandiarán and L. Seijo, *Can. J. Chem.*, 1992, **70**, 409.
29 Z. Barandiarán, L. Seijo and S. Huzinaga, *J. Chem. Phys.*, 1990, **93**, 5843.
30 T. X. T. Sayle, S. C. Parker and C. R. A. Catlow, *Surf. Sci.*, 1994, **316**, 329.
31 B. G. Dick and A. W. Overhauser, *Phys. Rev.*, 1958, **112**, 90.
32 M. Baudin, M. Wójcik and K. Hermansson, *Surf. Sci.*, 1997, **375**, 374.
33 D. P. Chong and S. R. Langhoff, *J. Chem. Phys.*, 1986, **84**, 5606.

Paper 9/04987H

Structure and reactivity of iron oxide surfaces

Sh. K. Shaikhutdinov,† Y. Joseph, C. Kuhrs, W. Ranke and W. Weiss*

Fritz-Haber-Institut der Max-Planck-Gesellschaft, Faradayweg 4-6, 14195 Berlin, Germany. E-mail:w_weiss@fhi-berlin.mpg.de

Received 1st April 1999

Epitaxial films of different iron oxide phases and of potassium iron oxide were grown onto Pt(111) substrates and used for studying structure–reactivity correlations. The film morphologies and their atomic surface structures were characterized by scanning tunneling microscopy and low energy electron diffraction including multiple scattering calculations. The adsorption of water, ethylbenzene, and styrene was investigated by temperature programmed desorption and photoelectron spectroscopy. A dissociative chemisorption of water and a molecular chemisorption of ethylbenzene and styrene is observed on all oxides that expose metal cations in their topmost layers, whereas purely oxygen-terminated FeO(111) monolayer films are chemically inert and only physisorption occurs. Regarding the technical styrene synthesis reaction, which is performed over iron oxide based catalysts, we find a decreasing chemisorption strength of the reaction product molecule styrene, if compared to ethylbenzene, when going from $Fe_3O_4(111)$ over α-$Fe_2O_3(0001)$ to $KFe_xO_y(111)$. Extrapolation of the adsorbate coverages to the technical styrene synthesis reaction conditions using the Langmuir isotherm for coadsorption suggests an increasing catalytic activity along the same direction. This result agrees with previous kinetic experiments performed at elevated gas pressures over the model systems studied here and over polycrystalline iron oxide catalyst samples. It indicates that the iron oxide surface chemistry does not change across the pressure gap and that the model systems simulate technical styrene synthesis catalysts in a realistic way.

1. Introduction

Our understanding of the surface chemical and catalytic properties of metal oxides is still not well developed at a fundamental level. In order to obtain direct evidence on the role of atomic scale structural elements for the surface chemistry of metal oxides, single crystalline model systems with defined chemical compositions and surface structures including non-ideal defects, are needed. A systematic investigation of structure–reactivity correlations becomes possible if these parameters can be varied in a controlled manner, which has proved to be very difficult.[1] Here, we first combine the structural characterization of oxide surfaces with adsorption studies under ultrahigh vacuum conditions. Secondly, the catalytic activities of the model systems are studied under real conditions. If the reaction kinetics at elevated gas pressures are in line with the elementary step kinetics determined under ultrahigh vacuum conditions, no major changes of the surface chemical properties can be assumed to occur across the pressure gap. In this case surface science experiments can reveal relevant information on reaction mechanisms also under real conditions.

---

† Permanent address: Boreskov Institute of Catalysis, Novosibirsk, 630090 Russia.

Iron oxides are used for catalyzing many reactions involving selective oxidations, dehydrogenations, oxo-dehydrogenations, and the water–gas shift reaction.[2] In particular, the dehydrogenation of ethylbenzene (EB) to styrene (St), a large scale sythesis reaction in chemical industry, is performed over iron oxide based catalysts in the presence of steam.[3,4] Kinetic experiments over polycrystalline samples revealed the promoter potassium to increase the catalyst activity by one order of magnitude,[5–7] and an active $KFeO_2$ phase was found to form on the catalyst surface under reaction conditions.[8,9] The reaction rate depends on the adsorption–desorption equilibrium of the educt and product molecules EB and St, where the stronger adsorption of St leads to a site-blocking effect. In this work we study the adsorption of water, EB and St onto single crystalline films of different iron oxide phases and of potassium iron oxide with ultraviolet photoelectron spectroscopy (UPS) and thermal desorption spectroscopy (TDS), after the corresponding surface structures were investigated by scanning tunneling microscopy (STM) and low energy electron diffraction (LEED). These studies give an insight into the styrene synthesis reaction mechanism and into structure–reactivity correlations of metal oxide materials in general.

Since under thermodynamic equilibrium the stoichiometry of binary oxides is determined by the oxygen gas phase pressure and temperature,[10] heteroepitaxial film growth onto chemically inert substrates allows good control over the oxide stoichiometry. Well ordered $Fe_3O_4(111)$ and α-$Fe_2O_3(0001)$ films were grown onto Al_2O_3 and MgO substrates by oxygen-plasma assisted molecular beam epitaxy using different oxygen partial pressures and growth rates.[11,12] In the present paper, the film growth is accomplished by repeated cycles of iron deposition and subsequent oxidation followed by final oxidation treatments in different oxygen gas pressures. Well ordered FeO(111) monolayer films and 100–200 Å thick $Fe_3O_4(111)$ and α-$Fe_2O_3(0001)$ films were prepared in this way. The growth and surface structures of these films was presented in previous work[13–21] and is briefly summarized in Section 3.1. For the first time, epitaxial potassium iron oxide films $KFe_xO_y(111)$ were prepared by depositing potassium onto $Fe_3O_4(111)$ films and subsequent annealing, as described in Section 3.2.

For EB adsorbed on $Fe_3O_4(111)$ a strong interaction between the π-electron system of the phenyl ring, with iron cations exposed on the regular surface areas, was observed in previous UPS studies, whereas no chemical interaction occurred on the oxygen-terminated FeO(111) surface.[22,23] Analogous behavior is observed for water adsorbed onto these surfaces, as discussed in Section 3.3, where chemisorption reactivity again is found to be related to the iron cations exposed on the regular oxide surface. In Section 3.4, a decreasing chemisorption strength of the reaction product St, as compared to EB, is observed when going from $Fe_3O_4(111)$ over α-Fe_2O_3 to $KFe_xO_y(111)$. The reasons are discussed why this result agrees with the catalytic activities of these model systems[24] and of polycrystalline iron oxide samples[5,6] observed in previous reaction experiments at elevated gas pressures.

2. Experimental

2.1. Instrumentation

The experiments were performed in three separate chambers described in detail in ref. 25, which all had base pressures below 1×10^{-10} mbar and contained standard facilities for sample cleaning. The STM and TDS chambers were equipped with backview LEED optics and cylindrical mirror analyzers for Auger electron spectroscopy (AES) measurements, which were carried out at 3 kV primary beam energy, a beam current of ~ 70 µA cm^{-2} and a modulation voltage of $V_{pp} = 10$ V. The UPS chamber contained a helium resonance lamp (He I line 21.2 eV, He II line 40.8 eV), a double pass cylindrical mirror analyzer and a high resolution spot profile analysis LEED system designed by Scheithauer et al.[26] For UPS measurements the direction of light incidence was normal to the analyser axis. As a result, the spectra were averaged over a large range of escape angles.

Each chamber contained a sample transfer system and a fully rotatable off-axis manipulator with identical sample heating–cooling stations, in which a Pt(111) crystal mounted onto a sample holder could be heated by electron bombardment from the back and cooled by a liquid nitrogen reservoir. A chromel–alumel thermocouple was spotwelded onto the edge of the crystal for temperature control. High pressure oxidation treatments were performed in preparation cells that could be completely separated from the analysis chamber by a gate valve after the sample transfer. Temperatures up to 1100 K could be reached in oxygen pressures up to 1 mbar using a 250 W halogen lamp radiation heater located in front of the sample.

The STM experiments were performed with a commercial instrument mounted horizontally on a 150 mm flange, into which the sample holder could be transferred from the off-axis manipulator. All measurements were done at room temperature in the constant current mode, using electrochemically etched tungsten tips cleaned *in situ* by electron bombardment. Calibration of the scanner was performed on a Si(111)-(7 × 7) surface. Image processing of the experimental data included background subtraction and smoothing procedures. All TDS experiments were performed with heating rates of 5 K s^{-1} after gas adsorption at sample temperatures of $T = 100$ K. The exposures are given in Langmuirs (1 L = 1.33×10^{-6} mbar s^{-1}) after correcting for the ionization probabilities of EB and St, which are about 7 times higher than for N_2.[27] High purity EB and St and triply distilled water were dosed *via* leak valves into the UHV chambers. Before that the EB and St were filled under N_2 atmosphere into a small glass bulb containing an activated molecular sieve to reduce possible contamination by water.

2.2. Preparation of iron oxide films

A clean Pt(111) surface is prepared by numerous cycles of argon ion bombardment (1 keV) and subsequent annealing to 1300 K, until it exhibited a sharp LEED pattern and no contamination signals in AES. Onto this surface, iron was evaporated with an electron beam assisted evaporator or by resistively heating a tungsten wire with an iron wire wrapped around it. As described in more detail in ref. 19, 1–2 monolayer thick FeO(111) films and 100–200 Å thick Fe_3O_4(111) films were prepared by repeated cycles of iron deposition at room temperature and subsequent oxidation for about 2 min at temperatures around 1000 K in 10^{-6} mbar oxygen partial pressure. The Fe_3O_4(111) films were transformed into α-Fe_2O_3(0001) films by an oxidation treatment for 10–15 min in oxygen partial pressures above 10^{-3} mbar at temperatures around $T = 1100$ K. Potassium iron oxide films were prepared by deposition of potassium from a getter source (SAES) onto Fe_3O_4(111) films at room temperature and subsequent annealing to 1000 K in vacuum or in 10^{-6} mbar oxygen.

3. Results and discussion

3.1. Growth and surface structures of iron oxide films

Fig. 1(a) displays a 55 × 55 Å^2 atomic resolution STM image of the first iron oxide layer grown onto the Pt(111) surface. A hexagonal lattice of protrusions with a 3.1 Å periodicity and a moiré superstructure with a 25 Å periodicity can be seen. STM image simulations performed by Galloway *et al.*[15] revealed the atomic protrusions to occur at oxygen atom positions located in the topmost layer of this film. Fig. 1(c) depicts a schematic model, which consists of a hexagonal close-packed iron–oxygen bilayer with oxygen located on top. This FeO(111) bilayer is laterally expanded to a lattice constant of $a_{FeO} = 3.11$ Å (the value in bulk FeO is 3.04 Å) and rotated by $\alpha = 1.3°$ against the Pt(111) surface lattice. As explained in detail in ref. 18 the 12% lattice mismatch creates the FeO(111) coincidence moiré superstructure visible in the STM image in Fig. 1(a). The LEED pattern shown in Fig. 1(b) exhibits the main diffraction spots that correspond to the lattice constant of $a_{FeO} = 3.1$ Å and that are surrounded by satellite spots, which are created by double diffraction between the Pt(111) surface and the FeO(111) overlayer. This FeO(111) bilayer structure terminated by an outermost close-packed oxygen layer was also deduced from photoelectron diffraction experiments. There, a strongly reduced iron–oxygen interlayer distance of 0.68 Å (the (111) interlayer distance in bulk FeO is 1.25 Å), which goes along with the lateral lattice expansion, was obtained.[28]

The first FeO(111) layer completely wets the platinum substrate. Depending on the exact oxidation temperature a second FeO(111) layer grows, followed by the formation of three-dimensional Fe_3O_4(111) islands.[19] These islands eventually coalesce and form closed Fe_3O_4(111) films that are at least 100 Å thick. They consist of hexagonally shaped crystallites with lateral dimensions of about 1000 Å as depicted in Fig. 6(a) (see later). These crystallites are atomically flat with a few

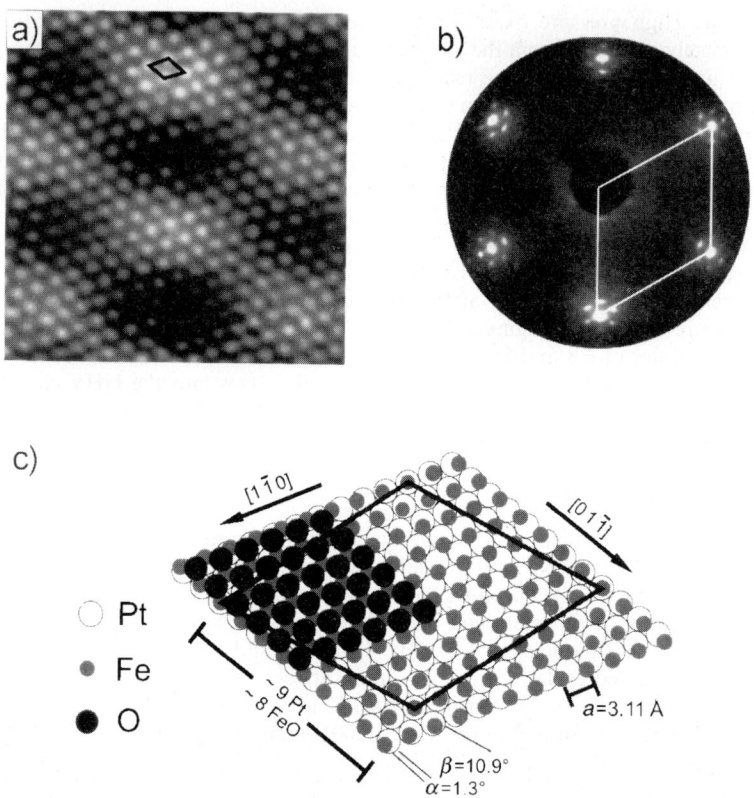

Fig. 1 (a) 55 × 55 Å² STM image of a FeO(111) monolayer film grown onto Pt(111). The protrusions correspond to surface oxygen atoms forming a hexagonal lattice with the indicated unit cell, which has a lattice constant of 3.1 Å. The large moiré superstructure is caused by the 12% lattice mismatch to the Pt(111) substrate. Tunneling voltage and current $U_t = 0.9$ V, $I_t = 0.3$ nA. (b) LEED pattern of the FeO(111) monolayer film at an electron energy of $E = 60$ eV. The main spots indicated by the reciprocal unit cell correspond to the 3.1 Å periodicity on the FeO(111)-(1 × 1) surface. The surrounding satellite spots are created by double diffraction between the platinum substrate and the FeO overlayer. (c) Model of the FeO(111) monolayer film. A hexagonal iron–oxygen bilayer with a lateral lattice constant of $a_{FeO} = 3.10$ Å is rotated by $\alpha = 1.3°$ against the platinum surface lattice. This results in the indicated large coincidence superstructure cell rotated by $\alpha + \beta$ against the FeO(111)-(1 × 1) unit cell, which is seen in the STM image in (a).

monoatomic steps and somewhat higher step densities at the crystallite edges. After a final oxidation treatment at 1000 K, the vertical roughness of the Fe_3O_4(111) films ranges between 30 and 100 Å on a length scale of 1 µm, and one defined surface structure is formed, which exhibits atomic resolution STM images as shown in Fig. 2(a), where a hexagonal lattice of protrusions with a 6 Å periodicity and randomly distributed missing protrusions can be seen. These images are observed both for positive and negative bias voltages applied to the sample. The LEED pattern of these films is shown in Fig. 2(b), it corresponds to a hexagonal surface unit cell with a lattice constant of 6 Å.

Fe_3O_4 magnetite crystallizes in the cubic inverse spinel structure where the oxygen anions form a close-packed fcc sublattice, with tetrahedrally and octahedrally coordinated Fe^{2+} and Fe^{3+} cations located in the interstitial sites.[29,30] The interatomic distance within the close-packed oxygen (111) planes is 2.96 Å. Since not all sites are occupied within the iron (111) planes, a two-dimensional hexagonal unit cell with a lattice constant of 5.92 Å is formed, as observed by STM and LEED. The Fe_3O_4(111) surface structure was determined by a full dynamical LEED intensity analysis and is displayed in a top view in Fig. 2(c).[16,21] It exposes $\frac{1}{4}$ monolayer (ML) of iron cations that would be tetrahedrally coordinated in the bulk, located over a close-packed oxygen layer underneath. The first four interlayer spacings were found to strongly relax. Based on

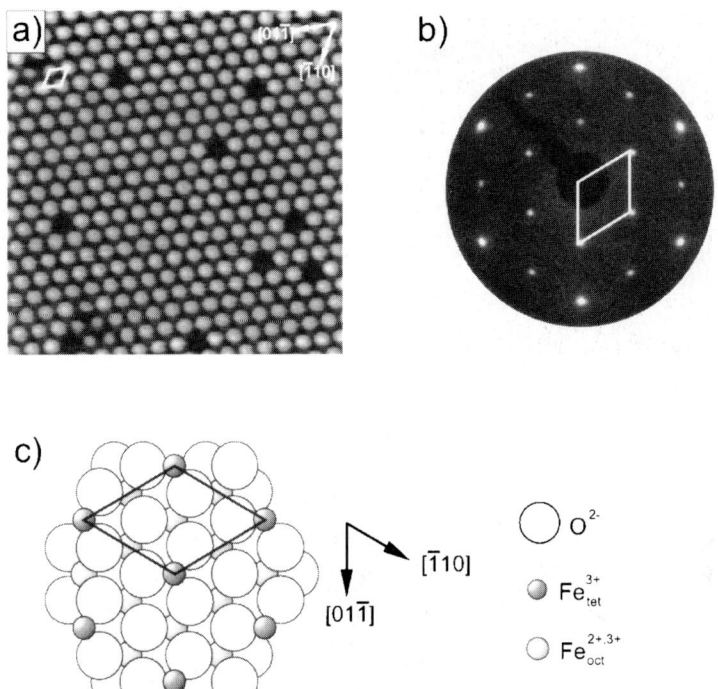

Fig. 2 (a) 90 × 90 Å² STM image of the Fe$_3$O$_4$(111) surface. The protrusions correspond to iron atoms in the topmost layer, which form a hexagonal lattice with the indicated unit cell 6 Å in size. $U_t = -0.9$ V, $I_t = 0.5$ nA. (b) LEED pattern taken at $E = 60$ eV. The indicated reciprocal unit cell corresponds to a surface periodicity of 6 Å. (c) Top view onto the Fe$_3$O$_4$(111) surface structure as obtained from a dynamical LEED intensity analysis[16] with the 6 Å unit cell indicated.

this result we assign the protrusions in the atomic resolution STM image to the topmost layer iron cation positions. This interpretation is supported by recent *ab initio* calculations performed for this surface termination. They revealed a high electron density of states located just above and below the Fermi level, which is related to Fe 3d orbitals of the top layer iron atoms.[31] The missing protrusions in the STM image in Fig. 2(a) represent the most abundant type of point defects on these surfaces, and they are attributed to iron vacancies in the topmost layer. This interpretation was confirmed by dynamical LEED calculations, where different types of vacancies were tested with a LEED program that simulates surface point defects in a random distribution. When analyzing LEED data of surfaces with high defect concentrations a considerable improvement of the R-factor (12%) was achieved only for iron vacancy defects.

α-Fe$_2$O$_3$(0001) films were prepared by oxidizing Fe$_3$O$_4$(111) films at oxygen pressures above 10^{-3} mbar. α-Fe$_2$O$_3$ hematite crystallizes in the corundum structure, where the oxygen anions form an hcp sublattice with octahedrally coordinated Fe^{3+} cations located in the interstitials.[30] Fig. 3(c) depicts a top view onto an unreconstructed α-Fe$_2$O$_3$(0001) surface terminated by an outermost iron layer. The oxygen anion positions within the close-packed (0001) planes slightly deviate from an ideal hexagonal arrangement, leading to an average oxygen–oxygen interatomic distance of 2.91 Å. Within the iron (0001) planes only 1/3 of all sites are occupied leading to a two-dimensional hexagonal unit cell with a lattice constant of 5.03 Å. Fig. 3(a) shows an atomic resolution STM image of a surface prepared in 10^{-3} mbar oxygen, where a hexagonal lattice of protrusions with a 5 Å periodicity can be seen. The LEED pattern in Fig. 3(b) corresponds to a hexagonal surface unit cell with a lattice constant of 5 Å. We attribute the protrusions in the STM image to the topmost iron atom positions on an α-Fe$_2$O$_3$(0001) surface, as depicted in Fig. 3(c). Recent *ab initio* calculations predicted an iron-terminated α-Fe$_2$O$_3$(0001) surface structure to be stable in low oxygen pressure environments and an oxygen-terminated surface structure to be

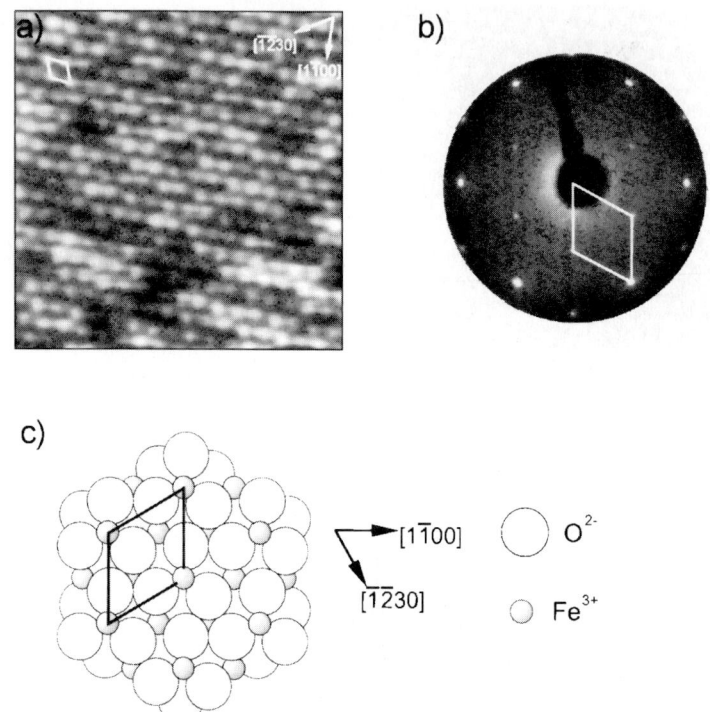

Fig. 3 (a) 120 × 120 Å² STM image of an α-Fe$_2$O$_3$(0001) surface terminated by an outermost iron layer. The protrusions correspond to iron atoms in the topmost layer, which form a hexagonal surface lattice with the indicated unit cell 5 Å in size. $U_t = 1.3$ V, $I_t = 1.25$ nA. (b) LEED pattern of this film taken at an electron energy of 60 eV, the reciprocal unit cell corresponds to a 5 Å periodicity. (c) Top view onto an iron-terminated α-Fe$_2$O$_3$(0001) surface with the indicated unit cell 5 Å in size.

stable in high oxygen pressures environments.[32] As described in ref. 20, we observe coexisting iron-terminated and oxygen-terminated surface domains after preparation in oxygen pressures between 10^{-4} and 10^{-1} mbar. When prepared in 10^{-5} mbar oxygen the iron-terminated surface structure depicted in Fig. 3(c) is formed over the entire sample. All adsorption experiments presented in this paper were performed on this surface.

3.2. Formation and surface structure of potassium iron oxide films

Potassium iron oxide films were prepared by deposition of approximately 5 ML of potassium at 300 K onto Fe$_3$O$_4$(111) films and subsequent annealing in a vacuum. We monitored the reaction between the deposited K and the oxide film by stepwise annealing to increasing temperatures for 10–15 s each. After cooling the sample, Auger spectra were taken and the surface composition was determined using AES sensitivity factors from ref. 33. The resulting temperature dependence of the surface composition is shown in Fig. 4(a). The K concentration decreases considerably between 300 and 450 K, accompanied by a corresponding increase of the Fe and O concentrations. This is due to thermal desorption of K within this temperature range as observed in the desorption trace depicted in (b), which was measured after room temperature deposition of the same amount of K as in (a). With further increasing temperature the K concentration decreases only slightly until it drops more rapidly beyond 700 K, but no further desorption of potassium is observed within this temperature range. From this we conclude that potassium diffuses into the bulk where it might undergo a solid state reaction with the iron oxide. At temperatures around 970 K a composition KFe$_x$O$_y$ with $x \approx 1$ and $y \approx 2$ is deduced from the Auger spectrum shown in Fig. 5, which is close to the stoichiometry of the KFeO$_2$ compound. At this point the film surface has become well-

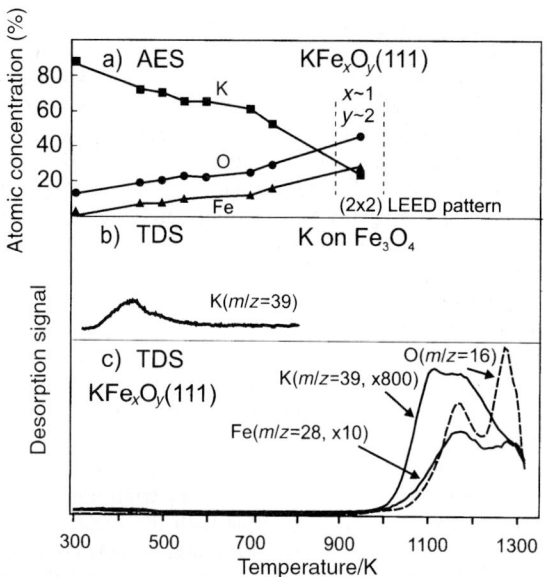

Fig. 4 (a) Surface composition (atomic concentration in percent) of an $Fe_3O_4(111)$ film after deposition of about 5 ML of potassium at 300 K and stepwise annealing as deduced from AES data. (b) Thermal desorption trace of K^+ after deposition of about 5 ML of potassium at 300 K on $Fe_3O_4(111)$. (c) Thermal desorption traces of K^+, Fe^{2+} and O^+ during heating up a KFe_xO_y (111) film, showing the thermal decomposition of the film.

ordered and displays the LEED pattern depicted in Fig. 7(a) (see later), which corresponds to a (2×2) superstructure with respect to $Fe_3O_4(111)$-(1×1).

Fig. 4(c) shows a thermal desorption trace measured for a $KFe_xO_y(111)$ film, which exhibits the characteristic (2×2) LEED pattern. The film is thermally stable up to 1000 K when potassium starts to desorb first. Three K desorption peaks located at at 1115 K, 1165 K and 1260 K can be seen, the last two coincide with maxima in the Fe and O desorption traces. For clean Fe_3O_4 and α-Fe_2O_3 films, only one broad desorption signal, with a maximum around 1300 K, is observed

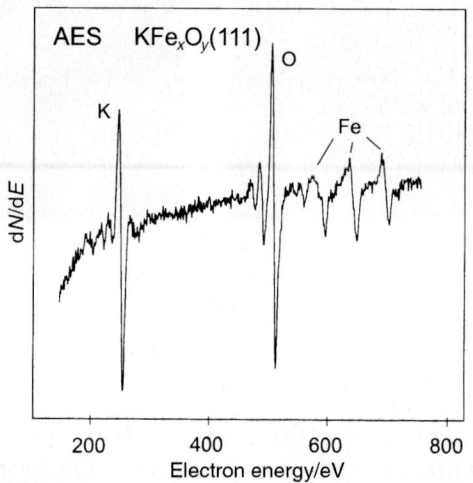

Fig. 5 Auger electron spectrum of an ordered $KFe_xO_y(111)$ film exhibiting a (2×2) LEED pattern with respect to $Fe_3O_4(111)$. It corresponds to a stoichiometry close to $KFeO_2$.

during thermal decomposition (not shown here). The initial desorption starting at 1000 K can be attributed to potassium at the surface and in the subsurface region, where it has accumulated and from where it can diffuse quickly to the surface at these temperatures. The depletion of K in the subsurface region might be accompanied by a decomposition of a ternary K–Fe–O compound into K and Fe_3O_4. This decomposition is not completed during the temperature ramp thereby leading to a common desorption maximum for potassium, iron and oxygen at 1165 K, which indicates decomposition and desorption of the ternary compound phase left in the subsurface region. The desorption maximum at 1260 K is dominated by iron and oxygen with only a small potassium contribution, which indicates the decomposition and desorption of the Fe_3O_4 bulk phase.

The surface structure and morphology of the $KFe_xO_y(111)$ film was studied by STM. Fig. 6 displays 5000 × 5000 Å^2 images of the original $Fe_3O_4(111)$ film (a) and of the $KFe_xO_y(111)$ film (b). The $KFe_xO_y(111)$ film forms small terraces with predominantly triangular shapes, and it is much rougher if compared to the $Fe_3O_4(111)$ film. This surface roughening is probably caused by the diffusion of K into the bulk of the film. However, the film surface is well ordered and exhibits the (2 × 2) LEED pattern depicted in Fig. 7(a), which corresponds to a hexagonal unit cell with a lattice constant of 12 Å. This unit cell is also resolved in the STM image shown in Fig. 7(b).

Since K diffuses into the bulk, the formation of a ternary compound seems quite possible. The only stable ternary K–Fe–O phase that is compatible with the surface composition measured by AES is $KFeO_2$.[34] Due to its orthorhombic crystal structure, $KFeO_2$ contains no hexagonal lattice planes. Therefore, only a strongly distorted $KFeO_2$ lattice could form the observed (2 × 2) superstructure with respect to $Fe_3O_4(111)$-(1 × 1), which seems unlikely. Potassium may also substitute iron atoms in the Fe_3O_4 lattice in variable concentrations. The strain produced in such a solid solution, while maintaining the pseudomorphic relationship to the $Fe_3O_4(111)$ substrate, may limit the maximum potassium concentration. Indeed, Auger measurements performed on films prepared after initial deposition of different amounts of K and different numbers of annealing cycles reveal varying KFe_xO_y compositions with $x \approx 1.1$–0.9 and $y \approx 1.3$–2. But since AES averages over the electron escape depth of 10–20 Å, these compositions may result from potassium concentrations that vary with the distance from the surface.

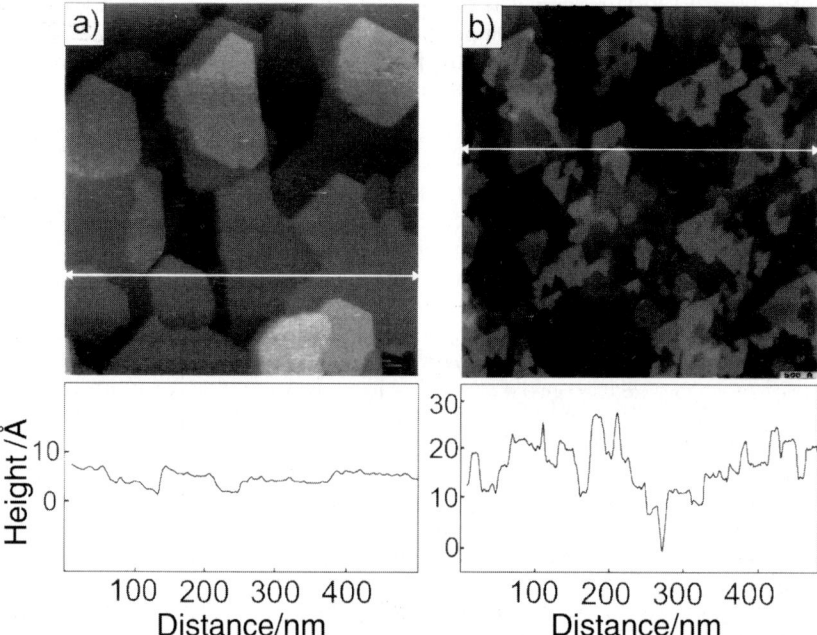

Fig. 6 5000 × 5000 Å^2 STM images of a clean $Fe_3O_4(111)$ (a) and of a $KFe_xO_y(111)$ film (b), showing the different surface morphologies. $U_t = 0.4$ V, $I_t = 0.6$ nA.

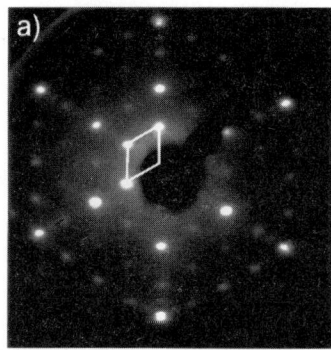 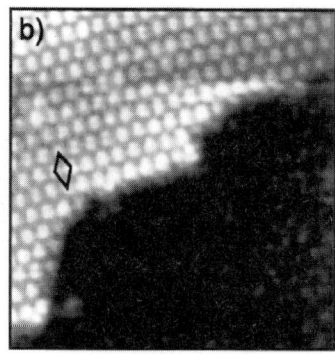

Fig. 7 (a) LEED pattern of the KFe$_x$O$_y$(111) film at an electron energy of $E = 60$ eV. The indicated reciprocal unit cell corresponds to a (2×2) superstructure referred to Fe$_3$O$_4$(111)-(1×1) with a surface periodicity of 12 Å. (b) 220×220 Å^2 STM image of the KFe$_x$O$_y$(111) surface. Two terraces separated by a 2.5 Å high step can be seen, which both exhibit hexagonal lattices of protrusions with a 12 Å periodicity. The protrusions on the upper terraces are wider than those on the lower terrace. $U_t = 1.4$ V, $I_t = 1.3$ nA.

The atomic resolution STM image in Fig. 7(b) reveals two different structures. Here they are separated by a step 2.5 Å high, but in other images they are also observed on the same level. Both exhibit hexagonal lattices of protrusions with a 12 Å periodicity, but the half-width of the protrusions on the upper terrace is about 6 Å, whereas that on the lower terrace is about 4 Å. Also, the upper terrace looks clean while the lower one is covered by adsorbate-like species, which suggests different chemical properties for these two surface structures. We assign the atomic protrusions to potassium atom positions in the topmost layer of this film. This is in line with other studies where alkali metals were found to form the uppermost surface layers.[35,36] Also, alkali metals adsorbed on metal and metal-oxide surfaces were imaged as protrusions with STM,[37–39] in agreement with theoretical calculations.[40] X-ray diffraction and ion scattering experiments are planned in order to further clarify the surface structure of these films, the interpretation of the STM images and the question about the formation of a ternary K–Fe–O compound phase.

Before and after performing adsorption and catalysis experiments the KFe$_x$O$_y$(111) films were cleaned by flashing to 900 K in 10^{-6} mbar oxygen. STM, LEED and AES showed no changes of the surface structure and composition after this treatment.

3.3. Active sites for water chemisorption

As discussed in Section 3.1, the surface of the FeO(111) film is oxygen-terminated whereas the surface of the Fe$_3$O$_4$(111) film exposes $\frac{1}{4}$ monolayer of iron cations over a close-packed oxygen layer underneath. Fig. 8 shows thermal desorption spectra after exposing both surfaces at 100 K to increasing amounts of water. On FeO(111) a desorption maximum at 170 K is observed first, which shifts to 164 K as the exposure increases to 0.3 L. This signal is attributed to physisorbed water (labeled β). With further increasing exposure, the desorption traces form a common leading edge, which is characteristic for zero-order desorption of condensed ice multilayers (labeled α). On Fe$_3$O$_4$(111) a desorption maximum around 280 K is observed first, which is attributed to chemisorbed water (labeled γ). With increasing exposure a physisorbed β signal at 190 K evolves, followed by the ice multilayer signal α. No species other than water desorb, especially no hydrogen. The absence of a chemisorption signal on FeO(111) but not on Fe$_3$O$_4$(111) indicates a specific chemical interaction related to the iron cations exposed on the Fe$_3$O$_4$(111) surface.

The TDS spectra were taken after exposing the sample to defined amounts of water at a temperature low enough to prevent desorption after pumping off the water vapor. Because of the low mobilities of the adsorbed species at this temperature equilibrium may not be reached, and different adsorption states may not be occupied strictly sequentially, or even incompletely in the case of activated adsorption. We have taken UP spectra under conditions of dynamic adsorption–desorption equilibrium. Water was admitted at a constant vapor pressure and spectra were taken at different sample temperatures after establishment of the corresponding equilibrium coverages,

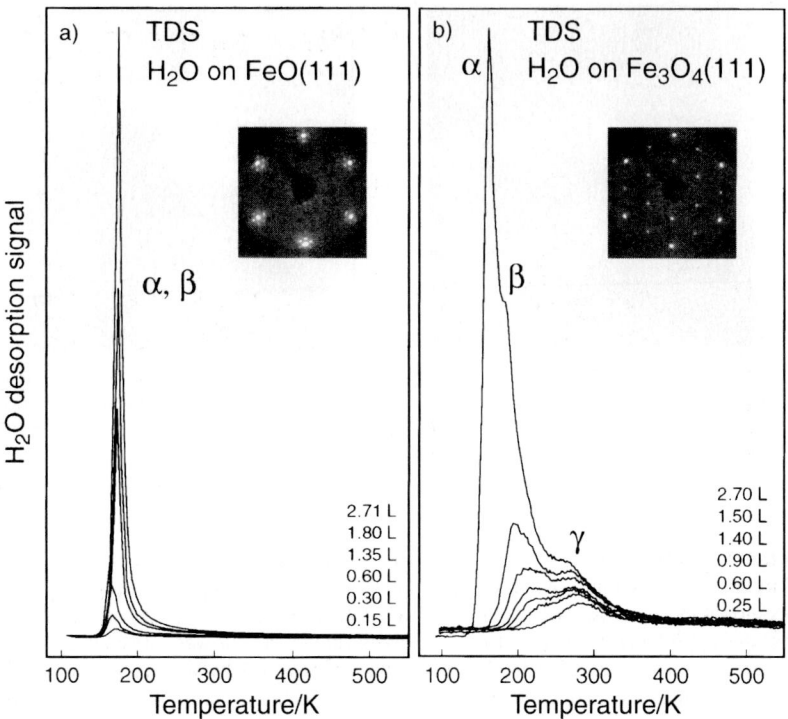

Fig. 8 Thermal desorption traces of water adsorbed onto FeO(111) (a) and $Fe_3O_4(111)$ (b). Exposures are indicated in Langmuir units. The inserts show LEED patterns of the respective clean oxide films.

which took up to 6 min at maximum for the chemisorbed species. Under such conditions the adsorption of the different species occurred strictly sequentially. UPS turned out to be a non-destructive technique and reveals both the nature of the adsorbate species and their coverages. It further allows the determination of work function changes, which yields information on the adsorbate orientation.

Fig. 9(a) presents a sequence of UP spectra taken from the FeO(111) monolayer film at a pressure $p(H_2O) = 1 \times 10^{-8}$ mbar upon a stepwise decrease of the sample temperature from 307 K (no adsorption) to 137 K (beginning condensation). Significant adsorption starts below about 180 K. Below -6 eV with respect to the Fermi level, the intensity increase due to adsorbate-induced emission dominates the spectral changes, whereas above -6 eV attenuation of the substrate emission dominates. In order to deduce the adsorbate coverage, Θ, we use the attenuation factor, AF, which was determined near the Fermi level, E_F, where no adsorbate emission features exist. According to the Lambert–Beer absorption law, the adsorbate thickness, d, in units of the electron escape depth, l_e, is given by $d/l_e = \ln(1/AF) \propto \Theta$. The escape depth is given by $l_e = \lambda_e \cos \alpha$ with λ_e the electron mean free path and α the mean escape angle (42° for the cylindrical mirror analyser used here). For $E_{kin} \approx 20$ eV (with respect to E_F) λ_e is between 3 and 10 Å[41] and the value $d/l_e = 1$ (corresponding to AF = 0.37) should correspond to an adsorbate thickness of 2.5–7 Å or 1–2 ML. This estimation reveals the coverage in Fig. 9 to increase from 0 to 3–5 ML. Precise coverages will be deduced from comparisons with adsorption models on the different oxide surfaces.

From incremental difference spectra (difference between subsequent spectra), changes in the adsorbate spectra can be seen which occur when an adsorbate species has saturated and a new species starts to form on top. This happens at coverages corresponding to the two dotted spectra (being the lower ones in the range below -6 eV) in Fig. 9(a). The highest dotted spectrum corresponds to condensation and does not saturate. The spectra in Fig. 9(b) represent the differences

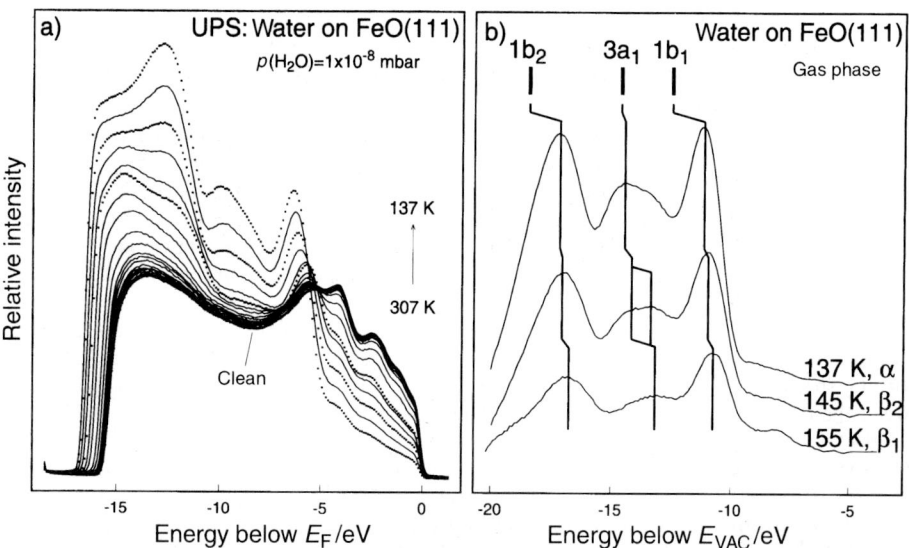

Fig. 9 (a) UP spectra of water adsorbed onto the FeO(111) surface, measured at constant pressure $p(H_2O) = 1 \times 10^{-8}$ mbar upon stepwise decreasing the sample temperature. The lower two dotted spectra represent saturation of two different physisorbed species β_1 and β_2, the upper dotted curve corresponds to condensation of ice α. (b) Adsorbate difference spectra corresponding to the dotted curves in (a). The energy axis refers to the vacuum level and the orbital positions for gas phase water[41] are indicated.

between the dotted spectra in Fig. 9(a) and the properly attenuated clean substrate spectrum. They reflect two saturated species labeled β_1 and β_2, which form sequentially, followed by the condensate spectrum α, which does not represent saturation. The work function change is given by the shift of the low energy onset of the spectra in Fig. 9(a). It decreases by 0.82 eV for saturation of β_1 and further by 0.33 eV upon saturation of β_2. Condensation does not change the work function further.

The spectra in Fig. 9(b) are referenced to the vacuum level E_{vac} so that comparison with the orbital positions of gas-phase water[42] is possible. The general spectral shapes of the three species are similar and correspond to the spectrum of gaseous water, shifted by different relaxation shifts $\Delta E_R = 1.8$, 1.5 and 1.3 eV for the β_1, β_2 and α species (taken from the average shift of the $1b_1$ and $1b_2$ orbitals). We therefore conclude that all species represent molecular water. The β_1 species saturates at a coverage corresponding to $d/l_e = 0.55 \pm 0.03$ and exhibits relative peak positions that agree well with those of gaseous water. Therefore, we ascribe it to physisorbed water monomers. From the work function decrease we conclude that the molecules are oriented with their oxygen atoms towards the substrate. Because they are weakly bound, they may be mobile and form a two-dimensional lattice gas as interpreted for similar species observed on other surfaces.[43,44] All orbitals of the β_2 species are shifted and broadened with respect to β_1, and the $3a_1$ emission contains contributions from two peaks as indicated by the bars in Fig. 9(b). This is typical for hydrogen-bonded aggregates with two kinds of water molecules in different bonding configurations,[43,45,46] where one acts as proton donor and the other as proton acceptor. The β_2 species saturates at $d/l_e = 1.3 \pm 0.2$, more than twice that of the β_1 species. We ascribe it to a physisorbed water bilayer as identified previously on several metal surfaces.[47] For the condensed α species the position of the $3a_1$ emission relative to the $1b_2$ and $1b_1$ orbitals is similar to that of the β_2 species. In the bulk of a well crystallized ice film, all molecules are equivalent and no splitting of the $3a_1$ orbital is expected. The broadening of the $3a_1$ signal in Fig. 9(b) is caused by differently coordinated water molecules at the surface and at the interface to the substrate, since the ice layer was only about two bilayers thick.

In contrast to the general tendency of adsorbed water to form aggregates at small coverages,[47] water monomers first physisorb on FeO(111) without aggregation. Although the interaction with the substrate is weak (≈ 50 kJ mol^{-1} desorption energy as determined from the TDS data), it appears to be strong enough to overbalance the energy gain from hydrogen bonding in aggregates. It saturates at less than half the coverage of the β_2 bilayer, which would be compatible with an $(\sqrt{3} \times \sqrt{3})R\,30°$ adsorbate structure referred to FeO(111)-(1 × 1). But no corresponding LEED pattern was observed, possibly due to the high mobility of the adsorbed species. The work function decrease implies adsorption with the oxygen atom and its lone pair orbital oriented towards the substrate, which is quite unexpected for an oxygen-terminated surface. The reason may be that the electronic structure of the FeO(111) monolayer film deviates from that of the corresponding bulk material. Indeed, the Fe–O interlayer distance is only 0.68 Å[28] (1.25 Å in bulk FeO), so that water molecules adsorbed onto threefold hollow sites of the outermost oxygen layer may "feel" the acidic character of the iron atoms underneath, resulting in the adsorption geometry we propose for the monomeric β_1 water species.

Fig. 10(a) presents UP spectra for water adsorbed on the Fe_3O_4(111) surface under the same equilibrium conditions as in Fig. 9(a). Here the adsorption starts at 350 K, and three sequentially adsorbing species can be distinguished. Their spectra in Fig. 10(b) are formed from the differences between the dotted saturation curves in (a). The first species (labeled γ) saturates at 225 K with $d/l_e = 0.43 \pm 0.03$ and exhibits two broad peaks. We attribute them to the 1π and 3σ orbitals of $OH^{\delta-}$ species formed by dissociation of water into $OH^{\delta-} + H^{\delta+}$,[48] as observed on many other metal-oxide surfaces.[49–54] The work function decreases by 0.75 eV, indicating that $OH^{\delta-}$ species are adsorbed with the oxygen atom oriented towards the surface. The features between -5 and -10 eV below E_{vac} are probably due to adsorbate-induced changes of the substrate emission and to emission from the coadsorbed atomic hydrogen.

The saturation coverage (in terms of d/l_e) of dissociatively chemisorbed water is about 1/3 of the water bilayer β_2 saturation coverage on FeO(111). This is slightly less than the monomeric β_1 water layer thereon, which is too high to be explained by defect-related chemisorption. Therefore, regular surface sites must be responsible for the chemisorption. On films with higher surface defect concentrations, as estimated from the spot widths and background intensities in the LEED pattern, we observe lower saturation coverages. Since the dominant type of point defects on Fe_3O_4(111) are iron vacancies, this observation clearly indicates that iron cations exposed on the surface are responsible for the dissociative chemisorption of water into $OH^{\delta-}$ and $H^{\delta+}$. The $OH^{\delta-}$ group is adsorbed onto an acidic Fe site and $H^{\delta+}$ onto a neighboring basic oxygen site. On low

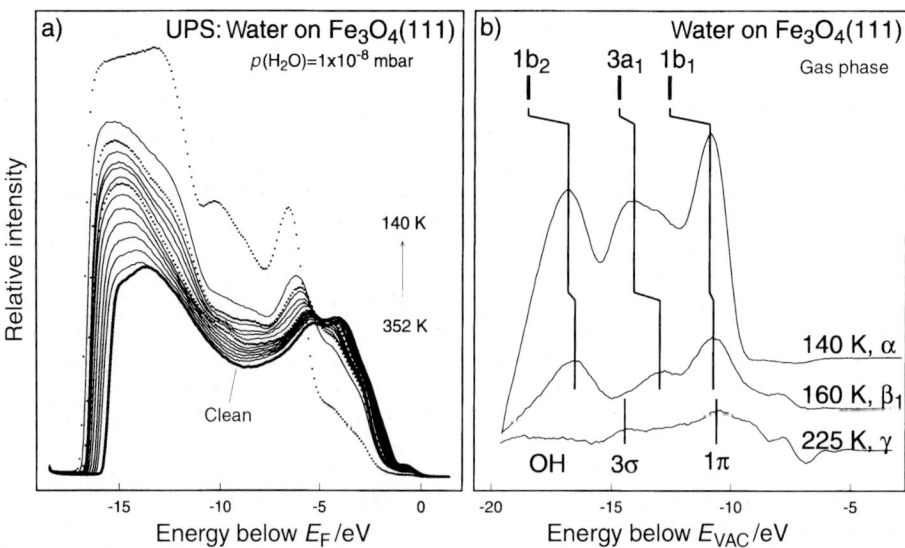

Fig. 10 (a) UP spectra of water adsorbed on the Fe_3O_4(111) surface, measured as in Fig. 9. The lower two dotted spectra represent saturation of the chemisorbed γ and physisorbed β_1 species, the upper dotted curve corresponds to condensation of ice α. (b) Incremental adsorbate spectra formed from the difference between the dotted curves in (a). The energy axis refers to the vacuum level and the orbital positions for gas phase water[42] are indicated.

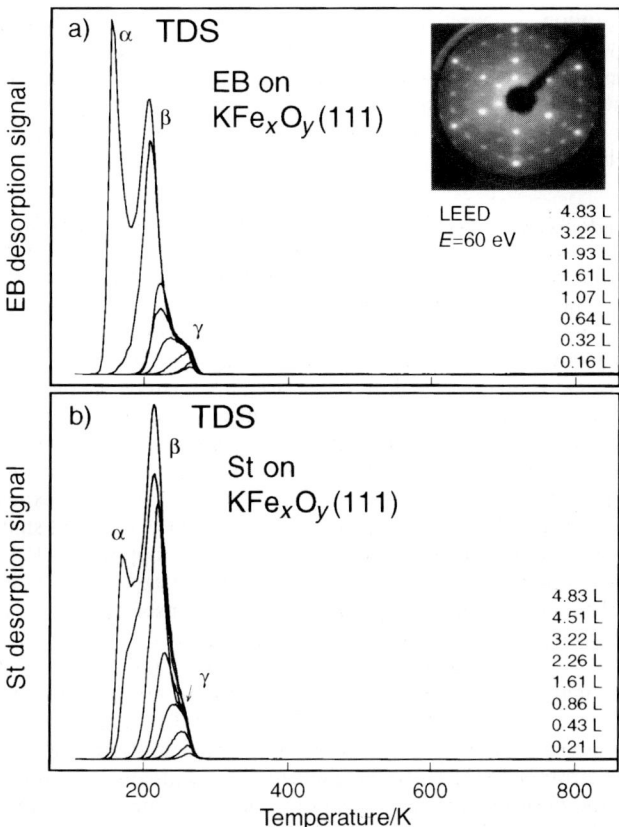

Fig. 11 Thermal desorption traces of ethylbenzene (a) and styrene (b) adsorbed onto the $KFe_xO_y(111)$ surface. The exposures are indicated in Langmuir. The insert in (a) shows the LEED pattern of the clean surface.

defect density films the saturation coverage of $OH^{\delta-}$ and $H^{\delta+}$ should be equal to the iron surface density, i.e., 3.3×10^{14} cm^{-2}. This corresponds to 31% of the density of water molecules in an ideal bilayer (10.6×10^{14} cm^{-2} for an adsorbed bilayer, slightly less than the value 11.2×10^{14} cm^{-2} for the bilayer density in thick ice[47]). Taking the experimental error into account, the observed ratio of the saturation coverages for the chemisorbed γ layer on $Fe_3O_4(111)$ ($d/l_e = 0.43 \pm 0.03$) and the physisorbed bilayer β_2 ($d/l_e = 1.3 \pm 0.2$) on FeO(111) is 0.33 ± 0.08, which represents a remarkable agreement and strongly supports the adsorption model.

The saturation temperature, spectral shape, and orbital binding energies of the second species formed on $Fe_3O_4(111)$ are similar to the β_1 species on FeO(111) and thus corresponds to monomeric physisorbed water. It saturates at a total coverage $d/l_e = 0.86 \pm 0.03$ (for $\gamma + \beta_1$), which is exactly twice that of the γ-species. This is consistent with a fully occupied first adsorbate layer consisting of equal amounts of the γ and β_1 species, i.e., of $OH^{\delta-}$, $H^{\delta+}$ and coadsorbed monomeric H_2O. At higher coverages water condenses on top of this $\gamma + \beta_1$ layer.

To summarize, on both surfaces a sequential adsorption of three different species is observed. Their saturation coverages correlate quantitatively with the respective surface structures. The oxygen-terminated FeO(111) surface is chemically inert and a physisorption of water monomers is observed, followed by the formation of a hydrogen bonded bilayer and finally by condensation of ice. On the iron-terminated $Fe_3O_4(111)$ surface a dissociative chemisorption of water occurs, which was also observed on the (0001) and (012) surfaces of α-Fe_2O_3.[55,56] On $Fe_3O_4(111)$ the chemisorption is not related to surface defects but occurs on regular surface areas. After saturation of the chemisorbed species, water monomers get coadsorbed followed by condensation of ice. The dissociative chemisorption is caused by the cooperative action of neighboring acidic iron and basic oxygen sites, which are exposed on the surface in a geometry that allows the adsorption of $OH^{\delta-}$ and $H^{\delta+}$ species after the dissociation of a water molecule. These findings agree with a previous study, where a specific chemical interaction between surface iron cations and the π-electron system of EB was observed[23] since EB is molecularly chemisorbed on the iron-terminated Fe_3O_4(111) surface but not on the oxygen-terminated FeO(111) surface.

3.4. Adsorption of ethylbenzene and styrene onto Fe_3O_4(111), α-Fe_2O_3(0001) and KFe_xO_y(111)

The dehydrogenation of ethylbenzene (EB) to styrene (St) is performed over iron oxide catalysts promoted with potassium. We studied the adsorption of the EB educt and St product molecules onto the Fe_3O_4(111), α-Fe_2O_3(0001) and KFe_xO_y(111) model catalyst films by TDS. As for water on Fe_3O_4(111), the sequential adsorption of chemisorbed γ species, physisorbed β species, and condensed α multilayers is observed on all three surfaces. This can be seen in Fig. 11, which shows TD spectra measured after exposing a KFe_xO_y(111) film to the indicated amounts of EB (a) and styrene (b) at 100 K. For the smallest exposure both chemisorbed γ-EB and γ-St desorb at 260 K. Their saturation exposures are about 1 L. The respective spectra for Fe_3O_4(111) and α-Fe_2O_3(0001) look similar and will be presented elsewhere, the saturation exposures thereon are also 1 L. The desorption peak for physisorbed β-EB appears at 230 K and shifts with increasing exposure to 206 K, then the condensed α-EB multilayer peak appears at 155 K. The peak for physisorbed β-styrene shifts from 238 K to 215 K, then the condensed α-St multilayer peak appears at 170 K. The desorption temperatures of the condensed species agree well with their respective heats of sublimation.[57]

An exposure of 1 L of EB or St at a gas temperature of 300 K corresponds to 1.96×10^{14} molecules cm^{-2} impinging onto the surface. The van der Waals area of EB and styrene molecules lying flat on the surface is approximately 50 Å^2, so that 1 ML corresponds to $N = 2 \times 10^{14}$ molecules cm^{-2}, assuming a densely packed layer. Since condensation occurs at 150 K for these molecules, a sticking probability of 1 can be assumed for a sample temperature of 100 K during adsorption, so that the observed saturation exposure of the chemisorbed γ species of 1 L corresponds to about 1 ML. Thus, the chemisorption sites do not represent a minority species as would be expected for surface defects, although surface defects can also contribute to the desorption signals.

Fig. 12 compares TD spectra after applying small exposures of EB and St to the three substrates, so that only the γ chemisorption states get populated. On Fe_3O_4(111), St is bound considerably more strongly than EB. On α-Fe_2O_3(0001), two chemisorbed species are observed for EB and St. A minority species labeled $γ_2$ creates desorption peaks around 370 K, for which saturation coverages of less than 0.1 ML were deduced. The desorption peaks of the main chemisorption species appear at lower temperatures and are labeled $γ_1$. Here again St is bound more strongly than EB, but the difference is smaller when compared to Fe_3O_4(111). On KFe_xO_y(111), the desorption temperatures of γ-EB and γ-St are equal and quite low. From the TDS data the desorption energies and frequency factors of the chemisorbed species were determined by a threshold analysis[58] (with the exception of the minority species $γ_2$ on Fe_2O_3). They are listed in Table 1. The desorption energies are quite reasonable and all frequency factors have values in the range 10^{11}–10^{13} s^{-1}.

Table 1 Desorption energies E_{des} and frequency prefactors $ν_{des}$ for ethylbenzene and styrene chemisorbed on Fe_3O_4(111), α-Fe_2O_3(0001) and KFe_xO_y(111) as determined from the TDS measurements

	Fe_3O_4(111)		α-Fe_2O_3(0001)		KFe_xO_y(111)	
	Ethylbenzene	Styrene	Ethylbenzene	Styrene	Ethylbenzene	Styrene
E_{des}/kJ mol^{-1}	86	118	64	73	65	65
$ν_{des}$/s^{-1}	1×10^{12}	3×10^{11}	1×10^{12}	5×10^{12}	2×10^{12}	1×10^{13}

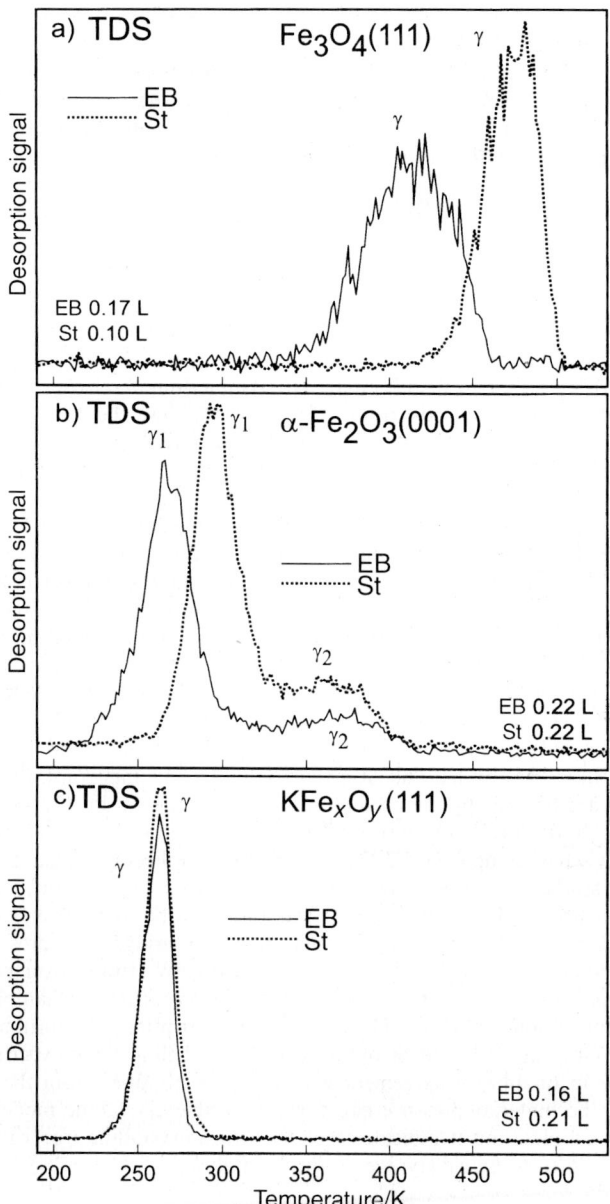

Fig. 12 Thermal desorption traces of EB and styrene chemisorbed on $Fe_3O_4(111)$ (a), $\alpha\text{-}Fe_2O_3(0001)$ (b), and $KFe_xO_y(111)$ (c), taken after small exposures.

The Langmuir isotherm gives the relative coverage Θ_r (number of occupied divided by the number of available adsorption sites) under adsorption–desorption equilibrium conditions:[59]

$$\Theta_r = bp/(1+bp)$$

with p being the gas pressure and b given by

$$b = \{s_0/(\nu_{des} N(2\pi mkT)^{0.5})\}\exp\{-(\Delta E_{ad} - \Delta E_{des})/kT\}.$$

Here, s_0 is the initial sticking coefficient, ν_{des} the frequency factor for desorption, N the adsorption site density which is 2×10^{14} cm^{-2} for EB and St, m the mass of the adsorbing molecule, k the

Boltzmann constant, T the temperature, and ΔE_{ad}, ΔE_{des} the activation energies for adsorption and desorption, respectively. The Langmuir formalism assumes that sticking occurs only on unoccupied sites without mobile precursor states. Although we observe precursor adsorption kinetics at $T = 100$ K,[22] the lifetime and diffusion length of the precursor species become very small at $T = 900$ K, so that one can assume Langmuir adsorption kinetics at such high temperatures. For competitive coadsorption of two different molecules A and B, we obtain

$$\Theta_{r,A} = b_A p_A / (1 + b_A p_A + b_B p_B), \quad \Theta_{r,B} = b_B p_B / (1 + b_A p_A + b_B p_B),$$

and

$$\Theta_r = \Theta_{r,A} + \Theta_{r,B}.$$

Here, p_A and p_B are the partial pressures of gases A and B, respectively.

For estimating the surface coverages of EB and St under the technical styrene synthesis reaction conditions ($T = 873$ K, p_{EB} or $p_{St} = 200$ mbar), we assume initial sticking coefficients $s_0 = 1$ for both molecules and non-activated adsorption ($\Delta E_{ad} \approx 0$). We use the ΔE_{des} and v_{des} values from Table 1. First, we consider the adsorption of only EB and only St (no coadsorption). It turns out that the physisorbed β states are occupied by less than 1%. The relative coverages Θ_r of the chemisorbed γ states of EB (St) would be 92% (100%) on $Fe_3O_4(111)$, 37% (29%) on α-$Fe_2O_3(0001)$ (majority species γ_1) and 25% (6%) on $KFe_xO_y(111)$. The more strongly bound γ_2 minority species on Fe_2O_3 are always fully occupied, but for the following discussion of the coadsorption of EB and St we neglect its contribution because the number of sites is only 9% of the γ_1 sites.

For the competitive coadsorption of EB and St we assume equal partial pressures of 100 mbar each. The Langmuir extrapolation yields, on $Fe_3O_4(111)$, $\Theta_r = \Theta_{r,EB} + \Theta_{r,St} = 100\%$ and $\Theta_{EB}/\Theta_{St} \approx 0.004$; on α-$Fe_2O_3(0001)$, $\Theta_r = 35\%$ and $\Theta_{EB}/\Theta_{St} \approx 1.3$, and on $KFe_xO_y(111)$, $\Theta_r = 18\%$ and $\Theta_{EB}/\Theta_{St} \approx 5$. Thus on $Fe_3O_4(111)$ all chemisorption sites are occupied, on $Fe_2O_3(0001)$ 35%, and on $KFe_xO_y(111)$ 18%. For the catalytic dehydrogenation reaction of EB to St the relative occupation of these sites by EB and St given by Θ_{EB}/Θ_{St} is important. On Fe_3O_4 almost all chemisorption sites are blocked by the product molecule styrene, on Fe_2O_3 43% of these sites are occupied by styrene, and on KFe_xO_y only 17% are occupied by styrene. This suggests an increasing catalytic activity when going from Fe_3O_4 over Fe_2O_3 to KFe_xO_y, which indeed was observed in recent reactivity studies performed over the unpromoted $Fe_3O_4(111)$ and α-$Fe_2O_3(0001)$ films at elevated gas pressures.[24] There, $Fe_3O_4(111)$ was found to be inactive and $Fe_2O_3(0001)$ films with high surface defect concentrations were active, which bridges the pressure gap for these model systems. Furthermore, kinetic experiments performed by Hirano over polycrystalline iron oxide catalyst samples in the mbar range revealed potassium-promoted samples to be 10 times more active than unpromoted Fe_2O_3.[5] The adsorption–desorption equilibrium constants for the educt and product molecules EB and St obtained from modelling these experiments with simple rate equations were in qualitative agreement with the ones obtained from the TDS data in this work. This bridges the material–pressure gap from the single crystalline model systems to polycrystalline catalyst samples, and it indicates that the single crystalline iron oxide films represent realistic model systems for technical styrene synthesis catalysts.

4. Conclusions

Well ordered epitaxial FeO(111), $Fe_3O_4(111)$, α-$Fe_2O_3(0001)$, and potassium iron oxide films with a surface composition close to $KFeO_2$ were grown onto Pt(111). On FeO(111) an oxygen-terminated surface structure is formed, the other three oxides form surface structures terminated by metal cations if prepared in oxygen partial pressures around 10^{-6} mbar. The adsorption studies clearly reveal a specific interaction of water with the iron-terminated $Fe_3O_4(111)$ surface, whereas the purely oxygen-terminated FeO(111) surface is chemically inert. The dissociative chemisorption of water becomes possible by the cooperative action of an acidic iron and a neighboring basic oxygen site. On $Fe_3O_4(111)$ these sites are located at a distance from each other that fits the geometry of the water molecule, so that $OH^{\delta-}$ can adsorb onto an Fe site and $H^{\delta+}$ onto a neighboring oxygen site immediately after the dissociation. The cooperative action of acidic and basic sites arranged in a way that fits the geometry of the reactant molecule most likely plays a key role for many surface reactions on metal oxides, *e.g.*, dehydrogenations and others. Regarding the catalytic dehydrogenation of ethylbenzene to styrene, the TDS experiments reveal desorption energies for the ethylbenzene educt and styrene product molecules that suggest an increasing catalytic activity when going from $Fe_3O_4(111)$ over α-$Fe_2O_3(0001)$ to $KFe_xO_y(111)$. This result is in line with kinetic experiments performed at elevated gas pressures over polycrystalline catalyst samples and over the epitaxial model systems presented here. It bridges the pressure–material gap and demonstrates that the single crystalline iron oxide films model technical styrene synthesis catalysts in a realistic way.

Acknowledgements

We thank Robert Schlögl for fruitful discussions and Manfred Swoboda for technical assistance. Y. Joseph acknowledges financial support from the Deutsche Forschungsgemeinschaft.

References

1. V. E. Henrich and P. A. Cox, *The Surface Science of Metal Oxides*, Cambridge University Press, Cambridge, 1994.
2. J. W. Geus, *Appl. Catal.*, 1986, **25**, 313.
3. E. H. Lee, *Catal. Rev.*, 1973, **8**, 285.
4. K. Kochloefl, in *Handbook of Heterogeneous Catalysis*, ed. G. Ertl, H. Knözinger and J. Weitkamp, Wiley-VCH, Weinheim, 1997, Vol. 5, p. 2151.
5. T. Hirano, *Appl. Catal.*, 1986, **26**, 65.
6. T. Hirano, *Appl. Catal.*, 1986, **28**, 119.
7. K. Coulter, D. W. Goodman and R. G. More, *Catal. Lett.*, 1995, **31**, 1.
8. M. Muhler, J. Schütze, M. Wesemann, T. Rayment, A. Dent, R. Schlögl and G. Ertl, *J. Catal.*, 1990, **126**, 339.
9. M. Muhler, R. Schlögl and G. Ertl, *J. Catal.*, 1992, **138**, 413.
10. D. Schmalzried, *Chemical Kinetics of Solids*, VCH, Weinheim, 1995.
11. Y. J. Kim, Y. Gao and S. A. Chambers, *Surf. Sci.*, 1997, **371**, 358.
12. Y. Gao, Y. K. Kim, S. Thevuthasan and S. A. Chambers, *J. Appl. Phys.*, 1997, **81**, 3253.
13. G. H. Vurens, M. Salmeron and G. A. Somorjai, *Surf. Sci.*, 1988, **201**, 129.
14. H. C. Galloway, J. J. Benitez and M. Salmeron, *Surf. Sci.*, 1993, **298**, 127.
15. H. C. Galloway, P. Sautet and M. Salmeron, *Phys. Rev. B: Condens. Matter*, 1996, **54**, R11145.
16. W. Weiss, A. Barbieri, M. A. VanHove and G. A. Somorjai, *Phys. Rev. Lett.*, 1993, **71**, 1848.
17. A. Barbieri, W. Weiss, M. A. Van Hove and G. A. Somorjai, *Surf. Sci.*, 1994, **302**, 259.
18. M. Ritter, W. Ranke and W. Weiss, *Phys. Rev. B: Condens. Matter*, 1998, **57**, 7240.
19. W. Weiss and M. Ritter, *Phys. Rev. B: Condens. Matter*, 1999, **50**, 5201.
20. Sh. K. Shaikhutdinov and W. Weiss, *Surf. Sci. Lett.*, 1999, **432**, L627.
21. M. Ritter and W. Weiss, *Surf. Sci.*, 1999, **432**, 81.
22. D. Zscherpel, W. Ranke, W. Weiss and R. Schlögl, *J. Chem. Phys.*, 1998, **108**, 9506.
23. W. Ranke and W. Weiss, *Surf. Sci.*, 1998, **414**, 238.
24. W. Weiss, D. Zscherpel and R. Schlögl, *Catal. Lett.*, 1998, **52**, 215.
25. W. Weiss, M. Ritter, D. Zscherpel, M. Swoboda and R. Schlögl, *J. Vac. Sci. Technol. A*, 1998, **16**, 21.
26. U. Scheithauer, G. Meyer and M. Henzler, *Surf. Sci.*, 1986, **178**, 441.
27. F. Nakao, *Vacuum*, 1975, **25**, 431.
28. Y. J. Kim, C. Westphal, R. X. Ynzunza, H. C. Galloway, M. Salmeron, M. A. Van Hove and C. S. Fadley, *Phys. Rev. B: Condens. Matter*, 1997, **55**, R13448.
29. R. W. G. Wyckoff, *Crystal Structures*, Interscience Publishers, New York, 2nd edn., 1982, Vol. I, p. 85.
30. R. M. Cornell and U. Schwertmann, *The Iron Oxides*, VCH, Weinheim, 1996.
31. X.-G. Wang and M. Scheffler, personal communication.
32. X.-G. Wang, W. Weiss, Sh. K. Shaikhutdinov, M. Ritter, M. Petersen, F. Wagner, R. Schlögl and M. Scheffler, *Phys. Rev. Lett.*, 1998, **81**, 1038.
33. L. E. Davis, N. C. MacDonald, P. W. Palmberg, P. E. Riach and R. E. Weber, *Handbook of Auger electron spectroscopy*, Physical Electronics Industries, Eden Prairie, MN, 1976.
34. Z. Tomkowicz and A. Szytuka, *J. Phys. Chem. Solids*, 1977, **38**, 1117.
35. R. D. Diehl and R. McGrath, *Surf. Sci. Rep.*, 1996, **23**, 43.
36. J. Nerlov, S. V. Hoffmann, M. Shimomura and P. J. Moeller, *Surf. Sci.*, 1998, **401**, 56.
37. R. Schuster, J. V. Barth, G. Ertl and R. J. Behm, *Phys. Rev. Lett.*, 1992, **69**, 2547.
38. P. W. Murray, D. Abrooks, F. M. Leibsle, R. D. Diehl and R. McGrath, *Surf. Sci.*, 1994, **314**, 307.
39. P. W. Murray, N. G. Condon and G. Thornton, *Surf. Sci.*, 1995, **323**, L281.

40 N. D. Lang, *Phys. Rev. Lett.*, 1986, **56**, 1164.
41 M. P. Seah and W. A. Dench, *SIA Surf. Interface Anal.*, 1979, **1**, 2.
42 D. W. Turner, A. D. Baker, C. Baker and C. R. Brundle, *Molecular Photoelectron Spectroscopy*, Wiley-Interscience, New York, 1970.
43 D. Schmeisser, F. J. Himpsel, G. Hollinger, B. Reihl and K. Jacobi, *Phys. Rev. B: Condens. Matter*, 1983, **27**, 3279.
44 S. Fölsch, A. Stock and M. Henzler, *Surf. Sci.*, 1992, **264**, 65.
45 K. Morokuma, *J. Chem. Phys.*, 1971, **55**, 1236.
46 H. Umeyama and K. Morokuma, *J. Am. Chem. Soc.*, 1977, **99**, 1316.
47 P. A. Thiel and T. E. Madey, *Surf. Sci. Rep.*, 1987, **7**, 211.
48 We write $OH^{\delta-}$ and $H^{\delta+}$ instead of OH^- and H^+ because a full elementary charge transfer upon chemisorption is unlikely.
49 R. L. Kurtz and V. E. Henrich, *Phys. Rev. B: Condens. Matter*, 1982, **26**, 6682.
50 R. L. Kurtz and V. E. Henrich, *Surf. Sci.*, 1983, **129**, 345.
51 P. B. Smith and S. L. Bernasek, *Surf. Sci.*, 1987, **188**, 241.
52 R. J. Lad and V. E. Henrich, *Surf. Sci.*, 1988, **193**, 81.
53 X. D. Peng and M. A. Barteau, *Surf. Sci.*, 1990, **233**, 283.
54 V. A. Gercher and D. F. Cox, *Surf. Sci.*, 1995, **322**, 177.
55 P. Liu, T. Kendelewicz, G. E. Brown, E. J. Nelson and S. A. Chambers, *Surf. Sci.*, 1998, **417**, 53.
56 M. A. Henderson, S. A. Joyce and J. R. Rustad, *Surf. Sci.*, 1998, **417**, 66.
57 *Landolt-Börnstein, Zahlenwerte und Funktionen*, 6th edn., ed. K. Schäfer and E. Lax, Springer, Berlin, 1960, Vol.II/2a.
58 E. Habenschaden and J. Küppers, *Surf. Sci.*, 1984, **138**, L147.
59 K. Christmann, *Surface Physical Chemistry*, Steinkopff, Darmstadt, 1991.

Paper 9/02633I

Atomistic simulation of oxide surfaces and their reactivity with water

S. C. Parker, N. H. de Leeuw and S. E. Redfern

Department of Chemistry, University of Bath, Claverton Down, Bath, UK BA2 7AY.
E-mail: s.c.parker@bath.ac.uk; Tel: +44(0)1225 826505; Fax: +44(0)1225 826231

Received 19th April 1999

Atomistic simulation is a valuable tool for interpreting and predicting surface structures. This paper describes our current work aimed at applying this approach to model oxide surfaces in contact with water. The atomistic simulation techniques used are energy minimisation and molecular dynamics, which are coupled with interatomic potentials. Energy minimisation allows us to evaluate the most stable surface configurations and molecular dynamics provides the effect of temperature on the surface. The use of interatomic potentials, which describe the forces between the atoms, allows the surface properties to be calculated rapidly hence enabling us to increase the complexity of the systems studied. We have extended our previous work in two ways, first by modelling the interaction of water with more complex materials such as magnesium silicate and iron oxide and secondly, by considering the initial stages of dissolution by evaluating the energies of replacing the surface cations with protons. We find that there is a strong interaction between the surfaces and water. The bonding of the surface to the water molecules is dominated by the cation–water interactions but is moderated by the area occupied by each water molecule, which is approximately 10 Å^2. In addition, as expected, the dissolution energies are highly dependent on cation coordination and the type of cation present, with Ca being energetically more favoured than Mg, and the surface structure as illustrated by Fe_2O_3.

1 Introduction

The aim of the work described in this paper is to model the surface structure and reactivity of oxides, particularly the reactivity with water. In earlier work[1-5] we have shown that simulation techniques can model the interaction of water with perfect surfaces. We are now attempting to extend this in two ways by, first, studying the energetics of removing ions into solution and forming defective surfaces to give insights into the initial steps of dissolution and, secondly, modelling the interaction of surface defects, which are likely to be present on real crystals, with water. As we want to compare the reactivity of different materials and for different surfaces of each material we need a technique which is material specific and is sufficiently fast to perform a large number of simulations rapidly. At present atomistic simulation techniques are the most appropriate. Atomistic simulation is a powerful but straightforward tool for generating models of surface structure and energies. The approach is based on the Born model of solids where simple parametrised analytical equations are used to describe the forces between atoms. The disadvantages are that great care must be taken when deriving the parameters used in these analytical equations and that these approaches do not model electron transfer. However, the benefit is that when the parameters are available this approach can be used to study many different surfaces with a high level of complexity and thus provide a useful complement to experimental structural techniques. There are now many examples in the literature, where there is good agreement between the structural models developed by the atomistic simulation methods and those observed by experiment. Such examples include the surfaces of rutile,[6] alumina[7] and tungsten oxide.[8] In addition to giving good structural information these techniques also provide detailed energetics and where there are available experimental data the energies of adsorption of water, for example on magnesia, calcia[1,9] and alumina,[5,10] are in good agreement.

The level agreement between simulation and experiment gives us sufficient confidence to apply these techniques to problems where there are much less experimental data. Thus after describing the simulation methods in a little more detail we shall describe results on firstly the simple binary oxides of MgO and CaO and then consider the more complex examples of iron oxides, Mg_2SiO_4 and $CaCO_3$.

2 Methodology

The atomistic simulation of the surfaces of polar solids was pioneered by the work of Tasker[11,12] and Mackrodt and Stewart.[13] Much of the early work was confined to modelling planar surfaces of the cubic rocksalt oxides MgO, CaO and NiO,[14,15] and has been extended only recently to more complicated materials such as $CaCO_3$[16,17] and $BaSO_4$.[18-20]

The low energy, and hence the most common, surfaces of a crystal are generally those of low Miller index. These planes are the closest packed and hence have the smallest surface area per unit cell. As the unit cell volume is fixed these will have the largest interplanar spacings and are hence most easily cleaved. However, in polar solids other factors apply. The crystal must not only be electrically neutral but also have no net dipole moment perpendicular to the surface.[12] Bertaut[21] demonstrated that when there is a dipole moment perpendicular to the surface, the surface energy diverges and increases with increasing size. These surfaces are therefore unstable, and cannot occur naturally without the adsorption of foreign atoms or surface roughening.[22] As a way of identifying these potentially unstable surfaces Tasker[11] characterized the surfaces in terms of the repeat unit, which when repeated into the bulk generates the crystal. He identified three types of surface: type I, in which the repeat unit is charge neutral stoichiometric layers; type II, which is comprised of charged layers but in such a way that there is no dipole moment perpendicular to the surface; and finally, type III, where there is such a dipole moment.

There are two strategies for modelling free surfaces. One is to use a two region approach. The crystal is considered as a stack of planes and assumed to be periodic in two dimensions, with the stack being divided into two regions: a region I adjacent to the interface where the ions are allowed to move independently; and a region II in which the ions are held fixed relative to each other but the region as a whole may move. The inclusion of a region II ensures that all the ions in region I experience the forces associated with the rest of the crystal and that the energies are fully converged. The second approach is to take only region I and assume three-dimensional periodicity. Thus the simulation cell comprises a slab of solid separated by a vacuum gap or fluid which is then repeated infinitely. The virtue of this approach is that it is possible to exploit the periodicity to ensure that the energies, particularly the electrostatic component, of large simulation cells can be calculated rapidly using efficient algorithms but the disadvantage is that, as a consequence of the long range nature of the electrostatic forces, the ions on one surface may be influenced by the behaviour of ions on the other.

Once the energies and forces are evaluated we next apply either energy minimisation or molecular dynamics simulation techniques. Energy minimisation is achieved by adjusting the atoms in region I until the total interaction energy is at a minimum. Molecular dynamics allows explicit treatment of temperature by giving all the ions in region I kinetic energy. Generally, we begin by using energy minimisation to evaluate the surface structure and energy of a range of surfaces and select suitable candidate surfaces for further study with molecular dynamics. Again we first investigate the surface structure and energy before considering dynamical properties, such as molecular transport. The simulations are all performed assuming constant area. The specific surface energy is defined as the energy per unit area required to form the crystal surface relative to the bulk. The surface energy is therefore given by;

$$\gamma = \frac{U_s - U_b}{A} \tag{1}$$

where U_s refers to the energy of the near surface region, U_b refers to the energy of an equivalent number of bulk atoms and A is the surface area. In addition to calculating the surface energy we calculate the hydration energy, i.e., the energy to adsorb water on the surface per water molecule, E_{hyd}, and is given by;

$$E_{hyd} = \frac{U_W - (U_S + nE_{H_2O})}{n} \tag{2}$$

where n is the number of water molecules, E_{H_2O} is the self-energy of water and U_W is the energy of the simulation cell including water. Finally, to begin to address the dissolution of the oxide, at least the initial step, which normally takes place under acidic conditions, we define a replacement energy, E_{rep}, which is the energy per proton for the following reaction;

$$nH_{(aq)}^+ + M_{(surface)}^{n+} \rightarrow M_{(aq)}^{n+} + nH_{(surface)}^+ \tag{3}$$

Thus the energy

$$E_{rep} = \frac{U_d - (U_s + E_{corr})}{n} \tag{4}$$

where U_d is the energy of the simulation cell containing protons bound to the cation vacancy and E_{corr} is the energy of the hydrated species relative to the ions infinitely separated. The latter is required because the energies from the simulations are essentially lattice energies, which are always relative to the separated ions in the gas phase rather than the atoms in their standard state.

We first applied these simulations to MgO and CaO before proceeding to more complex oxides.

3 Application

3.1 Dissolution of MgO and CaO

Both CaO and MgO have the rock-salt structure and in the bulk each cation and oxygen ion is six-coordinate. Perfect crystals show a cubic morphology with one dominant surface, namely the {100}, where all ions are five-coordinate. In our earlier work we modelled the hydration energies of water interacting with the flat and stepped surfaces of CaO and MgO.[1,9] We found, for example, the {100} surface with its five-coordinate cation sites did not dissociatively adsorb water but energetically preferred the water to remain intact. When cations with lower coordinations were introduced the energies favour dissociatively adsorbed water. This perhaps implies that hydroxy groups would not be observed on the {100} surface and that dissolution may not occur on this surface, both of which are unlikely. Furthermore, the differences in energies between the hydration of perfect MgO and CaO surfaces was not large and yet CaO is observed to be more reactive. We have attempted to study these issues in more detail by evaluating the replacement energies, i.e., dissolving the cation in solution and replacing it by two protons.

The lower-coordinated cations, namely those with four- and three-fold coordination were modelled by considering the micro-facetted {110} and {111} surfaces, respectively. Micro-facetting usually occurs on surfaces that are unstable in their planar form and have a high surface energy. For example, micro-facetting of the planar {110} surface leads to rows of four-coordinated ions with {100} planes inclined at 45° on both sides (Fig. 1). We considered two such facets, $(a_0\sqrt{2})/2$ and $a_0\sqrt{2}$, where a_0 is the lattice parameter and the label, e.g., $a_0\sqrt{2}$, indicates the depth from the topmost ion to the first fully coordinated plane of ions. The planar {111} surface is dipolar and, as discussed above, is stabilised by surface rearrangement, for example, by shifting one half of the surface atoms to the top of the unit cell[23] or by micro-facetting (Fig. 2), in which case the now non-dipolar micro-facetted surface becomes a collection of three-sided pyramids where the sides of the pyramids, inclined at 52.7°, consist of {100} planes. Both the micro-facetted {110} and {111}

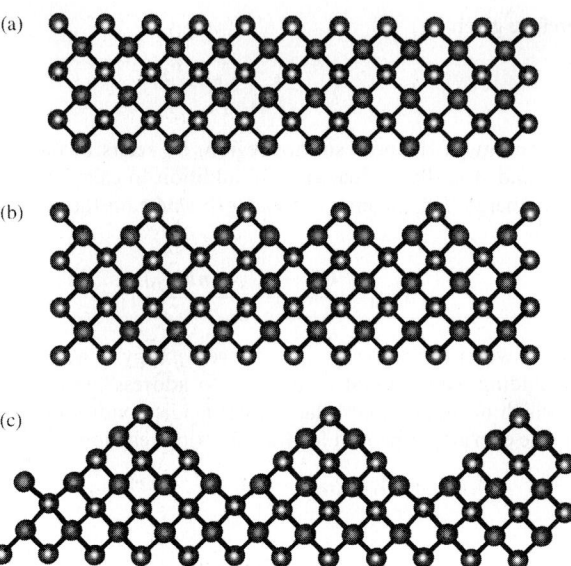

Fig. 1 Relaxed structures of the {110} surfaces of MgO: (a) unfacetted, (b) $(a_0\sqrt{2})/2$ facetted and (c) $a_0\sqrt{2}$ facetted (where $a_0\sqrt{2}$ refers to the distance between the top and bottom of the facet).

surfaces have been observed experimentally and were shown to be stable against repeated sputter–anneal cycles.[24,25] The surface energies of the surfaces considered are collected in Table 1, where a low positive value indicates a stable surface. It is clear that the {100} surface is the most stable surface and that facetting of the relatively unstable {110} and {111} surface exposing {100} planes[1,9] thereby increases their stability.

As noted above the crystals begin to dissolve in acidic conditions. By investigating the effect of dissolving metal units (M^{2+}) from surface sites of varying coordination number and replacing them with H^+ ions we can also evaluate whether the lower-coordinated atoms are the most likely initial sites of crystal dissolution.

Table 2 shows the energies per proton of replacing magnesium or calcium ions by two protons. The table shows that the removal of magnesium ions from the {100} surface is energetically

Fig. 2 Relaxed structure of the $a_0\sqrt{3}$ facetted {111} surface of MgO.

Table 1 Relaxed surface energies (J m^{-2}) of the planar and facetted surfaces of MgO and CaO

	Surface					
	{100}	{110}	$(a_0\sqrt{2})/2$	$a_0\sqrt{2}$	{111}	$a_0\sqrt{3}$
CaO	0.77	1.95	1.31	1.14	2.47	1.67
MgO	1.25	3.02	2.09	1.87	3.86	2.39

Table 2 Energies of replacement (kJ mol^{-1}) of one cation by two hydrogen ions

Surface	MgO				CaO			
	25%	50%	75%	100%	25%	50%	75%	100%
{100}	−64.0	−94.7	−84.8	−23.4	−134.7	−152.4	−146.4	−108.0
{110}	−212.8	−192.5	−168.6	−201.2	−279.6	−263.0	−242.1	−224.9
$(a_0\sqrt{2})/2$ edge	−85.9	−85.0	−109.9	−121.1	−173.2	−171.7	−186.2	−193.0
$a_0\sqrt{2}$ edge	—	−110.6	—	−138.5	—	−196.5	—	−197.1
{111} cation	—	—	—	−158.8	—	—	—	−235.1
$a_0\sqrt{3}$ ox.	—	—	—	−115.8	—	—	—	−213.9

favourable and is more favourable at low coverages than at 100%, which for MgO is only just energetically preferable. Dissolution of M^{2+} ions from the {100} surfaces leads to a surface of lattice oxygen ions, cation vacancies and hydrogen atoms that are coordinated to oxygen atoms both in the surface layer and in the next layer down resulting in an uneven plane.

The energies of removing the cations for the unfacetted {110} surfaces are very large at all coverages. This is not surprising as this surface is very unstable with a large surface energy and dissolving M^{2+} ions from this surface leads to the facetted surface. The energies favour removal of the cations in rows rather than isolated over the surface.

We have also considered removal of the four-coordinated cations from the facetted {110} surfaces. The energies are larger than for those at the five-coordinated sites (on the {100} surfaces). Unlike the unfacetted surface, the removal of those cations adjacent to a site where a cation has been removed are not energetically favoured. There is no particular site preference for removal of subsequent cations, with typical energy differences between sites of 6 kJ mol^{-1} we expect the process to be random. This may be because removing rows of cations does not immediately lead to a bigger facet as was the case for the unfacetted {110} surface. The formation of a larger facet requires second layer M^{2+} ions to be dissolved. One difference in the energies of removing cations between the facetted surfaces and the flat surface is that the average energy increases as the cations are removed. This may arise from the strong interaction energy between the incoming proton and the low-coordinated oxygen species on the step and by the binding energy between the proton and the oxygen adjacent (Fig. 3). The latter configuration is energetically more stable (by approximately 275 kJ mol^{-1}) than configurations where protons are bound to more distant oxygen atoms.

Following the above trend, the dissolution of M^{2+} ions from three-coordinated surface sites is even more energetically favoured. In contrast to hydroxylating the corner sites, which gave the same energies for both a three-coordinate cation or a three-coordinate oxygen atom, the dissolution energies for the two different corner atoms are very different because substitution of a cation at the apex of the pyramid is modelling the dissolution of a three-coordinate ion (Fig. 4), whereas dissolution at the oxygen-terminated pyramid removes a four-coordinated cation below

Fig. 3 MgO $a_0\sqrt{2}$ facetted {110} surface with 100% of edge MgO units dissolved, showing the energetically preferred positions of the adsorbed hydrogen atoms (Mg = blue, O = red, H = white).

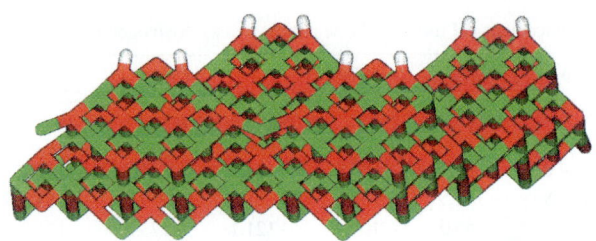

Fig. 4 Dissolution of CaO units from the $a_0\sqrt{3}$ facetted calcium-terminated {111} surface showing replacement of a three-coordinated Ca ion by two hydrogen atoms, the now three-coordinated oxygen atoms and hydrogen atoms are shown as balls (Ca = green, O = red, H = white).

the three-coordinated oxygen. Hence it is not surprising that in this case the energy is similar to the energies of removing cations from the four-coordinated sites found on the facetted {110} surfaces.

In all configurations on the above surfaces the two hydrogen atoms replacing the dissolved cation, tilt towards the empty cation site. This is a structural analogue to the hydrogarnet defect in grossular[26] (where four hydrogen atoms point towards an empty silicon site) and, as shown previously,[9] stabilises hydroxylation of the MgO {310} edge atoms where the oxygen atom of the hydroxy group adsorbes at an interstitial position and the hydrogen atom tilts towards the corresponding interstitial cation site.

Finally, we note that the energies of removal are significantly larger, in some cases by a factor of two, for CaO compared to MgO. This implies that CaO has a much faster dissolution rate, although we should be cautious about comparing dissolution rates based on the energetics of reacting the top cation layer only.

In summary, the energies of replacing cations at the corner sites, modelled by the facetted {111} surfaces, are generally larger than those of the edge sites, modelled by the facetted {110} surfaces, and both are larger than the energies of the sites on the flat {100} surfaces.

In addition to studying defects on the surfaces of simple oxides the second theme of this paper is to show that these techniques can be applied to modelling the interaction of water with more complex materials, first a transition metal oxide in this case hematite and secondly, replacing the oxide ion by a polyanion such as a silicate or carbonate.

3.2 Interaction of hematite (α-Fe$_2$O$_3$) surfaces with water

Hematite is the most stable of the ion oxides at atmospheric conditions.[27] It occurs mainly in sediments and is found in soils as a weathering product of iron bearing minerals and a regulator of pollutants and plant nutrients. In the laboratory, hematite is used in heterogeneous catalysis, such as dehydrogenation and oxidation, and as electrodes in photoelectrolysis.[28] Hematite is hexagonal with spacegroup $R\bar{3}c$. We used the unit cell of $a = b = 5.030$ Å and $c = 13.772$ Å, $\alpha = \beta = 90°$ and $\lambda = 120°$,[29] which relaxed to a minimum-energy configuration with $a = b = 5.063$ Å and $c = 13.3601$ Å, $\alpha = \beta = 90°$ and $\gamma = 120°$. The experimental morphologies are dominated by two surfaces, the {0001} basal plane and the {01$\bar{1}$2} surface. Studies of the adsorption of water to hematite surface conclude that water dissociatively adds to the surfaces, but the reactivity of the {0001} basal plane with water remains unclear. The interatomic potentials used were those derived by Lewis and Catlow[30] for the iron oxide, and for the water and hydroxide, those by de Leeuw and Parker[3] and Baram and Parker,[31] respectively.

The {01$\bar{1}$2} surface is a stable surface with five-coordinated cations. On relaxation the surface undergoes little relaxation thus leaving the surface ions close to their bulk terminated positions (Table 3). When associated water is added to the surface a well-ordered water layer is formed with the adsorbate molecules relaxing into rows across the surface as they follow the periodicity of the substrate below (Fig. 5(a)). The water molecules bind strongly to the surface cations with cation–water distances of around 2.0 Å. On allowing the water to dissociate the hydroxide layer on this surface again forms an ordered structure (Fig. 5(b)), which is very similar to the stable {10$\bar{1}$0} face of goethite (FeOOH), which might be expected as it is possible to transform between goethite and

Table 3 Surface energies of {0001} and {01$\bar{1}$2} surfaces of α-Fe$_2$O$_3$

Surface	Unrelaxed surface energy/J m^{-2}	Relaxed surface energy/J m^{-2}
{01$\bar{1}$2}	5.06	2.41
{0001}O	11.95	3.74
{0001}Fe	2.91	2.1

hematite. Comparison of the hydroxylation and hydration energies for the {01$\bar{1}$2} face show that they are very similar (−103 and −109 kJ mol^{-1}) (Table 4) and hence at a macroscopic scale we would expect to observe both species. In addition, Henderson et al.[27] reported a hydration energy of −118 kJ mol^{-1}, which agrees well with our prediction.

The {0001} surface has two possible terminations, one of which is non-dipolar and is terminated with Fe cations in three-fold coordination (type II). The other plane is a dipolar type III surface, which is terminated by a half filled surface layer of oxygen atoms and as a result there are four Fe sites with coordinations 3, 4, 5 and 6. The Fe terminated face is the lower in energy (Table 3) and

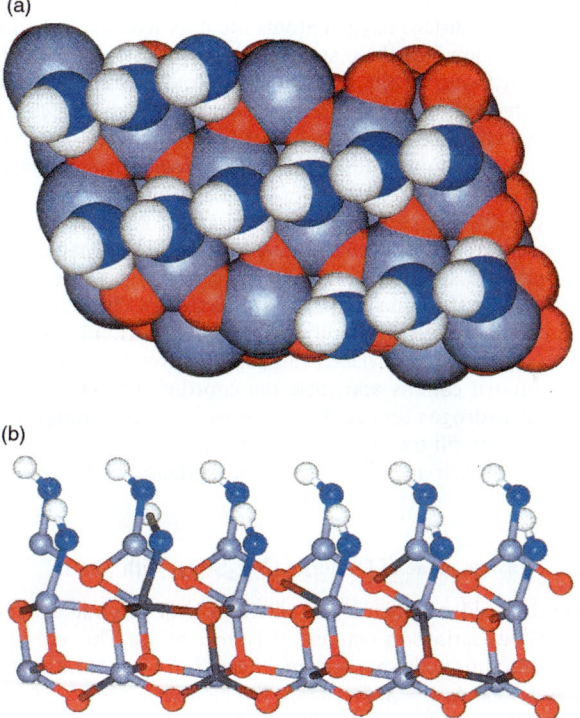

Fig. 5 (a) Associative adsorption of water in the hematite {01$\bar{1}$2} surface, showing coordination of the hydrogen atoms to oxygen atoms of neighbouring water molecules and lattice oxygens on the surface, and (b) hydroxylated {01$\bar{1}$2} surface (Fe = purple, O = red, O$_{water}$ = blue, O$_{hydroxyl}$ = blue, H = white).

Table 4 Energies of associative and dissociative adsorption of water on hematite surfaces

Surface	Energy$_{associative}$/kJ mol^{-1}	Energy$_{dissociative}$/kJ mol^{-1}
{01$\bar{1}$2}	−109.9	−103.2
{0001}O	−150.5	−298.1
{0001}Fe	−132.2	−146.7

is smooth, particularly after the surface iron atoms relax into the surface,[7] whereas the O terminated face is quite rough and has a much lower surface ion density. The hydroxylation of both of the {0001} terminations was found to be favoured over the molecular adsorption of water (Table 4). The hydroxylation energy of the oxygen terminated face was calculated to be very large because the effect of the reaction is to fill the empty anion sites forming a flat surface that is very stable. The highly stable surface formed is similar to that predicted for the basal plane of alumina, which is itself similar to the most stable surface of gibbsite, $Al(OH)_3$.[31]

We next calculated the replacement energies of Fe^{3+} ions with H^+ ions for the $\{01\bar{1}2\}$ and $\{0001\}$ surfaces to model the initial stages of the dissolution process. In each case we removed the lowest-coordinated species, which was five-fold for the $\{01\bar{1}2\}$ and three-fold for the $\{0001\}$ surface. Unlike the examples above, where the cation is replaced by two protons, on reacting with hematite the cation is replaced by three protons. We found that on all three surfaces studied, it is predicted that hematite begins to dissolve in acidic conditions.

In its dry form, the $\{01\bar{1}2\}$ face is more stable than the two $\{0001\}$ planes. It is quite smooth, with slight corrugations of the surface oxygen atoms. On dissolution, the resulting surface remains very smooth, with hydrogen atoms tilting towards the cation vacancies or lying along the corrugations. Our calculations predict the lowest energies per proton for removing Fe atoms are -42.5 kJ mol^{-1} when 12.5% of the surface Fe atoms are removed and -48.5 kJ mol^{-1} when 25% are removed. At this point all the surface oxygen atoms are fully hydroxylated and further removal of Fe would necessitate the removal of these surface hydroxys as well, in the form of water. These energies are closest to those found for the removal of five-fold coordinated cations in MgO.

Following the work on MgO and CaO we would expect that the energies of removing the three-coordinated cations would be between a factor of two or three more favourable than removing the five-fold coordinated cations. This is indeed the case with the removal of Fe from the Fe terminated {0001} surface releasing 113.9 and 111.0 kJ (mol H^+)$^{-1}$ for 50% and 100% of the cations removed, respectively. The oxygen-terminated {0001} face has by far the highest energies of removal, because we remove the three-coordinated cations and because it is the least stable and hence most reactive. Since the surface layer is made up of oxygen ions, the cation vacancies are created in the layer below and the initial stages of dissolution, when 12.5% and 25% of the cations are removed result in energies of -187.2 and -235.4 kJ (mol H^+)$^{-1}$, respectively. The latter, as with the $\{01\bar{1}2\}$ surface, results in the surface being fully hydroxylated. The effect of the dissolution is to remove low-coordinated cations and raise the coordination number of the surface oxygen atoms by the addition of hydrogen atoms. As before with the dissolution of MgO and CaO, the hydrogens replacing the cation tilt towards the vacancy site.

We next apply this approach to the most stable surfaces of the more complex oxides of Mg_2SiO_4 and $CaCO_3$.

3.3 Interaction of the cleavage plane of forsterite (Mg_2SiO_4) with water

Forsterite, the magnesium end member of the olivine group of minerals, consists of SiO_4 tetrahedra linked by magnesium cations in octahedral coordination. This prominent Mantle material has an orthorhombic structure with spacegroup *Pbnm*.[29] We used the unit cell of $a = 4.7560$ Å, $b = 10.2070$ Å and $c = 5.9800$ Å, $\alpha = \beta = \gamma = 90°$,[32] which relaxed to a minimum-energy structure with $a = 4.7898$ Å, $b = 10.2464$ Å and $c = 5.9863$ Å, $\alpha = \beta = \gamma = 90°$. We modelled the dissociative and associative adsorption of water on the {010} surface, which is the main cleavage plane of forsterite.

On associative adsorption we found that a maximum of one water molecule per 10 Å^2 could be accommodated before a second water layer was formed, which agreed with previous molecular dynamics simulations of MgO in liquid water.[3] In that study and also previous simulations of the hydration of calcite[4,33] and α-alumina surfaces[5] we found that water molecules often adsorb flat onto the surface, in agreement with experimental findings of hydrated neutral clay surfaces.[34] In addition, it is not unreasonable that one water molecule occupies 10 Å^2 if the experimental non-bonded hydrogen–oxygen distance in ice is 1.76–1.95 Å,[35] the effective area of water then becomes $\pi r^2 = 9.7$–11.9 Å^2.

Under dry conditions the non-dipolar termination of the {010} surface ({010}a) with a relaxed surface energy of $\gamma = 1.28$ J m^{-2} (Table 5) is more stable than the dipolar surface ($\gamma = 2.32$ J m^{-2})

Table 5 Surface energies (J m^{-2}) of {010} surfaces of forsterite

Surface	$\gamma_{unrelaxed}$	$\gamma_{relaxed}$	$\gamma_{associative}$	$\gamma_{dissociative}$	$\gamma_{ass.\,on\,hydrox.}$
{010}a	2.23	1.28	0.30	0.76	0.27
{010}b[a]	4.43	2.32	0.86	0.58	0.17

[a] Denotes dipolar plane.

({010}b) in agreement with previous calculations.[36] Both surfaces are amenable to associative adsorption of water and the relative stabilities of the two surfaces remain the same. However, when hydroxylated the stabilities are reversed with the dipolar plane becoming the more stable surface. In fact, the surface energy of the dipolar plane has decreased steadily from dry to associative adsorption to hydroxylation, but the surface energy of the non-dipolar {010} surface has increased from associative adsorption ($\gamma = 0.30$ J m^{-2}) to dissociative adsorption ($\gamma = 0.76$ J m^{-2}) (Table 5). In addition, the energy of adsorption of dissociated water molecules at -89 kJ mol^{-1} is less than for associative adsorption (-99 kJ mol^{-1}) (Table 6) and from both adsorption energies and surface energies we may assume that the non-dipolar {010} surface will prefer not to be hydroxylated.

Fig. 6 shows the unhydrated non-dipolar {010} surface and the hydroxylated dipolar {010} plane. The non-dipolar surface is a smooth plane terminated by O–Mg–O bridges. Hydroxylation of this plane leads to a much rougher surface. In a previous study on the stability of quartz surfaces[2] we found that the quartz {0001} surface, which is terminated by O–Si–O bridges similarly to the O–Mg–O bridges on the forsterite {010} surface, was not as amenable to dissociative adsorption of water as the other quartz surfaces, which had low-coordinated surface species. Furthermore, we adsorbed concentrated NaOH on the quartz surface and found that the formation of O–Na–O bridges had a large stabilising effect as well. Hydroxylation of the dipolar forsterite {010} surface on the other hand is energetically very favourable. The surface consists of a half-vacant magnesium plane. The hydroxy groups of the dissociated water molecules become bonded to two surface magnesium ions (Fig. 6(b)).

Finally, we adsorbed a layer of water molecules on the hydroxylated surface. Both hydroxylated surfaces have been further stabilised by associative adsorption with fairly low adsorption energies indicative of physisorption of the water molecules to the hydroxylated surfaces. Fubini et al.[37] in their TPD study of a range of hydrated oxides, found energies of between 50–70 kJ mol^{-1} for adsorption of water molecules to hydrogen atoms of surface hydroxy groups. The stabilities of the two {010} planes are still reversed with the dipolar plane the more stable of the two, although the difference in surface energies (0.1 J m^{-1}) is small enough that we may expect at the macroscopic scale to find a stepped surface containing both planes under aqueous conditions.

The greater stability of the hydroxylated dipolar {010} plane over the non-dipolar plane indicates that some dissolution of the magnesium ions may occur upon hydration, as is, for example, observed experimentally for wollastonite (CaSiO$_3$)[38] and can be induced for forsterite at low pH. Seyama et al.[39], for example, used X-ray photoelectron spectroscopy to investigate the removal of magnesium ions from forsterite surfaces. After acid dissolution they only found surface SiO$_2 \cdot$ H$_2$O species while micrographs showed an etched structure. In addition, high temperature vacuum experiments by Nagahara and Ozawa[40] and Young et al.[41] showed forsterite to evaporate in H$_2$ gas leaving pits and islands on the surface. We were therefore interested to see whether dissolution of surface magnesium ions, similar to the process in MgO and CaO, was energetically feasible for

Table 6 Energies (kJ mol^{-1}) of associative and dissociative adsorption of water onto forsterite {010} surfaces and energies of replacement of surface mg ions by two hydrogen ions

Surface	$E_{associative}$	$E_{dissociative}$	$E_{ass.\,on\,hydrox.}$	$E_{dissolution}/H^+$	$E_{ass.\,on\,dissol.}$
{010}a	-99.4	-89.2	-71.4	-115.2	-63.1
{010}b	-127.4	-300.1	-67.5	-203.8	-130.9

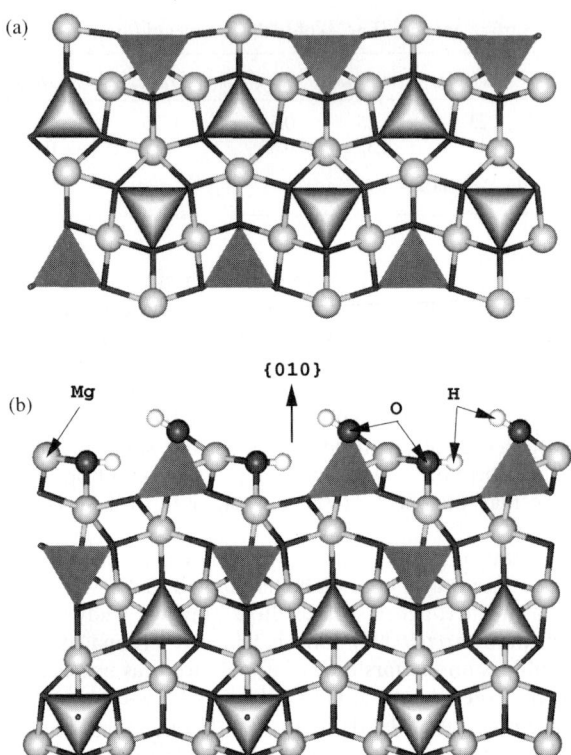

Fig. 6 (a) Relaxed unhydrated non-dipolar {010} surface showing a smooth plane of O–Mg–O bridges and (b) hydroxylated dipolar {010} surface showing extensive coordination between surface Mg ions and hydroxy groups (SiO$_4$ shown as tetrahedra).

both the dipolar and non-dipolar forsterite {010} surfaces. To this end we again replaced surface magnesium ions by two hydrogen ions each under acidic conditions.

The replacement energy for the non-dipolar plane was -41.9 kJ mol^{-1} per H$^+$ while the dissolution energy for the dipolar plane was -130.5 kJ mol^{-1}, clearly a more favourable process. Thus we find again that dissolution of magnesium ions from both surfaces is energetically favourble under acidic conditions, providing a route for obtaining the most stable surface under different conditions. The surfaces including defects were then hydrated by molecular water again, which released another 63.1 and 130.9 kJ mol^{-1} for the non-dipolar and dipolar surfaces, respectively.

3.4 Activation energies for dissolution of CaCO$_3$ from the cleavage plane of calcite

Calcite, the most stable form of calcium carbonate, is one of the most abundant minerals in the environment and of fundamental importance in many fields, both inorganic and biological. It has a rhombohedral crystal structure with space group $R\bar{3}c$ and $a = b = 4.990$ Å, $c = 17.061$ Å, $\alpha = \beta = 90°$ and $\gamma = 120°$.[29] On energy minimisation the structure relaxed to $a = b = 4.797$ Å, $c = 17.482$ Å, $\alpha = \beta = 90°$ and $\gamma = 120°$. The {10$\bar{1}$4} surface is by far the most stable plane of calcite and dominates the observed morphology. However, no experimental surface is truly planar and there are always defects present like steps and kinks. We have employed molecular dynamics simulations, using the DL_POLY code,[42] to investigate the formation of a point defect on the planar {10$\bar{1}$4} surface and the introduction of kink sites on two different monatomic steps on the {10$\bar{1}$4} surface, both under aqueous conditions. In effect, we have simulated and obtained an energy for the reaction given in the following equation:

$$n\text{CaCO}_3(\text{s}) \rightarrow (n-1)\text{CaCO}_3(\text{s}) + \text{Ca}^{2+}(\text{aq}) + \text{CO}_3^{2-}(\text{aq}) \tag{5}$$

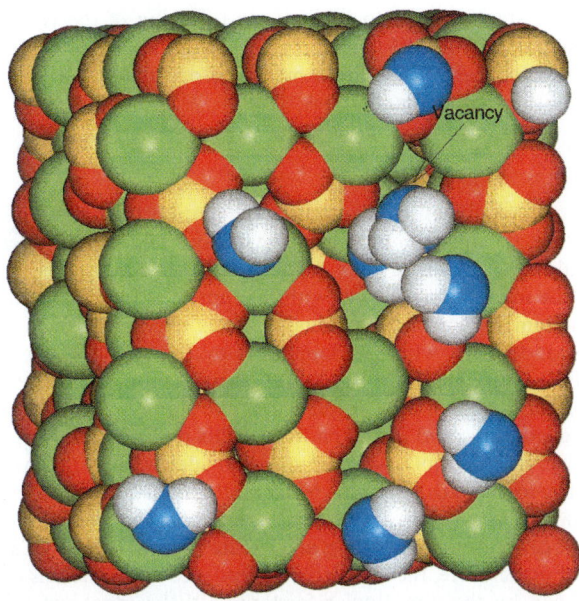

Fig. 7 Plan view of the calcite $\{10\bar{1}4\}$ surface ($T = 1000$ K) with surface $CaCO_3$ unit removed, showing clustering of water molecules round the defect (Ca = green, C = yellow, O = red, O_{water} = blue, H = white).

which we consider will be the major component of the activation energy for dissolution. We modelled the planar surface at two temperatures, 300 K and 1000 K. On the perfect surface the water molecules adsorb flat onto the surface in a regular herring-bone pattern, coordinating by their oxygen atom to surface calcium atoms. At 300 K the water molecules do not diffuse but once adsorbed stay on the surface. At 1000 K, however, the water molecules diffuse through the gap between the surfaces, randomly adsorbing and desorbing. When a calcium carbonate unit is removed from the surface the water molecules cluster in and around the defect. Fig. 7 is plan view of the $\{10\bar{1}4\}$ surface with defect at 1000 K clearly showing the clustering of the water molecules around the vacancy. At 300 K, the energy of dissolution of a $CaCO_3$ unit from the planar surface is found to be $+328.0$ kJ mol^{-1}. On raising the temperature to 1000 K the dissolution energy decreases to $+135.1$ kJ mol^{-1}, which, although much lower, still does not fall in the thermal energy region.

In view of the higher reactivity of the low-coordinated sites on MgO and CaO reported in Section 3.1 we modelled the dissolution of $CaCO_3$ units from two step edges on the $\{10\bar{1}4\}$ surface to see whether dissolution from these edges is less endothermic. The sides of the two steps are also $\{10\bar{1}4\}$ surfaces and the step edges are identical to the two different edges of the calcite rhomb. One step is acute, *i.e.*, the carbonate groups on the edge of the step overhang the plane below the step and the angle between the step wall and plane is 80° on the relaxed surface (*cf.* exp. 78°).[43] The other step is obtuse, *i.e.*, the carbonate groups on the step edge lean back with respect to the plane below with an angle between the step wall and plane of 105° on the relaxed surface (exp. 102°).[43] These two types of step are found experimentally to form the dissolving edges of etch

Table 7 Energies of dissolution (kJ mol^{-1}) of $CaCO_3$ unit from $\{10\bar{1}4\}$ plane and step edges

	Surface		
	Planar $\{10\bar{1}4\}$	Acute step	Obtuse step
300 K	+328.0	+103.7	+45.8
1000 K	+135.1	—	—

pits[43,44] and the obtuse step is found to be the fastest moving of the two. Dissolving a $CaCO_3$ unit from the acute step costs 103.7 kJ mol^{-1} (Table 7), but dissolution from the obtuse step edge only costs 45.8 kJ mol^{-1}. Clearly, dissolution prefers to occur from steps rather than the planar surface in agreement with experimental findings.[45,46]

4 Conclusions

This study has shown the application of atomistic simulation to a number of interfaces in ionic crystals. We have extended our previous work in two ways: first, by considering the initial stages of dissolution by evaluating the energies of replacing the surface cations with protons; and secondly, by modelling the interaction of water with the more complex materials, iron oxide, magnesium silicate and calcium carbonate. We found that, as expected there is a strong interaction between the surfaces and water, which is dominated by the cation–water interactions although moderated by the fact that each water molecule occupies approximately 10 Å^2. The energies associated with the initial stages of dissolution, where the surface cations are replaced by protons, were all found to be exothermic and the magnitude of this replacement energy is highly dependent on cation coordination. For example, with the simple oxides we find that very approximately the energy to replace a five-coordinated cation is 60 kJ mol^{-1} of protons, four-coordinated is 150 and three-coordinated 180 kJ mol^{-1}. However, this is not the complete picture as the energies do depend on the type of cation replaced, as Mg and Fe are slightly less and Ca more than the crude energies just quoted. Furthermore, there is also a dependence on the details of the surface, which is dramatically exemplified by the non-dipolar forsterite {010} surface, where the Mg ions are in three-fold coordination. In summary, atomistic simulation is a powerful tool for modelling surface structure, stability and reactivity and one which is complementary to experimental surface analysis.

Acknowledgements

We thank EPSRC grant no. GR/L35577 and NERC grant no. GR3/11779 for support.

References

1. N. H. de Leeuw, G. W. Watson and S. C. Parker, *J. Chem. Soc., Faraday Trans.*, 1996, **92**, 2081.
2. N. H. de Leeuw, F. M. Higgins and S. C. Parker, *J. Phys. Chem. B*, 1998, **103**, 1270.
3. N. H. de Leeuw and S. C. Parker, *Phys. Rev. B: Condens. Matter*, 1998, **58**, 13901.
4. N. H. de Leeuw and S. C. Parker, *J. Phys. Chem. B*, 1998, **102**, 2914.
5. N. H. de Leeuw and S. C. Parker, *J. Am. Ceram. Soc.*, 1999, in the press.
6. P. M. Oliver, G. W. Watson, E. T. Kelsey and S. C. Parker, *J. Mater. Chem.*, 1997, **7**, 563.
7. W. C. Mackrodt, R. J. Davey, S. N. Black and R. Docherty, *J. Cryst. Growth*, 1987, **80**, 441.
8. P. M. Oliver, S. C. Parker, R. G. Egdell and F. H. Jones, *J. Chem. Soc., Faraday Trans.*, 1996, **92**, 2049.
9. N. H. de Leeuw, G. W. Watson and S. C. Parker, *J. Phys. Chem.*, 1995, **99**, 17219.
10. N. H. de Leeuw, S. E. Redfern and S. C. Parker, *Recent Res. Dev. Phys. Chem.*, 1998, **2**, 441.
11. P. W. Tasker, *Philos. Mag. A*, 1979, **39**, 119.
12. P. W. Tasker, *J. Phys. C: Solid State Phys.*, 1979, **12**, 4977.
13. W. C. Mackrodt and R. F. Stewart, *J. Phys. C: Solid State Phys.*, 1979, **12**, 431.
14. E. A. Colbourn, W. C. Mackrodt and P. W. Tasker, *J. Mater. Sci.*, 1983, **18**, 1917.
15. P. W. Tasker and D. M. Duffy, *Surf. Sci.*, 1984, **137**, 91.
16. J. O. Titiloye, S. C. Parker, D. J. Osguthorphe and S. Mann, *J. Chem. Soc., Chem. Commun.*, 1991, 1494.
17. S. C. Parker, J. O. Titiloye and G. W. Watson, *Philos. Trans. R. Soc. London, Ser. A*, 1993, **344**, 37.
18. S. C. Parker, E. T. Kelsey, P. M. Oliver, and J. O. Titiloye, *Faraday Discuss.*, 1993, **95**, 75.
19. N. L. Allen, A. L. Rohl, D. H. Gay, C. R. A. Catlow, R. J. Davey and W. C. Mackrodt, *Faraday Discuss.*, 1993, **95**, 273.
20. S. E. Redfern and S. C. Parker, *J. Chem. Soc., Faraday Trans.*, 1998, **94**, 1947.
21. F. Bertaut, *C. R. Hebd. Seances Acad. Sci.*, 1958, **246**, 3447.
22. G. L. Benson and K. S. Yon, *The Solid Gas Interface*, ed. E. A. Flood, Arnold, London, 1967.
23. P. M. Oliver, S. C. Parker and W. C. Mackrodt, *Model. Simul. Mater. Sci. Eng.*, 1993, **1**, 755.
24. V. E. Henrich, *Surf. Sci.*, 1976, **57**, 385.
25. H. Onishi, C. Egawa, T. Aruga and Y. Iwasawa, *Surf. Sci.*, 1987, **191**, 479.
26. J. Purton, R. Jones, M. Heggie, S. Oberg and C. R. A. Catlow, *Phys. Chem. Miner.*, 1992, **18**, 389.
27. M. A. Henderson, S. A. Joyce and J. R. Rustad, *Surf. Sci.*, 1998, **417**, 66.

28   J. H. Kennedy and K. W. Frese, Jr., *J. Electrochem. Soc.*, 1978, **125**, 709.
29   W. A. Deer, R. A. Howie and J. Zussman, *Introduction to the rock-forming minerals*, Longman, UK, 1992.
30   G. V. Lewis and C. R. A. Catlow, *J. Phys. C: Solid State Phys.*, 1985, **18**, 1149.
31   P. S. Baram and S. C. Parker, *Philos. Mag. B*, 1996, **73**, 49.
32   J. R. Smyth and R. M. Hazen, *Am. Mineral.*, 1973, **58**, 588.
33   N. H. de Leeuw and S. C. Parker, *J. Chem. Soc., Faraday Trans.*, 1997, **93**, 467.
34   F.-R. C. Chang, N. G. Skipper and G. Sposito, *Langmuir*, 1995, **11**, 2734.
35   B. Kamb, *Water and Aqueous Solutions*, ed. R. A. Horne, John Wiley & Sons Inc., New York, 1972.
36   G. W. Watson, P. M. Oliver and S. C. Parker, *Phys. Chem. Miner.*, 1997, **25**, 70.
37   B. Fubini, V. Bolis, M. Bailes and F. S. Stone, *Solid State Ionics*, 1989, **32**, 258.
38   K. H. Rao, 1998, personal communication.
39   H. Seyama, M. Soma and A. Tanaka, *Chem. Geol.*, 1996, **129**, 209.
40   H. Nagahara and K. Ozawa, *Geochim. Cosmochim. Acta*, 1996, **60**, 1445.
41   E. D. Young, H. Nagahara, B. O. Mysen and D. M. Audet, *Geochim. Cosmochim. Acta*, 1998, **62**, 3109.
42   T. R. Forester and W. Smith, *DL_POLY user manual*, CCLRC, Daresbury Laboratory, Daresbury, Warrington, 1995.
43   N.-S. Park, M.-W. Kim, S. C. Langford and J. T. Dickinson, *J. Appl. Phys.*, 1996, **80**, 2680.
44   Y. Liang, D. R. Baer, J. M. McCoy, J. E. Amonette and J. P. LaFemina, *Geochim. Cosmochim. Acta*, 1996, **60**, 4883.
45   A. J. Gratz, P. E. Hillner and P. K. Hansma, *Geochim. Cosmochim. Acta*, 1993, **57**, 491.
46   P. E. Hillner, S. Manne and P. K. Hansma, *Faraday Discuss.*, 1993, **95**, 191.

Paper 9/03111A

Theory of PbTiO$_3$, BaTiO$_3$, and SrTiO$_3$ surfaces

B. Meyer, J. Padilla, and David Vanderbilt*

Department of Physics and Astronomy, Rutgers University, Piscataway, NJ 0855-0849, USA

Received 16th April 1999

First-principles total-energy calculations are carried out for (001) surfaces of the cubic perovskite ATiO$_3$ compounds PbTiO$_3$, BaTiO$_3$, and SrTiO$_3$. Both AO-terminated and TiO$_2$-terminated surfaces are considered, and fully-relaxed atomic configurations are determined. In general, BaTiO$_3$ and SrTiO$_3$ are found to have a rather similar behavior, while PbTiO$_3$ is different in many respects because of the partially covalent character of the Pb–O bonds. PbTiO$_3$ and BaTiO$_3$ are ferroelectrics, and the influence of the surface upon the ferroelectric distortions is studied for a tetragonal ferroelectric distortion parallel to the surface. The surface relaxation energies are found to be substantial, *i.e.*, many times larger than the bulk ferroelectric well depth. Nevertheless, the influence of the surface upon the ferroelectric order parameter is modest, and is qualitatively as well as quantitatively different for the two materials. Surface energies and electronic properties are also computed. It is found that for BaTiO$_3$ and SrTiO$_3$ surfaces, both AO-terminated and TiO$_2$-terminated surfaces can be thermodynamically stable, whereas for PbTiO$_3$ only the PbO surface termination is stable.

1. Introduction

The surfaces of insulating cubic perovskite materials such as PbTiO$_3$, BaTiO$_3$, and SrTiO$_3$, are of interest from several points of view. First, some of these materials (notably SrTiO$_3$) are very widely used as substrates for growth of other oxide materials (*e.g.*, layered high-T_c superconductors and "colossal magnetoresistance" materials). Second, this class of materials is of enormous importance for actual and potential applications that make use of their unusual piezoelectric, ferroelectric, and dielectric properties (*e.g.*, for piezoelectric transducers, non-volatile memories, and wireless communications applications, respectively). Many of these applications are increasingly oriented towards thin-film geometries, where surface properties are of growing importance. Third, the bulk materials display a variety of structural phase transitions; the ferroelectric (FE) structural phases are of special interest, but antiferroelectric (AFE) or antiferrodistortive (AFD) transitions can also take place.[1] It is then of considerable fundamental interest to consider how these structural distortions couple to the surface, *e.g.*, whether the presence of the surface acts to enhance or to suppress the structural distortion. The ferroelectric properties are well known to degrade in thin-film[2] and particulate[3] geometries, and it is very important to understand whether such behavior is intrinsic to the presence of a surface, or whether it arises from extrinsic factors, such as compositional non-uniformities or structural defects in the surface region. Finally, the cubic perovskites can serve as model systems for the study of transition-metal oxide surfaces more generally.[4]

In the last decade, there has been a surge of activity in the application of first-principles computational methods based on density-functional theory (DFT) to the study of the bulk properties, and especially the ferroelectric transitions, in bulk perovskite oxides (for a recent review, see refs. 5 or 6). The importance of these methods was recently underlined by the award of the Nobel Prize in Chemistry to Walter Kohn, the primary originator of DFT. In the materials theory community, these methods have been widely used for two decades to predict properties of semiconductors and simple metals. However, recent advances in computational algorithms and computer power now allow these methods to be applied to more complex materials (*e.g.*, perovskites) and more complex geometries (*e.g.*, defects and surfaces). In particular, pioneering studies of $BaTiO_3$[7-9] and $SrTiO_3$[10-12] surfaces have recently appeared.

Experimental investigations of the surface structure of cubic perovskites have not been very extensive. Such studies are hindered by the difficulties of preparing clean and defect-free surfaces, and of overcoming charging effects associated with many experimental probes. Even for $SrTiO_3$, the best studied of these surfaces, there is a disappointing level of agreement among experimental results[13-16] and between experiment and theory.[11] We are not aware of comparable studies of $BaTiO_3$ and $PbTiO_3$ surfaces.

The purpose of the present contribution is to present new theoretical work on the structural properties of the $PbTiO_3$(001) surface, and to compare and contrast these results with the previous work of our group on $BaTiO_3$ and $SrTiO_3$ surfaces.[9,11] As regards bulk properties, lead-based compounds such as $PbTiO_3$ and $PbZrO_3$ are known to behave quite differently from alkaline-earth-based perovskites, such as $BaTiO_3$ and $SrTiO_3$. Previous theoretical work has shown that the FE distortion is typically larger and that Pb atoms participate much more strongly in (and sometimes even dominate) the FE distortion, compared with non-Pb perovskites.[17-21] Moreover, the Pb-based compounds are generally more susceptible to more complex AFD and AFE instabilities involving tilting of the oxygen octahedra,[20-23] and the ground-state structures often involve the formation of some quite short Pb–O bonds.[23-26] All of these effects point to a strong and active involvement of the Pb atoms in the bonding, most naturally interpreted in terms of the formation of partially covalent Pb–O bonds with the closest oxygen neighbors. Finally, a focus on Pb-based materials is motivated by the fact that these are the leading candidates for many practical piezoelectric and switching applications, especially in the form of solid solutions, such as PZT ($PbZr_x Ti_{1-x} O_3$), PMN ($PbMg_{1/3} Nb_{2/3} O_3$), and PZN ($PbZn_{1/3} Nb_{2/3} O_3$).

The manuscript is organized as follows. Section II contains a brief account of the technical details of the work, including the theoretical methods used, the slab geometries studied, and the formulation of the surface energy. In Section III we present the computed structural relaxations of the $PbTiO_3$ surfaces, and compare these to the previous results on $BaTiO_3$ and $SrTiO_3$ surfaces. Additionally, we discuss the surface energetics (surface energies and surface relaxation energies), and point out some characteristic differences in the surface electronic structure of the three compounds. Finally, the paper ends with a summary in Section IV.

II. Preliminaries

II.A. Theoretical methods

We carried out self-consistent plane-wave pseudopotential calculations within Kohn–Sham density-functional theory using a conjugate-gradient technique.[19] Exchange and correlation were treated using the Ceperley–Alder form.[27] Vanderbilt ultrasoft pseudopotentials were employed,[28] with semicore Pb 5d, Ba 5s and 5p, Sr 4s and 4p, and Ti 3s and 3p orbitals included as valence states. A plane-wave cutoff of 25 Ry has been used throughout. Relaxations of atomic coordinates are iterated until the forces are less than 0.01 eV Å^{-1}. Justification of the convergence and accuracy of this approach can be found in the previously published work.[9,11,19]

II.B. Surface and slab geometries

In this work we consider only II–IV cubic perovskites, *i.e.*, ABO_3 perovskites in which atoms A and B are divalent and tetravalent, respectively. In this case, two non-polar (001) surface terminations are possible: the AO–terminated surface, and the BO_2–terminated surface.

We have studied both types of surface termination for all three materials ($PbTiO_3$, $BaTiO_3$, and $SrTiO_3$) using a repeated slab geometry. The slabs are symmetrically terminated and typically

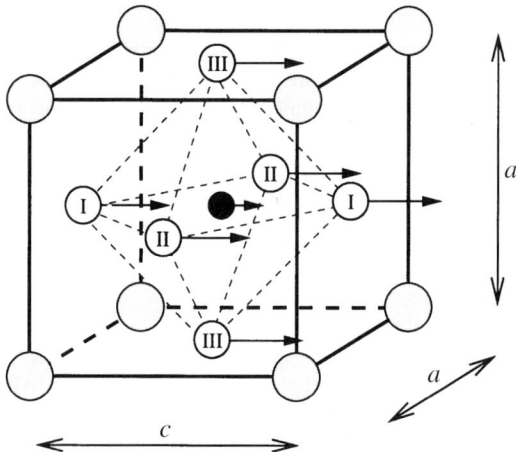

Fig. 1 Structure of the cubic perovskite compounds $ATiO_3$. Atoms A, Ti and O are represented by shaded, solid and open circles, and O_I, O_{II} and O_{III} are the oxygen atoms lying along the $\hat{x}$, $\hat{y}$ and $\hat{z}$-direction from the Ti atom, respectively. Arrows indicate the displacements of the Ti and O atoms relative to the A atoms in the case of the tetragonal phase of $PbTiO_3$.

contain seven layers (17 or 18 atoms), as illustrated in Fig. 2. The vacuum region was chosen to be two lattice constants thick. The calculations were done using a (4,4,2) Monkhorst–Pack mesh,[29] corresponding to three or four k-points in the irreducible Brillouin zone for cubic and tetragonal surfaces, respectively. The convergence of the calculations has been very carefully checked for

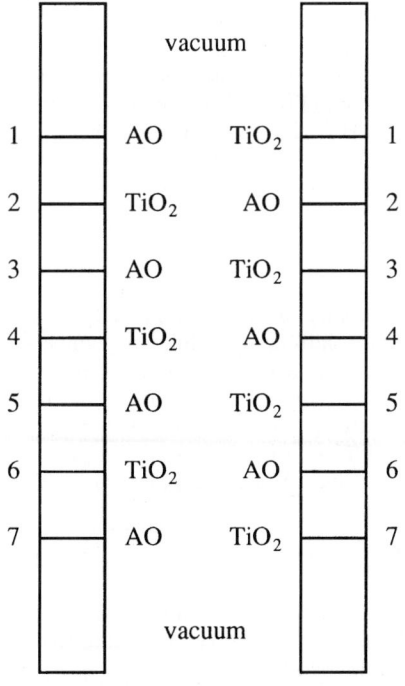

Fig. 2 Schematic illustration of the supercell geometries for the two differently terminated $ATiO_3$(001) surfaces.

PbTiO$_3$ by repeating some of the calculations with asymmetrically terminated eight-layer slabs and symmetrically terminated nine-layer slabs. Additionally, we have enlarged the vacuum region to a thickness of three lattice constants, and we have checked the convergence of the Brillouin zone integration by going to a (6,6,2) k-point mesh. In all cases, the results for the structural properties of the surfaces given in the Tables 1–5 (see below) change by less than 0.2%.

For all three materials, we first computed the relaxations for the "cubic" surface, i.e., for the case where there is no symmetry lowering relative to a slab of ideal cubic material. In this case we preserved M_x, M_y, and M_z mirror symmetries relative to the center of the slab, and set the lattice constants in the $\hat{x}$ and $\hat{y}$ directions equal to those computed theoretically for the corresponding bulk material (3.89 Å, 3.95 Å, and 3.86 Å for PbTiO$_3$, BaTiO$_3$, and SrTiO$_3$, respectively). The symmetry-allowed displacements of the atoms in the z (surface-normal) direction were then fully relaxed.

Each of the three materials studied displays a different sequence of structural phase transitions from the cubic para-electric phase as the temperature is lowered.[1] PbTiO$_3$ undergoes a single transition into a tetragonal ferroelectric (FE) phase at 763 K and then remains in this structure down to zero temperature. BaTiO$_3$ displays a series of three transitions to tetragonal, orthorhombic, and rhombohedral FE phases at 403 K, 278 K, and 183 K, respectively. SrTiO$_3$ remains cubic down to 105 K, at which point it undergoes an antiferrodistortive transition involving rotation of the oxygen octahedra and doubling of the unit cell. The material nearly goes ferroelectric at about $T = 30$ K, but is evidently prevented from doing so by quantum zero-point fluctuations.[30]

Because we are primarily interested in the room-temperature structures of these materials and their surfaces, we have chosen to focus on the tetragonal FE phases of PbTiO$_3$ and BaTiO$_3$ for our surface studies. We consider only the case of the tetragonal c axis (i.e., polarization) lying *parallel* to the surface, since polarization normal to the surface is strongly suppressed by the depolarization fields that would arise from the accumulated charge at the surfaces.[31] We take the tetragonal axis to lie along $\hat{x}$, and relax the M_x symmetry while retaining the M_y and M_z symmetries with respect to the center of the slab. For PbTiO$_3$, which is tetragonal at $T = 0$, this will indeed be the ground-state structure of the slab. For BaTiO$_3$, on the other hand, the M_y symmetry is artificially imposed so that the theoretical $T = 0$ calculation will mimic the experimental room-temperature surface structure. In both cases, the slab lattice constants in the $\hat{x}$ and $\hat{y}$ directions were set equal to the corresponding theoretical equilibrium lattice constants computed for the bulk tetragonal phase: $c = 4.04$ Å and $a = 3.86$ Å for PbTiO$_3$, and $c = 3.99$ Å and $a = 3.94$ Å for BaTiO$_3$.

II.C. Surface energies

A comparison of the relative stability of the AO and TiO$_2$ surface terminations is problematic because the corresponding surface slabs contain different numbers of AO and TiO$_2$ formula subunits. We treat this problem by introducing chemical potentials μ_{AO} and μ_{TiO_2} for these subunits, defined in such a way that $\mu_{AO} = 0$ and $\mu_{TiO_2} = 0$ correspond to thermal equilibrium with bulk crystalline AO and TiO$_2$, respectively. We have computed the cohesive energies E_{AO} and E_{TiO_2} of crystalline PbO, BaO, SrO, and TiO$_2$ using the same first-principles pseudopotential method in order to provide these reference values. The grand potential for a given surface structure can then be computed as

$$F_{surf} = \tfrac{1}{2}[E_{slab} - N_{TiO_2}(E_{TiO_2} + \mu_{TiO_2}) - N_{AO}(E_{AO} + \mu_{AO})], \qquad (1)$$

where N is the number of formula subunits contained in the slab, and the factor of 1/2 accounts for the fact that each slab contains two surfaces. Assuming that the surface of the ATiO$_3$ is in equilibrium with its own bulk, it follows that

$$\mu_{AO} + \mu_{TiO_2} = -E_f, \qquad (2)$$

where E_f is the heat of formation of bulk ATiO$_3$ from bulk AO and bulk TiO$_2$. The two chemical potentials are thus not independent, and we choose to treat μ_{TiO_2} as the independent variable when presenting our results. Accordingly, μ_{TiO_2} is allowed to vary over the range

$$-E_f \leq \mu_{TiO_2} \leq 0, \qquad (3)$$

the lower and upper limit corresponding to the precipitation of particulates of AO and TiO_2 on the surface, respectively.

III. Results and discussion

III.A. Structural relaxations

We begin by presenting our new results on the structural properties of the $PbTiO_3$ (001) surfaces. The equilibrium atomic positions for both surface terminations in the two phases were obtained by starting from the ideal structures of the surfaces and then relaxing the atomic positions while preserving the symmetries described in Section II.B. The results for the fully relaxed geometries are summarized in Table 1 and 2. By symmetry, there are no forces along $\hat{x}$ and $\hat{y}$ for the cubic surface, and no forces along $\hat{y}$ for the tetragonal surface.

Tables 1 and 2 show for both surfaces a substantial inward contraction towards the bulk for the uppermost surface layers, whereas for the second layers we find an outward relaxation of the atoms relative to the positions of the atoms on the ideal surface. Generally, the metal and the oxygen atoms move in the same direction, but the relaxations of the metal atoms are much larger, leading to a rumpling of the layers. The single exception is the surface layer of the tetragonal TiO_2-terminated surface, where one of the two oxygen atoms moves in the opposite direction to the metal atom. Therefore we can see here a significant asymmetry between the O atoms with respect to their positions perpendicular to the surface. This asymmetry between the oxygen atoms in the topmost surface layer of the tetragonal TiO_2-terminated surface was also found for $BaTiO_3$ but with a much smaller amplitude. As expected, we find the largest relaxations for the surface-layer atoms, but the displacement of the Pb atom in the second layer of the TiO_2-terminated surface is of the same magnitude.

Table 1 Atomic relaxations (relative to ideal atomic positions) of the PbO-terminated surface in the cubic (C) and tetragonal (T) phases[a]

Atom	$\delta_z(C)$	$\delta_x(T)$	$\delta_z(T)$
Pb(1)	−4.36	−3.44	−2.38
O_{III}(1)	−0.46	+11.85	−1.17
Ti(2)	+2.39	+3.62	+1.15
O_I(2)	+1.21	+9.27	+0.81
O_{II}(2)	+1.21	+11.45	+0.06
Pb(3)	−1.37	+0.00	−0.81
O_{III}(3)	−0.20	+11.14	−0.17
Ti(4)	0	+3.86	0
O_I(4)	0	+9.60	0
O_{II}	0	+10.98	0

[a] The relaxations perpendicular (δ_z) and parallel (δ_x) to the surface are given in percent of the lattice constants a and c, respectively. For reference, the theoretical δ_x values in the bulk ferroelectric phase, relative to the Pb atoms, are δ_x(Ti) = 3.45, $\delta_x(O_I)$ = 9.26 and $\delta_x(O_{II})$ = $\delta_x(O_{III})$ = 10.44. Atom labels refer to Figs. 1 and 2; results are only given for the top half of the slab, since the bottom half is equivalent by M_z mirror symmetry.

Table 2 Atomic relaxations of the TiO_2-terminated surface in the cubic (C) and tetragonal (T) phases (notation is the same as in Table 1)

Atom	δ_z(C)	δ_x(T)	δ_z(T)
Ti(1)	−3.40	+3.62	−3.47
O_I(1)	−0.34	+9.27	−1.60
O_{II}(1)	−0.34	+11.45	+0.79
Pb(2)	+4.53	+0.00	+4.06
O_{III}(2)	+0.43	+11.14	+0.17
Ti(3)	−0.92	+3.86	−0.79
O_I(3)	−0.27	+9.60	−0.03
O_{II}(3)	−0.27	+10.98	−0.06
Pb(4)	0	−3.44	0
O_{III}(4)	0	+11.85	0

In order to compare these results with previous calculations for $SrTiO_3$ and $BaTiO_3$, we have calculated the changes in the interlayer distances Δd_{ij} and the amplitudes of the rumpling η_i of the layers in the surface slabs for all three perovskites. The results for both surface terminations and the different phases are given in the Tables 3 and 4. We denote the change in the z-position of a metal atom relative to the ideal unrelaxed structure as δ_z(M), and δ_z(O) is the same for the oxygen atom in the same layer (defined as $[\delta_z(O_I) + \delta_z(O_{II})]/2$ for a TiO_2 layer). We then define the change of the interlayer distance Δd_{ij} as the difference between the averaged atomic displacements $[\delta_z(M) + \delta_z(O)]/2$ of layer i and j, and the rumpling η_i is defined as the amplitude of these displacements

Table 3 Change of the interlayer distance Δd_{ij} and layer rumpling η_i (in percent of the lattice constant a) for the relaxed AO-terminated surface of the three perovskites in the cubic and tetragonal phases

	$SrTiO_3$	$BaTiO_3$		$PbTiO_3$	
	Cubic	Cubic	Tetrag	Cubic	Tetrag
Δd_{12}	−3.4	−2.8	−2.8	−4.2	−2.6
Δd_{23}	+1.2	+1.1	+1.1	+2.6	+1.3
Δd_{34}	−0.6	−0.4	−0.4	−0.8	−0.5
η_1	5.8	1.4	1.5	3.9	1.2
η_2	1.2	0.4	0.5	1.2	0.7
η_3	1.1	0.3	0.4	1.2	0.6

Table 4 Change of the interlayer distance Δd_{ij} and layer rumpling η_i (in percent of the lattice constant a) for the relaxed TiO_2-terminated surface of the three perovskites in the cubic and tetragonal phases

	$SrTiO_3$	$BaTiO_3$		$PbTiO_3$	
	Cubic	Cubic	Tetrag	Cubic	Tetrag
Δd_{12}	−3.5	−3.1	−2.9	−4.4	−4.1
Δd_{23}	+1.6	+0.9	+1.2	+3.1	+2.5
Δd_{34}	−0.6	−0.6	−0.4	−0.6	−0.4
η_1	1.8	2.3	2.5	3.1	3.1
η_2	3.0	1.9	2.1	4.1	3.9
η_3	0.2	0.4	0.4	0.7	0.8

$|\delta_z(M) - \delta_z(O)|$. From Tables 3 and 4 we can see that, for all three perovskites and for both terminations, the surfaces display a similar oscillating relaxation pattern with a reduction of the interlayer distance d_{12}, an expansion of d_{23} and again a reduction for d_{34}. However, compared to BaTiO$_3$ and SrTiO$_3$, the amplitudes of the relaxations in PbTiO$_3$ are significantly increased.

The second interesting feature of Tables 3 and 4 is that for BaTiO$_3$, there is almost no difference in the relaxations of the surface layers between the cubic and the tetragonal phase. The same is true for the TiO$_2$–terminated surface of PbTiO$_3$. For the PbO–terminated surface, in contrast, the changes in the interlayer distances and the layer rumplings are strongly reduced in the tetragonal phase. We will come back to this point at the end of the next subsection.

III.B. Influence of the surface upon ferroelectricity

We turn now to the question of whether the presence of the surface has a strong effect upon the near–surface ferroelectricity. To analyze whether the ferroelectric order is enhanced or suppressed near the surface, we introduce average ferroelectric distortions δ_{FE} for each layer of the surface slabs:

$$\delta_{FE} = |\delta_x(A) - \delta_x(O_{III})| \quad \text{for AO planes and}$$

$$\delta_{FE} = |\delta_x(Ti) - [\delta_x(O_I) + \delta_x(O_{II})]/2| \quad \text{for TiO}_2 \text{ planes.} \quad (4)$$

The calculated values of δ_{FE} for BaTiO$_3$ and PbTiO$_3$ are given in Table 5; the last row of the table gives the bulk values for reference.

For the PbO–terminated surface of PbTiO$_3$, one can see a clear increase in the average ferroelectric distortions δ_{FE} when going from the bulk values to the surface layer. On the other hand, for the TiO$_2$–terminated surface, the average distortions are slightly decreased at the surface. Surprisingly, this is just the opposite of what one observes for BaTiO$_3$, where one sees a reduction of the ferroelectric distortions for the BaO–terminated surface and a moderate enhancement for the TiO$_2$–terminated surface. (Of course, the distortions are also much smaller for BaTiO$_3$ surfaces, as they are in the bulk, compared to PbTiO$_3$.) These results tend to confirm that Pb is a much more active constituent in PbTiO$_3$ than is Ba in BaTiO$_3$, presumably because of the partially covalent nature of the Pb–O bonds as discussed in Section I.

In any case, the present results again confirm that the presence of the surface does not lead to any drastic suppression of the ferroelectric order near the surface, supporting the view that extrinsic effects must be responsible for degradation of ferroelectricity in thin-film geometries.

Finally, we note that there are interesting signs of interplay between the relaxations parallel and perpendicular to the surface for PbTiO$_3$. In particular, the relaxations perpendicular to the surface are substantially reduced (by a factor of ≈ 3) on the PbO-terminated surface when going from the cubic to the tetragonal case. This can be rationalized as follows. Because of the partial covalency of the Pb–O bonds, there is a tendency to reduce the Pb–O bond length (this length is 2.75, 2.51, and 2.30 Å in cubic PbTiO$_3$, tetragonal PbTiO$_3$, and PbO, respectively). For the cubic surface, by symmetry, the only possibility to shorten this bond length is by a strong movement of the Pb atom towards the bulk and a strong movement upwards of the O atoms in the second

Table 5 Average layer-by-layer ferroelectric distortions δ_{FE} of the relaxed slabs, in percent of the lattice constant c (last row shows the theoretical bulk values for reference)

	AO-terminated				TiO$_2$-terminated			
	BaTiO$_3$		PbTiO$_3$		BaTiO$_3$		PbTiO$_3$	
Layer	δ_{FE}(BaO)	δ_{FE}(TiO$_2$)	δ_{FE}(PbO)	δ_{FE}(TiO$_2$)	δ_{FE}(BaO)	δ_{FE}(TiO$_2$)	δ_{FE}(PbO)	δ_{FE}(TiO$_2$)
1	1.6		15.3		4.4		5.7	
2		1.8		6.8	1.4		7.0	
3	1.3		11.1			3.4		6.3
4		2.6		6.4	1.7		9.7	
Bulk	1.5	3.2	10.4	6.4	1.5	3.2	10.4	6.4

layer. This leads to the strong rumpling and the decrease of d_{12}. But in the tetragonal phase there is also the possibility to enlarge the ferroelectric distortion in order to shorten the Pb–O bond length. Evidently, the enlargement of the ferroelectric distortion is preferred to the relaxation perpendicular to the surface.

III.C. Surface energies

In this section we discuss the surface energetics of the three perovskite compounds. In order to compare the relative stability of the AO– and TiO_2–terminated surfaces, we have calculated the grand thermodynamic potential F_{surf} (as introduced in Section II.C.) for the different surfaces as a function of the chemical potential μ_{TiO_2}. The results for the tetragonal surfaces of $BaTiO_3$ and $PbTiO_3$ are shown in Fig. 3. The graphs of the grand thermodynamic potentials for the $SrTiO_3$ surfaces are very similar to those of $BaTiO_3$ and are therefore not shown separately.

Fig. 3 shows a very different behavior for the $BaTiO_3$ and $PbTiO_3$ surfaces. First of all, the formation energy E_f of $PbTiO_3$ (when formed from bulk PbO and TiO_2) is 0.36 eV, much lower than the formation energies of $SrTiO_3$ and $BaTiO_3$ which are about 3.2 eV. This leads to a much smaller range for the chemical potential μ_{TiO_2} for which $PbTiO_3$ surfaces can grow in thermodynamic equilibrium. Second, for $BaTiO_3$ the two different surfaces have a comparable range of thermodynamic stability, indicating that either BaO–terminated surfaces or TiO_2–terminated surfaces could be formed depending on whether growth occurs in Ba–rich or Ti–rich conditions. In contrast, for $PbTiO_3$ only the PbO–terminated surface can be obtained in thermodynamic equilibrium.

To get a quantity describing the surface energetics that is independent of the chemical potential μ_{TiO_2} and therefore allows a more direct comparison of the three compounds, we define the average surface energy per surface unit cell

$$E_{surf} = \tfrac{1}{4}(E_{slab}^{AO} + E_{slab}^{TiO_2} - 7E_{bulk}), \qquad (5)$$

which is equal to the average of the grand thermodynamic potential F_{surf} for the two kinds of surfaces. Again, the results for E_{surf} shown in Table 6 are very similar for $SrTiO_3$ and $BaTiO_3$, whereas the value for $PbTiO_3$ is significantly lower.

Finally we have computed the average relaxation energy E_{relax} of the three perovskite compounds. E_{relax} is defined as the difference between the average surface energy E_{surf} of the ideal surface without relaxation of the atoms, and the fully relaxed surfaces. The largest and smallest value for E_{relax} (see Table 6) were found for $PbTiO_3$ and $BaTiO_3$, respectively, which is in agree-

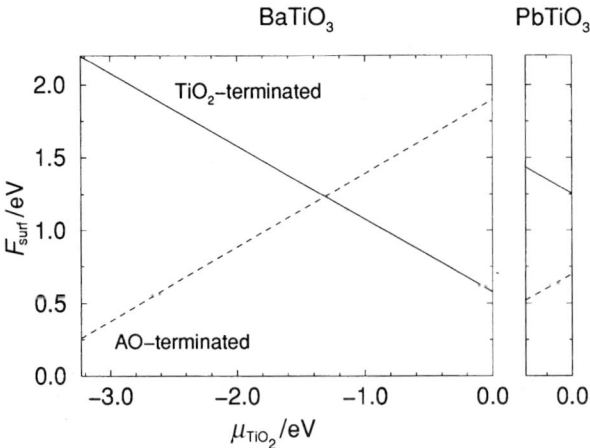

Fig. 3 Grand thermodynamic potential F_{surf} as a function of the chemical potential μ_{TiO_2} for the two types of surfaces of $BaTiO_3$ (left) and $PbTiO_3$ (right), in the tetragonal phase. Dashed and solid lines correspond to AO-terminated and TiO_2-terminated surfaces, respectively.

Table 6 Formation energy, E_f, average surface energy, E_{surf}, and average relaxation energy, E_{relax} (in eV (unit cell)$^{-1}$) for the three perovskites in the cubic and tetragonal phases

	SrTiO$_3$	BaTiO$_3$		PbTiO$_3$	
	Cubic	Cubic	Tetrag	Cubic	Tetrag
E_f	3.2	3.20	3.23	0.30	0.36
E_{surf}	1.26	1.24	1.24	0.97	0.97
E_{relax}	0.18	0.13		0.21	0.22

ment with the observation that the atomic relaxations are largest in PbTiO$_3$ and smallest in BaTiO$_3$.

For all three compounds the average relaxation energy E_{relax} is many times larger than a typical bulk ferroelectric well depth, which is approximately 0.03 eV for BaTiO$_3$ and 0.05 eV for PbTiO$_3$. This would indicate that the surface is capable of acting as a strong perturbation on the ferroelectric order. As we have shown in Section III.B., this is not the case for BaTiO$_3$ and PbTiO$_3$. One reason why the ferroelectric order is not as strongly affected by the surface as one might have thought has been pointed out in ref. 9: the soft phonon eigenmode, which is responsible for the ferroelectric distortion, is only one of three zone center modes having the same symmetry. By looking at how strongly the surface relaxations are related to each of these zone center modes it has turned out that the distortions induced by the presence of the surface are to a large extent of non–ferroelectric character.

III.D. Surface band structure

For all three perovskite compounds we have carried out LDA (local density approximation) calculations of the bulk and the surface electronic structure for our various surface slabs. It is well known that the LDA is quantitatively unreliable regarding excitation properties such as band gaps. Since we are in the following only looking at differences between band structures, we think that our conclusions drawn from the LDA results are nevertheless qualitatively correct.

As has already been shown in ref. 19, the bulk band structures of SrTiO$_3$ and BaTiO$_3$ are very similar, whereas PbTiO$_3$ shows some significant differences. In SrTiO$_3$ and BaTiO$_3$ the upper edge of the valence band is very flat throughout the Brillouin zone. On the other hand, in PbTiO$_3$ the shallow 6s semi-core states of the Pb atoms hybridize with the 2p states of the O atoms, leading to a lifting of the upper valence band states near the X point of the Brillouin zone.

This fact is responsible for a different behavior of the PbTiO$_3$ surface band structure compared to SrTiO$_3$ and BaTiO$_3$. If we look at the calculated band gaps in Table 7, we see that for TiO$_2$-terminated surfaces the band gap is significantly reduced for SrTiO$_3$ and BaTiO$_3$, whereas for PbTiO$_3$ the band gap is almost unchanged. The reduction of the band gap in SrTiO$_3$ and BaTiO$_3$ is mainly due to an upward intrusion of the upper valence band states near the M point into the lower part of the band gap (as pointed out in ref. 9, this is caused by the suppression of the hybridization of certain O 2p and Ti 3d orbitals in the surface layer). In PbTiO$_3$ we find the same upward movement of the upper valence band states near the M point, but these states stay just below the highest valence states at the X point, and so the band gap is almost unchanged.

Table 7 Calculated band gaps (in eV) for the relaxed cubic and tetragonal surface slabs

	SrTiO$_3$	BaTiO$_3$		PbTiO$_3$	
	Cubic	Cubic	Tetrag	Cubic	Tetrag
AO-term.	1.86	1.80	2.01	1.53	2.12
TiO$_2$-term	1.13	0.84	1.18	1.61	1.79
Bulk	1.85	1.79	1.80	1.54	1.56

On the other hand, for the AO–terminated surfaces we see no reduction of the band gap for any of the three perovskite compounds. Even here, however, there is a subtle difference between $PbTiO_3$ and the other materials, this time concerning the conduction band edge. According to our calculations, the Pb 6p states overlap the Ti 3d states to some degree in bulk $PbTiO_3$, and this effect is accentuated at the Γ point of the surface Brillouin zone on the Pb–O terminated surface, where the lowest Pb 6p state falls just below the lowest Ti 3d state. We thus suggest that the conduction band minimum may actually have Pb 6p character at this surface, although the effect is too small to affect the band gaps in Table 7 substantially. This might be an interesting target of investigation for future spectroscopic experimental studies.

IV. Summary

In summary, we have calculated structural and electronic properties of $PbTiO_3$(001) surfaces using a first-principles density functional approach. The results are compared and contrasted with corresponding previous calculations on $BaTiO_3$ and $SrTiO_3$ surfaces. We observe qualitatively different behavior of the $PbTiO_3$ surfaces in several respects. First, within the narrow range of PbO and TiO_2 chemical potentials permitted by bulk thermodynamics, we find that the TiO_2-terminated surface is never thermodynamically stable. Thus, the PbO-terminated surface is expected to be the one observed experimentally. Second, the interaction between the ferroelectric distortion and the presence of the surface is quite different for $PbTiO_3$, compared to $BaTiO_3$. In particular, the ferroelectricity is strongly enhanced at the AO-terminated surface and suppressed at the TiO_2-terminated surface, just the opposite of the behavior found for $BaTiO_3$. Moreover, the ferroelectric distortion at the surface allows for a drastic reduction of the rumpling of the surface layer on the PbO-terminated surface, an effect which is not seen on the BaO-terminated surface of $BaTiO_3$. Third, the surface electronic band structure is qualitatively modified in the case of $PbTiO_3$ by the presence of Pb 6s and 6p states in the upper valence and lower conduction regions.

Acknowledgements

This work was supported by the ONR grant N00014-97-1-0048.

References

1. M. E. Lines and A. M. Glass, *Principles and Applications of Ferroelectrics and Related Materials*, Clarendon Press, Oxford, 1977; F. Jona and G. Shirane, *Ferroelectric Crystals*, Dover Publications, New York, 1993.
2. F. Tsai and J. M. Cowley, *Appl. Phys. Lett.*, 1994, **65**, 1906.
3. J. C. Niepce, *Surface and Interfaces of Ceramic Materials*, ed. L. C. Dufour, Kluwer Academic Publishers, Dordrecht, 1989, p. 521.
4. V. E. Henrich and P. A. Cox, *The Surface Science of Metal Oxides*, Cambridge University Press, New York, 1994.
5. D. Vanderbilt, *Curr. Opin. Solid State Mater. Sci.*, 1997, **2**, 701.
6. D. Vanderbilt, *J. Korean Phys. Soc.*, 1998, **32**, S103.
7. R. E. Cohen, *J. Phys. Chem. Solids*, 1996, **57**, 1393.
8. R. E. Cohen, *Ferroelectrics*, 1997, **194**, 323.
9. J. Padilla and D. Vanderbilt, *Phys. Rev. B: Condens. Matter*, 1997, **56**, 1625.
10. S. Kimura, J. Yamauchi, M. Tsukada and S. Watanabe, *Phys. Rev. B: Condens. Matter*, 1995, **51**, 11049.
11. J. Padilla and D. Vanderbilt, *Surf. Sci.*, 1998, **418**, 64.
12. Z.-Q. Li, J.-L. Zhu, C. Q. Wu, Z. Tang and Y. Kawazoe, *Phys. Rev. B: Condens. Matter*, 1998, **58**, 8075.
13. T. Hikita, T. Hanada and M. Kudo, *Surf. Sci.*, 1993, **287/288**, 377.
14. N. Bickel, G. Schmidt, K. Heinz and K. Muller, *Vacuum*, 1990, **41**, 46; *Phys. Rev. Lett.*, 1989, **62**, 2009.
15. M. Naito and H. Sato, *Physica C*, 1994, **229**, 1.
16. K. Kitahama, Q. R. Feng, T. Kawai and S. Kawai, *Bull. Fall Meeting Jpn. Appl. Phys. Soc.*, 1992, **2**, 494.
17. R. E. Cohen, *Nature (London)*, 1992, **358**, 136.
18. R. E. Cohen and H. Krakauer, *Ferroelectrics*, 1992, **136**, 65.
19. R. D. King-Smith and D. Vanderbilt, *Phys. Rev. B: Condens. Matter*, 1994, **49**, 5828.
20. A. García and D. Vanderbilt, *Phys. Rev. B: Condens. Matter*, 1996, **54**, 3817.
21. U. V. Waghmare and K. M. Rabe, *Ferroelectrics*, 1997, **194**, 135.
22. W. Zhong and D. Vanderbilt, *Phys. Rev. Lett.*, 1995, **74**, 2587.
23. D. J. Singh, *Phys. Rev. B: Condens. Matter*, 1995, **52**, 12559.

24 L. Bellaiche, J. Padilla and D. Vanderbilt, in *First-Principles Calculations for Ferroelectrics: Fifth Williamsburg Workshop*, ed. R. E. Cohen, AIP, Woodbury, New York, 1998, p. 11.
25 L. Bellaiche, J. Padilla and D. Vanderbilt, *Phys. Rev.: Condens. Matter*, 1999, **59**, 1834.
26 T. Egami, W. Domowski, M. Akbas and P. K. Davies, in *First-Principles Calculations for Ferroelectrics: Fifth Williamsburg Workshop*, ed. R. E. Cohen, AIP, Woodbury, New York, 1998, p. 1.
27 D. M. Ceperley and B. J. Alder, *Phys. Rev. Lett.*, 1980, **45**, 566.
28 D. Vanderbilt, *Phys. Rev. B: Condens. Matter*, 1990, **41**, 7892.
29 H. J. Monkhorst and J. D. Pack, *Phys. Rev. B: Condens. Matter*, 1976, **13**, 5188.
30 W. Zhong and D. Vanderbilt, *Phys. Rev. B: Condens. Matter*, 1996, **53**, 5047.
31 W. Zhong, R. D. King-Smith and D. Vanderbilt, *Phys. Rev. Lett.*, 1994, **72**, 3618.

Paper 9/03029H

Electronic structure and surface reactivity of $La_{1-x}Sr_xCoO_3$

Wendy R. Flavell,*[a] Andrew G. Thomas,[a] Jane Hollingworth,[a] Samantha Warren,[a]
Sarah C. Grice,[a] Patricia M. Dunwoody,[a] Caroline E. J. Mitchell,[a] Peter G. D. Marr,[a]
David Teehan,[b] Stuart Downes,[b] Elaine A. Seddon,[b] Vinod R. Dhanak,[b]
Kichizo Asai,[c] Yoshihiko Koboyashi[c] and Nobuyoshi Yamada[c]

[a] *Department of Physics, UMIST, P.O. Box 88, Manchester, UK M60 1QD.*
E-mail: wendy.flavell@umist.ac.uk
[b] *CLRC, Daresbury Laboratory, Warrington, Cheshire, UK WA4 5AD*
[c] *The University of Electro-Communications, 1-5-1, Chofugaoka, Chofu, Tokyo 182-8585, Japan*

Received 25th March 1999

Resonant photoemission performed at the SRS Daresbury is used to investigate the temperature- and doping-dependent spin transitions in single crystal samples of the $La_{1-x}Sr_xCoO_3$ system ($x = 0$, $x = 0.1$). For $LaCoO_3$, the measurements taken over a wide temperature range are interpreted in the light of current models for the spin transitions. The Sr-doped compound is found to be in the same spin state at all temperatures. The temperature dependence of low binding energy features primarily associated with Co in the low spin state is investigated in detail. In particular, the intensity and onset energy of the Co $3p \rightarrow 3d$ resonance associated with these features is measured at small binding energy intervals. This is used to locate the separate contributions of the low spin and higher spin states to these signals. In addition, the surface reactivity of the $LaCoO_3$ (111) surface is investigated using H_2O as a probe molecule, and the results are used to interpret changes which may occur in the spectra as a function of time in UHV.

1. Introduction

$LaCoO_3$ is of great technological interest, for example as a solid fuel cell and a catalytic material.[1,2] These properties may be further enhanced by the substitution of the La ion by Sr.[2] In addition to the technological interest there is also much investigation of the unusual magnetic and electronic properties of the doped and undoped $La_{1-x}Sr_xCoO_3$ series of compounds, in particular, the large magnetoresistive anomalies exhibited by the Sr-doped materials.[3–5]

In the undoped compound (*i.e.*, $LaCoO_3$) there is a temperature-induced transition at *ca.* 90 K from a non-magnetic semiconductor to a magnetic state.[3,6,7] The crystal field splitting, $10\ Dq$, is just large enough to overcome the intra-atomic exchange interaction, J, which favours aligned spins and therefore a high spin (HS) state, $(t_{2g}^4 e_g^2 : {}^5T_2)$ so that in fact $LaCoO_3$ is in the low spin (LS) state $(t_{2g}^6 e_g^0 : {}^1A_1)$ below this temperature. It was originally thought that this transition involved a gradual low to high spin transition as a function of temperature. However, the occurrence of an electronic transition at *ca.* 500 K has led to much debate about, and study of, the exact nature of the spin state in $LaCoO_3$.[8–15] Abbate *et al.* concluded from X-ray absorption and X-ray photoelectron spectroscopies on single crystal samples that $LaCoO_3$ is in a highly covalent

LS state up to 300 K on the grounds that neither the XAS and XPS spectra undergo a change in the temperature range 80–300 K.[9] Below 100 K, direct magnetic measurements have confirmed the compound exists in a low spin state.[3,6,7] However, a study of $SrCoO_3$[16] and recent calculations on the occupation of states in $LaCoO_3$ suggest the transition at 100 K is to an intermediate spin state, $^3T_1$, with a mixture of $t_{2g}^5 e_g^1$ and $t_{2g}^5 e_g^2 \underline{L}$ states (where $\underline{L}$ indicates a ligand hole state) stabilised by strong hybridisation of Co 3d and O 2p.[8,15] Further support for the intermediate spin state comes from the fact that oxides with unusually high valence state transition metals tend to undergo a ground state ligand to metal electron transfer mechanism thus reducing the valence of the metal.[14,16] Korotin and co-workers[15] suggest that in the case of $LaCoO_3$ this corresponds to d^6 to $d^7\underline{L}$, giving exactly the intermediate state described above. Furthermore these authors suggest that this state would be of lower energy than the high spin state, thus giving a possible explanation for the low spin-intermediate spin transition. More recent calculations using the self-energy corrected Hartree–Fock approximation have suggested that the IS and HS states have an energy around 1 eV higher than the LS phase.[17]

The intermediate spin state would be expected to be metallic. In fact, $LaCoO_3$ is insulating below ca. 500 K. It has been suggested that this is due to magnetic orbital ordering, which leads to the states being pulled back from the Fermi level.[15] At 500 K there is a second transition to a metallic state, which Korotin and co-workers suggest is due to the loss of orbital ordering allowing the broad e_g bands to cross the Fermi level.[15] These authors suggest the high spin state is at such a high energy above the intermediate state it cannot be reached until temperatures well in excess of 500 K. More recently it has been shown *via* neutron scattering experiments that the 500 K transition entails an anomalous lattice expansion which implies a second spin-state transition from an IS to HS state.[7] The results of this study suggest that the energy gap between the IS and HS states decreases as the temperature is raised, eventually leading to the formation of a mixed IS–HS state.

In contrast to the undoped compound, $La_{1-x}Sr_xCoO_3$ (where $x \geqslant 0.08$) was originally thought to exist in a high spin state at all temperatures due to lattice expansion induced by the dopant Sr ions.[13] However, it has been suggested more recently from cluster calculations that this compound exists in an IS state.[18]

In this work we have studied both single crystal $LaCoO_3$ (111) and single crystal $La_{0.9}Sr_{0.1}CoO_3$ (111) by resonance photoemission, over a wide temperature range. This complements previous low temperature studies of $LaCoO_3$.[19] In particular, the Co 3p → 3d resonance onset energy, and the way in which this changes with temperature is studied for both materials. We investigate in detail the resonance behaviour of valence band features as a function of binding energy, in order to make comparison with calculations for the different possible spin states.[17,18] The data are discussed in the light of recent models for the spin transitions, including the two-stage spin-state transition model involving an IS state.[7] In addition, the surface reactivity of the $LaCoO_3$ (111) surface is studied using H_2O as a probe molecule, and these data are used to interpret the degradation of the surface, which is sometimes seen to occur as a function of time inside the UHV spectrometer. This aids us in distinguishing changes due to spin-state transitions from any effects due to surface reactions.

2. Experimental

The single crystals were grown by a lamp image floating zone method. Laué back reflection from the $LaCoO_3$ crystal showed the crystal to be slightly twinned as a result of the rhombohedral distortion of the perovskite lattice ($a = 90.6°$).[13] Simulations of the expected Laué patterns performed at the CLRC Daresbury laboratory showed the best agreement with the (111) face. Clean samples were prepared by cleaving using a VSW crystal cleaving anvil or scraping with a clean diamond file in UHV at a pressure of better than 3×10^{-10} mbar. Sample cleanliness was checked by monitoring the evolution of contamination peaks at around 9.2 eV [18] and 5.1 eV. The origin of these features is discussed further in sections 3.1 and 3.4. The sample was recleaved or rescraped at regular intervals.

The photoemission measurements were performed using three stations at the Synchrotron Radiation Source (SRS) at the CLRC Daresbury Laboratory. These were the grazing incidence monochromator (20 eV $\leqslant h\nu \leqslant$ 280 eV) on station 6.1, the toroidal grazing monochromator (15 eV ⩽ hv ⩽ 90 eV) on station 6.2 and the spherical grating monochromator (15 eV ⩽ hv ⩽ 220 eV) on station 4.1. Photoemitted electrons were analysed using a µMott detector in a spin-integrated mode (essentially an HA50 rotatable hemispherical analyser) station 6.1 and using an ADES 400 rotatable hemispherical analyser fitted with a 1 mm aperture (angular acceptance ±2°) on station 6.2. On station 4.1, a Scienta 200 mm fixed hemispherical analyser was used. The resolution (monochromator + analyser) in energy distribution curve mode was 0.16 eV for the experimental set-up on station 6.2, approximately 0.25 eV for station 6.1 and 0.20 eV for station 4.1. For $La_{0.9}Sr_{0.1}CoO_3$, data were accumulated at temperatures down to 25 K, using a VG liquid He-cooled sample manipulator. However, for the undoped compound, significant sample charging effects, with shifting and broadening of the spectral features was observed below 130 K. No data are therefore presented at lower temperatures for $LaCoO_3$.

In dosing experiments, doubly distilled, freeze-degassed water was admitted to the vacuum chamber via a leak valve. The exposure in langmuirs (1 L = 10^{-6} Torr s) was estimated from the chamber ion gauge. The $LaCoO_3$ crystal was cooled to 150 K before dosing, and then warmed to room temperature. Spectra were recorded at intervals during this process.

3. Results and discussion

3.1. General features of the valence band spectra and the effect of temperature

All spectra are normalised to the photon flux, by means of a tungsten grid placed in the beamline, and aligned on the binding energy scale to a Fermi edge spectrum recorded from the clean sample plate, unless otherwise indicated. Fig. 1 shows energy distribution curve (EDC) spectra of cleaved $LaCoO_3$ compared to the calculations of Saitoh et al.[8] and the scraped 10% Sr doped samples.

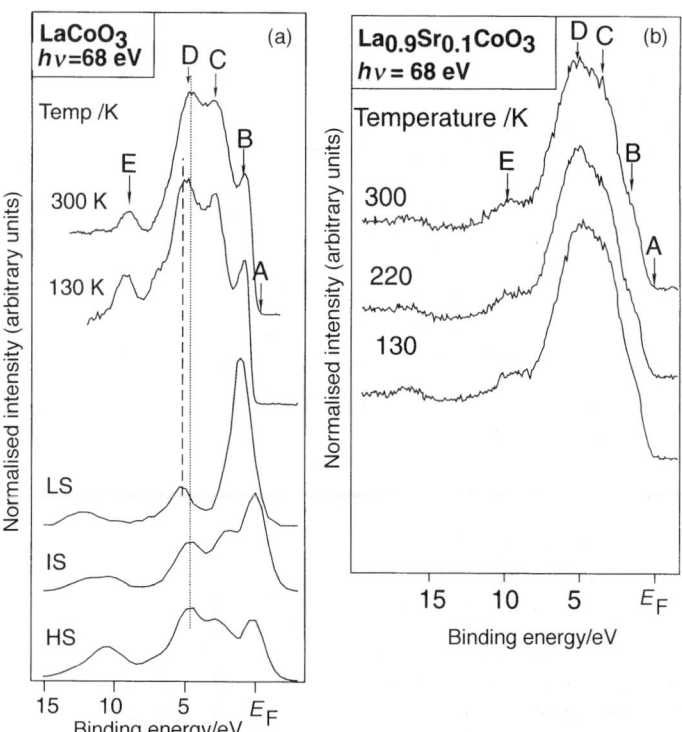

Fig. 1 Valence band EDCs recorded from (a) cleaved $LaCoO_3$ (111) and (b) scraped $La_{0.9}Sr_{0.1}CoO_3$ (111). Note the decreased intensity of the 1 eV binding energy feature in the Sr-doped compound. The results of cluster calculations[8] for the Co 3d density of states from Co in low spin (LS), intermediate spin (IS) and high spin (HS) states are shown in the lower part of Fig. 1(a). Features A–E are discussed in the text.

The spectra are generally in good agreement with those recorded from polycrystalline samples.[8,9,20] Spectra of the undoped compound show a number of features, which may be related to the cluster calculations of Saitoh et al.,[8] in the region of the main valence band. The Co 3d photoemission spectra calculated by these authors for different spin states (neglecting the O 2p contribution)[8] are shown in the lower part of Fig. 1, and the comparison is discussed in more detail below. The large structure observed at ca. 3 eV in the experimental data (feature C, Fig. 2) is not present in the calculations as this is derived mainly from the O 2p states.[8] A comparison with the simulated O 2p partial densities of states derived by Saitoh et al.[8] from the He I spectrum of $LaCoO_3$ indicates that this probably also contributes to the broadening of feature D observed in the experimental data (Fig. 2). A clearly resolved feature can be observed at a binding energy (E_b) of 1 eV in the $LaCoO_3$ spectrum. Although this region has been associated with Co in a low spin, $^1A_1$ state, cluster calculations suggest there are some contributions from the IS and HS states in this part of the valence band[8] (Fig. 1). The sharp cut-off of these states just below the Fermi energy is in agreement with recent LDA + U calculations where orbital ordering is thought to occur resulting in states being pulled back from the Fermi energy.[15] The rest of the valence band (to higher binding energy) is dominated by contributions from other spin states (intermediate state, IS and high spin, HS are possible) and O 2p states.[8]

The spectral quality from the 10% Sr doped sample is rather poorer than that from $LaCoO_3$, but it is nevertheless clear that the feature at a E_b of 1 eV is largely absent [Fig. 1(b)]. Sr doping beyond a level of approximately 8% leads to a lattice expansion suppressing the transition to a low spin state at low temperatures. We thus expect the $La_{0.9}Sr_{0.1}CoO_3$ sample to be in a high or intermediate spin state at all temperatures studied in this work,[7] and the dominant low spin contribution at around 1 eV should be largely absent.

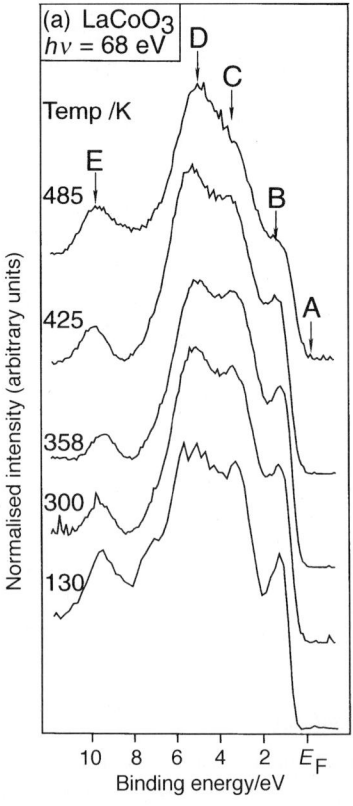

Fig. 2 Effect of temperature on the valence band EDC spectrum of scraped $LaCoO_3$ (111) recorded at a photon energy of 68 eV.

A number of changes are observed in the spectra of $LaCoO_3$ as a function of temperature. As can be seen from the spectra in Fig. 1(a) there is a decrease in the intensity of the peak at 1 eV E_b as the temperature is increased. This is likely to be caused by depopulation of the LS state at the higher temperature relative to the lower temperature. We also observe a small shift in the binding energy of the feature at a binding energy of ca. 6 eV between the spectra recorded at 130 and 300 K in direct agreement with the calculated DOS (density of states) for $LaCoO_3$ as we move from the LS to IS or HS states [Fig. 1(a)]. There appears to be very little, if any, change in the spectra recorded from the Sr-doped compound between these temperatures [Fig. 1(b)], suggesting that there are no substantial variations in spin state across the temperature range studied. This is presumably because the 10% Sr-doped compound does not exhibit the low spin state.

In both the doped and undoped crystal spectra we see a peak at a binding energy of 9.5 eV. An analogous peak is observed in valence band spectra of CoO (Co^{2+}, rather than Co^{3+}, as in $LaCoO_3$), where it has been assigned to a d^6 final state satellite with the valence band assigned to $d^7\underline{L}$ final states. The peak undergoes a resonance at the Co 3p to 3d resonance energy,[21] confirming its Co character. The peak has also been observed in earlier $LaCoO_3$ valence band spectra[22] and again is seen to undergo a Co resonance. In this work, we also observe a resonance in this feature at the Co threshold, confirming that there is some Co character in this part of the spectral profile [data are shown at Fig. 4(b) (see later), and are discussed in the next section]. Nevertheless, it is possible that this peak arises as a result of defects at the surface following cleaving or possibly contamination, given its similarity to the notorious '9.5 eV' peak observed in spectra of the related cuprate superconductors.[23] On some occasions, perhaps associated with a poor cleave, we have observed changes in spectral shape as a function of time in UHV. This has prompted us to carry out a study of the reactivity of the surface to probe molecules, and the results of water adsorption experiments are described in section 3.4. For the present purposes, we show in Fig. 3 the spectrum recorded from a freshly cleaved surface and the same surface after 2 h in UHV (1×10^{-10} mbar) at 300 K. Spectra are recorded at a lower photon energy than in Figs. 1 and 2 (33 eV) in order to enhance the photoionisation cross-section of any oxygen states involved in the degradation process. It is clear that in the event of a poor cleave, contamination or surface changes occur with time but that this contamination leads to an increase in the spectral intensity of both features D and E. The assignment of the difference spectrum shown in Fig. 3 is discussed further in section 3.4. For the purposes of deconvolution of these effects from any effects due to spin changes, we note that surface degradation is characterised by a uniform increase in the intensity of *both* features D and E.

Fig. 2 shows a more detailed investigation of the effect of temperature on the valence band spectra of scraped $LaCoO_3$ recorded at a photon energy of 68 eV. It was not possible to record meaningful spectra below 130 K due to charging of the sample. The slight difference in the spectral quality between the spectra presented in Figs. 1 and 2 is believed to be caused by the different methods of cleaning the surfaces. However, the general features and intensity changes of the two data sets are consistent. It is clear that feature B, largely associated with Co in a low spin state decreases in intensity with increasing temperature, indicating that the low spin state is becoming depopulated. At the same time there is an increase in the spectral weight in the rest of the valence band as other spin states become more dominant. The data recorded at 425 K show a clear increase in intensity of the feature D at a binding energy of 5 eV, in accord with the cluster calculations of Saitoh et al.,[8] where this peak is predicted to show the maximum spectral weight in the valence band region for the high spin state. At 485 K this enhancement of feature D is much clearer and is coupled with a further decrease in the intensity of feature C. We note that cluster calculations (Fig. 1 [8]) show that the Co DOS is stronger at feature D than at C for the HS state, but the converse is true for the IS state. It is clear that the observed increases in intensity in feature D are *not* accompanied by increases in intensity in feature E. This differs from the pattern for surface decomposition shown in Fig. 3, where we show that as the surface degrades, both features are enhanced. We therefore attribute the intensity changes in feature D in Fig. 2 to changes in the spin state of the sample. The relative changes in intensity in features B, C and D appear consistent with a three state spin model involving the IS state. This possibility is discussed further in the light of resonance experiments described in section 3.2. It is clear that the transition to the high spin state is a gradual one, and this is in agreement with the prediction that the energy gap between the IS and HS states decreases gradually on moving through the transition.[7] Neutron

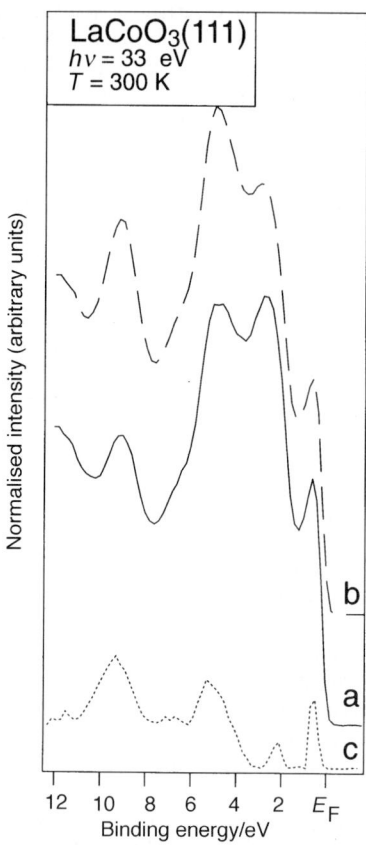

Fig. 3 EDCs from LaCoO$_3$ (111) recorded immediately after cleaving (a), and after approximately 2 h in UHV at a vacuum of 1×10^{-10} mbar (b). The difference spectrum (b)—(a) is also shown (c). Spectra are recorded at 33 eV (see text).

scattering results[7] suggest that the IS state is never the only state occupied, *i.e.*, at all temperatures sampled there is a contribution from the LS or the HS state, and this appears consistent with our data. However, we should add that we have not been able satisfactorily to simulate the observed spectra using the proportions of LS, IS and HS states determined from these experiments,[7] in combination with the cluster calculation results.[8] We believe this may be associated in part with difficulties in determining the O 2p partial DOS to be convoluted into the Co DOS from the cluster calculation. In order to investigate the spin state more thoroughly, we have therefore used resonant photoemission as described below; this allows us to comment further on the reasons for difficulties in simulating the observed spectra in section 3.3.

3.2. Resonant photoemission experiments

Fig. 4 shows constant initial state (CIS) spectra recorded over the Co 3p → 3d resonance for LaCoO$_3$, at 130 K, 300 K, 425 K and 485 K. The binding energy positions chosen for the CIS spectra are labelled in Fig. 2. All of the CIS spectra show a small resonance at a photon energy of *ca.* 59 eV followed by a larger main resonance at energies between 61 and 64 eV with the exception of point A which shows only the 59 eV feature. This is in agreement with CIS spectra recorded from CoO where the double-peaked structure is assigned to transitions from different spin orbital components of the Co 3p initial state.[21] However, the observation of only the small 59 eV feature at point A may suggest that it is not an intrinsic part of the resonance structure. Our discussion therefore centres on the main resonance at *ca.* 62 eV. It is clear that the main resonances at position B for the low and room temperature data are at approximately 3 eV higher

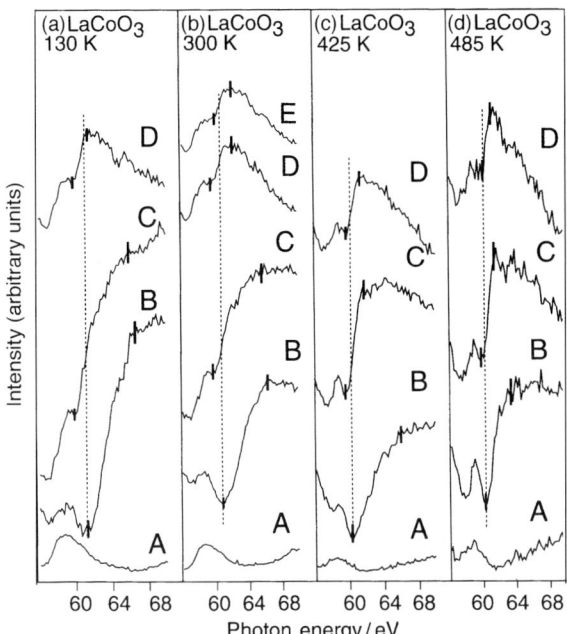

Fig. 4 CIS spectra recorded from LaCoO$_3$ (111) at temperatures of (a) 130 K (scraped), (b) 300 K (cleaved), (c) 425 K (scraped) and (d) 485 K (scraped). The dashed lines originating at the resonance minimum of feature B are drawn to emphasise the *ca.* 3 eV shift in resonance position between B and C–E at low temperature. Spectra are normalised to the incident photon flux.

energy than for the other features. If the major part of the intensity of the resonance at B derives from low spin Co, then this would be expected as in the low spin state all of the t_{2g} states are full, thus only the e_g states (which, according to calculations, lie approximately 3 eV above the t_{2g} states[15]) can contribute to the spectrum leading to a delayed resonance onset. In the higher binding energy parts of the valence band (features C and D), which correspond to Co in intermediate ($t_{2g}^5 e_g^1$ and $t_{2g}^5 e_g^2 \underline{L}$, where $\underline{L}$ represents a ligand hole), and high spin ($t_{2g}^4 e_g^2$) states we now have empty t_{2g} states for the Co 3p electrons to be excited into thus the resonance onset is at a lower energy at these points.

The data recorded at 130 K show the resonance intensity of feature B to be enhanced relative to the other valence band features, presumably as at this temperature the ratio of LS to IS–HS is the highest investigated in this work. This change in the intensity of the resonance edge is shown more clearly in Fig. 5(a) where the relative intensities of the resonances of features B and C are plotted against temperature. The spectra were normalised to the intensity of the 59 eV resonance of point A in order to remove any differences in the analyser performance/sample positions during data acquisition. The 'before' and 'after' resonance intensities were taken at the points indicated by the markers in Fig. 4. The gradual decrease in intensity I_B/I_C is suggestive of a gradual transition from the low spin state into other spin states as the temperature increases. Fig. 5(b) shows the same information for peaks C and D, where it can be seen that there is also a gradual decrease in the ratio I_C/I_D as the temperature is raised. Given the relative intensities of these features predicted by cluster calculations [discussed above and shown in Fig. 1(a)], this change may suggest a further gradual transition from the intermediate spin to a high spin state as the temperature is increased.

Fig. 6 shows the effect of temperature on the energy of the resonance onset for features B and C (taken as the mid-point of the markers in Fig. 4). The resonance onset energy of B is clearly falling as the temperature is increased because as the low spin state is depopulated in favour of other spin states (IS or HS), empty t_{2g} states become available into which the resonance can occur. It is clear that the total drop in energy over the range 130–500 K is approximately 3 eV, which, as stated previously, approximately corresponds to the crystal field stabilization energy. It should be noted

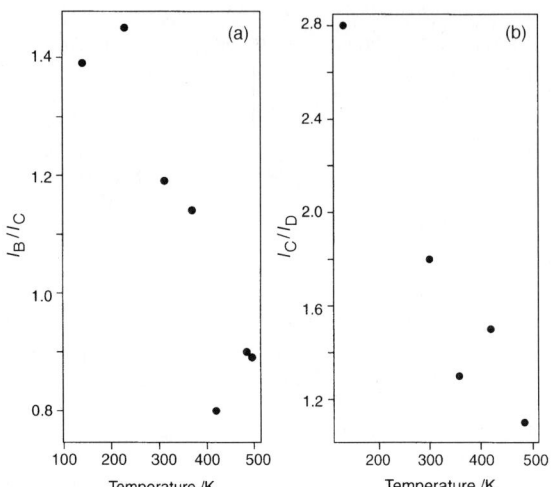

Fig. 5 (a) The ratio of intensities of resonance onset, I_B/I_C for features B and C and (b) for features C and D. The resonance edges were normalised to the intensity of feature A to remove any possible discrepancies due to different analysers or sample positions.

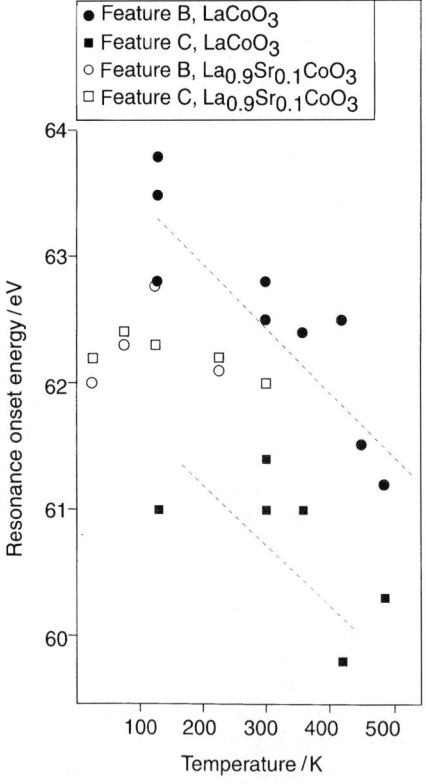

Fig. 6 Resonance energy onset for features B and C for $LaCoO_3$ (filled markers) and $La_{0.9}Sr_{0.1}CoO_3$ (open markers) plotted against temperature to show the decrease in resonance onset energy as t_{2g} states become available to the 3p → 3d resonance process. The dashed lines are drawn for heuristic purposes only, see text.

that there will also be a contribution from the energy differences in electron correlation and charge–transfer effects, depending on the electronic configuration changes occurring between the different spin states (*i.e.*, to account for the fact that in the region of LS the CIS experiment excites an electron into the empty e_g state, whereas in the IS case an electron is excited into a t_{2g} state already containing five electrons and the e_g also contains one or two electrons), thus the measured energy gap we present here may not be purely the crystal field splitting (10 Dq). It should also be remembered that there are some contributions from the IS and HS states under the main feature at B and it is these that allow us to monitor the reduction in excitation energy of B. The dashed lines in Fig. 6 are for heuristic purposes only as given the discussion above, we clearly have no reason to expect the changes in resonance energies with temperature to be linear. It can be seen from the cluster calculations shown in Fig. 1(a)[8] that all three spin states contribute to some extent to the resonance measured at feature B, and this will complicate the temperature dependence; in principle, we expect a delayed onset only for the LS state, and not for HS or IS. The calculations predict that the intensity maximum at B for LS is slightly displaced from that for IS and HS. Thus CIS recorded at small binding energy increments across feature B may show different onset behaviour at slightly different binding energy positions. Experiments designed to probe this possibility are described in the next section (3.3).

Fig. 7 shows the CIS spectra recorded from valence band features A, B, C and D [labelled in Fig. 1(b)] in $La_{0.9}Sr_{0.1}CoO_3$ at temperatures of 25, 125 and 300 K. There is little change as a function of temperature. More importantly we see no systematic shift in the energy of the resonance onset for feature B (plotted in Fig. 6), even at 25 K, thus indicating the Sr-doped compound

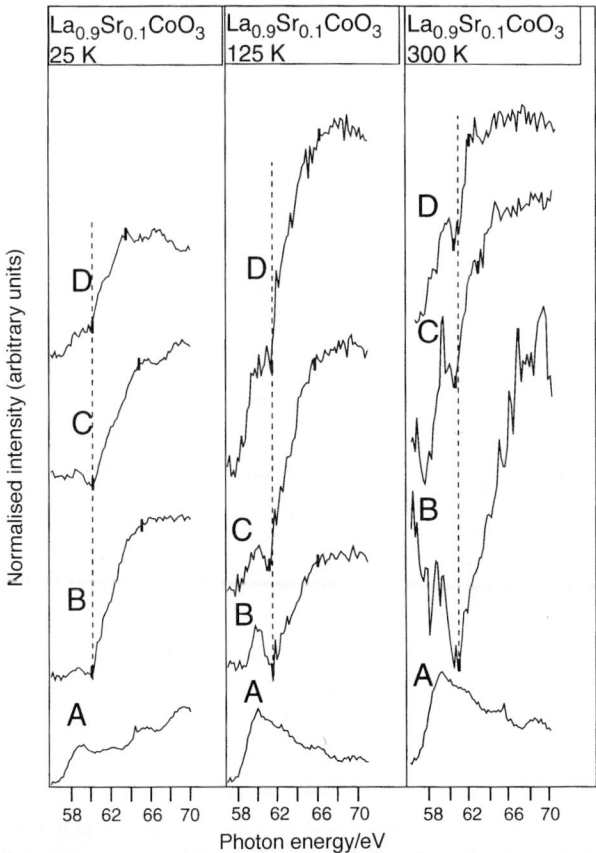

Fig. 7 CIS spectra of scraped $La_{0.9}Sr_{0.1}CoO_3$ (111) at 25 K, 125 K and 300 K. Dashed lines are drawn from the resonance minimum of feature B to guide the eye. There is no resonance onset delay for any of the features even at 25 K.

is in the same spin state at all temperatures sampled. This combined with the absence of a strong feature at B in the spectra, strongly suggests that the Sr-doped compound is in a constant spin state at all temperatures, which in the light of recent data[7] we assign to an IS state. The resonance onset energies of the Sr-doped compound in Fig. 6 are at around the same energy as those for room temperature $LaCoO_3$ further suggesting a majority IS spin state for the Sr-doped compound at the temperatures investigated in this work.

3.3. Detailed CIS investigation of feature B as a function of binding energy

In order to separate the contributions of the different spin states to feature B, station 4.1 at the SRS was used in combination with a Scienta 200 mm analyser to investigate CIS profiles taken at small binding energy increments around feature B, as a function of temperature. Fig. 8 shows these data, at temperatures of 200 K, 300 K and 450 K, in the binding energy range 0.2–2.2 eV (centred on feature B at around 1.2 eV). The spectra taken at 450 K are relatively straightforward; for binding energies in the range 1.2–2.2 eV, all resonances have the same onset energy of *ca.* 61.8 eV, and there is no significant resonance for binding energies less than 1.2 eV. However, at both 200 K and 300 K, the resonance onset energy changes with binding energy, from *ca.* 62 eV at 2.2 eV binding energy to around 64 eV at 0.7 eV binding energy. Below 0.7 eV, there is no significant resonance. We attribute the changes as a function of temperature to the decreasing contribution of the LS state to feature B as the temperature is raised. The binding energy variation suggests that the maxima in the contributions of the LS and HS–IS states to B occur at slightly different binding energies, as implied by the cluster calculations [Fig. 1(a)[8]]. The data taken at 450 K show no delayed onset, and we conclude that the sample is therefore predominantly in higher spin states (HS, IS or mixed HS–IS), *i.e.*, there is no significant LS contribution. The intensities of the CIS spectra suggest that the maximum contribution to feature B from these states is centred at binding energies of around 1.7–2.2 eV. The presence of a delayed resonance onset at 200 K and 300 K indicates the presence of a significant proportion of LS states, centred in the binding energy range

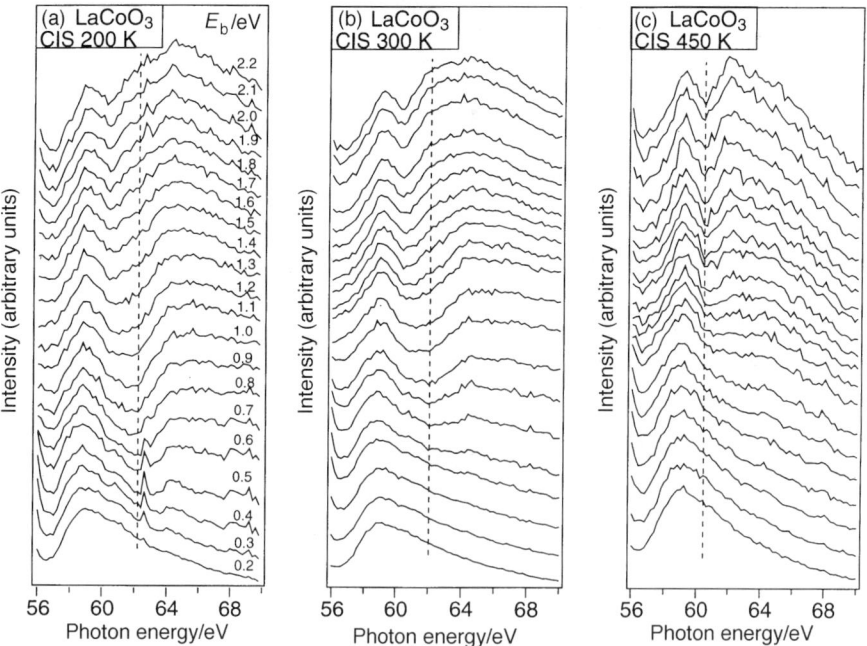

Fig. 8 CIS spectra recorded for small binding energy increments across feature B from scraped $LaCoO_3$ (111) at temperatures of (a) 200 K, (b) 300 K and (c) 450 K. The binding energy scale is as given in (a). Spectra are normalised to the incident photon flux. Dashed lines are drawn from the lowest binding energy resonance minimum in all cases. Delayed resonance onset over the lower part of the binding energy range is seen in (a) and (b), but is absent in (c).

0.7–1.2 eV, *i.e.*, at *lower* binding energy than the contribution from higher spin states. In the range 1.2–1.7 eV binding energy, we observe a transition from 'normal' to delayed onset. We note that these binding energies are in the *reverse* order to that expected from calculations [Fig. 1(a)[8]], where the HS–IS states give a contribution to B at lower binding energy than the LS states. To some extent this is expected, as the calculation is aligned arbitrarily by its authors, placing the maximum in the LS states at 1 eV.[8] One result is that it incorrectly shows a Fermi level crossing for the IS state, which is known not to occur in practice (*i.e.*, the compound is non-metallic to *ca.* 500 K, whereas the calculation as presented indicates it should be metallic). These difficulties have been acknowledged previously. The location of the relative binding energy of these states by our experiment does explain why we are unable satisfactorily to simulate the temperature dependence of our spectra using the cluster results.[8] Our result suggests a binding energy minimum for the LS state around 1 eV lower than any higher states (HS–IS). We note that this agrees well with the difference in ground state energies calculated by Takahashi and Igarashi,[17] although this comparison should be treated with caution, as our spectra are of course representative of the joint DOS of initial and final states, rather than the ground state alone.

3.4. Adsorption studies

Fig. 9 shows the effect of dosing a single crystal (111) surface of $LaCoO_3$ with H_2O at low temperature. The figure shows the spectrum obtained from a freshly scraped surface at 150 K and the spectrum obtained after dosing with 3 L H_2O. The effect on the spectra of warming the surface

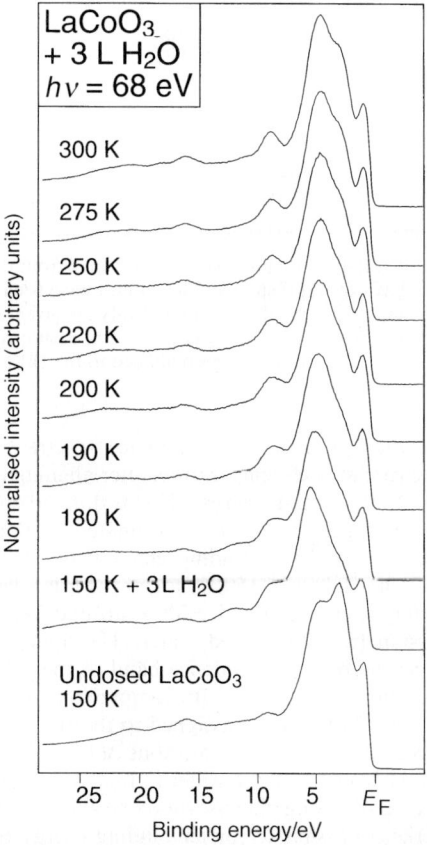

Fig. 9 EDC spectra recorded at 68 eV photon energy showing the effect of water adsorption and subsequent warming on $LaCoO_3$ (111). Spectra have been normalised to the beam monitor reading, and aligned on the binding energy scale to the La 5p core level intensity in the binding energy range 16–20 eV.

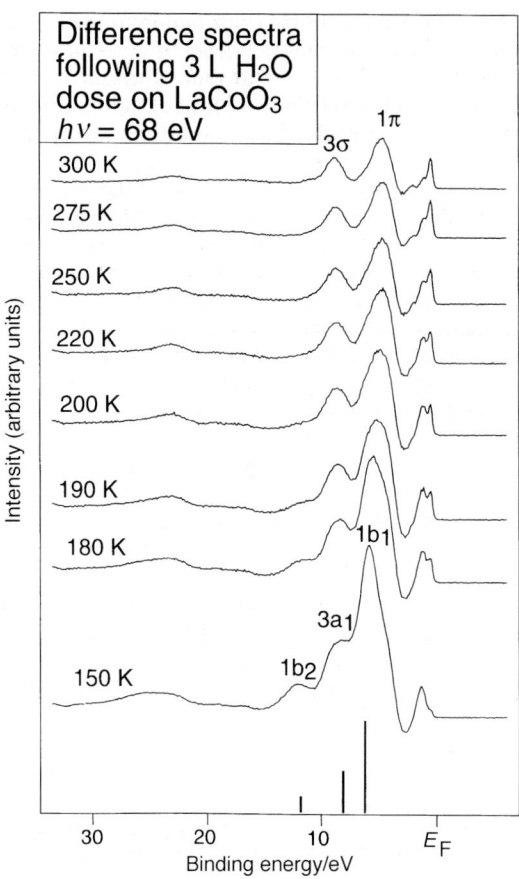

Fig. 10 Difference spectra (dosed surface − undosed surface) obtained from the data shown in Fig. 9, following water dosing and warming a LaCoO$_3$ (111) surface. The spectra have been normalised to feature C before subtraction. Assignments to the molecular orbitals of H$_2$O and OH are given. The molecular orbital positions for gas phase water are shown for comparison. The gas phase BE positions have been shifted so that the binding energy position of the gas phase 1b$_1$ peak has been aligned to the 1b$_1$ orbital of the adsorbed H$_2$O.

after dosing is also shown. Fig. 10 shows the difference spectra obtained by subtracting the undosed from the dosed spectrum at each temperature, after aligning the spectra at the La 5p core level signals in the range 16–20 eV binding energy. (This is done to correct for any changes in the workfunction of the sample on dosing; we were unfortunately unable to bias the sample during data recording.) In Fig. 10, three peaks at binding energies of 5.8 eV, 8.1 eV and 12.1 eV are evident in the difference spectrum taken at 150 K, together with a feature in the vicinity of B at around 1 eV. At higher binding energy, a broad peak at around 24–25 eV binding energy is also observed, due to O 2s emission from the dosed water. The appearance of three valence band adsorbate peaks, two of which lie below the valence band of the substrate is typical for nondissociative adsorption of water on metal oxides.[24] By comparison with data for other oxides,[23] the three features in the range ca. 6–12 eV may be assigned to the 1b$_1$, 3a$_1$ and 1b$_2$ molecular orbitals of undissociated water, respectively. The peak separations $\Delta(3a_1 - 1b_1)$ of 2.3 eV and $\Delta(1b_2 - 3a_1)$ of 4.0 eV compare well with those of gas phase water (2.1 eV and 3.8 eV, respectively[25]). It is now well established that in the case of strong chemisorption to the surface, the 3a$_1$ molecular orbital of the adsorbed water experiences a shift to higher binding energy relative to the other orbitals compared with its position in gas phase water.[23] This shift may be up to 1.3 eV.[26] The small shift here suggests that water is weakly chemisorbed to the surface at low temperature. Adsorption, followed by warming the surface causes a loss of spectral resolution, which may be associated with phonon broadening. This is particularly apparent for sharp features of the spectrum, such as feature B, which can be seen to be considerably broadened on warming (Fig. 9). This in turn leads to generation of artefacts when the difference spectra are obtained (Fig. 10, and also Fig. 3). Thus the features observed in the difference spectra in the vicinity of B are due to a combination of miscancellation due to broadening on adsorption, on heating, and changes in the spin state of the sample.

On warming the surface, further changes are seen in the valence band spectra. In the difference spectrum, the three features due to water adsorption are gradually replaced by two features at binding energies of 4.7 eV and 8.8 eV, and the absolute intensity of the difference spectrum decreases. We associate these features with the 1π and 3σ orbitals of adsorbed OH groups, which appear at 4.9 eV and 9.0 eV binding energy in the spectra of the related layered perovskite $La_{2-x}Sr_xCuO_4$.[27] Thus, the adsorbed water dissociates as the sample is warmed, leaving surface hydroxide, which is not completely desorbed even at room temperature. The temperature at which dissociation occurs is difficult to estimate from the difference plots, but intensity from the $1b_2$ molecular orbital of water disappears completely between 190 K and 200 K. Careful examination of the initial difference spectrum taken at 150 K shows that the $3a_1$ signal is rather broad, and the $1b_1$ signal has a low binding energy shoulder, suggesting that some OH may co-exist with H_2O on this surface even at low temperature. Recent elegant *ab initio* calculations have demonstrated that this may occur on oxide surfaces.[28] The binding energy positions and energy separation (4.1 eV) of the OH features are in very good agreement with the difference spectrum of the degraded $LaCoO_3$ surface (Fig. 3), where the two main features are also separated by 4.1 eV. We therefore conclude that the degradation of the surface sometimes observed as a function of time in UHV is due to the formation of surface OH, presumably through adsorption of H_2O from the residual vacuum. The dissociation of water to give OH may be enhanced at the step and defect sites associated with a poor cleave. Preferential dissociation on stepped oxide surfaces has been previously observed for related perovskite oxides, such as $SrTiO_3$.[29]

4. Conclusions

Clear temperature effects are seen in the valence band photoemission spectra of single crystal $LaCoO_3$, which may be attributed to changes in the spin state of the sample. The most obvious change is a gradual transition from the LS to other spin states as the temperature is raised. This is clearly identified by a delayed resonance onset in the CIS spectra. At low temperatures the CIS excitation energy for features associated with low spin states is found to be 3 eV higher than that for IS and HS features, roughly in accord with the 3 eV gap between the t_{2g} and e_g states calculated for $LaCoO_3$.[15] Our data clearly show that the lowest binding energy feature in the spectra (feature B) is associated predominantly with the LS state, but that it also contains contributions from higher spin states. CIS spectra recorded as a function of both binding energy and temperature have allowed us to identify the binding energy range 0.7–1.2 eV as predominantly LS, and the range 1.7–2.2 eV as predominantly higher spin states. The gross features of our spectra show features which correspond well with cluster calculations for the LS, HS and IS states, but importantly, our data show that the LS contribution is the *lowest* binding energy feature and not IS–HS as suggested by cluster calculation.[8] The binding energy difference between the LS and HS–IS states is around 1 eV.

While the LS state is clearly identified by its delayed resonance, neither HS nor IS states should show a delayed onset. It is therefore difficult to distinguish these two spin states on the basis of their CIS spectra. However, certain features of the EDC spectra are consistent with the existence of an IS state, and with a further gradual transition from it into a HS state as the temperature is raised. In particular, the relative changes in intensity in features B, C and D as a function of temperature appear consistent with a three state spin model involving the IS state.[7] In contrast the 10% Sr-doped compound is found to be in the same spin state at all temperatures, thought to be an IS spin state,[18] as indicated by the lack of resonance shifts and the unchanging valence band spectra observed as a function of temperature.

Adsorption measurements show that water is chemisorbed on the $LaCoO_3$ (111) surface at low temperatures (possibly with some dissociation). As the sample temperature is raised, the water dissociates completely when the temperature reaches 200 K. The resulting OH does not desorb completely at room temperature. The spectrum is strikingly similar to that from a poorly cleaved surface which has spent some time in UHV, and we therefore conclude that the surface degradation observed on poorly prepared surfaces is associated with hydroxylation of the surface by water in the residual vacuum. The identification of the signature of the degradation product allows us to distinguish these effects from effects due to spin changes in the EDC spectra as a function of temperature.

5. Acknowledgements

We would like to thank John Campbell of CLRC Daresbury Laboratory for the Laué simulations, and S. Murata at the University of Electrocommunications for growing the crystals. This work was funded by EPSRC (UK) and partly by scientific research grant/08640449 from the ministry of education, sports and culture, Japan.

References

1. M. Hrovat, N. Katsarakis, K. Reichmann, S. Bernik, D. Kuscer and J. Holc, *Solid State Ionics*, 1996, **83**, 99.
2. See, *e.g.*, T. Nitadori, S. Kurihara and M. Misono, *J. Catal.*, 1986, **98**, 221.
3. S. Yamaguchi, H. Taniguchi, H. Takagi, T. Arima and Y. Tokura, *J. Phys. Soc. Jpn.*, 1995, **64**, 1885.
4. G. Briceno, X.-D. Xiang, H. Chang, X. Sun and P. G. Schultz, *Science*, 1995, **270**, 273.
5. V. Golovanov, L. Mihaly and A. R. Moodenbaugh, *Phys. Rev. B: Condens. Matter*, 1996, **53**, 8207.
6. M. Itoh, M. Sagahara, I. Natori and K. Motoya, *J. Phys. Soc. Jpn.*, 1995, **64**, 3967.
7. K. Asai, A. Yoneda, O. Yokokura, J. M. Tranquada, G. Shirane and K. Kohn, *J. Phys. Soc. Jpn.*, 1998, **67**, 290.
8. T. Saitoh, T. Mizokawa, A. Fujimori, M. Abbate, Y. Takeda and M. Takano, *Phys. Rev. B: Condens. Matter*, 1997, **55**, 4257.
9. M. Abbate, J. C. Fuggle, A. Fujimori, L. H. Tjeng, C. T. Chen, R. Potze, G. A. Sawatzky, H. Eisaki and S. Uchida, *Phys. Rev. B: Condens. Matter*, 1993, **47**, 16124.
10. G. Thornton, B. C. Tolfield and A. W. Hewat, *J. Solid State Chem.*, 1986, **61**, 301.
11. G. Thornton, F. C. Morrison, S. Partington, B. C. Tolfield and D. E. Williams, *J. Phys. C: Solid State Phys.*, 1988, **21**, 2781.
12. G. Thornton, I. W. Owen and G. P. Diakun, *J. Phys.: Condens. Matter*, 1991, **3**, 417.
13. K. Asai, O. Yokokura, N. Nishimori, H. Chou, J. M. Tranquada, G. Shirane, S. Higuchi, Y. Okajima and K. Kohn, *Phys. Rev. B: Condens. Matter*, 1994, **50**, 3025.
14. J. B. Goodenough, *Mater. Res. Bull.*, 1971, **6**, 967.
15. M. A. Korotin, S. Yu. Ezhov, I. V. Solovyev, V. I. Anisimov, D. I. Khomski and G. A. Sawatzky, *Phys. Rev. B: Condens. Matter*, 1996, **54**, 5309.
16. R. Potze, G. A. Sawatzky and M. Abbate, *Phys. Rev. B: Condens. Matter*, 1995, **51**, 11501.
17. M. Takahashi and J.-I. Igarashi, *Phys. Rev. B: Condens. Matter*, 1997, **55**, 13557.
18. T. Saitoh, T. Mizokawa, A. Fujimori, M. Abbate, Y. Takeda and M. Takano, *Phys. Rev. B: Condens. Matter*, 1997, **56**, 1290.
19. Y. Taguchi, K. Ichikawa, T. Katsumi, K. Jouda, Y. Ohta, K. Soda, S. Kawamata, K. Okuda and O. Aita, *J. Phys.: Condens. Matter*, 1997, **9**, 6761.
20. A. Chainani, M. Mathew and D. D. Sarma, *Phys. Rev. B: Condens. Matter*, 1992, **46**, 9976.
21. Z.-X. Shen, J. W. Allen, P. A. P. Lindberg, D. S. Dessau, B. O. Wells, A. Borg, W. Ellis, J. S. Kang, S.-J. Oh, I. Lindau and W. E. Spicer, *Phys. Rev. B: Condens. Matter*, 1990, **42**, 1817.
22. S. Masuda, M. Aoki, Y. Harada, S. Hirohashi, Y. Watanabe, Y. Sakisaka and H. Koto, *Phys. Rev. Lett.*, 1993, **25**, 4214.
23. W. R. Flavell, J. H. Laverty, D. S.-L. Law, R. Linsay, C. A. Muryn, C. F. J. Flipse, G. N. Raiker, P. L. Wincott and G. Thornton, *Phys. Rev. B: Condens. Matter*, 1990, **41**, 11623.
24. S. Eriksen, P. D. Naylor and R. G. Egdell, *Spectrochim. Acta, Part A*, 1987, **43**, 1535.
25. K. Siegbahn, *J. Electron Spectrosc. Relat. Phenom.*, 1974, **5**, 1.
26. F. H. Potter and R. G. Egdell, *Surf. Sci.*, 1993, **297**, 286.
27. R. L. Kurtz, R. Stockbauer, T. E. Madey, D. Mueller, A. Shih and L. Toth, *Phys. Rev. B: Condens. Matter*, 1988, **37**, 7936.
28. P. J. D. Lindan, N. M. Harrison and M. J. Gillan, *Phys. Rev. Lett.*, 1998, **80**, 762.
29. N. B. Brookes, G. Thornton and F. M. Quinn, *Solid State Commun.*, 1987, **64**, 383.

Paper 9/02409C

QM investigations on perovskite-structured transition metal oxides: bulk, surfaces and interfaces

F. Corà* and C. R. A. Catlow

Davy-Faraday Research Laboratory, The Royal Institution of Great Britain, 21 Albemarle Street, London, UK W1X 4BS. E-mail: furio@ri.ac.uk

Received 7th June 1999

We present the results of electronic-structure, *ab initio* calculations on a set of perovskite-structured transition metal oxides, in which the transition metal ion has electronic configuration $d^{(0)}$. We perform an analysis of the QM solution for the bulk materials, based on a phenomenological, tight binding-like examination of the band structures in reciprocal space. This treatment allows us to understand the trends in the properties of bulk perovskites as a function of their chemical composition; a parameter is defined, easily calculated from the band structure of the cubic phase, that controls the extent of covalence in the M–O interaction. Ferroelectric-like distortions from the cubic phase are seeded by a symmetry breaking around either a M or an O ion of the structure; the electronic perturbation is then transferred to the neighbouring sites *via* a delocalisation of the π M–O bonding levels in the valence band. Investigations on the $\langle 001 \rangle$ surface termination of $BaTiO_3$ and WO_3 show that the electronic perturbation induced by the surface can couple with the ferro-/antiferro-electric (FE/AFE) distortional modes of the bulk materials in the surface and sub-surface regions. We have been able to attribute to different surface terminations an FE or AFE character. Finally, *via* the design of suitable perovskite/perovskite interfaces, we have combined materials with FE bulk behaviour with AFE surface terminations, and *vice versa* AFE bulk materials with FE surfaces. Our results show that the interface strain may significantly alter the behaviour of the support. The complexity of the systems investigated is already comparable to applications of FE materials in the field of computer memories.

1. Introduction

The perovskite structure is a common polymorph in the solid-state chemistry of transition metal oxides. A perovskite-structured material, hereafter referred to as perovskite (PV), is schematically displayed in Fig. 1; it has general stoichiometry AMO_3, and is formed by a network of corner-sharing MO_6 octahedra, centered on the transition metal ions M. All the oxygens are in topologically equivalent, 2-coordinate positions, bridging the two adjacent M sites. The dodecahedral interstices of the oxygen sublattice are occupied by a second cation A, and the cationic charge can be distributed in different ways between octahedral and dodecahedral sites. In the paper, we shall refer to $(n - n')$ perovskites as to those in which the octahedral cation has charge $+n$, and the dodecahedral $+n'$; for stoichiometry requirements, $n + n' = 6$. Examples are $KNbO_3$ (5-1), $CaTiO_3$ (4-2), $LaMnO_3$ (3-3). Under this classification, the binary oxides WO_3 and β-MoO_3,

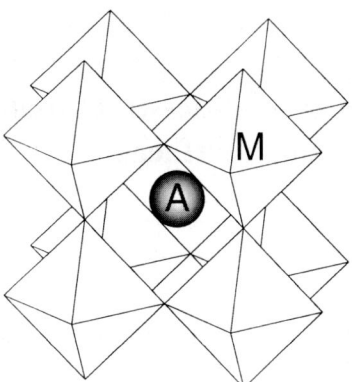

Fig. 1 Network of corner-sharing MO$_6$ octahedra in the perovskite structure.

normally referred to as ReO$_3$-structured, can be seen as a special case of perovskite, in which the dodecahedral sites are vacant. We refer to them as (6-0) perovskites.

PVs containing transition metal (TM) cations are of interest for applications in electronics; the properties of PVs range from the ferro- and piezoelectric behaviour of PbTiO$_3$ and KNbO$_3$, to magnetic (SrRuO$_3$[1]) and superconducting oxides (PV layers are one of the building blocks of the high-T_c oxides and of Sr$_2$RuO$_4$[2]), to the colossal magneto-resistance effect of doped LaMnO$_3$.[3] Furthermore, the conductivity properties of PVs range from metallic (NaWO$_3$ and SrNbO$_3$[4]) to semiconducting (KNbO$_3$ and WO$_3$[5]) and insulating (CaTiO$_3$[6] and LaGaO$_3$). The isostructural arrangement of ions in all the above materials makes them compatible and miscible with one another, opening the way to novel applications, based on the combination of two or more of the cited properties. Rationalising the physico-chemical behaviour of different PVs is crucial to the design of new materials suitable for desired applications, and a field where theoretical investigations can play an important rôle.

In this paper we focus first on the issue of ferroelectricity, and compare the behaviour of a set of PVs, in which the TM ion has formal electronic configuration d$^{(0)}$; and secondly on the surface properties of ferroelectric PVs. Ferroelectric (FE) materials have non-centrosymmetric crystal structures, in which the baricentre of positive and negative charges in each unit cell does not coincide. Such atomic configuration gives rise to a non-zero dipole moment in the unit cell, and to a macroscopic polarisation P_s of the sample if the FE state is long-range ordered. In PVs, ferroelectricity is linked to the off-centering of the TM ions in their coordination octahedra.

Furthermore, FE materials have two stable minima with opposite directions of P_s, between which the system can be switched by applying external electric fields. The presence of two stable states (which can be labelled + and −, or 0 and 1), recalls the Boolean algebra based on the binary (0 and 1) arithmetics on which computers are built. FE compounds are therefore researched for applications in computer memories, in which the single bit of information can be associated with the direction of polarisation of a single FE microcrystal.[7] To obtain gigabite memories, the components must be properly miniaturised: sub-micrometre, possibly even only a few nanometres thick, single crystals of FE materials need to be employed to store information, in conjunction with nanometre thin electrodes which provide a stable contact with the FE capacitor microcrystals. The latter are required to control (i.e., read and write) the polarisation verse of the FE unit.

Apart from the technological difficulty in manufacturing such devices, important questions remain open concerning the basic science underlying the materials employed, such as the minimum thickness of the FE layer which can still be technologically useful, and the finite-size effect on the materials properties.[7] Understanding the atomic-scale details of the metal/capacitor junction, to find combinations that do not induce stresses in the materials, and hence guarantee reliability and durability of the device, is another field where fundamental research is needed.

In the limit of very thin components, the influence of surfaces and interfaces on the overall properties is maximised. Advancing in the fields referred to above requires, therefore, a better understanding of the surface and interface properties of the materials employed. The required dimensions of the devices are such that state-of-the-art quantum-mechanical examinations of model materials are already directly comparable to the experimental applications.

PV-structured transition metal oxides include both FE materials, such as $KNbO_3$ and $BaTiO_3$, and metallic compounds ($SrNbO_3$, $BaNbO_3$, ReO_3). They represent therefore a suitable candidate for the above application in computer memories. Combining the properties of different PV-structured oxides, would allow an all-oxide technology, where, for instance, an isostructural conductor and FE capacitor (such as $BaNbO_3$ and $BaTiO_3$) could provide a smooth and stable interface. The properties of such an interface, in the limit of very thin layers for both metal and FE material, are however largely unknown.

In this contribution, we examine the above topics with *ab initio* Hartree-Fock (HF) calculations, to show how the FE behaviour of PVs depends on the chemical composition and on the *local* symmetry of both M and O ions of the structure.

2. Computational details

In the calculations reported, the solids are represented *via* periodic boundary conditions; the electronic distribution is described in terms of crystalline orbitals, obtained as linear combination of localised functions, or atomic orbitals (AOs), associated with lattice sites. To provide a better comparability of results, care has been devoted to select basis sets of similar quality for all the TM ions. We adopted a basis of at least split-valence quality for each ion, coupled with a Hay-Wadt effective core pseudopotential[8] for the TM ions: small-core for 3d and 4d metals (Ti, Nb, Mo) and large-core for the 5d elements (W, Re). The code employed is CRYSTAL.[9,10]

For the bulk materials, all the computational tolerances, such as the selection of integrals and the sampling of reciprocal space, have been fully converged; the calculations performed represent the best description within the Hamiltonian and basis-set chosen.

Surfaces and interfaces have been described with a slab model, a layer periodic in two dimensions and of finite thickness. The complexity and cost of the surface calculations increase rapidly compared to the bulk study. In the surface study, the computational tolerances have been eased (the two-electron integrals are calculated exactly if the overlap between the AOs involved is greater than 10^{-5} in the surfaces and 10^{-8} in the bulk oxides) to afford the simulation of slabs with increased thickness. The reference solution for the bulk materials, against which surfaces are compared in Sections 4 and 5, has been recalculated with the lower computational tolerances.

3. Results and discussion: bulk PVs

In order to investigate the properties of bulk PVs, we describe first the cubic phase, and then the phase transition from the cubic to the tetragonal ferro-or antiferro-electric (FE/AFE) structure. The latter is obtained by displacing the M ions towards one of their nearest neighbour oxygens.

3.1. Cubic PVs

Using periodic HF calculations, as described in Section 2, we have optimised the geometry and calculated the equilibrium electronic distribution for several cubic PVs. A summary of results is reported in Table 1.

The topology of the band structures in reciprocal space is similar for all the materials investigated; band-gaps and widths, instead, vary with the composition. In Fig. 2, we report the HF band-structure for two of the PV materials examined, the (6-0) WO_3 and the (4-2) $CaTiO_3$, which represent the two extremes in the set of perovskites studied, as will be clarified in the following discussion.

To analyse the *ab initio* solution, and highlight how the chemico-physical behaviour of PVs depends on the composition, we have performed a standard Mulliken population analysis of the electronic density; also we have constructed the band-structure of a cubic PV, based on a tight-binding (TB) interaction pattern between the frontier π AOs on nearest and next-nearest neighbour ions in the structure. The TB solution depends on few *effective* interaction parameters, which can be chosen with a direct physical or chemical meaning; it is therefore a useful tool to analyse the more complex solution obtained *via* fully *ab-initio* QM studies.

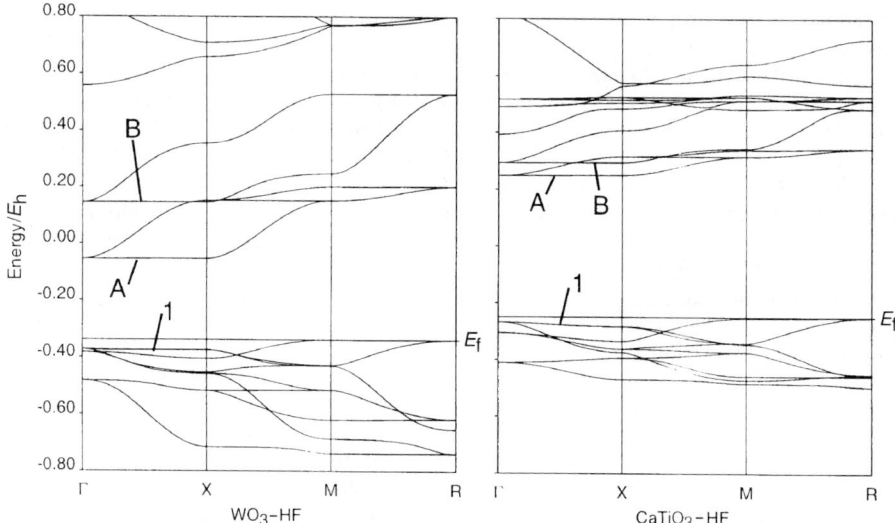

Fig. 2 HF band-structure for cubic WO_3 and $CaTiO_3$, along the Γ–X–M–R path in reciprocal space.

In the discussion, we make reference to the effective interaction parameters shown in Fig. 3(a): the diagonal elements, α; in particular α_d, relative to the M d(t_{2g}) AOs, and α_p for the O 2p(π) AOs (*i.e.* the 2p AOs on each oxygen, perpendicular to the M–O–M direction); β, between M d(t_{2g}) and O 2p(π) AOs on nearest neighbour ions; and γ, between the 2p(π) AOs on next-nearest neighbour oxygens. We further assume that the dodecahedral cation A behaves as a perfect ion in the crystal structure, whose only rôle is that of charge compensation; in other words, that there is no covalent interaction between A and the surrounding oxygens. For alkali or alkaline-earth A cations, the above assumption is justified; we shall not however examine PVs containing Pb as dodecahedral ion, such as $PbTiO_3$ or $PbZrO_3$.

By examining how the TB band-structure is modified upon a change of the parameters, and by comparing the result with the HF solution, we can understand the importance of the interaction represented by each TB parameter in the solid-state chemistry of PVs. Moreover, since the TB and HF solutions match closely, the TB parameters that best reproduce the HF band-structure can be used to analyse *quantitatively* the ab-initio solution. An extensive discussion of the TB derivation and of the TB-based phenomenological analysis of the HF solution can be found in ref. 11, and will be presented elsewhere; only the major conclusions are discussed here.

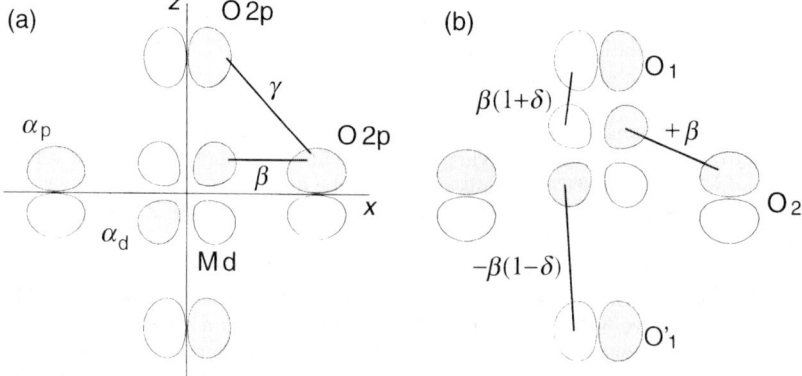

Fig. 3 Valence π AOs on metal and oxygens, and their TB interaction parameters for (a) the cubic and (b) the tetragonal-FE phases of bulk PVs.

(1) In the origin of reciprocal space, Γ, the TB Hamiltonian in the AO basis set is diagonal. The solutions coincide with the pure O $2p(\pi)$ and M $d(t_{2g})$ ionic orbitals, of energy α_p and α_d. No M–O covalence is present in Γ, and the M–O interaction in the crystal occurs *via* their ionic charge only. The splitting of levels, $\Delta\alpha = \alpha_d - \alpha_p$, is due only to the crystal (Coulomb) field.

The transition metal has two equidistant oxygen neighbours in each coordinate direction (x, y and z); similarly, each oxygen is exactly half the way between its two nearest M cations. The chemical contributions from the two neighbours have opposite sign ($+\beta$ and $-\beta$), and cancel reciprocally: covalence is symmetry forbidden. The same symmetry constraint applies whenever the *local* environment of M and O ions has the same symmetry features described here; conversely, when the symmetry is broken, either on the metal or on the oxygen site, M–O covalence is no longer forbidden.

(2) The solution along the Γ-X direction of reciprocal space consists of combinations of the M d and O 2p levels, of M–O bonding character in the valence band, whose levels are stabilised at lower energy compared to the solution in Γ (where $e = \alpha_p$), and M–O antibonding character in the conduction band, whose levels are shifted at higher energy. The Γ-X band-width, *i.e.*, the energy effectiveness of M–O covalence, equals $4\beta^2/\Delta\alpha$. For symmetry constraints, the two energy levels labelled 'A' and '1' in Fig. 2 remain non k-dependent and M–O non-bonding. They have been called *superdegenerate* in the literature.[12]

(3) The inter-oxygen parameter, γ, enters the TB solution *only* in the X-M direction of reciprocal space. Neglecting the O-O interaction ($\gamma = 0$), one of the energy solutions is still equal to α_p. The other eigenstates correspond to bonding and antibonding combinations of M d and O 2p states; the extent of hybridisation and the energy change due to M–O covalence in M is twice as large as in the X point, and equals $8\beta^2/\Delta\alpha$.

If $\gamma \neq 0$, the inter-oxygen interaction resolves the *superdegeneration* of having $e_1 = \alpha_p$ in the X-M direction of reciprocal space. In particular, an O-O repulsion causes a destabilisation of the O 2p levels at the top of the valence band.

3.1.1. Physical meaning of the TB parameters.

The parameters α_d and α_p represent the energy levels in a purely ionic solution. Since metal and oxygen ions are not isolated, but ordered in a regular lattice, the values of α_d and α_p will be shifted by an amount proportional to the intensity of the crystal field in the M and O positions, which is proportional to the net ionic charges. The value of $\Delta\alpha = \alpha_d - \alpha_p$ is therefore associated with the degree of ionicity in the solid.

The hybridisation of AOs, and their energy change due to M–O covalence is proportional to $\beta^2/\Delta\alpha$. The latter parameter represents the degree of covalence in the solid. We also see that $\Delta\alpha$ is a compatibility parameter that measures the tendency of the M d and O 2p AOs towards covalent bonding; the lower $\Delta\alpha$, the more effective covalence is in the M–O interaction. Covalence ($\beta^2/\Delta\alpha$) and ionicity ($\Delta\alpha$) are therefore complementary: the greater the ionicity of the solid, the higher the splitting of ionic levels due to the crystal field ($\Delta\alpha$), and the lower the importance of covalence, and *vice versa*.

The parameter γ is relative to the next-nearest-neighbour interactions. The number and relative orientation of next-nearest-neighbour oxygens will be different in different structures; we identify γ as a structural parameter of the solid.

Having available a calculated band structure for a cubic PV, the values of the ionicity parameter $\Delta\alpha$ can be derived from the energy spectrum in Γ, which does not depend on β and γ. Moreover, the k-dependence of the band structure along Γ-X is a function only of $\beta^2/\Delta\alpha$, which can be quantified from the valence band width along Γ-X. Finally, γ enters only in the dispersion of the one-electron energy levels along the X-M direction, which can be employed to evaluate the importance of the structural constraints, γ.

By careful analysis of the *ab-initio* band structure, calculated for a cubic PV-structured transition metal oxide, we can effectively separate the three TB parameters employed, but also the three components of bonding in the solid state chemistry. The TB solution, therefore, provides us with a mathematical framework, with which we can highlight and compare the importance of different forces in the solid-state chemistry of PV materials. Such a framework can be employed to analyse in a *quantitative* way the *ab-initio* results. In Table 1, we report a summary of results, relative to the HF calculations; the symbols used refer to the equilibrium lattice parameter (a in Å)

Table 1 HF results, concerning the electronic distribution of cubic PV materials at their optimised lattice parameter, a, and at a common lattice spacing, a_0

Material	a	Mulliken population analysis					Tight-binding analysis			
		$q(O)$	$q(e_g)$	$q(t_{2g})$	$q_b(MO)$	$q_b(OO)$	$\Delta\alpha$	β	γ	$\beta^2/\Delta\alpha$

Optimised lattice parameter—

					(6-0)					
MoO_3	3.78	−0.9978	0.660	0.397	0.031	−0.056	0.2995	−0.1263	−0.0162	0.0533
WO_3	3.77	−1.3625	0.676	0.403	−0.027	−0.039	0.3173	−0.1301	−0.0168	0.0533
					(5-1)					
$LiNbO_3$	3.96	−1.3851	0.551	0.248	0.025	−0.008	0.4125	−0.1166	−0.0161	0.0330
$NaNbO_3$	3.98	−1.3736	0.547	0.249	0.031	−0.008	0.3998	−0.1140	−0.0157	0.0325
$KNbO_3$	4.03	−1.3948	0.538	0.247	0.039	−0.007	0.3915	−0.1087	−0.0151	0.0302
					(4-2)					
$CaTiO_3$	3.88	−1.5820	0.332	0.108	0.009	−0.026	0.5148	−0.1205	−0.0204	0.0282
$SrTiO_3$	3.93	−1.6093	0.326	0.106	0.012	−0.022	0.5052	−0.1162	−0.0195	0.0267
$BaTiO_3$	4.02	−1.5714	0.313	0.103	0.021	−0.015	0.4748	−0.1017	−0.0188	0.0218

Common lattice parameter, $a_0 = 3.9225$ Å—

					(6-0)					
MoO_3		−1.0141	0.640	0.394	0.075	−0.036	0.2741	−0.1144	−0.0135	0.0477
WO_3		−1.3106	0.628	0.393	0.051	−0.024	0.2999	−0.1186	−0.0138	0.0469
					(5-1)					
$LiNbO_3$		−1.3822	0.559	0.252	0.016	−0.010	0.4156	−0.1195	−0.0167	0.0344
$NaNbO_3$		−1.3718	0.559	0.256	0.018	−0.011	0.4141	−0.1193	−0.0167	0.0344
$KNbO_3$		−1.3868	0.560	0.259	0.016	−0.011	0.4079	−0.1168	−0.0171	0.0334
					(4-2)					
$CaTiO_3$		−1.5880	0.325	0.105	0.014	−0.023	0.5092	−0.1168	−0.0194	0.0268
$SrTiO_3$		−1.6090	0.326	0.106	0.012	−0.022	0.5062	−0.1166	−0.0196	0.0267
$BaTiO_3$		−1.5639	0.326	0.110	0.011	−0.020	0.4837	−0.1062	−0.0211	0.0233

(symbols have the meaning and units defined in the text)

and charges derived from a Mulliken population analysis (q, in electrons). For the Mulliken charges, we report the net charge on the oxygens, $q(O)$, the bond-population between the transition metal and the nearest oxygens, $q_b(M-O)$, and between next nearest neighbour oxygens, $q_b(O-O)$. The three q values above have a direct correspondence with the three TB parameters $\Delta\alpha$, $\beta^2/\Delta\alpha$ and γ. In Table 1, we also report the values of the TB parameters ($\Delta\alpha$, $\beta^2/\Delta\alpha$ and γ, in atomic units), that best reproduce the HF band-structures.

The values reported in the upper part of Table 1 refer to the equilibrium geometry for the *ab-initio* Hamiltonian employed. On the other hand, the interaction parameters depend on the overlap between AOs, and will change by changing the lattice spacing of the material. The equilibrium lattice parameters of the PVs considered range from 3.77 to 4.03 Å, a wide variation, which can introduce appreciable changes in the relevant TB parameters. To make the comparison of the electronic distribution between different compounds more direct, we have repeated all the calculations at a common lattice spacing, $a_0 = 3.9225$ Å. The TB parameters fitted to the *ab-initio* band structure and the Mulliken charges at lattice parameter a_0 are collected in the bottom part of Table 1.

In the following discussion, we refer to the values calculated at lattice parameter a_0. To highlight the trends in the properties of PV materials, we imagine the solid as formed in two subsequent steps: first by creating an array of non interacting oxygen and metal ions in their crystalline positions; the crystal (Coulomb and exchange) field generated in each other's positions causes a splitting of the ionic energy levels. The actual energy levels obtained in this first step can be derived from the band-structure in the Γ point, whose solutions (the superdegenerate levels A and 1) correspond to the non-interacting ionic orbitals. The ionic energy levels provide a reference zero, according to which the intensity of the chemical (covalent) interactions between metal and oxygens can be measured. The second step consists in 'switching on' the M–O chemical interaction; the ground-state wave function of the system is hybridised by M–O covalence into a linear combination of the metal and oxygen AOs; this combination is of bonding character in the valence band, which is stabilised at lower energies, and antibonding in the conduction band, which is destabilised at higher energy values.

Before examining the results, it is worth recalling that HF Hamiltonians, such as the one employed in the present study, overestimate the energy of empty states, which will of course influence the calculated band gaps. We can, however, assume this error to be systematic when the computational conditions employed are identical for all the oxides investigated, so that comparisons of results for different compounds have quantitative meaning.

Let us now examine some of the trends:

(1) *$\Delta\alpha$ vs. formal charge of M*. As we see from Table 1, $\Delta\alpha$ increases considerably from (6-0) to (5-1) and (4-2) perovskites, of about 0.1 E_h (2.7 eV) in each step. In the sequence (6-0) → (5-1) → (4-2), the cationic charge is progressively displaced from the transition metal site to the dodecahedral interstices, modifying the Coulomb field and the relative energy of the superdegenerate levels A and 1. The energy of level A, which is a metal d AO, is more sensitive to the nuclear charge of M than is the energy of level 1, which is an O 2p state; $\Delta\alpha$ is therefore lowest in the (6-0) materials and highest in the (4-2). This comparison suggests that the value of the band gap in cubic PVs is controlled mostly by the electrostatic forces acting on the ions, that is by the value of the crystal field in the metal and oxygen positions; in particular it depends on the fraction of the total cationic charge which is located in the transition metal site.

(2) *$\Delta\alpha$ vs. principal quantum number of M*. Let us consider two materials with same partition of charge between octahedral and dodecahedral sites, but a different transition metal ion in the octahedral site, such as Mo in β-MoO$_3$ and W in WO$_3$. The behaviour is now differentiated by the principal quantum number of the d AOs: the orbitals involved are the 4d in Mo, the 5d in W, and $\Delta\alpha$ is lower in MoO$_3$ than it is in WO$_3$ (0.274 vs. 0.300 E_h in the present HF studies). The influence on $\Delta\alpha$ of the principal quantum number is lower than that of a different partition of the positive charge between the two cationic sites, A and M.

(3) *$\Delta\alpha$ vs. dodecahedral cation, A*. If we consider the lithium, sodium and potassium niobates, we see that at a common lattice parameter, the value of $\Delta\alpha$ is fairly constant (the change is of the order of the mE_h). This feature supports the argument that the only rôle of the dodecahedral cation is that of charge neutralising the structure.

(4) *Energy effectiveness of covalence, $\beta^2/\Delta\alpha$*. TM cations form a uniform class, in which the values of β are roughly constant. The importance of covalence ($\beta^2/\Delta\alpha$) follows therefore the changes in $\Delta\alpha$, and decreases in the sequence of (6-0) > (5-1) > (4-2) materials. It also decreases on increasing the principal quantum number of the transition metal ion occupying the octahedral sites of the lattice.

The net charge of the oxygen ions in the structure also depends on $\beta^2/\Delta\alpha$: in the ionic solution (the first in the two-step formation of the solid, corresponding to O^{2-} ions), the valence band is formed only by O 2p states; the more covalence is important in the bonding, the more the M d(t_{2g}) AOs are involved in the description of the electronic charge in the valence band, leading therefore to less ionic oxygens.

(5) *Inter-oxygen repulsion, γ*. The TB parameter γ represents the interaction between next-nearest-neighbour oxygens in the structure. We note in Table 1 that the absolute value of γ constantly increases when moving from (6-0) to (5-1) and (4-2) materials. As we have seen in the discussion above, the net ionic charge of the oxygen ions in the structure increases in the same order; the parallel between the two trends can be explained by the increase in the effective size of the oxygen ions with an increase in their net charge, which in turn causes an increase of their effective interaction parameter γ.

The latest finding suggests therefore that in the balance of forces present in the structure, the two-body repulsion among next-nearest-neighbours is not an independent effect, but reacts to changes in the balance between ionicity and covalence in the M–O bonding. In other words, γ depends on the value of $\Delta\alpha$. Although the absolute value of γ is lower than that of β, the multiplicity of next-nearest-neighbour interactions is higher than the multiplicity of the nearest-neighbour ones, and the inter-oxygen repulsion expressed by γ is a non-negligible force, that has to be taken properly into account in the overall description of the material.

The above comparisons show that in the sequence of (6-0) → (5-1) → (4-2) perovskites, we move from soft and highly covalent to hard and very ionic materials, in agreement with experiment. Changing the partition of positive charge between the two cationic sites A and M, we can modulate the value of $\Delta\alpha$, and hence the balance between covalence and ionicity in the solid.

The results above are qualitatively confirmed by a Mulliken population analysis of the electronic distribution, whose values are also summarised in Table 1. To make the analogy between the results of the TB-based and of the Mulliken population analysis more direct, in Fig. 4 we plot the variation of several quantities, relative to the strength of covalence in the M–O bonding: $\beta^2/\Delta\alpha$ from the phenomenological analysis; $2-|q(O)|$, $q_b(M-O)$, $q(e_g)$ and $q(t_{2g})$ from the Mulliken population analysis. The scale of the plot has been chosen in such a way that, for each of the quantities plotted, the value for the (6-0) materials has unitary value; the (5-1) and (4-2) values are scaled correspondingly. In each of the groups (6-0), (5-1) and (4-2), the value plotted is the average of all those reported in Table 1.

All the quantities reported in Fig. 4 have the same trend: the values derived from the Mulliken population analysis are subject to a higher uncertainty, due to the arbitrariness in the partition of charges; overall they are dispersed around the value derived from the TB-based analysis, which appears to provide the best reference, and the most accurate quantitative information to describe the trends in the physico-chemical behaviour of the set of materials examined.

3.2. FE/AFE distortions in bulk PVs

We now examine the electronic rearrangement that accompanies a displacement off-centre of the transition metal ion M in its octahedron, for instance towards one of its six nearest oxygens (that in the $+z$ direction). If the M displacements in the structure are long-range ordered, we obtain the FE and AFE phases: in an FE material all the M ions displace uniformly with respect to the O sublattice; in the AFE the M ions displace in alternate directions. The local environment of the M

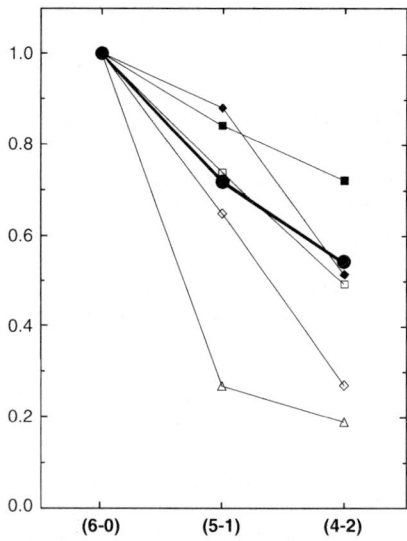

Fig. 4 Variation of parameters representing the strength of covalence in cubic $d^{(0)}$ perovskites, and derived from both the phenomenological and the Mulliken population analysis of the *ab initio* results, as a function of the partition of the nuclear charge in the materials. The large filled circles and the heavy line refer to the value of $\beta^2/\Delta\alpha$ obtained from the TB-based analysis; the thin lines to the values derived from the Mulliken population analysis: the filled and open squares to $1/q(O)$ and $2-|q(O)|$, respectively; the open triangles to $q_b(M-O)$; the filled and open diamonds to $q(e_g)$ and $q(t_{2g})$. All the above quantities measure the degree of M–O covalence in the solid.

ions, limited to their nearest oxygens, is the same in FE and AFE structures, and the electronic relaxation that occurs in the distortion is common to FE and AFE PV materials.

In the description of results, the labelling of the oxygen ions in the distorted structure is as follows: the displacement along z of the transition metal ion breaks the symmetry equivalence of its six nearest oxygens, and creates two sets of four equatorial (those in the xy plane) and two axial ions (along the z direction). We call O_1 the two axial oxygens, in particular O_1 the one towards which the transition metal moves and O_1' the oxygen in the opposite direction; and O_2 the four equatorial oxygens [see Fig. 3(b)].

Let us start from the TB solution for the cubic PV materials, and perform a first-order perturbative treatment. The TB parameters in the distorted system are shown in Fig. 3(b). We call ζ the displacement of the transition metal, in fractional coordinates, in the $+z$ direction. The interaction between the transition metal and its nearest neighbour oxygens varies under the M displacement; we label $\beta(1 + \delta)$ and $\beta(1 - \delta)$ the new effective Hamiltonian elements between the M d_{xz} AO and its two neighbour oxygens along z, O_1 and O_1' respectively, while we assume that the interaction with the four equatorial oxygens is unchanged, and its value is the same β as in the cubic phase.

The parameter δ can be taken as equal to the change in overlap between the relevant M d and O 2p AOs under the displacement of M; in first approximation, δ is linear in ζ.[12]

The M off-centering removes the symmetry constraints described for the cubic phase. The TB Hamiltonian in Γ contains off-diagonal terms in the block of M d and O_1 2p AOs, which can be expressed as:

$$H = \begin{vmatrix} \alpha_d & 2\beta\delta \\ 2\beta\delta & \alpha_p \end{vmatrix} = \begin{vmatrix} \alpha_d & 0 \\ 0 & \alpha_p \end{vmatrix} + 2\beta\delta \cdot \begin{vmatrix} 0 & 1 \\ 1 & 0 \end{vmatrix} = H_0 + \lambda \cdot H' \tag{1}$$

where H_0 is the Hamiltonian for the cubic phase.

In the distorted phase, covalence between M and O_1, its nearest oxygen along the direction of displacement z, is no longer symmetry forbidden. The superdegenerate levels, highlighted in Fig. 2 for the cubic materials, hybridise into a π M–O_1 bonding level (1, in the valence band), and a π M–O_1 antibonding level (A, in the conduction band). The electronic modification in the tetragonal distortion involves therefore the formation of a new π bond between the transition metal and its closest oxygen, O_1.

As clear from eqn. (1), in a first-order perturbative treatment, the energy effectiveness of the new covalent bond is proportional to λ^2 (more precisely it equals $4\beta^2 \delta^2/\Delta\alpha$), while the covalent mixing of M d and O_1 2p AOs in the crystalline orbitals is proportional to $\lambda = 2\beta\delta$. The extra splitting of the superdegenerate M d and O_1 2p energy levels due to M–O covalence obtained at the Γ point, $\pm 4\beta^2 \delta^2/\Delta\alpha$, is retained along the whole Γ-X direction.

In Fig. 5, we report the HF band-structure of WO_3 along the Γ-X direction of reciprocal space, corresponding to increasing ferroelectric (FE) displacements of the W ion, ζ, in fractional coordinates. We see there that the shift in energy of the superdegenerate levels A and 1 increases as a function of the metal displacement, and that the superdegenerate levels remain flat in k-space.

The energy effectiveness of the new π M–O bond formed upon the M off-centering is proportional to $\beta^2/\Delta\alpha$. The same trends highlighted for the latter parameter in the cubic phase (see Table 1) are observed here during the distortion; in particular, the stability of the M off-centering decreases as we move from the soft (6-0) PV materials to the hard (4-2) PVs.

The re-hybridisation of the frontier π levels causes a displacement of electronic charge from the O_1 2p towards the M d AOs, in the opposite direction with respect to the displacement of nuclear charge from which it originates. This effect causes the large values of effective (Born) charges measured experimentally for FE PVs. The values of Born-charges available in the literature for different PV materials (see, for instance, refs. 13-19), follow the same trend highlighted here for the TB parameter $\beta^2/\Delta\alpha$. It is important to add that the electronic rearrangement described is linear in δ and hence in the M off-centering, as suggested in ref. 14, *only* as a first-order perturbative treatment for eqn. (1), that is when $\beta^2 \ll \Delta\alpha$. This is the case for the (4-2) and (5-1) materials with highest $\Delta\alpha$, but not for the (6-0) PVs in which $\Delta\alpha$ is lowest.

FE-like distortions give rise to an alternation of strong and weak (or short and long) M–O bonds in each of the axial M–O–M chains that compose the tetragonal PV structure. Once one M

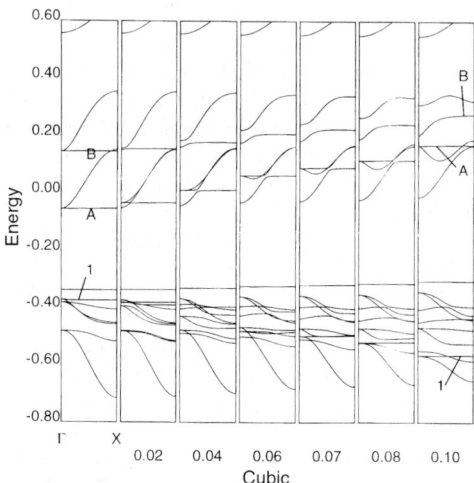

Fig. 5 Band structure of WO_3, corresponding to increasing FE displacements of W in fractional coordinates, as derived from the HF study.

ion is displaced off-centre, its covalent interaction polarises the electronic distribution of the neighbouring oxygens; the electronic perturbation is thus propagated towards the next M ions, favouring a long-range order of the M displacements. Each axial chain behaves as a *conjugated* π *system*, whose electronic delocalisation is proportional to the extent of covalence in the M–O interaction, *i.e.*, to $\beta^2/\Delta\alpha$. Furthermore, each chain has a polarisation vector, P_s, oriented parallel to the M off-centering, and whose electronic component has modulus proportional to $\beta^2/\Delta\alpha$.

3.2.1. FE, AFE and rotational modes. In principle, the distortion examined is energetically favourable for all the PVs in which the transition metal ion has electronic configuration $d^{(0)}$; in such cases, only the M–O bonding level in the valence band is populated. For those PVs in which the electronic configuration of M is greater than $d^{(0)}$, such as the conductor ReO_3, the population of the antibonding level in the conduction band opposes the distortion. The topic is examined in ref. 20.

The same rehybridisation of AOs occurs in both FE and AFE distortions of cubic PVs. The long-range order is, however, different in the two phases: in the AFE phase, nearest axial chains have polarisation vectors P_s anti-aligned; in the FE phase, instead, all the chains have P_s aligned. On purely electrostatic grounds we would expect an AFE arrangement of the distortion in nearest axial chains to be energetically stable. From experimental evidence we know, however, that, while WO_3, β-MoO_3 and $NaNbO_3$ are AFE, other PVs, such as $KNbO_3$ and $BaTiO_3$, are FE, and others still, such as $CaTiO_3$ and as limiting cases $MgSiO_3$ and $LaGaO_3$, undergo distortions that can be described as mutual rotations of the MO_6 octahedra. The three distortional modes (FE, AFE and rotational) are competitive, and are due to a subtle compromise of forces. For symmetry requirements in the AFE phase, no local dipole exists in the positions occupied by the equatorial oxygens and in the dodecahedral sites of the lattice. The AFE mode involves a displacement along z of the M and axial O ions *only*. Conversely, the FE ordering creates a local dipole moment in both O_2 and A sites; the FE distortional mode can, therefore, also have contributions from the equatorial oxygens and from the dodecahedral cations A. In the soft materials, those in which $\Delta\alpha$ is lowest, the energy contribution to the distortion arising from the M off-centering overcomes the components from A and O_2; these materials are stable in the AFE phase.

On increasing $\Delta\alpha$, the M off-centering becomes less energetically stable, and its contribution to the stability of the distortion is comparable to that due to the other ions of the structure (O_2 and A). In this case the materials are stable in the FE phase.

Increasing further $\Delta\alpha$, such as in the case of main group elements (Si, Ga) whose d AOs have very high energy, the M off-centering becomes unstable compared to a mutual rotation of the MO_6 octahedra. The latter reduces the interoxygen repulsion, which as seen for the cubic phase destabilises the O 2p levels at the top of the valence band in the X-M direction of reciprocal space.

4. Surfaces

In the previous section we have shown that the driving force for a FE-like distortional mode in PVs is represented by the formation of a new, π M–O bond, whose interaction was symmetry forbidden in higher symmetry phases. *Via* the analysis of the band-structure, we have been able to illustrate and isolate the contribution of covalence, ionicity and structural constraints in the solid-state chemistry of PVs. A privileged rôle in the electronic distribution is assumed by the parameter $\beta^2/\Delta\alpha$: for each M–O interaction, it is directly proportional to the degree of covalence in the M–O bonding. The *local* symmetry of M and O ions is also important: a local imbalance in $\beta^2/\Delta\alpha$ in the first coordination shell around either an oxygen or a transition metal ion in the PV structure will electronically polarise the surrounding lattice, according to the magnitude of the $\beta^2/\Delta\alpha$ imbalance, and seed an FE-like distortion, which then propagates in the structure *via* a delocalisation of the π M–O levels. The effectiveness of the latter is proportional to the value of $\beta^2/\Delta\alpha$ in the host oxide. We have in this way been able to define an electronic parameter, easily calculated from the *ab-initio* band structure of the pure oxides, which controls the ferroelectricity of perovskites.

As shown schematically in Fig. 6, local imbalances in $\beta^2/\Delta\alpha$ can be induced by intrinsic phonon vibrations, or introduced in the material by chemical doping, as shown experimentally by examining a Nb/Ta substitution in $KTaO_3$.[21,22] Compared to the case of chemical doping, in which a cation of the host lattice (say Ta^{5+}) is replaced by a similar ion (Nb^{5+}), surfaces represent a much stronger perturbation: one or more nearest neighbours of the surface ions are missing, inducing a large local difference in the value of $\beta^2/\Delta\alpha$ at the surface. It is reasonable to assume that such a change may couple with the FE distortion at the surface; moreover the electronic delocalisation of the perturbation may involve a large portion of material in the sub-surface region.

A general requirement for the stability of a surface is that it must be charge neutral, avoiding macroscopic separations of positive and negative charges in the direction perpendicular to the surface.[23] This charge-neutrality requirement has different consequences for PV materials, such as $BaTiO_3$ and WO_3, in which the cationic charge is distributed differently in the structure. The situation is exemplified in Fig. 7.

In the (4-2) PV materials, such as $BaTiO_3$ (left diagram in Fig. 7), the structure along the ⟨001⟩ direction is composed of alternating planes with composition BaO and TiO_2; each atomic layer is charge neutral. In such a case, the surface has the same periodicity as the bulk material, and the coordination of all the surface transition metal ions is equivalent: all are either 6-coordinate, if the crystal is cleaved along a BaO plane (upper surface in Fig. 7), or 5-coordinate when the cleaving plane is the TiO_2 layer (lower surface in Fig. 7). Similarly, all the oxygen ions directly exposed onto the surface have the same local environment.

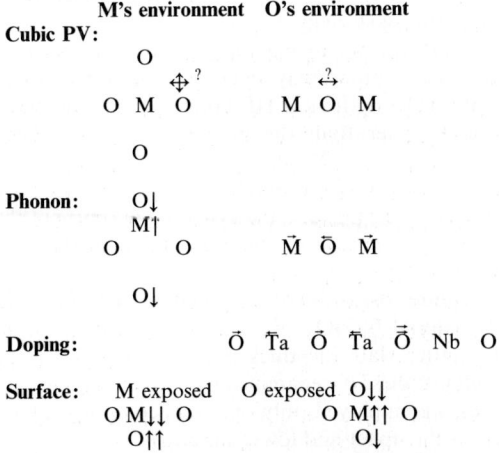

Fig. 6 Three local perturbations that may induce FE instabilities in PV-structured oxides. The top picture represents the symmetric local environment of the metal (left column) and oxygen (right column) ions in the cubic phase. The distortions are initiated, respectively, by a phonon vibration in which M displaces upwards, by chemical doping (Nb:Ta) and by a surface termination of the material.

BaO lay.	O	Ba	O	Ba	O	FE		1/2 O lay.			O				AFE
	Ti	O	Ti	O	Ti				W	O	W	O	W		
	O	Ba	O	Ba	O				O		O		O		
	Ti	O	Ti	O	Ti				W	O	W	O	W		
	O	Ba	O	Ba	O				O		O		O		
TiO$_2$ lay.	Ti	O	Ti	O	Ti	FE			W	O	W	O	W		
									O				O		AFE

Fig. 7 FE and AFE character of PV $\langle 001 \rangle$ surfaces: the (4-2) material BaTiO$_3$ (left diagram) is formed by charge neutral planes; not so the (6-0) material WO$_3$ (right diagram), in which the structure needs to reconstruct.

In the (6-0) PV material WO$_3$, the bulk structure along the $\langle 001 \rangle$ direction is instead formed by an alternation of planes with composition WO$_2$ and O (right diagram in Fig. 7). Neither atomic layer is charge-neutral, and cleavage along any $\langle 001 \rangle$ atomic layer would separate two surfaces with opposite charge. Such a process is obviously energetically unfavourable.

Charge neutral WO$_3$ $\langle 001 \rangle$ faces need to be obtained by appropriate surface reconstructions, in which the surface periodicity is a multiple of that of the underlying bulk. Since each oxygen in PV WO$_3$ is shared between two nearest W ions, stable terminations are provided by a cleavage along the O layer, which leaves half of the O ions on each side. In practice, this causes a doubling of the unit cell, and a $\sqrt{2} \times \sqrt{2}$ reconstruction of the $\langle 001 \rangle$ surface with respect to the bulk structure. The $\sqrt{2} \times \sqrt{2}$ doubling has been observed experimentally, by STM imaging of the WO$_3$ $\langle 001 \rangle$ faces.[24]

The unit cell of the reconstructed $\langle 001 \rangle$ surface contains two formula units, and two unequivalent surface W ions: one is 6-coordinate and has a terminal oxygen above, the other is 5-coordinate and is directly exposed onto the surface (see Fig. 7).

For both BaTiO$_3$ and WO$_3$ surface terminations, the M–O–M chains that compose the PV polymorph are oriented perpendicularly to the $\langle 001 \rangle$ surface.

In the light of the description summarised in Fig. 6, we expect that the different *local* structure in the $\langle 001 \rangle$ surface of BaTiO$_3$ and WO$_3$ may induce different electronic and geometrical relaxations. In particular, all the surface M and/or O ions on the BaTiO$_3$ $\langle 001 \rangle$ surface will relax in the same direction, inducing an FE behaviour in the sub-surface region, while in WO$_3$ $\langle 001 \rangle$, the 5 and 6-coordinate W ions will relax in opposite directions, inducing an AFE pattern in the sub-surface region.

Explicit QM calculations on the two types of surfaces described above, in which none of the expected features is imposed *a priori*, may be employed to validate or disprove the model developed to rationalise the FE behaviour of both bulk and surface PV materials. In the calculations, we have adopted the following strategy:

(1) In creating the surfaces, we disregarded the FE/AFE distortions calculated for the bulk materials, whose long-range order may impose structural constraints at the surface; for each material, we started from the optimised HF lattice parameter reported in Table 1. No structural or electronic constraint, other than the presence of the surface, is included in the starting geometry.

(2) We created a slab model of the unrelaxed $\langle 001 \rangle$ surfaces, by cleaving the bulk materials along a $\langle 001 \rangle$ plane. Reflection through the plane in the centre of the slab has been imposed to be a symmetry element in the calculations; the two surfaces at the upper and lower ends of the slab are symmetry equivalent.

The above choice requires us to employ slabs composed of an odd number of atomic layers. In the case of the (4-2) material BaTiO$_3$, we have used slabs whose thickness ranged from 7 to 15 atomic layers; in the latter slab, the thickness equals ≈ 30 Å, and represents the upper limit affordable with *ab-initio* studies. We examined both the BaO and TiO$_2$ terminations of BaTiO$_3$.

In the case of WO$_3$, we employed only one slab, composed of 11 atomic layers (counting as 1 both the 1/2 O layers at the upper and lower surfaces).

(3) For each material, we first considered surface unit cells with a $\sqrt{2} \times \sqrt{2}$ periodicity with respect to the bulk cubic cell. Although, as seen earlier in the discussion, the cell doubling is not required for BaTiO$_3$, in a surface with 1×1 periodicity all the Ti ions would be forced by symmetry to relax in the same vertical direction, and only an FE description of the displacements

could be achieved. If possible, after examining results for the thinner slabs, the thicker slabs have been studied with the 1 × 1 cell.

(4) We imposed an initial distortion of the ions at the surface, in a direction opposite to the one predicted earlier: AFE for $BaTiO_3$ $\langle 001 \rangle$; FE for WO_3 $\langle 001 \rangle$. The ions have been displaced by 0.01 Å ; according to the model presented in Fig. 6 (phonon), this movement is sufficient to break the symmetry around the M and/or O ions, and seed an FE/AFE distortional mode.

(5) The slab geometry has been fully relaxed with a numerical conjugate gradient method, allowing each ion (except those in the central layer, which are fixed according to the choice of point 2 above) to displace in the direction perpendicular to the surface. The central layer may be imagined as representing the bulk material, in the region unperturbed by the presence of the surface.

We shall now employ the results of the slab calculations in two ways. First, by examining the M–O bonding in each of the M–O–M chains perpendicular to the surface (axial chain), we shall check the predictions of the qualitative model of Fig. 6 regarding the FE/AFE character of the surface/subsurface distortion. Second, we shall examine the propagation of the perturbation introduced by the surface along the axial chains. For symmetry requirements, the central layer of the slab is unrelaxed, and the perturbation decays to zero at the slab centre. By comparing slabs with different thickness, and examining the convergence of the calculated properties, we can however estimate the range of propagation for the surface perturbation.

We now examine separately the results for the two types of surfaces defined earlier. In the following discussion, we define the direction perpendicular to the surface as z, and label with an index n the ions belonging to the nth atomic layer, starting from the surface layer, in which $n = 1$. We also define a positive direction for the ionic relaxations, which corresponds to ionic displacements directed from the centre of the slab towards the surface; conversely, a negative relaxation will correspond to displacements towards the centre of the slab. The zero of displacement for each ion coincides with the position it would occupy if the structure at the surface were identical to that of the bulk material (unrelaxed surface).

4.1 $BaTiO_3$ $\langle 001 \rangle$ surface

As mentioned earlier, we started the calculations for the $BaTiO_3$ $\langle 001 \rangle$ surfaces with a $\sqrt{2} \times \sqrt{2}$ double unit cell, and with an AFE relaxation pattern imposed onto the two columns of TiO ions perpendicular to the surface. Despite this initial constraint, geometry optimisation of 7-layer slabs, in either BaO or TiO_2 terminations, produces a final surface structure in which the non equivalent Ti and O ions in each layer are displaced along the same vertical direction, i.e., a structure with FE long-range order. We examine in detail the BaO termination of the $BaTiO_3$ slab. Only ionic displacements along z have been examined, which can be represented in a two-dimensional projec-

	Column a		Column b	
Starting geometry	$\uparrow O_{1a}^{+0.01}$	Ba_1^0	$O_{1b}^{-0.01} \downarrow$	AFE
	$\uparrow Ti_{2a}^{+0.01}$	O_2^0	$Ti_{2b}^{-0.01} \downarrow$	
	$O_{3a}^{+0.01}$	Ba_3^0	$O_{3b}^{-0.01}$	
Slab centre $\longrightarrow$	Ti_{4a}^0	O_4^0	Ti_{4b}^0	
	Column a		Column b	
Optimised geometry	$\downarrow\downarrow O_{1a}^{-0.046}$	$Ba_1^{-0.010}$	$O_{1b}^{-0.053} \downarrow\downarrow$	FE
	$\uparrow\uparrow Ti_{2a}^{+0.040}$	$O_2^{+0.050}$	$Ti_{2b}^{+0.037} \uparrow\uparrow$	
	$\downarrow O_{3a}^{-0.005}$	$Ba_3^{-0.004}$	$O_{3b}^{-0.008} \downarrow$	
	Ti_{4a}^0	O_4^0	Ti_{4b}^0	

Fig. 8 BaO termination of the $BaTiO_3$ $\langle 001 \rangle$ surface, represented by a 7-layer slab with a $\sqrt{2} \times \sqrt{2}$ doubling of the surface periodicity. The two non equivalent Ti–O axial columns are labelled with a and b. The upper diagram refers to the starting configuration; the lower diagram to the fully optimised geometry. The vertical relaxation of each ion is reported as exponent, in Å.

tion, reporting the z displacement for each symmetry-unique ion at its z coordinate in the system. The surface relaxation for the $BaTiO_3$ $\langle 001 \rangle$ slab is summarised in Fig. 8.

The AFE pattern of the two Ti-O axial columns (labelled a and b in Fig. 8) in the initial configuration, is represented in the upper diagram of Fig. 8; the optimised surface structure in the lower diagram. Comparison of the two shows that the surface relaxation has reversed the initial AFE into a final FE displacement pattern.

We now examine the electronic and geometric effect of the surface relaxation. The surface perturbation acts in the following way: the oxygen ion exposed on the surface (O_1) has only one nearest Ti ion, the Ti_2 species in the sub-surface layer. O_1 reacts to the coordination imbalance by forming a shorter and more covalent π bond with Ti_2. The features of the Coulomb field at the surface make the O_1–Ti_2 bond stronger than it would be in bulk $BaTiO_3$: the absence of the second Ti^{4+} ion in the coordination shell of the surface O_1 ion, makes the Coulomb field in the O_1 region less positive than in the bulk. The O_1 2p levels are therefore shifted to higher energy compared to the bulk, reducing the energy difference $\Delta\alpha$ with the 3d levels of the sub-surface Ti_2 ion. The Ti-O covalence, which is inversely proportional to the *local* value of $\Delta\alpha$, is correspondingly increased.

The electronic polarisation of the Ti_2 ion towards O_1, leaves in turn the oxygen in the third layer (O_3) in an asymmetric environment; O_3 reacts by polarising towards the Ti ion in the fourth layer (Ti_4). The latter is in the centre of the 7-layer slab; the propagation of the electronic perturbation has thus reached its nodal plane at this depth from the surface.

The above effect is purely electronic, and is already present in the unrelaxed surface. The Ti-O bond populations in the unrelaxed surface, for instance, equal 54, 16 and 21 $m|e|$ for the O_1–Ti_2, Ti_2–O_3 and O_3–Ti_4 bonds respectively, already showing the alternance of high and low values typical of the FE distortion in the M–O chains.

The surface relaxation follows the electronic polarisation, and causes a shortening of the O_1–Ti_2 bond, by 0.086 and 0.090 Å, respectively, in the two non-symmetry-equivalent columns a and b, the relaxation being slightly higher in the column b, in which the initial configuration had the correct direction of displacement. Such a shortening of the surface Ti-O bond is more than twice the value of 0.040 Å, calculated in the fully optimised FE phase of bulk $BaTiO_3$.

The weakened Ti_2–O_3 bond elongates by 0.045 in both columns a and b, while the O_3–Ti_4 bond, which shows an increase of electronic density, shortens by 0.005 and 0.008 Å in columns a and b, respectively.

The ionic displacements and electronic distribution in the two columns a and b are very similar, a feature not imposed in the calculations; the differences in columns a and b can be attributed to the numerical procedure employed in the geometry optimisation. Comparison of the behaviour along columns a and b confirms therefore the FE distortion pattern caused by the surface perturbation.

We may assume that this finding, although obtained for the slab with minimum thickness examined, has general validity, and will not be modified by increasing the number of layers composing the slab. In the following discussion, we shall therefore examine the BaO $\langle 001 \rangle$ surface termination of $BaTiO_3$ using surface cells with 1×1 periodicity only. The computational effort saved in reducing the lateral dimensions of the cell, can in this case be devoted to study slabs with increased vertical depth, up to 15 atomic layers in the case of $BaTiO_3$.

To focus on the equilibrium distances and electronic distribution along the Ti–O chains perpendicular to the surface, we introduce a tabular notation, as follows: for each symmetry independent Ti-O chain, in the n-th row of the table we report the calculated properties of the ion in the nth atomic layer, up to the slab centre. Atomic relaxation (Δ), bond distances (R) and bond populations (Q) can thus be easily tabulated.

Results for the BaO terminated, $BaTiO_3$ $\langle 001 \rangle$ slabs examined are summarised in Table 2. The values of the Ti–O overlap populations, Q, are calculated from a Mulliken population analysis on the equilibrium electronic density, and reported in units of $m|e|$; ionic displacements and bond distances are instead in Å. The reference values of R and Q for the Ti-O bond in the bulk material are reported in the bottom line of the table.

The alternation of shortened and lengthened Ti-O bonds along the perpendicular to the surface, typical of the FE distortional mode, is clear in the data of Table 2. Furthermore, the bond population of the short bonds increases compared with the bulk; that of the long bonds decreases. The

Table 2 Relaxation and electronic distribution in BaO terminated $\langle 001 \rangle$ surfaces of $BaTiO_3$[a]

N: ion(s)	7a			7b			9			15		
	Δ	R	Q	Δ	R	Q	Δ	R	Q	Δ	R	Q
O_1	−0.046			−0.053			−0.055			−0.034		
O_1–Ti_2		1.927	57		1.923	58		1.925	58		1.933	58
Ti_2	+0.040			+0.037			+0.033			+0.046		
Ti_2–O_3		2.057	13		2.058	13		2.055	12		2.048	12
O_3	−0.005			−0.008			−0.009			+0.011		
O_3–Ti_4		2.008	21		2.005	21		2.002	22		2.014	22
Ti_4	0			0			+0.002			+0.010		
Ti_4–O_5	—			—				2.015	19		2.019	20
O_5	—			—			0			+0.004		
O_5–Ti_6	—			—			—				2.015	20
Ti_6	—			—			—			+0.002		
Ti_6–O_7	—			—			—				2.016	20
O_7	—			—			—			−0.001		
O_7–Ti_8	—			—			—				2.012	20
Ti_8	—			—			—			0		
Bulk Ti–O											2.013	20

[a] N is the slab thickness, in number of atomic layers; for the 7-layer slab, 7a and 7b refer to the two columns indicated as a and b in Fig. 8. Δ is the atomic relaxation and R the Ti–O bond distance, in Å; Q the Ti–O bond population, in $m|e|$. The notation is explained in the text.

surface perturbation appears however to be short-ranged: the outermost O_1–Ti_2 bond is substantially modified by the surface, but only a smooth oscillation in the bond distances survives beyond the fourth atomic layer. The 9 and 15 layer slabs yield very similar results for the four outermost layers. We recall that the amount of delocalisation of the M–O π levels, in the model employed to rationalise ferroelectricity, is proportional to the value of $\beta^2/\Delta\alpha$ in the material examined; the (4-2) perovskites are those in which $\beta^2/\Delta\alpha$ is lowest, as is therefore the electronic propagation of the perturbation. From the data of Table 2, it appears that after the fourth Ti-O bond, the electronic distribution converges to the bulk values.

The situation is similar for the termination of the $BaTiO_3$ crystal in the TiO_2 plane. In this case, it is a surface Ti_1, and not an O_1 ion, that is unsaturated. The passage from 6- to 5-fold coordination of the surface Ti_1 has a less pronounced effect on the energy levels than the decrease of coordination number from 2 to 1 of the surface O_1 in the BaO termination. As such, the surface effect is less important for the TiO_2 termination of the solid than in the BaO case; the bond population of the terminal Ti_1–O_2 bond, for instance, is now 36 $m|e|$, while it was 58 $m|e|$ for the terminal O_1–Ti_2 bond at the BaO termination.

4.2. WO_3 $\langle 001 \rangle$ surface

The $\langle 001 \rangle$ surface of WO_3 has the second type of termination examined, with a $\sqrt{2} \times \sqrt{2}$ reconstruction, which creates two non-symmetry equivalent W–O–W columns. The first, which we label as a, terminates with a surface O_1 ion, while the second (labelled b) ends in a surface W_2 ion.

The surface relaxation is represented in Fig. 9, for the 11 layer slab employed. In the initial configuration we have imposed an FE pattern to the W ions; this is completely overcome in the geometry optimisation, which ends with an AFE state, evident in Fig. 9.

The AFE distortion of PV WO_3 is energetically very stable also in the bulk material (see ref. 25); in the optimised surface structure reported in Fig. 9 the effects of surface and bulk relaxations may therefore overlap. To highlight the surface effect, we have repeated the slab calculations, employing the geometry of the fully optimised tetragonal AFE phase of WO_3 as starting point.

The ionic relaxation in the bulk AFE phase of WO_3 involves a displacement along z of both the W and equatorial O_1 ions, and resolves the degeneracy of atomic coordinates along z. Each atomic layer of the cubic phase is split in the tetragonal: the WO_2 plane into a triplet of $\{W; O_2; W\}$ the O plane into $\{\frac{1}{2}O; \frac{1}{2}O\}$ with different z coordinates, as shown in Fig. 10.

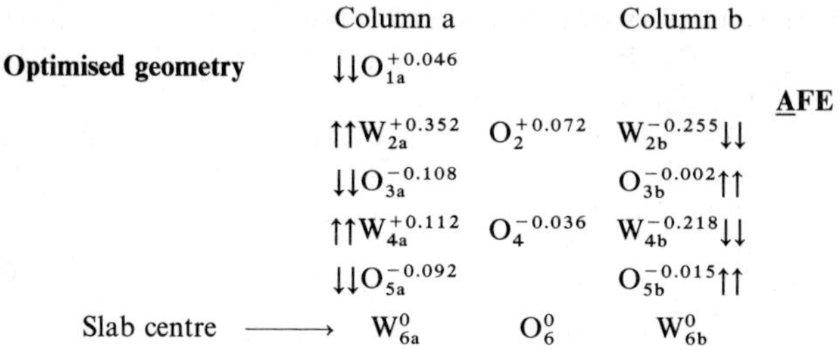

Column a
Column b
Optimised geometry
AFE
↓↓$O_{1a}^{+0.046}$
↑↑$W_{2a}^{+0.352}$ $O_2^{+0.072}$ $W_{2b}^{-0.255}$↓↓
↓↓$O_{3a}^{-0.108}$ $O_{3b}^{-0.002}$↑↑
↑↑$W_{4a}^{+0.112}$ $O_4^{-0.036}$ $W_{4b}^{-0.218}$↓↓
↓↓$O_{5a}^{-0.092}$ $O_{5b}^{-0.015}$↑↑
Slab centre → W_{6a}^{0} O_6^0 W_{6b}^{0}

Fig. 9 Structural relaxation of the $\sqrt{2} \times \sqrt{2}$ WO_3 ⟨001⟩ surface, represented with an 11-layer slab. The two non equivalent W–O columns are labelled with a and b. The vertical relaxation of each ion is reported as exponent, in Å.

The 11 layer slab of the cubic phase, examined earlier, contained 5 WO_3 formula units in each W–O–W column; in the tetragonal AFE phase it corresponds to a 25 atomic layer slab.

The symmetry of the AFE phase requires that reflection through the plane at the centre of the slab be no longer a symmetry operator (this is clear in the diagram at the right side of Fig. 10); the new geometry allows therefore a full columnar relaxation, that extends beyond the slab centre without nodal planes. To avoid geometrical interferences in the relaxation induced by the upper and lower surfaces of the slab, we have kept unrelaxed the whole central layer (5 atomic planes) of the slab. Electronic distribution and relaxation of the energy-minimised structure are reported in Table 3. We employ the same notation as used for the cubic phase, by attributing each ion in the tetragonal phase to the atomic layer it occupied in the cubic structure.

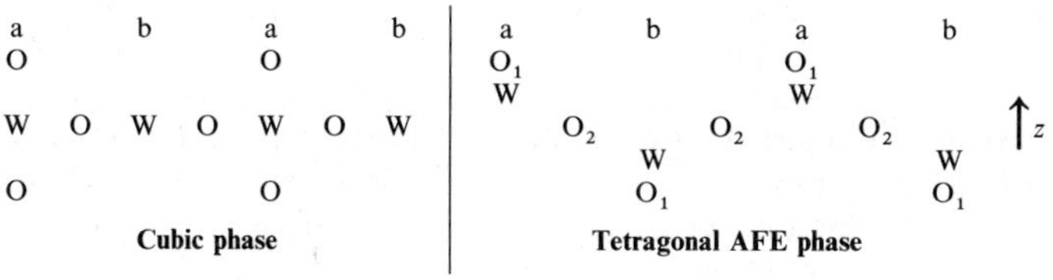

a b a b
O O
W O W O W O W
O O
Cubic phase
a b a b
O_1 O_1
W W
O_2 O_2 O_2
W W
O_1 O_1
Tetragonal AFE phase
z

Fig. 10 Splitting of the degeneracy along z of the atomic coordinates in the tetragonal AFE phase (right diagram) of WO_3 compared to the cubic (left diagram). The symbols a and b label the two WO columns along the ⟨001⟩ direction.

Table 3 Relaxation and electronic distribution in the AFE-WO_3 ⟨001⟩ surface[a]

	Column a			Column b		
Ion(s)	Δ	R	Q	Δ	R	Q
O_1	−0.023			—		
O_1–W_2		1.579	349		—	
W_2	+0.026			−0.009		
W_2–O_3		2.406	−1		1.628	195
O_3	−0.014			−0.009		
O_3–W_4		1.625	218		2.357	−7
W_4	−0.012			0.000		
W_4–O_5		2.355	−7		1.628	187
O_5	0			0		
O_5–W_6		1.628	184		2.366	−7
W_6	0			0		
Bulk W–O		1.628	187		2.366	−7
Bulk W–O		2.366	−7		1.628	187

[a] The labelling of columns a and b is the same as in Fig. 10. Symbols have the same meaning introduced for Table 2.

Despite being already considerably distorted in bulk (the W–O bond distances are 1.628 and 2.366 Å), the structure of AFE WO_3 reacts even further to the $\langle 001 \rangle$ surface. The overlap population in the outermost O_1–W_2 bond, for instance, increases from 187 $m|e|$ in bulk to 349 $m|e|$ on the surface. The perturbation is transferred to the subsurface layers in WO_3 much more effectively than in $BaTiO_3$; the O–W short bonds in column a have an overlap population of 218 $m|e|$ (O_3–W_4) and 184 $m|e|$ (O_5–W_6); the latter is unrelaxed, but still has a different overlap population from that in the bulk. The picture that emerges from the above analysis is consistent with WO_3 being the PV material examined with the highest value of $\beta^2/\Delta\alpha$, and hence in which the electronic delocalisation of any perturbation via the π W–O bands is highest.

The electronic distribution at the centre of the AFE WO_3 slab examined, appears not yet to have converged to the bulk value; it is however impossible to increase further the vertical dimension of the slab to confirm this conclusion; the surface calculations become quickly very demanding. We recall that the WO_3 $\langle 001 \rangle$ surface cannot be modelled with a 1×1 periodicity with respect to the bulk, and the 11-layer slab employed represents the current upper limit.

To conclude the present section, in Fig. 11 we report the electron density maps for the bulk and $\langle 001 \rangle$ surfaces of $BaTiO_3$ and WO_3, which confirm in a pictorial way the findings highlighted in the discussion.

The results presented confirm the validity of the model used for rationalising the coupling between the surface perturbation and the FE/AFE distortional mode of PV-structured materials (see Fig. 7); the FE/AFE relaxation of the bulk materials is considerably enhanced at the surface.

The unreconstructed surfaces, examined in Section 4.1, can therefore be associated with an FE character; the $\sqrt{2} \times \sqrt{2}$ reconstructed $\langle 001 \rangle$ surface of WO_3 can instead be associated with an AFE character.

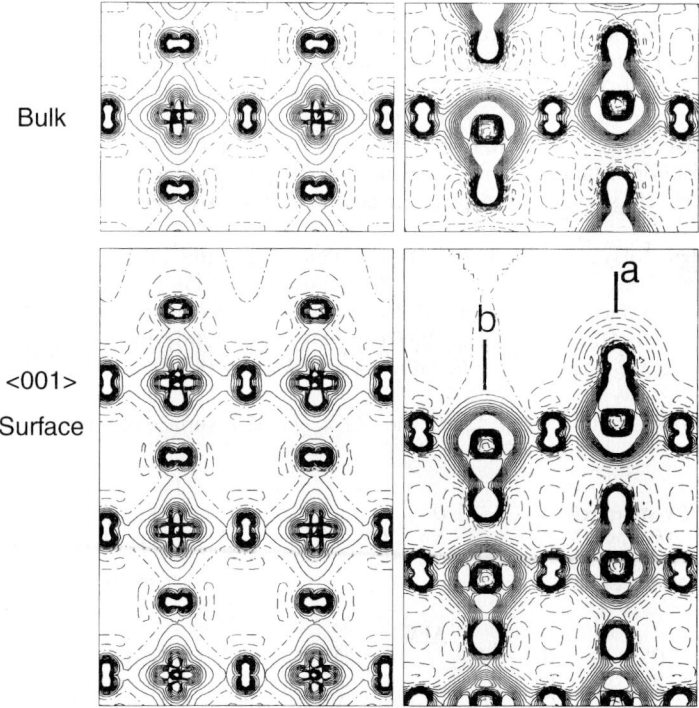

Fig. 11 Difference electron density plots (solid minus isolated formal ions) for $BaTiO_3$ (left column) and WO_3 (right column). The upper maps refer to the stable bulk phase of the materials: FE for $BaTiO_3$, AFE for WO_3. The lower plots are relative to the $\langle 001 \rangle$ surface; the BaO termination is shown for $BaTiO_3$. The surface is located at the upper end of the pictures. The labelling of columns a and b in the WO_3 surface is the same as in Fig. 10 and Table 3. Continuous, dashed and dot–dashed lines refer to positive, negative and zero difference densities; the interval between isodensity lines is 0.005 au ($|e| \cdot a_0^{-3}$). Note the increased electronic density of the bonds close to the surface.

It is possible that, in those materials where perturbations are long-ranged, the FE/AFE distortional pattern imposed by the surfaces, may propagate over extended sub-surface regions, and prevail there over the distortional modes calculated for the bulk systems. To investigate this topic, we shall now describe a model system, based on an atomic-level surface engineering of the PV $\langle 001 \rangle$ surfaces examined here.

5. Surface engineering: FE properties of oxide/oxide interfaces

In the following discussion, we shall examine the electronic structure of two model PV/PV interfaces. In ref. 26, we have shown that in oxide/oxide interfaces, when a component (overlayer) is deposited onto the other (support), in the limit of low deposition the structure of the overlayer is likely to be epitaxially related to that of the support. This feature has been widely investigated, and exploited, for applications in catalysis, but it is as yet unexplored in the field of materials science and electronics.

We shall now try to bridge the two fields, by transferring the knowledge gathered in catalysis towards the materials properties of oxide/oxide interfaces. We shall investigate two topics: (1) the stability of FE WO_3 under surface constraints; and (2) the FE properties of PV/PV interfaces under conditions similar to those researched for the application in computer memories.

The FE/AFE properties of the $\langle 001 \rangle$ surface of PV materials are schematically represented in Fig. 6. We have already examined the electronic properties of systems which couple an FE bulk material with an FE surface termination ($BaTiO_3$ $\langle 001 \rangle$) or an AFE bulk material with an AFE surface termination (WO_3 $\langle 001 \rangle$). Here we move a step further, and investigate systems that combine an FE bulk behaviour with AFE surfaces, and *vice versa*, i.e., an AFE bulk behaviour with FE surfaces.

No PV compound has the above properties in itself; to obtain the required combinations we shall create appropriate oxide/oxide interfaces, and deposit a PV-structured transition metal oxide monolayer onto the $BaTiO_3$ and WO_3 $\langle 001 \rangle$ surfaces examined in Section 4.

The process is shown schematically in Fig. 12. Let us consider, for instance, a PV-structured Nb_2O_5 monolayer [Fig. 12(a)]. Although Nb_2O_5 by itself is not PV-structured, Nb-based ternary perovskites exist (for instance, $KNbO_3$) in which the Nb–O framework has the required structure; as shown in ref. 26, single-layer deposition is then likely to provide epitaxial growth on the support.

The Nb_2O_5 monolayer can be built with one flat surface (this layer has composition NbO_2), and one with a $\sqrt{2} \times \sqrt{2}$ reconstruction (and composition 1/2 O). From the previous section, we recall that the flat surface is associated with FE character, the reconstructed with AFE.

Depositing the Nb_2O_5 monolayer on $BaTiO_3$ $\langle 001 \rangle$ [Fig. 12(b)], the flat FE surface of Nb_2O_5 will match the flat FE surface of the $BaTiO_3$ support, leaving the reconstructed AFE termination exposed.

Vice versa, on WO_3 $\langle 001 \rangle$ [Fig. 12(c)], the $\sqrt{2} \times \sqrt{2}$ AFE surface of the Nb_2O_5 overlayer will

(a) **PV-structured Nb_2O_5 ovl**

O		*O*		*O*	**AFE; ½ O**
Nb	*O*	*Nb*	*O*	*Nb*	**FE; NbO_2**

(b) **$Nb_2O_5/BaTiO_3$**

½ O					
NbO_2	*Nb*	*O*	*Nb*	*O*	*Nb*
BaO	O	Ba	O	Ba	O
	Ti	O	Ti	O	Ti
	O	Ba	O	Ba	O
TiO_2	Ti	O	Ti	O	Ti

AFE

(c) **Nb_2O_5/WO_3**

NbO_2	Nb	O	Nb	O	Nb	FE
½O + ½O	O		O		O
WO_2	W	O	W	O	W
	O		O		O
	W	O	W	O	W
	O		O		O

Fig. 12 FE/AFE character of PV/PV interfaces. (a) PV-structured Nb_2O_5 monolayer (ions in the Nb_2O_5 layer are represented with italic letters) constructed to have one flat FE side and one with AFE $\sqrt{2} \times \sqrt{2}$ periodicity. (b) $Nb_2O_5/BaTiO_3$ interface, which couples the FE support with an AFE surface termination; (c) Nb_2O_5/WO_3 interface, which couples the AFE support with an FE surface termination.

dock onto the substrate $\sqrt{2} \times \sqrt{2}$ periodicity, completing the O layer and leaving the flat FE side exposed.

Apart from the Nb_2O_5 monolayer, illustrated in Fig. 12, PV-structured monolayers of all the early transition metal cations with oxidation state +5 or +7 (such as V_2O_5, Ta_2O_5, Re_2O_7) provide the suitable combination of one FE and one AFE side. There are therefore several combinations of support and overlayer that can be examined. Here, we report on $Nb_2O_5/BaTiO_3$ and Re_2O_7/WO_3.

The computational procedure is the same as for the pure $\langle 001 \rangle$ surfaces: we have constructed a slab model with a double unit cell in the horizontal direction, and employed the thickest slab affordable in the calculations; in this case of 13 atomic layers (including both substrate and overlayer). Reflection through the central plane of the slab has again been imposed as a symmetry operation, and both upper and lower surfaces of the slab have been coated with the overlayer; we have then performed a full geometry optimisation of the slab, allowing each ion (excluding those in the central layer) to relax to minimum energy along the z direction. For each interface, we started from the optimised cubic bulk geometry of the support; the M and O ions of the overlayer have been located in the same position that would be occupied by the ions of the support, if the structure were extended of one atomic layer across the surface. The starting M–O bond lengths in the overlayer are therefore the same as in the support. We have then imposed a small initial shift to the overlayer ions, to break the symmetry between the two axial M–O columns across the slab.

As in Section 4, we label the two axial columns as a and b; for the AFE surface termination, column a is that terminated with an oxygen ion, column b the one terminating in a surface transition metal ion. In the FE surface, the labelling is arbitrary, as both columns terminate in the same ionic species.

We summarise the results of the geometry optimisation and of the calculated electronic structure in Table 4. Symbols and units are as in Table 2. We start by examining the $Nb_2O_5/BaTiO_3$ interface. The $BaTiO_3$ support, when in bulk, has a tendency to an FE distortion; as seen in Section 3, however, the value of $\beta^2/\Delta\alpha$ in $BaTiO_3$ is small, as is the energy gain associated with the cubic to FE phase transition. The distortional AFE pattern imposed by the surface is more intense and, at least for the slab thickness examined, it prevails over the FE order of the support. In Table 4, we note in fact the alternance of long and short Ti–O bonds in both columns, a and b, but with the reverse order typical of antiferroelectricity: in the layers where the Ti–O bond is short in column a, it is long in b, and *vice versa*. The oscillation of both bond-lengths and overlap population smoothens towards the slab centre, but less rapidly than in the case of the pure $BaTiO_3$

Table 4 Relaxation and electronic distribution[a] in the $Nb_2O_5/BaTiO_3$ and Re_2O_7/WO_3 interfaces, as a function of the depth from the surface. The overlayer ions compose the two outermost layers, and half of the O_3 ions in the case of Re_2O_7

	$Nb_2O_5/BaTiO_3$						Re_2O_7/WO_3					
	Column a			Column b			Column a			Column b		
Ion(s)	Δ	R	Q	Δ	R	Q	Δ	R	Q	Δ	R	Q
O_1 (ovl)	−0.093			—			−0.342			−0.340		
O_1–M_2		1.707	284		—			1.610	451		1.593	445
M_2 (ovl)	+0.487			+0.006			−0.069			−0.050		
M_2–O_3		2.580	0		1.922	72		1.859	−74		1.898	−98
O_3	−0.080			+0.097			−0.045			−0.065		
O_3–M_4		1.867	64		2.183	6		1.883	−124		1.826	−164
M_4	+0.065			−0.074			−0.046			−0.008		
M_4–O_5		2.111	7		1.921	33		1.843	−121		1.900	−113
O_5	−0.033			+0.018			−0.007			−0.026		
O_5–M_6		1.963	28		2.050	13		1.883	−109		1.864	−100
M_6	+0.017			−0.019			−0.007			−0.007		
M_6–O_7		2.029	18		1.994	21		1.876	−96		1.875	−116
O_7	0			0			0			0		
Bulk M–O		2.013	20		2.013	20		1.883	−110		1.883	−110

[a] Δ is the ionic relaxation, R the M–O bond distance, in Å; Q is the M–O bond population, in $m|e|$. The notation is the same employed in Table 2.

⟨001⟩ examined in Section 4.1. This result suggests that the perturbation caused by the interface is longer-ranged (and hence also more intense) than the perturbation created by the intrinsic $BaTiO_3$ ⟨001⟩ surface.

The AFE order of the $BaTiO_3$ substrate is clear in Fig. 13(a), where we show the calculated electron density across the slab for the $Nb_2O_5/BaTiO_3$ interface.

In the Re_2O_7/WO_3 interface, the electronic rearrangement at the surface is such as to form a very strong and short terminal O_1–Re_2 bond, as expected. The calculated bond-lengths in columns a and b are 1.610 and 1.593 Å, and the bond-populations 451 and 445 $m|e|$. The latter values are the highest bond-populations examined throughout this work: they are due to the unsaturation of the surface O_1 ions, but especially to the high oxidation state ($+7$) of the subsurface Re_2 cations. The electronic affinity of Re^{+7} is higher than that of W^{+6}; the energy difference, $\Delta\alpha$, between the O_1 2p and the Re_2 5d levels is therefore the lowest examined; as a consequence the O_1–Re_2 bond is the strongest. The very high electronic density in the terminal O_1–Re_2 bonds is clear in the electron density map of the Re_2O_7/WO_3 interface [Fig. 13(b)].

The propagation of the surface perturbation to the subsurface layers is different from that examined in the previous examples: the O_3 ion in the third atomic layer is now shared at the interface between the Re_2 and the W_4 ions. The Re_2 species is polarised away from O_3; in the other geometries examined, the electronic polarisation of Re_2 would be sufficient to transfer the perturbation to O_3 and cause an O_3–W_4 bond of increased strength. In the interface examined here, however, the two metal ions nearest neighbour of O_3 have a different charge; in particular, the ion electronically polarised away from O_3, i.e., Re_2, has higher nuclear charge than the W_4 at the opposite side. A closer examination of the data reported in Table 4 and in Fig. 13(b), suggests that O_3 is substantially unpolarised. The geometry and electronic distribution in the layers beyond the third is also very similar to that of cubic WO_3, if not for being slightly compressed along the $-z$ direction. The chemical difference between Re_2 and W_4 at the interface appears therefore to have absorbed completely the surface perturbation. The absence of the AFE distortion in the WO_3 substrate is, however, encouraging. The energy gained in the phase transition between cubic and AFE WO_3 is calculated as ≈ 0.8 eV per WO_3 formula unit.[25] Although the electronic relaxation imposed by the interface does not induce a long-range ordered FE state in the support, it completely cancels the driving force for the AFE distortion, typical of pure WO_3.

The almost undistorted structure of the WO_3 substrate appears to be a most favourable starting point for the exploitation of the Re_2O_7/WO_3 interface examined as a FE capacitor: the influence

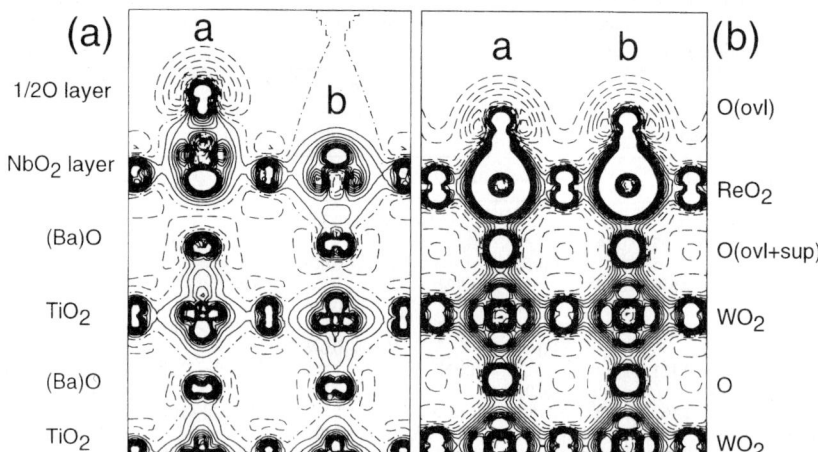

Fig. 13 Difference electron density plots (solid minus isolated formal ions) for the PV/PV interfaces examined. The maps refer to the $Nb_2O_5/BaTiO_3$ (a) and Re_2O_7/WO_3 (b) interfaces. The axial M–O columns are labelled here as in Table 4. The composition of each atomic layer is marked at the side of the pictures. Continuous, dashed and dot–dashed lines refer to positive, negative and zero difference densities; the interval between isodensity lines is 0.005 au ($|e| \cdot a_0^{-3}$).

of external applied fields would in this case be maximised, and even small intensities may be sufficient to pole the material in either direction, and to reverse its polarisation direction. We refer to the WO_3 phase obtained as *incipient ferroelectric*.

The results reported here confirm the importance of surfaces in determining the FE properties of PV materials, and suggest that surface engineering is a suitable route to provide novel materials with improved FE/AFE behaviour.

The theoretical predictions may in this case be validated by synthesising suitably coated particles with the same combinations of substrate and overlayer studied here, and examining experimentally how the FE properties depend on the particle size, up to the limit of nanometer-sized particles.

The interface between the WO_3 substrate and the Re-based overlayer examined here, can be imagined as being part of a more extended system

$$WO_3 \mid Re_2O_7(1 \text{ layer}) \mid ReO_3,$$

in which a Re_2O_7 monolayer has been inserted as buffer zone between a WO_3 capacitor and a ReO_3 conductor.

The above system provides a suitable combination of the features required for the application of FE materials in computer memories: the structure is continuous through the interface, which joins an isostructural conductor (ReO_3) with an incipient FE substrate (the undistorted WO_3 component); furthermore, the lattice parameters of ReO_3 and WO_3 are very similar (the mismatch is less than 2%), so that a ReO_3/WO_3 interface would have very little structural strain.

The results shown also suggest that the interface between an FE and a conducting PV oxide, as researched for application in the computer memories, may perturb significantly both the FE properties of the capacitor and the conductivity properties of the metallic oxide, and may therefore prevent, or favour, their practical application.

The thickness of the slabs employed in the simulations (≈ 25-30 Å) is already comparable with the thinnest FE layers synthesised experimentally. QM calculations of the kind described here may therefore play a relevant rôle to direct future experimental efforts in this field of research.

Finally, we would like to emphasise the symbiosis between the chemical and physical characterisation of materials, yielded by a common set of QM calculations. Such cooperation is obviously achievable also experimentally: overlayer oxides are widely investigated as catalysts, but so far their properties are unexplored for applications in electronics. The calculations performed here are the first examination of such a topic, and show its feasibility. While it is widely accepted that the supporting material can modify the chemical properties of the overlayer, the feature exploited in catalysis, using the ideas developed in the present paper, it is also reasonable to believe that the perturbation created by the surface (or interface) termination can modify the physical properties of the supporting material. We believe that the transfer of knowledge between the two areas of science of catalysis and materials for electronics will be particularly beneficial, and a coordinated effort will strengthen the ability to progress in both fields.

Acknowledgements

Financial support from EPSRC is gratefully acknowledged.

References

1. F. Fukunaga and N. Tsuda, *J. Phys. Soc. Jpn.*, 1994, **63**, 3798.
2. Y. Maeno, H. Hashimoto, K. Yoshida, S. Nishizaki, T. Fujita, J. D. Bednorz and F. Lichtenberg, *Nature (London)*, 1994, **372**, 532.
3. R. von Helmolt, J. Wecker, B. Holzapfel, L. Schultz and K. Samwer, *Phys. Rev. Lett.*, 1993, **71**, 2331; Y. Moritomo, A. Asamitsu, H. Kuwahara and Y. Tokura, *Nature (London)*, 1996, **380**, 141.
4. R. M. Bowman, D. O'Neill, M. McCurry and J. M. Gregg, *Appl. Phys. Lett.*, 1997, **70**, 2622.
5. C. G. Granqvist, *Handbook of Inorganic Electrochromic Materials*, Elsevier, Amsterdam, 1995.
6. S. Loridant, L. Abello, E. Siebert and G. Lucazeau, *Solid State Ionics*, 1995, **78**, 249.
7. O. Auciello, J. F. Scott and R. Ramesh, *Phys. Today*, 1998, July, p. 22–27.
8. P. J. Hay and W. R. Wadt, *J. Chem. Phys.*, 1985, **82**, 270; 284; 299.
9. C. Pisani, R. Dovesi and C. Roetti, *Lecture Notes in Chemistry*, Springer, Heidelberg, 1988, vol. 48.

10 R. Dovesi, V. R. Saunders, C. Roetti, M. Causà, N. M. Harrison, R. Orlando and E. Aprà, *CRYSTAL 95 User's manual*, University of Torino, Turin, 1996.
11 F. Corà, Ph.D. Thesis, The Royal Institution and the University of Portsmouth, 1999.
12 R. A. Wheeler, M.-H. Whangbo, T. Hughbanks, R. Hoffman, J. K. Burdett and T. A. Albright, *J. Am. Chem. Soc.*, 1986, **108**, 2222.
13 F. Detraux, Ph. Ghosez and X. Gonze, *Phys. Rev. B: Condens. Matter*, 1997, **56**, 983.
14 R. Resta, M. Posternak and A. Baldereschi, *Phys. Rev. Lett.*, 1993, **70**, 1010; M. Posternak, R. Resta and A. Baldereschi, *Phys. Rev. B: Condens. Matter*, 1994, **50**, 8911.
15 W. Zhong, R. D. King-Smith and D. Vanderbilt, *Phys. Rev. Lett.*, 1994, **72**, 3618.
16 R. Yu and H. Krakauer, *Phys. Rev. Lett.*, 1995, **74**, 4067.
17 C. Lasota, C.-Z. Wang, R. Yu and H. Krakauer, *Ferroelectrics*, 1997, **194**, 109.
18 J. D. Axe, *Phys. Rev.*, 1967, **157**, 429.
19 Ph. Ghosez, X. Gonze, Ph. Lambin and J. P. Michenaud, *Phys. Rev. B: Condens. Matter*, 1995, **51**, 6765; Ph. Ghosez, X. Gonze and J. P. Michenaud, *Ferroelectrics*, 1995, **164**, 113.
20 F. Corà, M. G. Stachiotti, C. R. A. Catlow and C. O. Rodriguez, *J. Phys. Chem. B*, 1997, **101**, 3945; M. G. Stachiotti, F. Corà, C. R. A. Catlow and C. O. Rodriguez, *Phys. Rev. B: Condens. Matter*, 1997, **55**, 7508.
21 J. J. van der Klink, S. Rod and A. Châtelain, *Phys. Rev. B*, 1986, **33**, 2084.
22 O. Bidault and M. Maglione, *J. Phys. I*, 1997, **7**, 543.
23 F. Bertaut, *C. R. Hebd. Seances Acad. Sci.*, 1958, **246**, 3447.
24 F. H. Jones, K. Rawlings, J. S. Foord, R. G. Egdell, J. B. Pethica, B. M. R. Wanklyn, S. C. Parker and P. M. Oliver, *Surf. Sci.*, 1996, **359**, 107.
25 F. Corà, A. Patel, N. M. Harrison, R. Dovesi and C. R. A. Catlow, *J. Am. Chem. Soc.*, 1996, **118**, 12174.
26 F. Corà and C. R. A. Catlow, *J. Mol. Catal. A: Chem.*, 1997, **119**, 57.

Paper 9/04517A

General Discussion

Prof. Catlow opened the discussion of Prof. Pettersson's paper: There are dangers in using potential models parametrised using static lattice techniques in dynamical simulations. In the present case, I consider that the parameter set you have employed is sufficiently robust for use in dynamical simulations. I would comment, however, that I consider that the best shell-model parameter set for CeO_2 is that recently derived by Grimes et al.[1] and I would recommend the use of these parameters in subsequent simulations (both static and dynamical) on CeO_2.

1   S. Vyas, R. W. Grimes, D. H. Gay and A. L. Holt, J. Chem. Soc., Faraday Trans., 1998, **94**, 427.

Prof. Pettersson responded: First, I would like to emphasize that what we have done is to use specific structures obtained from molecular dynamics simulations to illustrate some very fundamental principles concerning the chemistry of the O^{2-} anion that apply generally to all oxide materials. Statically derived interatomic potentials can introduce problems and limitations when used for dynamical simulations, something of which I am aware. I am glad that there is agreement that our choice of force field is reasonable. Since the O^{2-} anion is unstable in vacuum it exists only if stabilized by an external potential. From this it is clear that if one hypothetically increases the lattice parameter of a crystalline oxide material towards infinity, there is a point at which the Madelung potential is sufficiently weakened that the O^{2-} will donate an electron to anything in its surroundings or even auto-ionize. Thus, by expanding the lattice as in the sequence MgO–CaO–SrO–BaO the properties of the O^{2-} anion change drastically from a hard, neon-like closed-shell system with low oxygen affinity in MgO, to a soft ion with a strong tendency to form surface peroxo O_2^{2-} species, with a very balanced internal charge distribution, in the case of BaO. Temperature will have the same general effect. Increasing the temperature leads to a slight expansion of the lattice, but more importantly to a distribution of the ions around their most stabilized positions and thus a variation in the electrostatic stabilization of the charge states.

I should like to add that we have now tested the potentials suggested by Prof. Catlow in his comment and they indeed seem more stable in the sense that the computed internal forces are smaller which allows larger time steps to be used in the simulations. This is an advantage that we will exploit in our further simulations.

Dr Egdell asked: Could you give a qualitative explanation as to why Ca doping lowers the charge transfer energy, even in the static situation? Is this something that can be measured experimentally, for example by optical absorption spectroscopy.

Prof. Pettersson replied: Basically one Ce^{4+} and one O^{2-} are replaced by one Ca^{2+} ion at the Ce site. Although the total charge remains the same we have effectively reduced the Madelung potential. Assuming no change in the ion positions we could view the resulting potential as the original one (undoped system) plus the potential generated by a lattice built up from a -2 charge at each position of a removed cerium and a $+2$ charge at the sites where the oxygen vacancies are created. The latter potential counteracts the original one and results in a reduction of the electrostatic stabilization. This will be modified by the structural changes around the lattice positions induced by the doping, but the general effect should be this.

The experiments were done using nano-size doped particles and we have actually not considered investigating the effect on the optical absorption spectrum. It is a very good suggestion and I will discuss this with the experimental group.

Prof. Catlow added: Following from both your and Dr Egdell's remarks, I would add that the most important factor for reducing the calculated charge transfer energies is the stabilisation of the electron (idealised as a Ce^{3+} ion) by the neighbouring anion vacancy which has an effective positive charge.

Prof. Pettersson responded: This is certainly an interesting point, but it should be underlined that we have not found the charge-transfer (C-T) state, *i.e.* Ce^{3+}, to be energetically favourable in any of the dumps we have studied. The statistics we have are poor due to the small number of dumps that could be studied, but the indication is that the C-T state requires some excitation energy to be reached.

Prof. Joyner said: Just a comment to indicate that the doping of CeO_2 is complex. We have looked at the influence of a range of dopants on the catalytic reaction $CO + NO \rightarrow CO_2 + N_2$.[1] Several dopants enhance the rate of reaction significantly, including zirconium (4 +), neodymium, praseodymium and lanthanum (all 3 +). However yttrium has little effect.

In general, dopants may simply maintain surface area, or can also enhance the specific reactivity (rate per unit area of catalyst).

1   J. D. Burton, R. W. Joyner and C. J. Norman, in preparation.

Prof. Pettersson responded: Effects of doping on reactivities and the relation between dopants and the host lattice are certainly a very complex issue. As a matter of fact, in the experimental studies (see references in the paper) that the present investigation relates to, several different dopants and also reactions were studied. Among the dopants tested were Ca, Co, Mn, Nd and Pb and the effects on both reduction of SO_2 with CO and also total combustion of methane were studied; very varying effects have been found. In the case of, *e.g.*, Ca a very large enhancement of the SO_2 reaction, but no significant effect on the methane combustion, was found. For the Pb- and Co-doped systems the light-off temperature for SO_2 reduction was substantially increased. The CO oxidation was unaffected for the Co-doped system, but the light-off temperature was substantially increased for the Pb-doped CeO_2; in these two cases sulfide formation (PbS, CoS_2) at the surface was demonstrated. To make matters even more complicated there are strong hysteresis effects in the sense that when the temperature is lowered, *i.e.* scanning the experiment from higher to lower temperatures, the high conversion rate persists to lower temperatures than the initial light-off. For lead the 50% conversion of SO_2 persists to 30 °C lower temperature than for the undoped system, *i.e.* an enhancement of activity once the system has been activated. For Co- and Ca-doping the enhancement is even greater: 65 and 115 °C, respectively. I should finally add that the dopants and reactions studied here form only a small sample of what has been reported in the literature.

We have elected in the present work to study only the Ca-doped system since it seemed the most amenable to a theoretical treatment from an electronic structure point of view while still showing significant effects on the reactivity from the doping. Furthermore, it appears that a single phase is maintained at the levels of doping considered and that we may assume that no complicated chemistry with the dopant takes place (*c.f.* the formation of PbS and CoS_2 mentioned above); this would be a prerequisite for the type of atomistic modeling that we have used to obtain the structures for the quantum chemical modeling.

A final observation is that there will very likely always be many effects contributing to the observed reactivity. The O^{2-} chemistry and the coupling to the thermal motion that we discuss in our contribution does, however, apply generally. For any given system we cannot assert that this is the dominant effect, as is evidenced already for the Ca-doped CeO_2 discussed here, where already without dynamical effects the C-T energies are strongly reduced by the doping.

I believe the importance of our work is two-fold: first of all, in any chemical reaction involving a barrier it is the high-energy tail of the Maxwell–Boltzmann distribution that contributes to the rate. In this respect we introduce also the 'high-energy tail' of the lattice motion as a contributing factor. Secondly, if one can identify what specific phonon modes that produce the required destabilization then the field is open for computer simulations to predict the effects of a particular dopant. For example, for the oxygen abstraction from the lattice one needs to destabilize one oxygen and two cerium cations. If one finds a certain amplitude of this motion for the undoped system at the observed light-off temperature one can theoretically check a number of dopants to see whether a similar amplitude can be obtained at a lower or higher temperature. This is an exciting prospect that could lead to predictions for improved charge-transfer based catalysts.

Dr Shluger said: The thermal fluctuations which you describe are simply combinations of phonons. The lifetime of such fluctuations on a particular oxygen site is about characteristic vibrational times, *i.e.* about a picosecond. Your calculations of oxygen binding energies are static and correspond to particular snap shots of lattice structure, which is completely different after a picosecond. What is the physical meaning of the oxygen binding energy change in this time domain?

Prof. Pettersson replied: Our calculations represent an *ad hoc* combination of dynamical simulations and statical electronic structure calculations. Furthermore, the structure of the lattice before removal of the oxygen is fairly close to the optimum ground state structure, while the structure around the created defect can be expected to relax significantly. The latter effect is not included in our models and consequently our computed oxygen binding energies represent upper bounds. A closer and more consistent coupling between the description of the lattice dynamics and the chemical reaction is thus certainly desirable. At the present time it is difficult to see how this could be accomplished. A Car–Parinello type simulation involving also a coverage of CO molecules would be tempting, but there are several difficulties involved. One is the size of the unit cell for the doped system and the other is the requirement to describe also the highly localized 4f electron of the Ce(III) formed in the C-T process. It would be interesting if this can be done, but I believe it will be difficult.

The situation that we describe is a reaction involving a powdered material in a high-pressure environment. We assume that we always have a high flux of impinging CO molecules on the surface of the crystals such that reactions between CO and the lattice oxygens are attempted for all instantaneous distortions of the surface structure. We furthermore assume that the charge transfer from oxygen to cerium is sufficiently fast to be sensitive to the instantaneous surface structure. This is actually corroborated by the observation of a low-energy Urbach–Martienssen tail[1,2] in optical absorption spectra of non-metallic solids; this is due to coupling between the electronic structure and the phonon modes of the solid.

Our next assumption is that, once the bond begins to form and the adsorbate begins to become CO_2, the back-donation of charge becomes less favourable. With these assumptions the initial charge-transfer should be a reasonable measure of the expected reactivity of a given conformation. We do not have any immediate way to test this by direct computations at the moment. It would certainly be desirable.

Another aspect appears if one chooses to consider the surface as a large molecule and treats the chemical reaction theoretically in the normal way, *i.e.* by approaching the CO to the surface and allowing all degrees of freedom in adsorbate and substrate to relax in a search for the lowest energy transition state. The proposed coupling between the lattice phonons, which involve a large part of the lattice and can represent a significant inertia in the motion of the ions, and the chemical reaction is such that it is not clear to us whether this 0 K (and essentially local perturbation) approach is equivalent to the dynamical (and delocalized) picture presented. The full optimization of structural models of Ca-doped CeO_2 with adsorbed CO is unfortunately also beyond what can currently be treated.

1   F. Urbach, *Phys. Rev.*, 1953, **92**, 1424.
2   W. Martienssen, *J. Phys. Chem. Solids*, 1957, **2**, 257.

Prof. Hermann said: When you consider the charge transfer processes are there physical parameters other than energy that you can evaluate easily, such as, for example, the amount of transferred charge or spin?

Prof. Pettersson replied: In the model of the oxygen abstraction two electrons need to be accommodated after removal of a charge neutral oxygen atom. For the embedded models that we have studied it was always found that two cerium ions have been reduced to Ce(III). This is demonstrated by a 4f population of close to one electron on each center. Only in cases where one of the two cerium cations is not destabilized and where the charge-transfer thus is costly, do we find a significant population in the vacancy, *i.e.* an F_s^+-center. Regarding the charge-transfer energies we have computed these as the singlet–triplet energy difference for our cluster models. We always find one electron in the 4f shell and a hole in the oxygen 2p. The Mulliken population analysis that we have used here can be rather uncertain for the oxygen 2p populations, but owing to the local character of the 4f shell we expect the 4f population to be rather accurate.

Dr Gautier-Soyer opened the discussion of Dr Weiss's paper: I would like to come back to the structure of the thin FeO(111) film. You mention it is laterally expanded, and it seems that it is a more general feature, as we obtain a similar result when growing iron oxides on alumina by atomic oxygen assisted MBE. Could you comment on the way the film is stabilised? Have you ever observed the pressure of Fe^{3+} in this thin layer?

Dr Weiss responded: The first FeO(111) layers grown onto Pt(111) consist of hexagonal Fe–O bilayers which form polar surface structures terminated by close-packed oxygen layers.[1] Such surfaces would be unstable on bulk crystal samples because of infinite electrostatic surface energies. However, in the case of thin films these surfaces can be stabilized by lateral expansions which are accompanied by a reduction of the Fe–O interlayer spacings. The reduced interlayer spacings reduced the electric dipole moments perpendicular to the surface and therefore the electrostatic surface energies of these thin films. We find increased lateral expansions in the second FeO(111) layer if compared to the first layer. Since with each Fe–O bilayer the electrostatic dipole moment increases, the films have to relax more strongly with an increasing number of layers. This mechanism is able to stabilize the FeO(111) films on Pt(111) up to 2.5 ML thickness,[2] then a three-dimensional growth of unstrained Fe_3O_4(111) islands with a more stable surface structure begins.[3,4] This mechanism most likely can also stabilize thin films of other ionic materials grown along polar crystal directions, independent on the substrate material. It is very interesting that you observe the same phenomena when growing FeO(111) films onto alumina, in accordance with this mechanism. We never observed Fe^{3+} species, but always Fe^{2+} in our films on Pt(111). This can be explained by the fact that platinum represents a chemically more reducing substrate material than alumina.

1 H. C. Galloway, P. Sautet and M. Salmeron, *Phys. Rev. B*, 1996, **54**, R11145.
2 W. Ranke, M. Ritter and W. Weiss, *Phys. Rev. B*, 1999, **60**, 1527.
3 W. Weiss and M. Ritter, *Phys. Rev. B*, 1999, **59**, 5201.
4 M. Ritter and W. Weiss, *Surf. Sci.*, 1999, **432**, 81.

Prof. Kirschner asked: Suppose you wished to modify the substrate with respect to crystallographic surface orientation, material, or lattice constant. What are the essential requirements on the substrate to preserve good oxide film growth?

Dr Weiss replied: In the case of iron oxide the substrate overlayer lattice mismatch seems not to play an important role. On Pt(111) the mismatch is 12% and well ordered films are obtained, nevertheless. Onto Pt(100) surfaces with a square symmetry iron oxides also grow in (111) orientation, as on Pt(111).[1] Also on other substrates like alumina the lattice mismatch seems not to be important in the case of iron oxide overlayers. However, this might be different for other oxide materials. In any case, a small lattice mismatch is always good, and it is important that the substrate material is chemically inert. If it is inert against oxidation and if it does not alloy or react with the deposited metal, the best control over film thickness and stoichiometry can be expected. In that case it could even be possible to prepare different oxide phases by using different oxygen gas pressures during the oxidation treatments. In this way we can grow the Fe_3O_4 and Fe_2O_3 phases presented here in a controlled manner.

1 M. Ritter, H. Over and W. Weiss, *Surf. Sci.*, 1997, **371**, 245.

Prof. Joyner said: The addition of potassium is thought to influence the acidity of the surface, and the selectivity (as well as the rate) of the ethylbenzene to styrene reactions. Do you have evidence or comments on either of these points?
In general, the work that you present is both very impressive and convincing.

Dr Weiss answered: It is known that under reaction conditions an active $KFeO_2$ phase is formed on the catalyst surface.[1] For the dehydrogenation reaction a simultaneous action of Brønsted basic oxygen sites that deprotonate the ethyl group and of Lewis acidic Fe^{3+} sites that accept the electron charge is required, where the latter get reduced to Fe^{2+}. Potassium as an electron donor certainly enhances the Brønsted basicity of surface oxygen sites, and in the active Fe_2O_3 and $KFeO_2$ phases a higher number of acidic Fe^{3+} sites exists if compared to the inactive Fe_3O_4 phase. These two effects increase the activity and selectivity of the dehydrogenation of ethylbenzene to styrene, which we also observed in high pressure reaction studies on the single crystalline iron oxide model catalyst films.[2] Regarding an increased surface acidity upon potassium incorporation into the epitaxial Fe_3O_4 magnetite films, we do observe a strongly increased chemisorption of water on the KFe_xO_y films if compared to Fe_3O_4. This indeed indicates a stronger surface acidity, as usually the oxygen lone pair orbital of adsorbed water is bound to Lewis acidic sites on the substrate surface, and the bonding strength increases with the acidity of these sites. However, it is very interesting that the chemisorption strength for styrene considerably decreases on the KFe_xO_y films if compared to Fe_3O_4 and Fe_2O_3. This must be due to a very different chemical interaction in comparison to water, which is currently under investigation. The nature of this interaction is the key for understanding the reaction mechanism of the styrene synthesis and the promoting effect of potassium. Potassium promotion accelerates the desorption of the product molecule styrene, which creates a catalyst surface with low adsorbate coverage under reaction conditions. Thus, many active sites are empty on the time average leading to high conversions.

1 M. Muhler, R. Schlögl and G. Ertl, *J. Catal.*, 1992, **138**, 413.
2 W. Weiss, D. Zscherpel and R. Schlögl, *Catal. Lett.*, 1998, **52**, 215.

Prof. Madey said: My question concerns the stoichiometry of your $K Fe_xO_y$ films. To obtain $KFeO_2$ from $K + Fe_3O_4$, you either need to provide extra oxygen or to reduce Fe_3O_4. Based on your Auger sampling depth, the observed (average) stoichiometry extends rather deep, of order 10–20 Å. Please comment on your procedure for producing the film and indicate how the observed stoichiometry is achieved. Also, did you try depositing K on Fe_2O_3?

Dr Weiss answered: We obtain the same potassium iron oxide films with surface stoichiometries close to $KFeO_2$ and with identical surface structures when performing the annealing step after potassium deposition with oxygen (10^{-6} mbar) or without oxygen (as presented here). However, when annealing Fe_3O_4 films covered with larger amounts of K without oxygen we observe the formation of metallic iron on the surface. Thus, potassium reduces Fe^{2+} and Fe^{3+} species in Fe_3O_4 to metallic Fe^0. Since during the annealing step the metallic Fe^0 desorbs, the same $KFeO_2$ stoichiometry and the same STM and LEED patterns are obtained as after annealing with oxygen. We also deposited K onto Fe_2O_3 films. There, the very similar films with $KFeO_2$ stoichiometry but with more diffuse LEED patterns were obtained, indicating poorer crystalline quality.

Prof. Thornton asked: The behaviour you observe is very similar to what we have seen for $TiO_2(100)$.[1] In that case the data are consistent with K only at the surface. Can you rule out a similar situation here? The decrease in K AES signal above 500 K could be due to K diffusing to both surfaces of the film.

1 K. Prabhakaran, D. Purdie, R. Casanova, C. A. Muryn, P. Hardman, P. L. Wincott and G. Thornton, *Phys. Rev. B*, 1992, **45**, 6969.

Dr Weiss answered: We definitely can rule out that potassium is only sitting on the surface. We have performed AES measurements in combination with TDS. They clearly show that considerable amounts of potassium diffuse into the $Fe_3O_4(111)$ films, at temperatures between 500 and 900 K. However, we cannot rule out at this point that the potassium diffusing into the films accumulates at the substrate/overlayer interface as you suggest. We are going to perform grazing incidence X-ray scattering experiments to address this question.

Prof. Iwasawa said: From the TPD data you estimated the relative adsorption strength of ethylbenzene and styrene and extrapolated to get the relative value at 100 mbar and 900 K, similarly to the catalytic reaction conditions. You also conducted the catalytic performance of the Fe oxide sample in excess H_2O. The presenting H_2O can dissociatively adsorb on the Fe oxide surface. In these conditions the surface and morphology might be changed to those different from the surface characterised by STM and LEED.

Dr Weiss responded: You are absolutely right. The active model catalyst surface formed under reaction conditions at elevated pressures and temperatures might be different from the surfaces we have studied with STM and LEED under ultrahigh vacuum conditions. However, we observe different catalytic activities when running the reaction on the three different iron oxide films Fe_3O_4, Fe_2O_3 and KFe_xO_y. Therefore, the three films must have different surface properties also under reaction conditions. These properties are clearly related to the structures of the films we have characterized prior to the reaction experiments, as these films are transferred into the reactor without contact to air or other contaminants. This structure–reactivity correlation is further reflected by the different adsorption properties of the films observed with TDS, which can explain the observed catalytic activities when considering the extrapolated reactant surface coverages. This observation provides evidence that the surfaces do not change significantly under reaction conditions. However, we have to learn more about the active surface by real *in situ* and post reaction studies.

Prof. Campbell said: Beautiful paper! I have one minor technical issue. If I understand correctly, in extrapolating TPD measured desportion rate constants to 900 K, you have used the activation energies obtained with one set of prefactors (those in Table 1) together with an assumed prefactor of 10^{13} s^{-1}. A more accurate extrapolation would be found by using activation energies, obtained from the TPD rates but by assuming this same 10^{13} s^{-1}. This would ensure at least that your extrapolation curves go through the rate constants actually measured near the peak temperature. What you have done produces extrapolation curves which are several orders of magnitude different than the measured rates at the peak temperatures. One must use consistent activation energies and prefactors, to produce accurate rates.

Dr Weiss responded: This objection is correct, and therefore we have done two things. First, we repeated the TDS measurement and data evaluation for EB adsorbed on Fe_3O_4, where we originally had obtained a prefactor of 2×10^{15} s^{-1}, more than two orders of magnitude higher than all the others. It turned out that this high value was wrong, due to erroneous data analysis. Table 1 in our paper now lists all the correct values. Second, we repeated our surface coverage extrapolation using the actually evaluated activation energies and prefactors as listed in Table 1. The paper now contains the corrected coverage values. The tendency of decreasing site blocking by the product molecules styrene when going from Fe_3O_4 over Fe_2O_3 to KFe_xO_y has become even more evident.

Prof. Freund said: A simple measurement providing information as to whether K migrates into the iron oxide film could be work function measurements. Have you made such measurements?

Dr Weiss responded: We have not made such measurements yet, but we are going to do them soon. Furthermore, we are going to perform low energy ion scattering and grazing incidence X-ray scattering measurements in order to address the question of potassium segregation into the $Fe_3O_4(111)$ films, as well as the possible formation of a ternary compound phase like $KFeO_2$.

Prof. Catlow opened the discussion of Prof. Parker's paper: Eqn. (3) of your paper refers to a number of processes including hydration of H$^+$ and other cations. Could you explain how the energies associated with these processes have been evaluated?

Prof. Parker responded: The equation concerned, for a divalent cation, is

$$2H^+_{(aq)} + M^{2+}_{(surface)} \rightarrow M^{2+}_{(aq)} + 2H^+_{(surface)} \tag{a}$$

The approach is to take a simulation cell and calculate the total interaction energy. Then replace a divalent cation by two protons leaving the simulation cell charge neutral and recalculate the total interaction energy. The standard state is the ions separated to infinity. Thus the difference in interaction energies yields the energy of the following reaction:

$$2H^+_{(g)} + M^{2+}_{(surface)} \rightarrow M^{2+}_{(g)} + 2H^+_{(surface)} \qquad (b)$$

Thus to find the energy of eqn. (a) we add the energy of the reaction

$$2H^+_{(aq)} + M^{2+}_{(g)} \rightarrow M^{2+}_{(aq)} + 2H^+_{(g)} \qquad (c)$$

The components in eqn. (c) can be simulated, but we generally extract the reaction energy from the experimental heats of formation.

There is a final complication, namely, that on introducing the two protons, we replace an O^{2-} lattice ion by $2(OH)^-$ species. Thus in this case we modify eqn. (c) to consider the reaction:

$$2H^+_{(aq)} + 2MO_{(s)} \rightarrow M^{2+}_{(aq)} + M(OH)_{2(s)} \qquad (d)$$

where $MO_{(s)}$ and $M(OH)_{2(s)}$ are the bulk oxide and hydroxide structures. Thus the difference between the experimental reaction energy (d) and the calculated lattice energies of the two structures gives the correction energy for eqn. (a).

Dr Willock asked: Could you give us some more details regarding the potentials used in this work, in particular: (1) was a polarisable water model used, (2) how did the water potential perform for bulk water, (3) how was the oxide—H_2O interaction parametrised?
Was the water oxygen given the same van der Waals parameters as a lattice oxygen?

Prof. Parker responded: (1) This is a polarisable water model with 4 centres, one on each atom and in addition the oxygen atom has a shell. All atoms and the oxygen shell are free to move during the simulations. The potential parameters for the water molecule were empirically fitted to reproduce the experimental dipole moment, the OH bond length and the HOH angle of the water monomer and the structure of the water dimer and infra-red data. The constraint was that the fractional charges ($+0.4e$) for the hydrogen atoms were fixed, in line with the compatible hydroxide potential model developed by Baram and Parker.[1]

(2) We calculated the radial distribution functions, average energy, density, specific heat capacity, compressibility and the self-diffusion coefficient of water.[2] The agreement was adequate, at least as good as many of the potentials that are available, but for us this model has the advantage that we can use it with our solid–solid potentials. In general, the potential is probably slightly overbound resulting in a higher density. This is also indicated by the self-diffusion coefficient which was calculated to be 1.15×10^{-9} m^2 s^{-1} (2.3×10^{-9} m^2 s^{-1} at 298 K by experiment). This value is low compared to the experimental value at 298 K, but agrees with an experimental value of 1.17×10^{-9} m^2 s^{-1} for a water temperature of 275 K. The energy of vaporisation is calculated to be 43.0 kJ mol^{-1} which clearly agrees with the experimental value of 43.4 kJ mol^{-1} at 310 K.

(3) The potential model used for simulating the interactions between the water molecules and the solid surface follows the approach by Schroder et al.[3] The cation–oxygen interaction parameters, derived for cation–lattice oxygen interactions, are modified to compensate for the reduction in electrostatic interaction due to the lower charge on the water oxygen.

To answer your final question: no, we follow the view that the van der Waals interaction is related to the polarisabilities of the atoms concerned and since the water oxygen and lattice oxygen have different polarisabilities they have different polarisabilities.

1 P. S. Baram and S. C. Parker, *Philos. Mag. B*, 1996, **73**, 49.
2 N. H. de Leeuw and S. C. Parker, *Phys. Rev. B: Condens. Matter*, 1998, **58**, 13901.
3 K. P. Schroder, J. Sauer, M. Leslie, C. R. A. Catlow and J. M. Thomas, *Chem. Phys. Lett*, 1992, **188**, 320.

Prof. Jacobs asked: (1) You say in the text following eqn. (4) 'protons bound to the cation vacancy'—were the positions of the protons optimised? (2) Is this a shell model calculation, and if so, what potential are you using for OH^-?

Prof. Parker answered: (1) Yes, we check for convergence and find that we have to allow full relaxation, of all atoms within about 20 Å of an interface. If geometry optimisation is not included the energies from shell model calculations would be meaningless.

(2) This is a shell model calculation. The hydroxide potential was originally derived by Saul et al.[1] for NaOH, however, we modified it slightly when considering further hydroxides.[2]

1 P. Saul, C. R. A. Catlow and J. Kendrick, *Philos. Mag. B*, 1985, **51**, 107.
2 P. S. Baram and S. C. Parker, *Philos. Mag. B*, 1996, **73**, 49.

Dr Shluger said: You are using formal ionic charges in your simulations of low-coordinated surface sites. With appropriate choice of other parameters this may give reasonable lattice relaxations at corners and other surface sites. However, it is known that there is a significant electron density redistribution at low-coordinated surface sites. Therefore the electric field outside low-coordinated sites can be different from that predicted by your calculations. This could affect both the interaction of these sites with the solution and energies of dissolution of H^+ ions from low-coordinated sites. Could you possibly comment on these points?

Prof. Parker responded: We consider the potential model parameters to be robust and by including a shell model to give at least a very crude representation of the electronic relaxation we require the model to reproduce the relaxation and electric fields around defects at surfaces. Thus we would expect there to be reasonable accord with solution energies at low coordinated sites. However, the only way to be sure is to perform comparisons of these atomistic simulations with appropriate electronic structure calculations.

Dr Noguera said: I remember very nice results for the dissolution of calcite in water by McCoy and LaFemina[1] at the Les Houches meeting. How do your results compare with his?

1 J. M. McCoy and J. P. LaFemina, *Surf. Sci.*, 1997, **373**, 288.

Prof. Parker replied: The lovely work from the PNL group that McCoy and LaFemina described used kinetic Monte Carlo using a solid-on-solid model[1] to give energies for the creation and motion of kinks. The creation of a kink on the slow-moving (acute angled) step costs 0.69 eV (the exact value depends on which model is used to fit the data) and the creation of a kink on the fast-moving step costs 0.67 eV. Our calculated values are 103.7 kJ mol^{-1} (1.07 eV) for the slow step and 45.8 kJ mol^{-1} (0.47 eV) for the fast one. Our estimated error in the calculated values is $\pm$ 0.2 eV. The agreement is only fair but, given the errors, is reasonable.

1 J. M. McCoy and J. P. LaFemina, *Surf. Sci.*, 1997, **373**, 288.

Prof. Madey said: You indicate that MgO(111) is unstable and forms pyramidal microfacets with {100} facets upon annealing; you refer to early electron microscopy measurements by Henrich[1] to support this picture. There are recent careful measurements of MgO(111) by AFM and scanning electron microscopy by Gajdardziska-Josifovska et al.,[2] who demonstrated that the microfacets are not {100} but are vicinal surfaces inclined by $\approx 10°$ from (111).

1 V. E. Henrich, *Surf. Sci.*, 1976, **57**, 385.
2 R. Plass, J. Feller and M. Gajdardziska-Josifovska, *Surf. Sci.*, 1998, **414**, 26.

Prof. Parker said: This is interesting new work. As I understand it, one can view the pyramidal features as essentially {100} facets with steps and it is the addition of these steps that causes the change in incline. We have modelled much smaller features, perhaps best described as nanofacets, which do show the energetic preference for forming a faceted surface. It would be interesting to extend the atomistic simulation calculations to look at the stability of these vicinal surfaces.

Dr Taylor said: My question concerns the need to reconcile theory and experiment on the subject of the hydroxylation of MgO(100) terrace sites. As mentioned in your paper, the results of theoretical studies (*e.g.* ref. 1 and 2 below) are in agreement that the (100) surface does not dissociate water, energetically preferring it to remain intact, whilst defect sites are predicted to dissociatively chemisorb water. Equally, UHV experimental studies on annealed single crystal MgO(100) surfaces and on epitaxial MgO(100) films are in agreement that coverages of at least one OH^- per surface cation can be achieved (*e.g.* ref. 3–8 below). At room temperature the *rate* of this dissociative chemisorption seems to be controlled by the defect sites, but the build-up of such a high coverage implies that the *equilibrium* adsorption sites were, at least originally, (100) terrace sites. Do you have any ideas as to what is needed to reconcile theory and experiment? For example is it necessary to include entropy to refine predictions which at present are based on calculations which show that dissociation is endothermic? Alternatively, perhaps there is an adsorption-driven reconstruction of the (100) surface as yet not included in the calculations, although this would have to be minor or disordered since post-adsorption LEED shows only a slight broadening of the spots from a (1 × 1) (100) pattern.[7]

1 N. H. de Leeuw, G. W. Watson and S. C. Parker, *J. Phys. Chem.*, 1995, **99**, 17219.
2 C. Noguera, J. Goniakowski and S. Bouette-Russo, *Surf. Sci.*, 1993, **287/288**, 188.
3 X. D. Peng and M. A. Barteau, *Surf. Sci.*, 1990, **233**, 283.
4 M. A. Karolewski and R. G. Cavell, *Surf. Sci.*, 1992, **271**, 128.
5 M. C. Wu, C. A. Estrada, J. S. Corneille and D. W. Goodman, *J. Chem. Phys.*, 1992, **96**, 3892.
6 M. J. Stirniman, C. Huang, R. Scott-Smith, S. A. Joyce and B. D. Kay, *J. Chem. Phys.*, 1996, **105**, 1295.
7 P. Liu, T. Kendelewicz, G. E. Brown, Jr. and G. A. Parks, *Surf. Sci.*, 1998, **412/413**, 287.
8 J. I. Mmojieje and A. O. Taylor, poster presentation at this meeting, to be published.

Prof. Parker responded: I do not have any firm view as to how the theory and experiment can be reconciled. All calculations agree that at one monolayer coverage the associated water molecule is energetically preferred to the dissociated species. The energy difference is such that I consider that entropic factors are not going to change the order of stability at room temperature. Given the stability of the surface it is unlikely that there is a phase transition. This leaves only defect sites as sites for dissociation.

Dr Taylor added: Thank you, so it may be that dissociation is occurring at defect sites where a *local* high coverage is achieved and that these islands of high coverage gradually spread out over the surface maintaining the stable high coverage.

Dr Taylor commented: The failure to observe water dissociation on MgO smoke crystallites with low defect density[1] is sometimes cited as experimental evidence in support of the results of calculations which predict that such dissociation is thermodynamically unfavourable at most coverages.[2] Experimentally, we believe that the rate of water dissociation is controlled by defect sites, so the lack of observed reaction at the (100) facets of MgO smoke may simply reflect a kinetic limitation rather than a thermodynamic prohibition. Equally the experimentally observed high coverage on single crystal (100) planes may reflect a metastable state which very slowly desorbs at room temperature. There is some evidence of this from the work of Liu *et al.*[3]

1 C. F. Jones, R. A. Reeve, R. Rigg, R. L. Segall, R. St. C. Smart and P. S. Turner, *J. Chem. Soc., Faraday Trans. 1*, 1984, **80**, 2609.
2 N. H. de Leeuw, G. W. Watson and S. C. Parker, *J. Phys. Chem.*, 1995, **99**, 17219.
3 P. Liu, T. Kendelewicz, G. E. Brown, Jr. and G. A. Parks, *Surf. Sci.*, 1998, **412/413**, 287.

Dr Noguera addressed Prof. Parker: Coming back to the dissociation of water on MgO(100), it now seems well established in the calculations that dissociation probability is very different for a single molecule or for a full monolayer.[1]

1 L. Giordano, J. Goniakowski and J. Suzanne, *Phys. Rev. Lett.*, 1998, **81**, 1271.

Prof. Parker responded: Yes, in addition earlier atomistic simulations[1] support that view by virtue of a strong dependence of dissociation energy on coverage.

1 N. H. de Leeuw, G. W. Watson and S. C. Parker, *J. Phys. Chem.*, 1995, **99**, 17219.

Prof. Freund commented: Surface science in general has seen an enormous development over the last 30–35 years. There is a trend from studies of static properties towards the investigation of the dynamical behaviour of surfaces. Clearly for metal surfaces new experimental methods are being developed and also the analysis of more traditional methods is pushed in this direction. I would like to know whether you foresee a similar development in the studies on oxide surfaces?

Dr Shluger responded: With respect to the importance of dynamical issues noted by Prof. Freund, I would like to point out that the results presented by Lars Pettersson are a step in that direction. Thermal fluctuations are certainly important and their spectroscopic manifestations are well known. They are responsible, for instance, for the Urbach tail in the optical absorption of semiconductors and insulators, and for the dynamics of localisation of holes and excitons in insulators. However, one should be careful about the relative time-scale of thermal fluctuations and chemical processes at surfaces. In particular, the mobility of adsorbed molecules is not great and the effect of fluctuations on the charge transfer and chemical reactions studied by Pettersson can be important only at large coverages.

Dr Weiss said: The interaction of water vapour and liquid water with metal oxide surfaces plays a very important role in many fields like catalysis, environmental chemistry, electrochemistry and biology. Molecular or dissociative adsorption is observed on different oxides, but the role of atomic structural elements for the adsorption process is still not understood well. We have performed a comparative adsorption study for water on oxygen terminated FeO(111) films and iron terminated $Fe_3O_4(111)$ films. It clearly revealed the metal atoms exposed on $Fe_3O_4(111)$ to be responsible for the dissociative adsorption. I think that exposed metal sites are responsible for the chemical reactivity of oxides in general. Furthermore, we observe the coadsorption of molecular water species onto $Fe_3O_4(111)$ after the dissociated species gets separated, as predicted by recent calculations for water on MgO(100) and $TiO_2(110)$. Are there any other experimental observations of coadsorbed dissociated and molecular water species?

Prof. Flavell said: In support of this comment, I believe that in our experiments we also see co-adsorbed H_2O and OH on some perovskite surfaces even at quite low temperatures.

Dr Kantorovich said: It is not fair to say that dynamical issues in molecular adsorption have not been yet addressed. For example, MD simulations, notably of the *ab initio* Car–Parinello type, have already done a good job in this. I can give an example, Parinello study of H_2O dissociation on a MgO step, or even more sophisticated calculations of H_2 dissociation on a number of surfaces in which vibrational and rotational degrees of freedom in the quantum mechanical context have been shown to be absolutely essential in understanding the process.

Prof. Freund responded: I did not mean to imply that nothing has been done. Probably theory is here ahead of experiments. I intended to stimulate a discussion as to whether experiment and theory should move in this direction.

Dr Kawai opened the discussion of Prof. Vanderbilt's paper: (1) On TiO_2 terminated $SrTiO_3(100)$ surface, surface reconstructions have been reported, where they involve much larger surface unit cells. Since the results should strongly depend on the size of the cell and the symmetry calculated, I wonder what you would expect from the reconstructed surface that is observed experimentally.

(2) Concerning the application to nano-sized memory devices, knowledge of the interface between the metal electrodes should be of great interest. Prof. Jennison suggested the electrostatic interaction to operate between metals on oxides in the first session of this meeting, which may also be the case. What would you expect on the matter?

Prof. Vanderbilt responded: (1) The issue of surface reconstructions is an important one. Unfortunately we are in a situation with perovskites where the experimentalists tend to study reconstructed surfaces, while the theorists tend to study unreconstructed ones. I think it is important for both communities to make efforts to overcome this difficulty. That is, I believe that the theorists need to extend their studies to reconstructed surfaces, but also that the experimentalists should try to prepare and characterize (1 × 1) surfaces.

Theoretically, there is no obvious reason why a stoichiometric surface of a non-polar surface of a perovskite like $SrTiO_3(100)$ should be expected to reconstruct. Thus, I strongly suspect that the reconstructions result from deviations from ideal stoichiometry in the surface layer. For example, the reconstruction may correspond to some arrangement of oxygen vacancies. We should begin to test such models using the theoretical calculations. But it would also be extremely valuable if experimental information about the surface stoichiometry could somehow be obtained, or if surface stoichiometry could be controlled during the surface preparation. Another possible approach might be to try to obtain cleaved surfaces for study. However, I recognize that there are potential obstacles to such developments.

(2) A theoretical study of perovskite/metal interfaces would indeed be very valuable. We decided to focus first on the perovskite/vacuum interface as a natural first step. For the case of a metal, one has to choose one specific metal, and potentially deal with non-universal issues connected with lattice mismatch, *etc.* Alternatively, one could conceivably study an interface with a 'model metal' such as jellium. We will consider these possibilities when planning future work.

Prof. Thornton said: I believe it is possible to form a (1 × 1) termination of $SrTiO_3$(001) although it does not seem possible to image it with STM. We have preliminary room temperature SXRD data which points to an ordered surface with a structure which could be close to that which you have calculated.

Prof. Vanderbilt replied: It might be that the conditions of bulk stoichiometry needed to get a stoichiometric (1 × 1) surface are not compatible with STM imaging because of conductivity limitations. I look forward to learning more about your results.

Dr Castell addressed Prof. Thornton: We have found that where the $SrTiO_3$(001) surface displays a (1 × 1) LEED pattern, then STM shows this to be a rough surface. When preparing an atomically flat surface a c(2 × 4) reconstruction results. Are these observations similar to your results?

Prof. Thornton responded: Yes, our experience with STM is similar to yours, that is the (1 × 1) surface does appear rough, while the c(2 × 4) termination is relatively smooth. This may be an artefact associated with the difficulty of tunnelling into the SrO terminated terraces.[1] Our ability to record well defined SXRD data from the (1 × 1) termination gives this idea some credence.

1 Q. D. Jiang and J. Zegenhagen, *Surf. Sci.*, 1999, **425**, 343.

Dr Castell addressed Prof. Vanderbilt: For device applications and use as a substrate where flat surfaces are required, modelling reconstructed (001) surfaces of perovskites would seem a fruitful way forward.

Prof. Vanderbilt responded: Yes, the discussion with Dr Kawai has touched on this point earlier. From the theoretical point of view, we need a plausible model structure for the reconstructed geometry as a starting point for the calculations. In the absence of information about the surface stoichiometry, such models are poorly constrained. Any guidance that you can give us in this regard would be very useful.

Dr Noguera said: Could you comment on the fact that, in Table 7 of your paper, the values found for the gap in some of the slabs are larger than in the bulk?

Prof. Vanderbilt responded: This is an artifact of the use of slabs of finite thickness. In the limit that the slab thickness goes to infinity, one must either find that the gap becomes equal to the bulk gap, or smaller than it, the latter case corresponding to the intrusion of surface states into the gap.

Prof. Corà commented: In several transition metal oxides, the relaxation induced by the surface may extend over a relevant sub-surface region. When examining the calculated band-gap at the surface, we should take care that the slab employed has a sufficient thickness, to include the relevant geometric and electronic effects.

In my calculations (see Section 4 of the paper), I have used slabs up to 30 Å thick; slabs of thickness ≈ 20 Å often provided non-converged results.

Dr Noguera added: You are right that slabs must have a sufficient thickness, so that central layers are bulk-like. However, there is no magic number, defined once and for ever, which fixes the optimal slab thickness. It depends very much on the oxide upon consideration, and for a given oxide, it depends upon the surface orientation. For example, a good description of the very dense (100) surface of MgO requires very few layers: the sub-surface layer is nearly bulk-like. At variance, on the polar (111) surface, 7 to 10 layers are required. In the particular case of $TiO_2(110)$, we found, and this was in agreement with previous results,[1] that the slab should involve at least 5 Ti_2O_2 layers. This is the thickness that we now use, when studying the deposition of Na on $TiO_2(110)$.

1 S. P. Bates, G. Kresse and M. J. Gillan, *Surf. Sci.*, 1997, **385**, 386.

Prof. Harrison addressed Prof. Vanderbilt: Might relativistic effects alter the 6s6p splitting of the Pb orbitals and thus the covalent bonding. I believe that these effects have been seen in calculations on molecular complexes.

Prof. Vanderbilt responded: We included scalar relativistic effects in the usual way, *i.e.*, through the construction of a Pb pseudopotential based on a Dirac all-electron atomic calculation. (Of course, spin–orbit effects were not included.) Since we did not try the calculation without scalar-relativistic effects, I cannot tell you precisely what would have been the size of the errors introduced by omitting them, but I agree with you that they would probably be quite significant.

Prof. Finnis said: Without going into the technical detail, do you have any comment on how the ferroelectric displacement perpendicular to the surface will behave, and how the presence of a substrate might modify this displacement?

Prof. Vanderbilt replied: The case of a ferroelectric polarization perpendicular to the surface is a very interesting one that we have just begun to explore. To accommodate this case, one needs to provide and control the perpendicular electric field in the vacuum region that is connected with the presence of polarization-induced surface charges. We do this by placing an external dipole layer in the middle of the vacuum region. Our first idea was to adjust its magnitude such that the interior of the slab would be in a fully developed ferroelectric state with spontaneous polarization normal to the surfaces (*i.e.*, with vanishing macroscopic field inside the slab). However, for this case the electric field generated in the vacuum region is so large that electrons tend to get emitted from the negatively charged surface and accumulate in the vacuum region near the external dipole layer. We are now exploring the application of more modest external electric fields, and observe an interesting competition between the surface-related buckling and bulk-related ferroelectric distortion in the surface layers.

Dr Corà said: A second way to investigate the ferroelectric (FE) relaxation at the surface, that does not create a macroscopic polarisation in the direction perpendicular to the surface, consists in employing slabs with a mirror plane of symmetry in the centre. When the slab thickness is sufficient to avoid interferences in the relaxation induced by the upper and lower surfaces, we can assume the results represent with sufficient accuracy the FE-like relaxation perpendicular to the surface. A similar approach has been employed both by Prof. Vanderbilt and in my calculations (both studies have been presented at this meeting).

Although Prof. Vanderbilt examined a $BaTiO_3$ layer with a FE distortion parallel to the surface, the relaxed structure reproduces the same results as my calculations in the direction perpendicular to the surface. The results in Table 2 of Prof. Vanderbilt's paper, for instance, show that the outermost Ti–O bond [Ti(1)–O_{111}(2)] is shortened by $0.0383a$ (where a is the bulk lattice parameter); proceeding along the same column perpendicular to the surface, the O_{111}(2)–Ti(3) bond elongates by $0.0135a$ and Ti(3)–O_{111}(4) shortens by $0.0092a$. A very similar trend emerges in my calculations. According to eqn. (4) proposed in Prof. Vanderbilt's paper such a relaxation corresponds to a FE distortion perpendicular to the surface. The O_{111}(4) ion is in the centre of the slab, so that the FE mode induced by the surface cannot propagate further in the calculations performed in that study.

Prof. Vanderbilt responded: When considering a situation in which the bulk is paraelectric, I would suggest that the surface relaxation should not properly be described as 'ferroelectric' unless it involves a breaking of symmetry. After all, the surfaces of all oxides, including ones such as MgO that are not remotely ferroelectric, exhibit some surface rumpling that could otherwise be construed as 'ferroelectric' in the sense of your comment.

However, your point is otherwise well taken. I believe that one should expect such a surface distortion to decay exponentially into the bulk with a decay length that can be derived from the 'complex phonon band-structure' as the inverse of the imaginary wavevector at which a longitudinal mode frequency vanishes. It might be interesting to analyze our results and yours from this point of view.

Prof. Jennison said: I recall hearing of a rather complete treatment of ferroelectricity in a perovskite oxide carried out by the group of Krakauer, where using DFT they computed the phonon spectrum, identified the soft mode, modeled it with an effective Hamiltonian, and finally predicted rather well the phase transitions.[1]

My question then concerns whether, to understand the effects of a surface or finite film thickness, is it also necessary to compute the phonon spectrum, in addition to the relaxations and distortions presented in your paper?

1 H. Krakauer, R. Yu, C.-Z. Wang and C. LaSota, *Ferroelectrics*, 1998, **206**, 133.

Prof. Vanderbilt responded: Yes, we have also been heavily involved in this kind of calculation. See, for example, refs. 22 and 30 in our paper. This is what should be done to study the surface distortions at finite temperature. We have so far carried out only zero-temperature surface calculations. I believe it should be possible to extract from these zero-temperature calculations some modifications of the parameters of the effective Hamiltonian at the surface, such that the effective Hamiltonian method could be applied to the finite temperature surface problem. However, we have not pursued this approach, partly because of the point raised a little while ago by Dr Kawai to the effect that the perovskite/metal interface is the more interesting case from the point of view of applications.

Dr Egdell opened the discussion of Prof. Flavell's paper: Could you comment on why your experiments are carried out on (111) perovskite surfaces? The simplest considerations do not immediately suggest that this will be a stable surface.

Prof. Flavell responded: No, I agree that this surface is not obviously stable. Like the cuprate superconductors, these materials are very hard to grow. We work with very small crystals, where we are able to cleave or scape only parallel to an as-grown face. What is interesting about this is that the as-grown faces appear to be (111). I note that $LaCoO_3$ is not cubic, but significantly rhombohedrally distorted,[1] and I suspect that the distortion of the $Co-O_6$ octahedra entailed may have some role in stabilising the (111) surface. It would be very helpful if calculations (of any sort!) could be produced for this surface.

1 For example, R. Caciuffo, D. Rinaldi, G. Barucca, J. Mira, J. Rivas, M. A. Senaris-Rodriguez, P. G. Radaelli, D. Fiorani and J. B. Goodenough, *Phys. Rev. B*, 1999, **59**, 1068.

Dr Schindler said: Further indications for the stability of perovskite (111) surfaces come from the growth of hexagonal crystallites in ceramic $BaTiO_3$. In addition, it is easier to obtain well ordered $BaTiO_3$ (111) surfaces with a uniform superstructure than on the (100) surface.

Dr Noguera said: In relation to the question of stability of (111) faces of perovskites, we have recently performed calculations on the $SrTiO_3(111)$ surface.[1] We found that the planar stoichiometric, unreconstructed surface has a relatively low surface energy, despite the fact that it is polar. We assigned this effect to the fact that charge redistribution at the surface, necessary to cancel the macroscopic dipole moment, can take place without inducing metallisation of the surface, at variance with MgO(111).

1 A. Pojani, F. Finocchi and C. Noguera, *Surf. Sci.*, accepted.

Prof. Flavell responded: Yes, thank you, this is extremely helpful. I think there is an additional special feature of this surface, though, which is different to $SrTiO_3$, *viz.* the rhombohedral distortion mentioned earlier. I think this may play a role in the unusual stability of the (111) surface of $LaCoO_3$.

Prof. Thornton said: It seems curious that the surface is reactive to background gases and yet does not largely dissociate water at 150 K.

Prof. Flavell replied: My assumption here (as discussed in the paper) is that the surface degradation we observe is occurring at step and defect sites, whereas the majority water species which we see in Fig. 10 (of the paper) is adsorbed onto Co terrace sites. If one looks carefully at Fig. 10, it can be seen that the $3a_1$ feature appears rather broad at 150 K, and there is a low binding energy shoulder on the $1b_1$ feature. This strongly suggests that some water is dissociated to OH^- even at these low temperatures, probably at defect sites. Preferential dissociation on stepped oxide surfaces has been previously observed for related perovskite oxides, such as $SrTiO_3$.[1]

1 N. B. Brookes, G. Thornton and F. M. Quinn, *Solid State Commun.*, 1987, **64**, 383.

Dr Kawai said: $La_{1-x}Sr_xCoO_3$ is a correlated electron system and the electronic state strongly depends on the amount of doping. At surfaces, where symmetry breaks one would expect a quite different feature to the bulk. Does the reactivity with H_2O relate to the surface state?

Prof. Flavell responded: In principle, you are correct that there should be a very different electronic structure in the surface top-most layer from the bulk. This may well influence the surface reactivity. For this system, it is very interesting to compare photoemission results (which are merely surface sensitive) with MAES results (which are surface specific to only the top-most surface plane). A MAES study has been published for this surface,[1] which shows very strong enhancement of the feature at around 10 eV binding energy, and of one around 3.7 eV to lower binding energy. These features seem to us to be very similar to the difference spectrum of Fig. 3, and in our view may indicate very strong surface reactivity.

1 S. Masuda, M. Aoki, Y. Harada, H. Hirohashi, Y. Watanabe, Y. Sakisaka and H. Kato, *Phys. Rev. Lett.*, 1993, **71**, 4214.

Dr Lindsay asked: Have you been able to obtain LEED patterns from these samples?

Prof. Flavell replied: We are certainly not able to get good LEED patterns from these surfaces. The best we can obtain is a possible indication of the odd spot! The crystals are rather small, and we align them prior to an experiment by Laue back reflection. However, this tells us that the crystals are twinned, and we may be studying several (111) crystallites at the surface.

Prof. Catlow asked: Can you comment on the extent of dopant segregation at the surface in the Sr doped materials, and what effects will such segregation have on your measurements?

Prof. Flavell responded: Yes, Sr does appear to segregate in this and all related materials I know of, when the surfaces are equilibrated with their environment.[1] Interestingly, it segregates in both perovskite (ABO_3) and layered perovskites (A_2BO_4), over a range of compounds, including many systems studied by Egdell, Cox and others, such as $La_{1-x}Sr_xVO_3$,[2] $La_{1-x}Sr_xCoO_3$,[3] $La_{2-x}Sr_xCuO_4$,[4] $La_{2-x}Sr_xNiO_4$,[1] and the Fe-doped derivative,[1] as shown in Fig. 1.

All the available data appear to indicate that the segregation for all systems falls onto the same isotherm, shown in Fig. 1. Clearly, this is important in interpreting our own data for the Sr-doped crystal, as it means that the surface Sr content may be enhanced over the bulk doping level, and this is known to influence the spin-state of the sample (*e.g.* ref. 5). However, it should be noted that the data shown above refer to well established segregation in equilibrated ceramics; in our case, when we prepare a new crystal surface by cleaving or scraping, we study a composition which is more representative of the bulk sample.

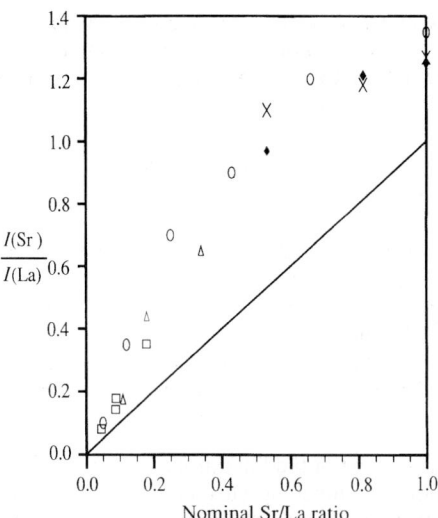

Fig. 1 The XPS Sr/La intensity ratio (corrected for ionization cross-sections) plotted against nominal bulk Sr/La ratio for perovskite and layered perovskite materials. Open circles $La_{1-x}Sr_xVO_3$, ref. 2; open triangles $La_{1-x}Sr_xCoO_3$, ref. 3; open squares $La_{2-x}Sr_xCuO_4$, ref. 4; crosses $La_{2-x}Sr_xNiO_4$, ref. 1; filled diamonds $La_{2-x}Sr_xNi_{1-y}Fe_yO_4$, ref. 1.

1   J. F. Howlett, W. R. Flavell, A. G. Thomas, J. Hollingworth, S. Warren, Z. Hashim, M. Mian, S. Squire, H. R. Aghabozorg, Md. M. Sarker, P. L. Wincott, D. Teehan, S. Downes, D. S.-L. Law and F. E. Hancock, *Faraday Discuss.*, 1996, **105**, 337.
2   R. G. Egdell, M. R. Harrison, M. D. Hill, L. Porte and G. Wall, *J. Phys. C: Solid State Phys.*, 1984, **17**, 2889.
3   J. P. Kemp, D. J. Beal and P. A. Cox, *J. Solid State Chem.*, 1990, **86**, 50.
4   R. G. Egdell, W. R. Flavell and M. S. Golden, *Supercond. Sci. Technol.*, 1990, **3**, 8.
5   R. Caciuffo, D. Rinaldi, G. Barucca, J. Mira, J. Rivas, M. A. Senaris-Rodriguez, P. G. Radaelli, D. Fiorani and J. B. Goodenough, *Phys. Rev. B*, 1999, **59**, 1068.

Prof. Freund asked: Because you mentioned that your samples exhibited weak LEED patterns, I wonder whether you have tried to measure angle resolved photoelectron spectra in order to perform a symmetry analysis of the valance band features?

Prof. Flavell responded: Yes, we have made angle-resolved photoemission measurements. We observed a very small amount of dispersion, but on the whole the behaviour appeared to be far more 'flat band' than calculations. While there may of course be good reasons for this, we felt that we were not sufficiently confident of the single-crystal nature of the surface to publish the data. We have evidence from Laue back reflection that the crystals are twinned, and feel that there is a good possibility that the experiment may have probed several different crystallites at the surface.

Prof. Harrison said: Perhaps the UPS difference spectra for water adsorption is best interpreted in terms of a dynamic model of OH, H and H_2O co-adsorption at the surface. First principles calculations demonstrate that co-adsorption is favourable and that thermal fluctuation are likely to be sufficient to explore a variety of 'molecular' and 'dissociated' states.

Prof. Flavell answered: Yes, I think this is perfectly possible. Your observation that water and OH^- may coexist on oxide surfaces, even in the absence of defects[1] shows us that adsorption is much more complex than we previously supposed, even on a perfect surface. As an experimentalist, it is very nice to see calculations which are now able to tackle the rather high 'real' doses of water which we use in our experiment, rather than single, isolated molecules.

1   P. J. D. Lindan, N. M. Harrison and M. J. Gillan, *Phys. Rev. Lett.*, 1998, **80**, 772.

Dr Egdell opened the discussion of Dr Corà's paper: It would be very interesting to use your surface calculation on WO_3 to help clarify whether STM images of the c(2 × 2) reconstruction described in your paper are dominated by electronic effects (which would lead to imaging W as maxima) or by structural effects (which would lead to imaging O as maxima).

Dr Corà responded: I didn't address explicitly this point in my calculations; I should therefore extend the study to provide a more accurate answer. By use of the Tersoff–Hamann approximation, the relevant information can be obtained with little extension to the functionality of the code employed. As I have shown in the paper, however, the (6-0) perovskites such as WO_3, are those where the hybridisation of the O 2p and M d atomic orbitals due to covalence is highest. The conduction band of WO_3, whose density of states is imaged in the STM micrographs, has therefore a relevant contribution from the O 2p states; in this case I would expect the tunnelling current to be more sensitive to the surface topology, leading to the imaging of the outermost oxygen ions (those labelled as O_1 in the paper) rather than the W ions in the second layer (labelled W_2 in the paper).

Prof. Hayden commented: In spite of the limitations in relying on the density of states extracted from the calculations to indicate tunnelling currents in STM, if the information is available it would provide an excellent starting point for the experimentalist in interpreting STM images.

Prof. Vanderbilt addressed Dr Corà: First, I would like to clarify a point on which a casual listener might have been misled by your presentation. It is, of course, not true that there is no hybridization between Ti 3d and oxygen 2p states in the undistorted cubic structure of $BaTiO_3$. As you correctly point out, such hybridization is absent at certain high-symmetry points in the Brillouin zone. However, as I'm sure you know, an integral over the whole Brillouin zone gives substantial hybridization even in the undistorted paraelectric structure.

Second, I wonder if you would care to comment on the relative suitability of Hartree–Fock vs. DFT for this class of systems, especially as regards the ability of Hartree–Fock to predict the ferroelectric distortions of perovskites, for which a considerable body of previous work has shown DFT to be rather successful.

Dr Corà responded: Concerning your first comment, you are absolutely right; the solution in the points of low symmetry of the reciprocal lattice includes important hybridisations of the O 2p and M d atomic orbitals. The perfectly ionic description of the M–O bonding in the cubic phase occurs only in special high-symmetry positions of the first Brillouin zone; the overall solution is due to a weighted average of the solution over the whole Brillouin zone, and gives rise to a non-negligible M–O covalence.

As I have shown in my paper, however, on reducing the symmetry of the crystal, for instance passing from the paraelectric cubic phase to the ferroelectric tetragonal phase, the spectrum of energy levels and their composition in terms of O and M atomic orbitals shows the biggest change in the Γ-X direction of reciprocal space. The description that I have given for this high-symmetry direction of the reciprocal space is therefore the relevant quantity to characterise how the electronic distribution changes during the ferroelectric distortion.

Answering your second question, I would first like to note that within the DFT we should distinguish between the different exchange and correlation functionals proposed. The instability of a high-symmetry phase towards structural distortions can be investigated by calculating its phonon spectrum; modes of imaginary frequency correspond to energetically stable distortions. In earlier work performed within the local density approximation (LDA), the structure at the optimised cubic geometry has no vibrational mode with imaginary frequency (*e.g.*, see ref. 1–3 given below).

The LDA lattice parameter is usually underestimated with respect to experiment; the accepted procedure has been to repeat the calculations at the experimental value of the lattice parameter. By imposing a larger interatomic distance, the dynamical matrix displays the correct, imaginary, frequency for the ferroelectric mode. We must nonetheless bear in mind that the investigations at a non-optimised geometry will concern the properties of a system which is not in thermodynamic equilibrium; whether the imaginary frequency should in this case be interpreted as a correct prediction of the LDA investigations, or as an artefact of the imposition of an unbalanced compromise between covalent, ionic and short-range repulsion contributions onto the ions in the structure, is not clear. This shortcoming is solved by the use of gradient corrected density functional Hamiltonians, in which the ferroelectric phonon has the correct imaginary frequency also at the optimised lattice parameter.

Results of Hartree–Fock calculations to describe the ferroelectric distortion in perovskites are less systematic. In the work of Dall'Olio et al. on $KNbO_3$,[4] the authors have found that HF calculations provide the correct order of stability of the cubic and ferroelectric phases. The results, however, depend on the basis set employed. In my calculations I have employed a basis set of split-valence quality, which does not include polarisation functions on the oxygens. As shown by Dall'Olio et al., this basis set underestimates the stability of the ferroelectric phase. Decreasing the basis set quality, however, has allowed me to increase the geometric complexity of the systems investigated. Even with the basis set employed here, both $BaTiO_3$ and $KNbO_3$ are calculated as stable in the ferroelectric phase; WO_3 and MoO_3 in the antiferroelectric, in agreement with experiment.[5] The latter results suggest that the HF Hamiltonian does correctly reproduce the ferroelectric properties of perovskites.

1 A. V. Postnikov, T. Neumann, G. Borstel and M. Methfessel, *Phys. Rev. B*, 1993, **48**, 5910.
2 A. V. Postnikov, T. Neumann and G. Borstel, *Phys. Rev. B*, 1994, **50**, 758.
3 D. J. Singh and L. L. Boyer, *Ferroelectrics*, 1992, **136**, 95.
4 S. Dall'Olio, R. Dovesi and R. Resta, *Phys. Rev. B*, 1997, **56**, 10105.
5 F. Corà, PhD Thesis, The Royal Institution of Great Britain and The University of Portsmouth, UK.

Prof. Harrison commented: Systematic studies of TiO_2 bulk and surfaces indicate that the HF, LDA and GGA approximations yield very similar geometries.[1] Surface formation energies are systematically lower in GGA than HF.

1 J. Muscat, PhD Thesis, University of Manchester, UK, 1999.

Dr Shluger commented: There have been several suggestions at this meeting that the surface density of electronic states calculated, for instance, by Dr Corà could be used for the interpretation of the STM data. I would like to point out that this is a very crude model of STM imaging and one should be very careful in using such data because they can be misleading. In many cases the STM tunneling current used for imaging is through the conduction band, which is not well described by existing computational methods. Secondly, even in the crudest Tersoff–Hamann model of STM one should use the surface density of states projected on the tip orbitals which are some 5–10 Å away from the surface. Finally, as the more accurate theories demonstrate, the Tersoff–Hamann model is in many cases two approximate to describe important details of STM images.

Prof. Diebold responded: While Dr Shluger's comment is well taken, it turns out that an evaluation of the spatial distribution of charge density is often a very good approximation for the observed contrast in STM images.[1] This is also our experience, at least for TiO_2. For example, calculations of charge densities of Cl adsorbed on TiO_2 [2] show that Cl should be imaged bright in empty-state images, in perfect agreement with experiment.

1 W. A. Hofer, G. Ritz, W. Hebenstreit, E. Platzgummer, M. Schmid and P. Varga, *Surf. Sci. Lett.*, 1998, **405**, L514.
2 D. Vogtenhuber et al., unpublished.

Dr Willock addressed Dr Corà: You have used the tight binding model to analyse your Hartree–Fock results. Do you think you have enough data to parametrise the tight binding (TB) model and use it to do structural relaxation? This would be particularly useful for examining large supercells to model surface reconstructions.

Dr Corà replied: The results of *ab initio* calculations can in general be used as a reference to parametrise simplified Hamiltonians, including that for tight binding (TB). The accuracy achievable with the parametrised Hamiltonian will, however, depend on how well the simplified method can reproduce the physics of the problem examined.

In the case of perovskites, the TB and HF (but also DFT, *e.g.*, see ref. 1 below) solutions match particularly well, suggesting that a TB-like Hamiltonian should be able to reproduce the solid-state chemistry of these materials as obtained from the *ab initio* calculations. To allow geometry optimisations with the TB scheme, I consider that the quantum mechanical part of the TB solution should be complemented by a set of pair potentials to represent the short-range repulsion between the ions of the structure.

1 F. Corà, M. G. Stachiotti, C. R. A. Carlow and C. O. Rodriguez, *J. Phys. Chem. B*, 1997, **101**, 3945.

Prof. Freund commented: In connection to a request for development in physics of more complex materials, I think there is a need for better samples. In surface science this has to do with the making of single crystals. There is a strong need for laboratories to study the crystal growth of complex materials.

Prof. Flavell responded: I would agree, I would add that the problem is so complex that it is possible for a crystal grower to spend an entire career growing just one system. It would be very good to have a European facility for single crystal growth. This film growth is very difficult for ternaries and quaternaries, so we really need high quality single crystals.

Prof. Thornton said: There is of course the alternative approach which involves growing thin films. Moreover, the type of growth methods employed in the classical semiconductor field could prove useful.

Prof. Campbell said: I support Prof. Freund's statement that the community needs crystal-growing facilities staffed with experts in the art. This would greatly facilitate microcalorimetry experiments as well, since free-standing crystals of only 1 µm thickness but 1 cm in diameter are needed there.

Concluding Remarks

T. E. Madey

Department of Physics and Astronomy, and Laboratory for Surface Modification, Rutgers University, Piscataway, NJ, 08854-8019, USA

Received 18th October 1999

I begin by thanking *Prof. Geoff Thornton* for giving me the opportunity to participate in this stimulating *Discussion* of the Surface Science of Oxides. When I accepted his invitation, I did not realize how challenging it would be to prepare a summary—particularly since these remarks may annoy more people than they please.

Thirty years ago, the scientists interested in oxide surfaces were mainly chemists who were driven by a desire to understand catalysis and corrosion. There has been an explosion of interest in oxide surfaces in the last few years driven by several factors. First, while catalysis and corrosion are still challenges, there are exciting opportunities arising from the discovery of new oxide materials that have useful and interesting properties for technology—from high T_c superconductors to colossal magnetoresistance and ferroelectric materials. Applications of oxide surfaces and interfaces also impact upon environmental clean-up, fuel cells, batteries, biomaterials—the list is almost endless. Second, there are significant scientific challenges in the surface studies of oxides. At the dawn of surface science, the main concern was 'can we generate a clean nickel or tungsten surface, free of impurities'? With oxides, the surface preparation is a more daunting challenge—not only impurities and cleanliness, but stoichiometry, termination (anions or cations), surface reconstruction, defect structure and phase purity are critical issues—all of this must be sorted out before one can even begin his/her experiment! The challenges to theorists are also enormous—although they have powerful new tools with the development of *ab initio* methods and high speed computers, they need to know whom to believe when faced with conflicting experimental results.

This meeting—with 25 oral presentations (approximately equal weight given to theory and experiment) and about 50 posters—provides vivid testimony to the excitement and vitality of the growing field. In my summary, I focus primarily on the oral presentations and discussions, with occasional reference to posters.

A friend suggested that I approach this presentation by borrowing *Hajo Freund's* transparencies from his excellent opening lecture, and giving his talk in the past tense. When we consider the topics that were discussed during the last three days, he proved to be remarkably prescient in his remarks.

In the following paragraphs, I summarize the key issues addressed in this *Discussion*, the range of oxide surfaces, the variety of experimental and theoretical approaches, and the topics that generated the liveliest discussion. I close with an outlook for the future.

A. Key issues

There were four recurrent themes that permeated the Discussion. First and foremost was the issue of the *geometrical structure of oxide surfaces*. It is here that theoreticians are having the greatest impact on our understanding. From the opening talks by *Freund, Finnis and Jennison* to the closing talks by *Vanderbilt, Flavell and Corà*, the questions of surface stoichiometry of crystalline oxides, terminations (anions or cations?), relaxation and reconstruction dominated the discussion. As long expected from Tasker's rules of surface stability in ionic materials, the polar surfaces of crystalline oxides exhibit the greatest richness of structure and the largest relaxations, while for non-polar surfaces, the bulk termination is a good start to our thinking. The issue of defects on otherwise perfect surfaces arose frequently (point defects: anion or cation vacancies; extended defects: steps, domain boundaries, phase boundaries; strain and dislocations at interfaces). The complexity of coexisting surface phases on oxides and the faceting of morphologically unstable polar surfaces adds to the richness of the discussion. *Harrison* proposed that lattice dynamics, mainly soft phonon modes with anharmonic character can influence the apparent atom positions in a static measurement of surface structure.

A second focus area included the *electronic properties* of oxide surfaces and the interplay between electronic and geometrical structure. Topics included surface charge densities and band structure, degree of ionicity (ionic *vs.* covalent), surface energy, and energetics of vacancy formation. Novel electronic properties of surfaces and films included ferroelectric distortions, spin state, and magnetism.

The third topic concentrated on *interfaces* and included nucleation and growth of metals on oxides and oxides on metals, the size and shape distributions of islands, and oxide/oxide interfaces.

The fourth topic focused on *reactivity and surface chemistry* of oxides, which included molecular adsorption (structure and kinetics), the identity of surface reaction pathways, the relative roles of kinetics and thermodynamics in determining pathways, and the energetics of metallic adsorption. Central to this issue was the role of defects of all types on reaction mechanisms, and the effects of surface termination on reactivity. The necessity of surface preparation and 'conditioning' to generate reproducible results was often mentioned in discussion.

In all cases, the focus was on the physical and chemical principles underlying these topics, but the motivating force was almost invariably one of the myriad applications of oxide materials—as evidenced also in the opening paragraph of each of the papers!

B. The materials studied

There were several binary oxides whose surfaces attracted attention of both theorists and experimentalists: TiO_2 (mostly the (110) surface); Al_2O_3 and alumina films; α-Fe_2O_3; NiO; MgO and Mg on MgO. Of these, by far the greatest emphasis was on TiO_2 with nine papers devoted to this material. Some may feel that there was too much discussion of TiO_2, but I believe the balance was appropriate. At this stage in our field, it is important to have a system which serves as a benchmark and a solid foundation of understanding. Pure single crystals of this reducible oxide are readily available to experimentalists (many interesting oxides are very scarce). TiO_2 exhibits a rich array of interesting surface structures, multiple phases, and multiple structures, and it displays beautiful and reproducible phenomena. It has attracted the attention of theorists, and an understanding of the physics and chemistry of this complex and varied system provides an excellent jumping off point for more complex systems.

Other materials studied by experimentalists included films of V_2O_3, Fe_3O_4, CoO and MoO_3. The only non-binary, multicomponent oxides for which experimental data were presented were a material with variable spin states, $La_{1-x}Sr_xCoO_3$, (*Flavell*), and catalytically active $KFeO_2$ (*Weiss*).

Theorists presented results for a wide range of interesting oxide surfaces, with occasional reference to data reported elsewhere but not presented here. These included V_2O_5, Ca-doped CeO_2, CaO, WO_3 and SiO_2 (the only amorphous oxide mentioned). Multicomponent oxides included ferroelectric titanates ($PbTiO_3$, $BaTiO_3$) by *Vanderbilt* and *Corà*. Other materials presented were $MgSiO_4$, $CaCO_3$ and interfaces of $Nb_2O_5/BaTiO_3$, and Re_2O_7/WO_3.

C. The approaches

It was readily apparent that there was excellent balance between theoretical and experimental presentations, and that theory is making significant impact in all areas of oxide surfaces, particularly surface geometrical structure and relaxations. A substantial fraction of the theoretical presentations were based on *ab initio* (first principles) methods, mostly density functional theory (with and without gradient corrections) and Hartree–Fock theory. There were a number of successes reported, in which theory is converging with experiment to varying degrees [*e.g.* geometrical structure of Al_2O_3 films (*Jennison*), electronic structure of V_2O_5 (*Hermann*), structure and ferroelectric properties of titanate surfaces (*Vanderbilt, Corà*), structure, electronic and defect properties of TiO_2 (*Harrison, Noguera*)]. There were two reports that incorporated dynamical effects, including the possible role of soft phonon modes on TiO_2 surface structure (*Harrison*) and a prediction that statistically infrequent large-amplitude anion displacements may play a significant role in the surface chemistry of CeO_2 (*Pettersson*). *Parker* used empirical potentials to provide insights into faceting and dissolution kinetics of oxide surfaces. Despite the successes, there are many open questions (*e.g.*, details of surface relaxation of $TiO_2(110)$ are still not completely understood (*Harrison*), and questions persist about the dissociation probability of H_2O on Al_2O_3 (*Jennison, Freund*).

In the experimental presentations, it was clear that scanning probe microscopy (SPM) methods, both scanning tunneling microscopy (STM) and non-contact atomic force microscopy (NC-AFM), are *sine qua non*—these insightful atomic-resolution methods are providing unprecedented views of a myriad of surface phenomena, (even on insulating oxides), and we were treated to one dazzling image after another by skilled experimentalists. Direct views were provided of a number of clean, anion- or cation-terminated surfaces and reconstructions, of static and mobile defects, and of surface adsorbates and reaction products (TiO_2: *Diebold, Bowker, Madix, Iwasawa, Goodman*; α-Fe_2O_3: *Weiss*; Al_2O_3: *Freund*, NiO and CoO: *Neddermeyer*). There remains, however, a significant challenge that was repeatedly presented to the theorists: we need help in determining the identity of the 'white spots' that comprise the measured STM images. Often the symmetry is obvious, but the atomic and electronic factors that influence image formation, tip–surface interactions, *etc.*, are not apparent.

Photoemission spectroscopy continues to be essential for determining electronic properties, with both laboratory photon sources and synchrotron radiation methods represented. Of particular note were *Weiss's* and *Flavell's* studies of molecular adsorption and decomposition, *Woodruff's* use of scanned photoelectron diffraction for quantitative determination of NO/NiO adsorption geometry, and *Flavell's* use of resonant photoemission in her study of $La_{1-x}Sr_xCoO_3$.

Diebold showed convincingly that LEED can mislead if one is not careful; both the perfect $TiO_2(110)$ termination and the oxidized (110) surface with 'rosette' structures as seen in STM gave good, high quality (1 × 1) images. Interestingly, despite the focus on structure in this Discussion, there were no new structural determinations reported based on LEED (I, V) measurements.

There was a considerable amount of new information provided by the application of novel methods. *Renaud* provided details of atom positions at metal/MgO interfaces using grazing incidence X-ray scattering (GIXS), and *Kantorovich* and *Kempter* used the high surface sensitivity of metastable impact excitation spectroscopy (MIES) to probe the defect properties of Mg on MgO. In a series of experiments with great significance to the understanding of adsorption energetics, *Campbell* described the use of microcalorimetry in measuring heats of adsorption of metals on oxides.

D. Stimulating discussions

As indicated above, there were stimulating discussions on many aspects of oxide surfaces. Here, at the risk of redundancy, I identify some of the areas that generated the most significant discussion.

A topic that arose time and time again concerned the role of preparation conditions of the oxide surfaces on the resultant structure, anion or cation termination, defect structure, reconstructions, faceting, *etc.* Not only were temperature and oxygen pressure found to be critical for termination and structure (Theory: *Finnis, Jennison, Parker, Vanderbilt*; experiment: *Diebold, Weiss, Bowker*) but also the state of reduction and the bulk defect structure of the substrate (TiO_2: *Diebold, Bowker, Madix, Henderson*) determined the surface structures and reconstructions of oxidized and/or reduced surfaces. While many O-terminated oxide surfaces are stable over a wide range of conditions, some O-terminations (α-Fe_2O_3, Al_2O_3, $BaTiO_3$) require high O_2 pressures.

Whereas a catalytic chemist may have been left hungry for more extensive consideration of complex reaction kinetics and mechanisms, there were, nonetheless, a number of important discussions of surface chemistry and reactivity. H_2O was the most frequently mentioned probe molecule and all aspects of its surface chemistry were explored: effects of termination on adsorption and dissociation (*Weiss*), role of point defects on H_2O surface chemistry (*Henderson*), structure and electronic properties of dissociation fragments (OH, H) (*Flavell, Thornton*), and dissolution of oxides in H_2O (*Parker*). The issues of hydroxy formation and its effect on metal bonding, nucleation and growth were explored by *Freund, Campbell and Jennison*. In addition to his studies of H_2O, *Weiss* showed that molecular chemisorption of ethylbenzene and styrene is observed on all Fe oxides that expose metal cations in their top-most layers, whereas purely oxygen-terminated FeO(111) monolayer films are chemically inert, and only physiorption occurs. *Goodman and Bowker* studied the effect of oxygen chemisorption and reaction on the structure and growth of metal particles on TiO_2 surfaces: *Goodman* showed that an oxygen-induced ripening of Ag particles affects cluster size and density, while *Bowker* found that oxygen-rich atmospheres promote a regrowth of TiO_2, around Pd particles, and can 'bury' the metal particles! The results of *Harrison* and *Pettersson* show that surface dynamical effects and large amplitude anion displacements (mentioned already above) may affect structure and reactivity of oxide surfaces stimulated lively discussion.

A recurrent issue in the above discussions concerned the relative roles of thermodynamics and kinetics in determining the structures of oxide surfaces, the reaction pathways on oxides, the growth of metal particles, and interface stability.

The issue of surface defects pervaded much of the discussion for the full three days, and the mechanism for vacancy formation was an important ingredient. In an important calculation, *Hermann* showed that oxygen cations are quite stable at V_2O_5, surfaces, but that exposure to atomic H reduces the binding energy of O dramatically; his calculations suggest that oxygen removal from the surface occurs preferentially by formation of surface H_2O, which is bound weakly enough to desorb and create oxygen vacancies. Point defects were shown to play important roles in the surface chemistry of H_2O (*Henderson*) and on metal nucleation and growth (*Jennison, Freund*); *Neddermeyer* showed a beautiful atomically resolved video sequence to illustrate the diffusion of defects. In a striking series of experiments, *Iwasawa* showed that extended geometrical defects, steps on $TiO_2(110)$, may play a major role in surface chemistry: he illustrated this by showing that steps suppress the formation of carbonate products in formic acid decomposition. In the lively discussion following *Iwasawa's* presentation, *Madix* emphasized the necessity of pre-conditioning the TiO_2 surface in order to obtain reproducible results. Several authors (*Diebold, Bowker, Henderson*) provided convincing evidence for the diffusion of cation interstitials to the surface of partially-reduced TiO_2; the reaction of cation interstitials with O_2 leads to a regrowth of the surface, with various complex structures (including *Diebold's* rosettes!).

There were extensive discussions of both oxide clusters and films, which were theoretically (and in some cases, experimentally) more tractable than bulk oxides. *Noguera* simulated TiO_2 clusters of various sizes, charges and stoichiometries; she focussed on the electron distribution in an attempt to understand the ionic and covalent character of the anion–cation bonding, and the screening properties. *Pacchioni* focused on SiO_2 clusters and identified preferred binding sites for Cu atoms and aggregates; *Hermann* determined vacancy formation energies in V_2O_5. *Mackrodt* computed the magnetic and electronic properties of supported NiO films, including both ground state and excited state properties (d → d and charge transfer excitonic states). The growth and characterization of NiO (CoO) and V_2O_3 films were reported by *Neddermeyer* and *Madix*; in particular, *Neddermeyer* showed that a fully developed (100) oxide film grown on Ag requires annealing at elevated temperatures.

Another topic that recurred throughout the meeting concerned metal growth and interface formation on oxide surfaces. In an elegant experiment, *Renaud* used GIXS to identify binding sites for Ag, Pd and Ni on MgO(001), and showed that the epitaxial site is invariably above the oxygen ions of the substrate. *Kantorovich* determined the electronic properties of Mg adatoms on MgO, while *Pacchioni* used DFT calculations to identify the stable binding sites for Cu on SiO_2. Many others (*Madix, Campbell, Goodman, Bowker, Jennison*) addressed various aspects of metal nucleation and growth, the reactivity of metal particles, and the energetics of interface formation.

Last (but not least) were the animated discussions of novel electronic effects in oxide materials. *Vanderbilt* computed the effect of the surface in ferroelectric distortions in the cubic perovskites $PbTiO_3$, $BaTiO_3$ and $SrTiO_3$; he finds system specific behavior, with both qualitative and quantitative differences that depend on surface termination. *Corà* also studied ferroelectric distortions in perovskites, and showed the effects of surface terminations and interface strain on ferroelectric properties. As indicated above, *Mackrodt* calculated the electronic and magnetic properties of NiO films supported on MgO. In the only experimental report of novel electronic effects, *Flavell* identified clear temperature effects in the valence band photoemission spectra of monocrystalline $LaCoO_3$, which may be attributed to changes in the spin state of the sample. *Flavell* closed with a plea for more crystal growers to prepare the samples needed for careful, systematic studies.

E. Outstanding issues

Many issues were explored in the three days of discussion and some were settled, but a large number of challenges remain, including the following.

It is often assumed and it has been occasionally demonstrated that an oxide film is a good model for a bulk crystal surface. More needs to be done here (both theoretically and experimentally) to define the conditions under which this is so, particularly for non-rocksalt oxides. As a warning, the vast range of surface phases and reconstructions seen for TiO_2 crystals have not yet been reported on thin oxide films.

The role of defects in surface chemistry and electronic properties of oxide surfaces is a continuing challenge: extended structural defects (steps, grain boundaries, phase boundaries), electronic defects, point defects, (vacancies, dopants), defects on particles and in pores. Vicinal (stepped) surfaces have barely been explored.

Theorists and experimentalists are converging on a few systems (TiO_2, MgO, Al_2O_3 ...) but we need more instances of theorists and experimentalists working on the same systems, particularly complex oxides. A quantity that is often calculated but seldom measured is surface energy—experimentalists take note! (In the poster session, *Castell* showed striking microscopic data illustrating the anisotropy of surface energy for UO_2.)

We have just 'scratched the surface' in studies of complex (multicomponent) oxides with interesting bulk properties. These include high-temperature superconducting oxides, ferroelectrics, giant magnetoresistive oxides, colossal magnetoresistance materials, *etc.*

Even for well studied oxides like TiO_2, almost all measurements have focussed on one phase, rutile. Other phases with interesting properties (anatase, brookite) need to be explored. The range of possible phases is enormous.

An area with important implications concerns electronic excited states and their effects on surface dynamics at oxide surfaces. Areas not discussed here, but with great challenges, include photochemistry at oxide surfaces (*e.g.*, the mechanisms for photolysis of H_2O) and DIET processes (desorption induced by electronic transitions).

There are numerous fundamental questions that arise in considering the many applications of oxides in technology. A young scientist could effectively 'carve a niche' in an area of enormous societal impact that involves oxide surfaces. A list of examples includes:

(i) catalysis, corrosion, (ii) gas and liquid sensors, (iii) environmental remediation, oxide/liquid interfaces, (iv) biomaterials and interfaces (implants, dental materials, . . .), (v) electronics—thin gate dielectrics, high-K materials; combinatorial methods.

In closing, it is clear that the opportunities are vast, and we have just begun!

List of Posters

On the mechanism of the selective oxidation of n-butane, but-1-ene and but-1,3-diene to maleic anhydride over a vanadyl pyrophosphate catalyst **K. C. Waugh** and **B. H. Sakakini**, *UMIST, Manchester, UK*, **Y. H. Taufiq-Yap**, *University of Putra, Malaysia*

Interaction of small molecules with oxide surfaces: CO_2 chemisorption on CaO films **P. Stracke, S. Krischok, D. Ochs, B. Braun, W. Maus-Friedrichs** and **V. Kempter**, *Physikalisches Institut der TU Clausthal, Germany*

Surface electronic states at cleavage surfaces of NiO and UO_2: the *ab initio* interpretation of STM images **M. R. Castell, S. L. Dudarev, A. P Sutton** and **G. A. D. Briggs**, *University of Oxford, UK*

Theoretical modelling of non-contact atomic force microscopy on insulators **A. L. Shluger, A. S. Foster, A. I. Livshits** and **L. N. Kantorovich**, *University College London, UK*

Initial stages of the oxidation of aluminium: A DFT study **P. W. M. Jacobs**, *University of Western Ontario, Canada*, **Y. F. Zhukovskii**, *University of Latvia, Latvia*, **M. Causa**, *University of Torino, Italy*

Surface structure of TiO_2(210). Atomically resolved STM and atomistic simulation **R. G. Egdell, A. Howard** and **C. E. J. Mitchell**, *University of Oxford, UK*, **S. C. Parker**, *University of Bath, UK*

Structure–reactivity-relationship in MoO_3-type oxide catalysts for selective partial oxidation **G. Mestl, R. Schlögl** and **R. Gottschall**, *Fritz-Haber-Institut der MPG, Berlin, Germany*, **Ch. Linsmeier**, *Max-Planck-Institut füur Plasmaphysik, Garching, Germany*

Spectroscopic characterisation and chemical reactivity of silicon monoxide layers deposited on Cu(100) **F. Yubero, A. Barranco, J. P. Espinos, J. A. Mejias** and **A. R. González-Elipe**, *Instituto de Ciencia de Materiales de Sevilla, Sevilla, Spain*

Vanadium oxide island structures on Pd(111): Structure and morphology of the metal/oxide interface **S. L. Surnev, F. P. Leisenberger, L. Vitali, M. G. Ramsey** and **F. P. Netzer**, *Karl-Franzens-Universität, Graz, Austria*

α-Alumina as a model fluorination catalyst: a periodic density functional theory study **D. J. Willock, G. W. Watson** and **G. J. Hutchings**, *University of Wales, Cardiff, UK*, **M. Bankhead**, *Liverpool University, UK*, **J. Scott**, *ICI Chemicals and Polymers, Runcorn, UK*

Oxidation reactions on thin-film oxides of Mo and mixed Co–Mo oxides **C. M. Friend**, *Harvard University, USA*

Towards surface chemical reactivity characterisation of oxides desorption of negative ions induced by low energy electron collisions **M. Bernheim**, *University of Paris-Sud, France*

Resonant photoemission, STM and IRAS studies on CeO_{2-x}/Pt(111) model catalysts **K. Schierbaum** and **U. Berner**, *Heinrich-Heine University, Düsseldorf, Germany*, **G. G. Jones, P. L. Wincott** and **G. Thornton**, *University of Manchester, UK*, **V. Dhanak**, *CLRC Daresbury Laboratory, Warrington, UK*, **S. Haq**, *University of Liverpool, UK*

Structural aspects of the adsorption of carboxylate species at TiO_2(110) **E. M. Williams**, *Liverpool University, UK*

Nucleation and growth of supported metal clusters at defect sites on MgO(001): The cases of Pd and Ag **J. A. Venables**, *Arizona State University, USA and University of Sussex, UK*, **G. Haas**, *Institut de Physique Experimental, Lausanne, Switzerland*, **J. H. Harding**, *University College London, UK*

Nanocrystalline SnO_2 gas-sensors: Influence of doping electronic structure **M. Sinner-Hettenbach, J. Kappler, N. Bârsan, U. Weimar** and **W. Göpel**, *Institute of Physical and Theoretical Chemistry and Center of Interface Analysis and Sensors, Tübingen, Germany*

A model for electron emission from triple oxide cathode materials **R. Devonshire** and **A. Buckley**, *Sheffield University, UK*

Microstructural changes to metal bond coatings on gas turbine alloys with time at high temperature **N. V. Russell, F. Wigley, J. Williamson**, *Imperial College London, UK*, **S. S. West**, *National Power plc, Swindon, UK*

Nanometric layers of magnetic iron oxides epitaxially grown on (-Al_2O_3 **M. Gautier-Soyer, S. Gota, E. Guiot** and **M. Henriot**, *CEA Saclay, Gif sur Yvette, France*

SPM study of rutile $TiO_2(111)$ surface **H. Onishi, H. Uetsuka, R. Sasahara, T. Sato** and **M. Iwatsuki**, *Kanagawa Academy of Science and Technology, Japan*

Force microscopy study of $SrTiO_3(100)$ surfaces with single atomic layer steps **M. Kawai, K. Iwahori** and **S. Watanabe**, *Institute of Physical and Chemical Research (RIKEN), Saitama, Japan*

Quantum chemical studies of carboxylic acids on oxide surfaces **P. Persson, S. Lunell**, *Uppsala University, Sweden*, **L. Ojamäe**, *University of Stockholm, Sweden*

A photoemission and STM study of the electronic structure of Nb-doped TiO_2 **D. Morris, Y. Dou, J. Rebane** and **R. G. Egdell**, *Oxford University, UK*, **D. S. L. Law, S. N. Downes** and **M. MacDonald**, *Daresbury Laboratory, Warrington, UK*

Oxygen pressure dependence of the α-$Fe_2O_3(0001)$ surface structure **W. Weiss, G. Ketteler** and **Sh. K. Shaikhutdinov**, *Fritz-Haber-Institut der MPG, Berlin, Germany*

Interaction of molecular oxygen with $TiO_2(110)$ surfaces. Influence of a metal deposit **H. Mostéfa-Sba, V. Blondeau-Patissier, B. Domenichini, A. Steinbrunn** and **S. Bourgeois**, *Université de Bourgogne, Dijon, France*

Characteristics of Pd deposition on the MgO(111) surface **J. Goniakowski**, *CNRS Campus de Luminy, Marseille, France*, **C. Noguera**, *Université Paris-Sud, France*

Facet formation of UO_2 crystals **M. R. Castell** and **G. A. D. Briggs**, *Oxford University, UK*, **D. D. Perovic**, *University of Toronto, Canada*, **G. A. Wood** and **D. T. Goddard**, *BNFL, Preston, UK*

Adsorption, diffusion and charge transfer of potassium on $Cr_2O_3(0001)/Cr(110)$ **M. Asscher, W. Zhao** and **G. Kerner**, *The Hebrew University, Jerusalem, Israel*, **M. Wilde, K. Al-Shamery** and **H.-J. Freund**, *Fritz-Haber-Institut der MPG, Berlin, Germany*, **V. Staemmler** and **M. Wieszbowska**, *Ruhr Universität Bochum, Germany*

Alkali metal reactions with Ni(110)–O surfaces **A. F. Carley** and **M. W. Roberts**, *Cardiff University, Wales, UK*, **S. D. Jackson**, *Synetix, Billingham, UK*, **J. N. O'Shea**, *Uppsala University, Sweden*

Atomistic simulation of Ti interstitial diffusion in $Ti_{1+x}O_2$ **A. J. Ramirez-Cuesta**, *Reading University, UK and Universidad Nacional de San Luis, Argentina*, **R. A. Bennett, P. Stone** and **M. Bowker**, *Reading University, UK*

Electron- and photon-stimulated desorption of sodium atoms from silicon dioxide surface: relevance to tenuous planetary atmospheres **B. V. Yakshinskiy** and **T. E. Madey**, *Rutgers, The State University of New Jersey, Piscataway, USA*

Reconstructions on $BaTiO_3(100)$ investigated by STM, ISS, XPS and LEED **Ch. Hagendorf, K.-M. Schindler** and **H. Neddermeyer**, *Martin-Luther-Universität, Halle-Wittenberg, Germany*

A cluster model analysis for the band gap structure of the $SnO_2(110)$ surface **K. Tabata, S. Shimomura** and **E. Suzuki**, *Research Institute of Innovative Technology for the Earth (RITE), Kyoto, Japan*

A first principles investigation of the CO, H_2O and H_2S chemisorption on the α-$Al_2O_3(0001)$ surface **M. Casarin** and **C. Maccato**, *Universita di Padova, Italy*, **A. Vittadini**, *Instituto di Chimica e Tecnologia dei Materiali Avanzati del CNR di Padova, Italy*

Formic acid adsorption on the TiO_2 anatase (101) surface by DFT calculations **A. Vittadini**, *CNR-CSSRCC, Padova, Italy*, **A. Selloni**, *University of Geneva, Switzerland*, **F. P. Rotzinger** and **M. Grätzel**, *EPFL, Lausanne, Switzerland*

The growth of Cu on $TiO_2(110)$ at room temperature and low temperature **G. Thornton, C. L. Pang, I. M. Brookes, S. A. Haycock, H. Raza** and **A. Limb**, *University of Manchester, UK*

Molecular beam and *in situ* cell measurements of the reactivity of MgO(100) surfaces towards simple molecules such as water, carbon dioxide and methane **J. Mmojieje** and **A. O. Taylor**, *Brunel University, UK*

The electronic structure of Na_xWO_3(001): The relationship between resonance photoemission spectra and STM images **F. H. Jones, R. A. Dixon, T. W. Fishlock** and **R. G. Egdell**, *Oxford University, UK*

Unoccupied electronic states of ZnO and MgO surfaces **S. A. Komolov** and **E. F. Lazneva**, *St. Petersburg University, Russia*, **P. J. Møller**, *University of Copenhagen, Denmark*

Modelling in-crystal anionic polarisability **P. W. Fowler** and **C. Domene**, *University of Exeter, UK*, **P. A. Madden** and **M. Wilson**, *Oxford University, UK*, **R. J. Wheatley**, *University of Nottingham, UK*

A computational study of bulk and surface properties of $LaCrO_3$ perovskite type **A. Damin, S. Bordiga** and **A. Zecchina**, *Universita di Torino, Italy*

The alignment of formate on TiO_2(110): Azimuthàl FT-RAIRS measurements **B. E. Hayden, A. King** and **M. A. Newton**, *University of Southampton, UK*

Two (1 × 2) reconstructions of TiO_2(110); surface rearrangement and reactivity studied using elevated temperature STM **P. Stone, R. A. Bennett** and **M. Bowker**, *University of Reading, UK*

Excitation and ionisation energies of defect states at the MgO(001) surface **P. V. Sushko** and **A. L. Shluger**, *University College London, UK*, **C. R. A. Catlow**, *The Royal Institution of Great Britain, UK*

Far-IR RAIRS studies of the adsorption of tin tetrachloride on silica surfaces using the buried metal layer approach **M. E. Pemble** and **P. Gardner**, *University of Salford, UK*, **M. J. Pilling** and **P. Gardner**, *UMIST, Manchester, UK*, **A. Awaluddin** and **M. E. Pemble**, *University of Salford, UK*, **M. Surmin**, *CCL RC Daresbury, Warrington, UK*

A TDS study: CO, NO and H_2O on NiO(100) and MgO(100) **R. Wichtendahl, M. Rodrigo-Rodriguez, U. Härtel, H. Kuhlenbeck** and **H.-J. Freund**, *Fritz-Haber-Institut der Max-Planck-Gesellschaft, Berlin, Germany*

List of Participants

Dr M. Alfredsson *The Royal Institution of Great Britain, UK*
Dr G. M. Allcock *The Royal Society of Chemistry, Cambridge, UK*
Prof. M. Asscher *The Hebrew University of Jerusalem, Israel*
Mr A. Awaluddin *University of Salford, UK*
Mr M. Baudin *Uppsala University, Sweden*
Dr R. Bennett *University of Reading, UK*
Mr U. Berner *University of Dusseldorf, Germany*
Dr M. Bernheim *Université Paris-Sud, Orsay, France*
Dr S. Bourgeois *Université de Bourgogne, Dijon, France*
Prof. M. Bowker *University of Reading, UK*
Mr I. M. Brookes *University of Manchester, UK*
Mr A. Buckley *University of Sheffield, UK*
Prof. C. T. Campbell *University of Washington, Seattle, USA*
Dr A. Carley *University of Cardiff, UK*
Prof. M. Casarin *University of Padova, Italy*
Dr M. Castell *University of Oxford, UK*
Mr D. J. M. Castle *University of Manchester, UK*
Prof. C. R. A. Catlow *The Royal Institution of Great Britain, UK*
Mr M. W. Collins *HMS Sultan, Gosport, UK*
Mr S. Colonna *ESRF, Grenoble, France*
Dr F. Corà *The Royal Institution of Great Britain, UK*
Dr A. Damin *University of Torino, Italy*
Prof. P. T. Dawson *McMaster University, Canada*
Dr N. H. de Leeuw *University of Bath, UK*
Dr R. Devonshire *University of Sheffield, UK*
Prof. U. Diebold *Tulane University, New Orleans, USA*
Miss C. Domene *University of Exeter, UK*
Dr R. G. Egdell *University of Oxford, UK*
Prof. M. W. Finnis *The Queen's University of Belfast, UK*
Prof. W. Flavell *UMIST, Manchester, UK*
Prof. P. W. Fowler *University of Exeter, UK*
Prof. H.-J. Freund *Fritz-Haber-Institut der Max-Planck-Gesellschaft, Berlin, Germany*
Prof. C. M. Friend *Harvard University, USA*
Dr M. Gautier-Soyer *CEA, Gif sur Yvette, France*
Dr D. A. Geeson *AWE, Reading, UK*
Prof. E. A. Gillet *Laboratory Sermec, Marseille, France*
Prof. M. F. A. Gillet *Laboratory Sermec, Marseille, France*
Dr J. Glascott *AWE, Reading, UK*
Dr J. Goniakowski *Universite d'Aix-Marseille, France*
Prof. D. W. Goodman *Texas A&M University, USA*
Prof. G. Granozzi *University of Padova, Italy*
Dr P. J. Hardman *University of Manchester, UK*
Prof. N. M. Harrison *Daresbury Laboratory, Warrington, UK*
Prof. B. E. Hayden *University of Southampton, UK*
Mr T. Heller *University of St Andrews, UK*
Dr M. A. Henderson *Pacific Northwest National Laboratory/Battelle, Richland, USA*
Prof. K. Hermann *Fritz-Haber-Institut der Max-Planck-Gesellschaft, Berlin, Germany*
Mr A. Howard *Oxford University, UK*
Mr K. Iwahori *The Institute of Physical & Chemical Research, Saitama, Japan*
Prof. Y. Iwasawa *University of Tokyo, Japan*
Prof. P. W. M. Jacobs *University of Western Ontario, Canada*
Prof. D. R. Jennison *Sandia National Laboratories, Albuquerque, USA*
Dr F. H. Jones *University of London, Reading, UK*
Miss I. Jones *Johnson Matthey Technology Centre, Reading, UK*
Mr R. Jones *University of Cardiff, UK*
Prof. R. W. Joyner *Nottingham Trent University, UK*
Dr L. Kantorovich *University College London, UK*

Mrs E. Karlsen *Norsk Hydro a.s., Porsgrunn, Norway*
Dr M. Kawai *The Institute of Physical & Chemical Research, Saitama, Japan*
Dr S. Kelling *Harvard University, USA*
Prof. V. Kempter *TU Clausthal, Clausthal-Zellerfeld, Germany*
Mr G. Ketteler *Fritz-Haber-Institut der Max-Planck-Gesellschaft, Berlin, Germany*
Prof. J. Kirschner *Max-Planck-Institut für Mikrostrukturphysik, Halle, Germany*
Mr M. Kittel *Fritz-Haber-Institut der Max-Planck-Gesellschaft, Berlin, Germany*
Dr M. Klaua *Max-Planck-Institut für Mikrostrukturphysik, Halle, Germany*
Mr H. KrishnankuttyRajamma *University of Cardiff, UK*
Mr A. J. Limb *University of Manchester, UK*
Dr R. Lindsay *University of Manchester, UK*
Mr P. Lo Pinto *UMIST, Manchester, UK*
Dr W. C. Mackrodt *University of St Andrews, UK*
Prof. T. E. Madey *Rutgers University, Piscataway, USA*
Prof. R. J. Madix *Stanford University, USA*
Prof. A. M. Masson *CNRS, Paris, France*
Dr G. Mestl *Fritz-Haber-Institut der Max-Planck-Gesellschaft, Berlin, Germany*
Dr C. E. J. Mitchell *University of Oxford, UK*
Mrs P. Mohamed *Royal Society of Chemistry, London, UK*
Prof. P. Møller *University of Copenhagen, Denmark*
Dr B. Montanari *Queen's University of Belfast, UK*
Mr P. G. Morrall *University of Manchester, UK*
Mr D. Morris *University of Oxford, UK*
Dr C. A. Muryn *University of Manchester, UK*
Prof. H. Neddermeyer *Martin-Luther-Universität, Halle, Germany*
Dr R. Nix *Queen Mary & Westfield College, London, UK*
Dr C. Noguera *Université de Paris-Sud, Orsay, France*
Dr L. Ojamäe *University of Stockholm, Sweden*
Dr R. Oldman *ICI Technology, Runcorn, UK*
Dr H. Onishi *Kanagawa Academy of Science and Technology, Kawasaki-shi, Japan*
Prof. G. Pacchioni *University of Milan, Italy*
Prof. S. C. Parker *University of Bath, UK*
Prof. M. Pemble *University of Salford, UK*
Mr P. Persson *Uppsala University, Sweden*
Prof. L. G. Pettersson *University of Stockholm, Sweden*
Dr S. Poulston *Johnson Matthey Technology Centre, Reading, UK*
Dr D. W. Price *University of Reading, UK*
Dr A. Ramirez-Cuesta *University of Reading, UK*
Dr G. Renaud *CEA-Grenoble/DRFMC/SCIB, Grenoble, France*
Miss S. Riaz *Royal Society of Chemistry, London, UK*
Dr N. V. Russell *Imperial College London, UK*
Dr B. F. Schedin *University of Manchester, UK*
Dr G. Scheying *Robert Bosch GmbH, Stuttgart, Germany*
Prof. K. Schierbaum *University of Dusseldorf, Germany*
Dr K.-M. Schindler *Martin-Luther-Universität Halle-Wittenberg, Halle/Saale, Germany*
Dr G. H. E. Scott *The Royal Society of Chemistry, Cambridge, UK*
Dr S. Shimomura *Research Institute of Innovative Technology for the Earth, Kyoto, Japan*
Dr A. Shluger *University College London, UK*
Mr M. Sinner-Hettenbach *University of Tübingen, Germany*
Mr P. Stone *University of Reading, UK*
Mr P. Stracke *TU Clausthal, Clausthal-Zellerfeld, Germany*
Dr S. Surnev *Karl-Franzens-Universität Graz, Austria*
Mr P. V. Sushko *The Royal Institution of Great Britain, UK*
Dr K. Tabata *Research Institute of Innovative Technology for the Earth, Kyoto, Japan*
Dr A. Taylor *Brunel University, Uxbridge, UK*
Prof. Sir John Meurig Thomas *The Royal Institution of Great Britain, UK*
Prof. G. Thornton *University of Manchester, UK*
Mr S. Tull *University of Cardiff, UK*

Prof. D. Vanderbilt   *Rutgers University, Piscataway, USA*
Dr J. Venables   *University of Sussex, Brighton, UK*
Dr A. Vittadini   *CNR – CSSRCC, Padova, Italy*
Dr S. Warren   *UMIST, Manchester, UK*
Dr S. Watanabe   *The Institute of Physical & Chemical Research, Saitama, Japan*
Prof. K. Waugh   *UMIST, Manchester, UK*
Dr W. Weiss   *Fritz-Haber-Institut der Max-Planck-Gesellschaft, Berlin, Germany*
Dr R. Whitelock   *The Royal Society of Chemistry, Cambridge, UK*
Dr R. Wichtendahl   *Fritz-Haber-Institut der Max-Planck-Gesellschaft, Berlin, Germany*
Dr E. M. Williams   *University of Liverpool, UK*
Dr D. J. Willock   *University of Cardiff, UK*
Prof. C. Wöll   *Ruhr-Universität Bochum, Germany*
Prof. D. P. Woodruff   *University of Warwick, Coventry, UK*
Dr F. Yubero   *ICMSE, Seville, Spain*

Index of Contributors*

Alavi, A., **33**
Albaret, T., **285**
Allan, N. L., **105**
Asai, K., **407**
Asscher, M., 102, 238
Bald, D. J., **195**
Barbier, A., **157**
Batyrev, I., **33**
Baudin, M., **351**
Baumgärtel, P., **141**
Bäumer, M., **67**
Bennett, R., **267**, 349
Bertrams, T., **129**
Biener, J., **67**
Bogicevic, A., **45**
Bowker, M., 86, 102, 230, **267**, 334, 335, 337, 343, 346
Bradshaw, A. M., **141**
Campbell, C. T., 93, 101, **195**, 225, 233, 238, 240, 241, 243, 334, 336, 347, 448, 460
Carley, A., 100, 230, 345
Castell, M., 225, 453
Castro, M. E., **313**
Catlow, C. R. A., 86, 87, 96, 241, **421**, 443, 448, 456
Corà, F., 97, 338, **421**, 453, 454, 458, 459
de Carolis, S., **351**
de Leeuw, N. H., **381**
Dhanak, V. R., **407**
Diebold, U., 94, 100, 225, 233, **245**, 331, 332, 333, 334, 335, 340, 347, 459
Dinger, A., **67**
Downes, S., **407**
Druzinic, R., **53**
Dunwoody, P. M., **407**
Egdell, R. G., 85, 92, 99, 224, 226, 342, 443, 455, 458
Finnis, M. W., **33**, 86, 87, 88, 90, 97, 231, 240, 340, 454
Finocchi, F., **285**
Flavell, W., 235, 341, **407**, 452, 455, 456, 457, 460
Freund, H.-J., **1**, 85, 86, 90, 94, 95, 99, 100, 101, 102, 224, 227, 228, 231, 238, 243, 343, 448, 451, 452, 457, 460
Friend, C. M., 91, 98, 228, 229, 335, 343, 344
Fukui, K.-i, **259**
Gautier-Soyer, M., 87, 88, 90, 446
Grice, S. C., **407**
Goodman, D. W., 101, **173**, 232, 240, 243, **279**, 335, 336, 337, 347
Günster, J., **173**
Harrison, N. M., 87, 88, 90, 91, 97, 224, **305**, 340, 341, 342, 454, 457, 459
Hayden, B. E., 227, 234, 240, 458
Hebenstreit, W., **245**
Henderson, M. A., 240, **245**, **313**, 332, 335, 341, 343, 344, 345, 346
Hermansson, K., **351**
Hermann, K., **53**, 95, 96, 97, 98, 236, 342, 445
Hollingworth, J., **407**
Iwasawa, Y., 86, 94, **259**, 331, 332, 333, 334, 337, 346, 447
Illas, F., **209**
Jacobs, P. W. M., 449
Jennison, D. R., **45**, 90, 91, 92, 93, 94, 95, 101, 223, 226, 229, 232, 239, 242, **245**, 331, 341, 343, 455
Joseph, Y., **363**
Joyner, R. W., 228, 237, 335, 444, 446
Kantorovich, L., 86, 98, **173**, 235, 236, 237, 238, 452
Kawai, M., 452, 456
Kempter, V., 96, 99, **173**, 237, 238, 242, 338
Kirschner, J. 241, 446
Koboyashi, Y., **407**
Kuhrs, C., **363**
Kulkarni, S., **141**
Lai, X., **279**
Li, M., **245**
Lindsay, R., **141**, 227, 332, 346, 456
Lopez, N., **209**
Mackrodt, W. C., **105**, 223, 224, 225
Madey, T. E., 93, 101, 225, 240, 334, 447, 450, **461**
Madix, R. J., **67**, 85, 92, 93, 99, 100, 101, 102, 103, 334, 344, 346
Marr, P. G. D., **407**
Meinel, K., **129**
Meyer, B., **395**
Mitchell, C. E. J., 336, **407**
Møller, P., 103
Muscat, J., **305**
Musgrove, J. E., **195**
Neddermeyer, H., **129**, 225, 226, 227
Noguera, C., 95, **105**, 224, 237, **285**, 337, 338, 340, 341, 450, 451, 453, 454, 455
Nygren, M. A., **351**
Onishi, H., **259**, 331, 349
Otero-Tapia, S., **313**
Pacchioni, G., **209**, 224, 236, 241, 242, 243
Padilla, J., **395**
Parker, S. C., **381**, 448, 449, 450, 451
Pettersson, L. G., 98, 227, 229, 237, 241, **351**, 443, 444, 445
Polcik, M., **141**
Ranke, W., **363**
Ranney, J. T., **195**
Redfern, S. E., **381**
Renaud, G., 90, 91, **157**, 231, 232, 233, 234, 342, 343
Robach, O., **157**
Sasaki, T., **259**
Sebastian, I., **129**
Schaff, O., **141**
Scheffler, M., **305**
Schindler, K.-M., 455
Seddon, E. A., **407**

* The page numbers in **bold** type indicate papers submitted for discussions.

Shaikhutdinov, Sh. K., **363**
Shluger, A., 92, 95, 100, **173**, 223, 226, 242, 340, 342, 445, 450, 452, 459
Starr, D. E., **195**
St.Clair, T. P., **279**
Stone, P., **267**
Stracke, P., **173**
Sushko, P. V., **173**
Suzuki, S., **259**
Taylor, A., 85, 450, 451
Teehan, D., **407**
Terborg, R., **141**
Thomas, A. G., **407**
Thomas, J. M., 85, 93, 100, 349
Thornton, G., 85, 90, 224, 226, 227, 232, 243, 331, 337, 341, 342, 345, 347, 350, 447, 453, 456, 460
Toomes, R. L., **141**
Triguero, L., **351**
Vanderbilt, D., 87, 90, 95, 331, 338, **395**, 452, 453, 454, 455, 458
Venables, J., 91, 227, 233, 241
Wang, X.-G., **305**
Warren, S., **407**
Waugh, K., 98, 345
Weiss, W., 89, 96, 226, **363**, 446, 447, 448, 452
Willock, D. J., 86, 341, 449, 459
Witko, M., **53**
Wójcik, M., **351**
Woodruff, D. P., **141**, 227, 228, 229, 230, 231
Yamada, N., **407**
Yubero, F., 102

General Discussions of the Faraday Society/Faraday Discussions of the Chemical Society

Date	Subject	Volume
1907	Osmotic Pressure	Trans. 3
1907	Hydrates in Solution	3
1910	The Constitution of Water	6
1911	High Temperature Work	7
1912	Magnetic Properties of Alloys	8
1913	Colloids and their Viscosity	9
1913	The Corrosion of Iron and Steel	9
1913	The Passivity of Metals	9
1914	Optical Rotary Power	10
1914	The Hardening of Metals	10
1915	The Transformation of Pure Iron	11
1916	Methods and Appliances for the Attainment of High Temperatures in a Laboratory	12
1916	Refractory Materials	12
1917	Training and Work of the Chemical Engineer	13
1917	Osmotic Pressure	13
1917	Pyrometers and Pyrometry	13
1918	The Setting of Cements and Plasters	14
1918	Electric Furnaces	14
1918	Co-ordination of Scientific Publication	14
1918	The Occlusion of Gases by Metals	14
1919	The Present Position of the Theory of Ionization	15
1919	The Examination of Materials by X-Rays	15
1920	The Microscope: Its Design, Construction and Applications	16
1920	Basic Slags: Their Production and Utilization in Agriculture	16
1920	Physics and Chemistry of Colloids	16
1920	Electrodeposition and Electroplating	16
1921	Capillarity	17
1921	The Failure of Metals under Internal and Prolonged Stress	17
1921	Physico-Chemical Problems Relating to the Soil	17
1921	Catalysis with special reference to Newer Theories of Chemical Action	17
1922	Some Properties of Powders with special reference to Grading by Elutriation	18
1922	The Generation and Utilization of Cold	18
1923	Alloys Resistant to Corrosion	19
1923	The Physical Chemistry of the Photographic Process	19
1923	The Electronic Theory of Valency	19
1923	Electrode Reactions and Equilibria	19
1923	Atmospheric Corrosion. First Report	19
1924	Investigation on Oppau Ammonium Sulphate-Nitrate	20
1924	Fluxes and Slags in Metal Melting and Working	20
1924	Physical and Physico-Chemical Problems relating to Textile Fibres	20
1924	The Physical Chemistry of Igneous Rock Formation	20
1924	Base Exchange in Soils	20
1925	The Physical Chemistry of Steel-Making Processes	21
1925	Photochemical Reactions of Liquids and Gases	21
1926	Explosive Reactions in Gaseous Media	22
1926	Physical Phenomena at Interfaces, with special reference to Molecular Orientation	22
1927	Atmospheric Corrosion, Second Report	23
1927	The Theory of Strong Electrolytes	23
1927	Cohesion and Related Problems	24
1928	Homogeneous Catalysis	24
1929	Crystal Structure and Chemical Constitution	25
1929	Atmospheric Corrosion of Metals, Third Report	25
1929	Molecular Spectra and Molecular Structure	26
1930	Colloid Science Applied to Biology	26
1931	Photochemical Processes	27
1932	The Adsorption of Gases by Solids	28
1932	The Colloid Aspect of Textile Materials	29
1933	Liquid Crystals and Anisotropic Melts	29
1933	Free Radicals	30
1934	Dipole Moments	30
1934	Colloidal Electrolytes	31

Date	Subject	Volume
1935	The Structure of Metallic Coatings, Films and Surfaces	31
1935	The Phenomena of Polymerization and Condensation	32
1936	Disperse Systems in Gases: Dust, Smoke and Fog	32
1936	Structure and Molecular Forces in (a) Pure Liquids, and (b) Solutions	33
1937	The Properties and Function of Membranes, Natural and Artificial	33
1937	Reaction Kinetics	34
1938	Chemical Reactions Involving Solids	34
1938	Luminescence	35
1939	Hydrocarbon Chemistry	35
1939	The Electrical Double Layer (owing to the outbreak of the war the meeting was abandoned, but the papers were printed in the *Transactions*)	35
1940	The Hydrogen Bond	36
1941	The Oil-Water Interface	37
1941	The Mechanism and Chemical Kinetics of Organic Reactions in Liquid Systems	37
1942	The Structure and Reactions of Rubber	38
1943	Modes of Drug Action	39
1944	Molecular Weight and Molecular Weight Distribution in High Polymers (Joint Meeting with the Plastics Group, Society of Chemical Industry)	40
1945	The Application of Infra-red Spectra to Chemical Problems	41
1945	Oxidation	42
1946	Dielectrics	42 A
1946	Swelling and Shrinking	42 B
1947	Electrode Processes	Disc. 1
1947	The Labile Molecule	2
1947	Surface Chemistry (Jointly with the Sociéitéi de Chimie Physique at Bordeaux Published by Butterworths Scientific Publications Ltd	
1947	Colloidal Electrolytes and Solutions	Trans. 43
1948	The Interaction of Water and Porous Materials	Disc. 3
1948	The Physical Chemistry of Process Metallurgy	4
1949	Crystal Growth	5*
1949	Lipo-proteins	6
1949	Chromatographic Analysis	7
1950	Heterogeneous Catalysis	8
1950	Physico-chemical Properties and Behaviour of Nuclear Acids	Trans. 46
1950	Spectroscopy and Molecular Structure and Optical Methods of Investigating Cell Structure	Disc. 9
1950	Electrical Double Layer	Trans. 47
1951	Hydrocarbons	Disc. 10
1951	The Size and Shape Factor in Colloidal Systems	11
1952	Radiation Chemistry	12
1952	The Physical Chemistry of Proteins	13
1952	The Reactivity of Free Radicals	14
1953	The Equilibrium Properties of Solutions on Non-electrolytes	15
1953	The Physical Chemistry of Dyeing and Tanning	16
1954	The Study of Fast Reactions	17
1954	Coagulation and Flocculation	18
1955	Microwave and Radio-frequency Spectroscopy	19
1955	Physical Chemistry of Enzymes	20
1956	Membrane Phenomena	21
1956	Physical Chemistry of Processes at High Pressures	22
1957	Molecular Mechanism of Rate Processes in Solids	23
1957	Interactions in Ionic Solutions	24
1958	Configurations and Interactions of Macromolecules and Liquid Crystals	25
1958	Ions of the Transition Elements	26
1959	Energy Transfer with special reference to Biological Systems	27
1959	Crystal Imperfections and the Chemical Reactivity of Solids	28
1960	Oxidation-Reduction Reactions in Ionizing Solvents	29
1960	The Physical Chemistry of Aerosols	30
1961	Radiation Effects in Inorganic Solids	31
1961	The Structure and Properties of Ionic Melts	32
1962	Inelastic Collisions of Atoms and Simple Molecules	33
1962	High Resolution Nuclear Magnetic Resonance	34
1963	The Structure of Electronically Excited Species in the Gas Phase	35
1963	Fundamental Processes in Radiation Chemistry	36
1964	Chemical Reactions in the Atmosphere	37
1964	Dislocations in Solids	38
1965	The Kinetics of Proton Transfer Processes	39
1965	Intermolecular Forces	40
1966	The Role of the Absorbed State in Heterogeneous Catalysis	41
1966	Colloid Stability in Aqueous and Non-aqueous Media	42
1967	The Structure and Properties of Liquids	43

Date	Subject	Volume
1967	Molecular Dynamics of the Chemical Reactions of Gases	44
1968	Electrode Reactions of Organic Compounds	45
1968	Homogeneous Catalysis with Special Reference to Hydrogenation and Oxidation	46
1969	Bonding in Metallo-organic Compounds	47
1969	Motions in Molecular Crystals	48
1970	Polymer Solutions	49
1970	The Vitreous State	50
1971	Electrical Conduction in Organic Solids	51
1971	Surface Chemistry of Oxides	52
1972	Reactions of Small Molecules in Excited States	53
1972	The Photoelectron Spectroscopy of Molecules	54
1973	Molecular Beam Scattering	55
1973	Intermediates in Electrochemical Reactions	56
1974	Gels and Gelling Processes	57
1974	Photo-effects in Adsorbed Species	58
1975	Physical Adsorption in Condensed Phases	59
1975	Electron Spectroscopy of Solids and Surfaces	60
1976	Precipitation	61
1977	Potential Energy Surfaces	62
1977	Radiation Effects in Liquids and Solids	63
1977	Ion–Ion and Ion–Solvent Interactions	64
1978	Colloid Stability	65
1978	Structures and Motion in Molecular Liquids	66
1979	Kinetics of State Selected Species	67
1979	Organization of Macromolecules in the Condensed Phase	68
1980	Phase Transitions in Molecular Solids	69
1980	Photoelectrochemistry	70
1981	High Resolution Spectroscopy	71
1981	Selectivity in Heterogeneous Catalysis	72
1982	Van der Waals Molecules	73
1982	Electron and Proton Transfer	74
1983	Intramolecular Kinetics	75
1983	Concentrated Colloidal Dispersions	76
1984	Interfacial Kinetics in Solution	77
1984	Radicals in Condensed Phases	78
1985	Polymer Liquid Crystals	79
1985	Physical Interactions and Energy Exchange at the Gas–Solid Interface	80
1986	Lipid Vesicles and Membranes	81
1986	Dynamics of Molecular Photofragmentation	82
1987	Brownian Motion	83
1987	Dynamics of Elementary Gas-phase Reactions	84
1988	Solvation	85
1988	Spectroscopy at Low Temperatures	86
1989	Catalysis by Well Characterised Materials	87
1989	Charge Transfer in Polymeric Systems	88
1990	Structure of Surfaces and Interfaces as studied using Synchrotron Radiation	89
1990	Colloidal Dispersions	90
1991	Structure and Dynamics of Reactive Transition States	91
1991	The Chemistry and Physics of Small Metallic Particles	92
1992	Structure and Activity of Enzymes	93
1992	The Liquid/Solid Interface at High Resolution	94
1993	Crystal Growth	95
1993	Dynamics at the Gas/Solid Interface	96
1994	Structure and Dynamics of Van der Waals Complexes	97
1994	Polymers at Surfaces and Interfaces	98*
1994	Vibrational Optical Activity: From Fundamentals to Biological Applications	99*
1995	Atmospheric Chemistry: Measurements, Mechanisms and Models	100*
1995	Gels	101*
1995	Unimolecular Reaction Dynamics	102*
1996	Hydration Processes in Biological and Macromolecular Systems	103*
1996	Complex Fluids at Interfaces	104*
1996	Catalysis and Surface Science at High Resolution	105
1997	Solid State Chemistry: New Opportunities from Computer Simulations	106*
1997	Interactions of Acoustic Waves with Thin Films and Interfaces	107*
1997	Dynamics of Electronically Excited States in Gaseous, Cluster and Condensed Media	108*
1998	Chemistry and Physics of Molecules and Grains in Space	109*
1998	Chemical Reaction Theory	110*
1998	Molecular Interactions of Biomembranes	111*
1999	Physical Chemistry in the Mesoscopic Regime	112*
1999	Stereochemistry and Control in Molecular Reaction Dynamics	113*

* *Available for purchase, for current information on prices* etc. *please contact the Sales and Promotion Department, The Royal Society of Chemistry, Thomas Graham House, Science Park, Milton Road, Cambridge, UK CB4 0WF.*